THE BERTRAND RUSSELL COLLECTION

Happy Hour Books

Published by:

HAPPY HOUR BOOKS

www.happyhoursbooks.com

email: happyhourbooks1@gmail.com

First published by Happy Hour Books 2023

Title : **The Bertrand Russell Collection**

Paperback ISBN **:** 9789358484472

Hardback ISBN : 9789358484441

THE PROBLEMS OF PHILOSOPHY

PREFACE

In the following pages I have confined myself in the main to those problems of philosophy in regard to which I thought it possible to say something positive and constructive, since merely negative criticism seemed out of place. For this reason, theory of knowledge occupies a larger space than metaphysics in the present volume, and some topics much discussed by philosophers are treated very briefly, if at all.

I have derived valuable assistance from unpublished writings of G. E. Moore and J. M. Keynes: from the former, as regards the relations of sense-data to physical objects, and from the latter as regards probability and induction. I have also profited greatly by the criticisms and suggestions of Professor Gilbert Murray.

1912

CHAPTER I.

APPEARANCE AND REALITY

Is there any knowledge in the world which is so certain that no reasonable man could doubt it? This question, which at first sight might not seem difficult, is really one of the most difficult that can be asked. When we have realized the obstacles in the way of a straightforward and confident answer, we shall be well launched on the study of philosophy—for philosophy is merely the attempt to answer such ultimate questions, not carelessly and dogmatically, as we do in ordinary life and even in the sciences, but critically, after exploring all that makes such questions puzzling, and after realizing all the vagueness and confusion that underlie our ordinary ideas.

In daily life, we assume as certain many things which, on a closer scrutiny, are found to be so full of apparent contradictions that only a great amount of thought enables us to know what it is that we really may believe. In the search for certainty, it is natural to begin with our present experiences, and in some sense, no doubt, knowledge is to be derived from them. But any statement as to what it is that our immediate experiences make us know is very likely to be wrong. It seems to me that I am now sitting in a chair, at a table of a certain shape, on which I see sheets of paper with writing or print. By turning my head, I see out of the window buildings and clouds and the sun. I believe that the sun is about ninety-three million miles from the earth; that it is a hot globe many times bigger than the earth; that, owing to the earth's rotation, it rises every morning, and will continue to do so for an indefinite time in the future. I believe that, if any other normal person comes into my room, he will see the same chairs and tables and books and papers as I see, and that the table which I see is the same as the table which I feel pressing against my arm. All this seems to be so evident as to be hardly worth stating, except in answer to a man who doubts whether I know anything. Yet all this may be reasonably doubted, and all of it requires much careful discussion before we can be sure that we have stated it in a form that is wholly true.

To make our difficulties plain, let us concentrate attention on the table. To the eye it is oblong, brown and shiny, to the touch it is smooth and cool and hard; when I tap it, it gives out a wooden sound. Anyone else who sees and feels and hears the table will agree with this description, so that it might seem as if no difficulty would arise; but as soon as we try to be more precise our troubles begin. Although I believe that the table is 'really' of the same colour all over, the parts that reflect the light look much brighter than the other parts, and some parts look white because of reflected light. I know that, if I move, the parts that reflect the light will be different, so that the apparent distribution of colours on the table will change. It follows that if several people are looking at the table at the same moment, no two of them will see exactly the same distribution of colours, because no two can see it from exactly the same point of view, and any change in the point of view makes some change in the way the light is reflected.

For most practical purposes these differences are unimportant, but to the painter they are all-important: the painter has to unlearn the habit of thinking that things seem to have the colour which common sense says they 'really' have, and to learn the habit of seeing things as they appear. Here we have already the beginning of one of the distinctions that cause most trouble in philosophy—the distinction between 'appearance' and 'reality', between what things seem to be and what they are. The painter wants to know what things seem to be, the practical man and the philosopher want to know what they are; but the philosopher's wish to know this is stronger than the practical man's, and is more troubled by knowledge as to the difficulties of answering the question.

To return to the table. It is evident from what we have found, that there is no colour which pre-eminently appears to be *the* colour of the table, or even of any one particular part of the table—it appears to be of different colours from different points of view, and there is no reason for regarding some of these as more really its colour than others. And we know that even from a given point of view the colour will seem different by artificial light, or to a colour-blind man, or to a man wearing blue spectacles, while in the dark there will be no colour at all, though to touch and hearing the table will be unchanged. This colour is not something which is inherent in the table, but something depending upon the table and the spectator and the way the light falls on the table. When, in ordinary life, we speak of *the* colour of the table, we only mean the sort of colour which it will seem to have to a normal spectator from an ordinary point of view under usual conditions of light. But the other colours which appear under other conditions have just as good a right to be considered real; and therefore, to avoid favouritism, we are compelled to deny that, in itself, the table has any one particular colour.

The same thing applies to the texture. With the naked eye one can see the grain, but otherwise the table looks smooth and even. If we looked at it through a microscope, we should see roughnesses and hills and valleys, and all sorts of differences that are imperceptible to the naked eye. Which of these is the 'real' table? We are naturally tempted to say that what we see through the microscope is more real, but that in turn would be changed by a still more powerful microscope. If, then, we cannot trust what we see with the naked eye, why should we trust what we see through a microscope? Thus, again, the confidence in our senses with which we began deserts us.

The shape of the table is no better. We are all in the habit of judging as to the 'real' shapes of things, and we do this so unreflectingly that we come to think we actually see the real shapes. But, in fact, as we all have to learn if we try to draw, a given thing looks different in shape from every different point of view. If our table is 'really' rectangular, it will look, from almost all points of view, as if it had two acute angles and two obtuse angles. If opposite sides are parallel, they will look as if they converged to a point away from the spectator; if they are of equal length, they will look as if the nearer side were longer. All these things are not commonly noticed in looking at a table, because experience has taught us to construct the 'real' shape from the apparent shape, and the 'real' shape is what interests us as practical men. But the 'real' shape is not what we see; it is something inferred from what we see. And what we see is constantly changing in shape as we move about the room; so that here again the senses seem not to give us the truth about the table itself, but only about the appearance of the table.

Similar difficulties arise when we consider the sense of touch. It is true that the table always gives us a sensation of hardness, and we feel that it resists pressure. But the sensation we obtain depends upon how hard we press the table and also upon what part of the body we press with; thus the various sensations due to various pressures or various parts of the body cannot be supposed to reveal *directly* any definite property of the table, but at most to be *signs* of some property which perhaps *causes* all the sensations, but is not actually apparent in any of them. And the same applies still more obviously to the sounds which can be elicited by rapping the table.

Thus it becomes evident that the real table, if there is one, is not the same as what we immediately experience by sight or touch or hearing. The real table, if there is one, is not *immediately* known to us at all, but must be an inference from what is immediately known. Hence, two very difficult questions at once arise; namely, (1) Is there a real table at all? (2) If so, what sort of object can it be?

It will help us in considering these questions to have a few simple terms of which the meaning is definite and clear. Let us give the name of 'sense-data' to the things that are immediately known in sensation: such things as colours, sounds, smells, hardnesses, roughnesses, and so on. We shall give the name 'sensation' to the experience of being immediately aware of these things. Thus, whenever we see a colour, we have a sensation *of* the colour, but the colour itself is a sense-datum, not a sensation. The colour is that *of* which we are immediately aware, and the awareness itself is the sensation. It is plain that if we are to know anything about the table, it must be by means of the sense-data—brown colour, oblong shape, smoothness, etc.—which we associate with the table; but, for the reasons which have been given, we cannot say that the table is the sense-data, or even that the sense-data are directly properties of the table. Thus a problem arises as to the relation of the sense-data to the real table, supposing there is such a thing.

The real table, if it exists, we will call a 'physical object'. Thus we have to consider the relation of sense-data to physical objects. The collection of all physical objects is called 'matter'. Thus our two questions may be re-stated as follows: (1) Is there any such thing as matter? (2) If so, what is its nature?

The philosopher who first brought prominently forward the reasons for regarding the immediate objects of our senses as not existing independently of us was Bishop Berkeley (1685-1753). His *Three Dialogues between Hylas and Philonous, in Opposition to Sceptics and Atheists*, undertake to prove that there is no such thing as matter at all, and that the world consists of nothing but minds and their ideas. Hylas has hitherto believed in matter, but he is no match for Philonous, who mercilessly drives him into contradictions and paradoxes, and makes his own denial of matter seem, in the end, as if it were almost common sense. The arguments employed are of very different value: some are important and sound, others are confused or quibbling. But Berkeley retains the merit of having shown that the existence of matter is capable of being denied without absurdity, and that if there are any things that exist independently of us they cannot be the immediate objects of our sensations.

There are two different questions involved when we ask whether matter exists, and it is important to keep them clear. We commonly mean by 'matter' something which is opposed to 'mind', something which we think of as occupying space and as radically incapable of any sort of thought or consciousness. It is chiefly in this sense that Berkeley denies matter; that is to say, he does not deny that the sense-data which we commonly take as signs of the existence of the table are really signs of the existence of *something* independent of us, but he does deny that this something is non-mental, that it is neither mind nor ideas entertained by some mind. He admits that there must be something which continues to exist when we go out of the room or shut our eyes, and that what we call seeing the table does really give us reason for believing in something which persists even when we are not seeing it. But he thinks that this something cannot be radically different in nature from what we see, and cannot be independent of seeing altogether, though it must be independent of *our* seeing. He is thus led to regard the 'real' table as an idea in the mind of God. Such an idea has the required permanence and independence of ourselves, without being—as matter would otherwise be—something quite unknowable, in the sense that we can only infer it, and can never be directly and immediately aware of it.

Other philosophers since Berkeley have also held that, although the table does not depend for its existence upon being seen by me, it does depend upon being seen (or otherwise apprehended in sensation) by *some* mind—not necessarily the mind of God, but more often the whole collective mind of the universe. This they hold, as Berkeley does, chiefly because they think there can be nothing real—or at any rate nothing known to be real except minds and their thoughts and feelings.

We might state the argument by which they support their view in some such way as this: 'Whatever can be thought of is an idea in the mind of the person thinking of it; therefore nothing can be thought of except ideas in minds; therefore anything else is inconceivable, and what is inconceivable cannot exist.'

Such an argument, in my opinion, is fallacious; and of course those who advance it do not put it so shortly or so crudely. But whether valid or not, the argument has been very widely advanced in one form or another; and very many philosophers, perhaps a majority, have held that there is nothing real except minds and their ideas. Such philosophers are called 'idealists'. When they come to explaining matter, they either say, like Berkeley, that matter is really nothing but a collection of ideas, or they say, like Leibniz (1646-1716), that what appears as matter is really a collection of more or less rudimentary minds.

But these philosophers, though they deny matter as opposed to mind, nevertheless, in another sense, admit matter. It will be remembered that we asked two questions; namely, (1) Is there a real table at all? (2) If so, what sort of object can it be? Now both Berkeley and Leibniz admit that there is a real table, but Berkeley says it is certain ideas in the mind of God, and Leibniz says it is a colony of souls. Thus both of them answer our first question in the affirmative, and only diverge from the views of ordinary mortals in their answer to our second question. In fact, almost all philosophers seem to be agreed that there is a real table: they almost all agree that, however much our sense-data—colour, shape, smoothness, etc.—may depend upon us, yet their occurrence is a sign of something existing independently of us, something differing, perhaps, completely from our sense-data, and yet to be regarded as causing those sense-data whenever we are in a suitable relation to the real table.

Now obviously this point in which the philosophers are agreed—the view that there *is* a real table, whatever its nature may be—is vitally important, and it will be worth while to consider what reasons there are for accepting this view before we go on to the further question as to the nature of the real table. Our next chapter, therefore, will be concerned with the reasons for supposing that there is a real table at all.

Before we go farther it will be well to consider for a moment what it is that we have discovered so far. It has appeared that, if we take any common object of the sort that is supposed to be known by the senses, what the senses *immediately* tell us is not the truth about the object as it is apart from us, but only the truth about certain sense-data which, so far as we can see, depend upon the relations between us and the object. Thus what we directly see and feel is merely 'appearance', which we believe to be a sign of some 'reality' behind. But if the reality is not what appears, have we any means of knowing whether there is any reality at all? And if so, have we any means of finding out what it is like?

Such questions are bewildering, and it is difficult to know that even the strangest hypotheses may not be true. Thus our familiar table, which has roused but the slightest thoughts in us hitherto, has become a problem full of surprising possibilities. The one thing we know about it is that it is not what it seems. Beyond this modest result, so far, we have the most complete liberty of conjecture. Leibniz tells us it is a community of souls: Berkeley tells us it is an idea in the mind of God; sober science, scarcely less wonderful, tells us it is a vast collection of electric charges in violent motion.

Among these surprising possibilities, doubt suggests that perhaps there is no table at all. Philosophy, if it cannot *answer* so many questions as we could wish, has at least the power of *asking* questions which increase the interest of the world, and show the strangeness and wonder lying just below the surface even in the commonest things of daily life.

CHAPTER II.

THE EXISTENCE OF MATTER

In this chapter we have to ask ourselves whether, in any sense at all, there is such a thing as matter. Is there a table which has a certain intrinsic nature, and continues to exist when I am not looking, or is the table merely a product of my imagination, a dream-table in a very prolonged dream? This question is of the greatest importance. For if we cannot be sure of the independent existence of objects, we cannot be sure of the independent existence of other people's bodies, and therefore still less of other people's minds, since we have no grounds for believing in their minds except such as are derived from observing their bodies. Thus, if we cannot be sure of the independent existence of objects, we shall be left alone in a desert—it may be that the whole outer world is nothing but a dream, and that we alone exist. This is an uncomfortable possibility; but although it cannot be strictly proved to be false, there is not the slightest reason to suppose that it is true. In this chapter we have to see why this is the case.

Before we embark upon doubtful matters, let us try to find some more or less fixed point from which to start. Although we are doubting the physical existence of the table, we are not doubting the existence of the sense-data which made us think there was a table; we are not doubting that, while we look, a certain colour and shape appear to us, and while we press, a certain sensation of hardness is experienced by us. All this, which is psychological, we are not calling in question. In fact, whatever else may be doubtful, some at least of our immediate experiences seem absolutely certain.

Descartes (1596-1650), the founder of modern philosophy, invented a method which may still be used with profit—the method of systematic doubt. He determined that he would believe nothing which he did not see quite clearly and distinctly to be true. Whatever he could bring himself to doubt, he would doubt, until he saw reason for not doubting it. By applying this method, he gradually became convinced that the only existence of which he could be *quite* certain was his own. He imagined a deceitful demon, who presented unreal things to his senses in a perpetual phantasmagoria; it might be very improbable that such a demon existed, but still it was possible, and therefore doubt concerning things perceived by the senses was possible.

But doubt concerning his own existence was not possible, for if he did not exist, no demon could deceive him. If he doubted, he must exist; if he had any experiences whatever, he must exist. Thus, his own existence was an absolute certainty to him. 'I think, therefore I am,' he said (*Cogito, ergo sum*); and on the basis of this certainty he set to work to build up again the world of knowledge which his doubt had laid in ruins. By inventing the method of doubt, and by showing that subjective things are the most certain, Descartes performed a great service to philosophy, and one which makes him still useful to all students of the subject.

But some care is needed in using Descartes' argument. 'I think, therefore I am' says rather more than is strictly certain. It might seem as though we were quite sure of being the same person to-day as we were yesterday, and this is no doubt true in some sense. But the real Self is as hard to arrive at as the real table, and does not seem to have that absolute, convincing certainty that belongs to particular experiences. When I look at my table and see a certain brown colour, what is quite certain at once is not '*I* am seeing a brown colour', but rather, 'a brown colour is being seen'. This of course involves something (or somebody) which (or who) sees the brown colour; but it does not of itself involve that more or less permanent person whom we call 'I'. So far as immediate certainty goes, it might be that the something which sees the brown colour is quite momentary, and not the same as the something which has some different experience the next moment.

Thus, it is our particular thoughts and feelings that have primitive certainty. And this applies to dreams and hallucinations as well as to normal perceptions: when we dream or see a ghost, we certainly do have the sensations we think we have, but for various reasons it is held that no physical object corresponds to these sensations. Thus, the certainty of our knowledge of our own experiences does not have to be limited in any way to allow for exceptional cases. Here, therefore, we have, for what it is worth, a solid basis from which to begin our pursuit of knowledge.

The problem we have to consider is this: Granted that we are certain of our own sense-data, have we any reason for regarding them as signs of the existence of something else, which we can call the physical object? When we have enumerated all the sense-data which we should naturally regard as connected with the table, have we said all there is to say about the table, or is there still something else—something not a sense-datum, something which persists when we go out of the room? Common sense unhesitatingly answers that there is. What can be bought and sold and pushed about and have a cloth laid on it, and so on, cannot be a *mere* collection of sense-data. If the cloth completely hides the table, we shall derive no sense-data from the table, and therefore, if the table were merely sense-data, it would have ceased to exist, and the cloth would be suspended in empty air, resting, by a miracle, in the place where the table formerly was. This seems plainly absurd; but whoever wishes to become a philosopher must learn not to be frightened by absurdities.

One great reason why it is felt that we must secure a physical object in addition to the sense-data, is that we want the same object for different people. When ten people are sitting round a dinner-table, it seems preposterous to maintain that they are not seeing the same tablecloth, the same knives and forks and spoons and glasses. But the sense-data are private to each separate person; what is immediately present to the sight of one is not immediately present to the sight of another: they all see things from slightly different points of view, and therefore see them slightly differently. Thus, if there are to be public neutral objects, which can be in some sense known to many different people, there must be something over and above the private and particular sense-data which appear to various people. What reason, then, have we for believing that there are such public neutral objects?

The first answer that naturally occurs to one is that, although different people may see the table slightly differently, still they all see more or less similar things when they look at the table, and the variations in what they see follow the laws of perspective and reflection of light, so that it is easy to arrive at a permanent object underlying all the different people's sense-data. I bought my table from the former occupant of my room; I could not buy *his* sense-data, which died when he went away, but I could and did buy the confident expectation of more or less similar sense-data. Thus it is the fact that different people have similar sense-data, and that one person in a given place at different times has similar sense-data, which makes us suppose that over and above the sense-data there is a permanent public object which underlies or causes the sense-data of various people at various times.

Now in so far as the above considerations depend upon supposing that there are other people besides ourselves, they beg the very question at issue. Other people are represented to me by certain sense-data, such as the sight of them or the sound of their voices, and if I had no reason to believe that there were physical objects independent of my sense-data, I should have no reason to believe that other people exist except as part of my dream. Thus, when we are trying to show that there must be objects independent of our own sense-data, we cannot appeal to the testimony of other people, since this testimony itself consists of sense-data, and does not reveal other people's experiences unless our own sense-data are signs of things existing independently of us. We must therefore, if possible, find, in our own purely private experiences, characteristics which show, or tend to show, that there are in the world things other than ourselves and our private experiences.

In one sense it must be admitted that we can never prove the existence of things other than ourselves and our experiences. No logical absurdity results from the hypothesis that the world consists of myself and my thoughts and feelings and sensations, and that everything else is mere fancy. In dreams a very complicated world may seem to be present, and yet on waking we find it was a delusion; that is to say, we find that the sense-data in the dream do not appear to have corresponded with such physical objects as we should naturally infer from our sense-data. (It is true that, when the physical world is assumed, it is possible to find physical causes for the sense-data in dreams: a door banging, for instance, may cause us to dream of a naval engagement. But although, in this case, there is a physical cause for the sense-data, there is not a physical object corresponding to the sense-data in the way in which an actual naval battle would correspond.) There is no logical impossibility in the supposition that the whole of life is a dream, in which we ourselves create all the objects that come before us. But although this is not logically impossible, there is no reason whatever to suppose that it is true; and it is, in fact, a less simple hypothesis, viewed as a means of accounting for the facts of our own life, than the common-sense hypothesis that there really are objects independent of us, whose action on us causes our sensations.

The way in which simplicity comes in from supposing that there really are physical objects is easily seen. If the cat appears at one moment in one part of the room, and at another in another part, it is natural to suppose that it has moved from the one to the other, passing over a series of intermediate positions. But if it is merely a set of sense-data, it cannot have ever been in any place where I did not see it; thus we shall have to suppose that it did not exist at all while I was not looking, but suddenly sprang into being in a new place. If the cat exists whether I see it or not, we can understand from our own experience how it gets hungry between one meal and the next; but if it does not exist when I am not seeing it, it seems odd that appetite should grow during non-existence as fast as during existence. And if the cat consists only of sense-data, it cannot be hungry, since no hunger but my own can be a sense-datum to me. Thus the behaviour of the sense-data which represent the cat to me, though it seems quite natural when regarded as an expression of hunger, becomes utterly inexplicable when regarded as mere movements and changes of patches of colour, which are as incapable of hunger as a triangle is of playing football.

But the difficulty in the case of the cat is nothing compared to the difficulty in the case of human beings. When human beings speak—that is, when we hear certain noises which we associate with ideas, and simultaneously see certain motions of lips and expressions of face—it is very difficult to suppose that what we hear is not the expression of a thought, as we know it would be if we emitted the same sounds. Of course similar things happen in dreams, where we are mistaken as to the existence of other people. But dreams are more or less suggested by what we call waking life, and are capable of being more or less accounted for on scientific principles if we assume that there really is a physical world. Thus every principle of simplicity urges us to adopt the natural view, that there really are objects other than ourselves and our sense-data which have an existence not dependent upon our perceiving them.

Of course it is not by argument that we originally come by our belief in an independent external world. We find this belief ready in ourselves as soon as we begin to reflect: it is what may be called an *instinctive* belief. We should never have been led to question this belief but for the fact that, at any rate in the case of sight, it seems as if the sense-datum itself were instinctively believed to be the independent object, whereas argument shows that the object cannot be identical with the sense-datum. This discovery, however—which is not at all paradoxical in the case of taste and smell and sound, and only slightly so in the case of touch—leaves undiminished our instinctive belief that there *are* objects *corresponding* to our sense-data. Since this belief does not lead to any difficulties, but on the contrary tends to simplify and systematize our account of our experiences, there seems no good reason for rejecting it. We may therefore admit—though with a slight doubt derived from dreams—that the external world does really exist, and is not wholly dependent for its existence upon our continuing to perceive it.

The argument which has led us to this conclusion is doubtless less strong than we could wish, but it is typical of many philosophical arguments, and it is therefore worth while to consider briefly its general character and validity. All knowledge, we find, must be built up upon our instinctive beliefs, and if these are rejected, nothing is left. But among our instinctive beliefs some are much stronger than others, while many have, by habit and association, become entangled with other beliefs, not really instinctive, but falsely supposed to be part of what is believed instinctively.

Philosophy should show us the hierarchy of our instinctive beliefs, beginning with those we hold most strongly, and presenting each as much isolated and as free from irrelevant additions as possible. It should take care to show that, in the form in which they are finally set forth, our instinctive beliefs do not clash, but form a harmonious system. There can never be any reason for rejecting one instinctive belief except that it clashes with others; thus, if they are found to harmonize, the whole system becomes worthy of acceptance.

It is of course *possible* that all or any of our beliefs may be mistaken, and therefore all ought to be held with at least some slight element of doubt. But we cannot have *reason* to reject a belief except on the ground of some other belief. Hence, by organizing our instinctive beliefs and their consequences, by considering which among them is most possible, if necessary, to modify or abandon, we can arrive, on the basis of accepting as our sole data what we instinctively believe, at an orderly systematic organization of our knowledge, in which, though the *possibility* of error remains, its likelihood is diminished by the interrelation of the parts and by the critical scrutiny which has preceded acquiescence.

This function, at least, philosophy can perform. Most philosophers, rightly or wrongly, believe that philosophy can do much more than this—that it can give us knowledge, not otherwise attainable, concerning the universe as a whole, and concerning the nature of ultimate reality. Whether this be the case or not, the more modest function we have spoken of can certainly be performed by philosophy, and certainly suffices, for those who have once begun to doubt the adequacy of common sense, to justify the arduous and difficult labours that philosophical problems involve.

CHAPTER III.

THE NATURE OF MATTER

In the preceding chapter we agreed, though without being able to find demonstrative reasons, that it is rational to believe that our sense-data—for example, those which we regard as associated with my table—are really signs of the existence of something independent of us and our perceptions. That is to say, over and above the sensations of colour, hardness, noise, and so on, which make up the appearance of the table to me, I assume that there is something else, of which these things are appearances. The colour ceases to exist if I shut my eyes, the sensation of hardness ceases to exist if I remove my arm from contact with the table, the sound ceases to exist if I cease to rap the table with my knuckles. But I do not believe that when all these things cease the table ceases. On the contrary, I believe that it is because the table exists continuously that all these sense-data will reappear when I open my eyes, replace my arm, and begin again to rap with my knuckles. The question we have to consider in this chapter is: What is the nature of this real table, which persists independently of my perception of it?

To this question physical science gives an answer, somewhat incomplete it is true, and in part still very hypothetical, but yet deserving of respect so far as it goes. Physical science, more or less unconsciously, has drifted into the view that all natural phenomena ought to be reduced to motions. Light and heat and sound are all due to wave-motions, which travel from the body emitting them to the person who sees light or feels heat or hears sound. That which has the wave-motion is either aether or 'gross matter', but in either case is what the philosopher would call matter. The only properties which science assigns to it are position in space, and the power of motion according to the laws of motion. Science does not deny that it *may* have other properties; but if so, such other properties are not useful to the man of science, and in no way assist him in explaining the phenomena.

It is sometimes said that 'light *is* a form of wave-motion', but this is misleading, for the light which we immediately see, which we know directly by means of our senses, is *not* a form of wave-motion, but something quite different—something which we all know if we are not blind, though we cannot describe it so as to convey our knowledge to a man who is blind. A wave-motion, on the contrary, could quite well be described to a blind man, since he can acquire a knowledge of space by the sense of touch; and he can experience a wave-motion by a sea voyage almost as well as we can. But this, which a blind man can understand, is not what we mean by *light*: we mean by *light* just that which a blind man can never understand, and which we can never describe to him.

Now this something, which all of us who are not blind know, is not, according to science, really to be found in the outer world: it is something caused by the action of certain waves upon the eyes and nerves and brain of the person who sees the light. When it is said that light *is* waves, what is really meant is that waves are the physical cause of our sensations of light. But light itself, the thing which seeing people experience and blind people do not, is not supposed by science to form any part of the world that is independent of us and our senses. And very similar remarks would apply to other kinds of sensations.

It is not only colours and sounds and so on that are absent from the scientific world of matter, but also *space* as we get it through sight or touch. It is essential to science that its matter should be in *a* space, but the space in which it is cannot be exactly the space we see or feel. To begin with, space as we see it is not the same as space as we get it by the sense of touch; it is only by experience in infancy that we learn how to touch things we see, or how to get a sight of things which we feel touching us. But the space of science is neutral as between touch and sight; thus it cannot be either the space of touch or the space of sight.

Again, different people see the same object as of different shapes, according to their point of view. A circular coin, for example, though we should always *judge* it to be circular, will *look* oval unless we are straight in front of it. When we judge that it *is* circular, we are judging that it has a real shape which is not its apparent shape, but belongs to it intrinsically apart from its appearance. But this real shape, which is what concerns science, must be in a real space, not the same as anybody's *apparent* space. The real space is public, the apparent space is private to the percipient. In different people's *private* spaces the same object seems to have different shapes; thus the real space, in which it has its real shape, must be different from the private spaces. The space of science, therefore, though *connected* with the spaces we see and feel, is not identical with them, and the manner of its connexion requires investigation.

We agreed provisionally that physical objects cannot be quite like our sense-data, but may be regarded as *causing* our sensations. These physical objects are in the space of science, which we may call 'physical' space. It is important to notice that, if our sensations are to be caused by physical objects, there must be a physical space containing these objects and our sense-organs and nerves and brain. We get a sensation of touch from an object when we are in contact with it; that is to say, when some part of our body occupies a place in physical space quite close to the space occupied by the object. We see an object (roughly speaking) when no opaque body is between the object and our eyes in physical space. Similarly, we only hear or smell or taste an object when we are sufficiently near to it, or when it touches the tongue, or has some suitable position in physical space relatively to our body. We cannot begin to state what different sensations we shall derive from a given object under different circumstances unless we regard the object and our body as both in one physical space, for it is mainly the relative positions of the object and our body that determine what sensations we shall derive from the object.

Now our sense-data are situated in our private spaces, either the space of sight or the space of touch or such vaguer spaces as other senses may give us. If, as science and common sense assume, there is one public all-embracing physical space in which physical objects are, the relative positions of physical objects in physical space must more or less correspond to the relative positions of sense-data in our private spaces. There is no difficulty in supposing this to be the case. If we see on a road one house nearer to us than another, our other senses will bear out the view that it is nearer; for example, it will be reached sooner if we walk along the road. Other people will agree that the house which looks nearer to us is nearer; the ordnance map will take the same view; and thus everything points to a spatial relation between the houses corresponding to the relation between the sense-data which we see when we look at the houses. Thus we may assume that there is a physical space in which physical objects have spatial relations corresponding to those which the corresponding sense-data have in our private spaces. It is this physical space which is dealt with in geometry and assumed in physics and astronomy.

Assuming that there is physical space, and that it does thus correspond to private spaces, what can we know about it? We can know *only* what is required in order to secure the correspondence. That is to say, we can know nothing of what it is like in itself, but we can know the sort of arrangement of physical objects which results from their spatial relations. We can know, for example, that the earth and moon and sun are in one straight line during an eclipse, though we cannot know what a physical straight line is in itself, as we know the look of a straight line in our visual space. Thus we come to know much more about the *relations* of distances in physical space than about the distances themselves; we may know that one distance is greater than another, or that it is along the same straight line as the other, but we cannot have that immediate acquaintance with physical distances that we have with distances in our private spaces, or with colours or sounds or other sense-data. We can know all those things about physical space which a man born blind might know through other people about the space of sight; but the kind of things which a man born blind could never know about the space of sight we also cannot know about physical space. We can know the properties of the relations required to preserve the correspondence with sense-data, but we cannot know the nature of the terms between which the relations hold.

With regard to time, our *feeling* of duration or of the lapse of time is notoriously an unsafe guide as to the time that has elapsed by the clock. Times when we are bored or suffering pain pass slowly, times when we are agreeably occupied pass quickly, and times when we are sleeping pass almost as if they did not exist. Thus, in so far as time is constituted by duration, there is the same necessity for distinguishing a public and a private time as there was in the case of space. But in so far as time consists in an *order* of before and after, there is no need to make such a distinction; the time-order which events seem to have is, so far as we can see, the same as the time-order which they do have. At any rate no reason can be given for supposing that the two orders are not the same. The same is usually true of space: if a regiment of men are marching along a road, the shape of the regiment will look different from different points of view, but the men will appear arranged in the same order from all points of view. Hence we regard the order as true also in physical space, whereas the shape is only supposed to correspond to the physical space so far as is required for the preservation of the order.

In saying that the time-order which events seem to have is the same as the time-order which they really have, it is necessary to guard against a possible misunderstanding. It must not be supposed that the various states of different physical objects have the same time-order as the sense-data which constitute the perceptions of those objects. Considered as physical objects, the thunder and lightning are simultaneous; that is to say, the lightning is simultaneous with the disturbance of the air in the place where the disturbance begins, namely, where the lightning is. But the sense-datum which we call hearing the thunder does not take place until the disturbance of the air has travelled as far as to where we are. Similarly, it takes about eight minutes for the sun's light to reach us; thus, when we see the sun we are seeing the sun of eight minutes ago. So far as our sense-data afford evidence as to the physical sun they afford evidence as to the physical sun of eight minutes ago; if the physical sun had ceased to exist within the last eight minutes, that would make no difference to the sense-data which we call 'seeing the sun'. This affords a fresh illustration of the necessity of distinguishing between sense-data and physical objects.

What we have found as regards space is much the same as what we find in relation to the correspondence of the sense-data with their physical counterparts. If one object looks blue and another red, we may reasonably presume that there is some corresponding difference between the physical objects; if two objects both look blue, we may presume a corresponding similarity. But we cannot hope to be acquainted directly with the quality in the physical object which makes it look blue or red. Science tells us that this quality is a certain sort of wave-motion, and this sounds familiar, because we think of wave-motions in the space we see. But the wave-motions must really be in physical space, with which we have no direct acquaintance; thus the real wave-motions have not that familiarity which we might have supposed them to have. And what holds for colours is closely similar to what holds for other sense-data. Thus we find that, although the *relations* of physical objects have all sorts of knowable properties, derived from their correspondence with the relations of sense-data, the physical objects themselves remain unknown in their intrinsic nature, so far at least as can be discovered by means of the senses. The question remains whether there is any other method of discovering the intrinsic nature of physical objects.

The most natural, though not ultimately the most defensible, hypothesis to adopt in the first instance, at any rate as regards visual sense-data, would be that, though physical objects cannot, for the reasons we have been considering, be *exactly* like sense-data, yet they may be more or less like. According to this view, physical objects will, for example, really have colours, and we might, by good luck, see an object as of the colour it really is. The colour which an object seems to have at any given moment will in general be very similar, though not quite the same, from many different points of view; we might thus suppose the 'real' colour to be a sort of medium colour, intermediate between the various shades which appear from the different points of view.

Such a theory is perhaps not capable of being definitely refuted, but it can be shown to be groundless. To begin with, it is plain that the colour we see depends only upon the nature of the light-waves that strike the eye, and is therefore modified by the medium intervening between us and the object, as well as by the manner in which light is reflected from the object in the direction of the eye. The intervening air alters colours unless it is perfectly clear, and any strong reflection will alter them completely. Thus the colour we see is a result of the ray as it reaches the eye, and not simply a property of the object from which the ray comes. Hence, also, provided certain waves reach the eye, we shall see a certain colour, whether the object from which the waves start has any colour or not. Thus it is quite gratuitous to suppose that physical objects have colours, and therefore there is no justification for making such a supposition. Exactly similar arguments will apply to other sense-data.

It remains to ask whether there are any general philosophical arguments enabling us to say that, if matter is real, it must be of such and such a nature. As explained above, very many philosophers, perhaps most, have held that whatever is real must be in some sense mental, or at any rate that whatever we can know anything about must be in some sense mental. Such philosophers are called 'idealists'. Idealists tell us that what appears as matter is really something mental; namely, either (as Leibniz held) more or less rudimentary minds, or (as Berkeley contended) ideas in the minds which, as we should commonly say, 'perceive' the matter. Thus idealists deny the existence of matter as something intrinsically different from mind, though they do not deny that our sense-data are signs of something which exists independently of our private sensations. In the following chapter we shall consider briefly the reasons—in my opinion fallacious—which idealists advance in favour of their theory.

CHAPTER IV

IDEALISM

The word 'idealism' is used by different philosophers in somewhat different senses. We shall understand by it the doctrine that whatever exists, or at any rate whatever can be known to exist, must be in some sense mental. This doctrine, which is very widely held among philosophers, has several forms, and is advocated on several different grounds. The doctrine is so widely held, and so interesting in itself, that even the briefest survey of philosophy must give some account of it.

Those who are unaccustomed to philosophical speculation may be inclined to dismiss such a doctrine as obviously absurd. There is no doubt that common sense regards tables and chairs and the sun and moon and material objects generally as something radically different from minds and the contents of minds, and as having an existence which might continue if minds ceased. We think of matter as having existed long before there were any minds, and it is hard to think of it as a mere product of mental activity. But whether true or false, idealism is not to be dismissed as obviously absurd.

We have seen that, even if physical objects do have an independent existence, they must differ very widely from sense-data, and can only have a *correspondence* with sense-data, in the same sort of way in which a catalogue has a correspondence with the things catalogued. Hence common sense leaves us completely in the dark as to the true intrinsic nature of physical objects, and if there were good reason to regard them as mental, we could not legitimately reject this opinion merely because it strikes us as strange. The truth about physical objects *must* be strange. It may be unattainable, but if any philosopher believes that he has attained it, the fact that what he offers as the truth is strange ought not to be made a ground of objection to his opinion.

The grounds on which idealism is advocated are generally grounds derived from the theory of knowledge, that is to say, from a discussion of the conditions which things must satisfy in order that we may be able to know them. The first serious attempt to establish idealism on such grounds was that of Bishop Berkeley. He proved first, by arguments which were largely valid, that our sense-data cannot be supposed to have an existence independent of us, but must be, in part at least, 'in' the mind, in the sense that their existence would not continue if there were no seeing or hearing or touching or smelling or tasting. So far, his contention was almost certainly valid, even if some of his arguments were not so. But he went on to argue that sense-data were the only things of whose existence our perceptions could assure us; and that to be known is to be 'in' a mind, and therefore to be mental. Hence he concluded that nothing can ever be known except what is in some mind, and that whatever is known without being in my mind must be in some other mind.

In order to understand his argument, it is necessary to understand his use of the word 'idea'. He gives the name 'idea' to anything which is *immediately* known, as, for example, sense-data are known. Thus a particular colour which we see is an idea; so is a voice which we hear, and so on. But the term is not wholly confined to sense-data. There will also be things remembered or imagined, for with such things also we have immediate acquaintance at the moment of remembering or imagining. All such immediate data he calls 'ideas'.

He then proceeds to consider common objects, such as a tree, for instance. He shows that all we know immediately when we 'perceive' the tree consists of ideas in his sense of the word, and he argues that there is not the slightest ground for supposing that there is anything real about the tree except what is perceived. Its being, he says, consists in being perceived: in the Latin of the schoolmen its '*esse*' is '*percipi*'. He fully admits that the tree must continue to exist even when we shut our eyes or when no human being is near it. But this continued existence, he says, is due to the fact that God continues to perceive it; the 'real' tree, which corresponds to what we called the physical object, consists of ideas in the mind of God, ideas more or less like those we have when we see the tree, but differing in the fact that they are permanent in God's mind so long as the tree continues to exist. All our perceptions, according to him, consist in a partial participation in God's perceptions, and it is because of this participation that different people see more or less the same tree. Thus apart from minds and their ideas there is nothing in the world, nor is it possible that anything else should ever be known, since whatever is known is necessarily an idea.

There are in this argument a good many fallacies which have been important in the history of philosophy, and which it will be as well to bring to light. In the first place, there is a confusion engendered by the use of the word 'idea'. We think of an idea as essentially something in somebody's mind, and thus when we are told that a tree consists entirely of ideas, it is natural to suppose that, if so, the tree must be entirely in minds. But the notion of being 'in' the mind is ambiguous. We speak of bearing a person in mind, not meaning that the person is in our minds, but that a thought of him is in our minds. When a man says that some business he had to arrange went clean out of his mind, he does not mean to imply that the business itself was ever in his mind, but only that a thought of the business was formerly in his mind, but afterwards ceased to be in his mind. And so when Berkeley says that the tree must be in our minds if we can know it, all that he really has a right to say is that a thought of the tree must be in our minds. To argue that the tree itself must be in our minds is like arguing that a person whom we bear in mind is himself in our minds. This confusion may seem too gross to have been really committed by any competent philosopher, but various attendant circumstances rendered it possible. In order to see how it was possible, we must go more deeply into the question as to the nature of ideas.

Before taking up the general question of the nature of ideas, we must disentangle two entirely separate questions which arise concerning sense-data and physical objects. We saw that, for various reasons of detail, Berkeley was right in treating the sense-data which constitute our perception of the tree as more or less subjective, in the sense that they depend upon us as much as upon the tree, and would not exist if the tree were not being perceived. But this is an entirely different point from the one by which Berkeley seeks to prove that whatever can be immediately known must be in a mind. For this purpose arguments of detail as to the dependence of sense-data upon us are useless. It is necessary to prove, generally, that by being known, things are shown to be mental. This is what Berkeley believes himself to have done. It is this question, and not our previous question as to the difference between sense-data and the physical object, that must now concern us.

Taking the word 'idea' in Berkeley's sense, there are two quite distinct things to be considered whenever an idea is before the mind. There is on the one hand the thing of which we are aware—say the colour of my table—and on the other hand the actual awareness itself, the mental act of apprehending the thing. The mental act is undoubtedly mental, but is there any reason to suppose that the thing apprehended is in any sense mental? Our previous arguments concerning the colour did not prove it to be mental; they only proved that its existence depends upon the relation of our sense organs to the physical object—in our case, the table. That is to say, they proved that a certain colour will exist, in a certain light, if a normal eye is placed at a certain point relatively to the table. They did not prove that the colour is in the mind of the percipient.

Berkeley's view, that obviously the colour must be in the mind, seems to depend for its plausibility upon confusing the thing apprehended with the act of apprehension. Either of these might be called an 'idea'; probably either would have been called an idea by Berkeley. The act is undoubtedly in the mind; hence, when we are thinking of the act, we readily assent to the view that ideas must be in the mind. Then, forgetting that this was only true when ideas were taken as acts of apprehension, we transfer the proposition that 'ideas are in the mind' to ideas in the other sense, i.e. to the things apprehended by our acts of apprehension. Thus, by an unconscious equivocation, we arrive at the conclusion that whatever we can apprehend must be in our minds. This seems to be the true analysis of Berkeley's argument, and the ultimate fallacy upon which it rests.

This question of the distinction between act and object in our apprehending of things is vitally important, since our whole power of acquiring knowledge is bound up with it. The faculty of being acquainted with things other than itself is the main characteristic of a mind. Acquaintance with objects essentially consists in a relation between the mind and something other than the mind; it is this that constitutes the mind's power of knowing things. If we say that the things known must be in the mind, we are either unduly limiting the mind's power of knowing, or we are uttering a mere tautology. We are uttering a mere tautology if we mean by '*in* the mind' the same as by '*before* the mind', i.e. if we mean merely being apprehended by the mind. But if we mean this, we shall have to admit that what, *in this sense*, is in the mind, may nevertheless be not mental. Thus when we realize the nature of knowledge, Berkeley's argument is seen to be wrong in substance as well as in form, and his grounds for supposing that 'ideas'—i.e. the objects apprehended—must be mental, are found to have no validity whatever. Hence his grounds in favour of idealism may be dismissed. It remains to see whether there are any other grounds.

It is often said, as though it were a self-evident truism, that we cannot know that anything exists which we do not know. It is inferred that whatever can in any way be relevant to our experience must be at least capable of being known by us; whence it follows that if matter were essentially something with which we could not become acquainted, matter would be something which we could not know to exist, and which could have for us no importance whatever. It is generally also implied, for reasons which remain obscure, that what can have no importance for us cannot be real, and that therefore matter, if it is not composed of minds or of mental ideas, is impossible and a mere chimaera.

To go into this argument fully at our present stage would be impossible, since it raises points requiring a considerable preliminary discussion; but certain reasons for rejecting the argument may be noticed at once. To begin at the end: there is no reason why what cannot have any *practical* importance for us should not be real. It is true that, if *theoretical* importance is included, everything real is of *some* importance to us, since, as persons desirous of knowing the truth about the universe, we have some interest in everything that the universe contains. But if this sort of interest is included, it is not the case that matter has no importance for us, provided it exists even if we cannot know that it exists. We can, obviously, suspect that it may exist, and wonder whether it does; hence it is connected with our desire for knowledge, and has the importance of either satisfying or thwarting this desire.

Again, it is by no means a truism, and is in fact false, that we cannot know that anything exists which we do not know. The word 'know' is here used in two different senses. (1) In its first use it is applicable to the sort of knowledge which is opposed to error, the sense in which what we know is *true*, the sense which applies to our beliefs and convictions, i.e. to what are called *judgements*. In this sense of the word we know *that* something is the case. This sort of knowledge may be described as knowledge of *truths*. (2) In the second use of the word 'know' above, the word applies to our knowledge of *things*, which we may call *acquaintance*. This is the sense in which we know sense-data. (The distinction involved is roughly that between *savoir* and *connaître* in French, or between *wissen* and *kennen* in German.)

Thus the statement which seemed like a truism becomes, when re-stated, the following: 'We can never truly judge that something with which we are not acquainted exists.' This is by no means a truism, but on the contrary a palpable falsehood. I have not the honour to be acquainted with the Emperor of China, but I truly judge that he exists. It may be said, of course, that I judge this because of other people's acquaintance with him. This, however, would be an irrelevant retort, since, if the principle were true, I could not know that any one else is acquainted with him. But further: there is no reason why I should not know of the existence of something with which nobody is acquainted. This point is important, and demands elucidation.

If I am acquainted with a thing which exists, my acquaintance gives me the knowledge that it exists. But it is not true that, conversely, whenever I can know that a thing of a certain sort exists, I or some one else must be acquainted with the thing. What happens, in cases where I have true judgement without acquaintance, is that the thing is known to me by *description*, and that, in virtue of some general principle, the existence of a thing answering to this description can be inferred from the existence of something with which I am acquainted. In order to understand this point fully, it will be well first to deal with the difference between knowledge by acquaintance and knowledge by description, and then to consider what knowledge of general principles, if any, has the same kind of certainty as our knowledge of the existence of our own experiences. These subjects will be dealt with in the following chapters.

CHAPTER V.

KNOWLEDGE BY ACQUAINTANCE AND KNOWLEDGE BY DESCRIPTION

In the preceding chapter we saw that there are two sorts of knowledge: knowledge of things, and knowledge of truths. In this chapter we shall be concerned exclusively with knowledge of things, of which in turn we shall have to distinguish two kinds. Knowledge of things, when it is of the kind we call knowledge by *acquaintance*, is essentially simpler than any knowledge of truths, and logically independent of knowledge of truths, though it would be rash to assume that human beings ever, in fact, have acquaintance with things without at the same time knowing some truth about them. Knowledge of things by *description*, on the contrary, always involves, as we shall find in the course of the present chapter, some knowledge of truths as its source and ground. But first of all we must make clear what we mean by 'acquaintance' and what we mean by 'description'.

We shall say that we have *acquaintance* with anything of which we are directly aware, without the intermediary of any process of inference or any knowledge of truths. Thus in the presence of my table I am acquainted with the sense-data that make up the appearance of my table—its colour, shape, hardness, smoothness, etc.; all these are things of which I am immediately conscious when I am seeing and touching my table. The particular shade of colour that I am seeing may have many things said about it—I may say that it is brown, that it is rather dark, and so on. But such statements, though they make me know truths about the colour, do not make me know the colour itself any better than I did before so far as concerns knowledge of the colour itself, as opposed to knowledge of truths about it, I know the colour perfectly and completely when I see it, and no further knowledge of it itself is even theoretically possible. Thus the sense-data which make up the appearance of my table are things with which I have acquaintance, things immediately known to me just as they are.

My knowledge of the table as a physical object, on the contrary, is not direct knowledge. Such as it is, it is obtained through acquaintance with the sense-data that make up the appearance of the table. We have seen that it is possible, without absurdity, to doubt whether there is a table at all, whereas it is not possible to doubt the sense-data. My knowledge of the table is of the kind which we shall call 'knowledge by description'. The table is 'the physical object which causes such-and-such sense-data'. This describes the table by means of the sense-data. In order to know anything at all about the table, we must know truths connecting it with things with which we have acquaintance: we must know that 'such-and-such sense-data are caused by a physical object'. There is no state of mind in which we are directly aware of the table; all our knowledge of the table is really knowledge of truths, and the actual thing which is the table is not, strictly speaking, known to us at all. We know a description, and we know that there is just one object to which this description applies, though the object itself is not directly known to us. In such a case, we say that our knowledge of the object is knowledge by description.

All our knowledge, both knowledge of things and knowledge of truths, rests upon acquaintance as its foundation. It is therefore important to consider what kinds of things there are with which we have acquaintance.

Sense-data, as we have already seen, are among the things with which we are acquainted; in fact, they supply the most obvious and striking example of knowledge by acquaintance. But if they were the sole example, our knowledge would be very much more restricted than it is. We should only know what is now present to our senses: we could not know anything about the past—not even that there was a past—nor could we know any truths about our sense-data, for all knowledge of truths, as we shall show, demands acquaintance with things which are of an essentially different character from sense-data, the things which are sometimes called 'abstract ideas', but which we shall call 'universals'. We have therefore to consider acquaintance with other things besides sense-data if we are to obtain any tolerably adequate analysis of our knowledge.

The first extension beyond sense-data to be considered is acquaintance by *memory*. It is obvious that we often remember what we have seen or heard or had otherwise present to our senses, and that in such cases we are still immediately aware of what we remember, in spite of the fact that it appears as past and not as present. This immediate knowledge by memory is the source of all our knowledge concerning the past: without it, there could be no knowledge of the past by inference, since we should never know that there was anything past to be inferred.

The next extension to be considered is acquaintance by *introspection*. We are not only aware of things, but we are often aware of being aware of them. When I see the sun, I am often aware of my seeing the sun; thus 'my seeing the sun' is an object with which I have acquaintance. When I desire food, I may be aware of my desire for food; thus 'my desiring food' is an object with which I am acquainted. Similarly we may be aware of our feeling pleasure or pain, and generally of the events which happen in our minds. This kind of acquaintance, which may be called self-consciousness, is the source of all our knowledge of mental things. It is obvious that it is only what goes on in our own minds that can be thus known immediately. What goes on in the minds of others is known to us through our perception of their bodies, that is, through the sense-data in us which are associated with their bodies. But for our acquaintance with the contents of our own minds, we should be unable to imagine the minds of others, and therefore we could never arrive at the knowledge that they have minds. It seems natural to suppose that self-consciousness is one of the things that distinguish men from animals: animals, we may suppose, though they have acquaintance with sense-data, never become aware of this acquaintance. I do not mean that they *doubt* whether they exist, but that they have never become conscious of the fact that they have sensations and feelings, nor therefore of the fact that they, the subjects of their sensations and feelings, exist.

We have spoken of acquaintance with the contents of our minds as *self*-consciousness, but it is not, of course, consciousness of our *self*: it is consciousness of particular thoughts and feelings. The question whether we are also acquainted with our bare selves, as opposed to particular thoughts and feelings, is a very difficult one, upon which it would be rash to speak positively. When we try to look into ourselves we always seem to come upon some particular thought or feeling, and not upon the 'I' which has the thought or feeling. Nevertheless there are some reasons for thinking that we are acquainted with the 'I', though the acquaintance is hard to disentangle from other things. To make clear what sort of reason there is, let us consider for a moment what our acquaintance with particular thoughts really involves.

When I am acquainted with 'my seeing the sun', it seems plain that I am acquainted with two different things in relation to each other. On the one hand there is the sense-datum which represents the sun to me, on the other hand there is that which sees this sense-datum. All acquaintance, such as my acquaintance with the sense-datum which represents the sun, seems obviously a relation between the person acquainted and the object with which the person is acquainted. When a case of acquaintance is one with which I can be acquainted (as I am acquainted with my acquaintance with the sense-datum representing the sun), it is plain that the person acquainted is myself. Thus, when I am acquainted with my seeing the sun, the whole fact with which I am acquainted is 'Self-acquainted-with-sense-datum'.

Further, we know the truth 'I am acquainted with this sense-datum'. It is hard to see how we could know this truth, or even understand what is meant by it, unless we were acquainted with something which we call 'I'. It does not seem necessary to suppose that we are acquainted with a more or less permanent person, the same to-day as yesterday, but it does seem as though we must be acquainted with that thing, whatever its nature, which sees the sun and has acquaintance with sense-data. Thus, in some sense it would seem we must be acquainted with our Selves as opposed to our particular experiences. But the question is difficult, and complicated arguments can be adduced on either side. Hence, although acquaintance with ourselves seems *probably* to occur, it is not wise to assert that it undoubtedly does occur.

We may therefore sum up as follows what has been said concerning acquaintance with things that exist. We have acquaintance in sensation with the data of the outer senses, and in introspection with the data of what may be called the inner sense—thoughts, feelings, desires, etc.; we have acquaintance in memory with things which have been data either of the outer senses or of the inner sense. Further, it is probable, though not certain, that we have acquaintance with Self, as that which is aware of things or has desires towards things.

In addition to our acquaintance with particular existing things, we also have acquaintance with what we shall call *universals*, that is to say, general ideas, such as *whiteness*, *diversity*, *brotherhood*, and so on. Every complete sentence must contain at least one word which stands for a universal, since all verbs have a meaning which is universal. We shall return to universals later on, in Chapter IX; for the present, it is only necessary to guard against the supposition that whatever we can be acquainted with must be something particular and existent. Awareness of universals is called *conceiving*, and a universal of which we are aware is called a *concept*.

It will be seen that among the objects with which we are acquainted are not included physical objects (as opposed to sense-data), nor other people's minds. These things are known to us by what I call 'knowledge by description', which we must now consider.

By a 'description' I mean any phrase of the form 'a so-and-so' or 'the so-and-so'. A phrase of the form 'a so-and-so' I shall call an 'ambiguous' description; a phrase of the form 'the so-and-so' (in the singular) I shall call a 'definite' description. Thus 'a man' is an ambiguous description, and 'the man with the iron mask' is a definite description. There are various problems connected with ambiguous descriptions, but I pass them by, since they do not directly concern the matter we are discussing, which is the nature of our knowledge concerning objects in cases where we know that there is an object answering to a definite description, though we are not acquainted with any such object. This is a matter which is concerned exclusively with definite descriptions. I shall therefore, in the sequel, speak simply of 'descriptions' when I mean 'definite descriptions'. Thus a description will mean any phrase of the form 'the so-and-so' in the singular.

We shall say that an object is 'known by description' when we know that it is 'the so-and-so', i.e. when we know that there is one object, and no more, having a certain property; and it will generally be implied that we do not have knowledge of the same object by acquaintance. We know that the man with the iron mask existed, and many propositions are known about him; but we do not know who he was. We know that the candidate who gets the most votes will be elected, and in this case we are very likely also acquainted (in the only sense in which one can be acquainted with some one else) with the man who is, in fact, the candidate who will get most votes; but we do not know which of the candidates he is, i.e. we do not know any proposition of the form 'A is the candidate who will get most votes' where A is one of the candidates by name. We shall say that we have 'merely descriptive knowledge' of the so-and-so when, although we know that the so-and-so exists, and although we may possibly be acquainted with the object which is, in fact, the so-and-so, yet we do not know any proposition '*a* is the so-and-so', where *a* is something with which we are acquainted.

When we say 'the so-and-so exists', we mean that there is just one object which is the so-and-so. The proposition '*a* is the so-and-so' means that *a* has the property so-and-so, and nothing else has. 'Mr. A. is the Unionist candidate for this constituency' means 'Mr. A. is a Unionist candidate for this constituency, and no one else is'. 'The Unionist candidate for this constituency exists' means 'some one is a Unionist candidate for this constituency, and no one else is'. Thus, when we are acquainted with an object which is the so-and-so, we know that the so-and-so exists; but we may know that the so-and-so exists when we are not acquainted with any object which we know to be the so-and-so, and even when we are not acquainted with any object which, in fact, is the so-and-so.

Common words, even proper names, are usually really descriptions. That is to say, the thought in the mind of a person using a proper name correctly can generally only be expressed explicitly if we replace the proper name by a description. Moreover, the description required to express the thought will vary for different people, or for the same person at different times. The only thing constant (so long as the name is rightly used) is the object to which the name applies. But so long as this remains constant, the particular description involved usually makes no difference to the truth or falsehood of the proposition in which the name appears.

Let us take some illustrations. Suppose some statement made about Bismarck. Assuming that there is such a thing as direct acquaintance with oneself, Bismarck himself might have used his name directly to designate the particular person with whom he was acquainted. In this case, if he made a judgement about himself, he himself might be a constituent of the judgement. Here the proper name has the direct use which it always wishes to have, as simply standing for a certain object, and not for a description of the object. But if a person who knew Bismarck made a judgement about him, the case is different. What this person was acquainted with were certain sense-data which he connected (rightly, we will suppose) with Bismarck's body. His body, as a physical object, and still more his mind, were only known as the body and the mind connected with these sense-data. That is, they were known by description. It is, of course, very much a matter af chance which characteristics of a man's appearance will come into a friend's mind when he thinks of him; thus the description actually in the friend's mind is accidental. The essential point is that he knows that the various descriptions all apply to the same entity, in spite of not being acquainted with the entity in question.

When we, who did not know Bismarck, make a judgement about him, the description in our minds will probably be some more or less vague mass of historical knowledge—far more, in most cases, than is required to identify him. But, for the sake of illustration, let us assume that we think of him as 'the first Chancellor of the German Empire'. Here all the words are abstract except 'German'. The word 'German' will, again, have different meanings for different people. To some it will recall travels in Germany, to some the look of Germany on the map, and so on. But if we are to obtain a description which we know to be applicable, we shall be compelled, at some point, to bring in a reference to a particular with which we are acquainted. Such reference is involved in any mention of past, present, and future (as opposed to definite dates), or of here and there, or of what others have told us. Thus it would seem that, in some way or other, a description known to be applicable to a particular must involve some reference to a particular with which we are acquainted, if our knowledge about the thing described is not to be merely what follows *logically* from the description. For example, 'the most long-lived of men' is a description involving only universals, which must apply to some man, but we can make no judgements concerning this man which involve knowledge about him beyond what the description gives. If, however, we say, 'The first Chancellor of the German Empire was an astute diplomatist', we can only be assured of the truth of our judgement in virtue of something with which we are acquainted—usually a testimony heard or read. Apart from the information we convey to others, apart from the fact about the actual Bismarck, which gives importance to our judgement, the thought we really have contains the one or more particulars involved, and otherwise consists wholly of concepts.

All names of places—London, England, Europe, the Earth, the Solar System—similarly involve, when used, descriptions which start from some one or more particulars with which we are acquainted. I suspect that even the Universe, as considered by metaphysics, involves such a connexion with particulars. In logic, on the contrary, where we are concerned not merely with what does exist, but with whatever might or could exist or be, no reference to actual particulars is involved.

It would seem that, when we make a statement about something only known by description, we often *intend* to make our statement, not in the form involving the description, but about the actual thing described. That is to say, when we say anything about Bismarck, we should like, if we could, to make the judgement which Bismarck alone can make, namely, the judgement of which he himself is a constituent. In this we are necessarily defeated, since the actual Bismarck is unknown to us. But we know that there is an object B, called Bismarck, and that B was an astute diplomatist. We can thus *describe* the proposition we should like to affirm, namely, 'B was an astute diplomatist', where B is the object which was Bismarck. If we are describing Bismarck as 'the first Chancellor of the German Empire', the proposition we should like to affirm may be described as 'the proposition asserting, concerning the actual object which was the first Chancellor of the German Empire, that this object was an astute diplomatist'. What enables us to communicate in spite of the varying descriptions we employ is that we know there is a true proposition concerning the actual Bismarck, and that however we may vary the description (so long as the description is correct) the proposition described is still the same. This proposition, which is described and is known to be true, is what interests us; but we are not acquainted with the proposition itself, and do not know it, though we know it is true.

It will be seen that there are various stages in the removal from acquaintance with particulars: there is Bismarck to people who knew him; Bismarck to those who only know of him through history; the man with the iron mask; the longest-lived of men. These are progressively further removed from acquaintance with particulars; the first comes as near to acquaintance as is possible in regard to another person; in the second, we shall still be said to know 'who Bismarck was'; in the third, we do not know who was the man with the iron mask, though we can know many propositions about him which are not logically deducible from the fact that he wore an iron mask; in the fourth, finally, we know nothing beyond what is logically deducible from the definition of the man. There is a similar hierarchy in the region of universals. Many universals, like many particulars, are only known to us by description. But here, as in the case of particulars, knowledge concerning what is known by description is ultimately reducible to knowledge concerning what is known by acquaintance.

The fundamental principle in the analysis of propositions containing descriptions is this: *Every proposition which we can understand must be composed wholly of constituents with which we are acquainted.*

We shall not at this stage attempt to answer all the objections which may be urged against this fundamental principle. For the present, we shall merely point out that, in some way or other, it must be possible to meet these objections, for it is scarcely conceivable that we can make a judgement or entertain a supposition without knowing what it is that we are judging or supposing about. We must attach *some* meaning to the words we use, if we are to speak significantly and not utter mere noise; and the meaning we attach to our words must be something with which we are acquainted. Thus when, for example, we make a statement about Julius Caesar, it is plain that Julius Caesar himself is not before our minds, since we are not acquainted with him. We have in mind some description of Julius Caesar: 'the man who was assassinated on the Ides of March', 'the founder of the

Roman Empire', or, perhaps, merely 'the man whose name was *Julius Caesar*'. (In this last description, *Julius Caesar* is a noise or shape with which we are acquainted.) Thus our statement does not mean quite what it seems to mean, but means something involving, instead of Julius Caesar, some description of him which is composed wholly of particulars and universals with which we are acquainted.

The chief importance of knowledge by description is that it enables us to pass beyond the limits of our private experience. In spite of the fact that we can only know truths which are wholly composed of terms which we have experienced in acquaintance, we can yet have knowledge by description of things which we have never experienced. In view of the very narrow range of our immediate experience, this result is vital, and until it is understood, much of our knowledge must remain mysterious and therefore doubtful.

CHAPTER VI.

ON INDUCTION

In almost all our previous discussions we have been concerned in the attempt to get clear as to our data in the way of knowledge of existence. What things are there in the universe whose existence is known to us owing to our being acquainted with them? So far, our answer has been that we are acquainted with our sense-data, and, probably, with ourselves. These we know to exist. And past sense-data which are remembered are known to have existed in the past. This knowledge supplies our data.

But if we are to be able to draw inferences from these data—if we are to know of the existence of matter, of other people, of the past before our individual memory begins, or of the future, we must know general principles of some kind by means of which such inferences can be drawn. It must be known to us that the existence of some one sort of thing, A, is a sign of the existence of some other sort of thing, B, either at the same time as A or at some earlier or later time, as, for example, thunder is a sign of the earlier existence of lightning. If this were not known to us, we could never extend our knowledge beyond the sphere of our private experience; and this sphere, as we have seen, is exceedingly limited. The question we have now to consider is whether such an extension is possible, and if so, how it is effected.

Let us take as an illustration a matter about which none of us, in fact, feel the slightest doubt. We are all convinced that the sun will rise to-morrow. Why? Is this belief a mere blind outcome of past experience, or can it be justified as a reasonable belief? It is not easy to find a test by which to judge whether a belief of this kind is reasonable or not, but we can at least ascertain what sort of general beliefs would suffice, if true, to justify the judgement that the sun will rise to-morrow, and the many other similar judgements upon which our actions are based.

It is obvious that if we are asked why we believe that the sun will rise to-morrow, we shall naturally answer 'Because it always has risen every day'. We have a firm belief that it will rise in the future, because it has risen in the past. If we are challenged as to why we believe that it will continue to rise as heretofore, we may appeal to the laws of motion: the earth, we shall say, is a freely rotating body, and such bodies do not cease to rotate unless something interferes from outside, and there is nothing outside to interfere with the earth between now and to-morrow. Of course it might be doubted whether we are quite certain that there is nothing outside to interfere, but this is not the interesting doubt. The interesting doubt is as to whether the laws of motion will remain in operation until to-morrow. If this doubt is raised, we find ourselves in the same position as when the doubt about the sunrise was first raised.

The *only* reason for believing that the laws of motion will remain in operation is that they have operated hitherto, so far as our knowledge of the past enables us to judge. It is true that we have a greater body of evidence from the past in favour of the laws of motion than we have in favour of the sunrise, because the sunrise is merely a particular case of fulfilment of the laws of motion, and there are countless other particular cases. But the real question is: Do *any* number of cases of a law being fulfilled in the past afford evidence that it will be fulfilled in the future? If not, it becomes plain that we have no ground whatever for expecting the sun to rise to-morrow, or for expecting the bread we shall eat at our next meal not to poison us, or for any of the other scarcely conscious expectations that control our daily lives. It is to be observed that all such expectations are only *probable*; thus we have not to seek for a proof that they *must* be fulfilled, but only for some reason in favour of the view that they are *likely* to be fulfilled.

Now in dealing with this question we must, to begin with, make an important distinction, without which we should soon become involved in hopeless confusions. Experience has shown us that, hitherto, the frequent repetition of some uniform succession or coexistence has been a *cause* of our expecting the same succession or coexistence on the next occasion. Food that has a certain appearance generally has a certain taste, and it is a severe shock to our expectations when the familiar appearance is found to be associated with an unusual taste. Things which we see become associated, by habit, with certain tactile sensations which we expect if we touch them; one of the horrors of a ghost (in many ghost-stories) is that it fails to give us any sensations of touch. Uneducated people who go abroad for the first time are so surprised as to be incredulous when they find their native language not understood.

And this kind of association is not confined to men; in animals also it is very strong. A horse which has been often driven along a certain road resists the attempt to drive him in a different direction. Domestic animals expect food when they see the person who usually feeds them. We know that all these rather crude expectations of uniformity are liable to be misleading. The man who has fed the chicken every day throughout its life at last wrings its neck instead, showing that more refined views as to the uniformity of nature would have been useful to the chicken.

But in spite of the misleadingness of such expectations, they nevertheless exist. The mere fact that something has happened a certain number of times causes animals and men to expect that it will happen again. Thus our instincts certainly cause us to believe that the sun will rise to-morrow, but we may be in no better a position than the chicken which unexpectedly has its neck wrung. We have therefore to distinguish the fact that past uniformities *cause* expectations as to the future, from the question whether there is any reasonable ground for giving weight to such expectations after the question of their validity has been raised.

The problem we have to discuss is whether there is any reason for believing in what is called 'the uniformity of nature'. The belief in the uniformity of nature is the belief that everything that has happened or will happen is an instance of some general law to which there are no exceptions. The crude expectations which we have been considering are all subject to exceptions, and therefore liable to disappoint those who entertain them. But science habitually assumes, at least as a working hypothesis, that general rules which have exceptions can be replaced by general rules which have no exceptions. 'Unsupported bodies in air fall' is a general rule to which balloons and aeroplanes are exceptions. But the laws of motion and the law of gravitation, which account for the fact that most bodies fall, also account for the fact that balloons and aeroplanes can rise; thus the laws of motion and the law of gravitation are not subject to these exceptions.

The belief that the sun will rise to-morrow might be falsified if the earth came suddenly into contact with a large body which destroyed its rotation; but the laws of motion and the law of gravitation would not be infringed by such an event. The business of science is to find uniformities, such as the laws of motion and the law of gravitation, to which, so far as our experience extends, there are no exceptions. In this search science has been remarkably successful, and it may be conceded that such uniformities have held hitherto. This brings us back to the question: Have we any reason, assuming that they have always held in the past, to suppose that they will hold in the future?

It has been argued that we have reason to know that the future will resemble the past, because what was the future has constantly become the past, and has always been found to resemble the past, so that we really have experience of the future, namely of times which were formerly future, which we may call past futures. But such an argument really begs the very question at issue. We have experience of past futures, but not of future futures, and the question is: Will future futures resemble past futures? This question is not to be answered by an argument which starts from past futures alone. We have therefore still to seek for some principle which shall enable us to know that the future will follow the same laws as the past.

The reference to the future in this question is not essential. The same question arises when we apply the laws that work in our experience to past things of which we have no experience—as, for example, in geology, or in theories as to the origin of the Solar System. The question we really have to ask is: 'When two things have been found to be often associated, and no instance is known of the one occurring without the other, does the occurrence of one of the two, in a fresh instance, give any good ground for expecting the other?' On our answer to this question must depend the validity of the whole of our expectations as to the future, the whole of the results obtained by induction, and in fact practically all the beliefs upon which our daily life is based.

It must be conceded, to begin with, that the fact that two things have been found often together and never apart does not, by itself, suffice to *prove* demonstratively that they will be found together in the next case we examine. The most we can hope is that the oftener things are found together, the more probable it becomes that they will be found together another time, and that, if they have been found together often enough, the probability will amount *almost* to certainty. It can never quite reach certainty, because we know that in spite of frequent repetitions there sometimes is a failure at the last, as in the case of the chicken whose neck is wrung. Thus probability is all we ought to seek.

It might be urged, as against the view we are advocating, that we know all natural phenomena to be subject to the reign of law, and that sometimes, on the basis of observation, we can see that only one law can possibly fit the facts of the case. Now to this view there are two answers. The first is that, even if *some* law which has no exceptions applies to our case, we can never, in practice, be sure that we have discovered that law and not one to which there are exceptions. The second is that the reign of law would seem to be itself only probable, and that our belief that it will hold in the future, or in unexamined cases in the past, is itself based upon the very principle we are examining.

The principle we are examining may be called the *principle of induction*, and its two parts may be stated as follows:

(a) When a thing of a certain sort A has been found to be associated with a thing of a certain other sort B, and has never been found dissociated from a thing of the sort B, the greater the number of cases in which A and B have been associated, the greater is the probability that they will be associated in a fresh case in which one of them is known to be present;

(b) Under the same circumstances, a sufficient number of cases of association will make the probability of a fresh association nearly a certainty, and will make it approach certainty without limit.

As just stated, the principle applies only to the verification of our expectation in a single fresh instance. But we want also to know that there is a probability in favour of the general law that things of the sort A are *always* associated with things of the sort B, provided a sufficient number of cases of association are known, and no cases of failure of association are known. The probability of the general law is obviously less than the probability of the particular case, since if the general law is true, the particular case must also be true, whereas the particular case may be true without the general law being true. Nevertheless the probability of the general law is increased by repetitions, just as the probability of the particular case is. We may therefore repeat the two parts of our principle as regards the general law, thus:

(a) The greater the number of cases in which a thing of the sort A has been found associated with a thing of the sort B, the more probable it is (if no cases of failure of association are known) that A is always associated with B;

b) Under the same circumstances, a sufficient number of cases of the association of A with B will make it nearly certain that A is always associated with B, and will make this general law approach certainty without limit.

It should be noted that probability is always relative to certain data. In our case, the data are merely the known cases of coexistence of A and B. There may be other data, which *might* be taken into account, which would gravely alter the probability. For example, a man who had seen a great many white swans might argue, by our principle, that on the data it was *probable* that all swans were white, and this might be a perfectly sound argument. The argument is not disproved by the fact that some swans are black, because a thing may very well happen in spite of the fact that some data render it improbable. In the case of the swans, a man might know that colour is a very variable characteristic in many species of animals, and that, therefore, an induction as to colour is peculiarly liable to error. But this knowledge would be a fresh datum, by no means proving that the probability relatively to our previous data had been wrongly estimated. The fact, therefore, that things often fail to fulfil our expectations is no evidence that our expectations will not *probably* be fulfilled in a given case or a given class of cases. Thus our inductive principle is at any rate not capable of being *disproved* by an appeal to experience.

The inductive principle, however, is equally incapable of being *proved* by an appeal to experience. Experience might conceivably confirm the inductive principle as regards the cases that have been already examined; but as regards unexamined cases, it is the inductive principle alone that can justify any inference from what has been examined to what has not been examined. All arguments which, on the basis of experience, argue as to the future or the unexperienced parts of the past or present, assume the inductive principle; hence we can never use experience to prove the inductive principle without begging the question. Thus we must either accept the inductive principle on the ground of its intrinsic evidence, or forgo all justification of our expectations about the future. If the principle is unsound, we have no reason to expect the sun to rise to-morrow, to expect bread to be more nourishing than a stone, or to expect that if we throw ourselves off the roof we shall fall. When we see what looks like our best friend approaching us, we shall have no reason to suppose that his body is not inhabited by the mind of our worst enemy or of some total stranger. All our conduct is based upon associations which have worked in the past, and which we therefore regard as likely to work in the future; and this likelihood is dependent for its validity upon the inductive principle.

The general principles of science, such as the belief in the reign of law, and the belief that every event must have a cause, are as completely dependent upon the inductive principle as are the beliefs of daily life All such general principles are believed because mankind have found innumerable instances of their truth and no instances of their falsehood. But this affords no evidence for their truth in the future, unless the inductive principle is assumed.

Thus all knowledge which, on a basis of experience tells us something about what is not experienced, is based upon a belief which experience can neither confirm nor confute, yet which, at least in its more concrete applications, appears to be as firmly rooted in us as many of the facts of experience. The existence and justification of such beliefs—for the inductive principle, as we shall see, is not the only example—raises some of the most difficult and most debated problems of philosophy. We will, in the next chapter, consider briefly what may be said to account for such knowledge, and what is its scope and its degree of certainty.

CHAPTER VII.

ON OUR KNOWLEDGE OF GENERAL PRINCIPLES

We saw in the preceding chapter that the principle of induction, while necessary to the validity of all arguments based on experience, is itself not capable of being proved by experience, and yet is unhesitatingly believed by every one, at least in all its concrete applications. In these characteristics the principle of induction does not stand alone. There are a number of other principles which cannot be proved or disproved by experience, but are used in arguments which start from what is experienced.

Some of these principles have even greater evidence than the principle of induction, and the knowledge of them has the same degree of certainty as the knowledge of the existence of sense-data. They constitute the means of drawing inferences from what is given in sensation; and if what we infer is to be true, it is just as necessary that our principles of inference should be true as it is that our data should be true. The principles of inference are apt to be overlooked because of their very obviousness—the assumption involved is assented to without our realizing that it is an assumption. But it is very important to realize the use of principles of inference, if a correct theory of knowledge is to be obtained; for our knowledge of them raises interesting and difficult questions.

In all our knowledge of general principles, what actually happens is that first of all we realize some particular application of the principle, and then we realize that the particularity is irrelevant, and that there is a generality which may equally truly be affirmed. This is of course familiar in such matters as teaching arithmetic: 'two and two are four' is first learnt in the case of some particular pair of couples, and then in some other particular case, and so on, until at last it becomes possible to see that it is true of any pair of couples. The same thing happens with logical principles. Suppose two men are discussing what day of the month it is. One of them says, 'At least you will admit that *if* yesterday was the 15th to-day must be the 16th.' 'Yes', says the other, 'I admit that.' 'And you know', the first continues, 'that yesterday was the 15th, because you dined with Jones, and your diary will tell you that was on the 15th.' 'Yes', says the second; 'therefore to-day *is* the 16th.'

Now such an argument is not hard to follow; and if it is granted that its premisses are true in fact, no one will deny that the conclusion must also be true. But it depends for its truth upon an instance of a general logical principle. The logical principle is as follows: 'Suppose it known that *if* this is true, then that is true. Suppose it also known that this *is* true, then it follows that that is true.' When it is the case that if this is true, that is true, we shall say that this 'implies' that, and that that 'follows from' this. Thus our principle states that if this implies that, and this is true, then that is true. In other words, 'anything implied by a true proposition is true', or 'whatever follows from a true proposition is true'.

This principle is really involved—at least, concrete instances of it are involved—in all demonstrations. Whenever one thing which we believe is used to prove something else, which we consequently believe, this principle is relevant. If any one asks: 'Why should I accept the results of valid arguments based on true premisses?' we can only answer by appealing to our principle. In fact, the truth of the principle is impossible to doubt, and its obviousness is so great that at first sight it seems almost trivial. Such principles, however, are not trivial to the philosopher, for they show that we may have indubitable knowledge which is in no way derived from objects of sense.

The above principle is merely one of a certain number of self-evident logical principles. Some at least of these principles must be granted before any argument or proof becomes possible. When some of them have been granted, others can be proved, though these others, so long as they are simple, are just as obvious as the principles taken for granted. For no very good reason, three of these principles have been singled out by tradition under the name of 'Laws of Thought'.

They are as follows:

(1) *The law of identity*: 'Whatever is, is.'

(2) *The law of contradiction*: 'Nothing can both be and not be.'

(3) *The law of excluded middle*: 'Everything must either be or not be.'

These three laws are samples of self-evident logical principles, but are not really more fundamental or more self-evident than various other similar principles: for instance, the one we considered just now, which states that what follows from a true premiss is true. The name 'laws of thought' is also misleading, for what is important is not the fact that we think in accordance with these laws, but the fact that things behave in accordance with them; in other words, the fact that when we think in accordance with them we think *truly*. But this is a large question, to which we must return at a later stage.

In addition to the logical principles which enable us to prove from a given premiss that something is *certainly* true, there are other logical principles which enable us to prove, from a given premiss, that there is a greater or less probability that something is true. An example of such principles—perhaps the most important example is the inductive principle, which we considered in the preceding chapter.

One of the great historic controversies in philosophy is the controversy between the two schools called respectively 'empiricists' and 'rationalists'. The empiricists—who are best represented by the British philosophers, Locke, Berkeley, and Hume—maintained that all our knowledge is derived from experience; the rationalists—who are represented by the Continental philosophers of the seventeenth century, especially Descartes and Leibniz—maintained that, in addition to what we know by experience, there are certain 'innate ideas' and 'innate principles', which we know independently of experience. It has now become possible to decide with some confidence as to the truth or falsehood of these opposing schools. It must be admitted, for the reasons already stated, that logical principles are known to us, and cannot be themselves proved by experience, since all proof presupposes them. In this, therefore, which was the most important point of the controversy, the rationalists were in the right.

On the other hand, even that part of our knowledge which is *logically* independent of experience (in the sense that experience cannot prove it) is yet elicited and caused by experience. It is on occasion of particular experiences that we become aware of the general laws which their connexions exemplify. It would certainly be absurd to suppose that there are innate principles in the sense that babies are born with a knowledge of everything which men know and which cannot be deduced from what is experienced. For this reason, the word 'innate' would not now be employed to describe our knowledge of logical principles. The phrase '*a priori*' is less objectionable, and is more usual in modern writers. Thus, while admitting that all knowledge is elicited and caused by experience, we shall nevertheless hold that some knowledge is *a priori*, in the sense that the experience which makes us think of it does not suffice to prove it, but merely so directs our attention that we see its truth without requiring any proof from experience.

There is another point of great importance, in which the empiricists were in the right as against the rationalists. Nothing can be known to *exist* except by the help of experience. That is to say, if we wish to prove that something of which we have no direct experience exists, we must have among our premisses the existence of one or more things of which we have direct experience. Our belief that the Emperor of China exists, for example, rests upon testimony, and testimony consists, in the last analysis, of sense-data seen or heard in reading or being spoken to. Rationalists believed that, from general consideration as to what must be, they could deduce the existence of this or that in the actual world. In this belief they seem to have been mistaken. All the knowledge that we can acquire *a priori* concerning existence seems to be hypothetical: it tells us that if one thing exists, another must exist, or, more generally, that if one proposition is true, another must be true. This is exemplified by the principles we have already dealt with, such as '*if* this is true, and this implies that, then that is true', or '*if* this and that have been repeatedly found connected, they will probably be connected in the next instance in which one of them is found'. Thus the scope and power of *a priori* principles is strictly limited. All knowledge that something exists must be in part dependent on experience. When anything is known immediately, its existence is known by experience alone; when anything is proved to exist, without being known immediately, both experience and *a priori* principles must be required in the proof. Knowledge is called *empirical* when it rests wholly or partly upon experience. Thus all knowledge which asserts existence is empirical, and the only *a priori* knowledge concerning existence is hypothetical, giving connexions among things that exist or may exist, but not giving actual existence.

A priori knowledge is not all of the logical kind we have been hitherto considering. Perhaps the most important example of non-logical *a priori* knowledge is knowledge as to ethical value. I am not speaking of judgements as to what is useful or as to what is virtuous, for such judgements do require empirical premisses; I am speaking of judgements as to the intrinsic desirability of things. If something is useful, it must be useful because it secures some end; the end must, if we have gone far enough, be valuable on its own account, and not merely because it is useful for some further end. Thus all judgements as to what is useful depend upon judgements as to what has value on its own account.

We judge, for example, that happiness is more desirable than misery, knowledge than ignorance, goodwill than hatred, and so on. Such judgements must, in part at least, be immediate and *a priori*. Like our previous *a priori* judgements, they may be elicited by experience, and indeed they must be; for it seems not possible to judge whether anything is intrinsically valuable unless we have experienced something of the same kind. But it is fairly obvious that they cannot be proved by experience; for the fact that a thing exists or does not exist cannot prove either that it is good that it should exist or that it is bad. The pursuit of this subject belongs to ethics, where the impossibility of deducing what ought to be from what is has to be established. In the present connexion, it is only important to realize that knowledge as to what is intrinsically of value is *a priori* in the same sense in which logic is *a priori*, namely in the sense that the truth of such knowledge can be neither proved nor disproved by experience.

All pure mathematics is *a priori*, like logic. This was strenuously denied by the empirical philosophers, who maintained that experience was as much the source of our knowledge of arithmetic as of our knowledge of geography. They maintained that by the repeated experience of seeing two things and two other things, and finding that altogether they made four things, we were led by induction to the conclusion that two things and two other things would *always* make four things altogether. If, however, this were the source of our knowledge that two and two are four, we should proceed differently, in persuading ourselves of its truth, from the way in which we do actually proceed. In fact, a certain number of instances are needed to make us think of two abstractly, rather than of two coins or two books or two people, or two of any other specified kind. But as soon as we are able to divest our thoughts of irrelevant particularity, we become able to see the general principle that two and two are four; any one instance is seen to be *typical*, and the examination of other instances becomes unnecessary.

(1) Cf. A. N. Whitehead, *Introduction to Mathematics* (Home University Library).

The same thing is exemplified in geometry. If we want to prove some property of *all* triangles, we draw some one triangle and reason about it; but we can avoid making use of any property which it does not share with all other triangles, and thus, from our particular case, we obtain a general result. We do not, in fact, feel our certainty that two and two are four increased by fresh instances, because, as soon as we have seen the truth of this proposition, our certainty becomes so great as to be incapable of growing greater. Moreover, we feel some quality of necessity about the proposition 'two and two are four', which is absent from even the best attested empirical generalizations. Such generalizations always remain mere facts: we feel that there might be a world in which they were false, though in the actual world they happen to be true. In any possible world, on the contrary, we feel that two and two would be four: this is not a mere fact, but a necessity to which everything actual and possible must conform.

The case may be made clearer by considering a genuinely-empirical generalization, such as 'All men are mortal.' It is plain that we believe this proposition, in the first place, because there is no known instance of men living beyond a certain age, and in the second place because there seem to be physiological grounds for thinking that an organism such as a man's body must sooner or later wear out. Neglecting the second ground, and considering merely our experience of men's mortality, it is plain that we should not be content with one quite clearly understood instance of a man dying, whereas, in the case of 'two and two are four', one instance does suffice, when carefully considered, to persuade us that the same must happen in any other instance. Also we can be forced to admit, on reflection, that there may be some doubt, however slight, as to whether *all* men are mortal. This may be made plain by the attempt to imagine two different worlds, in one of which there are men who are not mortal, while in the other two and two make five. When Swift invites us to consider the race of Struldbugs who never die, we are able to acquiesce in imagination. But a world where two and two make five seems quite on a different level. We feel that such a world, if there were one, would upset the whole fabric of our knowledge and reduce us to utter doubt.

The fact is that, in simple mathematical judgements such as 'two and two are four', and also in many judgements of logic, we can know the general proposition without inferring it from instances, although some instance is usually necessary to make clear to us what the general proposition means. This is why there is real utility in the process of *deduction*, which goes from the general to the general, or from the general to the particular, as well as in the process of *induction*, which goes from the particular to the particular, or from the particular to the general. It is an old debate among philosophers whether deduction ever gives *new* knowledge. We can now see that in certain cases, at least, it does do so. If we already know that two and two always make four, and we know that Brown and Jones are two, and so are Robinson and Smith, we can deduce that Brown and Jones and Robinson and Smith are four. This is new knowledge, not contained in our premisses, because the general proposition, 'two and two are four', never told us there were such people as Brown and Jones and Robinson and Smith, and the particular premisses do not tell us that there were four of them, whereas the particular proposition deduced does tell us both these things.

But the newness of the knowledge is much less certain if we take the stock instance of deduction that is always given in books on logic, namely, 'All men are mortal; Socrates is a man, therefore Socrates is mortal.' In this case, what we really know beyond reasonable doubt is that certain men, A, B, C, were mortal, since, in fact, they have died. If Socrates is one of these men, it is foolish to go the roundabout way through 'all men are mortal' to arrive at the conclusion that *probably* Socrates is mortal. If Socrates is not one of the men on whom our induction is based, we shall still do better to argue straight from our A, B, C, to Socrates, than to go round by the general proposition, 'all men are mortal'. For the probability that Socrates is mortal is greater, on our data, than the probability that all men are mortal. (This is obvious, because if all men are mortal, so is Socrates; but if Socrates is mortal, it does not follow that all men are mortal.) Hence we shall reach the conclusion that Socrates is mortal with a greater approach to certainty if we make our argument purely inductive than if we go by way of 'all men are mortal' and then use deduction.

This illustrates the difference between general propositions known *a priori* such as 'two and two are four', and empirical generalizations such as 'all men are mortal'. In regard to the former, deduction is the right mode of argument, whereas in regard to the latter, induction is always theoretically preferable, and warrants a greater confidence in the truth of our conclusion, because all empirical generalizations are more uncertain than the instances of them.

We have now seen that there are propositions known *a priori*, and that among them are the propositions of logic and pure mathematics, as well as the fundamental propositions of ethics. The question which must next occupy us is this: How is it possible that there should be such knowledge? And more particularly, how can there be knowledge of general propositions in cases where we have not examined all the instances, and indeed never can examine them all, because their number is infinite? These questions, which were first brought prominently forward by the German philosopher Kant (1724-1804), are very difficult, and historically very important.

CHAPTER VIII.

HOW A PRIOR I KNOWLEDGE IS POSSIBLE

Immanuel Kant is generally regarded as the greatest of the modern philosophers. Though he lived through the Seven Years War and the French Revolution, he never interrupted his teaching of philosophy at Königsberg in East Prussia. His most distinctive contribution was the invention of what he called the 'critical' philosophy, which, assuming as a datum that there is knowledge of various kinds, inquired how such knowledge comes to be possible, and deduced, from the answer to this inquiry, many metaphysical results as to the nature of the world. Whether these results were valid may well be doubted. But Kant undoubtedly deserves credit for two things: first, for having perceived that we have *a priori* knowledge which is not purely 'analytic', i.e. such that the opposite would be self-contradictory, and secondly, for having made evident the philosophical importance of the theory of knowledge.

Before the time of Kant, it was generally held that whatever knowledge was *a priori* must be 'analytic'. What this word means will be best illustrated by examples. If I say, 'A bald man is a man', 'A plane figure is a figure', 'A bad poet is a poet', I make a purely analytic judgement: the subject spoken about is given as having at least two properties, of which one is singled out to be asserted of it. Such propositions as the above are trivial, and would never be enunciated in real life except by an orator preparing the way for a piece of sophistry. They are called 'analytic' because the predicate is obtained by merely analysing the subject. Before the time of Kant it was thought that all judgements of which we could be certain *a priori* were of this kind: that in all of them there was a predicate which was only part of the subject of which it was asserted. If this were so, we should be involved in a definite contradiction if we attempted to deny anything that could be known *a priori*. 'A bald man is not bald' would assert and deny baldness of the same man, and would therefore contradict itself. Thus according to the philosophers before Kant, the law of contradiction, which asserts that nothing can at the same time have and not have a certain property, sufficed to establish the truth of all *a priori* knowledge.

Hume (1711-76), who preceded Kant, accepting the usual view as to what makes knowledge *a priori*, discovered that, in many cases which had previously been supposed analytic, and notably in the case of cause and effect, the connexion was really synthetic. Before Hume, rationalists at least had supposed that the effect could be logically deduced from the cause, if only we had sufficient knowledge. Hume argued—correctly, as would now be generally admitted—that this could not be done. Hence he inferred the far more doubtful proposition that nothing could be known *a priori* about the connexion of cause and effect. Kant, who had been educated in the rationalist tradition, was much perturbed by Hume's scepticism, and endeavoured to find an answer to it. He perceived that not only the connexion of cause and effect, but all the propositions of arithmetic and geometry, are 'synthetic', i.e. not analytic: in all these propositions, no analysis of the subject will reveal the predicate. His stock instance was the proposition 7 + 5 = 12. He pointed out, quite truly, that 7 and 5 have to be put together to give 12: the idea of 12 is not contained in them, nor even in the idea of adding them together. Thus he was led to the conclusion that all pure mathematics, though *a priori*, is synthetic; and this conclusion raised a new problem of which he endeavoured to find the solution.

The question which Kant put at the beginning of his philosophy, namely 'How is pure mathematics possible?' is an interesting and difficult one, to which every philosophy which is not purely sceptical must find some answer. The answer of the pure empiricists, that our mathematical knowledge is derived by induction from particular instances, we have already seen to be inadequate, for two reasons: first, that the validity of the inductive principle itself cannot be proved by induction; secondly, that the general propositions of mathematics, such as 'two and two always make four', can obviously be known with certainty by consideration of a single instance, and gain nothing by enumeration of other cases in which they have been found to be true. Thus our knowledge of the general propositions of mathematics (and the same applies to logic) must be accounted for otherwise than our (merely probable) knowledge of empirical generalizations such as 'all men are mortal'.

The problem arises through the fact that such knowledge is general, whereas all experience is particular. It seems strange that we should apparently be able to know some truths in advance about particular things of which we have as yet no experience; but it cannot easily be doubted that logic and arithmetic will apply to such things. We do not know who will be the inhabitants of London a hundred years hence; but we know that any two of them and any other two of them will make four of them. This apparent power of anticipating facts about things of which we have no experience is certainly surprising. Kant's solution of the problem, though not valid in my opinion, is interesting. It is, however, very difficult, and is differently understood by different philosophers. We can, therefore, only give the merest outline of it, and even that will be thought misleading by many exponents of Kant's system.

What Kant maintained was that in all our experience there are two elements to be distinguished, the one due to the object (i.e. to what we have called the 'physical object'), the other due to our own nature. We saw, in discussing matter and sense-data, that the physical object is different from the associated sense-data, and that the sense-data are to be regarded as resulting from an interaction between the physical object and ourselves. So far, we are in agreement with Kant. But what is distinctive of Kant is the way in which he apportions the shares of ourselves and the physical object respectively. He considers that the crude material given in sensation—the colour, hardness, etc.—is due to the object, and that what we supply is the arrangement in space and time, and all the relations between sense-data which result from comparison or from considering one as the cause of the other or in any other way. His chief reason in favour of this view is that we seem to have *a priori* knowledge as to space and time and causality and comparison, but not as to the actual crude material of sensation. We can be sure, he says, that anything we shall ever experience must show the characteristics affirmed of it in our *a priori* knowledge, because these characteristics are due to our own nature, and therefore nothing can ever come into our experience without acquiring these characteristics.

The physical object, which he calls the 'thing in itself', he regards as essentially unknowable; what can be known is the object as we have it in experience, which he calls the 'phenomenon'. The phenomenon, being a joint product of us and the thing in itself, is sure to have those characteristics which are due to us, and is therefore sure to conform to our *a priori* knowledge. Hence this knowledge, though true of all actual and possible experience, must not be supposed to apply outside experience. Thus in spite of the existence of *a priori* knowledge, we cannot know anything about the thing in itself or about what is not an actual or possible object of experience. In this way he tries to reconcile and harmonize the contentions of the rationalists with the arguments of the empiricists.

(1) Kant's 'thing in itself' is identical in *definition* with the physical object, namely, it is the cause of sensations. In the properties deduced from the definition it is not identical, since Kant held (in spite of some inconsistency as regards cause) that we can know that none of the categories are applicable to the 'thing in itself'.

Apart from minor grounds on which Kant's philosophy may be criticized, there is one main objection which seems fatal to any attempt to deal with the problem of *a priori* knowledge by his method. The thing to be accounted for is our certainty that the facts must always conform to logic and arithmetic. To say that logic and arithmetic are contributed by us does not account for this. Our nature is as much a fact of the existing world as anything, and there can be no certainty that it will remain constant. It might happen, if Kant is right, that to-morrow our nature would so change as to make two and two become five. This possibility seems never to have occurred to him, yet it is one which utterly destroys the certainty and universality which he is anxious to vindicate for arithmetical propositions. It is true that this possibility, formally, is inconsistent with the Kantian view that time itself is a form imposed by the subject upon phenomena, so that our real Self is not in time and has no to-morrow. But he will still have to suppose that the time-order of phenomena is determined by characteristics of what is behind phenomena, and this suffices for the substance of our argument.

Reflection, moreover, seems to make it clear that, if there is any truth in our arithmetical beliefs, they must apply to things equally whether we think of them or not. Two physical objects and two other physical objects must make four physical objects, even if physical objects cannot be experienced. To assert this is certainly within the scope of what we mean when we state that two and two are four. Its truth is just as indubitable as the truth of the assertion that two phenomena and two other phenomena make four phenomena. Thus Kant's solution unduly limits the scope of *a priori* propositions, in addition to failing in the attempt at explaining their certainty.

Apart from the special doctrines advocated by Kant, it is very common among philosophers to regard what is *a priori* as in some sense mental, as concerned rather with the way we must think than with any fact of the outer world. We noted in the preceding chapter the three principles commonly called 'laws of thought'. The view which led to their being so named is a natural one, but there are strong reasons for thinking that it is erroneous. Let us take as an illustration the law of contradiction. This is commonly stated in the form 'Nothing can both be and not be', which is intended to express the fact that nothing can at once have and not have a given quality. Thus, for example, if a tree is a beech it cannot also be not a beech; if my table is rectangular it cannot also be not rectangular, and so on.

Now what makes it natural to call this principle a law of *thought* is that it is by thought rather than by outward observation that we persuade ourselves of its necessary truth. When we have seen that a tree is a beech, we do not need to look again in order to ascertain whether it is also not a beech; thought alone makes us know that this is impossible. But the conclusion that the law of contradiction is a law of *thought* is nevertheless erroneous. What we believe, when we believe the law of contradiction, is not that the mind is so made that it must believe the law of contradiction. *This* belief is a subsequent result of psychological reflection, which presupposes the belief in the law of contradiction. The belief in the law of contradiction is a belief about things, not only about thoughts. It is not, e.g., the belief that if we *think* a certain tree is a beech, we cannot at the same time *think* that it is not a beech; it is the belief that if the tree *is* a beech, it cannot at the same time *be* not a beech. Thus the law of contradiction is about things, and not merely about thoughts; and although belief in the law of contradiction is a thought, the law of contradiction itself is not a thought, but a fact concerning the things in the world. If this, which we believe when we believe the law of contradiction, were not true of the things in the world, the fact that we were compelled to *think* it true would not save the law of contradiction from being false; and this shows that the law is not a law of *thought*.

A similar argument applies to any other *a priori* judgement. When we judge that two and two are four, we are not making a judgement about our thoughts, but about all actual or possible couples. The fact that our minds are so constituted as to believe that two and two are four, though it is true, is emphatically not what we assert when we assert that two and two are four. And no fact about the constitution of our minds could make it *true* that two and two are four. Thus our *a priori* knowledge, if it is not erroneous, is not merely knowledge about the constitution of our minds, but is applicable to whatever the world may contain, both what is mental and what is non-mental.

The fact seems to be that all our *a priori* knowledge is concerned with entities which do not, properly speaking, *exist*, either in the mental or in the physical world. These entities are such as can be named by parts of speech which are not substantives; they are such entities as qualities and relations. Suppose, for instance, that I am in my room. I exist, and my room exists; but does 'in' exist? Yet obviously the word 'in' has a meaning; it denotes a relation which holds between me and my room. This relation is something, although we cannot say that it exists *in the same sense* in which I and my room exist. The relation 'in' is something which we can think about and understand, for, if we could not understand it, we could not understand the sentence 'I am in my room'. Many philosophers, following Kant, have maintained that relations are the work of the mind, that things in themselves have no relations, but that the mind brings them together in one act of thought and thus produces the relations which it judges them to have.

This view, however, seems open to objections similar to those which we urged before against Kant. It seems plain that it is not thought which produces the truth of the proposition 'I am in my room'. It may be true that an earwig is in my room, even if neither I nor the earwig nor any one else is aware of this truth; for this truth concerns only the earwig and the room, and does not depend upon anything else. Thus relations, as we shall see more fully in the next chapter, must be placed in a world which is neither mental nor physical. This world is of great importance to philosophy, and in particular to the problems of *a priori* knowledge. In the next chapter we shall proceed to develop its nature and its bearing upon the questions with which we have been dealing.

CHAPTER IX.

THE WORLD OF UNIVERSALS

At the end of the preceding chapter, we saw that such entities as relations appear to have a being which is in some way different from that of physical objects, and also different from that of minds and from that of sense-data. In the present chapter we have to consider what is the nature of this kind of being, and also what objects there are that have this kind of being. We will begin with the latter question.

The problem with which we are now concerned is a very old one, since it was brought into philosophy by Plato. Plato's 'theory of ideas' is an attempt to solve this very problem, and in my opinion, it is one of the most successful attempts hitherto made. The theory to be advocated in what follows is largely Plato's, with merely such modifications as time has shown to be necessary.

The way the problem arose for Plato was more or less as follows. Let us consider, say, such a notion as *justice*. If we ask ourselves what justice is, it is natural to proceed by considering this, that, and the other just act, with a view to discovering what they have in common. They must all, in some sense, partake of a common nature, which will be found in whatever is just and in nothing else. This common nature, in virtue of which they are all just, will be justice itself, the pure essence the admixture of which with facts of ordinary life produces the multiplicity of just acts. Similarly with any other word which may be applicable to common facts, such as 'whiteness' for example. The word will be applicable to a number of particular things because they all participate in a common nature or essence. This pure essence is what Plato calls an 'idea' or 'form'. (It must not be supposed that 'ideas', in his sense, exist in minds, though they may be apprehended by minds.) The 'idea' *justice* is not identical with anything that is just: it is something other than particular things, which particular things partake of. Not being particular, it cannot itself exist in the world of sense. Moreover it is not fleeting or changeable like the things of sense: it is eternally itself, immutable and indestructible.

Thus Plato is led to a supra-sensible world, more real than the common world of sense, the unchangeable world of ideas, which alone gives to the world of sense whatever pale reflection of reality may belong to it. The truly real world, for Plato, is the world of ideas; for whatever we may attempt to say about things in the world of sense, we can only succeed in saying that they participate in such and such ideas, which, therefore, constitute all their character. Hence it is easy to pass on into a mysticism. We may hope, in a mystic illumination, to see the ideas as we see objects of sense; and we may imagine that the ideas exist in heaven.

These mystical developments are very natural, but the basis of the theory is in logic, and it is as based in logic that we have to consider it.

The word 'idea' has acquired, in the course of time, many associations which are quite misleading when applied to Plato's 'ideas'. We shall therefore use the word 'universal' instead of the word 'idea', to describe what Plato meant. The essence of the sort of entity that Plato meant is that it is opposed to the particular things that are given in sensation. We speak of whatever is given in sensation, or is of the same nature as things given in sensation, as a *particular*; by opposition to this, a *universal* will be anything which may be shared by many particulars, and has those characteristics which, as we saw, distinguish justice and whiteness from just acts and white things.

When we examine common words, we find that, broadly speaking, proper names stand for particulars, while other substantives, adjectives, prepositions, and verbs stand for universals. Pronouns stand for particulars, but are ambiguous: it is only by the context or the circumstances that we know what particulars they stand for. The word 'now' stands for a particular, namely the present moment; but like pronouns, it stands for an ambiguous particular, because the present is always changing.

It will be seen that no sentence can be made up without at least one word which denotes a universal. The nearest approach would be some such statement as 'I like this'. But even here the word 'like' denotes a universal, for I may like other things, and other people may like things. Thus all truths involve universals, and all knowledge of truths involves acquaintance with universals.

Seeing that nearly all the words to be found in the dictionary stand for universals, it is strange that hardly anybody except students of philosophy ever realizes that there are such entities as universals. We do not naturally dwell upon those words in a sentence which do not stand for particulars; and if we are forced to dwell upon a word which stands for a universal, we naturally think of it as standing for some one of the particulars that come under the universal. When, for example, we hear the sentence, 'Charles I's head was cut off', we may naturally enough think of Charles I, of Charles I's head, and of the operation of cutting off *his* head, which are all particulars; but we do not naturally dwell upon what is meant by the word 'head' or the word 'cut', which is a universal: We feel such words to be incomplete and insubstantial; they seem to demand a context before anything can be done with them. Hence we succeed in avoiding all notice of universals as such, until the study of philosophy forces them upon our attention.

Even among philosophers, we may say, broadly, that only those universals which are named by adjectives or substantives have been much or often recognized, while those named by verbs and prepositions have been usually overlooked. This omission has had a very great effect upon philosophy; it is hardly too much to say that most metaphysics, since Spinoza, has been largely determined by it. The way this has occurred is, in outline, as follows: Speaking generally, adjectives and common nouns express qualities or properties of single things, whereas prepositions and verbs tend to express relations between two or more things. Thus the neglect of prepositions and verbs led to the belief that every proposition can be regarded as attributing a property to a single thing, rather than as expressing a relation between two or more things. Hence it was supposed that, ultimately, there can be no such entities as relations between things. Hence either there can be only one thing in the universe, or, if there are many things, they cannot possibly interact in any way, since any interaction would be a relation, and relations are impossible.

The first of these views, advocated by Spinoza and held in our own day by Bradley and many other philosophers, is called *monism*; the second, advocated by Leibniz but not very common nowadays, is called *monadism*, because each of the isolated things is called a *monad*. Both these opposing philosophies, interesting as they are, result, in my opinion, from an undue attention to one sort of universals, namely the sort represented by adjectives and substantives rather than by verbs and prepositions.

As a matter of fact, if any one were anxious to deny altogether that there are such things as universals, we should find that we cannot strictly prove that there are such entities as *qualities*, i.e. the universals represented by adjectives and substantives, whereas we can prove that there must be *relations*, i.e. the sort of universals generally represented by verbs and prepositions. Let us take in illustration the universal *whiteness*. If we believe that there is such a universal, we shall say that things are white because they have the quality of whiteness. This view, however, was strenuously denied by Berkeley and Hume, who have been followed in this by later empiricists. The form which their denial took was to deny that there are such things as 'abstract ideas '. When we want to think of whiteness, they said, we form an image of some particular white thing, and reason concerning this particular, taking care not to deduce anything concerning it which we cannot see to be equally true of any other white thing. As an account of our actual mental processes, this is no doubt largely true. In geometry, for example, when we wish to prove something about all triangles, we draw a particular triangle and reason about it, taking care not to use any characteristic which it does not share with other triangles. The beginner, in order to avoid error, often finds it useful to draw several triangles, as unlike each other as possible, in order to make sure that his reasoning is equally applicable to all of them. But a difficulty emerges as soon as we ask ourselves how we know that a thing is white or a triangle. If we wish to avoid the universals *whiteness* and *triangularity*, we shall choose some particular patch of white or some particular triangle, and say that anything is white or a triangle if it has the right sort of resemblance to our chosen particular. But then the resemblance required will have to be a universal. Since there are many white things, the resemblance must hold between many pairs of particular white things; and this is the characteristic of a universal. It will be useless to say that there is a different resemblance for each pair, for then we shall have to say that these resemblances resemble each other, and thus at last we shall be forced to admit resemblance as a universal. The relation of resemblance, therefore, must be a true universal. And having been forced to admit this universal, we find that it is no longer worth while to invent difficult and unplausible theories to avoid the admission of such universals as whiteness and triangularity.

Berkeley and Hume failed to perceive this refutation of their rejection of 'abstract ideas', because, like their adversaries, they only thought of *qualities*, and altogether ignored *relations* as universals. We have therefore here another respect in which the rationalists appear to have been in the right as against the empiricists, although, owing to the neglect or denial of relations, the deductions made by rationalists were, if anything, more apt to be mistaken than those made by empiricists.

Having now seen that there must be such entities as universals, the next point to be proved is that their being is not merely mental. By this is meant that whatever being belongs to them is independent of their being thought of or in any way apprehended by minds. We have already touched on this subject at the end of the preceding chapter, but we must now consider more fully what sort of being it is that belongs to universals.

Consider such a proposition as 'Edinburgh is north of London'. Here we have a relation between two places, and it seems plain that the relation subsists independently of our knowledge of it. When we come to know that Edinburgh is north of London, we come to know something which has to do only with Edinburgh and London: we do not cause the truth of the proposition by coming to know it, on the contrary we merely apprehend a fact which was there before we knew it. The part of the earth's surface where Edinburgh stands would be north of the part where London stands, even if there were no human being to know about north and south, and even if there were no minds at all in the universe. This is, of course, denied by many philosophers, either for Berkeley's reasons or for Kant's. But we have already considered these reasons, and decided that they are inadequate. We may therefore now assume it to be true that nothing mental is presupposed in the fact that Edinburgh is north of London. But this fact involves the relation 'north of', which is a universal; and it would be impossible for the whole fact to involve nothing mental if the relation 'north of', which is a constituent part of the fact, did involve anything mental. Hence we must admit that the relation, like the terms it relates, is not dependent upon thought, but belongs to the independent world which thought apprehends but does not create.

This conclusion, however, is met by the difficulty that the relation 'north of' does not seem to *exist* in the same sense in which Edinburgh and London exist. If we ask 'Where and when does this relation exist?' the answer must be 'Nowhere and nowhen'. There is no place or time where we can find the relation 'north of'. It does not exist in Edinburgh any more than in London, for it relates the two and is neutral as between them. Nor can we say that it exists at any particular time. Now everything that can be apprehended by the senses or by introspection exists at some particular time. Hence the relation 'north of' is radically different from such things. It is neither in space nor in time, neither material nor mental; yet it is something.

It is largely the very peculiar kind of being that belongs to universals which has led many people to suppose that they are really mental. We can think *of* a universal, and our thinking then exists in a perfectly ordinary sense, like any other mental act. Suppose, for example, that we are thinking of whiteness. Then *in one sense* it may be said that whiteness is 'in our mind'. We have here the same ambiguity as we noted in discussing Berkeley in Chapter IV. In the strict sense, it is not whiteness that is in our mind, but the act of thinking of whiteness. The connected ambiguity in the word 'idea', which we noted at the same time, also causes confusion here. In one sense of this word, namely the sense in which it denotes the *object* of an act of thought, whiteness is an 'idea'. Hence, if the ambiguity is not guarded against, we may come to think that whiteness is an 'idea' in the other sense, i.e. an act of thought; and thus we come to think that whiteness is mental. But in so thinking, we rob it of its essential quality of universality. One man's act of thought is necessarily a different thing from another man's; one man's act of thought at one time is necessarily a different thing from the same man's act of thought at another time. Hence, if whiteness were the thought as opposed to its object, no two different men could think of it, and no one man could think of it twice. That which many different thoughts of whiteness have in common is their *object*, and this object is different from all of them. Thus universals are not thoughts, though when known they are the objects of thoughts.

We shall find it convenient only to speak of things *existing* when they are in time, that is to say, when we can point to some time at which they exist (not excluding the possibility of their existing at all times). Thus thoughts and feelings, minds and physical objects exist. But universals do not exist in this sense; we shall say that they *subsist* or *have being*, where 'being' is opposed to 'existence' as being timeless. The world of universals, therefore, may also be described as the world of being. The world of being is unchangeable, rigid, exact, delightful to the mathematician, the logician, the builder of metaphysical systems, and all who love perfection more than life. The world of existence is fleeting, vague, without sharp boundaries, without any clear plan or arrangement, but it contains all thoughts and feelings, all the data of sense, and all physical objects, everything that can do either good or harm, everything that makes any difference to the value of life and the world. According to our temperaments, we shall prefer the contemplation of the one or of the other. The one we do not prefer will probably seem to us a pale shadow of the one we prefer, and hardly worthy to be regarded as in any sense real. But the truth is that both have the same claim on our impartial attention, both are real, and both are important to the metaphysician. Indeed no sooner have we distinguished the two worlds than it becomes necessary to consider their relations.

But first of all we must examine our knowledge of universals. This consideration will occupy us in the following chapter, where we shall find that it solves the problem of *a priori* knowledge, from which we were first led to consider universals.

CHAPTER X.

ON OUR KNOWLEDGE OF UNIVERSALS

In regard to one man's knowledge at a given time, universals, like particulars, may be divided into those known by acquaintance, those known only by description, and those not known either by acquaintance or by description.

Let us consider first the knowledge of universals by acquaintance. It is obvious, to begin with, that we are acquainted with such universals as white, red, black, sweet, sour, loud, hard, etc., i.e. with qualities which are exemplified in sense-data. When we see a white patch, we are acquainted, in the first instance, with the particular patch; but by seeing many white patches, we easily learn to abstract the whiteness which they all have in common, and in learning to do this we are learning to be acquainted with whiteness. A similar process will make us acquainted with any other universal of the same sort. Universals of this sort may be called 'sensible qualities'. They can be apprehended with less effort of abstraction than any others, and they seem less removed from particulars than other universals are.

We come next to relations. The easiest relations to apprehend are those which hold between the different parts of a single complex sense-datum. For example, I can see at a glance the whole of the page on which I am writing; thus the whole page is included in one sense-datum. But I perceive that some parts of the page are to the left of other parts, and some parts are above other parts. The process of abstraction in this case seems to proceed somewhat as follows: I see successively a number of sense-data in which one part is to the left of another; I perceive, as in the case of different white patches, that all these sense-data have something in common, and by abstraction I find that what they have in common is a certain relation between their parts, namely the relation which I call 'being to the left of'. In this way I become acquainted with the universal relation.

In like manner I become aware of the relation of before and after in time. Suppose I hear a chime of bells: when the last bell of the chime sounds, I can retain the whole chime before my mind, and I can perceive that the earlier bells came before the later ones. Also in memory I perceive that what I am remembering came before the present time. From either of these sources I can abstract the universal relation of before and after, just as I abstracted the universal relation 'being to the left of'. Thus time-relations, like space-relations, are among those with which we are acquainted.

Another relation with which we become acquainted in much the same way is resemblance. If I see simultaneously two shades of green, I can see that they resemble each other; if I also see a shade of red: at the same time, I can see that the two greens have more resemblance to each other than either has to the red. In this way I become acquainted with the universal *resemblance* or *similarity*.

Between universals, as between particulars, there are relations of which we may be immediately aware. We have just seen that we can perceive that the resemblance between two shades of green is greater than the resemblance between a shade of red and a shade of green. Here we are dealing with a relation, namely 'greater than', between two relations. Our knowledge of such relations, though it requires more power of abstraction than is required for perceiving the qualities of sense-data, appears to be equally immediate, and (at least in some cases) equally indubitable. Thus there is immediate knowledge concerning universals as well as concerning sense-data.

Returning now to the problem of *a priori* knowledge, which we left unsolved when we began the consideration of universals, we find ourselves in a position to deal with it in a much more satisfactory manner than was possible before. Let us revert to the proposition 'two and two are four'. It is fairly obvious, in view of what has been said, that this proposition states a relation between the universal 'two' and the universal 'four'. This suggests a proposition which we shall now endeavour to establish: namely, *All* a priori *knowledge deals exclusively with the relations of universals*. This proposition is of great importance, and goes a long way towards solving our previous difficulties concerning *a priori* knowledge.

The only case in which it might seem, at first sight, as if our proposition were untrue, is the case in which an *a priori* proposition states that *all* of one class of particulars belong to some other class, or (what comes to the same thing) that *all* particulars having some one property also have some other. In this case it might seem as though we were dealing with the particulars that have the property rather than with the property. The proposition 'two and two are four' is really a case in point, for this may be stated in the form 'any two and any other two are four', or 'any collection formed of two twos is a collection of four'. If we can show that such statements as this really deal only with universals, our proposition may be regarded as proved.

One way of discovering what a proposition deals with is to ask ourselves what words we must understand—in other words, what objects we must be acquainted with—in order to see what the proposition means. As soon as we see what the proposition means, even if we do not yet know whether it is true or false, it is evident that we must have acquaintance with whatever is really dealt with by the proposition. By applying this test, it appears that many propositions which might seem to be concerned with particulars are really concerned only with universals. In the special case of 'two and two are four', even when we interpret it as meaning 'any collection formed of two twos is a collection of four', it is plain that we can understand the proposition, i.e. we can see what it is that it asserts, as soon as we know what is meant by 'collection' and 'two' and 'four'. It is quite unnecessary to know all the couples in the world: if it were necessary, obviously we could never understand the proposition, since the couples are infinitely numerous and therefore cannot all be known to us. Thus although our general statement *implies* statements about particular couples, *as soon as we know that there are such particular couples*, yet it does not itself assert or imply that there are such particular couples, and thus fails to make any statement whatever about any actual particular couple. The statement made is about 'couple', the universal, and not about this or that couple.

Thus the statement 'two and two are four' deals exclusively with universals, and therefore may be known by anybody who is acquainted with the universals concerned and can perceive the relation between them which the statement asserts. It must be taken as a fact, discovered by reflecting upon our knowledge, that we have the power of sometimes perceiving such relations between universals, and therefore of sometimes knowing general *a priori* propositions such as those of arithmetic and logic. The thing that seemed mysterious, when we formerly considered such knowledge, was that it seemed to anticipate and control experience. This, however, we can now see to have been an error. *No* fact concerning anything capable of being experienced can be known independently of experience. We know *a priori* that two things and two other things together make four things, but we do *not* know *a priori* that if Brown and Jones are two, and Robinson and Smith are two, then Brown and Jones and Robinson and Smith are four. The reason is that this proposition cannot be understood at all unless we know that there are such people as Brown and Jones and Robinson and Smith, and this we can only know by experience. Hence, although our general proposition is *a priori*, all its applications to actual particulars involve experience and therefore contain an empirical element. In this way what seemed mysterious in our *a priori* knowledge is seen to have been based upon an error.

It will serve to make the point clearer if we contrast our genuine *a priori* judgement with an empirical generalization, such as 'all men are mortals'. Here as before, we can *understand* what the proposition means as soon as we understand the universals involved, namely *man* and *mortal*. It is obviously unnecessary to have an individual acquaintance with the whole human race in order to understand what our proposition means. Thus the difference between an *a priori* general proposition and an empirical generalization does not come in the *meaning* of the proposition; it comes in the nature of the *evidence* for it. In the empirical case, the evidence consists in the particular instances. We believe that all men are mortal because we know that there are innumerable instances of men dying, and no instances of their living beyond a certain age. We do not believe it because we see a connexion between the universal *man* and the universal *mortal*. It is true that if physiology can prove, assuming the general laws that govern living bodies, that no living organism can last for ever, that gives a connexion between *man* and *mortality* which would enable us to assert our proposition without appealing to the special evidence of *men* dying. But that only means that our generalization has been subsumed under a wider generalization, for which the evidence is still of the same kind, though more extensive. The progress of science is constantly producing such subsumptions, and therefore giving a constantly wider inductive basis for scientific generalizations. But although this gives a greater *degree* of certainty, it does not give a different *kind*: the ultimate ground remains inductive, i.e. derived from instances, and not an *a priori* connexion of universals such as we have in logic and arithmetic.

Two opposite points are to be observed concerning *a priori* general propositions. The first is that, if many particular instances are known, our general proposition may be arrived at in the first instance by induction, and the connexion of universals may be only subsequently perceived. For example, it is known that if we draw perpendiculars to the sides of a triangle from the opposite angles, all three perpendiculars meet in a point. It would be quite possible to be first led to this proposition by actually drawing perpendiculars in many cases, and finding that they always met in a point; this experience might lead us to look for the general proof and find it. Such cases are common in the experience of every mathematician.

The other point is more interesting, and of more philosophical importance. It is, that we may sometimes know a general proposition in cases where we do not know a single instance of it. Take such a case as the following: We know that any two numbers can be multiplied together, and will give a third called their *product*. We know that all pairs of integers the product of which is less than 100 have been actually multiplied together, and the value of the product recorded in the multiplication table. But we also know that the number of integers is infinite, and that only a finite number of pairs of integers ever have been or ever will be thought of by human beings. Hence it follows that there are pairs of integers which never have been and never will be thought of by human beings, and that all of them deal with integers the product of which is over 100. Hence we arrive at the proposition: 'All products of two integers, which never have been and never will be thought of by any human being, are over 100.' Here is a general proposition of which the truth is undeniable, and yet, from the very nature of the case, we can never give an instance; because any two numbers we may think of are excluded by the terms of the proposition.

This possibility, of knowledge of general propositions of which no instance can be given, is often denied, because it is not perceived that the knowledge of such propositions only requires a knowledge of the relations of universals, and does not require any knowledge of instances of the universals in question. Yet the knowledge of such general propositions is quite vital to a great deal of what is generally admitted to be known. For example, we saw, in our early chapters, that knowledge of physical objects, as opposed to sense-data, is only obtained by an inference, and that they are not things with which we are acquainted. Hence we can never know any proposition of the form 'this is a physical object', where 'this' is something immediately known. It follows that all our knowledge concerning physical objects is such that no actual instance can be given. We can give instances of the associated sense-data, but we cannot give instances of the actual physical objects. Hence our knowledge as to physical objects depends throughout upon this possibility of general knowledge where no instance can be given. And the same applies to our knowledge of other people's minds, or of any other class of things of which no instance is known to us by acquaintance.

We may now take a survey of the sources of our knowledge, as they have appeared in the course of our analysis. We have first to distinguish knowledge of things and knowledge of truths. In each there are two kinds, one immediate and one derivative. Our immediate knowledge of things, which we called *acquaintance*, consists of two sorts, according as the things known are particulars or universals. Among particulars, we have acquaintance with sense-data and (probably) with ourselves. Among universals, there seems to be no principle by which we can decide which can be known by acquaintance, but it is clear that among those that can be so known are sensible qualities, relations of space and time, similarity, and certain abstract logical universals. Our derivative knowledge of things, which we call knowledge by *description*, always involves both acquaintance with something and knowledge of truths. Our immediate knowledge of *truths* may be called *intuitive* knowledge, and the truths so known may be called *self-evident* truths. Among such truths are included those which merely state what is given in sense, and also certain abstract logical and arithmetical principles, and (though with less certainty) some ethical propositions. Our *derivative* knowledge of truths consists of everything that we can deduce from self-evident truths by the use of self-evident principles of deduction.

If the above account is correct, all our knowledge of truths depends upon our intuitive knowledge. It therefore becomes important to consider the nature and scope of intuitive knowledge, in much the same way as, at an earlier stage, we considered the nature and scope of knowledge by acquaintance. But knowledge of truths raises a further problem, which does not arise in regard to knowledge of things, namely the problem of *error*. Some of our beliefs turn out to be erroneous, and therefore it becomes necessary to consider how, if at all, we can distinguish knowledge from error. This problem does not arise with regard to knowledge by acquaintance, for, whatever may be the object of acquaintance, even in dreams and hallucinations, there is no error involved so long as we do not go beyond the immediate object: error can only arise when we regard the immediate object, i.e. the sense-datum, as the mark of some physical object. Thus the problems connected with knowledge of truths are more difficult than those connected with knowledge of things. As the first of the problems connected with knowledge of truths, let us examine the nature and scope of our intuitive judgements.

CHAPTER XI.

ON INTUITIVE KNOWLEDGE

There is a common impression that everything that we believe ought to be capable of proof, or at least of being shown to be highly probable. It is felt by many that a belief for which no reason can be given is an unreasonable belief. In the main, this view is just. Almost all our common beliefs are either inferred, or capable of being inferred, from other beliefs which may be regarded as giving the reason for them. As a rule, the reason has been forgotten, or has even never been consciously present to our minds. Few of us ever ask ourselves, for example, what reason there is to suppose the food we are just going to eat will not turn out to be poison. Yet we feel, when challenged, that a perfectly good reason could be found, even if we are not ready with it at the moment. And in this belief we are usually justified.

But let us imagine some insistent Socrates, who, whatever reason we give him, continues to demand a reason for the reason. We must sooner or later, and probably before very long, be driven to a point where we cannot find any further reason, and where it becomes almost certain that no further reason is even theoretically discoverable. Starting with the common beliefs of daily life, we can be driven back from point to point, until we come to some general principle, or some instance of a general principle, which seems luminously evident, and is not itself capable of being deduced from anything more evident. In most questions of daily life, such as whether our food is likely to be nourishing and not poisonous, we shall be driven back to the inductive principle, which we discussed in Chapter VI. But beyond that, there seems to be no further regress. The principle itself is constantly used in our reasoning, sometimes consciously, sometimes unconsciously; but there is no reasoning which, starting from some simpler self-evident principle, leads us to the principle of induction as its conclusion. And the same holds for other logical principles. Their truth is evident to us, and we employ them in constructing demonstrations; but they themselves, or at least some of them, are incapable of demonstration.

Self-evidence, however, is not confined to those among general principles which are incapable of proof. When a certain number of logical principles have been admitted, the rest can be deduced from them; but the propositions deduced are often just as self-evident as those that were assumed without proof. All arithmetic, moreover, can be deduced from the general principles of logic, yet the simple propositions of arithmetic, such as 'two and two are four', are just as self-evident as the principles of logic.

It would seem, also, though this is more disputable, that there are some self-evident ethical principles, such as 'we ought to pursue what is good'.

It should be observed that, in all cases of general principles, particular instances, dealing with familiar things, are more evident than the general principle. For example, the law of contradiction states that nothing can both have a certain property and not have it. This is evident as soon as it is understood, but it is not so evident as that a particular rose which we see cannot be both red and not red. (It is of course possible that parts of the rose may be red and parts not red, or that the rose may be of a shade of pink which we hardly know whether to call red or not; but in the former case it is plain that the rose as a whole is not red, while in the latter case the answer is theoretically definite as soon as we have decided on a precise definition of 'red'.) It is usually through particular instances that we come to be able to see the general principle. Only those who are practised in dealing with abstractions can readily grasp a general principle without the help of instances.

In addition to general principles, the other kind of self-evident truths are those immediately derived from sensation. We will call such truths 'truths of perception', and the judgements expressing them we will call 'judgements of perception'. But here a certain amount of care is required in getting at the precise nature of the truths that are self-evident. The actual sense-data are neither true nor false. A particular patch of colour which I see, for example, simply exists: it is not the sort of thing that is true or false. It is true that there is such a patch, true that it has a certain shape and degree of brightness, true that it is surrounded by certain other colours. But the patch itself, like everything else in the world of sense, is of a radically different kind from the things that are true or false, and therefore cannot properly be said to be *true*. Thus whatever self-evident truths may be obtained from our senses must be different from the sense-data from which they are obtained.

It would seem that there are two kinds of self-evident truths of perception, though perhaps in the last analysis the two kinds may coalesce. First, there is the kind which simply asserts the *existence* of the sense-datum, without in any way analysing it. We see a patch of red, and we judge 'there is such-and-such a patch of red', or more strictly 'there is that'; this is one kind of intuitive judgement of perception. The other kind arises when the object of sense is complex, and we subject it to some degree of analysis. If, for instance, we see a *round* patch of red, we may judge 'that patch of red is round'. This is again a judgement of perception, but it differs from our previous kind. In our present kind we have a single sense-datum which has both colour and shape: the colour is red and the shape is round. Our judgement analyses the datum into colour and shape, and then recombines them by stating that the red colour is round in shape. Another example of this kind of judgement is 'this is to the right of that', where 'this' and 'that' are seen simultaneously. In this kind of judgement the sense-datum contains constituents which have some relation to each other, and the judgement asserts that these constituents have this relation.

Another class of intuitive judgements, analogous to those of sense and yet quite distinct from them, are judgements of *memory*. There is some danger of confusion as to the nature of memory, owing to the fact that memory of an object is apt to be accompanied by an image of the object, and yet the image cannot be what constitutes memory. This is easily seen by merely noticing that the image is in the present, whereas what is remembered is known to be in the past. Moreover, we are certainly able to some extent to compare our image with the object remembered, so that we often know, within somewhat wide limits, how far our image is accurate; but this would be impossible, unless the object, as opposed to the image, were in some way before the mind. Thus the essence of memory is not constituted by the image, but by having immediately before the mind an object which is recognized as past. But for the fact of memory in this sense, we should not know that there ever was a past at all, nor should we be able to understand the word 'past', any more than a man born blind can understand the word 'light'. Thus there must be intuitive judgements of memory, and it is upon them, ultimately, that all our knowledge of the past depends.

The case of memory, however, raises a difficulty, for it is notoriously fallacious, and thus throws doubt on the trustworthiness of intuitive judgements in general. This difficulty is no light one. But let us first narrow its scope as far as possible. Broadly speaking, memory is trustworthy in proportion to the vividness of the experience and to its nearness in time. If the house next door was struck by lightning half a minute ago, my memory of what I saw and heard will be so reliable that it would be preposterous to doubt whether there had been a flash at all. And the same applies to less vivid experiences, so long as they are recent. I am absolutely certain that half a minute ago I was sitting in the same chair in which I am sitting now. Going backward over the day, I find things of which I am quite certain, other things of which I am almost certain, other things of which I can become certain by thought and by calling up attendant circumstances, and some things of which I am by no means certain. I am quite certain that I ate my breakfast this morning, but if I were as indifferent to my breakfast as a philosopher should be, I should be doubtful. As to the conversation at breakfast, I can recall some of it easily, some with an effort, some only with a large element of doubt, and some not at all. Thus there is a continual gradation in the degree of self-evidence of what I remember, and a corresponding gradation in the trustworthiness of my memory.

Thus the first answer to the difficulty of fallacious memory is to say that memory has degrees of self-evidence, and that these correspond to the degrees of its trustworthiness, reaching a limit of perfect self-evidence and perfect trustworthiness in our memory of events which are recent and vivid.

It would seem, however, that there are cases of very firm belief in a memory which is wholly false. It is probable that, in these cases, what is really remembered, in the sense of being immediately before the mind, is something other than what is falsely believed in, though something generally associated with it. George IV is said to have at last believed that he was at the battle of Waterloo, because he had so often said that he was. In this case, what was immediately remembered was his repeated assertion; the belief in what he was asserting (if it existed) would be produced by association with the remembered assertion, and would therefore not be a genuine case of memory. It would seem that cases of fallacious memory can probably all be dealt with in this way, i.e. they can be shown to be not cases of memory in the strict sense at all.

One important point about self-evidence is made clear by the case of memory, and that is, that self-evidence has degrees: it is not a quality which is simply present or absent, but a quality which may be more or less present, in gradations ranging from absolute certainty down to an almost imperceptible faintness. Truths of perception and some of the principles of logic have the very highest degree of self-evidence; truths of immediate memory have an almost equally high degree. The inductive principle has less self-evidence than some of the other principles of logic, such as 'what follows from a true premiss must be true'. Memories have a diminishing self-evidence as they become remoter and fainter; the truths of logic and mathematics have (broadly speaking) less self-evidence as they become more complicated. Judgements of intrinsic ethical or aesthetic value are apt to have some self-evidence, but not much.

Degrees of self-evidence are important in the theory of knowledge, since, if propositions may (as seems likely) have some degree of self-evidence without being true, it will not be necessary to abandon all connexion between self-evidence and truth, but merely to say that, where there is a conflict, the more self-evident proposition is to be retained and the less self-evident rejected.

It seems, however, highly probable that two different notions are combined in 'self-evidence' as above explained; that one of them, which corresponds to the highest degree of self-evidence, is really an infallible guarantee of truth, while the other, which corresponds to all the other degrees, does not give an infallible guarantee, but only a greater or less presumption. This, however, is only a suggestion, which we cannot as yet develop further. After we have dealt with the nature of truth, we shall return to the subject of self-evidence, in connexion with the distinction between knowledge and error.

CHAPTER XII.

TRUTH AND FALSEHOOD

Our knowledge of truths, unlike our knowledge of things, has an opposite, namely *error*. So far as things are concerned, we may know them or not know them, but there is no positive state of mind which can be described as erroneous knowledge of things, so long, at any rate, as we confine ourselves to knowledge by acquaintance. Whatever we are acquainted with must be something; we may draw wrong inferences from our acquaintance, but the acquaintance itself cannot be deceptive. Thus there is no dualism as regards acquaintance. But as regards knowledge of truths, there is a dualism. We may believe what is false as well as what is true. We know that on very many subjects different people hold different and incompatible opinions: hence some beliefs must be erroneous. Since erroneous beliefs are often held just as strongly as true beliefs, it becomes a difficult question how they are to be distinguished from true beliefs. How are we to know, in a given case, that our belief is not erroneous? This is a question of the very greatest difficulty, to which no completely satisfactory answer is possible. There is, however, a preliminary question which is rather less difficult, and that is: What do we *mean* by truth and falsehood? It is this preliminary question which is to be considered in this chapter. In this chapter we are not asking how we can know whether a belief is true or false: we are asking what is meant by the question whether a belief is true or false. It is to be hoped that a clear answer to this question may help us to obtain an answer to the question what beliefs are true, but for the present we ask only 'What is truth?' and 'What is falsehood?' not 'What beliefs are true?' and 'What beliefs are false?' It is very important to keep these different questions entirely separate, since any confusion between them is sure to produce an answer which is not really applicable to either.

There are three points to observe in the attempt to discover the nature of truth, three requisites which any theory must fulfil.

(1) Our theory of truth must be such as to admit of its opposite, falsehood. A good many philosophers have failed adequately to satisfy this condition: they have constructed theories according to which all our thinking ought to have been true, and have then had the greatest difficulty in finding a place for falsehood. In this respect our theory of belief must differ from our theory of acquaintance, since in the case of acquaintance it was not necessary to take account of any opposite.

(2) It seems fairly evident that if there were no beliefs there could be no falsehood, and no truth either, in the sense in which truth is correlative to falsehood. If we imagine a world of mere matter, there would be no room for

falsehood in such a world, and although it would contain what may be called 'facts', it would not contain any truths, in the sense in which truths are things of the same kind as falsehoods. In fact, truth and falsehood are properties of beliefs and statements: hence a world of mere matter, since it would contain no beliefs or statements, would also contain no truth or falsehood.

(3) But, as against what we have just said, it is to be observed that the truth or falsehood of a belief always depends upon something which lies outside the belief itself. If I believe that Charles I died on the scaffold, I believe truly, not because of any intrinsic quality of my belief, which could be discovered by merely examining the belief, but because of an historical event which happened two and a half centuries ago. If I believe that Charles I died in his bed, I believe falsely: no degree of vividness in my belief, or of care in arriving at it, prevents it from being false, again because of what happened long ago, and not because of any intrinsic property of my belief. Hence, although truth and falsehood are properties of beliefs, they are properties dependent upon the relations of the beliefs to other things, not upon any internal quality of the beliefs.

The third of the above requisites leads us to adopt the view—which has on the whole been commonest among philosophers—that truth consists in some form of correspondence between belief and fact. It is, however, by no means an easy matter to discover a form of correspondence to which there are no irrefutable objections. By this partly—and partly by the feeling that, if truth consists in a correspondence of thought with something outside thought, thought can never know when truth has been attained—many philosophers have been led to try to find some definition of truth which shall not consist in relation to something wholly outside belief. The most important attempt at a definition of this sort is the theory that truth consists in *coherence*. It is said that the mark of falsehood is failure to cohere in the body of our beliefs, and that it is the essence of a truth to form part of the completely rounded system which is The Truth.

There is, however, a great difficulty in this view, or rather two great difficulties. The first is that there is no reason to suppose that only *one* coherent body of beliefs is possible. It may be that, with sufficient imagination, a novelist might invent a past for the world that would perfectly fit on to what we know, and yet be quite different from the real past. In more scientific matters, it is certain that there are often two or more hypotheses which account for all the known facts on some subject, and although, in such cases, men of science endeavour to find facts which will rule out all the hypotheses except one, there is no reason why they should always succeed.

In philosophy, again, it seems not uncommon for two rival hypotheses to be both able to account for all the facts. Thus, for example, it is possible that life is one long dream, and that the outer world has only that degree of reality that the objects of dreams have; but although such a view does not seem inconsistent with known facts, there is no reason to prefer it to the common-sense view, according to which other people and things do really exist. Thus coherence as the definition of truth fails because there is no proof that there can be only one coherent system.

The other objection to this definition of truth is that it assumes the meaning of 'coherence' known, whereas, in fact, 'coherence' presupposes the truth of the laws of logic. Two propositions are coherent when both may be true, and are incoherent when one at least must be false. Now in order to know whether two propositions can both be true, we must know such truths as the law of contradiction. For example, the two propositions, 'this tree is a beech' and 'this tree is not a beech', are not coherent, because of the law of contradiction. But if the law of contradiction itself were subjected to the test of coherence, we should find that, if we choose to suppose it false, nothing will any longer be incoherent with anything else. Thus the laws of logic supply the skeleton or framework within which the test of coherence applies, and they themselves cannot be established by this test.

For the above two reasons, coherence cannot be accepted as giving the *meaning* of truth, though it is often a most important *test* of truth after a certain amount of truth has become known.

Hence we are driven back to *correspondence with fact* as constituting the nature of truth. It remains to define precisely what we mean by 'fact', and what is the nature of the correspondence which must subsist between belief and fact, in order that belief may be true.

In accordance with our three requisites, we have to seek a theory of truth which (1) allows truth to have an opposite, namely falsehood, (2) makes truth a property of beliefs, but (3) makes it a property wholly dependent upon the relation of the beliefs to outside things.

The necessity of allowing for falsehood makes it impossible to regard belief as a relation of the mind to a single object, which could be said to be what is believed. If belief were so regarded, we should find that, like acquaintance, it would not admit of the opposition of truth and falsehood, but would have to be always true. This may be made clear by examples. Othello believes falsely that Desdemona loves Cassio. We cannot say that this belief consists in a relation to a single object, 'Desdemona's love for Cassio', for if there were such an object, the belief would be true. There is in fact no such object, and therefore Othello cannot have any relation to such an object. Hence his belief cannot possibly consist in a relation to this object.

It might be said that his belief is a relation to a different object, namely 'that Desdemona loves Cassio'; but it is almost as difficult to suppose that there is such an object as this, when Desdemona does not love Cassio, as it was to suppose that there is 'Desdemona's love for Cassio'. Hence it will be better to seek for a theory of belief which does not make it consist in a relation of the mind to a single object.

It is common to think of relations as though they always held between two terms, but in fact this is not always the case. Some relations demand three terms, some four, and so on. Take, for instance, the relation 'between'. So long as only two terms come in, the relation 'between' is impossible: three terms are the smallest number that render it possible. York is between London and Edinburgh; but if London and Edinburgh were the only places in the world, there could be nothing which was between one place and another. Similarly *jealousy* requires three people: there can be no such relation that does not involve three at least. Such a proposition as 'A wishes B to promote C's marriage with D' involves a relation of four terms; that is to say, A and B and C and D all come in, and the relation involved cannot be expressed otherwise than in a form involving all four. Instances might be multiplied indefinitely, but enough has been said to show that there are relations which require more than two terms before they can occur.

The relation involved in *judging* or *believing* must, if falsehood is to be duly allowed for, be taken to be a relation between several terms, not between two. When Othello believes that Desdemona loves Cassio, he must not have before his mind a single object, 'Desdemona's love for Cassio', or 'that Desdemona loves Cassio ', for that would require that there should be objective falsehoods, which subsist independently of any minds; and this, though not logically refutable, is a theory to be avoided if possible. Thus it is easier to account for falsehood if we take judgement to be a relation in which the mind and the various objects concerned all occur severally; that is to say, Desdemona and loving and Cassio must all be terms in the relation which subsists when Othello believes that Desdemona loves Cassio. This relation, therefore, is a relation of four terms, since Othello also is one of the terms of the relation. When we say that it is a relation of four terms, we do not mean that Othello has a certain relation to Desdemona, and has the same relation to loving and also to Cassio. This may be true of some other relation than believing; but believing, plainly, is not a relation which Othello has to *each* of the three terms concerned, but to *all* of them together: there is only one example of the relation of believing involved, but this one example knits together four terms. Thus the actual occurrence, at the moment when Othello is entertaining his belief, is that the relation called 'believing' is knitting together into one complex whole the four terms Othello, Desdemona, loving, and Cassio. What is called belief or judgement is nothing but this relation of believing or judging, which relates a mind to several things other than itself. An *act* of belief or of judgement is the occurrence between certain terms at some particular time, of the relation of believing or judging.

We are now in a position to understand what it is that distinguishes a true judgement from a false one. For this purpose we will adopt certain definitions. In every act of judgement there is a mind which judges, and there are terms concerning which it judges. We will call the mind the *subject* in the judgement, and the remaining terms the *objects*. Thus, when Othello judges that Desdemona loves Cassio, Othello is the subject, while the objects are Desdemona and loving and Cassio. The subject and the objects together are called the *constituents* of the judgement. It will be observed that the relation of judging has what is called a 'sense' or 'direction'. We may say, metaphorically, that it puts its objects in a certain *order*, which we may indicate by means of the order of the words in the sentence. (In an inflected language, the same thing will be indicated by inflections, e.g. by the difference between nominative and accusative.) Othello's judgement that Cassio loves Desdemona differs from his judgement that Desdemona loves Cassio, in spite of the fact that it consists of the same constituents, because the relation of judging places the constituents in a different order in the two cases. Similarly, if Cassio judges that Desdemona loves Othello, the constituents of the judgement are still the same, but their order is different. This property of having a 'sense' or 'direction' is one which the relation of judging shares with all other relations. The 'sense' of relations is the ultimate source of order and series and a host of mathematical concepts; but we need not concern ourselves further with this aspect.

We spoke of the relation called 'judging' or 'believing' as knitting together into one complex whole the subject and the objects. In this respect, judging is exactly like every other relation. Whenever a relation holds between two or more terms, it unites the terms into a complex whole. If Othello loves Desdemona, there is such a complex whole as 'Othello's love for Desdemona'. The terms united by the relation may be themselves complex, or may be simple, but the whole which results from their being united must be complex. Wherever there is a relation which relates certain terms, there is a complex object formed of the union of those terms; and conversely, wherever there is a complex object, there is a relation which relates its constituents. When an act of believing occurs, there is a complex, in which 'believing' is the uniting relation, and subject and objects are arranged in a certain order by the 'sense' of the relation of believing. Among the objects, as we saw in considering 'Othello believes that Desdemona loves Cassio', one must be a relation—in this instance, the relation 'loving'. But this relation, as it occurs in the act of believing, is not the relation which creates the unity of the complex whole consisting of the subject and the objects. The relation 'loving', as it occurs in the act of believing, is one of the objects—it is a brick in the structure, not the cement. The cement is the relation 'believing'. When the belief is *true*, there is another complex unity, in which the relation which was one of the objects of the belief relates the other objects. Thus, e.g., if Othello believes *truly* that Desdemona loves Cassio, then there is a complex unity, 'Desdemona's love for

Cassio', which is composed exclusively of the *objects* of the belief, in the same order as they had in the belief, with the relation which was one of the objects occurring now as the cement that binds together the other objects of the belief. On the other hand, when a belief is *false*, there is no such complex unity composed only of the objects of the belief. If Othello believes *falsely* that Desdemona loves Cassio, then there is no such complex unity as 'Desdemona's love for Cassio'.

Thus a belief is *true* when it *corresponds* to a certain associated complex, and *false* when it does not. Assuming, for the sake of definiteness, that the objects of the belief are two terms and a relation, the terms being put in a certain order by the 'sense' of the believing, then if the two terms in that order are united by the relation into a complex, the belief is true; if not, it is false. This constitutes the definition of truth and falsehood that we were in search of. Judging or believing is a certain complex unity of which a mind is a constituent; if the remaining constituents, taken in the order which they have in the belief, form a complex unity, then the belief is true; if not, it is false.

Thus although truth and falsehood are properties of beliefs, yet they are in a sense extrinsic properties, for the condition of the truth of a belief is something not involving beliefs, or (in general) any mind at all, but only the *objects* of the belief. A mind, which believes, believes truly when there is a *corresponding* complex not involving the mind, but only its objects. This correspondence ensures truth, and its absence entails falsehood. Hence we account simultaneously for the two facts that beliefs (a) depend on minds for their *existence*, (b) do not depend on minds for their *truth*.

We may restate our theory as follows: If we take such a belief as 'Othello believes that Desdemona loves Cassio', we will call Desdemona and Cassio the *object-terms*, and loving the *object-relation*. If there is a complex unity 'Desdemona's love for Cassio', consisting of the object-terms related by the object-relation in the same order as they have in the belief, then this complex unity is called the *fact corresponding to the belief*. Thus a belief is true when there is a corresponding fact, and is false when there is no corresponding fact.

It will be seen that minds do not *create* truth or falsehood. They create beliefs, but when once the beliefs are created, the mind cannot make them true or false, except in the special case where they concern future things which are within the power of the person believing, such as catching trains. What makes a belief true is a *fact*, and this fact does not (except in exceptional cases) in any way involve the mind of the person who has the belief.

Having now decided what we *mean* by truth and falsehood, we have next to consider what ways there are of knowing whether this or that belief is true or false. This consideration will occupy the next chapter.

CHAPTER XIII.

KNOWLEDGE, ERROR, AND PROBABLE OPINION

The question as to what we mean by truth and falsehood, which we considered in the preceding chapter, is of much less interest than the question as to how we can know what is true and what is false. This question will occupy us in the present chapter. There can be no doubt that *some* of our beliefs are erroneous; thus we are led to inquire what certainty we can ever have that such and such a belief is not erroneous. In other words, can we ever *know* anything at all, or do we merely sometimes by good luck believe what is true? Before we can attack this question, we must, however, first decide what we mean by 'knowing', and this question is not so easy as might be supposed.

At first sight we might imagine that knowledge could be defined as 'true belief'. When what we believe is true, it might be supposed that we had achieved a knowledge of what we believe. But this would not accord with the way in which the word is commonly used. To take a very trivial instance: If a man believes that the late Prime Minister's last name began with a B, he believes what is true, since the late Prime Minister was Sir Henry Campbell Bannerman. But if he believes that Mr. Balfour was the late Prime Minister, he will still believe that the late Prime Minister's last name began with a B, yet this belief, though true, would not be thought to constitute knowledge. If a newspaper, by an intelligent anticipation, announces the result of a battle before any telegram giving the result has been received, it may by good fortune announce what afterwards turns out to be the right result, and it may produce belief in some of its less experienced readers. But in spite of the truth of their belief, they cannot be said to have knowledge. Thus it is clear that a true belief is not knowledge when it is deduced from a false belief.

In like manner, a true belief cannot be called knowledge when it is deduced by a fallacious process of reasoning, even if the premisses from which it is deduced are true. If I know that all Greeks are men and that Socrates was a man, and I infer that Socrates was a Greek, I cannot be said to *know* that Socrates was a Greek, because, although my premisses and my conclusion are true, the conclusion does not follow from the premisses.

But are we to say that nothing is knowledge except what is validly deduced from true premisses? Obviously we cannot say this. Such a definition is at once too wide and too narrow. In the first place, it is too wide, because it is not enough that our premisses should be *true*, they must also be *known*. The man who believes that Mr. Balfour was the late Prime Minister may proceed to draw valid deductions from the true premiss that the late Prime Minister's name began with a B, but he cannot be said to *know* the conclusions reached by these deductions. Thus we shall have to amend our definition by saying that knowledge is what is validly deduced from *known* premisses. This, however, is a circular definition: it assumes that we already know what is meant by 'known premisses'. It can, therefore, at best define one sort of knowledge, the sort we call derivative, as opposed to intuitive knowledge. We may say: '*Derivative* knowledge is what is validly deduced from premisses known intuitively'. In this statement there is no formal defect, but it leaves the definition of *intuitive* knowledge still to seek.

Leaving on one side, for the moment, the question of intuitive knowledge, let us consider the above suggested definition of derivative knowledge. The chief objection to it is that it unduly limits knowledge. It constantly happens that people entertain a true belief, which has grown up in them because of some piece of intuitive knowledge from which it is capable of being validly inferred, but from which it has not, as a matter of fact, been inferred by any logical process.

Take, for example, the beliefs produced by reading. If the newspapers announce the death of the King, we are fairly well justified in believing that the King is dead, since this is the sort of announcement which would not be made if it were false. And we are quite amply justified in believing that the newspaper asserts that the King is dead. But here the intuitive knowledge upon which our belief is based is knowledge of the existence of sense-data derived from looking at the print which gives the news. This knowledge scarcely rises into consciousness, except in a person who cannot read easily. A child may be aware of the shapes of the letters, and pass gradually and painfully to a realization of their meaning. But anybody accustomed to reading passes at once to what the letters mean, and is not aware, except on reflection, that he has derived this knowledge from the sense-data called seeing the printed letters. Thus although a valid inference from the-letters to their meaning is possible, and *could* be performed by the reader, it is not in fact performed, since he does not in fact perform any operation which can be called logical inference. Yet it would be absurd to say that the reader does not *know* that the newspaper announces the King's death.

We must, therefore, admit as derivative knowledge whatever is the result of intuitive knowledge even if by mere association, provided there *is* a valid logical connexion, and the person in question could become aware of this connexion by reflection. There are in fact many ways, besides logical inference, by which we pass from one belief to another: the passage from the print to its meaning illustrates these ways. These ways may be called 'psychological inference'. We shall, then, admit such psychological inference as a means of obtaining derivative knowledge, provided there is a discoverable logical inference which runs parallel to the psychological inference. This renders our definition of derivative knowledge less precise than we could wish, since the word 'discoverable' is vague: it does not tell us how much reflection may be needed in order to make the discovery. But in fact 'knowledge' is not a precise conception: it merges into 'probable opinion', as we shall see more fully in the course of the present chapter. A very precise definition, therefore, should not be sought, since any such definition must be more or less misleading.

The chief difficulty in regard to knowledge, however, does not arise over derivative knowledge, but over intuitive knowledge. So long as we are dealing with derivative knowledge, we have the test of intuitive knowledge to fall back upon. But in regard to intuitive beliefs, it is by no means easy to discover any criterion by which to distinguish some as true and others as erroneous. In this question it is scarcely possible to reach any very precise result: all our knowledge of truths is infected with some degree of doubt, and a theory which ignored this fact would be plainly wrong. Something may be done, however, to mitigate the difficulties of the question.

Our theory of truth, to begin with, supplies the possibility of distinguishing certain truths as *self-evident* in a sense which ensures infallibility. When a belief is true, we said, there is a corresponding fact, in which the several objects of the belief form a single complex. The belief is said to constitute *knowledge* of this fact, provided it fulfils those further somewhat vague conditions which we have been considering in the present chapter. But in regard to any fact, besides the knowledge constituted by belief, we may also have the kind of knowledge constituted by *perception* (taking this word in its widest possible sense). For example, if you know the hour of the sunset, you can at that hour know the fact that the sun is setting: this is knowledge of the fact by way of knowledge of *truths*; but you can also, if the weather is fine, look to the west and actually see the setting sun: you then know the same fact by the way of knowledge of *things*.

Thus in regard to any complex fact, there are, theoretically, two ways in which it may be known: (1) by means of a judgement, in which its several parts are judged to be related as they are in fact related; (2) by means of *acquaintance* with the complex fact itself, which may (in a large sense) be called perception, though it is by no means confined to objects of the senses. Now it will be observed that the second way of knowing a complex fact, the way of acquaintance, is only possible when there really is such a fact, while the first way, like all judgement, is liable to error. The second way gives us the complex whole, and is therefore only possible when its parts do actually have that relation which makes them combine to form such a complex. The first way, on the contrary, gives us the parts and the relation severally, and demands only the reality of the parts and the relation: the relation may not relate those parts in that way, and yet the judgement may occur.

It will be remembered that at the end of Chapter XI we suggested that there might be two kinds of self-evidence, one giving an absolute guarantee of truth, the other only a partial guarantee. These two kinds can now be distinguished.

We may say that a truth is self-evident, in the first and most absolute sense, when we have acquaintance with the fact which corresponds to the truth. When Othello believes that Desdemona loves Cassio, the corresponding fact, if his belief were true, would be 'Desdemona's love for Cassio'. This would be a fact with which no one could have acquaintance except Desdemona; hence in the sense of self-evidence that we are considering, the truth that Desdemona loves Cassio (if it were a truth) could only be self-evident to Desdemona. All mental facts, and all facts concerning sense-data, have this same privacy: there is only one person to whom they can be self-evident in our present sense, since there is only one person who can be acquainted with the mental things or the sense-data concerned. Thus no fact about any particular existing thing can be self-evident to more than one person. On the other hand, facts about universals do not have this privacy. Many minds may be acquainted with the same universals; hence a relation between universals may be known by acquaintance to many different people. In all cases where we know by acquaintance a complex fact consisting of certain terms in a certain relation, we say that the truth that these terms are so related has the first or absolute kind of self-evidence, and in these cases the judgement that the terms are so related *must* be true. Thus this sort of self-evidence is an absolute guarantee of truth.

But although this sort of self-evidence is an absolute guarantee of truth, it does not enable us to be *absolutely* certain, in the case of any given judgement, that the judgement in question is true. Suppose we first perceive the sun shining, which is a complex fact, and thence proceed to make the judgement 'the sun is shining'. In passing from the perception to the judgement, it is necessary to analyse the given complex fact: we have to separate out 'the sun' and 'shining' as constituents of the fact. In this process it is possible to commit an error; hence even where a *fact* has the first or absolute kind of self-evidence, a judgement believed to correspond to the fact is not absolutely infallible, because it may not really correspond to the fact. But if it does correspond (in the sense explained in the preceding chapter), then it *must* be true.

The second sort of self-evidence will be that which belongs to judgements in the first instance, and is not derived from direct perception of a fact as a single complex whole. This second kind of self-evidence will have degrees, from the very highest degree down to a bare inclination in favour of the belief. Take, for example, the case of a horse trotting away from us along a hard road. At first our certainty that we hear the hoofs is complete; gradually, if we listen intently, there comes a moment when we think perhaps it was imagination or the blind upstairs or our own heartbeats; at last we become doubtful whether there was any noise at all; then we *think* we no longer hear anything, and at last we *know* we no longer hear anything. In this process, there is a continual gradation of self-evidence, from the highest degree to the least, not in the sense-data themselves, but in the judgements based on them.

Or again: Suppose we are comparing two shades of colour, one blue and one green. We can be quite sure they are different shades of colour; but if the green colour is gradually altered to be more and more like the blue, becoming first a blue-green, then a greeny-blue, then blue, there will come a moment when we are doubtful whether we can see any difference, and then a moment when we know that we cannot see any difference. The same thing happens in tuning a musical instrument, or in any other case where there is a continuous gradation. Thus self-evidence of this sort is a matter of degree; and it seems plain that the higher degrees are more to be trusted than the lower degrees.

In derivative knowledge our ultimate premisses must have some degree of self-evidence, and so must their connexion with the conclusions deduced from them. Take for example a piece of reasoning in geometry. It is not enough that the axioms from which we start should be self-evident: it is necessary also that, at each step in the reasoning, the connexion of premiss and conclusion should be self-evident. In difficult reasoning, this connexion has often only a very small degree of self-evidence; hence errors of reasoning are not improbable where the difficulty is great.

From what has been said it is evident that, both as regards intuitive knowledge and as regards derivative knowledge, if we assume that intuitive knowledge is trustworthy in proportion to the degree of its self-evidence, there will be a gradation in trustworthiness, from the existence of noteworthy sense-data and the simpler truths of logic and arithmetic, which may be taken as quite certain, down to judgements which seem only just more probable than their opposites. What we firmly believe, if it is true, is called *knowledge*, provided it is either intuitive or inferred (logically or psychologically) from intuitive knowledge from which it follows logically. What we firmly believe, if it is not true, is called *error*. What we firmly believe, if it is neither knowledge nor error, and also what we believe hesitatingly, because it is, or is derived from, something which has not the highest degree of self-evidence, may be called *probable opinion*. Thus the greater part of what would commonly pass as knowledge is more or less probable opinion.

In regard to probable opinion, we can derive great assistance from *coherence*, which we rejected as the *definition* of truth, but may often use as a *criterion*. A body of individually probable opinions, if they are mutually coherent, become more probable than any one of them would be individually. It is in this way that many scientific hypotheses acquire their probability. They fit into a coherent system of probable opinions, and thus become more probable than they would be in isolation. The same thing applies to general philosophical hypotheses. Often in a single case such hypotheses may seem highly doubtful, while yet, when we consider the order and coherence which they introduce into a mass of probable opinion, they become pretty nearly certain. This applies, in particular, to such matters as the distinction between dreams and waking life. If our dreams, night after night, were as coherent one with another as our days, we should hardly know whether to believe the dreams or the waking life. As it is, the test of coherence condemns the dreams and confirms the waking life. But this test, though it increases probability where it is successful, never gives absolute certainty, unless there is certainty already at some point in the coherent system. Thus the mere organization of probable opinion will never, by itself, transform it into indubitable knowledge.

CHAPTER XIV.

THE LIMITS OF PHILOSOPHICAL KNOWLEDGE

In all that we have said hitherto concerning philosophy, we have scarcely touched on many matters that occupy a great space in the writings of most philosophers. Most philosophers—or, at any rate, very many—profess to be able to prove, by *a priori* metaphysical reasoning, such things as the fundamental dogmas of religion, the essential rationality of the universe, the illusoriness of matter, the unreality of all evil, and so on. There can be no doubt that the hope of finding reason to believe such theses as these has been the chief inspiration of many life-long students of philosophy. This hope, I believe, is vain. It would seem that knowledge concerning the universe as a whole is not to be obtained by metaphysics, and that the proposed proofs that, in virtue of the laws of logic such and such things *must* exist and such and such others cannot, are not capable of surviving a critical scrutiny. In this chapter we shall briefly consider the kind of way in which such reasoning is attempted, with a view to discovering whether we can hope that it may be valid.

The great representative, in modern times, of the kind of view which we wish to examine, was Hegel (1770-1831). Hegel's philosophy is very difficult, and commentators differ as to the true interpretation of it. According to the interpretation I shall adopt, which is that of many, if not most, of the commentators and has the merit of giving an interesting and important type of philosophy, his main thesis is that everything short of the Whole is obviously fragmentary, and obviously incapable of existing without the complement supplied by the rest of the world. Just as a comparative anatomist, from a single bone, sees what kind of animal the whole must have been, so the metaphysician, according to Hegel, sees, from any one piece of reality, what the whole of reality must be—at least in its large outlines. Every apparently separate piece of reality has, as it were, hooks which grapple it to the next piece; the next piece, in turn, has fresh hooks, and so on, until the whole universe is reconstructed. This essential incompleteness appears, according to Hegel, equally in the world of thought and in the world of things. In the world of thought, if we take any idea which is abstract or incomplete, we find, on examination, that if we forget its incompleteness, we become involved in contradictions; these contradictions turn the idea in question into its opposite, or antithesis; and in order to escape, we have to find a new, less incomplete idea, which is the synthesis of our original idea and its antithesis. This new idea, though less incomplete than the idea we started with, will be found, nevertheless, to be still not wholly complete, but to pass into its antithesis, with which it must be combined in a new synthesis. In this way Hegel advances until he reaches the 'Absolute Idea', which, according to him, has no incompleteness, no opposite, and no need of further development. The Absolute Idea, therefore, is adequate to describe Absolute Reality; but all lower ideas only describe reality as it appears to a partial view, not as it is to one who simultaneously surveys the Whole. Thus Hegel reaches the conclusion that Absolute Reality forms one single harmonious system, not in space or time, not in any degree evil, wholly rational, and wholly spiritual. Any appearance to the contrary, in the world we know, can be proved logically—so he believes—to be entirely due to our fragmentary piecemeal view of the universe. If we saw the universe whole, as we may suppose God sees it, space and time and matter and evil and all striving and struggling would disappear, and we should see instead an eternal perfect unchanging spiritual unity.

In this conception, there is undeniably something sublime, something to which we could wish to yield assent. Nevertheless, when the arguments in support of it are carefully examined, they appear to involve much confusion and many unwarrantable assumptions. The fundamental tenet upon which the system is built up is that what is incomplete must be not self-subsistent, but must need the support of other things before it can exist. It is held that whatever has relations to things outside itself must contain some reference to those outside things in its own nature, and could not, therefore, be what it is if those outside things did not exist. A man's nature, for example, is constituted by his memories and the rest of his knowledge, by his loves and hatreds, and so on; thus, but for the objects which he knows or loves or hates, he could not be what he is. He is essentially and obviously a fragment: taken as the sum-total of reality he would be self-contradictory.

This whole point of view, however, turns upon the notion of the 'nature' of a thing, which seems to mean 'all the truths about the thing'. It is of course the case that a truth which connects one thing with another thing could not subsist if the other thing did not subsist. But a truth about a thing is not part of the thing itself, although it must, according to the above usage, be part of the 'nature' of the thing. If we mean by a thing's 'nature' all the truths about the thing, then plainly we cannot know a thing's 'nature' unless we know all the thing's relations to all the other things in the universe. But if the word 'nature' is used in this sense, we shall have to hold that the thing may be known when its 'nature' is not known, or at any rate is not known completely. There is a confusion, when this use of the word 'nature' is employed, between knowledge of things and knowledge of truths. We may have knowledge of a thing by acquaintance even if we know very few propositions about it—theoretically we need not know any propositions about it. Thus, acquaintance with a thing does not involve knowledge of its 'nature' in the above sense. And although acquaintance with a thing is involved in our knowing any one proposition about a thing, knowledge of its 'nature', in the above sense, is not involved. Hence, (1) acquaintance with a thing does not logically involve a knowledge of its relations, and (2) a knowledge of some of its relations does not involve a knowledge of all of its relations nor a knowledge of its 'nature' in the above sense. I may be acquainted, for example, with my toothache, and this knowledge may be as complete as knowledge by acquaintance ever can be, without knowing all that the dentist (who is not acquainted with it) can tell me about its cause, and without therefore knowing its 'nature' in the above sense. Thus the fact that a thing has relations does not prove that its relations are logically necessary. That is to say, from the mere fact that it is the thing it is we cannot deduce that it must have the various relations which in fact it has. This only *seems* to follow because we know it already.

It follows that we cannot prove that the universe as a whole forms a single harmonious system such as Hegel believes that it forms. And if we cannot prove this, we also cannot prove the unreality of space and time and matter and evil, for this is deduced by Hegel from the fragmentary and relational character of these things. Thus we are left to the piecemeal investigation of the world, and are unable to know the characters of those parts of the universe that are remote from our experience. This result, disappointing as it is to those whose hopes have been raised by the systems of philosophers, is in harmony with the inductive and scientific temper of our age, and is borne out by the whole examination of human knowledge which has occupied our previous chapters.

Most of the great ambitious attempts of metaphysicians have proceeded by the attempt to prove that such and such apparent features of the actual world were self-contradictory, and therefore could not be real. The whole tendency of modern thought, however, is more and more in the direction of showing that the supposed contradictions were illusory, and that very little can be proved *a priori* from considerations of what *must* be. A good illustration of this is afforded by space and time. Space and time appear to be infinite in extent, and infinitely divisible. If we travel along a straight line in either direction, it is difficult to believe that we shall finally reach a last point, beyond which there is nothing, not even empty space. Similarly, if in imagination we travel backwards or forwards in time, it is difficult to believe that we shall reach a first or last time, with not even empty time beyond it. Thus space and time appear to be infinite in extent.

Again, if we take any two points on a line, it seems evident that there must be other points between them however small the distance between them may be: every distance can be halved, and the halves can be halved again, and so on *ad infinitum*. In time, similarly, however little time may elapse between two moments, it seems evident that there will be other moments between them. Thus space and time appear to be infinitely divisible. But as against these apparent facts—infinite extent and infinite divisibility—philosophers have advanced arguments tending to show that there could be no infinite collections of things, and that therefore the number of points in space, or of instants in time, must be finite. Thus a contradiction emerged between the apparent nature of space and time and the supposed impossibility of infinite collections.

Kant, who first emphasized this contradiction, deduced the impossibility of space and time, which he declared to be merely subjective; and since his time very many philosophers have believed that space and time are mere appearance, not characteristic of the world as it really is. Now, however, owing to the labours of the mathematicians, notably Georg Cantor, it has appeared that the impossibility of infinite collections was a mistake. They are not in fact self-contradictory, but only contradictory of certain rather obstinate mental prejudices. Hence the reasons for regarding space and time as unreal have become inoperative, and one of the great sources of metaphysical constructions is dried up.

The mathematicians, however, have not been content with showing that space as it is commonly supposed to be is possible; they have shown also that many other forms of space are equally possible, so far as logic can show. Some of Euclid's axioms, which appear to common sense to be necessary, and were formerly supposed to be necessary by philosophers, are now known to derive their appearance of necessity from our mere familiarity with actual space, and not from any *a priori* logical foundation. By imagining worlds in which these axioms are false, the mathematicians have used logic to loosen the prejudices of common sense, and to show the possibility of spaces differing—some more, some less—from that in which we live. And some of these spaces differ so little from Euclidean space, where distances such as we can measure are concerned, that it is impossible to discover by observation whether our actual space is strictly Euclidean or of one of these other kinds. Thus the position is completely reversed. Formerly it appeared that experience left only one kind of space to logic, and logic showed this one kind to be impossible. Now, logic presents many kinds of space as possible apart from experience, and experience only partially decides between them. Thus, while our knowledge of what is has become less than it was formerly supposed to be, our knowledge of what may be is enormously increased. Instead of being shut in within narrow walls, of which every nook and cranny could be explored, we find ourselves in an open world of free possibilities, where much remains unknown because there is so much to know.

What has happened in the case of space and time has happened, to some extent, in other directions as well. The attempt to prescribe to the universe by means of *a priori* principles has broken down; logic, instead of being, as formerly, the bar to possibilities, has become the great liberator of the imagination, presenting innumerable alternatives which are closed to unreflective common sense, and leaving to experience the task of deciding, where decision is possible, between the many worlds which logic offers for our choice. Thus knowledge as to what exists becomes limited to what we can learn from experience—not to what we can actually experience, for, as we have seen, there is much knowledge by description concerning things of which we have no direct experience. But in all cases of knowledge by description, we need some connexion of universals, enabling us, from such and such a datum, to infer an object of a certain sort as implied by our datum. Thus in regard to physical objects, for example, the principle that sense-data are signs of physical objects is itself a connexion of universals; and it is only in virtue of this principle that experience enables us to acquire knowledge concerning physical objects. The same applies to the law of causality, or, to descend to what is less general, to such principles as the law of gravitation.

Principles such as the law of gravitation are proved, or rather are rendered highly probable, by a combination of experience with some wholly *a priori* principle, such as the principle of induction. Thus our intuitive knowledge, which is the source of all our other knowledge of truths, is of two sorts: pure empirical knowledge, which tells us of the existence and some of the properties of particular things with which we are acquainted, and pure *a priori* knowledge, which gives us connexions between universals, and enables us to draw inferences from the particular facts given in empirical knowledge. Our derivative knowledge always depends upon some pure *a priori* knowledge and usually also depends upon some pure empirical knowledge.

Philosophical knowledge, if what has been said above is true, does not differ essentially from scientific knowledge; there is no special source of wisdom which is open to philosophy but not to science, and the results obtained by philosophy are not radically different from those obtained from science. The essential characteristic of philosophy, which makes it a study distinct from science, is criticism. It examines critically the principles employed in science and in daily life; it searches out any inconsistencies there may be in these principles, and it only accepts them when, as the result of a critical inquiry, no reason for rejecting them has appeared. If, as many philosophers have believed, the principles underlying the sciences were capable, when disengaged from irrelevant detail, of giving us knowledge concerning the universe as a whole, such knowledge would have the same claim on our belief as scientific knowledge has; but our inquiry has not revealed any such knowledge, and therefore, as regards the special doctrines of the bolder metaphysicians, has had a mainly negative result. But as regards what would be commonly accepted as knowledge, our result is in the main positive: we have seldom found reason to reject such knowledge as the result of our criticism, and we have seen no reason to suppose man incapable of the kind of knowledge which he is generally believed to possess.

When, however, we speak of philosophy as a *criticism* of knowledge, it is necessary to impose a certain limitation. If we adopt the attitude of the complete sceptic, placing ourselves wholly outside all knowledge, and asking, from this outside position, to be compelled to return within the circle of knowledge, we are demanding what is impossible, and our scepticism can never be refuted. For all refutation must begin with some piece of knowledge which the disputants share; from blank doubt, no argument can begin. Hence the criticism of knowledge which philosophy employs must not be of this destructive kind, if any result is to be achieved. Against this absolute scepticism, no *logical* argument can be advanced. But it is not difficult to see that scepticism of this kind is unreasonable. Descartes' 'methodical doubt', with which modern philosophy began, is not of this kind, but is rather the kind of criticism which we are asserting to be the essence of philosophy. His 'methodical doubt' consisted in doubting whatever seemed doubtful; in pausing, with each apparent piece of knowledge, to ask himself whether, on reflection, he could feel certain that he really knew it. This is the kind of criticism which constitutes philosophy. Some knowledge, such as knowledge of the existence of our sense-data, appears quite indubitable, however calmly and thoroughly we reflect upon it. In regard to such knowledge, philosophical criticism does not require that we should abstain from belief. But there are beliefs—such, for example, as the belief that physical objects exactly resemble our sense-data—which are entertained until we begin to reflect, but are found to melt away when subjected to a close inquiry. Such beliefs philosophy will bid us reject, unless some new line of argument is found to support them. But to reject the beliefs which do not appear open to any objections, however closely we examine them, is not reasonable, and is not what philosophy advocates.

The criticism aimed at, in a word, is not that which, without reason, determines to reject, but that which considers each piece of apparent knowledge on its merits, and retains whatever still appears to be knowledge when this consideration is completed. That some risk of error remains must be admitted, since human beings are fallible. Philosophy may claim justly that it diminishes the risk of error, and that in some cases it renders the risk so small as to be practically negligible. To do more than this is not possible in a world where mistakes must occur; and more than this no prudent advocate of philosophy would claim to have performed.

CHAPTER XV.

THE VALUE OF PHILOSOPHY

Having now come to the end of our brief and very incomplete review of the problems of philosophy, it will be well to consider, in conclusion, what is the value of philosophy and why it ought to be studied. It is the more necessary to consider this question, in view of the fact that many men, under the influence of science or of practical affairs, are inclined to doubt whether philosophy is anything better than innocent but useless trifling, hair-splitting distinctions, and controversies on matters concerning which knowledge is impossible.

This view of philosophy appears to result, partly from a wrong conception of the ends of life, partly from a wrong conception of the kind of goods which philosophy strives to achieve. Physical science, through the medium of inventions, is useful to innumerable people who are wholly ignorant of it; thus the study of physical science is to be recommended, not only, or primarily, because of the effect on the student, but rather because of the effect on mankind in general. Thus, utility does not belong to philosophy. If the study of philosophy has any value at all for others than students of philosophy, it must be only indirectly, through its effects upon the lives of those who study it. It is in these effects, therefore, if anywhere, that the value of philosophy must be primarily sought.

But further, if we are not to fail in our endeavour to determine the value of philosophy, we must first free our minds from the prejudices of what are wrongly called 'practical' men. The 'practical' man, as this word is often used, is one who recognizes only material needs, who realizes that men must have food for the body, but is oblivious of the necessity of providing food for the mind. If all men were well off, if poverty and disease had been reduced to their lowest possible point, there would still remain much to be done to produce a valuable society; and even in the existing world the goods of the mind are at least as important as the goods of the body. It is exclusively among the goods of the mind that the value of philosophy is to be found; and only those who are not indifferent to these goods can be persuaded that the study of philosophy is not a waste of time.

Philosophy, like all other studies, aims primarily at knowledge. The knowledge it aims at is the kind of knowledge which gives unity and system to the body of the sciences, and the kind which results from a critical examination of the grounds of our convictions, prejudices, and beliefs. But it cannot be maintained that philosophy has had any very great measure of success in its attempts to provide definite answers to its questions. If you ask a mathematician, a mineralogist, a historian, or any other man of learning, what definite body of truths has been ascertained by his science, his answer will last as long as you are willing to listen. But if you put the same question to a philosopher, he will, if he is candid, have to confess that his study has not achieved positive results such as have been achieved by other sciences. It is true that this is partly accounted for by the fact that, as soon as definite knowledge concerning any subject becomes possible, this subject ceases to be called philosophy, and becomes a separate science. The whole study of the heavens, which now belongs to astronomy, was once included in philosophy; Newton's great work was called 'the mathematical principles of natural philosophy'. Similarly, the study of the human mind, which was a part of philosophy, has now been separated from philosophy and has become the science of psychology. Thus, to a great extent, the uncertainty of philosophy is more apparent than real: those questions which are already capable of definite answers are placed in the sciences, while those only to which, at present, no definite answer can be given, remain to form the residue which is called philosophy.

This is, however, only a part of the truth concerning the uncertainty of philosophy. There are many questions—and among them those that are of the profoundest interest to our spiritual life—which, so far as we can see, must remain insoluble to the human intellect unless its powers become of quite a different order from what they are now. Has the universe any unity of plan or purpose, or is it a fortuitous concourse of atoms? Is consciousness a permanent part of the universe, giving hope of indefinite growth in wisdom, or is it a transitory accident on a small planet on which life must ultimately become impossible? Are good and evil of importance to the universe or only to man? Such questions are asked by philosophy, and variously answered by various philosophers. But it would seem that, whether answers be otherwise discoverable or not, the answers suggested by philosophy are none of them demonstrably true. Yet, however slight may be the hope of discovering an answer, it is part of the business of philosophy to continue the consideration of such questions, to make us aware of their importance, to examine all the approaches to them, and to keep alive that speculative interest in the universe which is apt to be killed by confining ourselves to definitely ascertainable knowledge.

Many philosophers, it is true, have held that philosophy could establish the truth of certain answers to such fundamental questions. They have supposed that what is of most importance in religious beliefs could be proved by strict demonstration to be true. In order to judge of such attempts, it is necessary to take a survey of human knowledge, and to form an opinion as to its methods and its limitations. On such a subject it would be unwise to pronounce dogmatically; but if the investigations of our previous chapters have not led us astray, we shall be compelled to renounce the hope of finding philosophical proofs of religious beliefs. We cannot, therefore, include as part of the value of philosophy any definite set of answers to such questions. Hence, once more, the value of philosophy must not depend upon any supposed body of definitely ascertainable knowledge to be acquired by those who study it.

The value of philosophy is, in fact, to be sought largely in its very uncertainty. The man who has no tincture of philosophy goes through life imprisoned in the prejudices derived from common sense, from the habitual beliefs of his age or his nation, and from convictions which have grown up in his mind without the co-operation or consent of his deliberate reason. To such a man the world tends to become definite, finite, obvious; common objects rouse no questions, and unfamiliar possibilities are contemptuously rejected. As soon as we begin to philosophize, on the contrary, we find, as we saw in our opening chapters, that even the most everyday things lead to problems to which only very incomplete answers can be given. Philosophy, though unable to tell us with certainty what is the true answer to the doubts which it raises, is able to suggest many possibilities which enlarge our thoughts and free them from the tyranny of custom. Thus, while diminishing our feeling of certainty as to what things are, it greatly increases our knowledge as to what they may be; it removes the somewhat arrogant dogmatism of those who have never travelled into the region of liberating doubt, and it keeps alive our sense of wonder by showing familiar things in an unfamiliar aspect.

Apart from its utility in showing unsuspected possibilities, philosophy has a value—perhaps its chief value—through the greatness of the objects which it contemplates, and the freedom from narrow and personal aims resulting from this contemplation. The life of the instinctive man is shut up within the circle of his private interests: family and friends may be included, but the outer world is not regarded except as it may help or hinder what comes within the circle of instinctive wishes. In such a life there is something feverish and confined, in comparison with which the philosophic life is calm and free. The private world of instinctive interests is a small one, set in the midst of a great and powerful world which must, sooner or later, lay our private world in ruins. Unless we can so enlarge our interests as to include the whole outer world, we remain like a garrison in a beleagured fortress, knowing that the enemy prevents escape and that ultimate surrender is inevitable. In such a life there is no peace, but a constant strife between the insistence of desire and the powerlessness of will. In one way or another, if our life is to be great and free, we must escape this prison and this strife.

One way of escape is by philosophic contemplation. Philosophic contemplation does not, in its widest survey, divide the universe into two hostile camps—friends and foes, helpful and hostile, good and bad—it views the whole impartially. Philosophic contemplation, when it is unalloyed, does not aim at proving that the rest of the universe is akin to man. All acquisition of knowledge is an enlargement of the Self, but this enlargement is best attained when it is not directly sought. It is obtained when the desire for knowledge is alone operative, by a study which does not wish in advance that its objects should have this or that character, but adapts the Self to the characters which it finds in its objects. This enlargement of Self is not obtained when, taking the Self as it is, we try to show that the world is so similar to this Self that knowledge of it is possible without any admission of what seems alien. The desire to prove this is a form of self-assertion and, like all self-assertion, it is an obstacle to the growth of Self which it desires, and of which the Self knows that it is capable. Self-assertion, in philosophic speculation as elsewhere, views the world as a means to its own ends; thus it makes the world of less account than Self, and the Self sets bounds to the greatness of its goods. In contemplation, on the contrary, we start from the not-Self, and through its greatness the boundaries of Self are enlarged; through the infinity of the universe the mind which contemplates it achieves some share in infinity.

For this reason greatness of soul is not fostered by those philosophies which assimilate the universe to Man. Knowledge is a form of union of Self and not-Self; like all union, it is impaired by dominion, and therefore by any attempt to force the universe into conformity with what we find in ourselves. There is a widespread philosophical tendency towards the view which tells us that Man is the measure of all things, that truth is man-made, that space and time and the world of universals are properties of the mind, and that, if there be anything not created by the mind, it is unknowable and of no account for us. This view, if our previous discussions were correct, is untrue; but in addition to being untrue, it has the effect of robbing philosophic contemplation of all that gives it value, since it fetters contemplation to Self. What it calls knowledge is not a union with the not-Self, but a set of prejudices, habits, and desires, making an impenetrable veil between us and the world beyond. The man who finds pleasure in such a theory of knowledge is like the man who never leaves the domestic circle for fear his word might not be law.

The true philosophic contemplation, on the contrary, finds its satisfaction in every enlargement of the not-Self, in everything that magnifies the objects contemplated, and thereby the subject contemplating. Everything, in contemplation, that is personal or private, everything that depends upon habit, self-interest, or desire, distorts the object, and hence impairs the union which the intellect seeks. By thus making a barrier between subject and object, such personal and private things become a prison to the intellect. The free intellect will see as God might see, without a *here* and *now*, without hopes and fears, without the trammels of customary beliefs and traditional prejudices, calmly, dispassionately, in the sole and exclusive desire of knowledge—knowledge as impersonal, as purely contemplative, as it is possible for man to attain. Hence also the free intellect will value more the abstract and universal knowledge into which the accidents of private history do not enter, than the knowledge brought by the senses, and dependent, as such knowledge must be, upon an exclusive and personal point of view and a body whose sense-organs distort as much as they reveal.

The mind which has become accustomed to the freedom and impartiality of philosophic contemplation will preserve something of the same freedom and impartiality in the world of action and emotion. It will view its purposes and desires as parts of the whole, with the absence of insistence that results from seeing them as infinitesimal fragments in a world of which all the rest is unaffected by any one man's deeds. The impartiality which, in contemplation, is the unalloyed desire for truth, is the very same quality of mind which, in action, is justice, and in emotion is that universal love which can be given to all, and not only to those who are judged useful or admirable. Thus contemplation enlarges not only the objects of our thoughts, but also the objects of our actions and our affections: it makes us citizens of the universe, not only of one walled city at war with all the rest. In this citizenship of the universe consists man's true freedom, and his liberation from the thraldom of narrow hopes and fears.

Thus, to sum up our discussion of the value of philosophy; Philosophy is to be studied, not for the sake of any definite answers to its questions, since no definite answers can, as a rule, be known to be true, but rather for the sake of the questions themselves; because these questions enlarge our conception of what is possible, enrich our intellectual imagination and diminish the dogmatic assurance which closes the mind against speculation; but above all because, through the greatness of the universe which philosophy contemplates, the mind also is rendered great, and becomes capable of that union with the universe which constitutes its highest good.

OUR KNOWLEDGE OF THE EXTERNAL WORLD

THE following lectures are an attempt to show, by means of examples, the nature, capacity, and limitations of the logical-analytic method in philosophy. This method, of which the first complete example is to be found in the writings of Frege, has gradually, in the course of actual research, increasingly forced itself upon me as something perfectly definite, capable of embodiment in maxims, and adequate, in all branches of philosophy, to yield whatever objective scientific knowledge it is possible to obtain. Most of the methods hitherto practised have professed to lead to more ambitious results than any that logical analysis can claim to reach, but unfortunately these results have always been such as many competent philosophers considered inadmissible. Regarded merely as hypotheses and as aids to imagination, the great systems of the past serve a very useful purpose, and are abundantly worthy of study. But something different is required if philosophy is to become a science, and to aim at results independent of the tastes and temperament of the philosopher who advocates them. In what follows, I have endeavoured to show, however imperfectly, the way by which I believe that this *desideratum* is to be found.

The central problem by which I have sought to illustrate method is the problem of the relation between the crude data of sense and the space, time, and matter of mathematical physics. I have been made aware of the importance of this problem by my friend and collaborator Dr Whitehead, to whom are due almost all the differences between the views advocated here and those suggested in *The Problems of Philosophy*. I owe to him the definition of points, the suggestion for the treatment of instants and "things," and the whole conception of the world of physics as a *construction* rather than an *inference*. What is said on these topics here is, in fact, a rough preliminary account of the more precise results which he is giving in the fourth volume of our *Principia Mathematica*. It will be seen that if his way of dealing with these topics is capable of being successfully carried through, a wholly new light is thrown on the time-honoured controversies of realists and idealists, and a method is obtained of solving all that is soluble in their problem.

The speculations of the past as to the reality or unreality of the world of physics were baffled, at the outset, by the absence of any satisfactory theory of the mathematical infinite. This difficulty has been removed by the work of Georg Cantor. But the positive and detailed solution of the problem by means of mathematical constructions based upon sensible objects as data has only been rendered possible by the growth of mathematical logic, without which it is practically impossible to manipulate ideas of the requisite abstractness and complexity. This aspect, which is somewhat obscured in a merely popular outline such as is contained in the following lectures, will become plain as soon as Dr Whitehead's work is published. In pure logic, which, however, will be very briefly discussed in these lectures, I have had the benefit of vitally important discoveries, not yet published, by my friend Mr Ludwig Wittgenstein.

Since my purpose was to illustrate method, I have included much that is tentative and incomplete, for it is not by the study of finished structures alone that the manner of construction can be learnt. Except in regard to such matters as Cantor's theory of infinity, no finality is claimed for the theories suggested; but I believe that where they are found to require modification, this will be discovered by substantially the same method as that which at present makes them appear probable, and it is on this ground that I ask the reader to be tolerant of their incompleteness.

CAMBRIDGE,
June 1914.

LECTURE I
CURRENT TENDENCIES

PHILOSOPHY, from the earliest times, has made greater claims, and achieved fewer results, than any other branch of learning. Ever since Thales said that all is water, philosophers have been ready with glib assertions about the sum-total of things; and equally glib denials have come from other philosophers ever since Thales was contradicted by Anaximander. I believe that the time has now arrived when this unsatisfactory state of things can be brought to an end. In the following course of lectures I shall try, chiefly by taking certain special problems as examples, to indicate wherein the claims of philosophers have been excessive, and why their achievements have not been greater. The problems and the method of philosophy have, I believe, been misconceived by all schools, many of its traditional problems being insoluble with our means of knowledge, while other more neglected but not less important problems can, by a more patient and more adequate method, be solved with all the precision and certainty to which the most advanced sciences have attained.

Among present-day philosophies, we may distinguish three principal types, often combined in varying proportions by a single philosopher, but in essence and tendency distinct. The first of these, which I shall call the classical tradition, descends in the main from Kant and Hegel; it represents the attempt to adapt to present needs the methods and results of the great constructive philosophers from Plato downwards. The second type, which may be called evolutionism, derived its predominance from Darwin, and must be reckoned as having had Herbert Spencer for its first philosophical representative; but in recent times it has become, chiefly through William James and M. Bergson, far bolder and far more searching in its innovations than it was in the hands of Herbert Spencer. The third type, which may be called "logical atomism" for want of a better name, has gradually crept into philosophy through the critical scrutiny of mathematics. This type of philosophy, which is the one that I wish to advocate, has not as yet many whole-hearted adherents, but the "new realism" which owes its inception to Harvard is very largely impregnated with its spirit. It represents, I believe, the same kind of advance as was introduced into physics by Galileo: the substitution of piecemeal, detailed, and verifiable results for large untested generalities recommended only by a certain appeal to imagination. But before we can understand the changes advocated by this new philosophy, we must briefly examine and criticise the other two types with which it has to contend.

A. The Classical Tradition

Twenty years ago, the classical tradition, having vanquished the opposing tradition of the English empiricists, held almost unquestioned sway in all Anglo-Saxon universities. At the present day, though it is losing ground, many of the most prominent teachers still adhere to it. In academic France, in spite of M. Bergson, it is far stronger than all its opponents combined; and in Germany it has many vigorous advocates. Nevertheless, it represents on the whole a decaying force, and it has failed to adapt itself to the temper of the age. Its advocates are, in the main, those whose extra-philosophical knowledge is literary, rather than those who have felt the inspiration of science. There are, apart from reasoned arguments, certain general intellectual forces against it—the same general forces which are breaking down the other great syntheses of the past, and making our age one of bewildered groping where our ancestors walked in the clear daylight of unquestioning certainty.

The original impulse out of which the classical tradition developed was the naïve faith of the Greek philosophers in the omnipotence of reasoning. The discovery of geometry had intoxicated them, and its *a priori* deductive method appeared capable of universal application. They would prove, for instance, that all reality is one, that there is no such thing as change, that the world of sense is a world of mere illusion; and the strangeness of their results gave them no qualms because they believed in the correctness of their reasoning. Thus it came to be thought that by mere thinking the most surprising and important truths concerning the whole of reality could be established with a certainty which no contrary observations could shake. As the vital impulse of the early philosophers died away, its place was taken by authority and tradition, reinforced, in the Middle Ages and almost to our own day, by systematic theology. Modern philosophy, from Descartes onwards, though not bound by authority like that of the Middle Ages, still accepted more or less uncritically the Aristotelian logic. Moreover, it still believed, except in Great Britain, that *a priori* reasoning could reveal otherwise undiscoverable secrets about the universe, and could prove reality to be quite different from what, to direct observation, it appears to be. It is this belief, rather than any particular tenets resulting from it, that I regard as the distinguishing characteristic of the classical tradition, and as hitherto the main obstacle to a scientific attitude in philosophy.

The nature of the philosophy embodied in the classical tradition may be made clearer by taking a particular exponent as an illustration. For this purpose, let us consider for a moment the doctrines of Mr Bradley, who is probably the most distinguished living representative of this school. Mr Bradley's *Appearance and Reality* is a book consisting of two parts, the first called *Appearance*, the second *Reality*. The first part examines and condemns almost all that makes up our everyday world: things and qualities, relations, space and time, change, causation, activity, the self. All these, though in some sense facts which qualify reality, are not real as they appear. What is real is one single, indivisible, timeless whole, called the Absolute, which is in some sense spiritual, but does not consist of souls, or of thought and will as we know them. And all this is established by abstract logical reasoning professing to find self-contradictions in the categories condemned as mere appearance, and to leave no tenable alternative to the kind of Absolute which is finally affirmed to be real.

One brief example may suffice to illustrate Mr Bradley's method. The world appears to be full of many things with various relations to each other—right and left, before and after, father and son, and so on. But relations, according to Mr Bradley, are found on examination to be self-contradictory and therefore impossible. He first argues that, if there are relations, there must be qualities between which they hold. This part of his argument need not detain us. He then proceeds:

"But how the relation can stand to the qualities is, on the other side, unintelligible. If it is nothing to the qualities, then they are not related at all; and, if so, as we saw, they have ceased to be qualities, and their relation is a nonentity. But if it is to be something to them, then clearly we shall require a *new* connecting relation. For the relation hardly can be the mere adjective of one or both of its terms; or, at least, as such it seems indefensible. And, being something itself, if it does not itself bear a relation to the terms, in what intelligible way will it succeed in being anything to them? But here again we are hurried off into the eddy of a hopeless process, since we are forced to go on finding new relations without end. The links are united by a link, and this bond of union is a link which also has two ends; and these require each a fresh link to connect them with the old. The problem is to find how the relation can stand to its qualities, and this problem is insoluble."

I do not propose to examine this argument in detail, or to show the exact points where, in my opinion, it is fallacious. I have quoted it only as an example of method. Most people will admit, I think, that it is calculated to produce bewilderment rather than conviction, because there is more likelihood of error in a very subtle, abstract, and difficult argument than in so patent a fact as the interrelatedness of the things in the world. To the early Greeks, to whom geometry was practically the only known science, it was possible to follow reasoning with assent even when it led to the strangest conclusions. But to us, with our methods of experiment and observation, our knowledge of the long history of *a priori* errors refuted by empirical science, it has become natural to suspect a fallacy in any deduction of which the conclusion appears to contradict patent facts. It is easy to carry such suspicion too far, and it is very desirable, if possible, actually to discover the exact nature of the error when it exists. But there is no doubt that what we may call the empirical outlook has become part of most educated people's habit of mind; and it is this, rather than any definite argument, that has diminished the hold of the classical tradition upon students of philosophy and the instructed public generally.

The function of logic in philosophy, as I shall try to show at a later stage, is all-important; but I do not think its function is that which it has in the classical tradition. In that tradition, logic becomes constructive through negation. Where a number of alternatives seem, at first sight, to be equally possible, logic is made to condemn all of them except one, and that one is then pronounced to be realised in the actual world. Thus the world is constructed by means of logic, with little or no appeal to concrete experience. The true function of logic is, in my opinion, exactly the opposite of this. As applied to matters of experience, it is analytic rather than constructive; taken *a priori*, it shows the possibility of hitherto unsuspected alternatives more often than the impossibility of alternatives which seemed *primâ facie* possible. Thus, while it liberates imagination as to what the world *may* be, it refuses to legislate as to what the world *is*. This change, which has been brought about by an internal revolution in logic, has swept away the ambitious constructions of traditional metaphysics, even for those whose faith in logic is greatest; while to the many who regard logic as a chimera the paradoxical systems to which it has given rise do not seem worthy even of refutation. Thus on all sides these systems have ceased to attract, and even the philosophical world tends more and more to pass them by.

One or two of the favourite doctrines of the school we are considering may be mentioned to illustrate the nature of its claims. The universe, it tells us, is an "organic unity," like an animal or a perfect work of art. By this it means, roughly speaking, that all the different parts fit together and co-operate, and are what they are because of their place in the whole. This belief is sometimes advanced dogmatically, while at other times it is defended by certain logical arguments. If it is true, every part of the universe is a microcosm, a miniature reflection of the whole. If we knew ourselves thoroughly, according to this doctrine, we should know everything. Common sense would naturally object that there are people—say in China—with whom our relations are so indirect and trivial that we cannot infer anything important as to them from any fact about ourselves. If there are living beings in Mars or in more distant parts of the universe, the same argument becomes even stronger. But further, perhaps the whole contents of the space and time in which we live form only one of many universes, each seeming to itself complete. And thus the conception of the necessary unity of all that is resolves itself into the poverty of imagination, and a freer logic emancipates us from the strait-waistcoated benevolent institution which idealism palms off as the totality of being.

Another very important doctrine held by most, though not all, of the school we are examining is the doctrine that all reality is what is called "mental" or "spiritual," or that, at any rate, all reality is dependent for its existence upon what is mental. This view is often particularised into the form which states that the relation of knower and known is fundamental, and that nothing can exist unless it either knows or is known. Here again the same legislative function is ascribed to *a priori* argumentation: it is thought that there are contradictions in an unknown reality. Again, if I am not mistaken, the argument is fallacious, and a better logic will show that no limits can be set to the extent and nature of the unknown. And when I speak of the unknown, I do not mean merely what we personally do not know, but what is not known to any mind. Here as elsewhere, while the older logic shut out possibilities and imprisoned imagination within the walls of the familiar, the newer logic shows rather what may happen, and refuses to decide as to what *must* happen.

The classical tradition in philosophy is the last surviving child of two very diverse parents: the Greek belief in reason, and the mediæval belief in the tidiness of the universe. To the schoolmen, who lived amid wars, massacres, and pestilences, nothing appeared so delightful as safety and order. In their idealising dreams, it was safety and order that they sought: the universe of Thomas Aquinas or Dante is as small and neat as a Dutch interior. To us, to whom safety has become monotony, to whom the primeval savageries of nature are so remote as to become a mere pleasing condiment to our ordered routine, the world of dreams is very different from what it was amid the wars of Guelf and Ghibelline. Hence William James's protest against what he calls the "block universe" of the classical tradition; hence Nietzsche's worship of force; hence the verbal bloodthirstiness of many quiet literary men. The barbaric substratum of human nature, unsatisfied in action, finds an outlet in imagination. In philosophy, as elsewhere, this tendency is visible; and it is this, rather than formal argument, that has thrust aside the classical tradition for a philosophy which fancies itself more virile and more vital.

B. Evolutionism

Evolutionism, in one form or another, is the prevailing creed of our time. It dominates our politics, our literature, and not least our philosophy. Nietzsche, pragmatism, Bergson, are phases in its philosophic development, and their popularity far beyond the circles of professional philosophers shows its consonance with the spirit of the age. It believes itself firmly based on science, a liberator of hopes, an inspirer of an invigorating faith in human power, a sure antidote to the ratiocinative authority of the Greeks and the dogmatic authority of mediæval systems. Against so fashionable and so agreeable a creed it may seem useless to raise a protest; and with much of its spirit every modern man must be in sympathy. But I think that, in the intoxication of a quick success, much that is important and vital to a true understanding of the universe has been forgotten. Something of Hellenism must be combined with the new spirit before it can emerge from the ardour of youth into the wisdom of manhood. And it is time to remember that biology is neither the only science, nor yet the model to which all other sciences must adapt themselves. Evolutionism, as I shall try to show, is not a truly scientific philosophy, either in its method or in the problems which it considers. The true scientific philosophy is something more arduous and more aloof, appealing to less mundane hopes, and requiring a severer discipline for its successful practice.

Darwin's *Origin of Species* persuaded the world that the difference between different species of animals and plants is not the fixed, immutable difference that it appears to be. The doctrine of natural kinds, which had rendered classification easy and definite, which was enshrined in the Aristotelian tradition, and protected by its supposed necessity for orthodox dogma, was suddenly swept away for ever out of the biological world. The difference between man and the lower animals, which to our human conceit appears enormous, was shown to be a gradual achievement, involving intermediate beings who could not with certainty be placed either within or without the human family. The sun and planets had already been shown by Laplace to be very probably derived from a primitive more or less undifferentiated nebula. Thus the old fixed landmarks became wavering and indistinct, and all sharp outlines were blurred. Things and species lost their boundaries, and none could say where they began or where they ended.

But if human conceit was staggered for a moment by its kinship with the ape, it soon found a way to reassert itself, and that way is the "philosophy" of evolution. A process which led from the amœba to man appeared to the philosophers to be obviously a progress—though whether the amœba would agree with this opinion is not known. Hence the cycle of changes which science had shown to be the probable history of the past was welcomed as revealing a law of development towards good in the universe—an evolution or unfolding of an ideal slowly embodying itself in the actual. But such a view, though it might satisfy Spencer and those whom we may call Hegelian evolutionists, could not be accepted as adequate by the more whole-hearted votaries of change. An ideal to which the world continuously approaches is, to these minds, too dead and static to be inspiring. Not only the aspirations, but the ideal too, must change and develop with the course of evolution; there must be no fixed goal, but a continual fashioning of fresh needs by the impulse which is life and which alone gives unity to the process.

Ever since the seventeenth century, those whom William James described as the "tender-minded" have been engaged in a desperate struggle with the mechanical view of the course of nature which physical science seems to impose. A great part of the attractiveness of the classical tradition was due to the partial escape from mechanism which it provided. But now, with the influence of biology, the "tender-minded" believe that a more radical escape is possible, sweeping aside not merely the laws of physics, but the whole apparently immutable apparatus of logic, with its fixed concepts, its general principles, and its reasonings which seem able to compel even the most unwilling assent. The older kind of teleology, therefore, which regarded the End as a fixed goal, already partially visible, towards which we were gradually approaching, is rejected by M. Bergson as not allowing enough for the absolute dominion of change. After explaining why he does not accept mechanism, he proceeds:

"But radical finalism is quite as unacceptable, and for the same reason. The doctrine of teleology, in its extreme form, as we find it in Leibniz for example, implies that things and beings merely realise a programme previously arranged. But if there is nothing unforeseen, no invention or creation in the universe, time is useless again. As in the mechanistic hypothesis, here again it is supposed that *all is given*. Finalism thus understood is only inverted mechanism. It springs from the same postulate, with this sole difference, that in the movement of our finite intellects along successive things, whose successiveness is reduced to a mere appearance, it holds in front of us the light with which it claims to guide us, instead of putting it behind. It substitutes the attraction of the future for the impulsion of the past. But succession remains none the less a mere appearance, as indeed does movement itself. In the doctrine of Leibniz, time is reduced to a confused perception, relative to the human standpoint, a perception which would vanish, like a rising mist, for a mind seated at the centre of things.

"Yet finalism is not, like mechanism, a doctrine with fixed rigid outlines. It admits of as many inflections as we like. The mechanistic philosophy is to be taken or left: it must be left if the least grain of dust, by straying from the path foreseen by mechanics, should show the slightest trace of spontaneity. The doctrine of final causes, on the contrary, will never be definitively refuted. If one form of it be put aside, it will take another. Its principle, which is essentially psychological, is very flexible. It is so extensible, and thereby so comprehensive, that one accepts something of it as soon as one rejects pure mechanism. The theory we shall put forward in this book will therefore necessarily partake of finalism to a certain extent."

M. Bergson's form of finalism depends upon his conception of life. Life, in his philosophy, is a continuous stream, in which all divisions are artificial and unreal. Separate things, beginnings and endings, are mere convenient fictions: there is only smooth, unbroken transition. The beliefs of to-day may count as true to-day, if they carry us along the stream; but to-morrow they will be false, and must be replaced by new beliefs to meet the new situation. All our thinking consists of convenient fictions, imaginary congealings of the stream: reality flows on in spite of all our fictions, and though it can be lived, it cannot be conceived in thought. Somehow, without explicit statement, the assurance is slipped in that the future, though we cannot foresee it, will be better than the past or the present: the reader is like the child who expects a sweet because it has been told to open its mouth and shut its eyes. Logic, mathematics, physics disappear in this philosophy, because they are too "static"; what is real is an impulse and movement towards a goal which, like the rainbow, recedes as we advance, and makes every place different when we reach it from what it appeared to be at a distance.

Now I do not propose at present to enter upon a technical examination of this philosophy. At present I wish to make only two criticisms of it—first, that its truth does not follow from what science has rendered probable concerning the facts of evolution, and secondly, that the motives and interests which inspire it are so exclusively practical, and the problems with which it deals are so special, that it can hardly be regarded as really touching any of the questions that to my mind constitute genuine philosophy.

(1) What biology has rendered probable is that the diverse species arose by adaptation from a less differentiated ancestry. This fact is in itself exceedingly interesting, but it is not the kind of fact from which philosophical consequences follow. Philosophy is general, and takes an impartial interest in all that exists. The changes suffered by minute portions of matter on the earth's surface are very important to us as active sentient beings; but to us as philosophers they have no greater interest than other changes in portions of matter elsewhere. And if the changes on the earth's surface during the last few millions of years appear to our present ethical notions to be in the nature of a progress, that gives no ground for believing that progress is a general law of the universe. Except under the influence of desire, no one would admit for a moment so crude a generalisation from such a tiny selection of facts. What does result, not specially from biology, but from all the sciences which deal with what exists, is that we cannot understand the world unless we can understand change and continuity. This is even more evident in physics than it is in biology. But the analysis of change and continuity is not a problem upon which either physics or biology throws any light: it is a problem of a new kind, belonging to a different kind of study. The question whether evolutionism offers a true or a false answer to this problem is not, therefore, a question to be solved by appeals to particular facts, such as biology and physics reveal. In assuming dogmatically a certain answer to this question, evolutionism ceases to be scientific, yet it is only in touching on this question that evolutionism reaches the subject-matter of philosophy. Evolutionism thus consists of two parts: one not philosophical, but only a hasty generalisation of the kind which the special sciences might hereafter confirm or confute; the other not scientific, but a mere unsupported dogma, belonging to

philosophy by its subject-matter, but in no way deducible from the facts upon which evolution relies.

(2) The predominant interest of evolutionism is in the question of human destiny, or at least of the destiny of Life. It is more interested in morality and happiness than in knowledge for its own sake. It must be admitted that the same may be said of many other philosophies, and that a desire for the kind of knowledge which philosophy really can give is very rare. But if philosophy is to become scientific—and it is our object to discover how this can be achieved—it is necessary first and foremost that philosophers should acquire the disinterested intellectual curiosity which characterises the genuine man of science. Knowledge concerning the future—which is the kind of knowledge that must be sought if we are to know about human destiny—is possible within certain narrow limits. It is impossible to say how much the limits may be enlarged with the progress of science. But what is evident is that any proposition about the future belongs by its subject-matter to some particular science, and is to be ascertained, if at all, by the methods of that science. Philosophy is not a short cut to the same kind of results as those of the other sciences: if it is to be a genuine study, it must have a province of its own, and aim at results which the other sciences can neither prove nor disprove.

The consideration that philosophy, if there is such a study, must consist of propositions which could not occur in the other sciences, is one which has very far-reaching consequences. All the questions which have what is called a human interest—such, for example, as the question of a future life—belong, at least in theory, to special sciences, and are capable, at least in theory, of being decided by empirical evidence. Philosophers have too often, in the past, permitted themselves to pronounce on empirical questions, and found themselves, as a result, in disastrous conflict with well-attested facts. We must, therefore, renounce the hope that philosophy can promise satisfaction to our mundane desires. What it can do, when it is purified from all practical taint, is to help us to understand the general aspects of the world and the logical analysis of familiar but complex things. Through this achievement, by the suggestion of fruitful hypotheses, it may be indirectly useful in other sciences, notably mathematics, physics, and psychology. But a genuinely scientific philosophy cannot hope to appeal to any except those who have the wish to understand, to escape from intellectual bewilderment. It offers, in its own domain, the kind of satisfaction which the other sciences offer. But it does not offer, or attempt to offer, a solution of the problem of human destiny, or of the destiny of the universe.

Evolutionism, if what has been said is true, is to be regarded as a hasty generalisation from certain rather special facts, accompanied by a dogmatic rejection of all attempts at analysis, and inspired by interests which are practical rather than theoretical. In spite, therefore, of its appeal to detailed results in various sciences, it cannot be regarded as any more genuinely scientific than the classical tradition which it has replaced. How philosophy is to be rendered scientific, and what is the true subject-matter of philosophy, I shall try to show first by examples of certain achieved results, and then more generally. We will begin with the problem of the physical conceptions of space and time and matter, which, as we have seen, are challenged by the contentions of the evolutionists. That these conceptions stand in need of reconstruction will be admitted, and is indeed increasingly urged by physicists themselves. It will also be admitted that the reconstruction must take more account of change and the universal flux than is done in the older mechanics with its fundamental conception of an indestructible matter. But I do not think the reconstruction required is on Bergsonian lines, nor do I think that his rejection of logic can be anything but harmful. I shall not, however, adopt the method of explicit controversy, but rather the method of independent inquiry, starting from what, in a pre-philosophic stage, appear to be facts, and keeping always as close to these initial data as the requirements of consistency will permit.

Although explicit controversy is almost always fruitless in philosophy, owing to the fact that no two philosophers ever understand one another, yet it seems necessary to say something at the outset in justification of the scientific as against the mystical attitude. Metaphysics, from the first, has been developed by the union or the conflict of these two attitudes. Among the earliest Greek philosophers, the Ionians were more scientific and the Sicilians more mystical.But among the latter, Pythagoras, for example, was in himself a curious mixture of the two tendencies: the scientific attitude led him to his proposition on right-angled triangles, while his mystic insight showed him that it is wicked to eat beans. Naturally enough, his followers divided into two sects, the lovers of right-angled triangles and the abhorrers of beans; but the former sect died out, leaving, however, a haunting flavour of mysticism over much Greek mathematical speculation, and in particular over Plato's views on mathematics. Plato, of course, embodies both the scientific and the mystical attitudes in a higher form than his predecessors, but the mystical attitude is distinctly the stronger of the two, and secures ultimate victory whenever the conflict is sharp. Plato, moreover, adopted from the Eleatics the device of using logic to defeat common sense, and thus to leave the field clear for mysticism—a device still employed in our own day by the adherents of the classical tradition.

The logic used in defence of mysticism seems to me faulty as logic, and in a later lecture I shall criticise it on this ground. But the more thorough-going mystics do not employ logic, which they despise: they appeal instead directly to the immediate deliverance of their insight. Now, although fully developed mysticism is rare in the West, some tincture of it colours the thoughts of many people, particularly as regards matters on which they have strong convictions not based on evidence. In all who seek passionately for the fugitive and difficult goods, the conviction is almost irresistible that there is in the world something deeper, more significant, than the multiplicity of little facts chronicled and classified by science. Behind the veil of these mundane things, they feel, something quite different obscurely shimmers, shining forth clearly in the great moments of illumination, which alone give anything worthy to be called real knowledge of truth. To seek such moments, therefore, is to them the way of wisdom, rather than, like the man of science, to observe coolly, to analyse without emotion, and to accept without question the equal reality of the trivial and the important.

Of the reality or unreality of the mystic's world I know nothing. I have no wish to deny it, nor even to declare that the insight which reveals it is not a genuine insight. What I do wish to maintain—and it is here that the scientific attitude becomes imperative—is that insight, untested and unsupported, is an insufficient guarantee of truth, in spite of the fact that much of the most important truth is first suggested by its means. It is common to speak of an opposition between instinct and reason; in the eighteenth century, the opposition was drawn in favour of reason, but under the influence of Rousseau and the romantic movement instinct was given the preference, first by those who rebelled against artificial forms of government and thought, and then, as the purely rationalistic defence of traditional theology became increasingly difficult, by all who felt in science a menace to creeds which they associated with a spiritual outlook on life and the world. Bergson, under the name of "intuition," has raised instinct to the position of sole arbiter of metaphysical truth. But in fact the opposition of instinct and reason is mainly illusory. Instinct, intuition, or insight is what first leads to the beliefs which subsequent reason confirms or confutes; but the confirmation, where it is possible, consists, in the last analysis, of agreement with other beliefs no less instinctive. Reason is a harmonising, controlling force rather than a creative one. Even in the most purely logical realms, it is insight that first arrives at what is new.

Where instinct and reason do sometimes conflict is in regard to single beliefs, held instinctively, and held with such determination that no degree of inconsistency with other beliefs leads to their abandonment. Instinct, like all human faculties, is liable to error. Those in whom reason is weak are often unwilling to admit this as regards themselves, though all admit it in regard to others. Where instinct is least liable to error is in practical matters as to which right judgment is a help to survival; friendship and hostility in others, for instance, are often felt with extraordinary discrimination through very careful disguises. But even in such matters a wrong impression may be given by reserve or flattery; and in matters less directly practical, such as philosophy deals with, very strong instinctive beliefs may be wholly mistaken, as we may come to know through their perceived inconsistency with other equally strong beliefs. It is such considerations that necessitate the harmonising mediation of reason, which tests our beliefs by their mutual compatibility, and examines, in doubtful cases, the possible sources of error on the one side and on the other. In this there is no opposition to instinct as a whole, but only to blind reliance upon some one interesting aspect of instinct to the exclusion of other more commonplace but not less trustworthy aspects. It is such onesidedness, not instinct itself, that reason aims at correcting.

These more or less trite maxims may be illustrated by application to Bergson's advocacy of "intuition" as against "intellect." There are, he says, "two profoundly different ways of knowing a thing. The first implies that we move round the object; the second that we enter into it. The first depends on the point of view at which we are placed and on the symbols by which we express ourselves. The second neither depends on a point of view nor relies on any symbol. The first kind of knowledge may be said to stop at the *relative*; the second, in those cases where it is possible, to attain the *absolute*."The second of these, which is intuition, is, he says, "the kind of intellectual *sympathy* by which one places oneself within an object in order to coincide with what is unique in it and therefore inexpressible". In illustration, he mentions self-knowledge: "there is one reality, at least, which we all seize from within, by intuition and not by simple analysis. It is our own personality in its flowing through time—our self which endures". The rest of Bergson's philosophy consists in reporting, through the imperfect medium of words, the knowledge gained by intuition, and the consequent complete condemnation of all the pretended knowledge derived from science and common sense.

This procedure, since it takes sides in a conflict of instinctive beliefs, stands in need of justification by proving the greater trustworthiness of the beliefs on one side than of those on the other. Bergson attempts this justification in two ways—first, by explaining that intellect is a purely practical faculty designed to secure biological success; secondly, by mentioning remarkable feats of instinct in animals, and by pointing out characteristics of the world which, though intuition can apprehend them, are baffling to intellect as he interprets it.

Of Bergson's theory that intellect is a purely practical faculty developed in the struggle for survival, and not a source of true beliefs, we may say, first, that it is only through intellect that we know of the struggle for survival and of the biological ancestry of man: if the intellect is misleading, the whole of this merely inferred history is presumably untrue. If, on the other hand, we agree with M. Bergson in thinking that evolution took place as Darwin believed, then it is not only intellect, but all our faculties, that have been developed under the stress of practical utility. Intuition is seen at its best where it is directly useful—for example, in regard to other people's characters and dispositions. Bergson apparently holds that capacity for this kind of knowledge is less explicable by the struggle for existence than, for example, capacity for pure mathematics. Yet the savage deceived by false friendship is likely to pay for his mistake with his life; whereas even in the most civilised societies men are not put to death for mathematical incompetence. All the most striking of his instances of intuition in animals have a very direct survival value. The fact is, of course, that both intuition and intellect have been developed because they are useful, and that, speaking broadly, they are useful when they give truth and become harmful when they give falsehood. Intellect, in civilised man, like artistic capacity, has occasionally been developed beyond the point where it is useful to the individual; intuition, on the other hand, seems on the whole to diminish as civilisation increases. Speaking broadly, it is greater in children than in adults, in the uneducated than in the educated. Probably in dogs it exceeds anything to be found in human beings. But those who find in these facts a recommendation of intuition ought to return to running wild in the woods, dyeing themselves with woad and living on hips and haws.

Let us next examine whether intuition possesses any such infallibility as Bergson claims for it. The best instance of it, according to him, is our acquaintance with ourselves; yet self-knowledge is proverbially rare and difficult. Most men, for example, have in their nature meannesses, vanities, and envies of which they are quite unconscious, though even their best friends can perceive them without any difficulty. It is true that intuition has a convincingness which is lacking to intellect: while it is present, it is almost impossible to doubt its truth. But if it should appear, on examination, to be at least as fallible as intellect, its greater subjective certainty becomes a demerit, making it only the more irresistibly deceptive. Apart from self-knowledge, one of the most notable examples of intuition is the knowledge people believe themselves to possess of those with whom they are in love: the wall between different personalities seems to become transparent, and people think they see into another soul as into their own. Yet deception in such cases is constantly practised with success; and even where there is no intentional deception, experience gradually proves, as a rule, that the supposed insight was illusory, and that the slower, more groping methods of the intellect are in the long run more reliable.

Bergson maintains that intellect can only deal with things in so far as they resemble what has been experienced in the past, while intuition has the power of apprehending the uniqueness and novelty that always belong to each fresh moment. That there is something unique and new at every moment, is certainly true; it is also true that this cannot be fully expressed by means of intellectual concepts. Only direct acquaintance can give knowledge of what is unique and new. But direct acquaintance of this kind is given fully in sensation, and does not require, so far as I can see, any special faculty of intuition for its apprehension.

It is neither intellect nor intuition, but sensation, that supplies new data; but when the data are new in any remarkable manner, intellect is much more capable of dealing with them than intuition would be. The hen with a brood of ducklings no doubt has intuitions which seem to place her inside them, and not merely to know them analytically; but when the ducklings take to the water, the whole apparent intuition is seen to be illusory, and the hen is left helpless on the shore. Intuition, in fact, is an aspect and development of instinct, and, like all instinct, is admirable in those customary surroundings which have moulded the habits of the animal in question, but totally incompetent as soon as the surroundings are changed in a way which demands some non-habitual mode of action.

The theoretical understanding of the world, which is the aim of philosophy, is not a matter of great practical importance to animals, or to savages, or even to most civilised men. It is hardly to be supposed, therefore, that the rapid, rough and ready methods of instinct or intuition will find in this field a favourable ground for their application. It is the older kinds of activity, which bring out our kinship with remote generations of animal and semi-human ancestors, that show intuition at its best. In such matters as self-preservation and love, intuition will act sometimes (though not always) with a swiftness and precision which are astonishing to the critical intellect. But philosophy is not one of the pursuits which illustrate our affinity with the past: it is a highly refined, highly civilised pursuit, demanding, for its success, a certain liberation from the life of instinct, and even, at times, a certain aloofness from all mundane hopes and fears. It is not in philosophy, therefore, that we can hope to see intuition at its best. On the contrary, since the true objects of philosophy, and the habits of thought demanded for their apprehension, are strange, unusual, and remote, it is here, more almost than anywhere else, that intellect proves superior to intuition, and that quick unanalysed convictions are least deserving of uncritical acceptance.

Before embarking upon the somewhat difficult and abstract discussions which lie before us, it will be well to take a survey of the hopes we may retain and the hopes we must abandon. The hope of satisfaction to our more human desires—the hope of demonstrating that the world has this or that desirable ethical characteristic—is not one which, so far as I can see, philosophy can do anything whatever to satisfy. The difference between a good world and a bad one is a difference in the particular characteristics of the particular things that exist in these worlds: it is not a sufficiently abstract difference to come within the province of philosophy. Love and hate, for example, are ethical opposites, but to philosophy they are closely analogous attitudes towards objects. The general form and structure of those attitudes towards objects which constitute mental phenomena is a problem for philosophy; but the difference between love and hate is not a difference of form or structure, and therefore belongs rather to the special science of psychology than to philosophy. Thus the ethical interests which have often inspired philosophers must remain in the background: some kind of ethical interest may inspire the whole study, but none must obtrude in the detail or be expected in the special results which are sought.

If this view seems at first sight disappointing, we may remind ourselves that a similar change has been found necessary in all the other sciences. The physicist or chemist is not now required to prove the ethical importance of his ions or atoms; the biologist is not expected to prove the utility of the plants or animals which he dissects. In pre-scientific ages this was not the case. Astronomy, for example, was studied because men believed in astrology: it was thought that the movements of the planets had the most direct and important bearing upon the lives of human beings. Presumably, when this belief decayed and the disinterested study of astronomy began, many who had found astrology absorbingly interesting decided that astronomy had too little human interest to be worthy of study. Physics, as it appears in Plato's *Timæus* for example, is full of ethical notions: it is an essential part of its purpose to show that the earth is worthy of admiration. The modern physicist, on the contrary, though he has no wish to deny that the earth is admirable, is not concerned, as physicist, with its ethical attributes: he is merely concerned to find out facts, not to consider whether they are good or bad. In psychology, the scientific attitude is even more recent and more difficult than in the physical sciences: it is natural to consider that human nature is either good or bad, and to suppose that the difference between good and bad, so all-important in practice, must be important in theory also. It is only during the last century that an ethically neutral science of psychology has grown up; and here too ethical neutrality has been essential to scientific success.

In philosophy, hitherto, ethical neutrality has been seldom sought and hardly ever achieved. Men have remembered their wishes, and have judged philosophies in relation to their wishes. Driven from the particular sciences, the belief that the notions of good and evil must afford a key to the understanding of the world has sought a refuge in philosophy. But even from this last refuge, if philosophy is not to remain a set of pleasing dreams, this belief must be driven forth. It is a commonplace that happiness is not best achieved by those who seek it directly; and it would seem that the same is true of the good. In thought, at any rate, those who forget good and evil and seek only to know the facts are more likely to achieve good than those who view the world through the distorting medium of their own desires.

The immense extension of our knowledge of facts in recent times has had, as it had in the Renaissance, two effects upon the general intellectual outlook. On the one hand, it has made men distrustful of the truth of wide, ambitious systems: theories come and go swiftly, each serving, for a moment, to classify known facts and promote the search for new ones, but each in turn proving inadequate to deal with the new facts when they have been found. Even those who invent the theories do not, in science, regard them as anything but a temporary makeshift. The ideal of an all-embracing synthesis, such as the Middle Ages believed themselves to have attained, recedes further and further beyond the limits of what seems feasible. In such a world, as in the world of Montaigne, nothing seems worth while except the discovery of more and more facts, each in turn the deathblow to some cherished theory; the ordering intellect grows weary, and becomes slovenly through despair.

On the other hand, the new facts have brought new powers; man's physical control over natural forces has been increasing with unexampled rapidity, and promises to increase in the future beyond all easily assignable limits. Thus alongside of despair as regards ultimate theory there is an immense optimism as regards practice: what man can *do* seems almost boundless. The old fixed limits of human power, such as death, or the dependence of the race on an equilibrium of cosmic forces, are forgotten, and no hard facts are allowed to break in upon the dream of omnipotence. No philosophy is tolerated which sets bounds to man's capacity of gratifying his wishes; and thus the very despair of theory is invoked to silence every whisper of doubt as regards the possibilities of practical achievement.

In the welcoming of new fact, and in the suspicion of dogmatism as regards the universe at large, the modern spirit should, I think, be accepted as wholly an advance. But both in its practical pretensions and in its theoretical despair it seems to me to go too far. Most of what is greatest in man is called forth in response to the thwarting of his hopes by immutable natural obstacles; by the pretence of omnipotence, he becomes trivial and a little absurd. And on the theoretical side, ultimate metaphysical truth, though less all-embracing and harder of attainment than it appeared to some philosophers in the past, can, I believe, be discovered by those who are willing to combine the hopefulness, patience, and open-mindedness of science with something of the Greek feeling for beauty in the abstract world of logic and for the ultimate intrinsic value in the contemplation of truth.

The philosophy, therefore, which is to be genuinely inspired by the scientific spirit, must deal with somewhat dry and abstract matters, and must not hope to find an answer to the practical problems of life. To those who wish to understand much of what has in the past been most difficult and obscure in the constitution of the universe, it has great rewards to offer—triumphs as noteworthy as those of Newton and Darwin, and as important in the long run, for the moulding of our mental habits. And it brings with it—as a new and powerful method of investigation always does—a sense of power and a hope of progress more reliable and better grounded than any that rests on hasty and fallacious generalisation as to the nature of the universe at large. Many hopes which inspired philosophers in the past it cannot claim to fulfil; but other hopes, more purely intellectual, it can satisfy more fully than former ages could have deemed possible for human minds.

LECTURE II
LOGIC AS THE ESSENCE OF PHILOSOPHY

THE topics we discussed in our first lecture, and the topics we shall discuss later, all reduce themselves, in so far as they are genuinely philosophical, to problems of logic. This is not due to any accident, but to the fact that every philosophical problem, when it is subjected to the necessary analysis and purification, is found either to be not really philosophical at all, or else to be, in the sense in which we are using the word, logical. But as the word "logic" is never used in the same sense by two different philosophers, some explanation of what I mean by the word is indispensable at the outset.

Logic, in the Middle Ages, and down to the present day in teaching, meant no more than a scholastic collection of technical terms and rules of syllogistic inference. Aristotle had spoken, and it was the part of humbler men merely to repeat the lesson after him. The trivial nonsense embodied in this tradition is still set in examinations, and defended by eminent authorities as an excellent "propædeutic," *i.e.* a training in those habits of solemn humbug which are so great a help in later life. But it is not this that I mean to praise in saying that all philosophy is logic. Ever since the beginning of the seventeenth century, all vigorous minds that have concerned themselves with inference have abandoned the mediæval tradition, and in one way or other have widened the scope of logic.

The first extension was the introduction of the inductive method by Bacon and Galileo—by the former in a theoretical and largely mistaken form, by the latter in actual use in establishing the foundations of modern physics and astronomy. This is probably the only extension of the old logic which has become familiar to the general educated public. But induction, important as it is when regarded as a method of investigation, does not seem to remain when its work is done: in the final form of a perfected science, it would seem that everything ought to be deductive. If induction remains at all, which is a difficult question, it will remain merely as one of the principles according to which deductions are effected. Thus the ultimate result of the introduction of the inductive method seems not the creation of a new kind of non-deductive reasoning, but rather the widening of the scope of deduction by pointing out a way of deducing which is certainly not syllogistic, and does not fit into the medieval scheme.

The question of the scope and validity of induction is of great difficulty, and of great importance to our knowledge. Take such a question as, "Will the sun rise to-morrow?" Our first instinctive feeling is that we have abundant reason for saying that it will, because it has risen on so many previous mornings. Now, I do not myself know whether this does afford a ground or not, but I am willing to suppose that it does. The question which then arises is: What is the principle of inference by which we pass from past sunrises to future ones? The answer given by Mill is that the inference depends upon the law of causation. Let us suppose this to be true; then what is the reason for believing in the law of causation? There are broadly three possible answers: (1) that it is itself known *a priori*; (2) that it is a postulate; (3) that it is an empirical generalisation from past instances in which it has been found to hold. The theory that causation is known *a priori* cannot be definitely refuted, but it can be rendered very unplausible by the mere process of formulating the law exactly, and thereby showing that it is immensely more complicated and less obvious than is generally supposed. The theory that causation is a postulate, *i.e.* that it is something which we choose to assert although we know that it is very likely false, is also incapable of refutation; but it is plainly also incapable of justifying any use of the law in inference. We are thus brought to the theory that the law is an empirical generalisation, which is the view held by Mill.

But if so, how are empirical generalisations to be justified? The evidence in their favour cannot be empirical, since we wish to argue from what has been observed to what has not been observed, which can only be done by means of some known relation of the observed and the unobserved; but the unobserved, by definition, is not known empirically, and therefore its relation to the observed, if known at all, must be known independently of empirical evidence. Let us see what Mill says on this subject.

According to Mill, the law of causation is proved by an admittedly fallible process called "induction by simple enumeration." This process, he says, "consists in ascribing the nature of general truths to all propositions which are true in every instance that we happen to know of."As regards its fallibility, he asserts that "the precariousness of the method of simple enumeration is in an inverse ratio to the largeness of the generalisation. The process is delusive and insufficient, exactly in proportion as the subject-matter of the observation is special and limited in extent. As the sphere widens, this unscientific method becomes less and less liable to mislead; and the most universal class of truths, the law of causation for instance, and the principles of number and of geometry, are duly and satisfactorily proved by that method alone, nor are they susceptible of any other proof."

In the above statement, there are two obvious lacunæ: (1) How is the method of simple enumeration itself justified? (2) What logical principle, if any, covers the same ground as this method, without being liable to its failures? Let us take the second question first.

A method of proof which, when used as directed, gives sometimes truth and sometimes falsehood—as the method of simple enumeration does—is obviously not a valid method, for validity demands invariable truth. Thus, if simple enumeration is to be rendered valid, it must not be stated as Mill states it. We shall have to say, at most, that the data render the result *probable*. Causation holds, we shall say, in every instance we have been able to test; therefore it *probably* holds in untested instances. There are terrible difficulties in the notion of probability, but we may ignore them at present. We thus have what at least *may* be a logical principle, since it is without exception. If a proposition is true in every instance that we happen to know of, and if the instances are very numerous, then, we shall say, it becomes very probable, on the data, that it will be true in any further instance. This is not refuted by the fact that what we declare to be probable does not always happen, for an event may be probable on the data and yet not occur. It is, however, obviously capable of further analysis, and of more exact statement. We shall have to say something like this: that every instance of a proposition being true increases the probability of its being true in a fresh instance, and that a sufficient number of favourable instances will, in the absence of instances to the contrary, make the probability of the truth of a fresh instance approach indefinitely near to certainty. Some such principle as this is required if the method of simple enumeration is to be valid.

But this brings us to our other question, namely, how is our principle known to be true? Obviously, since it is required to justify induction, it cannot be proved by induction; since it goes beyond the empirical data, it cannot be proved by them alone; since it is required to justify all inferences from empirical data to what goes beyond them, it cannot itself be even rendered in any degree probable by such data. Hence, *if* it is known, it is not known by experience, but independently of experience. I do not say that any such principle is known: I only say that it is required to justify the inferences from experience which empiricists allow, and that it cannot itself be justified empirically.

A similar conclusion can be proved by similar arguments concerning any other logical principle. Thus logical knowledge is not derivable from experience alone, and the empiricist's philosophy can therefore not be accepted in its entirety, in spite of its excellence in many matters which lie outside logic.

Hegel and his followers widened the scope of logic in quite a different way—a way which I believe to be fallacious, but which requires discussion if only to show how their conception of logic differs from the conception which I wish to advocate. In their writings, logic is practically identical with metaphysics. In broad outline, the way this came about is as follows. Hegel believed that, by means of *a priori* reasoning, it could be shown that the world *must* have various important and interesting characteristics, since any world without these characteristics would be impossible and self-contradictory. Thus what he calls "logic" is an investigation of the nature of the universe, in so far as this can be inferred merely from the principle that the universe must be logically self-consistent. I do not myself believe that from this principle alone anything of importance can be inferred as regards the existing universe. But, however that may be, I should not regard Hegel's reasoning, even if it were valid, as properly belonging to logic: it would rather be an application of logic to the actual world. Logic itself would be concerned rather with such questions as what self-consistency is, which Hegel, so far as I know, does not discuss. And though he criticises the traditional logic, and professes to replace it by an improved logic of his own, there is some sense in which the traditional logic, with all its faults, is uncritically and unconsciously assumed throughout his reasoning. It is not in the direction advocated by him, it seems to me, that the reform of logic is to be sought, but by a more fundamental, more patient, and less ambitious investigation into the presuppositions which his system shares with those of most other philosophers.

The way in which, as it seems to me, Hegel's system assumes the ordinary logic which it subsequently criticises, is exemplified by the general conception of "categories" with which he operates throughout. This conception is, I think, essentially a product of logical confusion, but it seems in some way to stand for the conception of "qualities of Reality as a whole." Mr Bradley has worked out a theory according to which, in all judgment, we are ascribing a predicate to Reality as a whole; and this theory is derived from Hegel. Now the traditional logic holds that every proposition ascribes a predicate to a subject, and from this it easily follows that there can be only one subject, the Absolute, for if there were two, the proposition that there were two would not ascribe a predicate to either. Thus Hegel's doctrine, that philosophical propositions must be of the form, "the Absolute is such-and-such," depends upon the traditional belief in the universality of the subject-predicate form. This belief, being traditional, scarcely self-conscious, and not supposed to be important, operates underground, and is assumed in arguments which, like the refutation of relations, appear at first sight such as to establish its truth. This is the most important respect in which Hegel uncritically assumes the traditional logic. Other less important respects—though important enough to be the source of such essentially Hegelian conceptions as the "concrete universal" and the "union of identity in difference"—will be found where he explicitly deals with formal logic.

There is quite another direction in which a large technical development of logic has taken place: I mean the direction of what is called logistic or mathematical logic. This kind of logic is mathematical in two different senses: it is itself a branch of mathematics, and it is the logic which is specially applicable to other more traditional branches of mathematics. Historically, it began as *merely* a branch of mathematics: its special applicability to other branches is a more recent development. In both respects, it is the fulfilment of a hope which Leibniz cherished throughout his life, and pursued with all the ardour of his amazing intellectual energy. Much of his work on this subject has been published recently, since his discoveries have been remade by others; but none was published by him, because his results persisted in contradicting certain points in the traditional doctrine of the syllogism. We now know that on these points the traditional doctrine is wrong, but respect for Aristotle prevented Leibniz from realising that this was possible.

The modern development of mathematical logic dates from Boole's *Laws of Thought* (1854). But in him and his successors, before Peano and Frege, the only thing really achieved, apart from certain details, was the invention of a mathematical symbolism for deducing consequences from the premisses which the newer methods shared with those of Aristotle. This subject has considerable interest as an independent branch of mathematics, but it has very little to do with real logic. The first serious advance in real logic since the time of the Greeks was made independently by Peano and Frege—both mathematicians. They both arrived at their logical results by an analysis of mathematics. Traditional logic regarded the two propositions, "Socrates is mortal" and "All men are mortal," as being of the same form;Peano and Frege showed that they are utterly different in form. The philosophical importance of logic may be illustrated by the fact that this confusion—which is still committed by most writers—obscured not only the whole study of the forms of judgment and inference, but also the relations of things to their qualities, of concrete existence to abstract concepts, and of the world of sense to the world of Platonic ideas. Peano and Frege, who pointed out the error, did so for technical reasons, and applied their logic mainly to technical developments; but the philosophical importance of the advance which they made is impossible to exaggerate.

Mathematical logic, even in its most modern form, is not *directly* of philosophical importance except in its beginnings. After the beginnings, it belongs rather to mathematics than to philosophy. Of its beginnings, which are the only part of it that can properly be called *philosophical* logic, I shall speak shortly. But even the later developments, though not directly philosophical, will be found of great indirect use in philosophising. They enable us to deal easily with more abstract conceptions than merely verbal reasoning can enumerate; they suggest fruitful hypotheses which otherwise could hardly be thought of; and they enable us to see quickly what is the smallest store of materials with which a given logical or scientific edifice can be constructed. Not only Frege's theory of number, which we shall deal with in Lecture VII., but the whole theory of physical concepts which will be outlined in our next two lectures, is inspired by mathematical logic, and could never have been imagined without it.

In both these cases, and in many others, we shall appeal to a certain principle called "the principle of abstraction." This principle, which might equally well be called "the principle which dispenses with abstraction," and is one which clears away incredible accumulations of metaphysical lumber, was directly suggested by mathematical logic, and could hardly have been proved or practically used without its help. The principle will be explained in our fourth lecture, but its use may be briefly indicated in advance. When a group of objects have that kind of similarity which we are inclined to attribute to possession of a common quality, the principle in question shows that membership of the group will serve all the purposes of the supposed common quality, and that therefore, unless some common quality is actually known, the group or class of similar objects may be used to replace the common quality, which need not be assumed to exist. In this and other ways, the indirect uses of even the later parts of mathematical logic are very great; but it is now time to turn our attention to its philosophical foundations.

In every proposition and in every inference there is, besides the particular subject-matter concerned, a certain *form*, a way in which the constituents of the proposition or inference are put together. If I say, "Socrates is mortal," "Jones is angry," "The sun is hot," there is something in common in these three cases, something indicated by the word "is." What is in common is the *form* of the proposition, not an actual constituent. If I say a number of things about Socrates—that he was an Athenian, that he married Xantippe, that he drank the hemlock—there is a common constituent, namely Socrates, in all the propositions I enunciate, but they have diverse forms. If, on the other hand, I take any one of these propositions and replace its constituents, one at a time, by other constituents, the form remains constant, but no constituent remains. Take (say) the series of propositions, "Socrates drank the hemlock," "Coleridge drank the hemlock," "Coleridge drank opium," "Coleridge ate opium." The form remains unchanged throughout this series, but all the constituents are altered. Thus form is not another constituent, but is the way the constituents are put together. It is forms, in this sense, that are the proper object of philosophical logic.

It is obvious that the knowledge of logical forms is something quite different from knowledge of existing things. The form of "Socrates drank the hemlock" is not an existing thing like Socrates or the hemlock, nor does it even have that close relation to existing things that drinking has. It is something altogether more abstract and remote. We might understand all the separate words of a sentence without understanding the sentence: if a sentence is long and complicated, this is apt to happen. In such a case we have knowledge of the constituents, but not of the form. We may also have knowledge of the form without having knowledge of the constituents. If I say, "Rorarius drank the hemlock," those among you who have never heard of Rorarius (supposing there are any) will understand the form, without having knowledge of all the constituents. In order to understand a sentence, it is necessary to have knowledge both of the constituents and of the particular instance of the form. It is in this way that a sentence conveys information, since it tells us that certain known objects are related according to a certain known form. Thus some kind of knowledge of logical forms, though with most people it is not explicit, is involved in all understanding of discourse. It is the business of philosophical logic to extract this knowledge from its concrete integuments, and to render it explicit and pure.

In all inference, form alone is essential: the particular subject-matter is irrelevant except as securing the truth of the premisses. This is one reason for the great importance of logical form. When I say, "Socrates was a man, all men are mortal, therefore Socrates was mortal," the connection of premisses and conclusion does not in any way depend upon its being Socrates and man and mortality that I am mentioning. The general form of the inference may be expressed in some such words as, "If a thing has a certain property, and whatever has this property has a certain other property, then the thing in question also has that other property." Here no particular things or properties are mentioned: the proposition is absolutely general. All inferences, when stated fully, are instances of propositions having this kind of generality. If they seem to depend upon the subject-matter otherwise than as regards the truth of the premisses, that is because the premisses have not been all explicitly stated. In logic, it is a waste of time to deal with inferences concerning particular cases: we deal throughout with completely general and purely formal implications, leaving it to other sciences to discover when the hypotheses are verified and when they are not.

But the forms of propositions giving rise to inferences are not the simplest forms: they are always hypothetical, stating that if one proposition is true, then so is another. Before considering inference, therefore, logic must consider those simpler forms which inference presupposes. Here the traditional logic failed completely: it believed that there was only one form of simple proposition (*i.e.* of proposition not stating a relation between two or more other propositions), namely, the form which ascribes a predicate to a subject. This is the appropriate form in assigning the qualities of a given thing—we may say "this thing is round, and red, and so on." Grammar favours this form, but philosophically it is so far from universal that it is not even very common. If we say "this thing is bigger than that," we are not assigning a mere quality of "this," but a relation of "this" and "that." We might express the same fact by saying "that thing is smaller than this," where grammatically the subject is changed. Thus propositions stating that two things have a certain relation have a different form from subject-predicate propositions, and the failure to perceive this difference or to allow for it has been the source of many errors in traditional metaphysics.

The belief or unconscious conviction that all propositions are of the subject-predicate form—in other words, that every fact consists in some thing having some quality—has rendered most philosophers incapable of giving any account of the world of science and daily life. If they had been honestly anxious to give such an account, they would probably have discovered their error very quickly; but most of them were less anxious to understand the world of science and daily life, than to convict it of unreality in the interests of a super-sensible "real" world. Belief in the unreality of the world of sense arises with irresistible force in certain moods—moods which, I imagine, have some simple physiological basis, but are none the less powerfully persuasive. The conviction born of these moods is the source of most mysticism and of most metaphysics. When the emotional intensity of such a mood subsides, a man who is in the habit of reasoning will search for logical reasons in favour of the belief which he finds in himself. But since the belief already exists, he will be very hospitable to any reason that suggests itself. The paradoxes apparently proved by his logic are really the paradoxes of mysticism, and are the goal which he feels his logic must reach if it is to be in accordance with insight. It is in this way that logic has been pursued by those of the great philosophers who were mystics—notably Plato, Spinoza, and Hegel. But since they usually took for granted the supposed insight of the mystic emotion, their logical doctrines were presented with a certain dryness, and were believed by their disciples to be quite independent of the sudden illumination from which they sprang. Nevertheless their origin clung to them, and they remained—to borrow a useful word from Mr Santayana—"malicious" in regard to the world of science and common sense. It is only so that we can account for the complacency with which philosophers have accepted the inconsistency of their doctrines with all the common and scientific facts which seem best established and most worthy of belief.

The logic of mysticism shows, as is natural, the defects which are inherent in anything malicious. While the mystic mood is dominant, the need of logic is not felt; as the mood fades, the impulse to logic reasserts itself, but with a desire to retain the vanishing insight, or at least to prove that it *was* insight, and that what seems to contradict it is illusion. The logic which thus arises is not quite disinterested or candid, and is inspired by a certain hatred of the daily world to which it is to be applied. Such an attitude naturally does not tend to the best results. Everyone knows that to read an author simply in order to refute him is not the way to understand him; and to read the book of Nature with a conviction that it is all illusion is just as unlikely to lead to understanding. If our logic is to find the common world intelligible, it must not be hostile, but must be inspired by a genuine acceptance such as is not usually to be found among metaphysicians.

Traditional logic, since it holds that all propositions have the subject-predicate form, is unable to admit the reality of relations: all relations, it maintains, must be reduced to properties of the apparently related terms. There are many ways of refuting this opinion; one of the easiest is derived from the consideration of what are called "asymmetrical" relations. In order to explain this, I will first explain two independent ways of classifying relations.

Some relations, when they hold between A and B, also hold between B and A. Such, for example, is the relation "brother or sister." If A is a brother or sister of B, then B is a brother or sister of A. Such again is any kind of similarity, say similarity of colour. Any kind of dissimilarity is also of this kind: if the colour of A is unlike the colour of B, then the colour of B is unlike the colour of A. Relations of this sort are called *symmetrical*. Thus a relation is symmetrical if, whenever it holds between A and B, it also holds between B and A.

All relations that are not symmetrical are called *non-symmetrical*. Thus "brother" is non-symmetrical, because, if A is a brother of B, it may happen that B is a *sister* of A.

A relation is called *asymmetrical* when, if it holds between A and B, it *never* holds between B and A. Thus husband, father, grandfather, etc., are asymmetrical relations. So are *before*, *after*, *greater*, *above*, *to the right of*, etc. All the relations that give rise to series are of this kind.

Classification into symmetrical, asymmetrical, and merely non-symmetrical relations is the first of the two classifications we had to consider. The second is into transitive, intransitive, and merely non-transitive relations, which are defined as follows.

A relation is said to be *transitive*, if, whenever it holds between A and B and also between B and C, it holds between A and C. Thus *before*, *after*, *greater*, *above* are transitive. All relations giving rise to series are transitive, but so are many others. The transitive relations just mentioned were asymmetrical, but many transitive relations are symmetrical—for instance, equality in any respect, exact identity of colour, being equally numerous (as applied to collections), and so on.

A relation is said to be *non-transitive* whenever it is not transitive. Thus "brother" is non-transitive, because a brother of one's brother may be oneself. All kinds of dissimilarity are non-transitive.

A relation is said to be *intransitive* when, if A has the relation to B, and B to C, A never has it to C. Thus "father" is intransitive. So is such a relation as "one inch taller" or "one year later."

Let us now, in the light of this classification, return to the question whether all relations can be reduced to predications.

In the case of symmetrical relations—*i.e.* relations which, if they hold between A and B, also hold between B and A—some kind of plausibility can be given to this doctrine. A symmetrical relation which is transitive, such as equality, can be regarded as expressing possession of some common property, while one which is not transitive, such as inequality, can be regarded as expressing possession of different properties. But when we come to asymmetrical relations, such as before and after, greater and less, etc., the attempt to reduce them to properties becomes obviously impossible. When, for example, two things are merely known to be unequal, without our knowing which is greater, we may say that the inequality results from their having different magnitudes, because inequality is a symmetrical relation; but to say that when one thing is *greater* than another, and not merely unequal to it, that means that they have different magnitudes, is formally incapable of explaining the facts. For if the other thing had been greater than the one, the magnitudes would also have been different, though the fact to be explained would not have been the same. Thus mere *difference* of magnitude is not *all* that is involved, since, if it were, there would be no difference between one thing being greater than another, and the other being greater than the one. We shall have to say that the one magnitude is *greater* than the other, and thus we shall have failed to get rid of the relation "greater." In short, both possession of the same property and possession of different properties are *symmetrical* relations, and therefore cannot account for the existence of *asymmetrical* relations.

Asymmetrical relations are involved in all series—in space and time, greater and less, whole and part, and many others of the most important characteristics of the actual world. All these aspects, therefore, the logic which reduces everything to subjects and predicates is compelled to condemn as error and mere appearance. To those whose logic is not malicious, such a wholesale condemnation appears impossible. And in fact there is no reason except prejudice, so far as I can discover, for denying the reality of relations. When once their reality is admitted, all *logical* grounds for supposing the world of sense to be illusory disappear. If this is to be supposed, it must be frankly and simply on the ground of mystic insight unsupported by argument. It is impossible to argue against what professes to be insight, so long as it does not argue in its own favour. As logicians, therefore, we may admit the possibility of the mystic's world, while yet, so long as we do not have his insight, we must continue to study the everyday world with which we are familiar. But when he contends that our world is impossible, then our logic is ready to repel his attack. And the first step in creating the logic which is to perform this service is the recognition of the reality of relations.

Relations which have two terms are only one kind of relations. A relation may have three terms, or four, or any number. Relations of two terms, being the simplest, have received more attention than the others, and have generally been alone considered by philosophers, both those who accepted and those who denied the reality of relations. But other relations have their importance, and are indispensable in the solution of certain problems. Jealousy, for example, is a relation between three people. Professor Royce mentions the relation "giving": when A gives B to C, that is a relation of three terms. When a man says to his wife: "My dear, I wish you could induce Angelina to accept Edwin," his wish constitutes a relation between four people, himself, his wife, Angelina, and Edwin. Thus such relations are by no means recondite or rare. But in order to explain exactly how they differ from relations of two terms, we must embark upon a classification of the logical forms of facts, which is the first business of logic, and the business in which the traditional logic has been most deficient.

The existing world consists of many things with many qualities and relations. A complete description of the existing world would require not only a catalogue of the things, but also a mention of all their qualities and relations. We should have to know not only this, that, and the other thing, but also which was red, which yellow, which was earlier than which, which was between which two others, and so on. When I speak of a "fact," I do not mean one of the simple things in the world; I mean that a certain thing has a certain quality, or that certain things have a certain relation. Thus, for example, I should not call Napoleon a fact, but I should call it a fact that he was ambitious, or that he married Josephine. Now a fact, in this sense, is never simple, but always has two or more constituents. When it simply assigns a quality to a thing, it has only two constituents, the thing and the quality. When it consists of a relation between two things, it has three constituents, the things and the relation. When it consists of a relation between three things, it has four constituents, and so on. The constituents of facts, in the sense in which we are using the word "fact," are not other facts, but are things and qualities or relations. When we say that there are relations of more than two terms, we mean that there are single facts consisting of a single relation and more than two things. I do not mean that one relation of two terms may hold between A and B, and also between A and C, as, for example, a man is the son of his father and also the son of his mother. This constitutes two distinct facts: if we choose to treat it as one fact, it is a fact which has facts for its constituents. But the facts I am speaking of have no facts among their constituents, but only things and relations. For example, when A is jealous of B on account of C, there is only one fact, involving three people; there are not two instances of jealousy, but only one. It is in such cases that I speak of a relation of three terms, where the simplest possible fact in which the relation occurs is one involving three things in addition to the relation. And the same applies to relations of four terms or five or any other number. All such relations must be admitted in our inventory of the logical forms of facts: two facts involving the same number of things have the same form, and two which involve different numbers of things have different forms.

Given any fact, there is an assertion which expresses the fact. The fact itself is objective, and independent of our thought or opinion about it; but the assertion is something which involves thought, and may be either true or false. An assertion may be positive or negative: we may assert that Charles I. was executed, or that he did *not* die in his bed. A negative assertion may be said to be a *denial*. Given a form of words which must be either true or false, such as "Charles I. died in his bed," we may either assert or deny this form of words: in the one case we have a positive assertion, in the other a negative one. A form of words which must be either true or false I shall call a *proposition*. Thus a proposition is the same as what may be significantly asserted or denied. A proposition which expresses what we have called a fact, *i.e.* which, when asserted, asserts that a certain thing has a certain quality, or that certain things have a certain relation, will be called an atomic proposition, because, as we shall see immediately, there are other propositions into which atomic propositions enter in a way analogous to that in which atoms enter into molecules. Atomic propositions, although, like facts, they may have any one of an infinite number of forms, are only one kind of propositions. All other kinds are more complicated. In order to preserve the parallelism in language as regards facts and propositions, we shall give the name "atomic facts" to the facts we have hitherto been considering. Thus atomic facts are what determine whether atomic propositions are to be asserted or denied.

Whether an atomic proposition, such as "this is red," or "this is before that," is to be asserted or denied can only be known empirically. Perhaps one atomic fact may sometimes be capable of being inferred from another, though this seems very doubtful; but in any case it cannot be inferred from premisses no one of which is an atomic fact. It follows that, if atomic facts are to be known at all, some at least must be known without inference. The atomic facts which we come to know in this way are the facts of sense-perception; at any rate, the facts of sense-perception are those which we most obviously and certainly come to know in this way. If we knew all atomic facts, and also knew that there were none except those we knew, we should, theoretically, be able to infer all truths of whatever form.Thus logic would then supply us with the whole of the apparatus required. But in the first acquisition of knowledge concerning atomic facts, logic is useless. In pure logic, no atomic fact is ever mentioned: we confine ourselves wholly to forms, without asking ourselves what objects can fill the forms. Thus pure logic is independent of atomic facts; but conversely, they are, in a sense, independent of logic. Pure logic and atomic facts are the two poles, the wholly *a priori* and the wholly empirical. But between the two lies a vast intermediate region, which we must now briefly explore.

"Molecular" propositions are such as contain conjunctions—*if*, *or*, *and*, *unless*, etc.—and such words are the marks of a molecular proposition. Consider such an assertion as, "If it rains, I shall bring my umbrella." This assertion is just as capable of truth or falsehood as the assertion of an atomic proposition, but it is obvious that either the corresponding fact, or the nature of the correspondence with fact, must be quite different from what it is in the case of an atomic proposition. Whether it rains, and whether I bring my umbrella, are each severally matters of atomic fact, ascertainable by observation. But the connection of the two involved in saying that *if* the one happens, *then* the other will happen, is something radically different from either of the two separately. It does not require for its truth that it should actually rain, or that I should actually bring my umbrella; even if the weather is cloudless, it may still be true that I should have brought my umbrella if the weather had been different. Thus we have here a connection of two propositions, which does not depend upon whether they are to be asserted or denied, but only upon the second being inferable from the first. Such propositions, therefore, have a form which is different from that of any atomic proposition.

Such propositions are important to logic, because all inference depends upon them. If I have told you that if it rains I shall bring my umbrella, and if you see that there is a steady downpour, you can infer that I shall bring my umbrella. There can be no inference except where propositions are connected in some such way, so that from the truth or falsehood of the one something follows as to the truth or falsehood of the other. It seems to be the case that we can sometimes know molecular propositions, as in the above instance of the umbrella, when we do not know whether the component atomic propositions are true or false. The *practical* utility of inference rests upon this fact.

The next kind of propositions we have to consider are *general* propositions, such as "all men are mortal," "all equilateral triangles are equiangular." And with these belong propositions in which the word "some" occurs, such as "some men are philosophers" or "some philosophers are not wise." These are the denials of general propositions, namely (in the above instances), of "all men are non-philosophers" and "all philosophers are wise." We will call propositions containing the word "some" *negative* general propositions, and those containing the word "all" *positive* general propositions. These propositions, it will be seen, begin to have the appearance of the propositions in logical text-books. But their peculiarity and complexity are not known to the text-books, and the problems which they raise are only discussed in the most superficial manner.

When we were discussing atomic facts, we saw that we should be able, theoretically, to infer all other truths by logic if we knew all atomic facts and also knew that there were no other atomic facts besides those we knew. The knowledge that there are no other atomic facts is positive general knowledge; it is the knowledge that "all atomic facts are known to me," or at least "all atomic facts are in this collection"—however the collection may be given. It is easy to see that general propositions, such as "all men are mortal," cannot be known by inference from atomic facts alone. If we could know each individual man, and know that he was mortal, that would not enable us to know that all men are mortal, unless we *knew* that those were all the men there are, which is a general proposition. If we knew every other existing thing throughout the universe, and knew that each separate thing was not an immortal man, that would not give us our result unless we *knew* that we had explored the whole universe, *i.e.* unless we knew "all things belong to this collection of things I have examined." Thus general truths cannot be inferred from particular truths alone, but must, if they are to be known, be either self-evident, or inferred from premisses of which at least one is a general truth. But all *empirical* evidence is of *particular* truths. Hence, if there is any knowledge of general truths at all, there must be *some* knowledge of general truths which is independent of empirical evidence, *i.e.* does not depend upon the data of sense.

The above conclusion, of which we had an instance in the case of the inductive principle, is important, since it affords a refutation of the older empiricists. They believed that all our knowledge is derived from the senses and dependent upon them. We see that, if this view is to be maintained, we must refuse to admit that we know any general propositions. It is perfectly possible logically that this should be the case, but it does not appear to be so in fact, and indeed no one would dream of maintaining such a view except a theorist at the last extremity. We must therefore admit that there is general knowledge not derived from sense, and that some of this knowledge is not obtained by inference but is primitive.

Such general knowledge is to be found in logic. Whether there is any such knowledge not derived from logic, I do not know; but in logic, at any rate, we have such knowledge. It will be remembered that we excluded from pure logic such propositions as, "Socrates is a man, all men are mortal, therefore Socrates is mortal," because Socrates and *man* and *mortal* are empirical terms, only to be understood through particular experience. The corresponding proposition in pure logic is: "If anything has a certain property, and whatever has this property has a certain other property, then the thing in question has the other property." This proposition is absolutely general: it applies to all things and all properties. And it is quite self-evident. Thus in such propositions of pure logic we have the self-evident general propositions of which we were in search.

A proposition such as, "If Socrates is a man, and all men are mortal, then Socrates is mortal," is true in virtue of its *form* alone. Its truth, in this hypothetical form, does not depend upon whether Socrates actually is a man, nor upon whether in fact all men are mortal; thus it is equally true when we substitute other terms for Socrates and *man* and *mortal*. The general truth of which it is an instance is purely formal, and belongs to logic. Since it does not mention any particular thing, or even any particular quality or relation, it is wholly independent of the accidental facts of the existent world, and can be known, theoretically, without any experience of particular things or their qualities and relations.

Logic, we may say, consists of two parts. The first part investigates what propositions are and what forms they may have; this part enumerates the different kinds of atomic propositions, of molecular propositions, of general propositions, and so on. The second part consists of certain supremely general propositions, which assert the truth of all propositions of certain forms. This second part merges into pure mathematics, whose propositions all turn out, on analysis, to be such general formal truths. The first part, which merely enumerates forms, is the more difficult, and philosophically the more important; and it is the recent progress in this first part, more than anything else, that has rendered a truly scientific discussion of many philosophical problems possible.

The problem of the nature of judgment or belief may be taken as an example of a problem whose solution depends upon an adequate inventory of logical forms. We have already seen how the supposed universality of the subject-predicate form made it impossible to give a right analysis of serial order, and therefore made space and time unintelligible. But in this case it was only necessary to admit relations of two terms. The case of judgment demands the admission of more complicated forms. If all judgments were true, we might suppose that a judgment consisted in apprehension of a *fact*, and that the apprehension was a relation of a mind to the fact. From poverty in the logical inventory, this view has often been held. But it leads to absolutely insoluble difficulties in the case of error. Suppose I believe that Charles I. died in his bed. There is no objective fact "Charles I.'s death in his bed" to which I can have a relation of apprehension. Charles I. and death and his bed are objective, but they are not, except in my thought, put together as my false belief supposes. It is therefore necessary, in analysing a belief, to look for some other logical form than a two-term relation. Failure to realise this necessity has, in my opinion, vitiated almost everything that has hitherto been written on the theory of knowledge, making the problem of error insoluble and the difference between belief and perception inexplicable.

Modern logic, as I hope is now evident, has the effect of enlarging our abstract imagination, and providing an infinite number of possible hypotheses to be applied in the analysis of any complex fact. In this respect it is the exact opposite of the logic practised by the classical tradition. In that logic, hypotheses which seem *primâ facie* possible are professedly proved impossible, and it is decreed in advance that reality must have a certain special character. In modern logic, on the contrary, while the *primâ facie* hypotheses as a rule remain admissible, others, which only logic would have suggested, are added to our stock, and are very often found to be indispensable if a right analysis of the facts is to be obtained. The old logic put thought in fetters, while the new logic gives it wings. It has, in my opinion, introduced the same kind of advance into philosophy as Galileo introduced into physics, making it possible at last to see what kinds of problems may be capable of solution, and what kinds must be abandoned as beyond human powers. And where a solution appears possible, the new logic provides a method which enables us to obtain results that do not merely embody personal idiosyncrasies, but must command the assent of all who are competent to form an opinion.

LECTURE III
ON OUR KNOWLEDGE OF THE EXTERNAL WORLD

PHILOSOPHY may be approached by many roads, but one of the oldest and most travelled is the road which leads through doubt as to the reality of the world of sense. In Indian mysticism, in Greek and modern monistic philosophy from Parmenides onward, in Berkeley, in modern physics, we find sensible appearance criticised and condemned for a bewildering variety of motives. The mystic condemns it on the ground of immediate knowledge of a more real and significant world behind the veil; Parmenides and Plato condemn it because its continual flux is thought inconsistent with the unchanging nature of the abstract entities revealed by logical analysis; Berkeley brings several weapons, but his chief is the subjectivity of sense-data, their dependence upon the organisation and point of view of the spectator; while modern physics, on the basis of sensible evidence itself, maintains a mad dance of electrons which has, superficially at least, very little resemblance to the immediate objects of sight or touch.

Every one of these lines of attack raises vital and interesting problems.

The mystic, so long as he merely reports a positive revelation, cannot be refuted; but when he *denies* reality to objects of sense, he may be questioned as to what he means by "reality," and may be asked how their unreality follows from the supposed reality of his super-sensible world. In answering these questions, he is led to a logic which merges into that of Parmenides and Plato and the idealist tradition.

The logic of the idealist tradition has gradually grown very complex and very abstruse, as may be seen from the Bradleian sample considered in our first lecture. If we attempted to deal fully with this logic, we should not have time to reach any other aspect of our subject; we will therefore, while acknowledging that it deserves a long discussion, pass by its central doctrines with only such occasional criticism as may serve to exemplify other topics, and concentrate our attention on such matters as its objections to the continuity of motion and the infinity of space and time—objections which have been fully answered by modern mathematicians in a manner constituting an abiding triumph for the method of logical analysis in philosophy. These objections and the modern answers to them will occupy our fifth, sixth, and seventh lectures.

Berkeley's attack, as reinforced by the physiology of the sense-organs and nerves and brain, is very powerful. I think it must be admitted as probable that the immediate objects of sense depend for their existence upon physiological conditions in ourselves, and that, for example, the coloured surfaces which we see cease to exist when we shut our eyes. But it would be a mistake to infer that they are dependent upon mind, not real while we see them, or not the sole basis for our knowledge of the external world. This line of argument will be developed in the present lecture.

The discrepancy between the world of physics and the world of sense, which we shall consider in our fourth lecture, will be found to be more apparent than real, and it will be shown that whatever there is reason to believe in physics can probably be interpreted in terms of sense.

The instrument of discovery throughout is modern logic, a very different science from the logic of the text-books and also from the logic of idealism. Our second lecture has given a short account of modern logic and of its points of divergence from the various traditional kinds of logic.

In our last lecture, after a discussion of causality and free will, we shall try to reach a general account of the logical-analytic method of scientific philosophy, and a tentative estimate of the hopes of philosophical progress which it allows us to entertain.

In this lecture, I wish to apply the logical-analytic method to one of the oldest problems of philosophy, namely, the problem of our knowledge of the external world. What I have to say on this problem does not amount to an answer of a definite and dogmatic kind; it amounts only to an analysis and statement of the questions involved, with an indication of the directions in which evidence may be sought. But although not yet a definite solution, what can be said at present seems to me to throw a completely new light on the problem, and to be indispensable, not only in seeking the answer, but also in the preliminary question as to what parts of our problem may possibly have an ascertainable answer.

In every philosophical problem, our investigation starts from what may be called "data," by which I mean matters of common knowledge, vague, complex, inexact, as common knowledge always is, but yet somehow commanding our assent as on the whole and in some interpretation pretty certainly true. In the case of our present problem, the common knowledge involved is of various kinds. There is first our acquaintance with particular objects of daily life—furniture, houses, towns, other people, and so on. Then there is the extension of such particular knowledge to particular things outside our personal experience, through history and geography, newspapers, etc. And lastly, there is the systematisation of all this knowledge of particulars by means of physical science, which derives immense persuasive force from its astonishing power of foretelling the future. We are quite willing to admit that there may be errors of detail in this knowledge, but we believe them to be discoverable and corrigible by the methods which have given rise to our beliefs, and we do not, as practical men, entertain for a moment the hypothesis that the whole edifice may be built on insecure foundations. In the main, therefore, and without absolute dogmatism as to this or that special portion, we may accept this mass of common knowledge as affording data for our philosophical analysis.

It may be said—and this is an objection which must be met at the outset—that it is the duty of the philosopher to call in question the admittedly fallible beliefs of daily life, and to replace them by something more solid and irrefragable. In a sense this is true, and in a sense it is effected in the course of analysis. But in another sense, and a very important one, it is quite impossible. While admitting that doubt is possible with regard to all our common knowledge, we must nevertheless accept that knowledge in the main if philosophy is to be possible at all. There is not any superfine brand of knowledge, obtainable by the philosopher, which can give us a standpoint from which to criticise the whole of the knowledge of daily life. The most that can be done is to examine and purify our common knowledge by an internal scrutiny, assuming the canons by which it has been obtained, and applying them with more care and with more precision. Philosophy cannot boast of having achieved such a degree of certainty that it can have authority to condemn the facts of experience and the laws of science. The philosophic scrutiny, therefore, though sceptical in regard to every detail, is not sceptical as regards the whole. That is to say, its criticism of details will only be based upon their relation to other details, not upon some external criterion which can be applied to all the details equally. The reason for this abstention from a universal criticism is not any dogmatic confidence, but its exact opposite; it is not that common knowledge *must* be true, but that we possess no radically different kind of knowledge derived from some other source. Universal scepticism, though logically irrefutable, is practically barren; it can only, therefore, give a certain flavour of hesitancy to our beliefs, and cannot be used to substitute other beliefs for them.

Although data can only be criticised by other data, not by an outside standard, yet we may distinguish different grades of certainty in the different kinds of common knowledge which we enumerated just now. What does not go beyond our own personal sensible acquaintance must be for us the most certain: the "evidence of the senses" is proverbially the least open to question. What depends on testimony, like the facts of history and geography which are learnt from books, has varying degrees of certainty according to the nature and extent of the testimony. Doubts as to the existence of Napoleon can only be maintained for a joke, whereas the historicity of Agamemnon is a legitimate subject of debate. In science, again, we find all grades of certainty short of the highest. The law of gravitation, at least as an approximate truth, has acquired by this time the same kind of certainty as the existence of Napoleon, whereas the latest speculations concerning the constitution of matter would be universally acknowledged to have as yet only a rather slight probability in their favour. These varying degrees of certainty attaching to different data may be regarded as themselves forming part of our data; they, along with the other data, lie within the vague, complex, inexact body of knowledge which it is the business of the philosopher to analyse.

The first thing that appears when we begin to analyse our common knowledge is that some of it is derivative, while some is primitive; that is to say, there is some that we only believe because of something else from which it has been inferred in some sense, though not necessarily in a strict logical sense, while other parts are believed on their own account, without the support of any outside evidence. It is obvious that the senses give knowledge of the latter kind: the immediate facts perceived by sight or touch or hearing do not need to be proved by argument, but are completely self-evident. Psychologists, however, have made us aware that what is actually given in sense is much less than most people would naturally suppose, and that much of what at first sight seems to be given is really inferred. This applies especially in regard to our space-perceptions. For instance, we instinctively infer the "real" size and shape of a visible object from its apparent size and shape, according to its distance and our point of view. When we hear a person speaking, our actual sensations usually miss a great deal of what he says, and we supply its place by unconscious inference; in a foreign language, where this process is more difficult, we find ourselves apparently grown deaf, requiring, for example, to be much nearer the stage at a theatre than would be necessary in our own country. Thus the first step in the analysis of data, namely, the discovery of what is really given in sense, is full of difficulty. We will, however, not linger on this point; so long as its existence is realised, the exact outcome does not make any very great difference in our main problem.

The next step in our analysis must be the consideration of how the derivative parts of our common knowledge arise. Here we become involved in a somewhat puzzling entanglement of logic and psychology. Psychologically, a belief may be called derivative whenever it is caused by one or more other beliefs, or by some fact of sense which is not simply what the belief asserts. Derivative beliefs in this sense constantly arise without any process of logical inference, merely by association of ideas or some equally extra-logical process. From the expression of a man's face, we judge as to what he is feeling: we say we *see* that he is angry, when in fact we only see a frown. We do not judge as to his state of mind by any logical process: the judgment grows up, often without our being able to say what physical mark of emotion we actually saw. In such a case, the knowledge is derivative psychologically; but logically it is in a sense primitive, since it is not the result of any logical deduction. There may or may not be a possible deduction leading to the same result, but whether there is or not, we certainly do not employ it. If we call a belief "logically primitive" when it is not actually arrived at by a logical inference, then innumerable beliefs are logically primitive which psychologically are derivative. The separation of these two kinds of primitiveness is vitally important to our present discussion.

When we reflect upon the beliefs which are logically but not psychologically primitive, we find that, unless they can on reflection be deduced by a logical process from beliefs which are also psychologically primitive, our confidence in their truth tends to diminish the more we think about them. We naturally believe, for example, that tables and chairs, trees and mountains, are still there when we turn our backs upon them. I do not wish for a moment to maintain that this is certainly not the case, but I do maintain that the question whether it is the case is not to be settled off-hand on any supposed ground of obviousness. The belief that they persist is, in all men except a few philosophers, logically primitive, but it is not psychologically primitive; psychologically, it arises only through our having seen those tables and chairs, trees and mountains. As soon as the question is seriously raised whether, because we have seen them, we have a right to suppose that they are there still, we feel that some kind of argument must be produced, and that if none is forthcoming, our belief can be no more than a pious opinion. We do not feel this as regards the immediate objects of sense: there they are, and as far as their momentary existence is concerned, no further argument is required. There is accordingly more need of justifying our psychologically derivative beliefs than of justifying those that are primitive.

We are thus led to a somewhat vague distinction between what we may call "hard" data and "soft" data. This distinction is a matter of degree, and must not be pressed; but if not taken too seriously it may help to make the situation clear. I mean by "hard" data those which resist the solvent influence of critical reflection, and by "soft" data those which, under the operation of this process, become to our minds more or less doubtful. The hardest of hard data are of two sorts: the particular facts of sense, and the general truths of logic. The more we reflect upon these, the more we realise exactly what they are, and exactly what a doubt concerning them really means, the more luminously certain do they become. *Verbal* doubt concerning even these is possible, but verbal doubt may occur when what is nominally being doubted is not really in our thoughts, and only words are actually present to our minds. Real doubt, in these two cases, would, I think, be pathological. At any rate, to me they seem quite certain, and I shall assume that you agree with me in this. Without this assumption, we are in danger of falling into that universal scepticism which, as we saw, is as barren as it is irrefutable. If we are to continue philosophising, we must make our bow to the sceptical hypothesis, and, while admitting the elegant terseness of its philosophy, proceed to the consideration of other hypotheses which, though perhaps not certain, have at least as good a right to our respect as the hypothesis of the sceptic.

Applying our distinction of "hard" and "soft" data to psychologically derivative but logically primitive beliefs, we shall find that most, if not all, are to be classed as soft data. They may be found, on reflection, to be capable of logical proof, and they then again become believed, but no longer as data. As data, though entitled to a certain limited respect, they cannot be placed on a level with the facts of sense or the laws of logic. The kind of respect which they deserve seems to me such as to warrant us in hoping, though not too confidently, that the hard data may prove them to be at least probable. Also, if the hard data are found to throw no light whatever upon their truth or falsehood, we are justified, I think, in giving rather more weight to the hypothesis of their truth than to the hypothesis of their falsehood. For the present, however, let us confine ourselves to the hard data, with a view to discovering what sort of world can be constructed by their means alone.

Our data now are primarily the facts of sense (*i.e.* of *our own* sense-data) and the laws of logic. But even the severest scrutiny will allow some additions to this slender stock. Some facts of memory—especially of recent memory—seem to have the highest degree of certainty. Some introspective facts are as certain as any facts of sense. And facts of sense themselves must, for our present purposes, be interpreted with a certain latitude. Spatial and temporal relations must sometimes be included, for example in the case of a swift motion falling wholly within the specious present. And some facts of comparison, such as the likeness or unlikeness of two shades of colour, are certainly to be included among hard data. Also we must remember that the distinction of hard and soft data is psychological and subjective, so that, if there are other minds than our own—which at our present stage must be held doubtful—the catalogue of hard data may be different for them from what it is for us.

Certain common beliefs are undoubtedly excluded from hard data. Such is the belief which led us to introduce the distinction, namely, that sensible objects in general persist when we are not perceiving them. Such also is the belief in other people's minds: this belief is obviously derivative from our perception of their bodies, and is felt to demand logical justification as soon as we become aware of its derivativeness. Belief in what is reported by the testimony of others, including all that we learn from books, is of course involved in the doubt as to whether other people have minds at all. Thus the world from which our reconstruction is to begin is very fragmentary. The best we can say for it is that it is slightly more extensive than the world at which Descartes arrived by a similar process, since that world contained nothing except himself and his thoughts.

We are now in a position to understand and state the problem of our knowledge of the external world, and to remove various misunderstandings which have obscured the meaning of the problem. The problem really is: Can the existence of anything other than our own hard data be inferred from the existence of those data? But before considering this problem, let us briefly consider what the problem is *not*.

When we speak of the "external" world in this discussion, we must not mean "spatially external," unless "space" is interpreted in a peculiar and recondite manner. The immediate objects of sight, the coloured surfaces which make up the visible world, are spatially external in the natural meaning of this phrase. We feel them to be "there" as opposed to "here"; without making any assumption of an existence other than hard data, we can more or less estimate the distance of a coloured surface. It seems probable that distances, provided they are not too great, are actually given more or less roughly in sight; but whether this is the case or not, ordinary distances can certainly be estimated approximately by means of the data of sense alone. The immediately given world is spatial, and is further not wholly contained within our own bodies. Thus our knowledge of what is external in this sense is not open to doubt.

Another form in which the question is often put is: "Can we know of the existence of any reality which is independent of ourselves?" This form of the question suffers from the ambiguity of the two words "independent" and "self." To take the Self first: the question as to what is to be reckoned part of the Self and what is not, is a very difficult one. Among many other things which we may mean by the Self, two may be selected as specially important, namely, (1) the bare subject which thinks and is aware of objects, (2) the whole assemblage of things that would necessarily cease to exist if our lives came to an end. The bare subject, if it exists at all, is an inference, and is not part of the data; therefore this meaning of Self may be ignored in our present inquiry. The second meaning is difficult to make precise, since we hardly know what things depend upon our lives for their existence. And in this form, the definition of Self introduces the word "depend," which raises the same questions as are raised by the word "independent." Let us therefore take up the word "independent," and return to the Self later.

When we say that one thing is "independent" of another, we may mean either that it is logically possible for the one to exist without the other, or that there is no causal relation between the two such that the one only occurs as the effect of the other. The only way, so far as I know, in which one thing can be *logically* dependent upon another is when the other is *part* of the one. The existence of a book, for example, is logically dependent upon that of its pages: without the pages there would be no book. Thus in this sense the question, "Can we know of the existence of any reality which is independent of ourselves?" reduces to the question, "Can we know of the existence of any reality of which our Self is not part?" In this form, the question brings us back to the problem of defining the Self; but I think, however the Self may be defined, even when it is taken as the bare subject, it cannot be supposed to be part of the immediate object of sense; thus in this form of the question we must admit that we can know of the existence of realities independent of ourselves.

The question of causal dependence is much more difficult. To know that one kind of thing is causally independent of another, we must know that it actually occurs without the other. Now it is fairly obvious that, whatever legitimate meaning we give to the Self, our thoughts and feelings are causally dependent upon ourselves, *i.e.* do not occur when there is no Self for them to belong to. But in the case of objects of sense this is not obvious; indeed, as we saw, the common-sense view is that such objects persist in the absence of any percipient. If this is the case, then they are causally independent of ourselves; if not, not. Thus in this form the question reduces to the question whether we can know that objects of sense, or any other objects not our own thoughts and feelings, exist at times when we are not perceiving them. This form, in which the difficult word "independent" no longer occurs, is the form in which we stated the problem a minute ago.

Our question in the above form raises two distinct problems, which it is important to keep separate. First, can we know that objects of sense, or very similar objects, exist at times when we are not perceiving them? Secondly, if this cannot be known, can we know that other objects, inferable from objects of sense but not necessarily resembling them, exist either when we are perceiving the objects of sense or at any other time? This latter problem arises in philosophy as the problem of the "thing in itself," and in science as the problem of matter as assumed in physics. We will consider this latter problem first.

Owing to the fact that we feel passive in sensation, we naturally suppose that our sensations have outside causes. Now it is necessary here first of all to distinguish between (1) our sensation, which is a mental event consisting in our being aware of a sensible object, and (2) the sensible object of which we are aware in sensation. When I speak of the sensible object, it must be understood that I do not mean such a thing as a table, which is both visible and tangible, can be seen by many people at once, and is more or less permanent. What I mean is just that patch of colour which is momentarily seen when we look at the table, or just that particular hardness which is felt when we press it, or just that particular sound which is heard when we rap it. Each of these I call a sensible object, and our awareness of it I call a sensation. Now our sense of passivity, if it really afforded any argument, would only tend to show that the *sensation* has an outside cause; this cause we should naturally seek in the sensible object. Thus there is no good reason, so far, for supposing that sensible objects must have outside causes. But both the thing-in-itself of philosophy and the matter of physics present themselves as outside causes of the sensible object as much as of the sensation. What are the grounds for this common opinion?

In each case, I think, the opinion has resulted from the combination of a belief that *something* which can persist independently of our consciousness makes itself known in sensation, with the fact that our sensations often change in ways which seem to depend upon us rather than upon anything which would be supposed to persist independently of us. At first, we believe unreflectingly that everything is as it seems to be, and that, if we shut our eyes, the objects we had been seeing remain as they were though we no longer see them. But there are arguments against this view, which have generally been thought conclusive. It is extraordinarily difficult to see just what the arguments prove; but if we are to make any progress with the problem of the external world, we must try to make up our minds as to these arguments.

A table viewed from one place presents a different appearance from that which it presents from another place. This is the language of common sense, but this language already assumes that there is a real table of which we see the appearances. Let us try to state what is known in terms of sensible objects alone, without any element of hypothesis. We find that as we walk round the table, we perceive a series of gradually changing visible objects. But in speaking of "walking round the table," we have still retained the hypothesis that there is a single table connected with all the appearances. What we ought to say is that, while we have those muscular and other sensations which make us say we are walking, our visual sensations change in a continuous way, so that, for example, a striking patch of colour is not suddenly replaced by something wholly different, but is replaced by an insensible gradation of slightly different colours with slightly different shapes. This is what we really know by experience, when we have freed our minds from the assumption of permanent "things" with changing appearances. What is really known is a correlation of muscular and other bodily sensations with changes in visual sensations.

But walking round the table is not the only way of altering its appearance. We can shut one eye, or put on blue spectacles, or look through a microscope. All these operations, in various ways, alter the visual appearance which we call that of the table. More distant objects will also alter their appearance if (as we say) the state of the atmosphere changes—if there is fog or rain or sunshine. Physiological changes also alter the appearances of things. If we assume the world of common sense, all these changes, including those attributed to physiological causes, are changes in the intervening medium. It is not quite so easy as in the former case to reduce this set of facts to a form in which nothing is assumed beyond sensible objects. Anything intervening between ourselves and what we see must be invisible: our view in every direction is bounded by the nearest visible object. It might be objected that a dirty pane of glass, for example, is visible although we can see things through it. But in this case we really see a spotted patchwork: the dirtier specks in the glass are visible, while the cleaner parts are invisible and allow us to see what is beyond. Thus the discovery that the intervening medium affects the appearances of things cannot be made by means of the sense of sight alone.

Let us take the case of the blue spectacles, which is the simplest, but may serve as a type for the others. The frame of the spectacles is of course visible, but the blue glass, if it is clean, is not visible. The blueness, which we say is in the glass, appears as being in the objects seen through the glass. The glass itself is known by means of the sense of touch. In order to know that it is between us and the objects seen through it, we must know how to correlate the space of touch with the space of sight. This correlation itself, when stated in terms of the data of sense alone, is by no means a simple matter. But it presents no difficulties of principle, and may therefore be supposed accomplished. When it has been accomplished, it becomes possible to attach a meaning to the statement that the blue glass, which we can touch, is between us and the object seen, as we say, "through" it.

But we have still not reduced our statement completely to what is actually given in sense. We have fallen into the assumption that the object of which we are conscious when we touch the blue spectacles still exists after we have ceased to touch them. So long as we are touching them, nothing except our finger can be seen through the part touched, which is the only part where we immediately know that there is something. If we are to account for the blue appearance of objects other than the spectacles, when seen through them, it might seem as if we must assume that the spectacles still exist when we are not touching them; and if this assumption really is necessary, our main problem is answered: we have means of knowing of the present existence of objects not given in sense, though of the same kind as objects formerly given in sense.

It may be questioned, however, whether this assumption is actually unavoidable, though it is unquestionably the most natural one to make. We may say that the object of which we become aware when we touch the spectacles continues to have effects afterwards, though perhaps it no longer exists. In this view, the supposed continued existence of sensible objects after they have ceased to be sensible will be a fallacious inference from the fact that they still have effects. It is often supposed that nothing which has ceased to exist can continue to have effects, but this is a mere prejudice, due to a wrong conception of causality. We cannot, therefore, dismiss our present hypothesis on the ground of *a priori* impossibility, but must examine further whether it can really account for the facts.

It may be said that our hypothesis is useless in the case when the blue glass is never touched at all. How, in that case, are we to account for the blue appearance of objects? And more generally, what are we to make of the hypothetical sensations of touch which we associate with untouched visible objects, which we know would be verified if we chose, though in fact we do not verify them? Must not these be attributed to permanent possession, by the objects, of the properties which touch would reveal?

Let us consider the more general question first. Experience has taught us that where we see certain kinds of coloured surfaces we can, by touch, obtain certain expected sensations of hardness or softness, tactile shape, and so on. This leads us to believe that what is seen is usually tangible, and that it has, whether we touch it or not, the hardness or softness which we should expect to feel if we touched it. But the mere fact that we are able to infer what our tactile sensations would be shows that it is not logically necessary to assume tactile qualities before they are felt. All that is really known is that the visual appearance in question, together with touch, will lead to certain sensations, which can necessarily be determined in terms of the visual appearance, since otherwise they could not be inferred from it.

We can now give a statement of the experienced facts concerning the blue spectacles, which will supply an interpretation of common-sense beliefs without assuming anything beyond the existence of sensible objects at the times when they are sensible. By experience of the correlation of touch and sight sensations, we become able to associate a certain place in touch-space with a certain corresponding place in sight-space. Sometimes, namely in the case of transparent things, we find that there is a tangible object in a touch-place without there being any visible object in the corresponding sight-place. But in such a case as that of the blue spectacles, we find that whatever object is visible beyond the empty sight-place in the same line of sight has a different colour from what it has when there is no tangible object in the intervening touch-place; and as we move the tangible object in touch-space, the blue patch moves in sight-space. If now we find a blue patch moving in this way in sight-space, when we have no sensible experience of an intervening tangible object, we nevertheless infer that, if we put our hand at a certain place in touch-space, we should experience a certain touch-sensation. If we are to avoid non-sensible objects, this must be taken as the whole of our meaning when we say that the blue spectacles are in a certain place, though we have not touched them, and have only seen other things rendered blue by their interposition.

I think it may be laid down quite generally that, *in so far* as physics or common sense is verifiable, it must be capable of interpretation in terms of actual sense-data alone. The reason for this is simple. Verification consists always in the occurrence of an expected sense-datum. Astronomers tell us there will be an eclipse of the moon: we look at the moon, and find the earth's shadow biting into it, that is to say, we see an appearance quite different from that of the usual full moon. Now if an expected sense-datum constitutes a verification, what was asserted must have been about sense-data; or, at any rate, if part of what was asserted was not about sense-data, then only the other part has been verified. There is in fact a certain regularity or conformity to law about the occurrence of sense-data, but the sense-data that occur at one time are often causally connected with those that occur at quite other times, and not, or at least not very closely, with those that occur at neighbouring times. If I look at the moon and immediately afterwards hear a train coming, there is no very close causal connection between my two sense-data; but if I look at the moon on two nights a week apart, there is a very close causal connection between the two sense-data. The simplest, or at least the easiest, statement of the connection is obtained by imagining a "real" moon which goes on whether I look at it or not, providing a series of *possible* sense-data of which only those are actual which belong to moments when I choose to look at the moon.

But the degree of verification obtainable in this way is very small. It must be remembered that, at our present level of doubt, we are not at liberty to accept testimony. When we hear certain noises, which are those we should utter if we wished to express a certain thought, we assume that that thought, or one very like it, has been in another mind, and has given rise to the expression which we hear. If at the same time we see a body resembling our own, moving its lips as we move ours when we speak, we cannot resist the belief that it is alive, and that the feelings inside it continue when we are not looking at it. When we see our friend drop a weight upon his toe, and hear him say—what we should say in similar circumstances, the phenomena *can* no doubt be explained without assuming that he is anything but a series of shapes and noises seen and heard by us, but practically no man is so infected with philosophy as not to be quite certain that his friend has felt the same kind of pain as he himself would feel. We will consider the legitimacy of this belief presently; for the moment, I only wish to point out that it needs the same kind of justification as our belief that the moon exists when we do not see it, and that, without it, testimony heard or read is reduced to noises and shapes, and cannot be regarded as evidence of the facts which it reports. The verification of physics which is possible at our present level is, therefore, only that degree of verification which is possible by one man's unaided observations, which will not carry us very far towards the establishment of a whole science.

Before proceeding further, let us summarise the argument so far as it has gone. The problem is: "Can the existence of anything other than our own hard data be inferred from these data?" It is a mistake to state the problem in the form: "Can we know of the existence of anything other than ourselves and our states?" or: "Can we know of the existence of anything independent of ourselves?" because of the extreme difficulty of defining "self" and "independent" precisely. The felt passivity of sensation is irrelevant, since, even if it proved anything, it could only prove that sensations are caused by sensible objects. The natural *naïve* belief is that things seen persist, when unseen, exactly or approximately as they appeared when seen; but this belief tends to be dispelled by the fact that what common sense regards as the appearance of one object changes with what common sense regards as changes in the point of view and in the intervening medium, including in the latter our own sense-organs and nerves and brain. This fact, as just stated, assumes, however, the common-sense world of stable objects which it professes to call in question; hence, before we can discover its precise bearing on our problem, we must find a way of stating it which does not involve any of the assumptions which it is designed to render doubtful. What we then find, as the bare outcome of experience, is that gradual changes in certain sense-data are correlated with gradual changes in certain others, or (in the case of bodily motions) with the other sense-data themselves.

The assumption that sensible objects persist after they have ceased to be sensible—for example, that the hardness of a visible body, which has been discovered by touch, continues when the body is no longer touched—may be replaced by the statement that the *effects* of sensible objects persist, *i.e.* that what happens now can only be accounted for, in many cases, by taking account of what happened at an earlier time. Everything that one man, by his own personal experience, can verify in the account of the world given by common sense and physics, will be explicable by some such means, since verification consists merely in the occurrence of an expected sense-datum. But what depends upon testimony, whether heard or read, cannot be explained in this way, since testimony depends upon the existence of minds other than our own, and thus requires a knowledge of something not given in sense. But before examining the question of our knowledge of other minds, let us return to the question of the thing-in-itself, namely, to the theory that what exists at times when we are not perceiving a given sensible object is something quite unlike that object, something which, together with us and our sense-organs, causes our sensations, but is never itself given in sensation.

The thing-in-itself, when we start from common-sense assumptions, is a fairly natural outcome of the difficulties due to the changing appearances of what is supposed to be one object. It is supposed that the table (for example) causes our sense-data of sight and touch, but must, since these are altered by the point of view and the intervening medium, be quite different from the sense-data to which it gives rise. There is, in this theory, a tendency to a confusion from which it derives some of its plausibility, namely, the confusion between a sensation as a psychical occurrence and its object. A patch of colour, even if it only exists when it is seen, is still something quite different from the seeing of it: the seeing of it is mental, but the patch of colour is not. This confusion, however, can be avoided without our necessarily abandoning the theory we are examining. The objection to it, I think, lies in its failure to realise the radical nature of the reconstruction demanded by the difficulties to which it points. We cannot speak legitimately of changes in the point of view and the intervening medium until we have already constructed some world more stable than that of momentary sensation. Our discussion of the blue spectacles and the walk round the table has, I hope, made this clear. But what remains far from clear is the nature of the reconstruction required.

Although we cannot rest content with the above theory, in the terms in which it is stated, we must nevertheless treat it with a certain respect, for it is in outline the theory upon which physical science and physiology are built, and it must, therefore, be susceptible of a true interpretation. Let us see how this is to be done.

The first thing to realise is that there are no such things as "illusions of sense." Objects of sense, even when they occur in dreams, are the most indubitably real objects known to us. What, then, makes us call them unreal in dreams? Merely the unusual nature of their connection with other objects of sense. I dream that I am in America, but I wake up and find myself in England without those intervening days on the Atlantic which, alas! are inseparably connected with a "real" visit to America. Objects of sense are called "real" when they have the kind of connection with other objects of sense which experience has led us to regard as normal; when they fail in this, they are called "illusions." But what is illusory is only the inferences to which they give rise; in themselves, they are every bit as real as the objects of waking life. And conversely, the sensible objects of waking life must not be expected to have any more intrinsic reality than those of dreams. Dreams and waking life, in our first efforts at construction, must be treated with equal respect; it is only by some reality not *merely* sensible that dreams can be condemned.

Accepting the indubitable momentary reality of objects of sense, the next thing to notice is the confusion underlying objections derived from their changeableness. As we walk round the table, its aspect changes; but it is thought impossible to maintain either that the table changes, or that its various aspects can all "really" exist in the same place. If we press one eyeball, we shall see two tables; but it is thought preposterous to maintain that there are "really" two tables. Such arguments, however, seem to involve the assumption that there can be something more real than objects of sense. If we see two tables, then there *are* two visual tables. It is perfectly true that, at the same moment, we may discover by touch that there is only one tactile table. This makes us declare the two visual tables an illusion, because usually one visual object corresponds to one tactile object. But all that we are warranted in saying is that, in this case, the manner of correlation of touch and sight is unusual. Again, when the aspect of the table changes as we walk round it, and we are told there cannot be so many different aspects in the same place, the answer is simple: what does the critic of the table mean by "the same place"? The use of such a phrase presupposes that all our difficulties have been solved; as yet, we have no right to speak of a "place" except with reference to one given set of momentary sense-data. When all are changed by a bodily movement, no place remains the same as it was. Thus the difficulty, if it exists, has at least not been rightly stated.

We will now make a new start, adopting a different method. Instead of inquiring what is the minimum of assumption by which we can explain the world of sense, we will, in order to have a model hypothesis as a help for the imagination, construct one possible (not necessary) explanation of the facts. It may perhaps then be possible to pare away what is superfluous in our hypothesis, leaving a residue which may be regarded as the abstract answer to our problem.

Let us imagine that each mind looks out upon the world, as in Leibniz's monadology, from a point of view peculiar to itself; and for the sake of simplicity let us confine ourselves to the sense of sight, ignoring minds which are devoid of this sense. Each mind sees at each moment an immensely complex three-dimensional world; but there is absolutely nothing which is seen by two minds simultaneously. When we say that two people see the same thing, we always find that, owing to difference of point of view, there are differences, however slight, between their immediate sensible objects. (I am here assuming the validity of testimony, but as we are only constructing a *possible* theory, that is a legitimate assumption.) The three-dimensional world seen by one mind therefore contains no place in common with that seen by another, for places can only be constituted by the things in or around them. Hence we may suppose, in spite of the differences between the different worlds, that each exists entire exactly as it is perceived, and might be exactly as it is even if it were not perceived. We may further suppose that there are an infinite number of such worlds which are in fact unperceived. If two men are sitting in a room, two somewhat similar worlds are perceived by them; if a third man enters and sits between them, a third world, intermediate between the two previous worlds, begins to be perceived. It is true that we cannot reasonably suppose just this world to have existed before, because it is conditioned by the sense-organs, nerves, and brain of the newly arrived man; but we can reasonably suppose that *some* aspect of the universe existed from that point of view, though no one was perceiving it. The system consisting of all views of the universe perceived and unperceived, I shall call the system of "perspectives"; I shall confine the expression "private worlds" to such views of the universe as are actually perceived. Thus a "private world" is a perceived "perspective"; but there may be any number of unperceived perspectives.

Two men are sometimes found to perceive very similar perspectives, so similar that they can use the same words to describe them. They say they see the same table, because the differences between the two tables they see are slight and not practically important. Thus it is possible, sometimes, to establish a correlation by similarity between a great many of the things of one perspective, and a great many of the things of another. In case the similarity is very great, we say the points of view of the two perspectives are near together in space; but this space in which they are near together is totally different from the spaces inside the two perspectives. It is a relation between the perspectives, and is not in either of them; no one can perceive it, and if it is to be known it can be only by inference. Between two perceived perspectives which are similar, we can imagine a whole series of other perspectives, some at least unperceived, and such that between any two, however similar, there are others still more similar. In this way the space which consists of relations between perspectives can be rendered continuous, and (if we choose) three-dimensional.

We can now define the momentary common-sense "thing," as opposed to its momentary appearances. By the similarity of neighbouring perspectives, many objects in the one can be correlated with objects in the other, namely, with the similar objects. Given an object in one perspective, form the system of all the objects correlated with it in all the perspectives; that system may be identified with the momentary common-sense "thing." Thus an aspect of a "thing" is a member of the system of aspects which *is* the "thing" at that moment. (The correlation of the times of different perspectives raises certain complications, of the kind considered in the theory of relativity; but we may ignore these at present.) All the aspects of a thing are real, whereas the thing is a mere logical construction. It has, however, the merit of being neutral as between different points of view, and of being visible to more than one person, in the only sense in which it can ever be visible, namely, in the sense that each sees one of its aspects.

It will be observed that, while each perspective contains its own space, there is only one space in which the perspectives themselves are the elements. There are as many private spaces as there are perspectives; there are therefore at least as many as there are percipients, and there may be any number of others which have a merely material existence and are not seen by anyone. But there is only one perspective-space, whose elements are single perspectives, each with its own private space. We have now to explain how the private space of a single perspective is correlated with part of the one all-embracing perspective space.

Perspective space is the system of "points of view" of private spaces (perspectives), or, since "points of view" have not been defined, we may say it is the system of the private spaces themselves. These private spaces will each count as one point, or at any rate as one element, in perspective space. They are ordered by means of their similarities. Suppose, for example, that we start from one which contains the appearance of a circular disc, such as would be called a penny, and suppose this appearance, in the perspective in question, is circular, not elliptic. We can then form a whole series of perspectives containing a graduated series of circular aspects of varying sizes: for this purpose we only have to move (as we say) towards the penny or away from it. The perspectives in which the penny looks circular will be said to lie on a straight line in perspective space, and their order on this line will be that of the sizes of the circular aspects. Moreover—though this statement must be noticed and subsequently examined—the perspectives in which the penny looks big will be said to be nearer to the penny than those in which it looks small. It is to be remarked also that any other "thing" than our penny might have been chosen to define the relations of our perspectives in perspective space, and that experience shows that the same spatial order of perspectives would have resulted.

In order to explain the correlation of private spaces with perspective space, we have first to explain what is meant by "the place (in perspective space) where a thing is." For this purpose, let us again consider the penny which appears in many perspectives. We formed a straight line of perspectives in which the penny looked circular, and we agreed that those in which it looked larger were to be considered as nearer to the penny. We can form another straight line of perspectives in which the penny is seen end-on and looks like a straight line of a certain thickness. These two lines will meet in a certain place in perspective space, *i.e.* in a certain perspective, which may be defined as "the place (in perspective space) where the penny is." It is true that, in order to prolong our lines until they reach this place, we shall have to make use of other things besides the penny, because, so far as experience goes, the penny ceases to present any appearance after we have come so near to it that it touches the eye. But this raises no real difficulty, because the spatial order of perspectives is found empirically to be independent of the particular "things" chosen for defining the order. We can, for example, remove our penny and prolong each of our two straight lines up to their intersection by placing other pennies further off in such a way that the aspects of the one are circular where those of our original penny were circular, and the aspects of the other are straight where those of our original penny were straight. There will then be just one perspective in which one of the new pennies looks circular and the other straight. This will be, by definition, the place where the original penny was in perspective space.

The above is, of course, only a first rough sketch of the way in which our definition is to be reached. It neglects the size of the penny, and it assumes that we can remove the penny without being disturbed by any simultaneous changes in the positions of other things. But it is plain that such niceties cannot affect the principle, and can only introduce complications in its application.

Having now defined the perspective which is the place where a given thing is, we can understand what is meant by saying that the perspectives in which a thing looks large are nearer to the thing than those in which it looks small: they are, in fact, nearer to the perspective which is the place where the thing is.

We can now also explain the correlation between a private space and parts of perspective space. If there is an aspect of a given thing in a certain private space, then we correlate the place where this aspect is in the private space with the place where the thing is in perspective space.

We may define "here" as the place, in perspective space, which is occupied by our private world. Thus we can now understand what is meant by speaking of a thing as near to or far from "here." A thing is near to "here" if the place where it is is near to my private world. We can also understand what is meant by saying that our private world is inside our head; for our private world is a place in perspective space, and may be part of the place where our head is.

It will be observed that *two* places in perspective space are associated with every aspect of a thing: namely, the place where the thing is, and the place which is the perspective of which the aspect in question forms part. Every aspect of a thing is a member of two different classes of aspects, namely: (1) the various aspects of the thing, of which at most one appears in any given perspective; (2) the perspective of which the given aspect is a member, *i.e.* that in which the thing has the given aspect. The physicist naturally classifies aspects in the first way, the psychologist in the second. The two places associated with a single aspect correspond to the two ways of classifying it. We may distinguish the two places as that *at* which, and that *from* which, the aspect appears. The "place at which" is the place of the thing to which the aspect belongs; the "place from which" is the place of the perspective to which the aspect belongs.

Let us now endeavour to state the fact that the aspect which a thing presents at a given place is affected by the intervening medium. The aspects of a thing in different perspectives are to be conceived as spreading outwards from the place where the thing is, and undergoing various changes as they get further away from this place. The laws according to which they change cannot be stated if we only take account of the aspects that are near the thing, but require that we should also take account of the things that are at the places from which these aspects appear. This empirical fact can, therefore, be interpreted in terms of our construction.

We have now constructed a largely hypothetical picture of the world, which contains and places the experienced facts, including those derived from testimony. The world we have constructed can, with a certain amount of trouble, be used to interpret the crude facts of sense, the facts of physics, and the facts of physiology. It is therefore a world which *may* be actual. It fits the facts, and there is no empirical evidence against it; it also is free from logical impossibilities. But have we any good reason to suppose that it is real? This brings us back to our original problem, as to the grounds for believing in the existence of anything outside my private world. What we have derived from our hypothetical construction is that there are no grounds *against* the truth of this belief, but we have not derived any positive grounds in its favour. We will resume this inquiry by taking up again the question of testimony and the evidence for the existence of other minds.

It must be conceded to begin with that the argument in favour of the existence of other people's minds cannot be conclusive. A phantasm of our dreams will appear to have a mind—a mind to be annoying, as a rule. It will give unexpected answers, refuse to conform to our desires, and show all those other signs of intelligence to which we are accustomed in the acquaintances of our waking hours. And yet, when we are awake, we do not believe that the phantasm was, like the appearances of people in waking life, representative of a private world to which we have no direct access. If we are to believe this of the people we meet when we are awake, it must be on some ground short of demonstration, since it is obviously possible that what we call waking life may be only an unusually persistent and recurrent nightmare. It may be that our imagination brings forth all that other people seem to say to us, all that we read in books, all the daily, weekly, monthly, and quarterly journals that distract our thoughts, all the advertisements of soap and all the speeches of politicians. This *may* be true, since it cannot be shown to be false, yet no one can really believe it. Is there any *logical* ground for regarding this possibility as improbable? Or is there nothing beyond habit and prejudice?

The minds of other people are among our data, in the very wide sense in which we used the word at first. That is to say, when we first begin to reflect, we find ourselves already believing in them, not because of any argument, but because the belief is natural to us. It is, however, a psychologically derivative belief, since it results from observation of people's bodies; and along with other such beliefs, it does not belong to the hardest of hard data, but becomes, under the influence of philosophic reflection, just sufficiently questionable to make us desire some argument connecting it with the facts of sense.

The obvious argument is, of course, derived from analogy. Other people's bodies behave as ours do when we have certain thoughts and feelings; hence, by analogy, it is natural to suppose that such behaviour is connected with thoughts and feelings like our own. Someone says, “Look out!” and we find we are on the point of being killed by a motor-car; we therefore attribute the words we heard to the person in question having seen the motor-car first, in which case there are existing things of which we are not directly conscious. But this whole scene, with our inference, may occur in a dream, in which case the inference is generally considered to be mistaken. Is there anything to make the argument from analogy more cogent when we are (as we think) awake?

The analogy in waking life is only to be preferred to that in dreams on the ground of its greater extent and consistency. If a man were to dream every night about a set of people whom he never met by day, who had consistent characters and grew older with the lapse of years, he might, like the man in Calderon's play, find it difficult to decide which was the dream-world and which was the so-called "real" world. It is only the failure of our dreams to form a consistent whole either with each other or with waking life that makes us condemn them. Certain uniformities are observed in waking life, while dreams seem quite erratic. The natural hypothesis would be that demons and the spirits of the dead visit us while we sleep; but the modern mind, as a rule, refuses to entertain this view, though it is hard to see what could be said against it. On the other hand, the mystic, in moments of illumination, seems to awaken from a sleep which has filled all his mundane life: the whole world of sense becomes phantasmal, and he sees, with the clarity and convincingness that belongs to our morning realisation after dreams, a world utterly different from that of our daily cares and troubles. Who shall condemn him? Who shall justify him? Or who shall justify the seeming solidity of the common objects among which we suppose ourselves to live?

The hypothesis that other people have minds must, I think, be allowed to be not susceptible of any very strong support from the analogical argument. At the same time, it is a hypothesis which systematises a vast body of facts and never leads to any consequences which there is reason to think false. There is therefore nothing to be said against its truth, and good reason to use it as a working hypothesis. When once it is admitted, it enables us to extend our knowledge of the sensible world by testimony, and thus leads to the system of private worlds which we assumed in our hypothetical construction. In actual fact, whatever we may try to think as philosophers, we cannot help believing in the minds of other people, so that the question whether our belief is justified has a merely speculative interest. And if it is justified, then there is no further difficulty of principle in that vast extension of our knowledge, beyond our own private data, which we find in science and common sense.

This somewhat meagre conclusion must not be regarded as the whole outcome of our long discussion. The problem of the connection of sense with objective reality has commonly been dealt with from a standpoint which did not carry initial doubt so far as we have carried it; most writers, consciously or unconsciously, have assumed that the testimony of others is to be admitted, and therefore (at least by implication) that others have minds. Their difficulties have arisen after this admission, from the differences in the appearance which one physical object presents to two people at the same time, or to one person at two times between which it cannot be supposed to have changed. Such difficulties have made people doubtful how far objective reality could be known by sense at all, and have made them suppose that there were positive arguments against the view that it can be so known. Our hypothetical construction meets these arguments, and shows that the account of the world given by common sense and physical science can be interpreted in a way which is logically unobjectionable, and finds a place for all the data, both hard and soft. It is this hypothetical construction, with its reconciliation of psychology and physics, which is the chief outcome of our discussion. Probably the construction is only in part necessary as an initial assumption, and can be obtained from more slender materials by the logical methods of which we shall have an example in the definitions of points, instants, and particles; but I do not yet know to what lengths this diminution in our initial assumptions can be carried.

LECTURE IV
THE WORLD OF PHYSICS AND THE WORLD OF SENSE

AMONG the objections to the reality of objects of sense, there is one which is derived from the apparent difference between matter as it appears in physics and things as they appear in sensation. Men of science, for the most part, are willing to condemn immediate data as "merely subjective," while yet maintaining the truth of the physics inferred from those data. But such an attitude, though it may be *capable* of justification, obviously stands in need of it; and the only justification possible must be one which exhibits matter as a logical construction from sense-data—unless, indeed, there were some wholly *a priori* principle by which unknown entities could be inferred from such as are known. It is therefore necessary to find some way of bridging the gulf between the world of physics and the world of sense, and it is this problem which will occupy us in the present lecture. Physicists appear to be unconscious of the gulf, while psychologists, who are conscious of it, have not the mathematical knowledge required for spanning it. The problem is difficult, and I do not know its solution in detail. All that I can hope to do is to make the problem felt, and to indicate the kind of methods by which a solution is to be sought.

Let us begin by a brief description of the two contrasted worlds. We will take first the world of physics, for, though the other world is given while the physical world is inferred, to us now the world of physics is the more familiar, the world of pure sense having become strange and difficult to rediscover. Physics started from the common-sense belief in fairly permanent and fairly rigid bodies—tables and chairs, stones, mountains, the earth and moon and sun. This common-sense belief, it should be noticed, is a piece of audacious metaphysical theorising; objects are not continually present to sensation, and it may be doubted whether they are there when they are not seen or felt. This problem, which has been acute since the time of Berkeley, is ignored by common sense, and has therefore hitherto been ignored by physicists. We have thus here a first departure from the immediate data of sensation, though it is a departure merely by way of extension, and was probably made by our savage ancestors in some very remote prehistoric epoch.

But tables and chairs, stones and mountains, are not *quite* permanent or *quite* rigid. Tables and chairs lose their legs, stones are split by frost, and mountains are cleft by earthquakes and eruptions. Then there are other things, which seem material, and yet present almost no permanence or rigidity. Breath, smoke, clouds, are examples of such things—so, in a lesser degree, are ice and snow; and rivers and seas, though fairly permanent, are not in any degree rigid.

Breath, smoke, clouds, and generally things that can be seen but not touched, were thought to be hardly real; to this day the usual mark of a ghost is that it can be seen but not touched. Such objects were peculiar in the fact that they seemed to disappear completely, not merely to be transformed into something else. Ice and snow, when they disappear, are replaced by water; and it required no great theoretical effort to invent the hypothesis that the water was the same thing as the ice and snow, but in a new form. Solid bodies, when they break, break into parts which are practically the same in shape and size as they were before. A stone can be hammered into a powder, but the powder consists of grains which retain the character they had before the pounding. Thus the ideal of absolutely rigid and absolutely permanent bodies, which early physicists pursued throughout the changing appearances, seemed attainable by supposing ordinary bodies to be composed of a vast number of tiny atoms. This billiard-ball view of matter dominated the imagination of physicists until quite modern times, until, in fact, it was replaced by the electromagnetic theory, which in its turn is developing into a new atomism. Apart from the special form of the atomic theory which was invented for the needs of chemistry, some kind of atomism dominated the whole of traditional dynamics, and was implied in every statement of its laws and axioms.

The pictorial accounts which physicists give of the material world as they conceive it undergo violent changes under the influence of modifications in theory which are much slighter than the layman might suppose from the alterations of the description. Certain features, however, have remained fairly stable. It is always assumed that there is *something* indestructible which is capable of motion in space; what is indestructible is always very small, but does not always occupy a mere point in space. There is supposed to be one all-embracing space in which the motion takes place, and until lately we might have assumed one all-embracing time also. But the principle of relativity has given prominence to the conception of "local time," and has somewhat diminished men's confidence in the one even-flowing stream of time. Without dogmatising as to the ultimate outcome of the principle of relativity, however, we may safely say, I think, that it does not destroy the possibility of correlating different local times, and does not therefore have such far-reaching philosophical consequences as is sometimes supposed. In fact, in spite of difficulties as to measurement, the one all-embracing time still, I think, underlies all that physics has to say about motion. We thus have still in physics, as we had in Newton's time, a set of indestructible entities which may be called particles, moving relatively to each other in a single space and a single time.

The world of immediate data is quite different from this. Nothing is permanent; even the things that we think are fairly permanent, such as mountains, only become data when we see them, and are not immediately given as existing at other moments. So far from one all-embracing space being given, there are several spaces for each person, according to the different senses which give relations that may be called spatial. Experience teaches us to obtain one space from these by correlation, and experience, together with instinctive theorising, teaches us to correlate our spaces with those which we believe to exist in the sensible worlds of other people. The construction of a single time offers less difficulty so long as we confine ourselves to one person's private world, but the correlation of one private time with another is a matter of great difficulty. Thus, apart from any of the fluctuating hypotheses of physics, three main problems arise in connecting the world of physics with the world of sense, namely (1) the construction of permanent "things," (2) the construction of a single space, and (3) the construction of a single time. We will consider these three problems in succession.

(1) The belief in indestructible "things" very early took the form of atomism. The underlying motive in atomism was not, I think, any empirical success in interpreting phenomena, but rather an instinctive belief that beneath all the changes of the sensible world there must be something permanent and unchanging. This belief was, no doubt, fostered and nourished by its practical successes, culminating in the conservation of mass; but it was not produced by these successes. On the contrary, they were produced by it. Philosophical writers on physics sometimes speak as though the conservation of something or other were essential to the possibility of science, but this, I believe, is an entirely erroneous opinion. If the *a priori* belief in permanence had not existed, the same laws which are now formulated in terms of this belief might just as well have been formulated without it. Why should we suppose that, when ice melts, the water which replaces it is the same thing in a new form? Merely because this supposition enables us to state the phenomena in a way which is consonant with our prejudices. What we really know is that, under certain conditions of temperature, the appearance we call ice is replaced by the appearance we call water. We can give laws according to which the one appearance will be succeeded by the other, but there is no reason except prejudice for regarding both as appearances of the same substance.

One task, if what has just been said is correct, which confronts us in trying to connect the world of sense with the world of physics, is the task of reconstructing the conception of matter without the *a priori* beliefs which historically gave rise to it. In spite of the revolutionary results of modern physics, the empirical successes of the conception of matter show that there must be some legitimate conception which fulfils roughly the same functions. The time has hardly come when we can state precisely what this legitimate conception is, but we can see in a general way what it must be like. For this purpose, it is only necessary to take our ordinary common-sense statements and reword them without the assumption of permanent substance. We say, for example, that things change gradually—sometimes very quickly, but not without passing through a continuous series of intermediate states. What this means is that, given any sensible appearance, there will usually be, *if we watch*, a continuous series of appearances connected with the given one, leading on by imperceptible gradations to the new appearances which common-sense regards as those of the same thing. Thus a thing may be defined as a certain series of appearances, connected with each other by continuity and by certain causal laws. In the case of slowly changing things, this is easily seen. Consider, say, a wall-paper which fades in the course of years. It is an effort not to conceive of it as one "thing" whose colour is slightly different at one time from what it is at another. But what do we really *know* about it? We know that under suitable circumstances—*i.e.* when we are, as is said, "in the room"—we perceive certain colours in a certain pattern: not always precisely the same colours, but sufficiently similar to feel familiar. If we can state the laws according to which the colour varies, we can state all that is empirically verifiable; the assumption that there is a constant entity, the wall-paper, which "has" these various colours at various times, is a piece of gratuitous metaphysics. We may, if we like, *define* the wall-paper as the series of its aspects. These are collected together by the same motives which led us to regard the wall-paper as one thing, namely a combination of sensible continuity and causal connection. More generally, a "thing" will be defined as a certain series of aspects, namely those which would commonly be said to be *of* the thing. To say that a certain aspect is an aspect *of* a certain thing will merely mean that it is one of those which, taken serially, *are* the thing. Everything will then proceed as before: whatever was verifiable is unchanged, but our language is so interpreted as to avoid an unnecessary metaphysical assumption of permanence.

The above extrusion of permanent things affords an example of the maxim which inspires all scientific philosophising, namely "Occam's razor": *Entities are not to be multiplied without necessity.* In other words, in dealing with any subject-matter, find out what entities are undeniably involved, and state everything in terms of these entities. Very often the resulting statement is more complicated and difficult than one which, like common sense and most philosophy, assumes hypothetical entities whose existence there is no good reason to believe in. We find it easier to imagine a wall-paper with changing colours than to think merely of the series of colours; but it is a mistake to suppose that what is easy and natural in thought is what is most free from unwarrantable assumptions, as the case of "things" very aptly illustrates.

The above summary account of the genesis of "things," though it may be correct in outline, has omitted some serious difficulties which it is necessary briefly to consider. Starting from a world of helter-skelter sense-data, we wish to collect them into series, each of which can be regarded as consisting of the successive appearances of one "thing." There is, to begin with, some conflict between what common sense regards as one thing, and what physics regards an unchanging collection of particles. To common sense, a human body is one thing, but to science the matter composing it is continually changing. This conflict, however, is not very serious, and may, for our rough preliminary purpose, be largely ignored. The problem is: by what principles shall we select certain data from the chaos, and call them all appearances of the same thing?

A rough and approximate answer to this question is not very difficult. There are certain fairly stable collections of appearances, such as landscapes, the furniture of rooms, the faces of acquaintances. In these cases, we have little hesitation in regarding them on successive occasions as appearances of one thing or collection of things. But, as the *Comedy of Errors* illustrates, we may be led astray if we judge by mere resemblance. This shows that something more is involved, for two different things may have any degree of likeness up to exact similarity.

Another insufficient criterion of one thing is *continuity*. As we have already seen, if we watch what we regard as one changing thing, we usually find its changes to be continuous so far as our senses can perceive. We are thus led to assume that, if we see two finitely different appearances at two different times, and if we have reason to regard them as belonging to the same thing, then there was a continuous series of intermediate states of that thing during the time when we were not observing it. And so it comes to be thought that continuity of change is necessary and sufficient to constitute one thing. But in fact it is neither. It is not *necessary*, because the unobserved states, in the case where our attention has not been concentrated on the thing throughout, are purely hypothetical, and cannot possibly be our ground for supposing the earlier and later appearances to belong to the same thing; on the contrary, it is because we suppose this that we assume intermediate unobserved states. Continuity is also not sufficient, since we can, for example, pass by sensibly continuous gradations from any one drop of the sea to any other drop. The utmost we can say is that discontinuity during uninterrupted observation is as a rule a mark of difference between things, though even this cannot be said in such cases as sudden explosions.

The assumption of continuity is, however, successfully made in physics. This proves something, though not anything of very obvious utility to our present problem: it proves that nothing in the known world is inconsistent with the hypothesis that all changes are really continuous, though from too great rapidity or from our lack of observation they may not always appear continuous. In this hypothetical sense, continuity may be allowed to be a *necessary* condition if two appearances are to be classed as appearances of the same thing. But it is not a *sufficient* condition, as appears from the instance of the drops in the sea. Thus something more must be sought before we can give even the roughest definition of a "thing."

What is wanted further seems to be something in the nature of fulfilment of causal laws. This statement, as it stands, is very vague, but we will endeavour to give it precision. When I speak of "causal laws," I mean any laws which connect events at different times, or even, as a limiting case, events at the same time provided the connection is not logically demonstrable. In this very general sense, the laws of dynamics are causal laws, and so are the laws correlating the simultaneous appearances of one "thing" to different senses. The question is: How do such laws help in the definition of a "thing"?

To answer this question, we must consider what it is that is proved by the empirical success of physics. What is proved is that its hypotheses, though unverifiable where they go beyond sense-data, are at no point in contradiction with sense-data, but, on the contrary, are ideally such as to render all sense-data calculable from a sufficient collection of data all belonging to a given period of time. Now physics has found it empirically possible to collect sense-data into series, each series being regarded as belonging to one "thing," and behaving, with regard to the laws of physics, in a way in which series not belonging to one thing would in general not behave. If it is to be unambiguous whether two appearances belong to the same thing or not, there must be only one way of grouping appearances so that the resulting things obey the laws of physics. It would be very difficult to prove that this is the case, but for our present purposes we may let this point pass, and assume that there is only one way. We must include in our definition of a "thing" those of its aspects, if any, which are not observed. Thus we may lay down the following definition: *Things are those series of aspects which obey the laws of physics.* That such series exist is an empirical fact, which constitutes the verifiability of physics.

It may still be objected that the "matter" of physics is something other than series of sense-data. Sense-data, it may be said, belong to psychology and are, at any rate in some sense, subjective, whereas physics is quite independent of psychological considerations, and does not assume that its matter only exists when it is perceived.

To this objection there are two answers, both of some importance.

(*a*) We have been considering, in the above account, the question of the *verifiability* of physics. Now verifiability is by no means the same thing as truth; it is, in fact, something far more subjective and psychological. For a proposition to be verifiable, it is not enough that it should be true, but it must also be such as we can *discover* to be true. Thus verifiability depends upon our capacity for acquiring knowledge, and not only upon the objective truth. In physics, as ordinarily set forth, there is much that is unverifiable: there are hypotheses as to (α) how things would appear to a spectator in a place where, as it happens, there is no spectator; (β) how things would appear at times when, in fact, they are not appearing to anyone; (γ) things which never appear at all. All these are introduced to simplify the statement of the causal laws, but none of them form an integral part of what is *known* to be true in physics. This brings us to our second answer.

(*b*) If physics is to consist wholly of propositions known to be true, or at least capable of being proved or disproved, the three kinds of hypothetical entities we have just enumerated must all be capable of being exhibited as logical functions of sense-data. In order to show how this might possibly be done, let us recall the hypothetical Leibnizian universe of Lecture III. In that universe, we had a number of perspectives, two of which never had any entity in common, but often contained entities which could be sufficiently correlated to be regarded as belonging to the same thing. We will call one of these an "actual" private world when there is an actual spectator to which it appears, and "ideal" when it is merely constructed on principles of continuity. A physical thing consists, at each instant, of the whole set of its aspects at that instant, in all the different worlds; thus a momentary state of a thing is a whole set of aspects. An "ideal" appearance will be an aspect merely calculated, but not actually perceived by any spectator. An "ideal" state of a thing will be a state at a moment when all its appearances are ideal. An ideal thing will be one whose states at all times are ideal. Ideal appearances, states, and things, since they are calculated, must be functions of actual appearances, states, and things; in fact, ultimately, they must be functions of actual appearances. Thus it is unnecessary, for the enunciation of the laws of physics, to assign any reality to ideal elements: it is enough to accept them as logical constructions, provided we have means of knowing how to determine when they become actual. This, in fact, we have with some degree of approximation; the starry heaven, for instance, becomes actual whenever we choose to look at it. It is open to us to believe that the ideal elements exist, and there can be no reason for *dis*believing this; but unless in virtue of some *a priori* law we cannot *know* it, for empirical knowledge is confined to what we actually observe.

(2) The three main conceptions of physics are space, time, and matter. Some of the problems raised by the conception of matter have been indicated in the above discussion of "things." But space and time also raise difficult problems of much the same kind, namely, difficulties in reducing the haphazard untidy world of immediate sensation to the smooth orderly world of geometry and kinematics. Let us begin with the consideration of space.

People who have never read any psychology seldom realise how much mental labour has gone into the construction of the one all-embracing space into which all sensible objects are supposed to fit. Kant, who was unusually ignorant of psychology, described space as "an infinite given whole," whereas a moment's psychological reflection shows that a space which is infinite is not given, while a space which can be called given is not infinite. What the nature of "given" space really is, is a difficult question, upon which psychologists are by no means agreed. But some general remarks may be made, which will suffice to show the problems, without taking sides on any psychological issue still in debate.

The first thing to notice is that different senses have different spaces. The space of sight is quite different from the space of touch: it is only by experience in infancy that we learn to correlate them. In later life, when we see an object within reach, we know how to touch it, and more or less what it will feel like; if we touch an object with our eyes shut, we know where we should have to look for it, and more or less what it would look like. But this knowledge is derived from early experience of the correlation of certain kinds of touch-sensations with certain kinds of sight-sensations. The one space into which both kinds of sensations fit is an intellectual construction, not a datum. And besides touch and sight, there are other kinds of sensation which give other, though less important spaces: these also have to be fitted into the one space by means of experienced correlations. And as in the case of things, so here: the one all-embracing space, though convenient as a way of speaking, need not be supposed really to exist. All that experience makes certain is the several spaces of the several senses, correlated by empirically discovered laws. The one space may turn out to be valid as a logical construction, compounded of the several spaces, but there is no good reason to assume its independent metaphysical reality.

Another respect in which the spaces of immediate experience differ from the space of geometry and physics is in regard to *points*. The space of geometry and physics consists of an infinite number of points, but no one has ever seen or touched a point. If there are points in a sensible space, they must be an inference. It is not easy to see any way in which, as independent entities, they could be validly inferred from the data; thus here again, we shall have, if possible, to find some logical construction, some complex assemblage of immediately given objects, which will have the geometrical properties required of points. It is customary to think of points as simple and infinitely small, but geometry in no way demands that we should think of them in this way. All that is necessary for geometry is that they should have mutual relations possessing certain enumerated abstract properties, and it may be that an assemblage of data of sensation will serve this purpose. Exactly how this is to be done, I do not yet know, but it seems fairly certain that it can be done.

The following illustrative method, simplified so as to be easily manipulated, has been invented by Dr Whitehead for the purpose of showing how points might be manufactured from sense-data. We have first of all to observe that there are no infinitesimal sense-data: any surface we can see, for example, must be of some finite extent. But what at first appears as one undivided whole is often found, under the influence of attention, to split up into parts contained within the whole. Thus one spatial object may be contained within another, and entirely enclosed by the other. This relation of enclosure, by the help of some very natural hypotheses, will enable us to define a "point" as a certain class of spatial objects, namely all those (as it will turn out in the end) which would naturally be said to contain the point. In order to obtain a definition of a "point" in this way, we proceed as follows:

Given any set of volumes or surfaces, they will not in general converge into one point. But if they get smaller and smaller, while of any two of the set there is always one that encloses the other, then we begin to have the kind of conditions which would enable us to treat them as having a point for their limit. The hypotheses required for the relation of enclosure are that (1) it must be transitive; (2) of two *different* spatial objects, it is impossible for each to enclose the other, but a single spatial object always encloses itself; (3) any set of spatial objects such that there is at least one spatial object enclosed by them all has a lower limit or minimum, *i.e.* an object enclosed by all of them and enclosing all objects which are enclosed by all of them; (4) to prevent trivial exceptions, we must add that there are to be instances of enclosure, *i.e.* there are really to be objects of which one encloses the other. When an enclosure-relation has these properties, we will call it a "point-producer." Given any relation of enclosure, we will call a set of objects an "enclosure-series" if, of any two of them, one is contained in the other. We require a condition which shall secure that an enclosure-series converges to a point, and this is obtained as follows: Let our enclosure-series be such that, given any other enclosure-series of which there are members enclosed in any arbitrarily chosen member of our first series, then there are members of our first series enclosed in any arbitrarily chosen member of our second series. In this case, our first enclosure-series may be called a "punctual enclosure-series." Then a "point" is all the objects which enclose members of a given punctual enclosure-series. In order to ensure infinite divisibility, we require one further property to be added to those defining point-producers, namely that any object which encloses itself also encloses an object other than itself. The "points" generated by point-producers with this property will be found to be such as geometry requires.

(3) The question of time, so long as we confine ourselves to one private world, is rather less complicated than that of space, and we can see pretty clearly how it might be dealt with by such methods as we have been considering. Events of which we are conscious do not last merely for a mathematical instant, but always for some finite time, however short. Even if there be a physical world such as the mathematical theory of motion supposes, impressions on our sense-organs

produce sensations which are not merely and strictly instantaneous, and therefore the objects of sense of which we are immediately conscious are not strictly instantaneous. Instants, therefore, are not among the data of experience, and, if legitimate, must be either inferred or constructed. It is difficult to see how they can be validly inferred; thus we are left with the alternative that they must be constructed. How is this to be done?

Immediate experience provides us with two time-relations among events: they may be simultaneous, or one may be earlier and the other later. These two are both part of the crude data; it is not the case that only the events are given, and their time-order is added by our subjective activity. The time-order, within certain limits, is as much given as the events. In any story of adventure you will find such passages as the following: "With a cynical smile he pointed the revolver at the breast of the dauntless youth. 'At the word *three* I shall fire,' he said. The words one and two had already been spoken with a cool and deliberate distinctness. The word *three* was forming on his lips. At this moment a blinding flash of lightning rent the air." Here we have simultaneity—not due, as Kant would have us believe, to the subjective mental apparatus of the dauntless youth, but given as objectively as the revolver and the lightning. And it is equally given in immediate experience that the words *one* and *two* come earlier than the flash. These time-relations hold between events which are not strictly instantaneous. Thus one event may begin sooner than another, and therefore be before it, but may continue after the other has begun, and therefore be also simultaneous with it. If it persists after the other is over, it will also be later than the other. Earlier, simultaneous, and later, are not inconsistent with each other when we are concerned with events which last for a finite time, however short; they only become inconsistent when we are dealing with something instantaneous.

It is to be observed that we cannot give what may be called *absolute* dates, but only dates determined by events. We cannot point to a time itself, but only to some event occurring at that time. There is therefore no reason in experience to suppose that there are times as opposed to events: the events, ordered by the relations of simultaneity and succession, are all that experience provides. Hence, unless we are to introduce superfluous metaphysical entities, we must, in defining what mathematical physics can regard as an instant, proceed by means of some construction which assumes nothing beyond events and their temporal relations.

If we wish to assign a date exactly by means of events, how shall we proceed? If we take any one event, we cannot assign our date exactly, because the event is not instantaneous, that is to say, it may be simultaneous with two events which are not simultaneous with each other. In order to assign a date exactly, we must be able, theoretically, to determine whether any given event is before, at, or after this date, and we must know that any other date is either before or after this date, but not simultaneous with it. Suppose, now, instead of taking one event A, we take two events A and B, and suppose A and B partly overlap, but B ends before A ends. Then an event which is simultaneous with both A and B must exist during the time when A and B overlap; thus we have come rather nearer to a precise date than when we considered A and B alone. Let C be an event which is simultaneous with both A and B, but which ends before either A or B has ended. Then an event which is simultaneous with A and B and C must exist during the time when all three overlap, which is a still shorter time. Proceeding in this way, by taking more and more events, a new event which is dated as simultaneous with all of them becomes gradually more and more accurately dated. This suggests a way by which a completely accurate date can be defined.

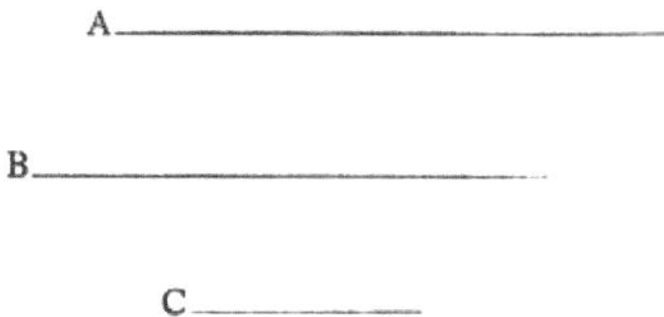

Let us take a group of events of which any two overlap, so that there is some time, however short, when they all exist. If there is any other event which is simultaneous with all of these, let us add it to the group; let us go on until we have constructed a group such that no event outside the group is simultaneous with all of them, but all the events inside the group are simultaneous with each other. Let us define this whole group as an instant of time. It remains to show that it has the properties we expect of an instant.

What are the properties we expect of instants? First, they must form a series: of any two, one must be before the other, and the other must be not before the one; if one is before another, and the other before a third, the first must be before the third. Secondly, every event must be at a certain number of instants; two events are simultaneous if they are at the same instant, and one is before the other if there is an instant, at which the one is, which is earlier than some instant at which the other is. Thirdly, if we assume that there is always some change going on somewhere during the time when any given event persists, the series of instants ought to be compact, *i.e.* given any two instants, there ought to be other instants between them. Do instants, as we have defined them, have these properties?

We shall say that an event is "at" an instant when it is a member of the group by which the instant is constituted; and we shall say that one instant is before another if the group which is the one instant contains an event which is earlier than, but not simultaneous with, some event in the group which is the other instant. When one event is earlier than, but not simultaneous with another, we shall say that it "wholly precedes" the other. Now we know that of two events which are not simultaneous, there must be one which wholly precedes the other, and in that case the other cannot also wholly precede the one; we also know that, if one event wholly precedes another, and the other wholly precedes a third, then the first wholly precedes the third. From these facts it is easy to deduce that the instants as we have defined them form a series.

We have next to show that every event is "at" at least one instant, *i.e.* that, given any event, there is at least one class, such as we used in defining instants, of which it is a member. For this purpose, consider all the events which are simultaneous with a given event, and do not begin later, *i.e.* are not wholly after anything simultaneous with it. We will call these the "initial contemporaries" of the given event. It will be found that this class of events is the first instant at which the given event exists, provided every event wholly after some contemporary of the given event is wholly after some *initial* contemporary of it.

Finally, the series of instants will be compact if, given any two events of which one wholly precedes the other, there are events wholly after the one and simultaneous with something wholly before the other. Whether this is the case or not, is an empirical question; but if it is not, there is no reason to expect the time-series to be compact.

Thus our definition of instants secures all that mathematics requires, without having to assume the existence of any disputable metaphysical entities.

Instants may also be defined by means of the enclosure-relation, exactly as was done in the case of points. One object will be temporally enclosed by another when it is simultaneous with the other, but not before or after it. Whatever encloses temporally or is enclosed temporally we shall call an "event." In order that the relation of temporal enclosure may be a "point-producer," we require (1) that it should be transitive, *i.e.* that if one event encloses another, and the other a third, then the first encloses the third; (2) that every event encloses itself, but if one event encloses another different event, then the other does not enclose the one; (3) that given any set of events such that there is at least one event enclosed by all of them, then there is an event enclosing all that they all enclose, and itself enclosed by all of them; (4) that there is at least one event. To ensure infinite divisibility, we require also that every event should enclose events other than itself. Assuming these characteristics, temporal enclosure is an infinitely divisible point-producer. We can now form an "enclosure-series" of events, by choosing a group of events such that of any two there is one which encloses the other; this will be a "punctual enclosure-series" if, given any other enclosure-series such that every member of our first series encloses some member of our second, then every member of our second series encloses some member of our first. Then an "instant" is the class of all events which enclose members of a given punctual enclosure-series.

The correlation of the times of different private worlds so as to produce the one all-embracing time of physics is a more difficult matter. We saw, in Lecture III., that different private worlds often contain correlated appearances, such as common sense would regard as appearances of the same "thing." When two appearances in different worlds are so correlated as to belong to one momentary "state" of a thing, it would be natural to regard them as simultaneous, and as thus affording a simple means of correlating different private times. But this can only be regarded as a first approximation. What we call one sound will be heard sooner by people near the source of the sound than by people further from it, and the same applies, though in a less degree, to light. Thus two correlated appearances in different worlds are not necessarily to be regarded as occurring at the same date in physical time, though they will be parts of one momentary state of a thing. The correlation of different private times is regulated by the desire to secure the simplest possible statement of the laws of physics, and thus raises rather complicated technical problems; but from the point of view of philosophical theory, there is no very serious difficulty of principle involved.

The above brief outline must not be regarded as more than tentative and suggestive. It is intended merely to show the kind of way in which, given a world with the kind of properties that psychologists find in the world of sense, it may be possible, by means of purely logical constructions, to make it amenable to mathematical treatment by defining series or classes of sense-data which can be called respectively particles, points, and instants. If such constructions are possible, then mathematical physics is applicable to the real world, in spite of the fact that its particles, points, and instants are not to be found among actually existing entities.

The problem which the above considerations are intended to elucidate is one whose importance and even existence has been concealed by the unfortunate separation of different studies which prevails throughout the civilised world. Physicists, ignorant and contemptuous of philosophy, have been content to assume their particles, points, and instants in practice, while conceding, with ironical politeness, that their concepts laid no claim to metaphysical validity. Metaphysicians, obsessed by the idealistic opinion that only mind is real, and the Parmenidean belief that the real is unchanging, repeated one after another the supposed contradictions in the notions of matter, space, and time, and therefore naturally made no endeavour to invent a tenable theory of particles, points, and instants. Psychologists, who have done invaluable work in bringing to light the chaotic nature of the crude materials supplied by unmanipulated sensation, have been ignorant of mathematics and modern logic, and have therefore been content to say that matter, space, and time are "intellectual constructions," without making any attempt to show in detail either how the intellect can construct them, or what secures the practical validity which physics shows them to possess. Philosophers, it is to be hoped, will come to recognise that they cannot achieve any solid success in such problems without some slight knowledge of logic, mathematics, and physics; meanwhile, for want of students with the necessary equipment, this vital problem remains unattempted and unknown.

There are, it is true, two authors, both physicists, who have done something, though not much, to bring about a recognition of the problem as one demanding study. These two authors are Poincaré and Mach, Poincaré especially in his *Science and Hypothesis*, Mach especially in his *Analysis of Sensations*. Both of them, however, admirable as their work is, seem to me to suffer from a general philosophical bias. Poincaré is Kantian, while Mach is ultra-empiricist; with Poincaré almost all the mathematical part of physics is merely conventional, while with Mach the sensation as a mental event is identified with its object as a part of the physical world. Nevertheless, both these authors, and especially Mach, deserve mention as having made serious contributions to the consideration of our problem.

When a point or an instant is defined as a class of sensible qualities, the first impression produced is likely to be one of wild and wilful paradox. Certain considerations apply here, however, which will again be relevant when we come to the definition of numbers. There is a whole type of problems which can be solved by such definitions, and almost always there will be at first an effect of paradox. Given a set of objects any two of which have a relation of the sort called "symmetrical and transitive," it is almost certain that we shall come to regard them as all having some common quality, or as all having the same relation to some one object outside the set. This kind of case is important, and I shall therefore try to make it clear even at the cost of some repetition of previous definitions.

A relation is said to be "symmetrical" when, if one term has this relation to another, then the other also has it to the one. Thus "brother or sister" is a "symmetrical" relation: if one person is a brother or a sister of another, then the other is a brother or sister of the one. Simultaneity, again, is a symmetrical relation; so is equality in size. A relation is said to be "transitive" when, if one term has this relation to another, and the other to a third, then the one has it to the third. The symmetrical relations mentioned just now are also transitive—provided, in the case of "brother or sister," we allow a person to be counted as his or her own brother or sister, and provided, in the case of simultaneity, we mean complete simultaneity, *i.e.* beginning and ending together.

But many relations are transitive without being symmetrical—for instance, such relations as "greater," "earlier," "to the right of," "ancestor of," in fact all such relations as give rise to series. Other relations are symmetrical without being transitive—for example, difference in any respect. If A is of a different age from B, and B of a different age from C, it does not follow that A is of a different age from C. Simultaneity, again, in the case of events which last for a finite time, will not necessarily be transitive if it only means that the times of the two events overlap. If A ends just after B has begun, and B ends just after C has begun, A and B will be simultaneous in this sense, and so will B and C, but A and C may well not be simultaneous.

All the relations which can naturally be represented as equality in any respect, or as possession of a common property, are transitive and symmetrical—this applies, for example, to such relations as being of the same height or weight or colour. Owing to the fact that possession of a common property gives rise to a transitive symmetrical relation, we come to imagine that wherever such a relation occurs it must be due to a common property. "Being equally numerous" is a transitive symmetrical relation of two collections; hence we imagine that both have a common property, called their number. "Existing at a given instant" (in the sense in which we defined an instant) is a transitive symmetrical relation; hence we come to think that there really is an instant which confers a common property on all the things existing at that instant. "Being states of a given thing" is a transitive symmetrical relation; hence we come to imagine that there really is a thing, other than the series of states, which accounts for the transitive symmetrical relation. In all such cases, the class of terms that have the given transitive symmetrical relation to a given term will fulfil all the formal requisites of a common property of all the members of the class. Since there certainly is the class, while any other common property may be illusory, it is prudent, in order to avoid needless assumptions, to substitute the class for the common property which would be ordinarily assumed. This is the reason for the definitions we have adopted, and this is the source of the apparent paradoxes. No harm is done if there are such common properties as language assumes, since we do not deny them, but merely abstain from asserting them. But if there are not such common properties in any given case, then our method has secured us against error. In the absence of special knowledge, therefore, the method we have adopted is the only one which is safe, and which avoids the risk of introducing fictitious metaphysical entities.

LECTURE V
THE THEORY OF CONTINUITY

THE theory of continuity, with which we shall be occupied in the present lecture, is, in most of its refinements and developments, a purely mathematical subject—very beautiful, very important, and very delightful, but not, strictly speaking, a part of philosophy. The logical basis of the theory alone belongs to philosophy, and alone will occupy us to-night. The way the problem of continuity enters into philosophy is, broadly speaking, the following: Space and time are treated by mathematicians as consisting of points and instants, but they also have a property, easier to feel than to define, which is called continuity, and is thought by many philosophers to be destroyed when they are resolved into points and instants. Zeno, as we shall see, proved that analysis into points and instants was impossible if we adhered to the view that the number of points or instants in a finite space or time must be finite. Later philosophers, believing infinite number to be self-contradictory, have found here an antinomy: Spaces and times could not consist of a *finite* number of points and instants, for such reasons as Zeno's; they could not consist of an *infinite* number of points and instants, because infinite numbers were supposed to be self-contradictory. Therefore spaces and times, if real at all, must not be regarded as composed of points and instants.

But even when points and instants, as independent entities, are discarded, as they were by the theory advocated in our last lecture, the problems of continuity, as I shall try to show presently, remain, in a practically unchanged form. Let us therefore, to begin with, admit points and instants, and consider the problems in connection with this simpler or at least more familiar hypothesis.

The argument against continuity, in so far as it rests upon the supposed difficulties of infinite numbers, has been disposed of by the positive theory of the infinite, which will be considered in Lecture VII. But there remains a feeling—of the kind that led Zeno to the contention that the arrow in its flight is at rest—which suggests that points and instants, even if they are infinitely numerous, can only give a jerky motion, a succession of different immobilities, not the smooth transitions with which the senses have made us familiar. This feeling is due, I believe, to a failure to realise imaginatively, as well as abstractly, the nature of continuous series as they appear in mathematics. When a theory has been apprehended logically, there is often a long and serious labour still required in order to *feel* it: it is necessary to dwell upon it, to thrust out from the mind, one by one, the misleading suggestions of false but more familiar theories, to acquire the kind of intimacy which, in the case of a foreign language, would enable us to think and dream in it, not merely to construct laborious sentences by the help of grammar and dictionary. It is, I believe, the absence of this kind of intimacy which makes many philosophers regard the mathematical doctrine of continuity as an inadequate explanation of the continuity which we experience in the world of sense.

In the present lecture, I shall first try to explain in outline what the mathematical theory of continuity is in its philosophically important essentials. The application to actual space and time will not be in question to begin with. I do not see any reason to suppose that the points and instants which mathematicians introduce in dealing with space and time are actual physically existing entities, but I do see reason to suppose that the continuity of actual space and time may be more or less analogous to mathematical continuity. The theory of mathematical continuity is an abstract logical theory, not dependent for its validity upon any properties of actual space and time. What is claimed for it is that, when it is understood, certain characteristics of space and time, previously very hard to analyse, are found not to present any logical difficulty. What we know empirically about space and time is insufficient to enable us to decide between various mathematically possible alternatives, but these alternatives are all fully intelligible and fully adequate to the observed facts. For the present, however, it will be well to forget space and time and the continuity of sensible change, in order to return to these topics equipped with the weapons provided by the abstract theory of continuity.

Continuity, in mathematics, is a property only possible to a *series* of terms, *i.e.* to terms arranged in an order, so that we can say of any two that one comes *before* the other. Numbers in order of magnitude, the points on a line from left to right, the moments of time from earlier to later, are instances of series. The notion of order, which is here introduced, is one which is not required in the theory of cardinal number. It is possible to know that two classes have the same number of terms without knowing any order in which they are to be taken. We have an instance of this in such a case as English husbands and English wives: we can see that there must be the same number of husbands as of wives, without having to arrange them in a series. But continuity, which we are now to consider, is essentially a property of an order: it does not belong to a set of terms in themselves, but only to a set in a certain order. A set of terms which can be arranged in one order can always also be arranged in other orders, and a set of terms which can be arranged in a continuous order can always also be arranged in orders which are not continuous. Thus the essence of continuity must not be sought in the nature of the set of terms, but in the nature of their arrangement in a series.

Mathematicians have distinguished different degrees of continuity, and have confined the word "continuous," for technical purposes, to series having a certain high degree of continuity. But for philosophical purposes, all that is important in continuity is introduced by the lowest degree of continuity, which is called "compactness." A series is called "compact" when no two terms are consecutive, but between any two there are others. One of the simplest examples of a compact series is the series of fractions in order of magnitude. Given any two fractions, however near together, there are other fractions greater than the one and smaller than the other, and therefore no two fractions are consecutive. There is no fraction, for example, which is next after ½: if we choose some fraction which is very little greater than ½, say $^{51}/_{100}$ we can find others, such as $^{101}/_{200}$, which are nearer to ½. Thus between any two fractions, however little they differ, there are an infinite number of other fractions. Mathematical space and time also have this property of compactness, though whether actual space and time have it is a further question, dependent upon empirical evidence, and probably incapable of being answered with certainty.

In the case of abstract objects such as fractions, it is perhaps not very difficult to realise the logical possibility of their forming a compact series. The difficulties that might be felt are those of infinity, for in a compact series the number of terms between any two given terms must be infinite. But when these difficulties have been solved, the mere compactness in itself offers no great obstacle to the imagination. In more concrete cases, however, such as motion, compactness becomes much more repugnant to our habits of thought. It will therefore be desirable to consider explicitly the mathematical account of motion, with a view to making its logical possibility felt. The mathematical account of motion is perhaps artificially simplified when regarded as describing what actually occurs in the physical world; but what actually occurs must be capable, by a certain amount of logical manipulation, of being brought within the scope of the mathematical account, and must, in its analysis, raise just such problems as are raised in their simplest form by this account. Neglecting, therefore, for the present, the question of its physical adequacy, let us devote ourselves merely to considering its possibility as a formal statement of the nature of motion.

In order to simplify our problem as much as possible, let us imagine a tiny speck of light moving along a scale. What do we mean by saying that the motion is continuous? It is not necessary for our purposes to consider the whole of what the mathematician means by this statement: only part of what he means is philosophically important. One part of what he means is that, if we consider any two positions of the speck occupied at any two instants, there will be other intermediate positions occupied at intermediate instants. However near together we take the two positions, the speck will not jump suddenly from the one to the other, but will pass through an infinite number of other positions on the way. Every distance, however small, is traversed by passing through all the infinite series of positions between the two ends of the distance.

But at this point imagination suggests that we may describe the continuity of motion by saying that the speck always passes from one position at one instant to *the next* position at *the next* instant. As soon as we say this or imagine it, we fall into error, because there is no *next* point or *next* instant. If there were, we should find Zeno's paradoxes, in some form, unavoidable, as will appear in our next lecture. One simple paradox may serve as an illustration. If our speck is in motion along the scale throughout the whole of a certain time, it cannot be at the same point at two consecutive instants. But it cannot, from one instant to the next, travel further than from one point to the next, for if it did, there would be no instant at which it was in the positions intermediate between that at the first instant and that at the next, and we agreed that the continuity of motion excludes the possibility of such sudden jumps. It follows that our speck must, so long as it moves, pass from one point at one instant to the next point at the next instant. Thus there will be just one perfectly definite velocity with which all motions must take place: no motion can be faster than this, and no motion can be slower. Since this conclusion is false, we must reject the hypothesis upon which it is based, namely that there are consecutive points and instants.Hence the continuity of motion must not be supposed to consist in a body's occupying consecutive positions at consecutive times.

The difficulty to imagination lies chiefly, I think, in keeping out the suggestion of *infinitesimal* distances and times. Suppose we halve a given distance, and then halve the half, and so on, we can continue the process as long as we please, and the longer we continue it, the smaller the resulting distance becomes. This infinite divisibility seems, at first sight, to imply that there are infinitesimal distances, *i.e.* distances so small that any finite fraction of an inch would be greater. This, however, is an error. The continued bisection of our distance, though it gives us continually smaller distances, gives us always *finite* distances. If our original distance was an inch, we reach successively half an inch, a quarter of an inch, an eighth, a sixteenth, and so on; but every one of this infinite series of diminishing distances is finite. "But," it may be said, "*in the end* the distance will grow infinitesimal." No, because there is no end. The process of bisection is one which can, theoretically, be carried on for ever, without any last term being attained. Thus infinite divisibility of distances, which must be admitted, does not imply that there are distances so small that any finite distance would be larger.

It is easy, in this kind of question, to fall into an elementary logical blunder. Given any finite distance, we can find a smaller distance; this may be expressed in the ambiguous form "there is a distance smaller than any finite distance." But if this is then interpreted as meaning "there is a distance such that, whatever finite distance may be chosen, the distance in question is smaller," then the statement is false. Common language is ill adapted to expressing matters of this kind, and philosophers who have been dependent on it have frequently been misled by it.

In a continuous motion, then, we shall say that at any given instant the moving body occupies a certain position, and at other instants it occupies other positions; the interval between any two instants and between any two positions is always finite, but the continuity of the motion is shown in the fact that, however near together we take the two positions and the two instants, there are an infinite number of positions still nearer together, which are occupied at instants that are also still nearer together. The moving body never jumps from one position to another, but always passes by a gradual transition through an infinite number of intermediaries. At a given instant, it is where it is, like Zeno's arrow;but we cannot say that it is at rest at the instant, since the instant does not last for a finite time, and there is not a beginning and end of the instant with an interval between them. Rest consists in being in the same position at all the instants throughout a certain finite period, however short; it does not consist simply in a body's being where it is at a given instant. This whole theory, as is obvious, depends upon the nature of compact series, and demands, for its full comprehension, that compact series should have become familiar and easy to the imagination as well as to deliberate thought.

What is required may be expressed in mathematical language by saying that the position of a moving body must be a continuous function of the time. To define accurately what this means, we proceed as follows. Consider a particle which, at the moment t, is at the point P. Choose now any small portion P_1P_2 of the path of the particle, this portion being one which contains P. We say then that, if the motion of the particle is continuous at the time t, it must be possible to find two instants t_1, t_2, one earlier than t and one later, such that throughout the whole time from t_1 to t_2 (both included), the particle lies between P_1 and P_2. And we say that this must still hold however small we make the portion P_1P_2. When this is the case, we say that the motion is continuous at the time t; and when the motion is continuous at all times, we say that the motion as a whole is continuous. It is obvious that if the particle were to jump suddenly from P to some other point Q, our definition would fail for all intervals P_1P_2 which were too small to include Q. Thus our definition affords an analysis of the continuity of motion, while admitting points and instants and denying infinitesimal distances in space or periods in time.

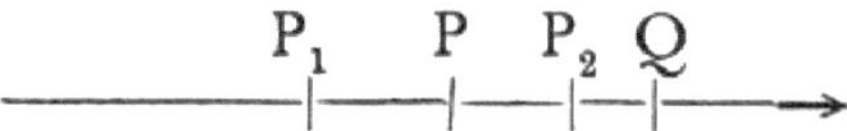

Philosophers, mostly in ignorance of the mathematician's analysis, have adopted other and more heroic methods of dealing with the *primâ facie* difficulties of continuous motion. A typical and recent example of philosophic theories of motion is afforded by Bergson, whose views on this subject I have examined elsewhere.

Apart from definite arguments, there are certain feelings, rather than reasons, which stand in the way of an acceptance of the mathematical account of motion. To begin with, if a body is moving at all fast, we *see* its motion just as we see its colour. A *slow* motion, like that of the hour-hand of a watch, is only known in the way which mathematics would lead us to expect, namely by observing a change of position after a lapse of time; but, when we observe the motion of the second-hand, we do not merely see first one position and then another—we see something as directly sensible as colour. What is this something that we see, and that we call visible motion? Whatever it is, it is *not* the successive occupation of successive positions: something beyond the mathematical theory of motion is required to account for it. Opponents of the mathematical theory emphasise this fact. "Your theory," they say, "may be very logical, and might apply admirably to some other world; but in this actual world, actual motions are quite different from what your theory would declare them to be, and require, therefore, some different philosophy from yours for their adequate explanation."

The objection thus raised is one which I have no wish to underrate, but I believe it can be fully answered without departing from the methods and the outlook which have led to the mathematical theory of motion. Let us, however, first try to state the objection more fully.

If the mathematical theory is adequate, nothing happens when a body moves except that it is in different places at different times. But in this sense the hour-hand and the second-hand are equally in motion, yet in the second-hand there is something perceptible to our senses which is absent in the hour-hand. We can see, at each moment, that the second-hand *is moving*, which is different from seeing it first in one place and then in another. This seems to involve our seeing it simultaneously in a number of places, although it must also involve our seeing that it is in some of these places earlier than in others. If, for example, I move my hand quickly from left to right, you seem to see the whole movement at once, in spite of the fact that you know it begins at the left and ends at the right. It is this kind of consideration, I think, which leads Bergson and many others to regard a movement as really one indivisible whole, not the series of separate states imagined by the mathematician.

To this objection there are three supplementary answers, physiological, psychological, and logical. We will consider them successively.

(1) The physiological answer merely shows that, if the physical world is what the mathematician supposes, its sensible appearance may nevertheless be expected to be what it is. The aim of this answer is thus the modest one of showing that the mathematical account is not impossible as applied to the physical world; it does not even attempt to show that this account is necessary, or that an analogous account applies in psychology.

When any nerve is stimulated, so as to cause a sensation, the sensation does not cease instantaneously with the cessation of the stimulus, but dies away in a short finite time. A flash of lightning, brief as it is to our sight, is briefer still as a physical phenomenon: we continue to see it for a few moments after the light-waves have ceased to strike the eye. Thus in the case of a physical motion, if it is sufficiently swift, we shall actually at one instant see the moving body throughout a finite portion of its course, and not only at the exact spot where it is at that instant. Sensations, however, as they die away, grow gradually fainter; thus the sensation due to a stimulus which is recently past is not exactly like the sensation due to a present stimulus. It follows from this that, when we see a rapid motion, we shall not only see a number of positions of the moving body simultaneously, but we shall see them with different degrees of intensity—the present position most vividly, and the others with diminishing vividness, until sensation fades away into immediate memory. This state of things accounts fully for the perception of motion. A motion is *perceived*, not merely *inferred*, when it is sufficiently swift for many positions to be sensible at one time; and the earlier and later parts of one perceived motion are distinguished by the less and greater vividness of the sensations.

This answer shows that physiology can account for our perception of motion. But physiology, in speaking of stimulus and sense-organs and a physical motion distinct from the immediate object of sense, is assuming the truth of physics, and is thus only capable of showing the physical account to be possible, not of showing it to be *necessary*. This consideration brings us to the psychological answer.

(2) The psychological answer to our difficulty about motion is part of a vast theory, not yet worked out, and only capable, at present, of being vaguely outlined. We considered this theory in the third and fourth lectures; for the present, a mere sketch of its application to our present problem must suffice. The world of physics, which was assumed in the physiological answer, is obviously inferred from what is given in sensation; yet as soon as we seriously consider what is actually given in sensation, we find it apparently very different from the world of physics. The question is thus forced upon us: Is the inference from sense to physics a valid one? I believe the answer to be affirmative, for reasons which I suggested in the third and fourth lectures; but the answer cannot be either short or easy. It consists, broadly speaking, in showing that, although the particles, points, and instants with which physics operates are not themselves given in experience, and are very likely not actually existing things, yet, out of the materials provided in sensation, it is possible to make logical constructions having the mathematical properties which physics assigns to particles, points, and instants. If this can be done, then all the propositions of physics can be translated, by a sort of dictionary, into propositions about the kinds of objects which are given in sensation.

Applying these general considerations to the case of motion, we find that, even within the sphere of immediate sense-data, it is necessary, or at any rate more consonant with the facts than any other equally simple view, to distinguish instantaneous states of objects, and to regard such states as forming a compact series. Let us consider a body which is moving swiftly enough for its motion to be perceptible, and long enough for its motion to be not wholly comprised in one sensation. Then, in spite of the fact that we see a finite extent of the motion at one instant, the extent which we see at one instant is different from that which we see at another. Thus we are brought back, after all, to a series of momentary views of the moving body, and this series will be compact, like the former physical series of points. In fact, though the *terms* of the series seem different, the mathematical character of the series is unchanged, and the whole mathematical theory of motion will apply to it *verbatim*.

When we are considering the actual data of sensation in this connection, it is important to realise that two sense-data may be, and *must* sometimes be, really different when we cannot perceive any difference between them. An old but conclusive reason for believing this was emphasised by Poincaré. In all cases of sense-data capable of gradual change, we may find one sense-datum indistinguishable from another, and that other indistinguishable from a third, while yet the first and third are quite easily distinguishable. Suppose, for example, a person with his eyes shut is holding a weight in his hand, and someone noiselessly adds a small extra weight. If the extra weight is small enough, no difference will be perceived in the sensation. After a time, another small extra weight may be added, and still no change will be perceived; but if both extra weights had been added at once, it may be that the change would be quite easily perceptible. Or, again, take shades of colour. It would be easy to find three stuffs of such closely similar shades that no difference could be perceived between the first and second, nor yet between the second and third, while yet the first and third would be distinguishable. In such a case, the second shade cannot be the same as the first, or it would be distinguishable from the third; nor the same as the third, or it would be distinguishable from the first. It must, therefore, though indistinguishable from both, be really intermediate between them.

Such considerations as the above show that, although we cannot distinguish sense-data unless they differ by more than a certain amount, it is perfectly reasonable to suppose that sense-data of a given kind, such as weights or colours, really form a compact series. The objections which may be brought from a psychological point of view against the mathematical theory of motion are not, therefore, objections to this theory properly understood, but only to a quite unnecessary assumption of simplicity in the momentary object of sense. Of the immediate object of sense, in the case of a visible motion, we may say that at each instant it is in all the positions which remain sensible at that instant; but this set of positions changes continuously from moment to moment, and is amenable to exactly the same mathematical treatment as if it were a mere point. When we assert that some mathematical account of phenomena is correct, all that we primarily assert is that *something* definable in terms of the crude phenomena satisfies our formulæ; and in this sense the mathematical theory of motion is applicable to the data of sensation as well as to the supposed particles of abstract physics.

There are a number of distinct questions which are apt to be confused when the mathematical continuum is said to be inadequate to the facts of sense. We may state these, in order of diminishing generality, as follows:—

(*a*) Are series possessing mathematical continuity logically possible?

(*b*) Assuming that they are possible logically, are they not impossible as applied to actual sense-data, because, among actual sense-data, there are no such fixed mutually external terms as are to be found, *e.g.*, in the series of fractions?

(*c*) Does not the assumption of points and instants make the whole mathematical account fictitious?

(*d*) Finally, assuming that all these objections have been answered, is there, in actual empirical fact, any sufficient reason to believe the world of sense continuous?

Let us consider these questions in succession.

(*a*) The question of the logical possibility of the mathematical continuum turns partly on the elementary misunderstandings we considered at the beginning of the present lecture, partly on the possibility of the mathematical infinite, which will occupy our next two lectures, and partly on the logical form of the answer to the Bergsonian objection which we stated a few minutes ago. I shall say no more on this topic at present, since it is desirable first to complete the psychological answer.

(*b*) The question whether sense-data are composed of mutually external units is not one which can be decided by empirical evidence. It is often urged that, as a matter of immediate experience, the sensible flux is devoid of divisions, and is falsified by the dissections of the intellect. Now I have no wish to argue that this view is *contrary* to immediate experience: I wish only to maintain that it is essentially incapable of being *proved* by immediate experience. As we saw, there must be among sense-data differences so slight as to be imperceptible: the fact that sense-data are immediately given does not mean that their differences also *must* be immediately given (though they *may* be). Suppose, for example, a coloured surface on which the colour changes gradually—so gradually that the difference of colour in two very neighbouring portions is imperceptible, while the difference between more widely separated portions is quite noticeable. The effect produced, in such a case, will be precisely that of "interpenetration," of transition which is not a matter of discrete units. And since it tends to be supposed that the colours, being immediate data, must *appear* different if they *are* different, it seems easily to follow that "interpenetration" must be the ultimately right account. But this does not follow. It is unconsciously assumed, as a premiss for a *reductio ad absurdum* of the analytic view, that, if A and B are immediate data, and A differs from B, then the fact that they differ must also be an immediate datum. It is difficult to say how this assumption arose, but I think it is to be connected with the confusion between "acquaintance" and "knowledge about." Acquaintance, which is what we derive from sense, does not, theoretically at least, imply even the smallest "knowledge about," *i.e.* it does not imply knowledge of any proposition concerning the object with which we are acquainted. It is a mistake to speak as if acquaintance had degrees: there is merely acquaintance and non-acquaintance. When we speak of becoming "better acquainted," as for instance with a person, what we must mean is, becoming acquainted with more parts of a certain whole; but the acquaintance with each part is either complete or nonexistent. Thus it is a mistake to say that if we were perfectly acquainted with an object we should know all about it. "Knowledge about" is knowledge of propositions, which is not involved necessarily in acquaintance with the constituents of the propositions. To know that two shades of colour are different is knowledge about them; hence acquaintance with the two shades does not in any way necessitate the knowledge that they are different.

From what has just been said it follows that the nature of sense-data cannot be validly used to prove that they are not composed of mutually external units. It may be admitted, on the other hand, that nothing in their empirical character specially necessitates the view that they are composed of mutually external units. This view, if it is held, must be held on logical, not on empirical, grounds. I believe that the logical grounds are adequate to the conclusion. They rest, at bottom, upon the impossibility of explaining complexity without assuming constituents. It is undeniable that the visual field, for example, is complex; and so far as I can see, there is always self-contradiction in the theories which, while admitting this complexity, attempt to deny that it results from a combination of mutually external units. But to pursue this topic would lead us too far from our theme, and I shall therefore say no more about it at present.

(*c*) It is sometimes urged that the mathematical account of motion is rendered fictitious by its assumption of points and instants. Now there are here two different questions to be distinguished. There is the question of absolute or relative space and time, and there is the question whether what occupies space and time must be composed of elements which have no extension or duration. And each of these questions in turn may take two forms, namely: (α) is the hypothesis *consistent* with the facts and with logic? (β) is it *necessitated* by the facts or by logic? I wish to answer, in each case, yes to the first form of the question, and no to the second. But in any case the mathematical account of motion will not be fictitious, provided a right interpretation is given to the words "point" and "instant." A few words on each alternative will serve to make this clear.

Formally, mathematics adopts an absolute theory of space and time, *i.e.* it assumes that, besides the things which are in space and time, there are also entities, called "points" and "instants," which are occupied by things. This view, however, though advocated by Newton, has long been regarded by mathematicians as merely a convenient fiction. There is, so far as I can see, no conceivable evidence either for or against it. It is logically possible, and it is consistent with the facts. But the facts are also consistent with the denial of spatial and temporal entities over and above things with spatial and temporal relations. Hence, in accordance with Occam's razor, we shall do well to abstain from either assuming or denying points and instants. This means, so far as practical working out is concerned, that we adopt the relational theory; for in practice the refusal to assume points and instants has the same effect as the denial of them. But in strict theory the two are quite different, since the denial introduces an element of unverifiable dogma which is wholly absent when we merely refrain from the assertion. Thus, although we shall derive points and instants from things, we shall leave the bare possibility open that they may also have an independent existence as simple entities.

We come now to the question whether the things in space and time are to be conceived as composed of elements without extension or duration, *i.e.* of elements which only occupy a point and an instant. Physics, formally, assumes in its differential equations that things consist of elements which occupy only a point at each instant, but persist throughout time. For reasons explained in Lecture IV., the persistence of things through time is to be regarded as the formal result of a logical construction, not as necessarily implying any actual persistence. The same motives, in fact, which lead to the division of things into point-particles, ought presumably to lead to their division into instant-particles, so that the ultimate *formal* constituent of the matter in physics will be a point-instant-particle. But such objects, as well as the particles of physics, are not data. The same economy of hypothesis, which dictates the practical adoption of a relative rather than an absolute space and time, also dictates the practical adoption of material elements which have a finite extension and duration. Since, as we saw in Lecture IV., points and instants can be constructed as logical functions of such elements, the mathematical account of motion, in which a particle passes continuously through a continuous series of points, can be interpreted in a form which assumes only elements which agree with our actual data in having a finite extension and duration. Thus, so far as the use of points and instants is concerned, the mathematical account of motion can be freed from the charge of employing fictions.

(*d*) But we must now face the question: Is there, in actual empirical fact, any sufficient reason to believe the world of sense continuous? The answer here must, I think, be in the negative. We may say that the hypothesis of continuity is perfectly consistent with the facts and with logic, and that it is technically simpler than any other tenable hypothesis. But since our powers of discrimination among very similar sensible objects are not infinitely precise, it is quite impossible to decide between different theories which only differ in regard to what is below the margin of discrimination. If, for example, a coloured surface which we see consists of a finite number of very small surfaces, and if a motion which we see consists, like a cinematograph, of a large finite number of successive positions, there will be nothing empirically discoverable to show that objects of sense are not continuous. In what is called *experienced* continuity, such as is said to be given in sense, there is a large negative element: absence of perception of difference occurs in cases which are *thought* to give perception of absence of difference. When, for example, we cannot distinguish a colour A from a colour B, nor a colour B from a colour C, but can distinguish A from C, the indistinguishability is a purely negative fact, namely, that we do not *perceive* a difference. Even in regard to immediate data, this is no reason for denying that there is a difference. Thus, if we see a coloured surface whose colour changes gradually, its sensible appearance if the change is continuous will be indistinguishable from what it would be if the change were by small finite jumps. If this is true, as it seems to be, it follows that there can never be any empirical evidence to demonstrate that the sensible world is continuous, and not a collection of a very large finite number of elements of which each differs from its neighbour in a finite though very small degree. The continuity of space and time, the infinite number of different shades in the spectrum, and so on, are all in the nature of unverifiable hypotheses—perfectly possible logically, perfectly consistent with the known facts, and simpler technically than any other tenable hypotheses, but not the sole hypotheses which are logically and empirically adequate.

If a relational theory of instants is constructed, in which an "instant" is defined as a group of events simultaneous with each other and not all simultaneous with any event outside the group, then if our resulting series of instants is to be compact, it must be possible, if x wholly precedes y, to find an event z, simultaneous with part of x, which wholly precedes some event which wholly precedes y. Now this requires that the number of events concerned should be infinite in any finite period of time. If this is to be the case in the world of one man's sense-data, and if each sense-datum is to have not less than a certain finite temporal extension, it will be necessary to assume that we always have an infinite number of sense-data simultaneous with any given sense-datum. Applying similar considerations to space, and assuming that sense-data are to have not less than a certain spatial extension, it will be necessary to suppose that an infinite number of sense-data overlap spatially with any given sense-datum. This hypothesis is possible, if we suppose a single sense-datum, *e.g.* in sight, to be a finite surface, enclosing other surfaces which are also single sense-data. But there are difficulties in such a hypothesis, and I do not know whether these difficulties could be successfully met. If they cannot, we must do one of two things: either declare that the world of one man's sense-data is not continuous, or else refuse to admit that there is any lower limit to the duration and extension of a single sense-datum. I do not know what is the right course to adopt as regards these alternatives. The logical analysis we have been considering provides the apparatus for dealing with the various hypotheses, and the empirical decision between them is a problem for the psychologist.

(3) We have now to consider the *logical* answer to the alleged difficulties of the mathematical theory of motion, or rather to the positive theory which is urged on the other side. The view urged explicitly by Bergson, and implied in the doctrines of many philosophers, is, that a motion is something indivisible, not validly analysable into a series of states. This is part of a much more general doctrine, which holds that analysis always falsifies, because the parts of a complex whole are different, as combined in that whole, from what they would otherwise be. It is very difficult to state this doctrine in any form which has a precise meaning. Often arguments are used which have no bearing whatever upon the question. It is urged, for example, that when a man becomes a father, his nature is altered by the new relation in which he finds himself, so that he is not strictly identical with the man who was previously not a father. This may be true, but it is a causal psychological fact, not a logical fact. The doctrine would require that a man who is a father cannot be strictly identical with a man who is a son, because he is modified in one way by the relation of fatherhood and in another by that of sonship. In fact, we may give a precise statement of the doctrine we are combating in the form: *There can never be two facts concerning the same thing.* A fact concerning a thing always is or involves a relation to one or more entities; thus two facts concerning the same thing would involve two relations of the same thing. But the doctrine in question holds that a thing is so modified by its relations that it cannot be the same in one relation as in another. Hence, if this doctrine is true, there can never be more than one fact concerning any one thing. I do not think the philosophers in question have realised that this is the precise statement of the view they advocate, because in this form the view is so contrary to plain truth that its falsehood is evident as soon as it is stated. The discussion of this question, however, involves so many logical subtleties, and is so beset with difficulties, that I shall not pursue it further at present.

When once the above general doctrine is rejected, it is obvious that, where there is change, there must be a succession of states. There cannot be change—and motion is only a particular case of change—unless there is something different at one time from what there is at some other time. Change, therefore, must involve relations and complexity, and must demand analysis. So long as our analysis has only gone as far as other smaller changes, it is not complete; if it is to be complete, it must end with terms that are not changes, but are related by a relation of earlier and later. In the case of changes which appear continuous, such as motions, it seems to be impossible to find anything other than change so long as we deal with finite periods of time, however short. We are thus driven back, by the logical necessities of the case, to the conception of instants without duration, or at any rate without any duration which even the most delicate instruments can reveal. This conception, though it can be made to seem difficult, is really easier than any other that the facts allow. It is a kind of logical framework into which any tenable theory must fit—not necessarily itself the statement of the crude facts, but a form in which statements which are true of the crude facts can be made by a suitable interpretation. The direct consideration of the crude facts of the physical world has been undertaken in earlier lectures; in the present lecture, we have only been concerned to show that nothing in the crude facts is inconsistent with the mathematical doctrine of continuity, or demands a continuity of a radically different kind from that of mathematical motion.

LECTURE VI
THE PROBLEM OF INFINITY CONSIDERED HISTORICALLY

IT will be remembered that, when we enumerated the grounds upon which the reality of the sensible world has been questioned, one of those mentioned was the supposed impossibility of infinity and continuity. In view of our earlier discussion of physics, it would seem that no *conclusive* empirical evidence exists in favour of infinity or continuity in objects of sense or in matter. Nevertheless, the explanation which assumes infinity and continuity remains incomparably easier and more natural, from a scientific point of view, than any other, and since Georg Cantor has shown that the supposed contradictions are illusory, there is no longer any reason to struggle after a finitist explanation of the world.

The supposed difficulties of continuity all have their source in the fact that a continuous series must have an infinite number of terms, and are in fact difficulties concerning infinity. Hence, in freeing the infinite from contradiction, we are at the same time showing the logical possibility of continuity as assumed in science.

The kind of way in which infinity has been used to discredit the world of sense may be illustrated by Kant's first two antinomies. In the first, the thesis states: "The world has a beginning in time, and as regards space is enclosed within limits"; the antithesis states: "The world has no beginning and no limits in space, but is infinite in respect of both time and space." Kant professes to prove both these propositions, whereas, if what we have said on modern logic has any truth, it must be impossible to prove either. In order, however, to rescue the world of sense, it is enough to destroy the proof of *one* of the two. For our present purpose, it is the proof that the world is *finite* that interests us. Kant's argument as regards space here rests upon his argument as regards time. We need therefore only examine the argument as regards time. What he says is as follows:

"For let us assume that the world has no beginning as regards time, so that up to every given instant an eternity has elapsed, and therefore an infinite series of successive states of the things in the world has passed by. But the infinity of a series consists just in this, that it can never be completed by successive synthesis. Therefore an infinite past world-series is impossible, and accordingly a beginning of the world is a necessary condition of its existence; which was the first thing to be proved."

Many different criticisms might be passed on this argument, but we will content ourselves with a bare minimum. To begin with, it is a mistake to define the infinity of a series as "impossibility of completion by successive synthesis." The notion of infinity, as we shall see in the next lecture, is primarily a property of *classes*, and only derivatively applicable to series; classes which are infinite are given all at once by the defining property of their members, so that there is no question of "completion" or of "successive synthesis." And the word "synthesis," by suggesting the mental activity of synthesising, introduces, more or less surreptitiously, that reference to mind by which all Kant's philosophy was infected. In the second place, when Kant says that an infinite series can "never" be completed by successive synthesis, all that he has even conceivably a right to say is that it cannot be completed *in a finite time*. Thus what he really proves is, at most, that if the world had no beginning, it must have already existed for an infinite time. This, however, is a very poor conclusion, by no means suitable for his purposes. And with this result we might, if we chose, take leave of the first antinomy.

It is worth while, however, to consider how Kant came to make such an elementary blunder. What happened in his imagination was obviously something like this: Starting from the present and going backwards in time, we have, if the world had no beginning, an infinite series of events. As we see from the word "synthesis," he imagined a mind trying to grasp these successively, *in the reverse order* to that in which they had occurred, *i.e.* going from the present backwards. *This* series is obviously one which has no end. But the series of events up to the present has an end, since it ends with the present. Owing to the inveterate subjectivism of his mental habits, he failed to notice that he had reversed the sense of the series by substituting backward synthesis for forward happening, and thus he supposed that it was necessary to identify the mental series, which had no end, with the physical series, which had an end but no beginning. It was this mistake, I think, which, operating unconsciously, led him to attribute validity to a singularly flimsy piece of fallacious reasoning.

The second antinomy illustrates the dependence of the problem of continuity upon that of infinity. The thesis states: "Every complex substance in the world consists of simple parts, and there exists everywhere nothing but the simple or what is composed of it." The antithesis states: "No complex thing in the world consists of simple parts, and everywhere in it there exists nothing simple." Here, as before, the proofs of both thesis and antithesis are open to criticism, but for the purpose of vindicating physics and the world of sense it is enough to find a fallacy in *one* of the proofs. We will choose for this purpose the proof of the antithesis, which begins as follows:

"Assume that a complex thing (as substance) consists of simple parts. Since all external relation, and therefore all composition out of substances, is only possible in space, the space occupied by a complex thing must consist of as many parts as the thing consists of. Now space does not consist of simple parts, but of spaces."

The rest of his argument need not concern us, for the nerve of the proof lies in the one statement: "Space does not consist of simple parts, but of spaces." This is like Bergson's objection to "the absurd proposition that motion is made up of immobilities." Kant does not tell us why he holds that a space must consist of spaces rather than of simple parts. Geometry regards space as made up of points, which are simple; and although, as we have seen, this view is not scientifically or logically *necessary*, it remains *primâ facie* possible, and its mere possibility is enough to vitiate Kant's argument. For, if his proof of the thesis of the antinomy were valid, and if the antithesis could only be avoided by assuming points, then the antinomy itself would afford a conclusive reason in favour of points. Why, then, did Kant think it impossible that space should be composed of points?

I think two considerations probably influenced him. In the first place, the essential thing about space is spatial order, and mere points, by themselves, will not account for spatial order. It is obvious that his argument assumes absolute space; but it is spatial *relations* that are alone important, and they cannot be reduced to points. This ground for his view depends, therefore, upon his ignorance of the logical theory of order and his oscillations between absolute and relative space. But there is also another ground for his opinion, which is more relevant to our present topic. This is the ground derived from infinite divisibility. A space may be halved, and then halved again, and so on *ad infinitum*, and at every stage of the process the parts are still spaces, not points. In order to reach points by such a method, it would be necessary to come to the end of an unending process, which is impossible. But just as an infinite class can be given all at once by its defining concept, though it cannot be reached by successive enumeration, so an infinite set of points can be given all at once as making up a line or area or volume, though they can never be reached by the process of successive division. Thus the infinite divisibility of space gives no ground for denying that space is composed of points. Kant does not give his grounds for this denial, and we can therefore only conjecture what they were. But the above two grounds, which we have seen to be fallacious, seem sufficient to account for his opinion, and we may therefore conclude that the antithesis of the second antinomy is unproved.

The above illustration of Kant's antinomies has only been introduced in order to show the relevance of the problem of infinity to the problem of the reality of objects of sense. In the remainder of the present lecture, I wish to state and explain the problem of infinity, to show how it arose, and to show the irrelevance of all the solutions proposed by philosophers. In the following lecture, I shall try to explain the true solution, which has been discovered by the mathematicians, but nevertheless belongs essentially to philosophy. The solution is definitive, in the sense that it entirely satisfies and convinces all who study it carefully. For over two thousand years the human intellect was baffled by the problem; its many failures and its ultimate success make this problem peculiarly apt for the illustration of method.

The problem appears to have first arisen in some such way as the following.Pythagoras and his followers, who were interested, like Descartes, in the application of number to geometry, adopted in that science more arithmetical methods than those with which Euclid has made us familiar. They, or their contemporaries the atomists, believed, apparently, that space is composed of indivisible points, while time is composed of indivisible instants.[23] This belief would not, by itself, have raised the difficulties which they encountered, but it was presumably accompanied by another belief, that the number of points in any finite area or of instants in any finite period must be finite. I do not suppose that this latter belief was a conscious one, because probably no other possibility had occurred to them. But the belief nevertheless operated, and very soon brought them into conflict with facts which they themselves discovered. Before explaining how this occurred, however, it is necessary to say one word in explanation of the phrase "finite number." The *exact* explanation is a matter for our next lecture; for the present, it must suffice to say that I mean 0 and 1 and 2 and 3 and so on, for ever—in other words, any number that can be obtained by successively adding ones. This includes all the numbers that can be expressed by means of our ordinary numerals, and since such numbers can be made greater and greater, without ever reaching an unsurpassable maximum, it is easy to suppose that there are no other numbers. But this supposition, natural as it is, is mistaken.

Whether the Pythagoreans themselves believed space and time to be composed of indivisible points and instants is a debatable question.[24] It would seem that the distinction between space and matter had not yet been clearly made, and that therefore, when an atomistic view is expressed, it is difficult to decide whether particles of matter or points of space are intended. There is an interesting passage in Aristotle's *Physics*,where he says:

"The Pythagoreans all maintained the existence of the void, and said that it enters into the heaven itself from the boundless breath, inasmuch as the heaven breathes in the void also; and the void differentiates natures, as if it were a sort of separation of consecutives, and as if it were their differentiation; and that this also is what is first in numbers, for it is the void which differentiates them."

This seems to imply that they regarded matter as consisting of atoms with empty space in between. But if so, they must have thought space could be studied by only paying attention to the atoms, for otherwise it would be hard to account for their arithmetical methods in geometry, or for their statement that "things are numbers."

The difficulty which beset the Pythagoreans in their attempts to apply numbers arose through their discovery of incommensurables, and this, in turn, arose as follows. Pythagoras, as we all learnt in youth, discovered the proposition that the sum of the squares on the sides of a right-angled triangle is equal to the square on the hypotenuse. It is said that he sacrificed an ox when he discovered this theorem; if so, the ox was the first martyr to science. But the theorem, though it has remained his chief claim to immortality, was soon found to have a consequence fatal to his whole philosophy. Consider the case of a right-angled triangle whose two sides are equal, such a triangle as is formed by two sides of a square and a diagonal. Here, in virtue of the theorem, the square on the diagonal is double of the square on either of the sides. But Pythagoras or his early followers easily proved that the square of one whole number cannot be double of the square of another.Thus the length of the side and the length of the diagonal are incommensurable; that is to say, however small a unit of length you take, if it is contained an exact number of times in the side, it is not contained any exact number of times in the diagonal, and *vice versa.*

Now this fact might have been assimilated by some philosophies without any great difficulty, but to the philosophy of Pythagoras it was absolutely fatal. Pythagoras held that number is the constitutive essence of all things, yet no two numbers could express the ratio of the side of a square to the diagonal. It would seem probable that we may expand his difficulty, without departing from his thought, by assuming that he regarded the length of a line as determined by the number of atoms contained in it—a line two inches long would contain twice as many atoms as a line one inch long, and so on. But if this were the truth, then there must be a definite numerical ratio between any two finite lengths, because it was supposed that the number of atoms in each, however large, must be finite. Here there was an insoluble contradiction. The Pythagoreans, it is said, resolved to keep the existence of incommensurables a profound secret, revealed only to a few of the supreme heads of the sect; and one of their number, Hippasos of Metapontion, is even said to have been shipwrecked at sea for impiously disclosing the terrible discovery to their enemies. It must be remembered that Pythagoras was the founder of a new religion as well as the teacher of a new science: if the science came to be doubted, the disciples might fall into sin, and perhaps even eat beans, which according to Pythagoras is as bad as eating parents' bones.

The problem first raised by the discovery of incommensurables proved, as time went on, to be one of the most severe and at the same time most far-reaching problems that have confronted the human intellect in its endeavour to understand the world. It showed at once that numerical measurement of lengths, if it was to be made accurate, must require an arithmetic more advanced and more difficult than any that the ancients possessed. They therefore set to work to reconstruct geometry on a basis which did not assume the universal possibility of numerical measurement—a reconstruction which, as may be seen in Euclid, they effected with extraordinary skill and with great logical acumen. The moderns, under the influence of Cartesian geometry, have reasserted the universal possibility of numerical measurement, extending arithmetic, partly for that purpose, so as to include what are called "irrational" numbers, which give the ratios of incommensurable lengths. But although irrational numbers have long been used without a qualm, it is only in quite recent years that logically satisfactory definitions of them have been given. With these definitions, the first and most obvious form of the difficulty which confronted the Pythagoreans has been solved; but other forms of the difficulty remain to be considered, and it is these that introduce us to the problem of infinity in its pure form.

We saw that, accepting the view that a length is composed of points, the existence of incommensurables proves that every finite length must contain an infinite number of points. In other words, if we were to take away points one by one, we should never have taken away all the points, however long we continued the process. The number of points, therefore, cannot be *counted*, for counting is a process which enumerates things one by one. The property of being unable to be counted is characteristic of infinite collections, and is a source of many of their paradoxical qualities. So paradoxical are these qualities that until our own day they were thought to constitute logical contradictions. A long line of philosophers, from Zeno to M. Bergson, have based much of their metaphysics upon the supposed impossibility of infinite collections. Broadly speaking, the difficulties were stated by Zeno, and nothing material was added until we reach Bolzano's *Paradoxien des Unendlichen*, a little work written in 1847–8, and published posthumously in 1851. Intervening attempts to deal with the problem are futile and negligible. The definitive solution of the difficulties is due, not to Bolzano, but to Georg Cantor, whose work on this subject first appeared in 1882.

In order to understand Zeno, and to realise how little modern orthodox metaphysics has added to the achievements of the Greeks, we must consider for a moment his master Parmenides, in whose interest the paradoxes were invented.Parmenides expounded his views in a poem divided into two parts, called "the way of truth" and "the way of opinion"—like Mr Bradley's "Appearance" and "Reality," except that Parmenides tells us first about reality and then about appearance. "The way of opinion," in his philosophy, is, broadly speaking, Pythagoreanism; it begins with a warning: "Here I shall close my trustworthy speech and thought about the truth. Henceforward learn the opinions of mortals, giving ear to the deceptive ordering of my words." What has gone before has been revealed by a goddess, who tells him what really *is*. Reality, she says, is uncreated, indestructible, unchanging, indivisible; it is "immovable in the bonds of mighty chains, without beginning and without end; since coming into being and passing away have been driven afar, and true belief has cast them away." The fundamental principle of his inquiry is stated in a sentence which would not be out of place in Hegel:"Thou canst not know what is not—that is impossible—nor utter it; for it is the same thing that can be thought and that can be." And again: "It needs must be that what can be thought and spoken of is; for it is possible for it to be, and it is not possible for what is nothing to be." The impossibility of change follows from this principle; for what is past can be spoken of, and therefore, by the principle, still is.

The great conception of a reality behind the passing illusions of sense, a reality one, indivisible, and unchanging, was thus introduced into Western philosophy by Parmenides, not, it would seem, for mystical or religious reasons, but on the basis of a logical argument as to the impossibility of not-being. All the great metaphysical systems—notably those of Plato, Spinoza, and Hegel—are the outcome of this fundamental idea. It is difficult to disentangle the truth and the error in this view. The contention that time is unreal and that the world of sense is illusory must, I think, be regarded as based upon fallacious reasoning. Nevertheless, there is some sense—easier to feel than to state—in which time is an unimportant and superficial characteristic of reality. Past and future must be acknowledged to be as real as the present, and a certain emancipation from slavery to time is essential to philosophic thought. The importance of time is rather practical than theoretical, rather in relation to our desires than in relation to truth. A truer image of the world, I think, is obtained by picturing things as entering into the stream of time from an eternal world outside, than from a view which regards time as the devouring tyrant of all that is. Both in thought and in feeling, to realise the unimportance of time is the gate of wisdom. But unimportance is not unreality; and therefore what we shall have to say about Zeno's arguments in support of Parmenides must be mainly critical.

The relation of Zeno to Parmenides is explained by Plato in the dialogue in which Socrates, as a young man, learns logical acumen and philosophic disinterestedness from their dialectic. I quote from Jowett's translation:

"I see, Parmenides, said Socrates, that Zeno is your second self in his writings too; he puts what you say in another way, and would fain deceive us into believing that he is telling us what is new. For you, in your poems, say All is one, and of this you adduce excellent proofs; and he on the other hand says There is no Many; and on behalf of this he offers overwhelming evidence. To deceive the world, as you have done, by saying the same thing in different ways, one of you affirming the one, and the other denying the many, is a strain of art beyond the reach of most of us.

"Yes, Socrates, said Zeno. But although you are as keen as a Spartan hound in pursuing the track, you do not quite apprehend the true motive of the composition, which is not really such an ambitious work as you imagine; for what you speak of was an accident; I had no serious intention of deceiving the world. The truth is, that these writings of mine were meant to protect the arguments of Parmenides against those who scoff at him and show the many ridiculous and contradictory results which they suppose to follow from the affirmation of the one. My answer is an address to the partisans of the many, whose attack I return with interest by retorting upon them that their hypothesis of the being of the many if carried out appears in a still more ridiculous light than the hypothesis of the being of the one."

Zeno's four arguments against motion were intended to exhibit the contradictions that result from supposing that there is such a thing as change, and thus to support the Parmenidean doctrine that reality is unchanging.Unfortunately, we only know his arguments through Aristotle,who stated them in order to refute them. Those philosophers in the present day who have had their doctrines stated by opponents will realise that a just or adequate presentation of Zeno's position is hardly to be expected from Aristotle; but by some care in interpretation it seems possible to reconstruct the so-called "sophisms" which have been "refuted" by every tyro from that day to this.

Zeno's arguments would seem to be "ad hominem"; that is to say, they seem to assume premisses granted by his opponents, and to show that, granting these premisses, it is possible to deduce consequences which his opponents must deny. In order to decide whether they are valid arguments or "sophisms," it is necessary to guess at the tacit premisses, and to decide who was the "homo" at whom they were aimed. Some maintain that they were aimed at the Pythagoreans,while others have held that they were intended to refute the atomists.M. Evellin, on the contrary, holds that they constitute a refutation of infinite divisibility,while M. G. Noël, in the interests of Hegel, maintains that the first two arguments refute infinite divisibility, while the next two refute indivisibles.Amid such a bewildering variety of interpretations, we can at least not complain of any restrictions on our liberty of choice.

The historical questions raised by the above-mentioned discussions are no doubt largely insoluble, owing to the very scanty material from which our evidence is derived. The points which seem fairly clear are the following: (1) That, in spite of MM. Milhaud and Paul Tannery, Zeno is anxious to prove that motion is really impossible, and that he desires to prove this because he follows Parmenides in denying plurality;(2) that the third and fourth arguments proceed on the hypothesis of indivisibles, a hypothesis which, whether adopted by the Pythagoreans or not, was certainly much advocated, as may be seen from the treatise *On Indivisible Lines* attributed to Aristotle. As regards the first two arguments, they would seem to be valid on the hypothesis of indivisibles, and also, without this hypothesis, to be such as would be valid if the traditional contradictions in infinite numbers were insoluble, which they are not.

We may conclude, therefore, that Zeno's polemic is directed against the view that space and time consist of points and instants; and that as against the view that a finite stretch of space or time consists of a finite number of points and instants, his arguments are not sophisms, but perfectly valid.

The conclusion which Zeno wishes us to draw is that plurality is a delusion, and spaces and times are really indivisible. The other conclusion which is possible, namely, that the number of points and instants is infinite, was not tenable so long as the infinite was infected with contradictions. In a fragment which is not one of the four famous arguments against motion, Zeno says:

"If things are a many, they must be just as many as they are, and neither more nor less. Now, if they are as many as they are, they will be finite in number.

"If things are a many, they will be infinite in number; for there will always be other things between them, and others again between these. And so things are infinite in number."

This argument attempts to prove that, if there are many things, the number of them must be both finite and infinite, which is impossible; hence we are to conclude that there is only one thing. But the weak point in the argument is the phrase: "If they are just as many as they are, they will be finite in number." This phrase is not very clear, but it is plain that it assumes the impossibility of definite infinite numbers. Without this assumption, which is now known to be false, the arguments of Zeno, though they suffice (on certain very reasonable assumptions) to dispel the hypothesis of finite indivisibles, do not suffice to prove that motion and change and plurality are impossible. They are not, however, on any view, mere foolish quibbles: they are serious arguments, raising difficulties which it has taken two thousand years to answer, and which even now are fatal to the teachings of most philosophers.

The first of Zeno's arguments is the argument of the race-course, which is paraphrased by Burnet as follows:

"You cannot get to the end of a race-course. You cannot traverse an infinite number of points in a finite time. You must traverse the half of any given distance before you traverse the whole, and the half of that again before you can traverse it. This goes on *ad infinitum*, so that there are an infinite number of points in any given space, and you cannot touch an infinite number one by one in a finite time."

Zeno appeals here, in the first place, to the fact that any distance, however small, can be halved. From this it follows, of course, that there must be an infinite number of points in a line. But, Aristotle represents him as arguing, you cannot touch an infinite number of points *one by one* in a finite time. The words "one by one" are important. (1) If *all* the points touched are concerned, then, though you pass through them continuously, you do not touch them "one by one." That is to say, after touching one, there is not another which you touch next: no two points are next each other, but between any two there are always an infinite number of others, which cannot be enumerated one by one. (2) If, on the other hand, only the successive middle points are concerned, obtained by always halving what remains of the course, then the points are reached one by one, and, though they are infinite in number, they are in fact all reached in a finite time. His argument to the contrary may be supposed to appeal to the view that a finite time must consist of a finite number of instants, in which case what he says would be perfectly true on the assumption that the possibility of continued dichotomy is undeniable. If, on the other hand, we suppose the argument directed against the partisans of infinite divisibility, we must suppose it to proceed as follows:"The points given by successive halving of the distances still to be traversed are infinite in number, and are reached in succession, each being reached a finite time later than its predecessor; but the sum of an infinite number of finite times must be infinite, and therefore the process will never be completed." It is very possible that this is historically the right interpretation, but in this form the argument is invalid. If half the course takes half a minute, and the next quarter takes a quarter of a minute, and so on, the whole course will take a minute. The apparent force of the argument, on this interpretation, lies solely in the mistaken supposition that there cannot be anything beyond the whole of an infinite series, which can be seen to be false by observing that 1 is beyond the whole of the infinite series $\frac{1}{2}$, $\frac{3}{4}$, $\frac{7}{8}$, $\frac{15}{16}$, ...

The second of Zeno's arguments is the one concerning Achilles and the tortoise, which has achieved more notoriety than the others. It is paraphrased by Burnet as follows:

"Achilles will never overtake the tortoise. He must first reach the place from which the tortoise started. By that time the tortoise will have got some way ahead. Achilles must then make up that, and again the tortoise will be ahead. He is always coming nearer, but he never makes up to it."

This argument is essentially the same as the previous one. It shows that, if Achilles ever overtakes the tortoise, it must be after an infinite number of instants have elapsed since he started. This is in fact true; but the view that an infinite number of instants make up an infinitely long time is not true, and therefore the conclusion that Achilles will never overtake the tortoise does not follow.

The third argument,that of the arrow, is very interesting. The text has been questioned. Burnet accepts the alterations of Zeller, and paraphrases thus:

"The arrow in flight is at rest. For, if everything is at rest when it occupies a space equal to itself, and what is in flight at any given moment always occupies a space equal to itself, it cannot move."

But according to Prantl, the literal translation of the unemended text of Aristotle's statement of the argument is as follows: "If everything, when it is behaving in a uniform manner, is continually either moving or at rest, but what is moving is always in the *now*, then the moving arrow is motionless." This form of the argument brings out its force more clearly than Burnet's paraphrase.

Here, if not in the first two arguments, the view that a finite part of time consists of a finite series of successive instants seems to be assumed; at any rate the plausibility of the argument seems to depend upon supposing that there are consecutive instants. Throughout an instant, it is said, a moving body is where it is: it cannot move during the instant, for that would require that the instant should have parts. Thus, suppose we consider a period consisting of a thousand instants, and suppose the arrow is in flight throughout this period. At each of the thousand instants, the arrow is where it is, though at the next instant it is somewhere else. It is never moving, but in some miraculous way the change of position has to occur *between* the instants, that is to say, not at any time whatever. This is what M. Bergson calls the cinematographic representation of reality. The more the difficulty is meditated, the more real it becomes. The solution lies in the theory of continuous series: we find it hard to avoid supposing that, when the arrow is in flight, there is a *next* position occupied at the *next* moment; but in fact there is no next position and no next moment, and when once this is imaginatively realised, the difficulty is seen to disappear.

The fourth and last of Zeno's arguments is the argument of the stadium.

The argument as stated by Burnet is as follows:

First Position.	Second Position.
A	A . . . .
B	B . . . .
C	C . . . .

"Half the time may be equal to double the time. Let us suppose three rows of bodies, one of which (A) is at rest while the other two (B, C) are moving with equal velocity in opposite directions. By the time they are all in the same part of the course, B will have passed twice as many of the bodies in C as in A. Therefore the time which it takes to pass C is twice as long as the time it takes to pass A. But the time which B and C take to reach the position of A is the same. Therefore double the time is equal to the half."

Gaye devoted an interesting article to the interpretation of this argument. His translation of Aristotle's statement is as follows:

"The fourth argument is that concerning the two rows of bodies, each row being composed of an equal number of bodies of equal size, passing each other on a race-course as they proceed with equal velocity in opposite directions, the one row originally occupying the space between the goal and the middle point of the course, and the other that between the middle point and the starting-post. This, he thinks, involves the conclusion that half a given time is equal to double the time. The fallacy of the reasoning lies in the assumption that a body occupies an equal time in passing with equal velocity a body that is in motion and a body of equal size that is at rest, an assumption which is false. For instance (so runs the argument), let A A ... be the stationary bodies of equal size, B B ... the bodies, equal in number and in size to A A ..., originally occupying the half of the course from the starting-post to the middle of the A's, and C C ... those originally occupying the other half from the goal to the middle of the A's, equal in number, size, and velocity, to B B ... Then three consequences follow. First, as the B's and C's pass one another, the first B reaches the last C at the same moment at which the first C reaches the last B. Secondly, at this moment the first C has passed all the A's, whereas the first B has passed only half the A's and has consequently occupied only half the time occupied by the first C, since each of the two occupies an equal time in passing each A. Thirdly, at the same moment all the B's have passed all the C's: for the first C and the first B will simultaneously reach the opposite ends of the course, since (so says Zeno) the time occupied by the first C in passing each of the B's is equal to that occupied by it in passing each of the A's, because an equal time is occupied by both the first B and the first C in passing all the A's. This is the argument: but it presupposes the aforesaid fallacious assumption."

First Position.			Second Position.				
B	B′	B″			B	B′	B″
A	A′	A″		A	A′	A″	
C	C′	C″	C	C′	C″		

This argument is not quite easy to follow, and it is only valid as against the assumption that a finite time consists of a finite number of instants. We may re-state it in different language. Let us suppose three drill-sergeants, A, A′, and A″, standing in a row, while the two files of soldiers march past them in opposite directions. At the first moment which we consider, the three men B, B′, B″ in one row, and the three men C, C′, C″ in the other row, are respectively opposite to A, A′, and A″. At the very next moment, each row has moved on, and now B and C″ are opposite A′. Thus B and C″ are opposite each other. When, then, did B pass C′? It must have been somewhere between the two moments which we supposed consecutive, and therefore the two moments cannot really have been consecutive. It follows that there must be other moments between any two given moments, and therefore that there must be an infinite number of moments in any given interval of time.

The above difficulty, that B must have passed C′ at some time between two consecutive moments, is a genuine one, but is not precisely the difficulty raised by Zeno. What Zeno professes to prove is that "half of a given time is equal to double that time." The most intelligible explanation of the argument known to me is that of Gaye.Since, however, his explanation is not easy to set forth shortly, I will re-state what seems to me to be the logical essence of Zeno's contention. If we suppose that time consists of a series of consecutive instants, and that motion consists in passing through a series of consecutive points, then the fastest possible motion is one which, at each instant, is at a point consecutive to that at which it was at the previous instant. Any slower motion must be one which has intervals of rest interspersed, and any faster motion must wholly omit some points. All this is evident from the fact that we cannot have more than one event for each instant. But now, in the case of our A's and B's and C's, B is opposite a fresh A every instant, and therefore the number of A's passed gives the number of instants since the beginning of the motion. But during the motion B has passed twice as many C's, and yet cannot have passed more than one each instant. Hence the number of instants since the motion began is twice the number of A's passed, though we previously found it was equal to this number. From this result, Zeno's conclusion follows.

Zeno's arguments, in some form, have afforded grounds for almost all the theories of space and time and infinity which have been constructed from his day to our own. We have seen that all his arguments are valid (with certain reasonable hypotheses) on the assumption that finite spaces and times consist of a finite number of points and instants, and that the third and fourth almost certainly in fact proceeded on this assumption, while the first and second, which were perhaps intended to refute the opposite assumption, were in that case fallacious. We may therefore escape from his paradoxes either by maintaining that, though space and time do consist of points and instants, the number of them in any finite interval is infinite; or by denying that space and time consist of points and instants at all; or lastly, by denying the reality of space and time altogether. It would seem that Zeno himself, as a supporter of Parmenides, drew the last of these three possible deductions, at any rate in regard to time. In this a very large number of philosophers have followed him. Many others, like M. Bergson, have preferred to deny that space and time consist of points and instants. Either of these solutions will meet the difficulties in the form in which Zeno raised them. But, as we saw, the difficulties can also be met if infinite numbers are admissible. And on grounds which are independent of space and time, infinite numbers, and series in which no two terms are consecutive, must in any case be admitted. Consider, for example, all the fractions less than 1, arranged in order of magnitude. Between any two of them, there are others, for example, the arithmetical mean of the two. Thus no two fractions are consecutive, and the total number of them is infinite. It will be found that much of what Zeno says as regards the series of points on a line can be equally well applied to the series of fractions. And we cannot deny that there are fractions, so that two of the above ways of escape are closed to us. It follows that, if we are to solve the whole class of difficulties derivable from Zeno's by analogy, we must discover some tenable theory of infinite numbers. What, then, are the difficulties which, until the last thirty years, led philosophers to the belief that infinite numbers are impossible?

The difficulties of infinity are of two kinds, of which the first may be called sham, while the others involve, for their solution, a certain amount of new and not altogether easy thinking. The sham difficulties are those suggested by the etymology, and those suggested by confusion of the mathematical infinite with what philosophers impertinently call the "true" infinite. Etymologically, "infinite" should mean "having no end." But in fact some infinite series have ends, some have not; while some collections are infinite without being serial, and can therefore not properly be regarded as either endless or having ends. The series of instants from any earlier one to any later one (both included) is infinite, but has two ends; the series of instants from the beginning of time to the present moment has one end, but is infinite. Kant, in his first antinomy, seems to hold that it is harder for the past to be infinite than for the future to be so, on the ground that the past is now completed, and that nothing infinite can be completed. It is very difficult to see how he can have imagined that there was any sense in this remark; but it seems most probable that he was thinking of the infinite as the "unended." It is odd that he did not see that the future too has one end at the present, and is precisely on a level with the past. His regarding the two as different in this respect illustrates just that kind of slavery to time which, as we agreed in speaking of Parmenides, the true philosopher must learn to leave behind him.

The confusions introduced into the notions of philosophers by the so-called "true" infinite are curious. They see that this notion is not the same as the mathematical infinite, but they choose to believe that it is the notion which the mathematicians are vainly trying to reach. They therefore inform the mathematicians, kindly but firmly, that they are mistaken in adhering to the "false" infinite, since plainly the "true" infinite is something quite different. The reply to this is that what they call the "true" infinite is a notion totally irrelevant to the problem of the mathematical infinite, to which it has only a fanciful and verbal analogy. So remote is it that I do not propose to confuse the issue by even mentioning what the "true" infinite is. It is the "false" infinite that concerns us, and we have to show that the epithet "false" is undeserved.

There are, however, certain genuine difficulties in understanding the infinite, certain habits of mind derived from the consideration of finite numbers, and easily extended to infinite numbers under the mistaken notion that they represent logical necessities. For example, every number that we are accustomed to, except 0, has another number immediately before it, from which it results by adding 1; but the first infinite number does not have this property. The numbers before it form an infinite series, containing all the ordinary finite numbers, having no maximum, no last finite number, after which one little step would plunge us into the infinite. If it is assumed that the first infinite number is reached by a succession of small steps, it is easy to show that it is self-contradictory. The first infinite number is, in fact, beyond the whole unending series of finite numbers. "But," it will be said, "there cannot be anything beyond the whole of an unending series." This, we may point out, is the very principle upon which Zeno relies in the arguments of the race-course and the Achilles. Take the race-course: there is the moment when the runner still has half his distance to run, then the moment when he still has a quarter, then when he still has an eighth, and so on in a strictly unending series. Beyond the whole of this series is the moment when he reaches the goal. Thus there certainly can be something beyond the whole of an unending series. But it remains to show that this fact is only what might have been expected.

The difficulty, like most of the vaguer difficulties besetting the mathematical infinite, is derived, I think, from the more or less unconscious operation of the idea of *counting*. If you set to work to count the terms in an infinite collection, you will never have completed your task. Thus, in the case of the runner, if half, three-quarters, seven-eighths, and so on of the course were marked, and the runner was not allowed to pass any of the marks until the umpire said "Now," then Zeno's conclusion would be true in practice, and he would never reach the goal.

But it is not essential to the existence of a collection, or even to knowledge and reasoning concerning it, that we should be able to pass its terms in review one by one. This may be seen in the case of finite collections; we can speak of "mankind" or "the human race," though many of the individuals in this collection are not personally known to us. We can do this because we know of various characteristics which every individual has if he belongs to the collection, and not if he does not. And exactly the same happens in the case of infinite collections:

they may be known by their characteristics although their terms cannot be enumerated. In this sense, an unending series may nevertheless form a whole, and there may be new terms beyond the whole of it.

Some purely arithmetical peculiarities of infinite numbers have also caused perplexity. For instance, an infinite number is not increased by adding one to it, or by doubling it. Such peculiarities have seemed to many to contradict logic, but in fact they only contradict confirmed mental habits. The whole difficulty of the subject lies in the necessity of thinking in an unfamiliar way, and in realising that many properties which we have thought inherent in number are in fact peculiar to finite numbers. If this is remembered, the positive theory of infinity, which will occupy the next lecture, will not be found so difficult as it is to those who cling obstinately to the prejudices instilled by the arithmetic which is learnt in childhood.

LECTURE VII
THE POSITIVE THEORY OF INFINITY

THE positive theory of infinity, and the general theory of number to which it has given rise, are among the triumphs of scientific method in philosophy, and are therefore especially suitable for illustrating the logical-analytic character of that method. The work in this subject has been done by mathematicians, and its results can be expressed in mathematical symbolism. Why, then, it may be said, should the subject be regarded as philosophy rather than as mathematics? This raises a difficult question, partly concerned with the use of words, but partly also of real importance in understanding the function of philosophy. Every subject-matter, it would seem, can give rise to philosophical investigations as well as to the appropriate science, the difference between the two treatments being in the direction of movement and in the kind of truths which it is sought to establish. In the special sciences, when they have become fully developed, the movement is forward and synthetic, from the simpler to the more complex. But in philosophy we follow the inverse direction: from the complex and relatively concrete we proceed towards the simple and abstract by means of analysis, seeking, in the process, to eliminate the particularity of the original subject-matter, and to confine our attention entirely to the logical *form* of the facts concerned.

Between philosophy and pure mathematics there is a certain affinity, in the fact that both are general and *a priori*. Neither of them asserts propositions which, like those of history and geography, depend upon the actual concrete facts being just what they are. We may illustrate this characteristic by means of Leibniz's conception of many *possible* worlds, of which one only is *actual*. In all the many possible worlds, philosophy and mathematics will be the same; the differences will only be in respect of those particular facts which are chronicled by the descriptive sciences. Any quality, therefore, by which our actual world is distinguished from other abstractly possible worlds, must be ignored by mathematics and philosophy alike. Mathematics and philosophy differ, however, in their manner of treating the general properties in which all possible worlds agree; for while mathematics, starting from comparatively simple propositions, seeks to build up more and more complex results by deductive synthesis, philosophy, starting from data which are common knowledge, seeks to purify and generalise them into the simplest statements of abstract form that can be obtained from them by logical analysis.

The difference between philosophy and mathematics may be illustrated by our present problem, namely, the nature of number. Both start from certain facts about numbers which are evident to inspection. But mathematics uses these facts to deduce more and more complicated theorems, while philosophy seeks, by analysis, to go behind these facts to others, simpler, more fundamental, and inherently more fitted to form the premisses of the science of arithmetic. The question, "What is a number?" is the pre-eminent philosophic question in this subject, but it is one which the mathematician as such need not ask, provided he knows enough of the properties of numbers to enable him to deduce his theorems. We, since our object is philosophical, must grapple with the philosopher's question. The answer to the question, "What is a number?" which we shall reach in this lecture, will be found to give also, by implication, the answer to the difficulties of infinity which we considered in the previous lecture.

The question "What is a number?" is one which, until quite recent times, was never considered in the kind of way that is capable of yielding a precise answer. Philosophers were content with some vague dictum such as, "Number is unity in plurality." A typical definition of the kind that contented philosophers is the following from Sigwart's *Logic* (§ 66, section 3): "Every number is not merely a *plurality*, but a plurality thought *as held together and closed, and to that extent as a unity*." Now there is in such definitions a very elementary blunder, of the same kind that would be committed if we said "yellow is a flower" because some flowers are yellow. Take, for example, the number 3. A single collection of three things might conceivably be described as "a plurality thought as held together and closed, and to that extent as a unity"; but a collection of three things is not the number 3. The number 3 is something which all collections of three things have in common, but is not itself a collection of three things. The definition, therefore, apart from any other defects, has failed to reach the necessary degree of abstraction: the number 3 is something more abstract than any collection of three things.

Such vague philosophic definitions, however, remained inoperative because of their very vagueness. What most men who thought about numbers really had in mind was that numbers are the result of *counting*. "On the consciousness of the law of counting," says Sigwart at the beginning of his discussion of number, "rests the possibility of spontaneously prolonging the series of numbers *ad infinitum*." It is this view of number as generated by counting which has been the chief psychological obstacle to the understanding of infinite numbers. Counting, because it is familiar, is erroneously supposed to be simple, whereas it is in fact a highly complex process, which has no meaning unless the numbers reached in counting have some significance independent of the process by which they are reached. And infinite numbers cannot be reached at all in this way. The mistake is of the same kind as if cows were defined as what can be bought from a cattle-merchant. To a person who knew several cattle-merchants, but had never seen a cow, this might seem an admirable definition. But if in his travels he came across a herd of wild cows, he would have to declare that they were not cows at all, because no cattle-merchant could sell them. So infinite numbers were declared not to be numbers at all, because they could not be reached by counting.

It will be worth while to consider for a moment what counting actually is. We count a set of objects when we let our attention pass from one to another, until we have attended once to each, saying the names of the numbers in order with each successive act of attention. The last number named in this process is the number of the objects, and therefore counting is a method of finding out what the number of the objects is. But this operation is really a very complicated one, and those who imagine that it is the logical source of number show themselves remarkably incapable of analysis. In the first place, when we say "one, two, three ..." as we count, we cannot be said to be discovering the number of the objects counted unless we attach some meaning to the words one, two, three, ... A child may learn to know these words in order, and to repeat them correctly like the letters of the alphabet, without attaching any meaning to them. Such a child may count correctly from the point of view of a grown-up listener, without having any idea of numbers at all. The operation of counting, in fact, can only be intelligently performed by a person who already has some idea what the numbers are; and from this it follows that counting does not give the logical basis of number.

Again, how do we know that the last number reached in the process of counting is the number of the objects counted? This is just one of those facts that are too familiar for their significance to be realised; but those who wish to be logicians must acquire the habit of dwelling upon such facts. There are two propositions involved in this fact: first, that the number of numbers from 1 up to any given number is that given number—for instance, the number of numbers from 1 to 100 is a hundred; secondly, that if a set of numbers can be used as names of a set of objects, each number occurring only once, then the number of numbers used as names is the same as the number of objects. The first of these propositions is capable of an easy arithmetical proof so long as finite numbers are concerned; but with infinite numbers, after the first, it ceases to be true. The second proposition remains true, and is in fact, as we shall see, an immediate consequence of the definition of number. But owing to the falsehood of the first proposition where infinite numbers are concerned, counting, even if it were practically possible, would not be a valid method of discovering the number of terms in an infinite collection, and would in fact give different results according to the manner in which it was carried out.

There are two respects in which the infinite numbers that are known differ from finite numbers: first, infinite numbers have, while finite numbers have not, a property which I shall call *reflexiveness*; secondly, finite numbers have, while infinite numbers have not, a property which I shall call *inductiveness*. Let us consider these two properties successively.

(1) *Reflexiveness.*—A number is said to be *reflexive* when it is not increased by adding 1 to it. It follows at once that any finite number can be added to a reflexive number without increasing it. This property of infinite numbers was always thought, until recently, to be self-contradictory; but through the work of Georg Cantor it has come to be recognised that, though at first astonishing, it is no more self-contradictory than the fact that people at the antipodes do not tumble off. In virtue of this property, given any infinite collection of objects, any finite number of objects can be added or taken away without increasing or diminishing the number of the collection. Even an infinite number of objects may, under certain conditions, be added or taken away without altering the number. This may be made clearer by the help of some examples.

Imagine all the natural numbers 0, 1, 2, 3, ... to be written down in a row, and immediately beneath them write down the numbers 1, 2, 3, 4, ..., so that 1 is under 0, 2 is under 1, and so on. Then every number in the top row has a number directly under it in the bottom row, and no number occurs twice in either row. It follows that the number of numbers in the two rows must be the same. But all the numbers that occur in the bottom row also occur in the top row, and one more, namely 0; thus the number of terms in the top row is obtained by adding one to the number of the bottom row. So long, therefore, as it was supposed that a number must be increased by adding 1 to it, this state of things constituted a contradiction, and led to the denial that there are infinite numbers.

$$0, 1, 2, 3, \ldots n \ldots$$

$$1, 2, 3, 4, \ldots n+1 \ldots$$

The following example is even more surprising. Write the natural numbers 1, 2, 3, 4, ... in the top row, and the even numbers 2, 4, 6, 8, ... in the bottom row, so that under each number in the top row stands its double in the bottom row. Then, as before, the number of numbers in the two rows is the same, yet the second row results from taking away all the odd numbers—an infinite collection—from the top row. This example is given by Leibniz to prove that there can be no infinite numbers. He believed in infinite collections, but, since he thought that a number must always be increased when it is added to and diminished when it is subtracted from, he maintained that infinite collections do not have numbers. "The number of all numbers," he says, "implies a contradiction, which I show thus: To any number there is a corresponding number equal to its double. Therefore the number of all numbers is not greater than the number of even numbers, *i.e.* the whole is not greater than its part."In dealing with this argument, we ought to substitute "the number of all finite numbers" for "the number of all numbers"; we then obtain exactly the illustration given by our two rows, one containing all the finite numbers, the other only the even finite numbers. It will be seen that Leibniz regards it as self-contradictory to maintain that the whole is not greater than its part. But the word "greater" is one which is capable of many meanings; for our purpose, we must substitute the less ambiguous phrase "containing a greater number of terms." In this sense, it is not self-contradictory for whole and part to be equal; it is the realisation of this fact which has made the modern theory of infinity possible.

There is an interesting discussion of the reflexiveness of infinite wholes in the first of Galileo's Dialogues on Motion. I quote from a translation published in 1730.The personages in the dialogue are Salviati, Sagredo, and Simplicius, and they reason as follows:

"*Simp.* Here already arises a Doubt which I think is not to be resolv'd; and that is this: Since 'tis plain that one Line is given greater than another, and since both contain infinite Points, we must surely necessarily infer, that we have found in the same Species something greater than Infinite, since the Infinity of Points of the greater Line exceeds the Infinity of Points of the lesser. But now, to assign an Infinite greater than an Infinite, is what I can't possibly conceive.

"*Salv.* These are some of those Difficulties which arise from Discourses which our finite Understanding makes about Infinites, by ascribing to them Attributes which we give to Things finite and terminate, which I think most improper, because those Attributes of Majority, Minority, and Equality, agree not with Infinities, of which we can't say that one is greater than, less than, or equal to another. For Proof whereof I have something come into my Head, which (that I may be the better understood) I will propose by way of Interrogatories to *Simplicius*, who started this Difficulty. To begin then: I suppose you know which are square Numbers, and which not?

"*Simp.* I know very well that a square Number is that which arises from the Multiplication of any Number into itself; thus 4 and 9 are square Numbers, that arising from 2, and this from 3, multiplied by themselves.

"*Salv.* Very well; And you also know, that as the Products are call'd Squares, the Factors are call'd Roots: And that the other Numbers, which proceed not from Numbers multiplied into themselves, are not Squares. Whence taking in all Numbers, both Squares and Not Squares, if I should say, that the Not Squares are more than the Squares, should I not be in the right?

"*Simp.* Most certainly.

"*Salv.* If I go on with you then, and ask you, How many squar'd Numbers there are? you may truly answer, That there are as many as are their proper Roots, since every Square has its own Root, and every Root its own Square, and since no Square has more than one Root, nor any Root more than one Square.

"*Simp.* Very true.

"*Salv.* But now, if I should ask how many Roots there are, you can't deny but there are as many as there are Numbers, since there's no Number but what's the Root to some Square. And this being granted, we may likewise affirm, that there are as many square Numbers, as there are Numbers; for there are as many Squares as there are Roots, and as many Roots as Numbers. And yet in the Beginning of this, we said, there were many more Numbers than Squares, the greater Part of Numbers being not Squares: And tho' the Number of Squares decreases in a greater proportion, as we go on to bigger Numbers, for count to an Hundred you'll find 10 Squares, viz. 1, 4, 9, 16, 25, 36, 49, 64, 81, 100, which is the same as to say the 10th Part are Squares; in Ten thousand only the 100th Part are Squares; in a Million only the 1000th: And yet in an infinite Number, if we can but comprehend it, we may say the Squares are as many as all the Numbers taken together.

"*Sagr.* What must be determin'd then in this Case?

"*Salv.* I see no other way, but by saying that all Numbers are infinite; Squares are Infinite, their Roots Infinite, and that the Number of Squares is not less than the Number of Numbers, nor this less than that: and then by concluding that the Attributes or Terms of Equality, Majority, and Minority, have no Place in Infinites, but are confin'd to terminate Quantities."

The way in which the problem is expounded in the above discussion is worthy of Galileo, but the solution suggested is not the right one. It is actually the case that the number of square (finite) numbers is the same as the number of (finite) numbers. The fact that, so long as we confine ourselves to numbers less than some given finite number, the proportion of squares tends towards zero as the given finite number increases, does not contradict the fact that the number of all finite squares is the same as the number of all finite numbers. This is only an instance of the fact, now familiar to mathematicians, that the *limit* of a function as the variable *approaches* a given point may not be the same as its *value* when the variable actually *reaches* the given point. But although the infinite numbers which Galileo discusses are equal, Cantor has shown that what Simplicius could not conceive is true, namely, that there are an infinite number of different infinite numbers, and that the conception of *greater* and *less* can be perfectly well applied to them. The whole of Simplicius's difficulty comes, as is evident, from his belief that, if *greater* and *less* can be applied, a part of an infinite collection must have fewer terms than the whole; and when this is denied, all contradictions disappear. As regards greater and less lengths of lines, which is the problem from which the above discussion starts, that involves a meaning of *greater* and *less* which is not arithmetical. The number of points is the same in a long line and in a short one, being in fact the same as the number of points in all space. The *greater* and *less* of metrical geometry involves the new metrical conception of *congruence*, which cannot be developed out of arithmetical considerations alone. But this question has not the fundamental importance which belongs to the arithmetical theory of infinity.

(2) *Non-inductiveness.*—The second property by which infinite numbers are distinguished from finite numbers is the property of non-inductiveness. This will be best explained by defining the positive property of inductiveness which characterises the finite numbers, and which is named after the method of proof known as "mathematical induction."

Let us first consider what is meant by calling a property "hereditary" in a given series. Take such a property as being named Jones. If a man is named Jones, so is his son; we will therefore call the property of being called Jones hereditary with respect to the relation of father and son. If a man is called Jones, all his descendants in the direct male line are called Jones; this follows from the fact that the property is hereditary. Now, instead of the relation of father and son, consider the relation of a finite number to its immediate successor, that is, the relation which holds between 0 and 1, between 1 and 2, between 2 and 3, and so on. If a property of numbers is hereditary with respect to this relation, then if it belongs to (say) 100, it must belong also to all finite numbers greater than 100; for, being hereditary, it belongs to 101 because it belongs to 100, and it belongs to 102 because it belongs to 101, and so on—where the "and so on" will take us, sooner or later, to any finite number greater than 100. Thus, for example, the property of being greater than 99 is hereditary in the series of finite numbers; and generally, a property is hereditary in this series when, given any number that possesses the property, the next number must always also possess it.

It will be seen that a hereditary property, though it must belong to all the finite numbers greater than a given number possessing the property, need not belong to all the numbers less than this number. For example, the hereditary property of being greater than 99 belongs to 100 and all greater numbers, but not to any smaller number. Similarly, the hereditary property of being called Jones belongs to all the descendants (in the direct male line) of those who have this property, but not to all their ancestors, because we reach at last a first Jones, before whom the ancestors have no surname. It is obvious, however, that any hereditary property possessed by Adam must belong to all men; and similarly any hereditary property possessed by 0 must belong to all finite numbers. This is the principle of what is called "mathematical induction." It frequently happens, when we wish to prove that all finite numbers have some property, that we have first to prove that 0 has the property, and then that the property is hereditary, *i.e.* that, if it belongs to a given number, then it belongs to the next number. Owing to the fact that such proofs are called "inductive," I shall call the properties to which they are applicable "inductive" properties. Thus an inductive property of numbers is one which is hereditary and belongs to 0.

Taking any one of the natural numbers, say , it is easy to see that it must have all inductive properties. For since such properties belong to 0 and are hereditary, they belong to 1; therefore, since they are hereditary, they belong to 2, and so on; by twenty-nine repetitions of such arguments we show that they belong to 29. We may *define* the "inductive" numbers as *all those that possess all inductive properties*; they will be the same as what are called the "natural" numbers, *i.e.* the ordinary finite whole numbers. To all such numbers, proofs by mathematical induction can be validly applied. They are those numbers, we may loosely say, which can be reached from 0 by successive additions of 1; in other words, they are all the numbers that can be reached by counting.

But beyond all these numbers, there are the infinite numbers, and infinite numbers do not have all inductive properties. Such numbers, therefore, may be called non-inductive. All those properties of numbers which are proved by an imaginary step-by-step process from one number to the next are liable to fail when we come to infinite numbers. The first of the infinite numbers has no immediate predecessor, because there is no greatest finite number; thus no succession of steps from one number to the next will ever reach from a finite number to an infinite one, and the step-by-step method of proof fails. This is another reason for the supposed self-contradictions of infinite numbers. Many of the most familiar properties of numbers, which custom had led people to regard as logically necessary, are in fact only demonstrable by the step-by-step method, and fail to be true of infinite numbers. But so soon as we realise the necessity of proving such properties by mathematical induction, and the strictly limited scope of this method of proof, the supposed contradictions are seen to contradict, not logic, but only our prejudices and mental habits.

The property of being increased by the addition of 1—*i.e.* the property of non-reflexiveness—may serve to illustrate the limitations of mathematical induction. It is easy to prove that 0 is increased by the addition of 1, and that, if a given number is increased by the addition of 1, so is the next number, *i.e.* the number obtained by the addition of 1. It follows that each of the natural numbers is increased by the addition of 1. This follows generally from the general argument, and follows for each particular case by a sufficient number of applications of the argument. We first prove that 0 is not equal to 1; then, since the property of being increased by 1 is hereditary, it follows that 1 is not equal to 2; hence it follows that 2 is not equal to 3; if we wish to prove that 30,000 is not equal to 30,001, we can do so by repeating this reasoning 30,000 times. But we cannot prove in this way that *all* numbers are increased by the addition of 1; we can only prove that this holds of the numbers attainable by successive additions of 1 starting from 0. The reflexive numbers, which lie beyond all those attainable in this way, are as a matter of fact not increased by the addition of 1.

The two properties of reflexiveness and non-inductiveness, which we have considered as characteristics of infinite numbers, have not so far been proved to be always found together. It is known that all reflexive numbers are non-inductive, but it is not known that all non-inductive numbers are reflexive. Fallacious proofs of this proposition have been published by many writers, including myself, but up to the present no valid proof has been discovered. The infinite numbers actually known, however, are all reflexive as well as non-inductive; thus, in mathematical practice, if not in theory, the two properties are always associated. For our purposes, therefore, it will be convenient to ignore the bare possibility that there may be non-inductive non-reflexive numbers, since all known numbers are either inductive or reflexive.

When infinite numbers are first introduced to people, they are apt to refuse the name of numbers to them, because their behaviour is so different from that of finite numbers that it seems a wilful misuse of terms to call them numbers at all. In order to meet this feeling, we must now turn to the logical basis of arithmetic, and consider the logical definition of numbers.

The logical definition of numbers, though it seems an essential support to the theory of infinite numbers, was in fact discovered independently and by a different man. The theory of infinite numbers—that is to say, the arithmetical as opposed to the logical part of the theory—was discovered by Georg Cantor, and published by him in 1882–3.The definition of number was discovered about the same time by a man whose great genius has not received the recognition it deserves—I mean Gottlob Frege of Jena. His first work, *Begriffsschrift*, published in 1879, contained the very important theory of hereditary properties in a series to which I alluded in connection with inductiveness. His definition of number is contained in his second work, published in 1884, and entitled *Die Grundlagen der Arithmetik, eine logisch-mathematische Untersuchung über den Begriff der Zahl*.It is with this book that the logical theory of arithmetic begins, and it will repay us to consider Frege's analysis in some detail.

Frege begins by noting the increased desire for logical strictness in mathematical demonstrations which distinguishes modern mathematicians from their predecessors, and points out that this must lead to a critical investigation of the definition of number. He proceeds to show the inadequacy of previous philosophical theories, especially of the "synthetic *a priori*" theory of Kant and the empirical theory of Mill. This brings him to the question: What kind of object is it that number can properly be ascribed to? He points out that physical things may be regarded as one or many: for example, if a tree has a thousand leaves, they may be taken altogether as constituting its foliage, which would count as one, not as a thousand; and *one* pair of boots is the same object as *two* boots. It follows that physical things are not the subjects of which number is properly predicated; for when we have discovered the proper subjects, the number to be ascribed must be unambiguous. This leads to a discussion of the very prevalent view that number is really something psychological and subjective, a view which Frege emphatically rejects. "Number," he says, "is as little an object of psychology or an outcome of psychical processes as the North Sea.... The botanist wishes to state something which is just as much a fact when he gives the number of petals in a flower as when he gives its colour. The one depends as little as the other upon our caprice. There is therefore a certain similarity between number and colour; but this does not consist in the fact that both are sensibly perceptible in external things, but in the fact that both are objective" .

"I distinguish the objective," he continues, "from the palpable, the spatial, the actual. The earth's axis, the centre of mass of the solar system, are objective, but I should not call them actual, like the earth itself". He concludes that number is neither spatial and physical, nor subjective, but non-sensible and objective. This conclusion is important, since it applies to all the subject-matter of mathematics and logic. Most philosophers have thought that the physical and the mental between them exhausted the world of being. Some have argued that the objects of mathematics were obviously not subjective, and therefore must be physical and empirical; others have argued that they were obviously not physical, and therefore must be subjective and mental. Both sides were right in what they denied, and wrong in what they asserted; Frege has the merit of accepting both denials, and finding a third assertion by recognising the world of logic, which is neither mental nor physical.

The fact is, as Frege points out, that no number, not even 1, is applicable to physical things, but only to general terms or descriptions, such as "man," "satellite of the earth," "satellite of Venus." The general term "man" is applicable to a certain number of objects: there are in the world so and so many men. The unity which philosophers rightly feel to be necessary for the assertion of a number is the unity of the general term, and it is the general term which is the proper subject of number. And this applies equally when there is one object or none which falls under the general term. "Satellite of the earth" is a term only applicable to one object, namely, the moon. But "one" is not a property of the moon itself, which may equally well be regarded as many molecules: it is a property of the general term "earth's satellite." Similarly, 0 is a property of the general term "satellite of Venus," because Venus has no satellite. Here at last we have an intelligible theory of the number 0. This was impossible if numbers applied to physical objects, because obviously no physical object could have the number 0. Thus, in seeking our definition of number we have arrived so far at the result that numbers are properties of general terms or general descriptions, not of physical things or of mental occurrences.

Instead of speaking of a general term, such as "man," as the subject of which a number can be asserted, we may, without making any serious change, take the subject as the class or collection of objects—*i.e.* "mankind" in the above instance—to which the general term in question is applicable. Two general terms, such as "man" and "featherless biped," which are applicable to the same collection of objects, will obviously have the same number of instances; thus the number depends upon the class, not upon the selection of this or that general term to describe it, provided several general terms can be found to describe the same class. But some general term is always necessary in order to describe a class. Even when the terms are enumerated, as "this and that and the other," the collection is constituted by the general property of being either this, or that, or the other, and only so acquires the unity which enables us to speak of it as *one* collection. And in the case of an infinite class, enumeration is impossible, so that description by a general characteristic common and peculiar to the members of the class is the only possible description. Here, as we see, the theory of number to which Frege was led by purely logical considerations becomes of use in showing how infinite classes can be amenable to number in spite of being incapable of enumeration.

Frege next asks the question: When do two collections have the same number of terms? In ordinary life, we decide this question by counting; but counting, as we saw, is impossible in the case of infinite collections, and is not logically fundamental with finite collections. We want, therefore, a different method of answering our question. An illustration may help to make the method clear. I do not know how many married men there are in England, but I do know that the number is the same as the number of married women. The reason I know this is that the relation of husband and wife relates one man to one woman and one woman to one man. A relation of this sort is called a one-one relation. The relation of father to son is called a one-many relation, because a man can have only one father but may have many sons; conversely, the relation of son to father is called a many-one relation. But the relation of husband to wife (in Christian countries) is called one-one, because a man cannot have more than one wife, or a woman more than one husband. Now, whenever there is a one-one relation between all the terms of one collection and all the terms of another severally, as in the case of English husbands and English wives, the number of terms in the one collection is the same as the number in the other; but when there is not such a relation, the number is different. This is the answer to the question: When do two collections have the same number of terms?

We can now at last answer the question: What is meant by the number of terms in a given collection? When there is a one-one relation between all the terms of one collection and all the terms of another severally, we shall say that the two collections are "similar." We have just seen that two similar collections have the same number of terms. This leads us to define the number of a given collection as the class of all collections that are similar to it; that is to say, we set up the following formal definition:

"The number of terms in a given class" is defined as meaning "the class of all classes that are similar to the given class."

This definition, as Frege (expressing it in slightly different terms) showed, yields the usual arithmetical properties of numbers. It is applicable equally to finite and infinite numbers, and it does not require the admission of some new and mysterious set of metaphysical entities. It shows that it is not physical objects, but classes or the general terms by which they are defined, of which numbers can be asserted; and it applies to 0 and 1 without any of the difficulties which other theories find in dealing with these two special cases.

The above definition is sure to produce, at first sight, a feeling of oddity, which is liable to cause a certain dissatisfaction. It defines the number 2, for instance, as the class of all couples, and the number 3 as the class of all triads. This does not *seem* to be what we have hitherto been meaning when we spoke of 2 and 3, though it would be difficult to say *what* we had been meaning. The answer to a feeling cannot be a logical argument, but nevertheless the answer in this case is not without importance. In the first place, it will be found that when an idea which has grown familiar as an unanalysed whole is first resolved accurately into its component parts—which is what we do when we define it—there is almost always a feeling of unfamiliarity produced by the analysis, which tends to cause a protest against the definition. In the second place, it may be admitted that the definition, like all definitions, is to a certain extent arbitrary. In the case of the small finite numbers, such as 2 and 3, it would be possible to frame definitions more nearly in accordance with our unanalysed feeling of what we mean; but the method of such definitions would lack uniformity, and would be found to fail sooner or later—at latest when we reached infinite numbers.

In the third place, the real desideratum about such a definition as that of number is not that it should represent as nearly as possible the ideas of those who have not gone through the analysis required in order to reach a definition, but that it should give us objects having the requisite properties. Numbers, in fact, must satisfy the formulæ of arithmetic; any indubitable set of objects fulfilling this requirement may be called numbers. So far, the simplest set known to fulfil this requirement is the set introduced by the above definition. In comparison with this merit, the question whether the objects to which the definition applies are like or unlike the vague ideas of numbers entertained by those who cannot give a definition, is one of very little importance. All the important requirements are fulfilled by the above definition, and the sense of oddity which is at first unavoidable will be found to wear off very quickly with the growth of familiarity.

There is, however, a certain logical doctrine which may be thought to form an objection to the above definition of numbers as classes of classes—I mean the doctrine that there are no such objects as classes at all. It might be thought that this doctrine would make havoc of a theory which reduces numbers to classes, and of the many other theories in which we have made use of classes. This, however, would be a mistake: none of these theories are any the worse for the doctrine that classes are fictions. What the doctrine is, and why it is not destructive, I will try briefly to explain.

On account of certain rather complicated difficulties, culminating in definite contradictions, I was led to the view that nothing that can be said significantly about things, *i.e.* particulars, can be said significantly (*i.e.* either truly or falsely) about classes of things. That is to say, if, in any sentence in which a thing is mentioned, you substitute a class for the thing, you no longer have a sentence that has any meaning: the sentence is no longer either true or false, but a meaningless collection of words. Appearances to the contrary can be dispelled by a moment's reflection. For example, in the sentence, "Adam is fond of apples," you may substitute *mankind*, and say, "Mankind is fond of apples." But obviously you do not mean that there is one individual, called "mankind," which munches apples: you mean that the separate individuals who compose mankind are each severally fond of apples.

Now, if nothing that can be said significantly about a thing can be said significantly about a class of things, it follows that classes of things cannot have the same kind of reality as things have; for if they had, a class could be substituted for a thing in a proposition predicating the kind of reality which would be common to both. This view is really consonant to common sense. In the third or fourth century B.C. there lived a Chinese philosopher named Hui Tzŭ, who maintained that "a bay horse and a dun cow are three; because taken separately they are two, and taken together they are one: two and one make three."The author from whom I quote says that Hui Tzŭ "was particularly fond of the quibbles which so delighted the sophists or unsound reasoners of ancient Greece," and this no doubt represents the judgment of common sense upon such arguments. Yet if collections of things were things, his contention would be irrefragable. It is only because the bay horse and the dun cow taken together are not a new thing that we can escape the conclusion that there are three things wherever there are two.

When it is admitted that classes are not things, the question arises: What do we mean by statements which are nominally about classes? Take such a statement as, "The class of people interested in mathematical logic is not very numerous." Obviously this reduces itself to, "Not very many people are interested in mathematical logic." For the sake of definiteness, let us substitute some particular number, say 3, for "very many." Then our statement is, "Not three people are interested in mathematical logic." This may be expressed in the form: "If x is interested in mathematical logic, and also y is interested, and also z is interested, then x is identical with y, or x is identical with z, or y is identical with z." Here there is no longer any reference at all to a "class." In some such way, all statements nominally about a class can be reduced to statements about what follows from the hypothesis of anything's having the defining property of the class. All that is wanted, therefore, in order to render the *verbal* use of classes legitimate, is a uniform method of interpreting propositions in which such a use occurs, so as to obtain propositions in which there is no longer any such use. The definition of such a method is a technical matter, which Dr Whitehead and I have dealt with elsewhere, and which we need not enter into on this occasion.

If the theory that classes are merely symbolic is accepted, it follows that numbers are not actual entities, but that propositions in which numbers verbally occur have not really any constituents corresponding to numbers, but only a certain logical form which is not a part of propositions having this form. This is in fact the case with all the apparent objects of logic and mathematics. Such words as *or*, *not*, *if*, *there is*, *identity*, *greater*, *plus*, *nothing*, *everything*, *function*, and so on, are not names of definite objects, like "John" or "Jones," but are words which require a context in order to have meaning. All of them are *formal*, that is to say, their occurrence indicates a certain form of proposition, not a certain constituent. "Logical constants," in short, are not entities; the words expressing them are not names, and cannot significantly be made into logical subjects except when it is the words themselves, as opposed to their meanings, that are being discussed.This fact has a very important bearing on all logic and philosophy, since it shows how they differ from the special sciences. But the questions raised are so large and so difficult that it is impossible to pursue them further on this occasion.

LECTURE VIII
ON THE NOTION OF CAUSE, WITH APPLICATIONS TO THE FREE-WILL PROBLEM

THE nature of philosophic analysis, as illustrated in our previous lectures, can now be stated in general terms. We start from a body of common knowledge, which constitutes our data. On examination, the data are found to be complex, rather vague, and largely interdependent logically. By analysis we reduce them to propositions which are as nearly as possible simple and precise, and we arrange them in deductive chains, in which a certain number of initial propositions form a logical guarantee for all the rest. These initial propositions are *premisses* for the body of knowledge in question. Premisses are thus quite different from data—they are simpler, more precise, and less infected with logical redundancy. If the work of analysis has been performed completely, they will be wholly free from logical redundancy, wholly precise, and as simple as is logically compatible with their leading to the given body of knowledge. The discovery of these premisses belongs to philosophy; but the work of deducing the body of common knowledge from them belongs to mathematics, if "mathematics" is interpreted in a somewhat liberal sense.

But besides the logical analysis of the common knowledge which forms our data, there is the consideration of its degree of certainty. When we have arrived at its premisses, we may find that some of them seem open to doubt, and we may find further that this doubt extends to those of our original data which depend upon these doubtful premisses. In our third lecture, for example, we saw that the part of physics which depends upon testimony, and thus upon the existence of other minds than our own, does not seem so certain as the part which depends exclusively upon our own sense-data and the laws of logic. Similarly, it used to be felt that the parts of geometry which depend upon the axiom of parallels have less certainty than the parts which are independent of this premiss. We may say, generally, that what commonly passes as knowledge is not all equally certain, and that, when analysis into premisses has been effected, the degree of certainty of any consequence of the premisses will depend upon that of the most doubtful premiss employed in proving this consequence. Thus analysis into premisses serves not only a logical purpose, but also the purpose of facilitating an estimate as to the degree of certainty to be attached to this or that derivative belief. In view of the fallibility of all human beliefs, this service seems at least as important as the purely logical services rendered by philosophical analysis.

In the present lecture, I wish to apply the analytic method to the notion of "cause," and to illustrate the discussion by applying it to the problem of free will. For this purpose I shall inquire: I., what is meant by a causal law; II., what is the evidence that causal laws have held hitherto; III., what is the evidence that they will continue to hold in the future; IV., how the causality which is used in science differs from that of common sense and traditional philosophy; V., what new light is thrown on the question of free will by our analysis of the notion of "cause."

I. By a "causal law" I mean any general proposition in virtue of which it is possible to infer the existence of one thing or event from the existence of another or of a number of others. If you hear thunder without having seen lightning, you infer that there nevertheless was a flash, because of the general proposition, "All thunder is preceded by lightning." When Robinson Crusoe sees a footprint, he infers a human being, and he might justify his inference by the general proposition, "All marks in the ground shaped like a human foot are subsequent to a human being's standing where the marks are." When we see the sun set, we expect that it will rise again the next day. When we hear a man speaking, we infer that he has certain thoughts. All these inferences are due to causal laws.

A causal law, we said, allows us to infer the existence of one *thing* (or *event*) from the existence of one or more others. The word "thing" here is to be understood as only applying to particulars, *i.e.* as excluding such logical objects as numbers or classes or abstract properties and relations, and including sense-data, with whatever is logically of the same type as sense-data.In so far as a causal law is directly verifiable, the thing inferred and the thing from which it is inferred must both be data, though they need not both be data at the same time. In fact, a causal law which is being used to extend our knowledge of existence must be applied to what, at the moment, is not a datum; it is in the possibility of such application that the practical utility of a causal law consists. The important point, for our present purpose, however, is that what is inferred is a "thing," a "particular," an object having the kind of reality that belongs to objects of sense, not an abstract object such as virtue or the square root of two.

But we cannot become acquainted with a particular except by its being actually given. Hence the particular inferred by a causal law must be only *described* with more or less exactness; it cannot be *named* until the inference is verified. Moreover, since the causal law is *general*, and capable of applying to many cases, the given particular from which we infer must allow the inference in virtue of some general characteristic, not in virtue of its being just the particular that it is. This is obvious in all our previous instances: we infer the unperceived lightning from the thunder, not in virtue of any peculiarity of the thunder, but in virtue of its resemblance to other claps of thunder. Thus a causal law must state that the existence of a thing of a certain sort (or of a number of things of a number of assigned sorts) implies the existence of another thing having a relation to the first which remains invariable so long as the first is of the kind in question.

It is to be observed that what is constant in a causal law is not the object or objects given, nor yet the object inferred, both of which may vary within wide limits, but the *relation* between what is given and what is inferred. The principle, “same cause, same effect,” which is sometimes said to be the principle of causality, is much narrower in its scope than the principle which really occurs in science; indeed, if strictly interpreted, it has no scope at all, since the “same” cause never recurs exactly. We shall return to this point at a later stage of the discussion.

The particular which is inferred may be uniquely determined by the causal law, or may be only described in such general terms that many different particulars might satisfy the description. This depends upon whether the constant relation affirmed by the causal law is one which only one term can have to the data, or one which many terms may have. If many terms may have the relation in question, science will not be satisfied until it has found some more stringent law, which will enable us to determine the inferred things uniquely.

Since all known things are in time, a causal law must take account of temporal relations. It will be part of the causal law to state a relation of succession or coexistence between the thing given and the thing inferred. When we hear thunder and infer that there was lightning, the law states that the thing inferred is earlier than the thing given. Conversely, when we see lightning and wait expectantly for the thunder, the law states that the thing given is earlier than the thing inferred. When we infer a man's thoughts from his words, the law states that the two are (at least approximately) simultaneous.

If a causal law is to achieve the precision at which science aims, it must not be content with a vague *earlier* or *later*, but must state how much earlier or how much later. That is to say, the time-relation between the thing given and the thing inferred ought to be capable of exact statement; and usually the inference to be drawn is different according to the length and direction of the interval. “A quarter of an hour ago this man was alive; an hour hence he will be cold.” Such a statement involves two causal laws, one inferring from a datum something which existed a quarter of an hour ago, the other inferring from the same datum something which will exist an hour hence.

Often a causal law involves not one datum, but many, which need not be all simultaneous with each other, though their time-relations must be given. The general scheme of a causal law will be as follows:

“Whenever things occur in certain relations to each other (among which their time-relations must be included), then a thing having a fixed relation to these things will occur at a date fixed relatively to their dates.”

The things given will not, in practice, be things that only exist for an instant, for such things, if there are any, can never be data. The things given will each occupy some finite time. They may be not static things, but processes, especially motions. We have considered in an earlier lecture the sense in which a motion may be a datum, and need not now recur to this topic.

It is not essential to a causal law that the object inferred should be later than some or all of the data. It may equally well be earlier or at the same time. The only thing essential is that the law should be such as to enable us to infer the existence of an object which we can more or less accurately describe in terms of the data.

II. I come now to our second question, namely: What is the nature of the evidence that causal laws have held hitherto, at least in the observed portions of the past? This question must not be confused with the further question: Does this evidence warrant us in assuming the truth of causal laws in the future and in unobserved portions of the past? For the present, I am only asking what are the grounds which lead to a belief in causal laws, not whether these grounds are adequate to support the belief in universal causation.

The first step is the discovery of approximate unanalysed uniformities of sequence or coexistence. After lightning comes thunder, after a blow received comes pain, after approaching a fire comes warmth; again, there are uniformities of coexistence, for example between touch and sight, between certain sensations in the throat and the sound of one's own voice, and so on. Every such uniformity of sequence or coexistence, after it has been experienced a certain number of times, is followed by an expectation that it will be repeated on future occasions, *i.e.* that where one of the correlated events is found, the other will be found also. The connection of experienced past uniformity with expectation as to the future is just one of those uniformities of sequence which we have observed to be true hitherto. This affords a psychological account of what may be called the animal belief in causation, because it is something which can be observed in horses and dogs, and is rather a habit of acting than a real belief. So far, we have merely repeated Hume, who carried the discussion of cause up to this point, but did not, apparently, perceive how much remained to be said.

Is there, in fact, any characteristic, such as might be called causality or uniformity, which is found to hold throughout the observed past? And if so, how is it to be stated?

The particular uniformities which we mentioned before, such as lightning being followed by thunder, are not found to be free from exceptions. We sometimes see lightning without hearing thunder; and although, in such a case, we suppose that thunder might have been heard if we had been nearer to the lightning, that is a supposition based on theory, and therefore incapable of being invoked to support the theory. What does seem, however, to be shown by scientific experience is this: that where an observed uniformity fails, some wider uniformity can be found, embracing more circumstances, and subsuming both the successes and the failures of the previous uniformity. Unsupported bodies in air fall, unless they are balloons or aeroplanes; but the principles of mechanics give uniformities which apply to balloons and aeroplanes just as accurately as to bodies that fall. There is much that is hypothetical and more or less artificial in the uniformities affirmed by mechanics, because, when they cannot otherwise be made applicable, unobserved bodies are inferred in order to account for observed peculiarities. Still, it is an empirical fact that it is possible to preserve the laws by assuming such bodies, and that they never have to be assumed in circumstances in which they ought to be observable. Thus the empirical verification of mechanical laws may be admitted, although we must also admit that it is less complete and triumphant than is sometimes supposed.

Assuming now, what must be admitted to be doubtful, that the whole of the past has proceeded according to invariable laws, what can we say as to the nature of these laws? They will not be of the simple type which asserts that the same cause always produces the same effect. We may take the law of gravitation as a sample of the kind of law that appears to be verified without exception. In order to state this law in a form which observation can confirm, we will confine it to the solar system. It then states that the motions of planets and their satellites have at every instant an acceleration compounded of accelerations towards all the other bodies in the solar system, proportional to the masses of those bodies and inversely proportional to the squares of their distances. In virtue of this law, given the state of the solar system throughout any finite time, however short, its state at all earlier and later times is determinate except in so far as other forces than gravitation or other bodies than those in the solar system have to be taken into consideration. But other forces, so far as science can discover, appear to be equally regular, and equally capable of being summed up in single causal laws. If the mechanical account of matter were complete, the whole physical history of the universe, past and future, could be inferred from a sufficient number of data concerning an assigned finite time, however short.

In the mental world, the evidence for the universality of causal laws is less complete than in the physical world. Psychology cannot boast of any triumph comparable to gravitational astronomy. Nevertheless, the evidence is not very greatly less than in the physical world. The crude and approximate causal laws from which science starts are just as easy to discover in the mental sphere as in the physical. In the world of sense, there are to begin with the correlations of sight and touch and so on, and the facts which lead us to connect various kinds of sensations with eyes, ears, nose, tongue, etc. Then there are such facts as that our body moves in answer to our volitions. Exceptions exist, but are capable of being explained as easily as the exceptions to the rule that unsupported bodies in air fall. There is, in fact, just such a degree of evidence for causal laws in psychology as will warrant the psychologist in assuming them as a matter of course, though not such a degree as will suffice to remove all doubt from the mind of a sceptical inquirer. It should be observed that causal laws in which the given term is mental and the inferred term physical, or *vice versa*, are at least as easy to discover as causal laws in which both terms are mental.

It will be noticed that, although we have spoken of causal laws, we have not hitherto introduced the word "cause." At this stage, it will be well to say a few words on legitimate and illegitimate uses of this word. The word "cause," in the scientific account of the world, belongs only to the early stages, in which small preliminary, approximate generalisations are being ascertained with a view to subsequent larger and more invariable laws. We may say, "Arsenic causes death," so long as we are ignorant of the precise process by which the result is brought about. But in a sufficiently advanced science, the word "cause" will not occur in any statement of invariable laws. There is, however, a somewhat rough and loose use of the word "cause" which may be preserved. The approximate uniformities which lead to its pre-scientific employment may turn out to be true in all but very rare and exceptional circumstances, perhaps in all circumstances that actually occur. In such cases, it is convenient to be able to speak of the antecedent event as the "cause" and the subsequent event as the "effect." In this sense, provided it is realised that the sequence is not necessary and may have exceptions, it is still possible to employ the words "cause" and "effect." It is in this sense, and in this sense only, that we shall intend the words when we speak of one particular event "causing" another particular event, as we must sometimes do if we are to avoid intolerable circumlocution.

III. We come now to our third question, namely: What reason can be given for believing that causal laws will hold in future, or that they have held in unobserved portions of the past?

What we have said so far is that there have been hitherto certain observed causal laws, and that all the empirical evidence we possess is compatible with the view that everything, both mental and physical, so far as our observation has extended, has happened in accordance with causal laws. The law of universal causation, suggested by these facts, may be enunciated as follows:

"There are such invariable relations between different events at the same or different times that, given the state of the whole universe throughout any finite time, however short, every previous and subsequent event can theoretically be determined as a function of the given events during that time."

Have we any reason to believe this universal law? Or, to ask a more modest question, have we any reason to believe that a particular causal law, such as the law of gravitation, will continue to hold in the future?

Among observed causal laws is this, that observation of uniformities is followed by expectation of their recurrence. A horse who has been driven always along a certain road expects to be driven along that road again; a dog who is always fed at a certain hour expects food at that hour and not at any other. Such expectations, as Hume pointed out, explain only too well the common-sense belief in uniformities of sequence, but they afford absolutely no logical ground for beliefs as to the future, not even for the belief that we shall continue to expect the continuation of experienced uniformities, for that is precisely one of those causal laws for which a ground has to be sought. If Hume's account of causation is the last word, we have not only no reason to suppose that the sun will rise to-morrow, but no reason to suppose that five minutes hence we shall still expect it to rise to-morrow.

It may, of course, be said that all inferences as to the future are in fact invalid, and I do not see how such a view could be disproved. But, while admitting the legitimacy of such a view, we may nevertheless inquire: If inferences as to the future *are* valid, what principle must be involved in making them?

The principle involved is the principle of induction, which, if it is true, must be an *a priori* logical law, not capable of being proved or disproved by experience. It is a difficult question how this principle ought to be formulated; but if it is to warrant the inferences which we wish to make by its means, it must lead to the following proposition: "If, in a great number of instances, a thing of a certain kind is associated in a certain way with a thing of a certain other kind, it is probable that a thing of the one kind is always similarly associated with a thing of the other kind; and as the number of instances increases, the probability approaches indefinitely near to certainty." It may well be questioned whether this proposition is true; but if we admit it, we can infer that any characteristic of the whole of the observed past is likely to apply to the future and to the unobserved past. This proposition, therefore, if it is true, will warrant the inference that causal laws probably hold at all times, future as well as past; but without this principle, the observed cases of the truth of causal laws afford no presumption as to the unobserved cases, and therefore the existence of a thing not directly observed can never be validly inferred.

It is thus the principle of induction, rather than the law of causality, which is at the bottom of all inferences as to the existence of things not immediately given. With the principle of induction, all that is wanted for such inferences can be proved; without it, all such inferences are invalid. This principle has not received the attention which its great importance deserves. Those who were interested in deductive logic naturally enough ignored it, while those who emphasised the scope of induction wished to maintain that all logic is empirical, and therefore could not be expected to realise that induction itself, their own darling, required a logical principle which obviously could not be proved inductively, and must therefore be *a priori* if it could be known at all.

The view that the law of causality itself is *a priori* cannot, I think, be maintained by anyone who realises what a complicated principle it is. In the form which states that "every event has a cause" it looks simple; but on examination, "cause" is merged in "causal law," and the definition of a "causal law" is found to be far from simple. There must necessarily be *some a priori* principle involved in inference from the existence of one thing to that of another, if such inference is ever valid; but it would appear from the above analysis that the principle in question is induction, not causality. Whether inferences from past to future are valid depends wholly, if our discussion has been sound, upon the inductive principle: if it is true, such inferences are valid, and if it is false, they are invalid.

IV. I come now to the question how the conception of causal laws which we have arrived at is related to the traditional conception of cause as it occurs in philosophy and common sense.

Historically, the notion of cause has been bound up with that of human volition. The typical cause would be the fiat of a king. The cause is supposed to be "active," the effect "passive." From this it is easy to pass on to the suggestion that a "true" cause must contain some prevision of the effect; hence the effect becomes the "end" at which the cause aims, and teleology replaces causation in the explanation of nature. But all such ideas, as applied to physics, are mere anthropomorphic superstitions. It is as a reaction against these errors that Mach and others have urged a purely "descriptive" view of physics: physics, they say, does not aim at telling us "why" things happen, but only "how" they happen. And if the question "why?" means anything more than the search for a general law according to which a phenomenon occurs, then it is certainly the case that this question cannot be answered in physics and ought not to be asked. In this sense, the descriptive view is indubitably in the right. But in using causal laws to support inferences from the observed to the unobserved, physics ceases to be *purely* descriptive, and it is these laws which give the scientifically useful part of the traditional notion of "cause." There is therefore *something* to preserve in this notion, though it is a very tiny part of what is commonly assumed in orthodox metaphysics.

In order to understand the difference between the kind of cause which science uses and the kind which we naturally imagine, it is necessary to shut out, by an effort, everything that differentiates between past and future. This is an extraordinarily difficult thing to do, because our mental life is so intimately bound up with difference. Not only do memory and hope make a difference in our feelings as regards past and future, but almost our whole vocabulary is filled with the idea of activity, of things done now for the sake of their future effects.

All transitive verbs involve the notion of cause as activity, and would have to be replaced by some cumbrous periphrasis before this notion could be eliminated.

Consider such a statement as, “Brutus killed Cæsar.” On another occasion, Brutus and Cæsar might engage our attention, but for the present it is the killing that we have to study. We may say that to kill a person is to cause his death intentionally. This means that desire for a person's death causes a certain act, because it is believed that that act will cause the person's death; or more accurately, the desire and the belief jointly cause the act. Brutus desires that Cæsar should be dead, and believes that he will be dead if he is stabbed; Brutus therefore stabs him, and the stab causes Cæsar's death, as Brutus expected it would. Every act which realises a purpose involves two causal steps in this way: C is desired, and it is believed (truly if the purpose is achieved) that B will cause C; the desire and the belief together cause B, which in turn causes C. Thus we have first A, which is a desire for C and a belief that B (an act) will cause C; then we have B, the act caused by A, and believed to be a cause of C; then, if the belief was correct, we have C, caused by B, and if the belief was incorrect we have disappointment. Regarded purely scientifically, this series A, B, C may equally well be considered in the inverse order, as they would be at a coroner's inquest. But from the point of view of Brutus, the desire, which comes at the beginning, is what makes the whole series interesting. We feel that if his desires had been different, the effects which he in fact produced would not have occurred. This is true, and gives him a sense of power and freedom. It is equally true that if the effects had not occurred, his desires would have been different, since being what they were the effects did occur. Thus the desires are determined by their consequences just as much as the consequences by the desires; but as we cannot (in general) know in advance the consequences of our desires without knowing our desires, this form of inference is uninteresting as applied to our own acts, though quite vital as applied to those of others.

A cause, considered scientifically, has none of that analogy with volition which makes us imagine that the effect is *compelled* by it. A cause is an event or group of events, of some known general character, and having a known relation to some other event, called the effect; the relation being of such a kind that only one event, or at any rate only one well-defined sort of event, can have the relation to a given cause. It is customary only to give the name “effect” to an event which is later than the cause, but there is no kind of reason for this restriction. We shall do better to allow the effect to be before the cause or simultaneous with it, because nothing of any scientific importance depends upon its being after the cause.

If the inference from cause to effect is to be indubitable, it seems that the cause can hardly stop short of the whole universe. So long as anything is left out, something may be left out which alters the expected result. But for practical and scientific purposes, phenomena can be collected into groups which are causally self-contained, or nearly so. In the common notion of causation, the cause is a single event—we say the lightning causes the thunder, and so on. But it is difficult to know what we mean by a single event; and it generally appears that, in order to have anything approaching certainty concerning the effect, it is necessary to include many more circumstances in the cause than unscientific common sense would suppose. But often a probable causal connection, where the cause is fairly simple, is of more practical importance than a more indubitable connection in which the cause is so complex as to be hard to ascertain.

To sum up: the strict, certain, universal law of causation which philosophers advocate is an ideal, possibly true, but not *known* to be true in virtue of any available evidence. What is actually known, as a matter of empirical science, is that certain constant relations are observed to hold between the members of a group of events at certain times, and that when such relations fail, as they sometimes do, it is usually possible to discover a new, more constant relation by enlarging the group. Any such constant relation between events of specified kinds with given intervals of time between them is a "causal law." But all causal laws are liable to exceptions, if the cause is less than the whole state of the universe; we believe, on the basis of a good deal of experience, that such exceptions can be dealt with by enlarging the group we call the cause, but this belief, wherever it is still unverified, ought not to be regarded as certain, but only as suggesting a direction for further inquiry.

A very common causal group consists of volitions and the consequent bodily acts, though exceptions arise (for example) through sudden paralysis. Another very frequent connection (though here the exceptions are much more numerous) is between a bodily act and the realisation of the purpose which led to the act. These connections are patent, whereas the causes of desires are more obscure. Thus it is natural to begin causal series with desires, to suppose that all causes are analogous to desires, and that desires themselves arise spontaneously. Such a view, however, is not one which any serious psychologist would maintain. But this brings us to the question of the application of our analysis of cause to the problem of free will.

V. The problem of free will is so intimately bound up with the analysis of causation that, old as it is, we need not despair of obtaining new light on it by the help of new views on the notion of cause. The free-will problem has, at one time or another, stirred men's passions profoundly, and the fear that the will might not be free has been to some men a source of great unhappiness. I believe that, under the influence of a cool analysis, the doubtful questions involved will be found to have no such emotional importance as is sometimes thought, since the disagreeable consequences supposed to flow from a denial of free will do not flow from this denial in any form in which there is reason to make it. It is not, however, on this account chiefly that I wish to discuss this problem, but rather because it affords a good example of the clarifying effect of analysis and of the interminable controversies which may result from its neglect.

Let us first try to discover what it is we really desire when we desire free will. Some of our reasons for desiring free will are profound, some trivial. To begin with the former: we do not wish to feel ourselves in the hands of fate, so that, however much we may desire to will one thing, we may nevertheless be compelled by an outside force to will another. We do not wish to think that, however much we may desire to act well, heredity and surroundings may force us into acting ill. We wish to feel that, in cases of doubt, our choice is momentous and lies within our power. Besides these desires, which are worthy of all respect, we have, however, others not so respectable, which equally make us desire free will. We do not like to think that other people, if they knew enough, could predict our actions, though we know that we can often predict those of other people, especially if they are elderly. Much as we esteem the old gentleman who is our neighbour in the country, we know that when grouse are mentioned he will tell the story of the grouse in the gun-room. But we ourselves are not so mechanical: we never tell an anecdote to the same person twice, or even once unless he is sure to enjoy it; although we once met (say) Bismarck, we are quite capable of hearing him mentioned without relating the occasion when we met him. In this sense, everybody thinks that he himself has free will, though he knows that no one else has. The desire for this kind of free will seems to be no better than a form of vanity. I do not believe that this desire can be gratified with any certainty; but the other, more respectable desires are, I believe, not inconsistent with any tenable form of determinism.

We have thus two questions to consider: (1) Are human actions theoretically predictable from a sufficient number of antecedents? (2) Are human actions subject to an external compulsion? The two questions, as I shall try to show, are entirely distinct, and we may answer the first in the affirmative without therefore being forced to give an affirmative answer to the second.

(1) *Are human actions theoretically predictable from a sufficient number of antecedents?* Let us first endeavour to give precision to this question. We may state the question thus: Is there some constant relation between an act and a certain number of earlier events, such that, when the earlier events are given, only one act, or at most only acts with some well-marked character, can have this relation to the earlier events? If this is the case, then, as soon as the earlier events are known, it is theoretically possible to predict either the precise act, or at least the character necessary to its fulfilling the constant relation.

To this question, a negative answer has been given by Bergson, in a form which calls in question the general applicability of the law of causation. He maintains that every event, and more particularly every mental event, embodies so much of the past that it could not possibly have occurred at any earlier time, and is therefore necessarily quite different from all previous and subsequent events. If, for example, I read a certain poem many times, my experience on each occasion is modified by the previous readings, and my emotions are never repeated exactly. The principle of causation, according to him, asserts that the same cause, if repeated, will produce the same effect. But owing to memory, he contends, this principle does not apply to mental events. What is apparently the same cause, if repeated, is modified by the mere fact of repetition, and cannot produce the same effect. He infers that every mental event is a genuine novelty, not predictable from the past, because the past contains nothing exactly like it by which we could imagine it. And on this ground he regards the freedom of the will as unassailable.

Bergson's contention has undoubtedly a great deal of truth, and I have no wish to deny its importance. But I do not think its consequences are quite what he believes them to be. It is not necessary for the determinist to maintain that he can foresee the whole particularity of the act which will be performed. If he could foresee that A was going to murder B, his foresight would not be invalidated by the fact that he could not know all the infinite complexity of A's state of mind in committing the murder, nor whether the murder was to be performed with a knife or with a revolver. If the *kind* of act which will be performed can be foreseen within narrow limits, it is of little practical interest that there are fine shades which cannot be foreseen. No doubt every time the story of the grouse in the gun-room is told, there will be slight differences due to increasing habitualness, but they do not invalidate the prediction that the story will be told. And there is nothing in Bergson's argument to show that we can never predict what *kind* of act will be performed.

Again, his statement of the law of causation is inadequate. The law does not state merely that, if the *same* cause is repeated, the *same* effect will result. It states rather that there is a constant relation between causes of certain kinds and effects of certain kinds. For example, if a body falls freely, there is a constant relation between the height through which it falls and the time it takes in falling. It is not necessary to have a body fall through the *same* height which has been previously observed, in order to be able to foretell the length of time occupied in falling. If this were necessary, no prediction would be possible, since it would be impossible to make the height exactly the same on two occasions. Similarly, the attraction which the sun will exert on the earth is not only known at distances for which it has been observed, but at all distances, because it is known to vary as the inverse square of the distance. In fact, what is found to be repeated is always the *relation* of cause and effect, not the cause itself; all that is necessary as regards the cause is that it should be of the same *kind* (in the relevant respect) as earlier causes whose effects have been observed.

Another respect in which Bergson's statement of causation is inadequate is in its assumption that the cause must be *one* event, whereas it may be two or more events, or even some continuous process. The substantive question at issue is whether mental events are determined by the past. Now in such a case as the repeated reading of a poem, it is obvious that our feelings in reading the poem are most emphatically dependent upon the past, but not upon one single event in the past. All our previous readings of the poem must be included in the cause. But we easily perceive a certain law according to which the effect varies as the previous readings increase in number, and in fact Bergson himself tacitly assumes such a law. We decide at last not to read the poem again, because we know that this time the effect would be boredom. We may not know all the niceties and shades of the boredom we should feel, but we know enough to guide our decision, and the prophecy of boredom is none the less true for being more or less general. Thus the kinds of cases upon which Bergson relies are insufficient to show the impossibility of prediction in the only sense in which prediction has practical or emotional interest. We may therefore leave the consideration of his arguments and address ourselves to the problem directly.

The law of causation, according to which later events can theoretically be predicted by means of earlier events, has often been held to be *a priori*, a necessity of thought, a category without which science would be impossible. These claims seem to me excessive. In certain directions the law has been verified empirically, and in other directions there is no positive evidence against it. But science can use it where it has been found to be true, without being forced into any assumption as to its truth in other fields. We cannot, therefore, feel any *a priori* certainty that causation must apply to human volitions.

The question how far human volitions are subject to causal laws is a purely empirical one. Empirically it seems plain that the great majority of our volitions have causes, but it cannot, on this account, be held necessarily certain that all have causes. There are, however, precisely the same kinds of reasons for regarding it as probable that they all have causes as there are in the case of physical events.

We may suppose—though this is doubtful—that there are laws of correlation of the mental and the physical, in virtue of which, given the state of all the matter in the world, and therefore of all the brains and living organisms, the state of all the minds in the world could be inferred, while conversely the state of all the matter in the world could be inferred if the state of all the minds were given. It is obvious that there is *some* degree of correlation between brain and mind, and it is impossible to say how complete it may be. This, however, is not the point which I wish to elicit. What I wish to urge is that, even if we admit the most extreme claims of determinism and of correlation of mind and brain, still the consequences inimical to what is worth preserving in free will do not follow. The belief that they follow results, I think, entirely from the assimilation of causes to volitions, and from the notion that causes *compel* their effects in some sense analogous to that in which a human authority can compel a man to do what he would rather not do. This assimilation, as soon as the true nature of scientific causal laws is realised, is seen to be a sheer mistake. But this brings us to the second of the two questions which we raised in regard to free will, namely, whether, assuming determinism, our actions can be in any proper sense regarded as compelled by outside forces.

(2) *Are human actions subject to an external compulsion?* We have, in deliberation, a subjective sense of freedom, which is sometimes alleged against the view that volitions have causes. This sense of freedom, however, is only a sense that we can choose which we please of a number of alternatives: it does not show us that there is no causal connection between what we please to choose and our previous history. The supposed inconsistency of these two springs from the habit of conceiving causes as analogous to volitions—a habit which often survives unconsciously in those who intend to conceive causes in a more scientific manner. If a cause is analogous to a volition, outside causes will be analogous to an alien will, and acts predictable from outside causes will be subject to compulsion. But this view of cause is one to which science lends no countenance. Causes, we have seen, do not *compel* their effects, any more than effects *compel* their causes. There is a mutual relation, so that either can be inferred from the other. When the geologist infers the past state of the earth from its present state, we should not say that the present state *compels* the past state to have been what it was; yet it renders it necessary as a consequence of the data, in the only sense in which effects are rendered necessary by their causes. The difference which we *feel*, in this respect, between causes and effects is a mere confusion due to the fact that we remember past events but do not happen to have memory of the future.

The apparent indeterminateness of the future, upon which some advocates of free will rely, is merely a result of our ignorance. It is plain that no desirable kind of free will can be dependent simply upon our ignorance; for if that were the case, animals would be more free than men, and savages than civilised people. Free will in any valuable sense must be compatible with the fullest knowledge. Now, quite apart from any assumption as to causality, it is obvious that complete knowledge would embrace the future as well as the past. Our knowledge of the past is not wholly based upon causal inferences, but is partly derived from memory. It is a mere accident that we have no memory of the future. We might—as in the pretended visions of seers—see future events immediately, in the way in which we see past events. They certainly will be what they will be, and are in this sense just as determined as the past. If we saw future events in the same immediate way in which we see past events, what kind of free will would still be possible? Such a kind would be wholly independent of determinism: it could not be contrary to even the most entirely universal reign of causality. And such a kind must contain whatever is worth having in free will, since it is impossible to believe that mere ignorance can be the essential condition of any good thing. Let us therefore imagine a set of beings who know the whole future with absolute certainty, and let us ask ourselves whether they could have anything that we should call free will.

Such beings as we are imagining would not have to wait for the event in order to know what decision they were going to adopt on some future occasion. They would know now what their volitions were going to be. But would they have any reason to regret this knowledge? Surely not, unless the foreseen volitions were in themselves regrettable. And it is less likely that the foreseen volitions would be regrettable if the steps which would lead to them were also foreseen. It is difficult not to suppose that what is foreseen is fated, and must happen however much it may be dreaded. But human actions are the outcome of desire, and no foreseeing can be true unless it takes account of desire. A foreseen volition will have to be one which does not become odious through being foreseen. The beings we are imagining would easily come to know the causal connections of volitions, and therefore their volitions would be better calculated to satisfy their desires than ours are. Since volitions are the outcome of desires, a prevision of volitions contrary to desires could not be a true one. It must be remembered that the supposed prevision would not create the future any more than memory creates the past. We do not think we were necessarily not free in the past, merely because we can now remember our past volitions. Similarly, we might be free in the future, even if we could now see what our future volitions were going to be. Freedom, in short, in any valuable sense, demands only that our volitions shall be, as they are, the result of our own desires, not of an outside force compelling us to will what we would rather not will. Everything else is confusion of thought, due to the feeling that knowledge *compels* the happening of what it knows when this is future, though it is at once obvious that knowledge has no such power in regard to the past. Free will, therefore, is true in the only form which is important; and the desire for other forms is a mere effect of insufficient analysis.

What has been said on philosophical method in the foregoing lectures has been rather by means of illustrations in particular cases than by means of general precepts. Nothing of any value can be said on method except through examples; but now, at the end of our course, we may collect certain general maxims which may possibly be a help in acquiring a philosophical habit of mind and a guide in looking for solutions of philosophic problems.

Philosophy does not become scientific by making use of other sciences, in the kind of way in which (*e.g.*) Herbert Spencer does. Philosophy aims at what is *general*, and the special sciences, however they may *suggest* large generalisations, cannot make them certain. And a hasty generalisation, such as Spencer's generalisation of evolution, is none the less hasty because what is generalised is the latest scientific theory. Philosophy is a study apart from the other sciences: its results cannot be established by the other sciences, and conversely must not be such as some other science might conceivably contradict. Prophecies as to the future of the universe, for example, are not the business of philosophy; whether the universe is progressive, retrograde, or stationary, it is not for the philosopher to say.

In order to become a scientific philosopher, a certain peculiar mental discipline is required. There must be present, first of all, the desire to know philosophical truth, and this desire must be sufficiently strong to survive through years when there seems no hope of its finding any satisfaction. The desire to know philosophical truth is very rare—in its purity, it is not often found even among philosophers. It is obscured sometimes—particularly after long periods of fruitless search—by the desire to *think* we know. Some plausible opinion presents itself, and by turning our attention away from the objections to it, or merely by not making great efforts to find objections to it, we may obtain the comfort of believing it, although, if we had resisted the wish for comfort, we should have come to see that the opinion was false. Again the desire for unadulterated truth is often obscured, in professional philosophers, by love of system: the one little fact which will not come inside the philosopher's edifice has to be pushed and tortured until it seems to consent. Yet the one little fact is more likely to be important for the future than the system with which it is inconsistent. Pythagoras invented a system which fitted admirably with all the facts he knew, except the incommensurability of the diagonal of a square and the side; this one little fact stood out, and remained a fact even after Hippasos of Metapontion was drowned for revealing it. To us, the discovery of this fact is the chief claim of Pythagoras to immortality, while his system has become a matter of merely historical curiosity.Love of system, therefore, and the system-maker's vanity which becomes associated with it, are among the snares that the student of philosophy must guard against.

The desire to establish this or that result, or generally to discover evidence for agreeable results, of whatever kind, has of course been the chief obstacle to honest philosophising. So strangely perverted do men become by unrecognised passions, that a determination in advance to arrive at this or that conclusion is generally regarded as a mark of virtue, and those whose studies lead to an opposite conclusion are thought to be wicked. No doubt it is commoner to wish to arrive at an agreeable result than to wish to arrive at a true result. But only those in whom the desire to arrive at a *true* result is paramount can hope to serve any good purpose by the study of philosophy.

But even when the desire to know exists in the requisite strength, the mental vision by which abstract truth is recognised is hard to distinguish from vivid imaginability and consonance with mental habits. It is necessary to practise methodological doubt, like Descartes, in order to loosen the hold of mental habits; and it is necessary to cultivate logical imagination, in order to have a number of hypotheses at command, and not to be the slave of the one which common sense has rendered easy to imagine. These two processes, of doubting the familiar and imagining the unfamiliar, are correlative, and form the chief part of the mental training required for a philosopher.

The naïve beliefs which we find in ourselves when we first begin the process of philosophic reflection may turn out, in the end, to be almost all capable of a true interpretation; but they ought all, before being admitted into philosophy, to undergo the ordeal of sceptical criticism. Until they have gone through this ordeal, they are mere blind habits, ways of behaving rather than intellectual convictions. And although it may be that a majority will pass the test, we may be pretty sure that some will not, and that a serious readjustment of our outlook ought to result. In order to break the dominion of habit, we must do our best to doubt the senses, reason, morals, everything in short. In some directions, doubt will be found possible; in others, it will be checked by that direct vision of abstract truth upon which the possibility of philosophical knowledge depends.

At the same time, and as an essential aid to the direct perception of the truth, it is necessary to acquire fertility in imagining abstract hypotheses. This is, I think, what has most of all been lacking hitherto in philosophy. So meagre was the logical apparatus that all the hypotheses philosophers could imagine were found to be inconsistent with the facts. Too often this state of things led to the adoption of heroic measures, such as a wholesale denial of the facts, when an imagination better stocked with logical tools would have found a key to unlock the mystery. It is in this way that the study of logic becomes the central study in philosophy: it gives the method of research in philosophy, just as mathematics gives the method in physics. And as physics, which, from Plato to the Renaissance, was as unprogressive, dim, and superstitious as philosophy, became a science through Galileo's fresh observation of facts and subsequent mathematical manipulation, so philosophy, in our own day, is becoming scientific through the simultaneous acquisition of new facts and logical methods.

In spite, however, of the new possibility of progress in philosophy, the first effect, as in the case of physics, is to diminish very greatly the extent of what is thought to be known. Before Galileo, people believed themselves possessed of immense knowledge on all the most interesting questions in physics. He established certain facts as to the way in which bodies fall, not very interesting on their own account, but of quite immeasurable interest as examples of real knowledge and of a new method whose future fruitfulness he himself divined. But his few facts sufficed to destroy the whole vast system of supposed knowledge handed down from Aristotle, as even the palest morning sun suffices to extinguish the stars. So in philosophy: though some have believed one system, and others another, almost all have been of opinion that a great deal was known; but all this supposed knowledge in the traditional systems must be swept away, and a new beginning must be made, which we shall esteem fortunate indeed if it can attain results comparable to Galileo's law of falling bodies.

By the practice of methodological doubt, if it is genuine and prolonged, a certain humility as to our knowledge is induced: we become glad to know *anything* in philosophy, however seemingly trivial. Philosophy has suffered from the lack of this kind of modesty. It has made the mistake of attacking the interesting problems at once, instead of proceeding patiently and slowly, accumulating whatever solid knowledge was obtainable, and trusting the great problems to the future. Men of science are not ashamed of what is intrinsically trivial, if its consequences are likely to be important; the *immediate* outcome of an experiment is hardly ever interesting on its own account. So in philosophy, it is often desirable to expend time and care on matters which, judged alone, might seem frivolous, for it is often only through the consideration of such matters that the greater problems can be approached.

When our problem has been selected, and the necessary mental discipline has been acquired, the method to be pursued is fairly uniform. The big problems which provoke philosophical inquiry are found, on examination, to be complex, and to depend upon a number of component problems, usually more abstract than those of which they are the components. It will generally be found that all our initial data, all the facts that we seem to know to begin with, suffer from vagueness, confusion, and complexity. Current philosophical ideas share these defects; it is therefore necessary to create an apparatus of precise conceptions as general and as free from complexity as possible, before the data can be analysed into the kind of premisses which philosophy aims at discovering. In this process of analysis, the source of difficulty is tracked further and further back, growing at each stage more abstract, more refined, more difficult to apprehend. Usually it will be found that a number of these extraordinarily abstract questions underlie any one of the big obvious problems. When everything has been done that can be done by method, a stage is reached where only direct philosophic vision can carry matters further. Here only genius will avail. What is wanted, as a rule, is some new effort of logical imagination, some glimpse of a possibility never conceived before, and then the direct perception that this possibility is realised in the case in question. Failure to think of the right possibility leaves insoluble difficulties, balanced arguments pro and con, utter bewilderment and despair. But the right possibility, as a rule, when once conceived, justifies itself swiftly by its astonishing power of absorbing apparently conflicting facts. From this point onward, the work of the philosopher is synthetic and comparatively easy; it is in the very last stage of the analysis that the real difficulty consists.

Of the prospect of progress in philosophy, it would be rash to speak with confidence. Many of the traditional problems of philosophy, perhaps most of those which have interested a wider circle than that of technical students, do not appear to be soluble by scientific methods. Just as astronomy lost much of its human interest when it ceased to be astrology, so philosophy must lose in attractiveness as it grows less prodigal of promises. But to the large and still growing body of men engaged in the pursuit of science—men who hitherto, not without justification, have turned aside from philosophy with a certain contempt—the new method, successful already in such time-honoured problems as number, infinity, continuity, space and time, should make an appeal which the older methods have wholly failed to make. Physics, with its principle of relativity and its revolutionary investigations into the nature of matter, is feeling the need for that kind of novelty in fundamental hypotheses which scientific philosophy aims at facilitating. The one and only condition, I believe, which is necessary in order to secure for philosophy in the near future an achievement surpassing all that has hitherto been accomplished by philosophers, is the creation of a school of men with scientific training and philosophical interests, unhampered by the traditions of the past, and not misled by the literary methods of those who copy the ancients in all except their merits.

WHY MEN FIGHT

I
THE PRINCIPLE OF GROWTH

TO all who are capable of new impressions and fresh thought, some modification of former beliefs and hopes has been brought by the war. What the modification has been has depended, in each case, upon character and circumstance; but in one form or another it has been almost universal. To me, the chief thing to be learnt through the war has been a certain view of the springs of human action, what they are, and what we may legitimately hope that they will become. This view, if it is true, seems to afford a basis for political philosophy more capable of standing erect in a time of crisis than the philosophy of traditional Liberalism has shown itself to be. The following lectures, though only one of them will deal with war, all are inspired by a view of the springs of action[1] which has been suggested by the war. And all of them are informed by the hope of seeing such political institutions established in Europe as shall make men averse to war—a hope which I firmly believe to be realizable, though not without a great and fundamental reconstruction of economic and social life.

To one who stands outside the cycle of beliefs and passions which make the war seem necessary, an isolation, an almost unbearable separation from the general activity, becomes unavoidable. At the very moment when the universal disaster raises compassion in the highest degree, compassion itself compels aloofness from the impulse to self-destruction which has swept over Europe. The helpless longing to save men from the ruin towards which they are hastening makes it necessary to oppose the stream, to incur hostility, to be thought unfeeling, to lose for the moment the power of winning belief. It is impossible to prevent others from feeling hostile, but it is possible to avoid any reciprocal hostility on one's own part, by imaginative understanding and the sympathy which grows out of it. And without understanding and sympathy it is impossible to find a cure for the evil from which the world is suffering.

There are two views of the war neither of which seems to me adequate. The usual view in this country is that it is due to the wickedness of the Germans; the view of most pacifists is that it is due to the diplomatic tangle and to the ambitions of Governments. I think both these views fail to realize the extent to which war grows out of ordinary human nature. Germans, and also the men who compose Governments, are on the whole average human beings, actuated by the same passions that actuate others, not differing much from the rest of the world except in their circumstances. War is accepted by men who are neither Germans nor diplomatists with a readiness, an acquiescence in untrue and inadequate reasons, which would not be possible if any deep repugnance to war were widespread in other nations or classes. The untrue things which men believe, and the true things which they disbelieve, are an index to their impulses—not necessarily to individual impulses in each case (since beliefs are contagious), but to the general impulses of the community. We all believe many things which we have no good ground for believing, because, subconsciously, our nature craves certain kinds of action which these beliefs would render reasonable if they were true. Unfounded beliefs are the homage which impulse pays to reason; and thus it is with the beliefs which, opposite but similar, make men here and in Germany believe it their duty to prosecute the war.

The first thought which naturally occurs to one who accepts this view is that it would be well if men were more under the dominion of reason. War, to those who see that it must necessarily do untold harm to all the combatants, seems a mere madness, a collective insanity in which all that has been known in time of peace is forgotten. If impulses were more controlled, if thought were less dominated by passion, men would guard their minds against the approaches of war fever, and disputes would be adjusted amicably. This is true, but it is not by itself sufficient. It is only those in whom the desire to think truly is itself a passion who will find this desire adequate to control the passions of war. Only passion can control passion, and only a contrary impulse or desire can check impulse. Reason, as it is preached by traditional moralists, is too negative, too little living, to make a good life. It is not by reason alone that wars can be prevented, but by a positive life of impulses and passions antagonistic to those that lead to war. It is the life of impulse that needs to be changed, not only the life of conscious thought.

All human activity springs from two sources: impulse and desire. The part played by desire has always been sufficiently recognized. When men find themselves not fully contented, and not able instantly to procure what will cause content, imagination brings before their minds the thought of things which they believe would make them happy. All desire involves an interval of time between the consciousness of a need and the opportunity for satisfying it. The acts inspired by desire may be in themselves painful, the time before satisfaction can be achieved may be very long, the object desired may be something outside our own lives, and even after our own death. Will, as a directing force, consists mainly in following desires for more or less distant objects, in spite of the painfulness of the acts involved and the solicitations of incompatible but more immediate desires and impulses. All this is familiar, and political philosophy hitherto has been almost entirely based upon desire as the source of human actions.

But desire governs no more than a part of human activity, and that not the most important but only the more conscious, explicit, and civilized part.

In all the more instinctive part of our nature we are dominated by impulses to certain kinds of activity, not by desires for certain ends. Children run and shout, not because of any good which they expect to realize, but because of a direct impulse to running and shouting. Dogs bay the moon, not because they consider that it is to their advantage to do so, but because they feel an impulse to bark. It is not any purpose, but merely an impulse, that prompts such actions as eating, drinking, love-making, quarreling, boasting. Those who believe that man is a rational animal will say that people boast in order that others may have a good opinion of them; but most of us can recall occasions when we have boasted in spite of knowing that we should be despised for it. Instinctive acts normally achieve some result which is agreeable to the natural man, but they are not performed from desire for this result. They are performed from direct impulse, and the impulse is often strong even in cases in which the normal desirable result cannot follow. Grown men like to imagine themselves more rational than children and dogs, and unconsciously conceal from themselves how great a part impulse plays in their lives. This unconscious concealment always follows a certain general plan. When an impulse is not indulged in the moment in which it arises, there grows up a desire for the expected consequences of indulging the impulse. If some of the consequences which are reasonably to be expected are clearly disagreeable, a conflict between foresight and impulse arises. If the impulse is weak, foresight may conquer; this is what is called acting on reason. If the impulse is strong, either foresight will be falsified, and the disagreeable consequences will be forgotten, or, in men of a heroic mold, the consequences may be recklessly accepted. When Macbeth realizes that he is doomed to defeat, he does not shrink from the fight; he exclaims:—

Lay on, Macduff,
And damned be him that first cries, Hold, enough!

But such strength and recklessness of impulse is rare. Most men, when their impulse is strong, succeed in persuading themselves, usually by a subconscious selectiveness of attention, that agreeable consequences will follow from the indulgence of their impulse. Whole philosophies, whole systems of ethical valuation, spring up in this way: they are the embodiment of a kind of thought which is subservient to impulse, which aims at providing a quasi-rational ground for the indulgence of impulse. The only thought which is genuine is that which springs out of the intellectual impulse of curiosity, leading to the desire to know and understand. But most of what passes for thought is inspired by some non-intellectual impulse, and is merely a means of persuading ourselves that we shall not be disappointed or do harm if we indulge this impulse.

When an impulse is restrained, we feel discomfort or even violent pain. We may indulge the impulse in order to escape from this pain, and our action is then one which has a purpose. But the pain only exists because of the impulse, and the impulse itself is directed to an act, not to escaping from the pain of restraining the impulse. The impulse itself remains without a purpose, and the purpose of escaping from pain only arises when the impulse has been momentarily restrained.

Impulse is at the basis of our activity, much more than desire. Desire has its place, but not so large a place as it seemed to have. Impulses bring with them a whole train of subservient fictitious desires: they make men feel that they desire the results which will follow from indulging the impulses, and that they are acting for the sake of these results, when in fact their action has no motive outside itself. A man may write a book or paint a picture under the belief that he desires the praise which it will bring him; but as soon as it is finished, if his creative impulse is not exhausted, what he has done grows uninteresting to him, and he begins a new piece of work. What applies to artistic creation applies equally to all that is most vital in our lives: direct impulse is what moves us, and the desires which we think we have are a mere garment for the impulse.

Desire, as opposed to impulse, has, it is true, a large and increasing share in the regulation of men's lives. Impulse is erratic and anarchical, not easily fitted into a well-regulated system; it may be tolerated in children and artists, but it is not thought proper to men who hope to be taken seriously. Almost all paid work is done from desire, not from impulse: the work itself is more or less irksome, but the payment for it is desired. The serious activities that fill a man's working hours are, except in a few fortunate individuals, governed mainly by purposes, not by impulses towards those activities. In this hardly any one sees an evil, because the place of impulse in a satisfactory existence is not recognized.

An impulse, to one who does not share it actually or imaginatively, will always seem to be mad. All impulse is essentially blind, in the sense that it does not spring from any prevision of consequences. The man who does not share the impulse will form a different estimate as to what the consequences will be, and as to whether those that must ensue are desirable. This difference of opinion will seem to be ethical or intellectual, whereas its real basis is a difference of impulse. No genuine agreement will be reached, in such a case, so long as the difference of impulse persists. In all men who have any vigorous life, there are strong impulses such as may seem utterly unreasonable to others. Blind impulses sometimes lead to destruction and death, but at other times they lead to the best things the world contains. Blind impulse is the source of war, but it is also the source of science, and art, and love. It is not the weakening of impulse that is to be desired, but the direction of impulse towards life and growth rather than towards death and decay.

The complete control of impulse by will, which is sometimes preached by moralists, and often enforced by economic necessity, is not really desirable. A life governed by purposes and desires, to the exclusion of impulse, is a tiring life; it exhausts vitality, and leaves a man, in the end, indifferent to the very purposes which he has been trying to achieve. When a whole nation lives in this way, the whole nation tends to become feeble, without enough grasp to recognize and overcome the obstacles to its desires. Industrialism and organization are constantly forcing civilized nations to live more and more by purpose rather than impulse. In the long run such a mode of existence, if it does not dry up the springs of life, produces new impulses, not of the kind which the will has been in the habit of controlling or of which thought is conscious. These new impulses are apt to be worse in their effects than those that have been checked. Excessive discipline, especially when it is imposed from without, often issues in impulses of cruelty and destruction; this is one reason why militarism has a bad effect on national character. Either lack of vitality, or impulses which are oppressive and against life, will almost always result if the spontaneous impulses are not able to find an outlet. A man's impulses are not fixed from the beginning by his native disposition: within certain wide limits, they are profoundly modified by his circumstances and his way of life. The nature of these modifications ought to be studied, and the results of such study ought to be taken account of in judging the good or harm that is done by political and social institutions.

The war has grown, in the main, out of the life of impulse, not out of reason or desire. There is an impulse of aggression, and an impulse of resistance to aggression. Either may, on occasion, be in accordance with reason, but both are operative in many cases in which they are quite contrary to reason. Each impulse produces a whole harvest of attendant beliefs. The beliefs appropriate to the impulse of aggression may be seen in Bernhardi, or in the early Mohammedan conquerors, or, in full perfection, in the Book of Joshua. There is first of all a conviction of the superior excellence of one's own group, a certainty that they are in some sense the chosen people. This justifies the feeling that only the good and evil of one's own group is of real importance, and that the rest of the world is to be regarded merely as material for the triumph or salvation of the higher race. In modern politics this attitude is embodied in imperialism. Europe as a whole has this attitude towards Asia and Africa, and many Germans have this attitude towards the rest of Europe.

Correlative to the impulse of aggression is the impulse of resistance to aggression. This impulse is exemplified in the attitude of the Israelites to the Philistines or of medieval Europe to the Mohammedans. The beliefs which it produces are beliefs in the peculiar wickedness of those whose aggression is feared, and in the immense value of national customs which they might suppress if they were victorious. When the war broke out, all the reactionaries in England and France began to speak of the danger to democracy, although until that moment they had opposed democracy with all their strength. They were not insincere in so speaking: the impulse of resistance to Germany made them value whatever was endangered by the German attack. They loved democracy because they hated Germany; but they thought they hated Germany because they loved democracy.

The correlative impulses of aggression and resistance to aggression have both been operative in all the countries engaged in the war. Those who have not been dominated by one or other of these impulses may be roughly divided into three classes. There are, first, men whose national sentiment is antagonistic to the State to which they are subject. This class includes some Irish, Poles, Finns, Jews, and other members of oppressed nations. From our point of view, these men may be omitted from our consideration, since they have the same impulsive nature as those who fight, and differ merely in external circumstances.

The second class of men who have not been part of the force supporting the war have been those whose impulsive nature is more or less atrophied. Opponents of pacifism suppose that all pacifists belong to this class, except when they are in German pay. It is thought that pacifists are bloodless, men without passions, men who can look on and reason with cold detachment while their brothers are giving their lives for their country. Among those who are merely passively pacifist, and do no more than abstain from actively taking part in the war, there may be a certain proportion of whom this is true. I think the supporters of war would be right in decrying such men. In spite of all the destruction which is wrought by the impulses that lead to war, there is more hope for a nation which has these impulses than for a nation in which all impulse is dead. Impulse is the expression of life, and while it exists there is hope of its turning towards life instead of towards death; but lack of impulse is death, and out of death no new life will come.

The active pacifists, however, are not of this class: they are not men without impulsive force but men in whom some impulse to which war is hostile is strong enough to overcome the impulses that lead to war. It is not the act of a passionless man to throw himself athwart the whole movement of the national life, to urge an outwardly hopeless cause, to incur obloquy and to resist the contagion of collective emotion. The impulse to avoid the hostility of public opinion is one of the strongest in human nature, and can only be overcome by an unusual force of direct and uncalculating impulse; it is not cold reason alone that can prompt such an act.

Impulses may be divided into those that make for life and those that make for death. The impulses embodied in the war are among those that make for death. Any one of the impulses that make for life, if it is strong enough, will lead a man to stand out against the war. Some of these impulses are only strong in highly civilized men; some are part of common humanity. The impulses towards art and science are among the more civilized of those that make for life. Many artists have remained wholly untouched by the passions of the war, not from feebleness of feeling, but because the creative instinct, the pursuit of a vision, makes them critical of the assaults of national passion, and not responsive to the myth in which the impulse of pugnacity clothes itself. And the few men in whom the scientific impulse is dominant have noticed the rival myths of warring groups, and have been led through understanding to neutrality. But it is not out of such refined impulses that a popular force can be generated which shall be sufficient to transform the world.

There are three forces on the side of life which require no exceptional mental endowment, which are not very rare at present, and might be very common under better social institutions.They are love, the instinct of constructiveness, and the joy of life. All three are checked and enfeebled at present by the conditions under which men live—not only the less outwardly fortunate, but also the majority of the well-to-do. Our institutions rest upon injustice and authority: it is only by closing our hearts against sympathy and our minds against truth that we can endure the oppressions and unfairnesses by which we profit. The conventional conception of what constitutes success leads most men to live a life in which their most vital impulses are sacrificed, and the joy of life is lost in listless weariness. Our economic system compels almost all men to carry out the purposes of others rather than their own, making them feel impotent in action and only able to secure a certain modicum of passive pleasure. All these things destroy the vigor of the community, the expansive affections of individuals, and the power of viewing the world generously. All these things are unnecessary and can be ended by wisdom and courage. If they were ended, the impulsive life of men would become wholly different, and the human race might travel towards a new happiness and a new vigor. To urge this hope is the purpose of these lectures.

The impulses and desires of men and women, in so far as they are of real importance in their lives, are not detached one from another, but proceed from a central principle of growth, an instinctive urgency leading them in a certain direction, as trees seek the light. So long as this instinctive movement is not thwarted, whatever misfortunes may occur are not fundamental disasters, and do not produce those distortions which result from interference with natural growth. This intimate center in each human being is what imagination must apprehend if we are to understand him intuitively. It differs from man to man, and determines for each man the type of excellence of which he is capable. The utmost that social institutions can do for a man is to make his own growth free and vigorous: they cannot force him to grow according to the pattern of another man. There are in men some impulses and desires—for example, those towards drugs—which do not grow out of the central principle; such impulses, when they become strong enough to be harmful, have to be checked by self-discipline. Other impulses, though they may grow out of the central principle in the individual, may be injurious to the growth of others, and they need to be checked in the interest of others. But in the main, the impulses which are injurious to others tend to result from thwarted growth, and to be least in those who have been unimpeded in their instinctive development.

Men, like trees, require for their growth the right soil and a sufficient freedom from oppression. These can be helped or hindered by political institutions. But the soil and the freedom required for a man's growth are immeasurably more difficult to discover and to obtain than the soil and the freedom required for the growth of a tree. And the full growth which may be hoped for cannot be defined or demonstrated; it is subtle and complex, it can only be felt by a delicate intuition and dimly apprehended by imagination and respect. It depends not only or chiefly upon the physical environment, but upon beliefs and affections, upon opportunities for action, and upon the whole life of the community. The more developed and civilized the type of man the more elaborate are the conditions of his growth, and the more dependent they become upon the general state of the society in which he lives. A man's needs and desires are not confined to his own life. If his mind is comprehensive and his imagination vivid, the failures of the community to which he belongs are his failures, and its successes are his successes: according as his community succeeds or fails, his own growth is nourished or impeded.

In the modern world, the principle of growth in most men and women is hampered by institutions inherited from a simpler age. By the progress of thought and knowledge, and by the increase in command over the forces of the physical world, new possibilities of growth have come into existence, and have given rise to new claims which must be satisfied if those who make them are not to be thwarted. There is less acquiescence in limitations which are no longer unavoidable, and less possibility of a good life while those limitations remain. Institutions which give much greater opportunities to some classes than to others are no longer recognized as just by the less fortunate, though the more fortunate still defend them vehemently. Hence arises a universal strife, in which tradition and authority are arrayed against liberty and justice. Our professed morality, being traditional, loses its hold upon those who are in revolt. Coöperation between the defenders of the old and the champions of the new has become almost impossible.An intimate disunion has entered into almost all the relations of life in continually increasing measure. In the fight for freedom, men and women become increasingly unable to break down the walls of the Ego and achieve the growth which comes from a real and vital union.

All our institutions have their historic basis in Authority. The unquestioned authority of the Oriental despot found its religious expression in the omnipotent Creator, whose glory was the sole end of man, and against whom man had no rights. This authority descended to the Emperor and Pope, to the kings of the Middle Ages, to the nobles in the feudal hierarchy, and even to every husband

and father in his dealings with his wife and children. The Church was the direct embodiment of the Divine authority, the State and the law were constituted by the authority of the King, private property in land grew out of the authority of conquering barons, and the family was governed by the authority of the pater-familias.

The institutions of the Middle Ages permitted only a fortunate few to develop freely: the vast majority of mankind existed to minister to the few. But so long as authority was genuinely respected and acknowledged even by its least fortunate subjects, medieval society remained organic and not fundamentally hostile to life, since outward submission was compatible with inward freedom because it was voluntary. The institutions of Western Christendom embodied a theory which was really believed, as no theory by which our present institutions can be defended is now believed.

The medieval theory of life broke down through its failure to satisfy men's demands for justice and liberty. Under the stress of oppression, when rulers exceeded their theoretical powers, the victims were forced to realize that they themselves also had rights, and need not live merely to increase the glory of the few. Gradually it came to be seen that if men have power, they are likely to abuse it, and that authority in practice means tyranny. Because the claim to justice was resisted by the holders of power, men became more and more separate units, each fighting for his own rights, not a genuine community bound together by an organic common purpose. This absence of a common purpose has become a source of unhappiness. One of the reasons which led many men to welcome the outbreak of the present war was that it made each nation again a whole community with a single purpose. It did this by destroying, for the present, the beginnings of a single purpose in the civilized world as a whole; but these beginnings were as yet so feeble that few were much affected by their destruction. Men rejoiced in the new sense of unity with their compatriots more than they minded the increased separation from their enemies.

The hardening and separation of the individual in the course of the fight for freedom has been inevitable, and is not likely ever to be wholly undone. What is necessary, if an organic society is to grow up, is that our institutions should be so fundamentally changed as to embody that new respect for the individual and his rights which modern feeling demands. The medieval Empire and Church swept away the individual. There were heretics, but they were massacred relentlessly, without any of the qualms aroused by later persecutions. And they, like their persecutors, were persuaded that there ought to be one universal Church: they differed only as to what its creed should be. Among a few men of art and letters, the Renaissance undermined the medieval theory, without, however, replacing it by anything but skepticism and confusion. The first serious breach in this medieval theory was caused by Luther's assertion of the right of private judgment and the fallibility of General Councils. Out of this assertion grew inevitably, with time, the belief that a man's religion could not be determined for him by authority, but must be left to the free choice of each individual. It was in matters of religion that the battle for liberty began, and it is in matters of religion that it has come nearest to a complete victory.

The development through extreme individualism to strife, and thence, one hopes, to a new reintegration, is to be seen in almost every department of life. Claims are advanced in the name of justice, and resisted in the name of tradition and prescriptive right. Each side honestly believes that it deserves to triumph, because two theories of society exist side by side in our thought, and men choose, unconsciously, the theory which fits their case. Because the battle is long and arduous all general theory is gradually forgotten; in the end, nothing remains but self-assertion, and when the oppressed win freedom they are as oppressive as their former masters.

This is seen most crudely in the case of what is called nationalism. Nationalism, in theory, is the doctrine that men, by their sympathies and traditions, form natural groups, called "nations," each of which ought to be united under one central Government. In the main this doctrine may be conceded. But in practice the doctrine takes a more personal form. "I belong," the oppressed nationalist argues, "by sympathy and tradition to nation A, but I am subject to a government which is in the hands of nation B. This is an injustice, not only because of the general principle of nationalism, but because nation A is generous, progressive, and civilized, while nation B is oppressive, retrograde, and barbarous. Because this is so, nation A deserves to prosper, while nation B deserves to be abased." The inhabitants of nation B are naturally deaf to the claims of abstract justice, when they are accompanied by personal hostility and contempt. Presently, however, in the course of war, nation A acquires its freedom. The energy and pride which have achieved freedom generates a momentum which leads on, almost infallibly, to the attempt at foreign conquest, or to the refusal of liberty to some smaller nation. "What? You say that nation C, which forms part of our State, has the same rights against us as we had against nation A? But that is absurd. Nation C is swinish and turbulent, incapable of good government, needing a strong hand if it is not to be a menace and a disturbance to all its neighbors." So the English used to speak of the Irish, so the Germans and Russians speak of the Poles, so the Galician Poles speak of the Ruthenes, so the Austrians used to speak of the Magyars, so the Magyars speak of the South Slav sympathizers with Serbia, so the Serbs speak of the Macedonian Bulgars. In this way nationalism, unobjectionable in theory, leads by a natural movement to oppression and wars of conquest. No sooner was France free from the English, in the fifteenth century, than it embarked upon the conquest of Italy; no sooner was Spain freed from the Moors than it entered into more than a century of conflict with France for the supremacy in Europe. The case of Germany is very interesting in this respect. At the beginning of the eighteenth century German culture was French: French was the language of the Courts, the language in which Leibnitz wrote his philosophy, the universal language of polite letters and learning. National consciousness hardly existed. Then a series of great men created a self-respect in Germany by their achievements in poetry, music, philosophy, and science. But politically German nationalism was only created by Napoleon's oppression and the uprising of 1813. After centuries during which every disturbance of the peace of Europe began with a French or Swedish or Russian invasion of Germany, the Germans discovered that by sufficient effort and union they could keep foreign armies off their territory. But the effort required had been too great to cease when its purely defensive purpose had been achieved by the defeat of Napoleon. Now, a hundred years later, they are still engaged in the same movement, which has become one of aggression and conquest. Whether we are now seeing the end of the movement it is not yet possible to guess.

If men had any strong sense of a community of nations, nationalism would serve to define the boundaries of the various nations. But because men only feel community within their own nation, nothing but force is able to make them respect the rights of other nations, even when they are asserting exactly similar rights on their own behalf.

Analogous development is to be expected, with the course of time, in the conflict between capital and labor which has existed since the growth of the industrial system, and in the conflict between men and women, which is still in its infancy.

What is wanted, in these various conflicts, is some principle, genuinely believed, which will have justice for its outcome. The tug of war of mutual self-assertion can only result in justice through an accidental equality of force. It is no use to attempt any bolstering up of institutions based on authority, since all such institutions involve injustice, and injustice once realized cannot be perpetuated without fundamental damage both to those who uphold it and to those who resist it. The damage consists in the hardening of the walls of the Ego, making them a prison instead of a window. Unimpeded growth in the individual depends upon many contacts with other people, which must be of the nature of free coöperation, not of enforced service. While the belief in authority was alive, free coöperation was compatible with inequality and subjection, but now equality and mutual freedom are necessary. All institutions, if they are not to hamper individual growth, must be based as far as possible upon voluntary combination, rather than the force of the law or the traditional authority of the holders of power. None of our institutions can survive the application of this principle without great and fundamental changes; but these changes are imperatively necessary if the world is to be withheld from dissolving into hard separate units each at war with all the others.

The two chief sources of good relations between individuals are instinctive liking and a common purpose. Of these two, a common purpose might seem more important politically, but, in fact, it is often the outcome, not the cause, of instinctive liking, or of a common instinctive aversion. Biological groups, from the family to the nation, are constituted by a greater or less degree of instinctive liking, and build their common purposes on this foundation.

Instinctive liking is the feeling which makes us take pleasure in another person's company, find an exhilaration in his presence, wish to talk with him, work with him, play with him. The extreme form of it is being in love, but its fainter forms, and even the very faintest, have political importance. The presence of a person who is instinctively disliked tends to make any other person more likable. An anti-Semite will love any fellow-Christian when a Jew is present. In China, or the wilds of Africa, any white man would be welcomed with joy. A common aversion is one of the most frequent causes of mild instinctive liking.

Men differ enormously in the frequency and intensity of their instinctive likings, and the same man will differ greatly at different times. One may take Carlyle and Walt Whitman as opposite poles in this respect. To Carlyle, at any rate in later life, most men and women were repulsive; they inspired an instinctive aversion which made him find pleasure in imagining them under the guillotine or perishing in battle. This led him to belittle most men, finding satisfaction only in those who had been notably destructive of human life—Frederick the Great, Dr. Francia, and Governor Eyre. It led him to love war and violence, and to despise the weak and the oppressed—for example, the "thirty thousand distressed needlewomen," on whom he was never weary of venting his scorn. His morals and his politics, in later life, were inspired through and through by repugnance to almost the whole human race.

Walt Whitman, on the contrary, had a warm, expansive feeling towards the vast majority of men and women. His queer catalogues seemed to him interesting because each item came before his imagination as an object of delight. The sort of joy which most people feel only in those who are exceptionally beautiful or splendid Walt Whitman felt in almost everybody. Out of this universal liking grew optimism, a belief in democracy, and a conviction that it is easy for men to live together in peace and amity. His philosophy and politics, like Carlyle's, were based upon his instinctive attitude towards ordinary men and women.

There is no objective reason to be given to show that one of these attitudes is essentially more rational than the other. If a man finds people repulsive, no argument can prove to him that they are not so. But both his own desires and other people's are much more likely to find satisfaction if he resembles Walt

Whitman than if he resembles Carlyle. A world of Walt Whitmans would be happier and more capable of realizing its purposes than a world of Carlyles. For this reason, we shall desire, if we can, to increase the amount of instinctive liking in the world and diminish the amount of instinctive aversion. This is perhaps the most important of all the effects by which political institutions ought to be judged.

The other source of good relations between individuals is a common purpose, especially where that purpose cannot be achieved without knowing its cause. Economic organizations, such as unions and political parties are constituted almost wholly by a common purpose; whatever instinctive liking may come to be associated with them is the result of the common purpose, not its cause. Economic organizations, such as railway companies, subsist for a purpose, but this purpose need only actually exist in those who direct the organization: the ordinary wage-earner need have no purpose beyond earning his wages. This is a defect in economic organizations, and ought to be remedied. One of the objects of syndicalism is to remedy this defect.

Marriage is (or should be) based on instinctive liking, but as soon as there are children, or the wish for children, it acquires the additional strength of a common purpose. It is this chiefly which distinguishes it from an irregular connection not intended to lead to children. Often, in fact, the common purpose survives, and remains a strong tie, after the instinctive liking has faded.

A nation, when it is real and not artificial, is founded upon a faint degree of instinctive liking for compatriots and a common instinctive aversion from foreigners. When an Englishman returns to Dover or Folkestone after being on the Continent, he feels something friendly in the familiar ways: the casual porters, the shouting paper boys, the women serving bad tea, all warm his heart, and seem more "natural," more what human beings ought to be, than the foreigners with their strange habits of behavior. He is ready to believe that all English people are good souls, while many foreigners are full of designing wickedness. It is such feelings that make it easy to organize a nation into a governmental unit. And when that has happened, a common purpose is added, as in marriage. Foreigners would like to invade our country and lay it waste, to kill us in battle, to humble our pride. Those who coöperate with us in preventing this disaster are our friends, and their coöperation intensifies our instinctive liking. But common purposes do not constitute the whole source of our love of country: allies, even of long standing,do not call out the same feelings as are called out by our compatriots. Instinctive liking, resulting largely from similar habits and customs, is an essential element in patriotism, and, indeed, the foundation upon which the whole feeling rests.

If men's natural growth is to be promoted and not hindered by the environment, if as many as possible of their desires and needs are to be satisfied, political institutions must, as far as possible, embody common purposes and foster instinctive liking. These two objects are interconnected, for nothing is so destructive of instinctive liking as thwarted purposes and unsatisfied needs, and nothing facilitates coöperation for common purposes so much as instinctive liking. When a man's growth is unimpeded, his self-respect remains intact, and he is not inclined to regard others as his enemies. But when, for whatever reason, his growth is impeded, or he is compelled to grow into some twisted and unnatural shape, his instinct presents the environment as his enemy, and he becomes filled with hatred. The joy of life abandons him, and malevolence takes the place of friendliness. The malevolence of hunchbacks and cripples is proverbial; and a similar malevolence is to be found in those who have been crippled in less obvious ways. Real freedom, if it could be brought about, would go a long way towards destroying hatred.

There is a not uncommon belief that what is instinctive in us cannot be changed, but must be simply accepted and made the best of. This is by no means the case. No doubt we have a certain native disposition, different in different people, which coöperates with outside circumstances in producing a certain character. But even the instinctive part of our character is very malleable. It may be changed by beliefs, by material circumstances, by social circumstances, and by institutions. A Dutchman has probably much the same native disposition as a German, but his instincts in adult life are very different owing to the absence of militarism and of the pride of a Great Power. It is obvious that the instincts of celibates become profoundly different from those of other men and women. Almost any instinct is capable of many different forms according to the nature of the outlets which it finds. The same instinct which leads to artistic or intellectual creativeness may, under other circumstances, lead to love of war. The fact that an activity or belief is an outcome of instinct is therefore no reason for regarding it as unalterable.

This applies to people's instinctive likes and dislikes as well as to their other instincts. It is natural to men, as to other animals, to like some of their species and dislike others; but the proportion of like and dislike depends on circumstances, often on quite trivial circumstances. Most of Carlyle's misanthropy is attributable to dyspepsia; probably a suitable medical regimen would have given him a completely different outlook on the world. The defect of punishment, as a means of dealing with impulses which the community wishes to discourage, is that it does nothing to prevent the existence of the impulses, but merely endeavors to check their indulgence by an appeal to self-interest. This method, since it does not eradicate the impulses, probably only drives them to find other outlets even when it is successful in its immediate object; and if the impulses are strong, mere self-interest is not likely to curb them effectually, since it is not a very powerful motive except with unusually reasonable and rather passionless people. It is thought to be a stronger motive than it is, because our moods make us deceive ourselves as to our interest, and lead us to believe that it is consistent with the actions to which we are prompted by desire or impulse.

Thus the commonplace that human nature cannot be changed is untrue. We all know that our own characters and those of our acquaintance are greatly affected by circumstances; and what is true of individuals is true also of nations. The root causes of changes in average human nature are generally either purely material changes—for instance, of climate—or changes in the degree of man's control over the material world. We may ignore the purely material changes, since these do not much concern the politician. But the changes due to man's increased control over the material world, by inventions and science, are of profound present importance. Through the industrial revolution, they have radically altered the daily lives of men; and by creating huge economic organizations, they have altered the whole structure of society. The general beliefs of men, which are, in the main, a product of instinct and circumstance, have become very different from what they were in the eighteenth century. But our institutions are not yet suited either to the instincts developed by our new circumstances, or to our real beliefs. Institutions have a life of their own, and often outlast the circumstances which made them a fit garment for instinct. This applies, in varying degrees, to almost all the institutions which we have inherited from the past: the State, private property, the patriarchal family, the Churches, armies and navies. All of these have become in some degree oppressive, in some measures hostile to life.

In any serious attempt at political reconstruction, it is necessary to realize what are the vital needs of ordinary men and women. It is customary, in political thought, to assume that the only needs with which politics is concerned are economic needs. This view is quite inadequate to account for such an event as the present war, since any economic motives that may be assigned for it are to a great extent mythical, and its true causes must be sought for outside the economic sphere. Needs which are normally satisfied without conscious effort remain unrecognized, and this results in a working theory of human needs which is far too simple. Owing chiefly to industrialism, many needs which were formerly satisfied without effort now remain unsatisfied in most men and women. But the old unduly simple theory of human needs survives, making men overlook the source of the new lack of satisfaction, and invent quite false theories as to why they are dissatisfied. Socialism as a panacea seems to me to be mistaken in this way, since it is too ready to suppose that better economic conditions will of themselves make men happy. It is not only more material goods that men need, but more freedom, more self-direction, more outlet for creativeness, more opportunity for the joy of life, more voluntary coöperation, and less involuntary subservience to purposes not their own. All these things the institutions of the future must help to produce, if our increase of knowledge and power over Nature is to bear its full fruit in bringing about a good life.

II
THE STATE

UNDER the influence of socialism, most liberal thought in recent years has been in favor of increasing the power of the State, but more or less hostile to the power of private property. On the other hand, syndicalism has been hostile both to the State and to private property. I believe that syndicalism is more nearly right than socialism in this respect, that both private property and the State, which are the two most powerful institutions of the modern world, have become harmful to life through excess of power, and that both are hastening the loss of vitality from which the civilized world increasingly suffers. The two institutions are closely connected, but for the present I wish to consider only the State. I shall try to show how great, how unnecessary, how harmful, many of its powers are, and how enormously they might be diminished without loss of what is useful in its activity. But I shall admit that in certain directions its functions ought to be extended rather than curtailed.

Some of the functions of the State, such as the Post Office and elementary education, might be performed by private agencies, and are only undertaken by the State from motives of convenience. But other matters, such as the law, the police, the Army, and the Navy, belong more essentially to the State: so long as there is a State at all it is difficult to imagine these matters in private hands. The distinction between socialism and individualism turns on the nonessential functions of the State, which the socialist wishes to extend and the individualist to restrict. It is the essential functions, which are admitted by individualists and socialists alike, that I wish to criticize, since the others do not appear to me in themselves objectionable.

The essence of the State is that it is the repository of the collective force of its citizens. This force takes two forms, one internal and one external. The internal form is the law and the police; the external form is the power of waging war, as embodied in the Army and Navy. The State is constituted by the combination of all the inhabitants in a certain area using their united force in accordance with the commands of a Government. In a civilized State force is only employed against its own citizens in accordance with rules previously laid down, which constitute the criminal law. But the employment of force against foreigners is not regulated by any code of rules, and proceeds, with few exceptions, according to some real or fancied national interest.

There can be no doubt that force employed according to law is less pernicious than force employed capriciously. If international law could acquire sufficient hold on men's allegiance to regulate the relations of States, a very great advance on our present condition would have been made. The primitive anarchy which precedes law is worse than law. But I believe there is a possibility of a stage to some extent above law, where the advantages now secured by the law are secured without loss of freedom, and without the disadvantages which the law and the police render inevitable. Probably some repository of force in the background will remain necessary, but the actual employment of force may become very rare, and the degree of force required very small. The anarchy which precedes law gives freedom only to the strong; the condition to be aimed at will give freedom as nearly as possible to every one. It will do this, not by preventing altogether the existence of organized force, but by limiting the occasions for its employment to the greatest possible extent.

The power of the State is only limited internally by the fear of rebellion and externally by the fear of defeat in war. Subject to these restrictions, it is absolute. In practice, it can seize men's property through taxation, determine the law of marriage and inheritance, punish the expression of opinions which it dislikes, put men to death for wishing the region they inhabit to belong to a different State, and order all able-bodied males to risk their lives in battle whenever it considers war desirable. On many matters disagreement with the purposes and opinions of the State is criminal. Probably the freest States in the world, before the war, were America and England; yet in America no immigrant may land until he has professed disbelief in anarchism and polygamy, while in England men were sent to prison in recent years for expressing disagreement with the Christian religion or agreement with the teaching of Christ.In time of war, all criticism of the external policy of the State is criminal. Certain objects having appeared desirable to the majority, or to the effective holders of power, those who do not consider these objects desirable are exposed to pains and penalties not unlike those suffered by heretics in the past. The extent of the tyranny thus exercised is concealed by its very success: few men consider it worth while to incur a persecution which is almost certain to be thorough and effective.

Universal military service is perhaps the extreme example of the power of the State, and the supreme illustration of the difference between its attitude to its own citizens and its attitude to the citizens of other States. The State punishes, with impartial rigor, both those who kill their compatriots and those who refuse to kill foreigners. On the whole, the latter is considered the graver crime. The phenomenon of war is familiar, and men fail to realize its strangeness; to those who stand inside the cycle of instincts which lead to war it all seems natural and reasonable. But to those who stand outside the strangeness of it grows with familiarity. It is amazing that the vast majority of men should tolerate a system which compels them to submit to all the horrors of the battlefield at any moment when their Government commands them to do so. A French artist, indifferent to politics, attentive only to his painting, suddenly finds himself called upon to shoot Germans, who, his friends assure him, are a disgrace to the human race. A

German musician, equally unknowing, is called upon to shoot the perfidious Frenchman. Why cannot the two men declare a mutual neutrality? Why not leave war to those who like it and bring it on? Yet if the two men declared a mutual neutrality they would be shot by their compatriots. To avoid this fate they try to shoot each other. If the world loses the artist, not the musician, Germany rejoices; if the world loses the musician, not the artist, France rejoices. No one remembers the loss to civilization, which is equal whichever is killed.

This is the politics of Bedlam. If the artist and the musician had been allowed to stand aside from the war, nothing but unmitigated good to mankind would have resulted. The power of the State, which makes this impossible,is a wholly evil thing, quite as evil as the power of the Church which in former days put men to death for unorthodox thought. Yet if, even in time of peace, an international league were founded to consist of Frenchmen and Germans in equal numbers, all pledged not to take part in war, the French State and the German State would persecute it with equal ferocity. Blind obedience, unlimited willingness to kill and die are exacted of the modern citizens of a democracy as much as of the Janizaries of medieval sultans or the secret agents of Oriental despots.

The power of the State may be brought to bear, as it often is in England, through public opinion rather than through the laws. By oratory and the influence of the Press, public opinion is largely created by the State, and a tyrannous public opinion is as great an enemy to liberty as tyrannous laws. If the young man who will not fight finds that he is dismissed from his employment, insulted in the streets, cold-shouldered by his friends, and thrown over with scorn by any woman who may formerly have liked him, he will feel the penalty quite as hard to bear as a death sentence.A free community requires not only legal freedom, but a tolerant public opinion, an absence of that instinctive inquisition into our neighbors' affairs which, under the guise of upholding a high moral standard, enables good people to indulge unconsciously a disposition to cruelty and persecution. Thinking ill of others is not in itself a good reason for thinking well of ourselves. But so long as this is not recognized, and so long as the State can manufacture public opinion, except in the rare cases where it is revolutionary, public opinion must be reckoned as a definite part of the power of the State.

The power of the State outside its own borders is in the main derived from war or the threat of war. Some power is derived from the ability to persuade its citizens to lend money or not to lend it, but this is unimportant in comparison with the power derived from armies and navies. The external activity of the State—with exceptions so rare as to be negligible—is selfish. Sometimes selfishness is mitigated by the need of retaining the goodwill of other States, but this only modifies the methods employed, not the ends pursued. The ends pursued, apart from mere defense against other States, are, on the one hand, opportunities for successful exploitation of weak or uncivilized countries, on the other hand, power and prestige, which are considered more glorious and less material than money. In pursuit of these objects, no State hesitates to put to death innumerable foreigners whose happiness is not compatible with exploitation or subjection, or to devastate territories into which it is thought necessary to strike terror. Apart from the present war, such acts have been performed within the last twenty years by many minor States and by all the Great Powers except Austria; and in the case of Austria only the opportunity, not the will, was lacking.

Why do men acquiesce in the power of the State? There are many reasons, some traditional, some very present and pressing.

The traditional reason for obedience to the State is personal loyalty to the sovereign. European States grew up under the feudal system, and were originally the several territories owned by feudal chiefs. But this source of obedience has decayed, and probably now counts for little except in Japan, and to a lesser extent in Russia.

Tribal feeling, which always underlay loyalty to the sovereign, has remained as strong as it ever was, and is now the chief support for the power of the State. Almost every man finds it essential to his happiness to feel himself a member of a group, animated by common friendships and enmities and banded together for defense and attack. But such groups are of two kinds: there are those which are essentially enlargements of the family, and there are those which are based upon a conscious common purpose. Nations belong to the first kind, Churches to the second. At times when men are profoundly swayed by creeds national divisions tend to break down, as they did in the wars of religion after the Reformation. At such times a common creed is a stronger bond than a common nationality. To a much slighter extent, the same thing has occurred in the modern world with the rise of socialism. Men who disbelieve in private property, and feel the capitalist the real enemy, have a bond which transcends national divisions. It has not been found strong enough to resist the passions aroused by the present war, but it has made them less bitter among socialists than among others, and has kept alive the hope of a European community to be reconstructed when the war is over. In the main, however, the universal disbelief in creeds has left tribal feeling triumphant, and has made nationalism stronger than at any previous period of the world's history. A few sincere Christians, a few sincere socialists, have found in their creed a force capable of resisting the assaults of national passion, but they have been too few to influence the course of events or even to cause serious anxiety to the Governments.

It is chiefly tribal feeling that generates the unity of a national State, but it is not only tribal feeling that generates its strength. Its strength results principally from two fears, neither of which is unreasonable: the fear of crime and anarchy within, and the fear of aggression from without.

The internal orderliness of a civilized community is a great achievement, chiefly brought about by the increased authority of the State. It would be inconvenient if peaceable citizens were constantly in imminent risk of being robbed and murdered. Civilized life would become almost impossible if adventurous people could organize private armies for purposes of plunder. These conditions existed in the Middle Ages, and have not passed away without a great struggle. It is thought by many—especially by the rich, who derive the greatest advantage from law and order—that any diminution in the power of the State might bring back a condition of universal anarchy. They regard strikes as portents of dissolution. They are terrified by such organizations as the Confédération Générale du Travail and the International Workers of the World. They remember the French Revolution, and feel a not unnatural desire to keep their heads on their shoulders. They dread particularly any political theory which seems to excuse private crimes, such as sabotage and political assassination. Against these dangers they see no protection except the maintenance of the authority of the State, and the belief that all resistance to the State is wicked.

Fear of the danger within is enhanced by fear of the danger without. Every State is exposed at all times to the risk of foreign invasion. No means has hitherto been devised for minimizing this risk except the increase of armaments. But the armaments which are nominally intended to repel invasion may also be used to invade. And so the means adopted to diminish the external fear have the effect of increasing it, and of enormously enhancing the destructiveness of war when it does break out. In this way a reign of terror becomes universal,and the State acquires everywhere something of the character of the Comité du Salut Public.

The tribal feeling out of which the State develops is natural, and the fear by which the State is strengthened is reasonable under present circumstances. And in addition to these two, there is a third source of strength in a national State, namely patriotism in its religious aspect.

Patriotism is a very complex feeling, built up out of primitive instincts and highly intellectual convictions. There is love of home and family and friends, making us peculiarly anxious to preserve our own country from invasion. There is the mild instinctive liking for compatriots as against foreigners. There is pride, which is bound up with the success of the community to which we feel that we belong. There is a belief, suggested by pride but reinforced by history, that one's own nation represents a great tradition and stands for ideals that are important to the human race. But besides all these, there is another element, at once nobler and more open to attack, an element of worship, of willing sacrifice, of joyful merging of the individual life in the life of the nation. This religious element in patriotism is essential to the strength of the State, since it enlists the best that is in most men on the side of national sacrifice.

The religious element in patriotism is reinforced by education, especially by a knowledge of the history and literature of one's own country, provided it is not accompanied by much knowledge of the history and literature of other countries. In every civilized country all instruction of the young emphasizes the merits of their own nation and the faults of other nations. It comes to be universally believed that one's own nation, because of its superiority, deserves support in a quarrel, however the quarrel may have originated. This belief is so genuine and deep that it makes men endure patiently, almost gladly, the losses and hardships and sufferings entailed by war. Like all sincerely believed religions, it gives an outlook on life, based upon instinct but sublimating it, causing a devotion to an end greater than any personal end, but containing many personal ends as it were in solution.

Patriotism as a religion is unsatisfactory because of its lack of universality. The good at which it aims is a good for one's own nation only, not for all mankind. The desires which it inspires in an Englishman are not the same as the desires which it inspires in a German. A world full of patriots may be a world full of strife. The more intensely a nation believes in its patriotism, the more fanatically indifferent it will become to the damage suffered by other nations. When once men have learnt to subordinate their own good to the good of a larger whole, there can be no valid reason for stopping short of the human race. It is the admixture of national pride that makes it so easy in practice for men's impulses towards sacrifice to stop short at the frontiers of their own country. It is this admixture that poisons patriotism, and makes it inferior, as a religion, to beliefs which aim at the salvation of all mankind. We cannot avoid having more love for our own country than for other countries, and there is no reason why we should wish to avoid it, any more than we should wish to love all individual men and women equally. But any adequate religion will lead us to temper inequality of affection by love of justice, and to universalize our aims by realizing the common needs of man. This change was effected by Christianity in Judaism, and must be effected in any merely national religion before it can be purged of evil.

In practice, patriotism has many other enemies to contend with. Cosmopolitanism cannot fail to grow as men acquire more knowledge of foreign countries by education and travel. There is also a kind of individualism which is continually increasing, a realization that every man ought to be as nearly free as possible to choose his own ends, not compelled by a geographical accident to pursue ends forced upon him by the community. Socialism, syndicalism, and anti-capitalist movements generally, are against patriotism in their tendency, since they make men aware that the present State is largely concerned in defending the privileges of the rich, and that many of the conflicts between States have their origin in the financial interests of a few plutocrats. This kind of opposition is perhaps temporary, a mere incident in the struggle of labor to acquire power. Australia, where labor feels its triumph secure, is full of patriotism and militarism, based upon determination to prevent foreign labor from sharing the benefits of a privileged position. It is not unlikely that England might develop a similar nationalism if it became a socialist State. But it is probable that such nationalism would be purely defensive. Schemes of foreign aggression, entailing great loss of life and wealth in the nation which adopts them, would hardly be initiated except by those whose instincts of dominion have been sharpened through the power derived from private property and the institutions of the capitalist State.

The evil wrought in the modern world by the excessive power of the State is very great, and very little recognized.

The chief harm wrought by the State is promotion of efficiency in war. If all States increase their strength, the balance of power is unchanged, and no one State has a better chance of victory than before. And when the means of offense exist, even though their original purpose may have been defensive, the temptation to use them is likely, sooner or later, to prove overwhelming. In this way the very measures which promoted security within the borders of the State promote insecurity elsewhere. It is of the essence of the State to suppress violence within and to facilitate it without. The State makes an entirely artificial division of mankind and of our duties toward them: towards one group we are bound by the law, towards the other only by the prudence of highwaymen.The State is rendered evil by its exclusions, and by the fact that, whenever it embarks upon aggressive war, it becomes a combination of men for murder and robbery. The present system is irrational, since external and internal anarchy must be both right or both wrong. It is supported because, so long as others adopt it, it is thought the only road to safety, and because it secures the pleasures of triumph and dominion, which cannot be obtained in a good community. If these pleasures were no longer sought, or no longer possible to obtain, the problem of securing safety from invasion would not be difficult.

Apart from war, the modern great State is harmful from its vastness and the resulting sense of individual helplessness. The citizen who is out of sympathy with the aims of the State, unless he is a man of very rare gifts, cannot hope to persuade the State to adopt purposes which seem to him better. Even in a democracy, all questions except a very few are decided by a small number of officials and eminent men; and even the few questions which are left to the popular vote are decided by a diffused mass-psychology, not by individual initiative. This is especially noticeable in a country like the United States, where, in spite of democracy, most men have a sense of almost complete impotence in regard to all large issues. In so vast a country the popular will is like one of the forces of Nature, and seems nearly as much outside the control of any one man. This state of things leads, not only in America but in all large States, to something of the weariness and discouragement that we associate with the Roman Empire. Modern States, as opposed to the small city States of ancient Greece or medieval Italy, leave little room for initiative, and fail to develop in most men any sense of ability to control their political destinies. The few men who achieve power in such States are men of abnormal ambition and thirst for dominion, combined with skill in cajolery and subtlety in negotiation. All the rest are dwarfed by knowledge of their own impotence.

A curious survival from the old monarchical idea of the State is the belief that there is some peculiar wickedness in a wish to secede on the part of any section of the population. If Ireland or Poland desires independence, it is thought obvious that this desire must be strenuously resisted, and any attempt to secure it is condemned as "high treason." The only instance to the contrary that I can remember is the separation of Norway and Sweden, which was commended but not imitated. In other cases, nothing but defeat in war has induced States to part with territory: although this attitude is taken for granted, it is not one which would be adopted if the State had better ends in view. The reason for its adoption is that the chief end of almost all great States is power, especially power in war. And power in war is often increased by the inclusion of unwilling citizens. If the well-being of the citizens were the end in view, the question whether a certain area should be included, or should form a separate State, would be left freely to the decision of that area. If this principle were adopted, one of the main reasons for war would be obviated, and one of the most tyrannical elements in the State would be removed.

The principal source of the harm done by the State is the fact that power is its chief end. This is not the case in America, because America is safe against aggression;but in all other great nations the chief aim of the State is to possess the greatest possible amount of external force. To this end, the liberty of the citizens is curtailed, and anti-militarist propaganda is severely punished. This attitude is rooted in pride and fear: pride, which refuses to be conciliatory, and fear, which dreads the results of foreign pride conflicting with our own pride. It seems something of a historical accident that these two passions, which by no means exhaust the political passions of the ordinary man, should so completely determine the external policy of the State. Without pride, there would be no occasion for fear: fear on the part of one nation is due to the supposed pride of another nation. Pride of dominion, unwillingness to decide disputes otherwise than by force or the threat of force, is a habit of mind greatly encouraged by the possession of power. Those who have long been in the habit of exercising power become autocratic and quarrelsome, incapable of regarding an equal otherwise than as a rival. It is notorious that head masters' conferences are more liable to violent disagreements than most similar bodies: each head master tries to treat the others as he treats his own boys; they resent such treatment, and he resents their resentment. Men who have the habit of authority are peculiarly unfit for friendly negotiation; but the official relations of States are mainly in the hands of men with a great deal of authority in their own country. This is, of course, more particularly the case where there is a monarch who actually governs. If is less true where there is a governing oligarchy, and still less true where there is some approach to real democracy. But it is true to a considerable extent in all countries, because Prime Ministers and Foreign Secretaries are necessarily men in authority. The first step towards remedying this state of things is a genuine interest in foreign affairs on the part of the ordinary citizen, and an insistence that national pride shall not be allowed to jeopardize his other interests. During war, when he is roused, he is willing to sacrifice everything to pride; but in quiet times he will be far more ready than men in authority to realize that foreign affairs, like private concerns, ought to be settled amicably according to principles, not brutally by force or the threat of force.

The effect of personal bias in the men who actually compose the Government may be seen very clearly in labor disputes. French syndicalists affirm that the State is simply a product of capitalism, a part of the weapons which capital employs in its conflict with labor. Even in democratic States there is much to bear out this view. In strikes it is common to order out the soldiers to coerce the strikers; although the employers are much fewer, and much easier to coerce, the soldiers are never employed against them. When labor troubles paralyze the industry of a country, it is the men who are thought to be unpatriotic, not the masters, though clearly the responsibility belongs to both sides. The chief reason for this attitude on the part of Governments is that the men composing them belong, by their success if not by their origin, to the same class as the great employers of labor. Their bias and their associates combine to make them view strikes and lockouts from the standpoint of the rich. In a democracy public opinion and the need of conciliating political supporters partially correct these plutocratic influences, but the correction is always only partial. And the same influences which warp the views of Governments on labor questions also warp their views on foreign affairs, with the added disadvantage that the ordinary citizen has much fewer means of arriving at an independent judgment.

The excessive power of the State, partly through internal oppression, but principally through war and the fear of war, is one of the chief causes of misery in the modern world, and one of the main reasons for the discouragement which prevents men from growing to their full mental stature. Some means of curing this excessive power must be found if men are not to be organized into despair, as they were in the Roman Empire.

The State has one purpose which is on the whole good, namely, the substitution of law for force in the relations of men. But this purpose can only be fully achieved by a world-State, without which international relations cannot be made subject to law. And although law is better than force, law is still not the best way of settling disputes. Law is too static, too much on the side of what is decaying, too little on the side of what is growing. So long as law is in theory supreme, it will have to be tempered, from time to time, by internal revolution and external war. These can only be prevented by perpetual readiness to alter the law in accordance with the present balance of forces. If this is not done, the motives for appealing to force will sooner or later become irresistible. A world-State or federation of States, if it is to be successful, will have to decide questions, not by the legal maxims which would be applied by the Hague tribunal, but as far as possible in the same sense in which they would be decided by war. The function of authority should be to render the appeal to force unnecessary, not to give decisions contrary to those which would be reached by force.

This view may be thought by some to be immoral. It may be said that the object of civilization should be to secure justice, not to give the victory to the strong. But when this antithesis is allowed to pass, it is forgotten that love of justice may itself set force in motion. A Legislature which wishes to decide an issue in the same way as it would be decided if there were an appeal to force will necessarily take account of justice, provided justice is so flagrantly on one side that disinterested parties are willing to take up the quarrel. If a strong man assaults a weak man in the streets of London, the balance of force is on the side of the weak man, because, even if the police did not appear, casual passers-by would step in to defend him. It is sheer cant to speak of a contest of might against right, and at the same time to hope for a victory of the right. If the contest is really between might and right, that *means* that right will be beaten. What is obscurely intended, when this phrase is used, is that the stronger side is only rendered stronger by men's sense of right. But men's sense of right is very subjective, and is only one factor in deciding the preponderance of force. What is desirable in a Legislature is, not that it should decide by its personal sense of right, but that it should decide in a way which is felt to make an appeal to force unnecessary.

Having considered what the State ought not to do, I come now to what it ought to do.

Apart from war and the preservation of internal order, there are certain more positive functions which the State performs, and certain others which it ought to perform.

We may lay down two principles as regards these positive functions.

First: there are matters in which the welfare of the whole community depends upon the practically universal attainment of a certain minimum; in such cases the State has the right to insist upon this minimum being attained.

Secondly: there are ways in which, by insisting upon the maintenance of law, the State, if it does nothing further, renders possible various forms of injustice which would otherwise be prevented by the anger of their victims. Such injustices ought, as far as possible, to be prevented by the State.

The most obvious example of a matter where the general welfare depends upon a universal minimum is sanitation and the prevention of infectious diseases. A single case of plague, if it is neglected, may cause disaster to a whole community. No one can reasonably maintain, on general grounds of liberty, that a man suffering from plague ought to be left free to spread infection far and wide. Exactly similar considerations apply to drainage, notification of fevers, and kindred matters. The interference with liberty remains an evil, but in some cases it is clearly a smaller evil than the spread of disease which liberty would produce. The stamping out of malaria and yellow fever by destroying mosquitoes is perhaps the most striking example of the good which can be done in this way. But when the good is small or doubtful, and the interference with liberty is great, it becomes better to endure a certain amount of preventable disease rather than suffer a scientific tyranny.

Compulsory education comes under the same head as sanitation. The existence of ignorant masses in a population is a danger to the community; when a considerable percentage are illiterate, the whole machinery of government has to take account of the fact. Democracy in its modern form would be quite impossible in a nation where many men cannot read. But in this case there is not the same need of absolute universality as in the case of sanitary measures. The gipsies, whose mode of life has been rendered almost impossible by the education authorities, might well have been allowed to remain a picturesque exception. But apart from such rather unimportant exceptions, the argument for compulsory education is irresistible.

What the State does for the care of children at present is less than what ought to be done, not more. Children are not capable of looking after their own interests, and parental responsibility is in many ways inadequate. It is clear that the State alone can insist upon the children being provided with the minimum of knowledge and health which, for the time being, satisfies the conscience of the community.

The encouragement of scientific research is another matter which comes rightly within the powers of the State, because the benefits of discoveries accrue to the community, while the investigations are expensive and never individually certain of achieving any result. In this matter, Great Britain lags behind all other civilized countries.

The second kind of powers which the State ought to possess are those that aim at diminishing economic injustice. It is this kind that has been emphasized by socialists. The law creates or facilitates monopolies, and monopolies are able to exact a toll from the community. The most glaring example is the private ownership of land. Railways are at present controlled by the State, since rates are fixed by law; and it is clear that if they were uncontrolled, they would acquire a dangerous degree of power. Such considerations, if they stood alone, would justify complete socialism. But I think justice, by itself, is, like law, too static to be made a supreme political principle: it does not, when it has been achieved, contain any seeds of new life or any impetus to development. For this reason, when we wish to remedy an injustice, it is important to consider whether, in so doing, we shall be destroying the incentive to some form of vigorous action which is on the whole useful to the community. No such form of action, so far as I can see, is associated with private ownership of land or of any other source of economic rent; if this is the case, it follows that the State ought to be the primary recipient of rent.

If all these powers are allowed to the State, what becomes of the attempt to rescue individual liberty from its tyranny?

This is part of the general problem which confronts all those who still care for the ideals which inspired liberalism, namely the problem of combining liberty and personal initiative with organization. Politics and economics are more and more dominated by vast organizations, in face of which the individual is in danger of becoming powerless. The State is the greatest of these organizations, and the most serious menace to liberty. And yet it seems that many of its functions must be extended rather than curtailed.

There is one way by which organization and liberty can be combined, and that is, by securing power for voluntary organizations, consisting of men who have chosen to belong to them because they embody some purpose which all their members consider important, not a purpose imposed by accident or outside force. The State, being geographical, cannot be a wholly voluntary association, but for that very reason there is need of a strong public opinion to restrain it from a tyrannical use of its powers. This public opinion, in most matters, can only be secured by combinations of those who have certain interests or desires in common.

The positive purposes of the State, over and above the preservation of order, ought as far as possible to be carried out, not by the State itself, but by independent organizations, which should be left completely free so long as they satisfied the State that they were not falling below a necessary minimum. This occurs to a certain limited extent at present in regard to elementary education. The universities, also, may be regarded as acting for the State in the matter of higher education and research, except that in their case no minimum of achievement is exacted. In the economic sphere, the State ought to exercise control, but ought to leave initiative to others. There is every reason to multiply opportunities of initiative, and to give the greatest possible share of initiative to each individual, for if this is not done there will be a general sense of impotence and discouragement.There ought to be a constant endeavor to leave the more positive aspects of government in the hands of voluntary organizations, the purpose of the State being merely to exact efficiency and to secure an amicable settlement of disputes, whether within or without its own borders. And with this ought to be combined the greatest possible toleration of exceptions and the least possible insistence upon uniform system.

A good deal may be achieved through local government by trades as well as by areas. This is the most original idea in syndicalism, and it is valuable as a check upon the tyranny which the community may be tempted to exercise over certain classes of its members. All strong organizations which embody a sectional public opinion, such as trade unions, coöperative societies, professions, and universities, are to be welcomed as safeguards of liberty and opportunities for initiative. And there is need of a strong public opinion in favor of liberty itself. The old battles for freedom of thought and freedom of speech, which it was thought had been definitively won, will have to be fought all over again, since most men are only willing to accord freedom to opinions which happen to be popular.Institutions cannot preserve liberty unless men realize that liberty is precious and are willing to exert themselves to keep it alive.

There is a traditional objection to every *imperium in imperio*, but this is only the jealousy of the tyrant. In actual fact, the modern State contains many organizations which it cannot defeat, except perhaps on rare occasions when public opinion is roused against them. Mr. Lloyd George's long fight with the medical profession over the Insurance Act was full of Homeric fluctuations of fortune. The Welsh miners recently routed the whole power of the State, backed by an excited nation. As for the financiers, no Government would dream of a conflict with them. When all other classes are exhorted to patriotism, they are allowed their 4½ per cent. and an increase of interest on their consols. It is well understood on all sides that an appeal to *their* patriotism would show gross ignorance of the world. It is against the traditions of the State to extort their money by threatening to withdraw police protection. This is not due to the difficulty of such a measure, but only to the fact that great wealth wins genuine admiration from us all, and we cannot bear to think of a very rich man being treated with disrespect.

The existence of strong organizations within the State, such as trade unions, is not undesirable except from the point of view of the official who wishes to wield unlimited power, or of the rival organizations, such as federations of employers, which would prefer a disorganized adversary. In view of the vastness of the State, most men can find little political outlet for initiative except in subordinate organizations formed for specific purposes. Without an outlet for political initiative, men lose their social vigor and their interest in public affairs: they become a prey to corrupt wire-pullers, or to sensation-mongers who have the art of capturing a tired and vagrant attention. The cure for this is to increase rather than diminish the powers of voluntary organizations, to give every man a sphere of political activity small enough for his interest and his capacity, and to confine the functions of the State, as far as possible, to the maintenance of peace among rival interests. The essential merit of the State is that it prevents the internal use of force by private persons. Its essential demerits are that it promotes the external use of force, and that, by its great size, it makes each individual feel impotent even in a democracy. I shall return in a later lecture to the question of preventing war. The prevention of the sense of individual impotence cannot be achieved by a return to the small City State, which would be as reactionary as a return to the days before machinery. It must be achieved by a method which is in the direction of present tendencies. Such a method would be the increasing devolution of positive political initiative to bodies formed voluntarily for specific purposes, leaving the State rather in the position of a federal authority or a court of arbitration. The State will then confine itself to insisting upon *some* settlement of rival interests: its only principle in deciding what is the right settlement will be an attempt to find the measure most acceptable, on the whole, to all the parties concerned. This is the direction in which democratic States naturally tend, except in so far as they are turned aside by war or the fear of war. So long as war remains a daily imminent danger, the State will remain a Moloch, sacrificing sometimes the life of the individual, and always his unfettered development, to the barren struggle for mastery in the competition with other States. In internal as in external affairs, the worst enemy of freedom is war.

III
WAR AS AN INSTITUTION

IN spite of the fact that most nations at most times, are at peace, war is one of the permanent institutions of all free communities, just as Parliament is one of our permanent institutions in spite of the fact that it is not always sitting. It is war as a permanent institution that I wish to consider: why men tolerate it; why they ought not to tolerate it; what hope there is of their coming not to tolerate it; and how they could abolish it if they wished to do so.

War is a conflict between two groups, each of which attempts to kill and maim as many as possible of the other group in order to achieve some object which it desires. The object is generally either power or wealth. It is a pleasure to exercise authority over other men, and it is a pleasure to live on the produce of other men's labor. The victor in war can enjoy more of these delights than the vanquished. But war, like all other natural activities, is not so much prompted by the end which it has in view as by an impulse to the activity itself. Very often men desire an end, not on its own account, but because their nature demands the actions which will lead to the end. And so it is in this case: the ends to be achieved by war appear in prospect far more important than they will appear when they are realized, because war itself is a fulfilment of one side of our nature. If men's actions sprang from desires for what would in fact bring happiness, the purely rational arguments against war would have long ago put an end to it. What makes war difficult to suppress is that it springs from an impulse, rather than from a calculation of the advantages to be derived from war.

War differs from the employment of force by the police through the fact that the actions of the police are ordered by a neutral authority, whereas in war it is the parties to the dispute themselves who set force in motion. This distinction is not absolute, since the State is not always wholly neutral in internal disturbances. When strikers are shot down, the State is taking the side of the rich. When opinions adverse to the existing State are punished, the State is obviously one of the parties to the dispute. And from the suppression of individual opinion up to civil war all gradations are possible. But broadly speaking, force employed according to laws previously laid down by the community as a whole may be distinguished from force employed by one community against another on occasions of which the one community is the sole judge. I have dwelt upon this difference because I do not think the use of force by the police can be wholly eliminated, and I think a similar use of force in international affairs is the best hope of permanent peace. At present, international affairs are regulated by the principle that a nation must not intervene unless its interests are involved: diplomatic usage forbids intervention for the mere maintenance of international law. America may protest when American citizens are drowned by German submarines, but must not protest when no American citizens are involved. The case would be analogous in internal affairs if the police would only interfere with murder when it happened that a policeman had been killed. So long as this principle prevails in the relations of States, the power of neutrals cannot be effectively employed to prevent war.

In every civilized country two forces coöperate to produce war. In ordinary times some men—usually a small proportion of the population—are bellicose: they predict war, and obviously are not unhappy in the prospect. So long as war is not imminent, the bulk of the population pay little attention to these men, and do not actively either support or oppose them. But when war begins to seem very near, a war fever seizes hold of people, and those who were already bellicose find themselves enthusiastically supported by all but an insignificant minority. The impulses which inspire war fever are rather different from those which make some men bellicose in ordinary times. Only educated men are likely to be warlike at ordinary times, since they alone are vividly aware of other countries or of the part which their own nation might play in the affairs of the world. But it is only their knowledge, not their nature, that distinguishes them from their more ignorant compatriots.

To take the most obvious example, German policy, in recent years before the war, was not averse from war, and not friendly to England. It is worth while to try to understand the state of mind from which this policy sprang.

The men who direct German policy are, to begin with, patriotic to an extent which is almost unknown in France and England. The interests of Germany appear to them unquestionably the only interests they need take into account. What injury may, in pursuing those interests, be done to other nations, what destruction may be brought upon populations and cities, what irreparable damage may result to civilization, it is not for them to consider. If they can confer what they regard as benefits upon Germany, everything else is of no account.

The second noteworthy point about German policy is that its conception of national welfare is mainly competitive. It is not the *intrinsic* wealth of Germany, whether materially or mentally, that the rulers of Germany consider important: it is the *comparative* wealth in the competition with other civilized countries. For this reason the destruction of good things abroad appears to them almost as desirable as the creation of good things in Germany. In most parts of the world the French are regarded as the most civilized of nations: their art and their literature and their way of life have an attraction for foreigners which those of Germany do not have. The English have developed political liberty, and the art of maintaining an Empire with a minimum of coercion, in a way for which Germany, hitherto, has shown no aptitude. These are grounds for envy, and envy wishes to destroy what is good in other countries. German militarists, quite rightly, judged that what was best in France and England would probably be destroyed by a great war, even if France and England were not in the end defeated in the actual fighting. I have seen a list of young French writers killed on the battlefield; probably the German authorities have also seen it, and have reflected with joy that another year of such losses will destroy French literature for a generation—perhaps, through loss of tradition, for ever. Every outburst against liberty in our more bellicose newspapers, every incitement to persecution of defenseless Germans, every mark of growing ferocity in our attitude, must be read with delight by German patriots, as proving their success in robbing us of our best, and in forcing us to imitate whatever is worst in Prussia.

But what the rulers of Germany have envied us most was power and wealth—the power derived from command of the seas and the straits,the wealth derived from a century of industrial supremacy. In both these respects they feel that their deserts are higher than ours. They have devoted far more thought and skill to military and industrial organization. Their average of intelligence and knowledge is far superior; their capacity for pursuing an attainable end, unitedly and with forethought, is infinitely greater. Yet we, merely (as they think) because we had a start in the race, have achieved a vastly larger Empire than they have, and an enormously greater control of capital. All this is unbearable; yet nothing but a great war can alter it.

Besides all these feelings, there is in many Germans, especially in those who know us best, a hot hatred of us on account of our pride. Farinata degli Uberti surveyed Hell "*come avesse lo Inferno in gran dispitto*." Just so, by German accounts, English officer prisoners look round them among their captors—holding aloof, as though the enemy were noxious, unclean creatures, toads or slugs or centipedes, which a man does not touch willingly, and shakes off with loathing if he is forced to touch them for a moment. It is easy to imagine how the devils hated Farinata, and inflicted greater pains upon him than upon his neighbors, hoping to win recognition by some slight wincing on his part, driven to frenzy by his continuing to behave as if they did not exist. In just the same way the Germans are maddened by our spiritual immobility. At bottom we have regarded the Germans as one regards flies on a hot day: they are a nuisance, one has to brush them off, but it would not occur to one to be turned aside by them. Now that the initial certainty of victory has faded, we begin to be affected inwardly by the Germans. In time, if we continue to fail in our military enterprises, we shall realize that they are human beings, not just a tiresome circumstance. Then perhaps we shall hate them with a hatred which they will have no reason to resent. And from such a hatred it will be only a short journey to a genuine *rapprochement*.

The problem which must be solved, if the future of the world is to be less terrible than its present, is the problem of preventing nations from getting into the moods of England and Germany at the outbreak of the war. These two nations as they were at that moment might be taken as almost mythical representatives of pride and envy—cold pride and hot envy. Germany declaimed passionately: “You, England, swollen and decrepit, you overshadow my whole growth—your rotting branches keep the sun from shining upon me and the rain from nourishing me. Your spreading foliage must be lopped, your symmetrical beauty must be destroyed, that I too may have freedom to grow, that my young vigor may no longer be impeded by your decaying mass.” England, bored and aloof, unconscious of the claims of outside forces, attempted absent-mindedly to sweep away the upstart disturber of meditation; but the upstart was not swept away, and remains so far with every prospect of making good his claim. The claim and the resistance to it are alike folly. Germany had no good ground for envy; we had no good ground for resisting whatever in Germany’s demands was compatible with our continued existence. Is there any method of averting such reciprocal folly in the future?

I think if either the English or the Germans were capable of thinking in terms of individual welfare rather than national pride, they would have seen that, at every moment during the war the wisest course would have been to conclude peace at once, on the best terms that could have been obtained. This course, I am convinced, would have been the wisest for each separate nation, as well as for civilization in general. The utmost evil that the enemy could inflict through an unfavorable peace would be a trifle compared to the evil which all the nations inflict upon themselves by continuing to fight. What blinds us to this obvious fact is pride, the pride which makes the acknowledgment of defeat intolerable, and clothes itself in the garb of reason by suggesting all kinds of evils which are supposed to result from admitting defeat. But the only real evil of defeat is humiliation, and humiliation is subjective; we shall not feel humiliated if we become persuaded that it was a mistake to engage in the war, and that it is better to pursue other tasks not dependent upon world-dominion. If either the English or the Germans could admit this inwardly, any peace which did not destroy national independence could be accepted without real loss in the self-respect which is essential to a good life.

The mood in which Germany embarked upon the war was abominable, but it was a mood fostered by the habitual mood of England. We have prided ourselves upon our territory and our wealth; we have been ready at all times to defend by force of arms what we have conquered in India and Africa. If we had realized the futility of empire, and had shown a willingness to yield colonies to Germany without waiting for the threat of force, we might have been in a position to persuade the Germans that their ambitions were foolish, and that the respect of the world was not to be won by an imperialist policy. But by our resistance we showed that we shared their standards. We, being in possession, became enamored of the *status quo*. The Germans were willing to make war to upset the *status quo*; we were willing to make war to prevent its being upset in Germany's favor. So convinced were we of the sacredness of the *status quo* that we never realized how advantageous it was to us, or how, by insisting upon it, we shared the responsibility for the war. In a world where nations grow and decay, where forces change and populations become cramped, it is not possible or desirable to maintain the *status quo* for ever. If peace is to be preserved, nations must learn to accept unfavorable alterations of the map without feeling that they must first be defeated in war, or that in yielding they incur a humiliation.

It is the insistence of legalists and friends of peace upon the maintenance of the *status quo* that has driven Germany into militarism. Germany had as good a right to an Empire as any other Great Power, but could only acquire an Empire through war. Love of peace has been too much associated with a static conception of international relations. In economic disputes we all know that whatever is vigorous in the wage-earning classes is opposed to "industrial peace," because the existing distribution of wealth is felt to be unfair. Those who enjoy a privileged position endeavor to bolster up their claims by appealing to the desire for peace, and decrying those who promote strife between the classes. It never occurs to them that by opposing changes without considering whether they are just, the capitalists share the responsibility for the class war. And in exactly the same way England shares the responsibility for Germany's war. If actual war is ever to cease there will have to be political methods of achieving the results which now can only be achieved by successful fighting, and nations will have voluntarily to admit adverse claims which appear just in the judgment of neutrals.

It is only by some such admission, embodying itself in a Parliament of the nations with full power to alter the distribution of territory, that militarism can be permanently overcome. It may be that the present war will bring, in the Western nations, a change of mood and outlook sufficient to make such an institution possible. It may be that more wars and more destruction will be necessary before the majority of civilized men rebel against the brutality and futile destruction of modern war. But unless our standards of civilization and our powers of constructive thought are to be permanently lowered, I cannot doubt that, sooner or later, reason will conquer the blind impulses which now lead nations into war. And if a large majority of the Great Powers had a firm determination that peace should be preserved, there would be no difficulty in devising diplomatic machinery for the settlement of disputes, and in establishing educational systems which would implant in the minds of the young an ineradicable horror of the slaughter which they are now taught to admire.

Besides the conscious and deliberate forces leading to war, there are the inarticulate feelings of common men, which, in most civilized countries, are always ready to burst into war fever at the bidding of statesmen. If peace is to be secure, the readiness to catch war fever must be somehow diminished. Whoever wishes to succeed in this must first understand what war fever is and why it arises.

The men who have an important influence in the world, whether for good or evil, are dominated as a rule by a threefold desire: they desire, first, an activity which calls fully into play the faculties in which they feel that they excel; secondly, the sense of successfully overcoming resistance; thirdly, the respect of others on account of their success. The third of these desires is sometimes absent: some men who have been great have been without the "last infirmity," and have been content with their own sense of success, or merely with the joy of difficult effort. But as a rule all three are present. Some men's talents are specialized, so that their choice of activities is circumscribed by the nature of their faculties; other men have, in youth, such a wide range of possible aptitudes that their choice is chiefly determined by the varying degrees of respect which public opinion gives to different kinds of success.

The same desires, usually in a less marked degree, exist in men who have no exceptional talents. But such men cannot achieve anything very difficult by their individual efforts; for them, as units, it is impossible to acquire the sense of greatness or the triumph of strong resistance overcome. Their separate lives are unadventurous and dull. In the morning they go to the office or the plow, in the evening they return tired and silent, to the sober monotony of wife and children. Believing that security is the supreme good, they have insured against sickness and death, and have found an employment where they have little fear of dismissal and no hope of any great rise. But security, once achieved, brings a Nemesis of ennui. Adventure, imagination, risk, also have their claims; but how can these claims be satisfied by the ordinary wage-earner? Even if it were possible to satisfy them, the claims of wife and children have priority and must not be neglected.

To this victim of order and good organization the realization comes, in some moment of sudden crisis, that he belongs to a nation, that his nation may take risks, may engage in difficult enterprises, enjoy the hot passion of doubtful combat, stimulate adventure and imagination by military expeditions to Mount Sinai and the Garden of Eden. What his nation does, in some sense, he does; what his nation suffers, he suffers. The long years of private caution are avenged by a wild plunge into public madness. All the horrid duties of thrift and order and care which he has learnt to fulfil in private are thought not to apply to public affairs: it is patriotic and noble to be reckless for the nation, though it would be wicked to be reckless for oneself. The old primitive passions, which civilization has denied, surge up all the stronger for repression. In a moment imagination and instinct travel back through the centuries, and the wild man of the woods emerges from the mental prison in which he has been confined. This is the deeper part of the psychology of the war fever.

But besides the irrational and instinctive element in the war fever, there is always also, if only as a liberator of primitive impulse, a certain amount of quasi-rational calculation and what is euphemistically called "thought." The war fever very seldom seizes a nation unless it believes that it will be victorious. Undoubtedly, under the influence of excitement, men over-estimate their chances of success; but there is some proportion between what is hoped and what a rational man would expect. Holland, though quite as humane as England, had no impulse to go to war on behalf of Belgium, because the likelihood of disaster was so obviously overwhelming. The London populace, if they had known how the war was going to develop, would not have rejoiced as they did on that August Bank Holiday long ago. A nation which has had a recent experience of war, and has come to know that a war is almost always more painful than it is expected to be at the outset, becomes much less liable to war fever until a new generation grows up. The element of rationality in war fever is recognized by Governments and journalists who desire war, as may be seen by their invariably minimizing the perils of a war which they wish to provoke. At the beginning of the South African War Sir William Butler was dismissed, apparently for suggesting that sixty thousand men and three months might not suffice to subdue the Boer Republics. And when the war proved long and difficult, the nation turned against those who had made it. We may assume, I think, without attributing too great a share to reason in human affairs, that a nation would not suffer from war fever in a case where every sane man could see that defeat was very probable.

The importance of this lies in the fact that it would make aggressive war very unlikely if its chances of success were very small. If the peace-loving nations were sufficiently strong to be obviously capable of defeating the nations which were willing to wage aggressive war, the peace-loving nations might form an alliance and agree to fight jointly against any nation which refused to submit its claims to an International Council. Before the present war we might have reasonably hoped to secure the peace of the world in some such way; but the military strength of Germany has shown that such a scheme has no great chance of success at present. Perhaps at some not far distant date it may be made more feasible by developments of policy in America.

The economic and political forces which make for war could be easily curbed if the will to peace existed strongly in all civilized nations. But so long as the populations are liable to war fever, all work for peace must be precarious; and if war fever could not be aroused, political and economic forces would be powerless to produce any long or very destructive war. The fundamental problem for the pacifist is to prevent the impulse towards war which seizes whole communities from time to time. And this can only be done by far-reaching changes in education, in the economic structure of society, and in the moral code by which public opinion controls the lives of men and women.

A great many of the impulses which now lead nations to go to war are in themselves essential to any vigorous or progressive life. Without imagination and love of adventure a society soon becomes stagnant and begins to decay. Conflict, provided it is not destructive and brutal, is necessary in order to stimulate men's activities, and to secure the victory of what is living over what is dead or merely traditional. The wish for the triumph of one's cause, the sense of solidarity with large bodies of men, are not things which a wise man will wish to destroy. It is only the outcome in death and destruction and hatred that is evil. The problem is, to keep these impulses, without making war the outlet for them.

All Utopias that have hitherto been constructed are intolerably dull. Any man with any force in him would rather live in this world, with all its ghastly horrors, than in Plato's Republic or among Swift's Houyhnhnms. The men who make Utopias proceed upon a radically false assumption as to what constitutes a good life. They conceive that it is possible to imagine a certain state of society and a certain way of life which should be once for all recognized as good, and should then continue for ever and ever. They do not realize that much the greater part of a man's happiness depends upon activity, and only a very small remnant consists in passive enjoyment. Even the pleasures which do consist in enjoyment are only satisfactory, to most men, when they come in the intervals of activity. Social reformers, like inventors of Utopias, are apt to forget this very obvious fact of human nature. They aim rather at securing more leisure, and more opportunity for enjoying it, than at making work itself more satisfactory, more consonant with impulse, and a better outlet for creativeness and the desire to employ one's faculties. Work, in the modern world, is, to almost all who depend on earnings, mere work, not an embodiment of the desire for activity. Probably this is to a considerable extent inevitable. But in so far as it can be prevented something will be done to give a peaceful outlet to some of the impulses which lead to war.

It would, of course, be easy to bring about peace if there were no vigor in the world. The Roman Empire was pacific and unproductive; the Athens of Pericles was the most productive and almost the most warlike community known to history. The only form of production in which our own age excels is science, and in science Germany, the most warlike of Great Powers, is supreme. It is useless to multiply examples; but it is plain that the very same vital energy which produces all that is best also produces war and the love of war. This is the basis of the opposition to pacifism felt by many men whose aims and activities are by no means brutal. Pacifism, in practice, too often expresses merely lack of force, not the refusal to use force in thwarting others. Pacifism, if it is to be both victorious and beneficent, must find an outlet, compatible with humane feeling, for the vigor which now leads nations into war and destruction.

This problem was considered by William James in an admirable address on "The Moral Equivalent of War," delivered to a congress of pacifists during the Spanish-American War of 1898. His statement of the problem could not be bettered; and so far as I know, he is the only writer who has faced the problem adequately. But his solution is not adequate; perhaps no adequate solution is possible. The problem, however, is one of degree: every additional peaceful outlet for men's energies diminishes the force which urges nations towards war, and makes war less frequent and less fierce. And as a question of degree, it is capable of more or less partial solutions.

Every vigorous man needs some kind of contest, some sense of resistance overcome, in order to feel that he is exercising his faculties. Under the influence of economics, a theory has grown up that what men desire is wealth; this theory has tended to verify itself, because people's actions are more often determined by what they think they desire than by what they really desire. The less active members of a community often do in fact desire wealth, since it enables them to gratify a taste for passive enjoyment, and to secure respect without exertion. But the energetic men who make great fortunes seldom desire the actual money: they desire the sense of power through a contest, and the joy of successful activity. For this reason, those who are the most ruthless in making money are often the most willing to give it away; there are many notorious examples of this among American millionaires. The only element of truth in the economic theory that these men are actuated by desire for money is this: owing to the fact that money is what is *believed* to be desirable, the making of money is recognized as the test of success. What is desired is visible and indubitable success; but this can only be achieved by being one of the few who reach a goal which many men would wish to reach. For this reason, public opinion has a great influence in directing the activities of vigorous men. In America a millionaire is more respected than a great artist; this leads men who might become either the one or the other to choose to become millionaires. In Renaissance Italy great artists were more respected than millionaires, and the result was the opposite of what it is in America.

Some pacifists and all militarists deprecate social and political conflicts. In this the militarists are in the right, from their point of view; but the pacifists seem to me mistaken. Conflicts of party politics, conflicts between capital and labor, and generally all those conflicts of principle which do not involve war, serve many useful purposes, and do very little harm. They increase men's interest in public affairs, they afford a comparatively innocent outlet for the love of contest, and they help to alter laws and institutions, when changing conditions or greater knowledge create the wish for an alteration. Everything that intensifies political life tends to bring about a peaceful interest of the same kind as the interest which leads to desire for war. And in a democratic community political questions give every voter a sense of initiative and power and responsibility which relieves his life of something of its narrow unadventurousness. The object of the pacifist should be to give men more and more political control over their own lives, and in particular to introduce democracy into the management of industry, as the syndicalists advise.

The problem for the reflective pacifist is two-fold: how to keep his own country at peace, and how to preserve the peace of the world. It is impossible that the peace of the world should be preserved while nations are liable to the mood in which Germany entered upon the war—unless, indeed, one nation were so obviously stronger than all others combined as to make war unnecessary for that one and hopeless for all the others. As this war has dragged on its weary length, many people must have asked themselves whether national independence is worth the price that has to be paid for it. Would it not perhaps be better to secure universal peace by the supremacy of one Power? "To secure peace by a world federation"—so a submissive pacifist may argue—"would require some faint glimmerings of reason in rulers and peoples, and is therefore out of the question; but to secure it by allowing Germany to dictate terms to Europe would be easy, in view of Germany's amazing military success. Since there is no other way of ending war"—so our advocate of peace at any price would contend—"let us adopt this way, which happens at the moment to be open to us." It is worth while to consider this view more attentively than is commonly considered.

There is one great historic example of a long peace secured in this way; I mean the Roman Empire. We in England boast of the *Pax Britannica* which we have imposed, in this way, upon the warring races and religions in India. If we are right in boasting of this, if we have in fact conferred a benefit upon India by enforced peace, the Germans would be right in boasting if they could impose a *Pax Germanica* upon Europe. Before the war, men might have said that India and Europe are not analogous, because India is less civilized than Europe; but now, I hope, no one would have the effrontery to maintain anything so preposterous. Repeatedly in modern history there has been a chance of achieving European unity by the hegemony of a single State; but always England, in obedience to the doctrine of the Balance of Power, has prevented this consummation, and preserved what our statesmen have called the "liberties of Europe." It is this task upon which we are now engaged. But I do not think our statesmen, or any others among us, have made much effort to consider whether the task is worth what it costs.

In one case we were clearly wrong: in our resistance to revolutionary France. If revolutionary France could have conquered the Continent and Great Britain, the world would now be happier, more civilized, and more free, as well as more peaceful. But revolutionary France was a quite exceptional case, because its early conquests were made in the name of liberty, against tyrants, not against peoples; and everywhere the French armies were welcomed as liberators by all except rulers and bigots. In the case of Philip II we were as clearly right as we were wrong in 1793. But in both cases our action is not to be judged by some abstract diplomatic conception of the "liberties of Europe," but by the ideals of the Power seeking hegemony, and by the probable effect upon the welfare of ordinary men and women throughout Europe.

"Hegemony" is a very vague word, and everything turns upon the degree of interference with liberty which it involves. There is a degree of interference with liberty which is fatal to many forms of national life; for example, Italy in the seventeenth and eighteenth centuries was crushed by the supremacy of Spain and Austria. If the Germans were actually to annex French provinces, as they did in 1871, they would probably inflict a serious injury upon those provinces, and make them less fruitful for civilization in general. For such reasons national liberty is a matter of real importance, and a Europe actually governed by Germany would probably be very dead and unproductive. But if "hegemony" merely means increased weight in diplomatic questions, more coaling stations and possessions in Africa, more power of securing advantageous commercial treaties, then it can hardly be supposed that it would do any vital damage to other nations; certainly it would not do so much damage as the present war is doing. I cannot doubt that, before the war, a hegemony of this kind would have abundantly satisfied the Germans. But the effect of the war, so far, has been to increase immeasurably all the dangers which it was intended to avert. We have now only the choice between certain exhaustion of Europe in fighting Germany and possible damage to the national life of France by German tyranny. Stated in terms of civilization and human welfare, not in terms of national prestige, that is now in fact the issue.

Assuming that war is not ended by one State conquering all the others, the only way in which it can be permanently ended is by a world-federation. So long as there are many sovereign States, each with its own Army, there can be no security that there will not be war. There will have to be in the world only one Army and one Navy before there will be any reason to think that wars have ceased. This means that, so far as the military functions of the State are concerned, there will be only one State, which will be world-wide.

The civil functions of the State—legislative, administrative, and judicial—have no very essential connection with the military functions, and there is no reason why both kinds of functions should normally be exercised by the same State. There is, in fact, every reason why the civil State and the military State should be different. The greater modern States are already too large for most civil purposes, but for military purposes they are not large enough, since they are not world-wide. This difference as to the desirable area for the two kinds of State introduces a certain perplexity and hesitation, when it is not realized that the two functions have little necessary connection: one set of considerations points towards small States, the other towards continually larger States. Of course, if there were an international Army and Navy, there would have to be some international authority to set them in motion. But this authority need never concern itself with any of the internal affairs of national States: it need only declare the rules which should regulate their relations, and pronounce judicially when those rules have been so infringed as to call for the intervention of the international force. How easily the limit of the authority could be fixed may be seen by many actual examples.

The civil and military State are often different in practice, for many purposes. The South American Republics are sovereign for all purposes except their relations with Europe, in regard to which they are subject to the United States: in dealings with Europe, the Army and Navy of the United States are their Army and Navy. Our self-governing Dominions depend for their defense, not upon their own forces but upon our Navy. Most Governments, nowadays, do not aim at formal annexation of a country which they wish to incorporate, but only at a protectorate—that is, civil autonomy subject to military control. Such autonomy is, of course, in practice incomplete, because it does not enable the "protected" country to adopt measures which are vetoed by the Power in military control. But it may be very nearly complete, as in the case of our self-governing Dominions. At the other extreme, it may become a mere farce, as in Egypt. In the case of an alliance, there is complete autonomy of the separate allied countries, together with what is practically a combination of their military forces into one single force.

The great advantage of a large military State is that it increases the area over which internal war is not possible except by revolution. If England and Canada have a disagreement, it is taken as a matter of course that a settlement shall be arrived at by discussion, not by force. Still more is this the case if Manchester and Liverpool have a quarrel, in spite of the fact that each is autonomous for many local purposes. No one would have thought it reasonable that Liverpool should go to war to prevent the construction of the Manchester Ship Canal, although almost any two Great Powers would have gone to war over an issue of the same relative importance. England and Russia would probably have gone to war over Persia if they had not been allies; as it is, they arrived by diplomacy at much the same iniquitous result as they would otherwise have reached by fighting. Australia and Japan would probably fight if they were both completely independent; but both depend for their liberties upon the British Navy, and therefore they have to adjust their differences peaceably.

The chief disadvantage of a large military State is that, when external war occurs, the area affected is greater. The quadruple Entente forms, for the present, one military State; the result is that, because of a dispute between Austria and Serbia, Belgium is devastated and Australians are killed in the Dardanelles. Another disadvantage is that it facilitates oppression. A large military State is practically omnipotent against a small State, and can impose its will, as England and Russia did in Persia and as Austria-Hungary has been doing in Serbia. It is impossible to make sure of avoiding oppression by any purely mechanical guarantees; only a liberal and humane spirit can afford a real protection. It has been perfectly possible for England to oppress Ireland, in spite of democracy and the presence of Irish Members at Westminster. Nor has the presence of Poles in the Reichstag prevented the oppression of Prussian Poland. But democracy and representative government undoubtedly make oppression less probable: they afford a means by which those who might be oppressed can cause their wishes and grievances to be publicly known, they render it certain that only a minority can be oppressed, and then only if the majority are nearly unanimous in wishing to oppress them. Also the practice of oppression affords much more pleasure to the governing classes, who actually carry it out, than to the mass of the population. For this reason the mass of the population, where it has power, is likely to be less tyrannical than an oligarchy or a bureaucracy.

In order to prevent war and at the same time preserve liberty it is necessary that there should be only one military State in the world, and that when disputes between different countries arise, it should act according to the decision of a central authority. This is what would naturally result from a federation of the world, if such a thing ever came about. But the prospect is remote, and it is worth while to consider why it is so remote.

The unity of a nation is produced by similar habits, instinctive liking, a common history, and a common pride. The unity of a nation is partly due to intrinsic affinities between its citizens, but partly also to the pressure and contrast of the outside world: if a nation were isolated, it would not have the same cohesion or the same fervor of patriotism. When we come to alliances of nations, it is seldom anything except outside pressure that produces solidarity. England and America, to some extent, are drawn together by the same causes which often make national unity: a (more or less) common language, similar political institutions, similar aims in international politics. But England, France, and Russia were drawn together solely by fear of Germany; if Germany had been annihilated by a natural cataclysm, they would at once have begun to hate one another, as they did before Germany was strong. For this reason, the possibility of coöperation in the present alliance against Germany affords no ground whatever for hoping that all the nations of the world might coöperate permanently in a peaceful alliance. The present motive for cohesion, namely a common fear, would be gone, and could not be replaced by any other motive unless men's thoughts and purposes were very different from what they are now.

The ultimate fact from which war results is not economic or political, and does not rest upon any mechanical difficulty of inventing means for the peaceful settlement of international disputes. The ultimate fact from which war results is the fact that a large proportion of mankind have an impulse to conflict rather than harmony, and can only be brought to coöperate with others in resisting or attacking a common enemy. This is the case in private life as well as in the relations of States. Most men, when they feel themselves sufficiently strong, set to work to make themselves feared rather than loved; the wish to gain the good opinion of others is confined, as a rule, to those who have not acquired secure power. The impulse to quarreling and self-assertion, the pleasure of getting one's own way in spite of opposition, is native to most men. It is this impulse, rather than any motive of calculated self-interest, which produces war, and causes the difficulty of bringing about a World-State. And this impulse is not confined to one nation; it exists, in varying degrees, in all the vigorous nations of the world.

But although this impulse is strong, there is no reason why it should be allowed to lead to war. It was exactly the same impulse which led to duelling; yet now civilized men conduct their private quarrels without bloodshed. If political contest within a World-State were substituted for war, imagination would soon accustom itself to the new situation, as it has accustomed itself to absence of duelling. Through the influence of institutions and habits, without any fundamental change in human nature, men would learn to look back upon war as we look upon the burning of heretics or upon human sacrifice to heathen deities. If I were to buy a revolver costing several pounds, in order to shoot my friend with a view to stealing sixpence out of his pocket, I should be thought neither very wise nor very virtuous. But if I can get sixty-five million accomplices to join me in this criminal absurdity, I become one of a great and glorious nation, nobly sacrificing the cost of my revolver, perhaps even my life, in order to secure the sixpence for the honor of my country. Historians, who are almost invariably sycophants, will praise me and my accomplices if we are successful, and say that we are worthy successors of the heroes who overthrew the might of Imperial Rome. But if my opponents are victorious, if their sixpences are defended at the cost of many pounds each and the lives of a large proportion of the population, then historians will call me a brigand (as I am), and praise the spirit and self-sacrifice of those who resisted me.

War is surrounded with glamour, by tradition, by Homer and the Old Testament, by early education, by elaborate myths as to the importance of the issues involved, by the heroism and self-sacrifice, which these myths call out. Jephthah sacrificing his daughter is a heroic figure, but he would have let her live if he had not been deceived by a myth. Mothers sending their sons to the battlefield are heroic, but they are as much deceived as Jephthah. And, in both cases alike, the heroism which issues in cruelty would be dispelled if there were not some strain of barbarism in the imaginative outlook from which myths spring. A God who can be pleased by the sacrifice of an innocent girl could only be worshiped by men to whom the thought of receiving such a sacrifice is not wholly abhorrent. A nation which believes that its welfare can only be secured by suffering and inflicting hundreds of thousands of equally horrible sacrifices, is a nation which has no very spiritual conception of what constitutes national welfare. It would be better a hundredfold to forgo material comfort, power, pomp, and outward glory than to kill and be killed, to hate and be hated, to throw away in a mad moment of fury the bright heritage of the ages. We have learnt gradually to free our God from the savagery with which the primitive Israelites and the Fathers endowed Him: few of us now believe that it is His pleasure to torture most of the human race in an eternity of hell-fire. But we have not yet learnt to free our national ideals from the ancient taint. Devotion to the nation is perhaps the deepest and most widespread religion of the present age. Like the ancient religions, it demands its persecutions, its holocausts, its lurid heroic cruelties; like them, it is noble, primitive, brutal, and mad. Now, as in the past, religion, lagging behind private consciences through the weight of tradition, steels the hearts of men against mercy and their minds against truth. If the world is to be saved, men must learn to be noble without being cruel, to be filled with faith and yet open to truth, to be inspired by great purposes without hating those who try to thwart them. But before this can happen, men must first face the terrible realization that the gods before whom they have bowed down were false gods and the sacrifices, they have made were vain.

IV
PROPERTY

AMONG the many gloomy novelists of the realistic school, perhaps the most full of gloom is Gissing. In common with all his characters, he lives under the weight of a great oppression: the power of the fearful and yet adored idol of Money. One of his typical stories is "Eve's Ransom," where the heroine, with various discreditable subterfuges, throws over the poor man whom she loves in order to marry the rich man whose income she loves still better. The poor man, finding that the rich man's income has given her a fuller life and a better character than the poor man's love could have given her, decides that she has done quite right, and that he deserves to be punished for his lack of money. In this story, as in his other books, Gissing has set forth, quite accurately, the actual dominion of money, and the impersonal worship which it exacts from the great majority of civilized mankind.

Gissing's facts are undeniable, and yet his attitude produces a revolt in any reader who has vital passions and masterful desires. His worship of money is bound up with his consciousness of inward defeat. And in the modern world generally, it is the decay of life which has promoted the religion of material goods; and the religion of material goods, in its turn, has hastened the decay of life on which it thrives. The man who worships money has ceased to hope for happiness through his own efforts or in his own activities: he looks upon happiness as a passive enjoyment of pleasures derived from the outside world. The artist or the lover does not worship money in his moments of ardor, because his desires are specific, and directed towards objects which only he can create. And conversely, the worshiper of money can never achieve greatness as an artist or a lover.

Love of money has been denounced by moralists since the world began. I do not wish to add another to the moral denunciations, of which the efficacy in the past has not been encouraging. I wish to show how the worship of money is both an effect and a cause of diminishing vitality, and how our institutions might be changed so as to make the worship of money grow less and the general vitality grow more. It is not the desire for money as a means to definite ends that is in question. A struggling artist may desire money in order to have leisure for his art, but this desire is finite, and can be satisfied fully by a very modest sum. It is the *worship* of money that I wish to consider: the belief that all values may be measured in terms of money, and that money is the ultimate test of success in life. This belief is held in fact, if not in words, by multitudes of men and women, and yet it is not in harmony with human nature, since it ignores vital needs and the instinctive tendency towards some specific kind of growth. It makes men treat as unimportant those of their desires which run counter to the acquisition of money, and yet such desires are, as a rule, more important to well-being than any increase of income. It leads men to mutilate their own natures from a mistaken theory of what constitutes success, and to give admiration to enterprises which add nothing to human welfare. It promotes a dead uniformity of character and purpose, a diminution in the joy of life, and a stress and strain which leaves whole communities weary, discouraged, and disillusioned.

America, the pioneer of Western progress, is thought by many to display the worship of money in its most perfect form. A well-to-do American, who already has more than enough money to satisfy all reasonable requirements, almost always continues to work at his office with an assiduity which would only be pardonable if starvation were the alternative.

But England, except among a small minority, is almost as much given over to the worship of money as America. Love of money in England takes, as a rule, the form of snobbishly desiring to maintain a certain social status, rather than of striving after an indefinite increase of income. Men postpone marriage until they have an income enabling them to have as many rooms and servants in their house as they feel that their dignity requires. This makes it necessary for them while they are young to keep a watch upon their affections, lest they should be led into an imprudence: they acquire a cautious habit of mind, and a fear of "giving themselves away," which makes a free and vigorous life impossible. In acting as they do they imagine that they are being virtuous, since they would feel it a hardship for a woman to be asked to descend to a lower social status than that of her parents, and a degradation to themselves to marry a woman whose social status was not equal to their own. The things of nature are not valued in comparison with money. It is not thought a hardship for a woman to have to accept, as her only experience of love, the prudent and limited attentions of a man whose capacity for emotion has been lost during years of wise restraint or sordid relations with women whom he did not respect. The woman herself does not know that it is a hardship; for she, too, has been taught prudence for fear of a descent in the social scale, and from early youth she has had it instilled into her that strong feeling does not become a young woman. So the two unite to slip through life in ignorance of all that is worth knowing. Their ancestors were not restrained from passion by the fear of hell-fire, but they are restrained effectually by a worse fear, the fear of coming down in the world.

The same motives which lead men to marry late also lead them to limit their families. Professional men wish to send their sons to a public school, though the education they will obtain is no better than at a grammar school, and the companions with whom they will associate are more vicious. But snobdom has decided that public schools are best, and from its verdict there is no appeal. What makes them the best is that they are the most expensive. And the same social struggle, in varying forms, runs through all classes except the very highest and the very lowest. For this purpose men and women make great moral efforts, and show amazing powers of self-control; but all their efforts and all their self-control, being not used for any creative end, serve merely to dry up the well-spring of life within them, to make them feeble, listless, and trivial. It is not in such a soil that the passion which produces genius can be nourished. Men's souls have exchanged the wilderness for the drawing-room: they have become cramped and petty and deformed, like Chinese women's feet. Even the horrors of war have hardly awakened them from the smug somnambulism of respectability. And it is chiefly the worship of money that has brought about this deathlike slumber of all that makes men great.

In France the worship of money takes the form of thrift. It is not easy to make a fortune in France, but an inherited competence is very common, and where it exists the main purpose of life is to hand it on undiminished, if not increased. The French *rentier* is one of the great forces in international politics: it is he through whom France has been strengthened in diplomacy and weakened in war, by increasing the supply of French capital and diminishing the supply of French men. The necessity of providing a *dot* for daughters, and the subdivision of property by the law of inheritance, have made the family more powerful, as an institution, than in any other civilized country. In order that the family may prosper, it is kept small, and the individual members are often sacrificed to it. The desire for family continuity makes men timid and unadventurous: it is only in the organized proletariat that the daring spirit survives which made the Revolution and led the world in political thought and practice. Through the influence of money, the strength of the family has become a weakness to the nation by making the population remain stationary and even tend to decline. The same love of safety is beginning to produce the same effects elsewhere; but in this, as in many better things, France has led the way.

In Germany the worship of money is more recent than in France, England, and America;indeed, it hardly existed until after the Franco-Prussian War. But it has been adopted now with the same intensity and whole-heartedness which have always marked German beliefs. It is characteristic that, as in France the worship of money is associated with the family, so in Germany it is associated with the State. Liszt, in deliberate revolt against the English economists, taught his compatriots to think of economics in national terms, and the German who develops a business is felt, by others as well as by himself, to be performing a service to the State. Germans believe that England's greatness is due to industrialism and Empire, and that our success in these is due to an intense nationalism. The apparent internationalism of our Free Trade policy they regard as mere hypocrisy. They have set themselves to imitate what they believe we really are, with only the hypocrisy omitted. It must be admitted that their success has been amazing. But in the process they have destroyed almost all that made

Germany of value to the world, and they have not adopted whatever of good there may have been among us, since that was all swept aside in the wholesale condemnation of "hypocrisy." And in adopting our worst faults,they have made them far worse by a system, a thoroughness, and a unanimity of which we are happily incapable. Germany's religion is of great importance to the world, since Germans have a power of real belief, and have the energy to acquire the virtues and vices which their creed demands. For the sake of the world, as well as for the sake of Germany, we must hope that they will soon abandon the worship of wealth which they have unfortunately learnt from us.

Worship of money is no new thing, but it is a more harmful thing than it used to be, for several reasons. Industrialism has made work more wearisome and intense, less capable of affording pleasure and interest by the way to the man who has undertaken it for the sake of money. The power of limiting families has opened a new field for the operation of thrift. The general increase in education and self-discipline has made men more capable of pursuing a purpose consistently in spite of temptations, and when the purpose is against life it becomes more destructive with every increase of tenacity in those who adopt it. The greater productivity resulting from industrialism has enabled us to devote more labor and capital to armies and navies for the protection of our wealth from envious neighbors, and for the exploitation of inferior races, which are ruthlessly wasted by the capitalist régime. Through the fear of losing money, forethought and anxiety eat away men's power of happiness, and the dread of misfortune becomes a greater misfortune than the one which is dreaded. The happiest men and women, as we can all testify from our own experience, are those who are indifferent to money because they have some positive purpose which shuts it out. And yet all our political thought, whether imperialist, radical, or socialist, continues to occupy itself almost exclusively with men's economic desires, as though they alone had real importance.

In judging of an industrial system, whether the one under which we live or one proposed by reformers, there are four main tests which may be applied. We may consider whether the system secures (1) the maximum of production, or (2) justice in distribution, or (3) a tolerable existence for producers, or (4) the greatest possible freedom and stimulus to vitality and progress. We may say, broadly, that the present system aims only at the first of these objects, while socialism aims at the second and third.Some defenders of the present system contend that technical progress is better promoted by private enterprise than it would be if industry were in the hands of the State; to this extent they recognize the fourth of the objects we have enumerated. But they recognize it only on the side of the goods and the capitalist, not on the side of the wage-earner. I believe that the fourth is much the most important of the objects to be aimed at, that the present system is fatal to it, and that orthodox socialism might well prove equally fatal.

One of the least questioned assumptions of the capitalist system is, that production ought to be increased in amount by every possible means: by new kinds of machinery, by employment of women and boys, by making hours of labor as long as is compatible with efficiency. Central African natives, accustomed to living on raw fruits of the earth and defeating Manchester by dispensing with clothes, are compelled to work by a hut tax which they can only pay by taking employment under European capitalists. It is admitted that they are perfectly happy while they remain free from European influences, and that industrialism brings upon them, not only the unwonted misery of confinement,but also death from diseases to which white men have become partially immune. It is admitted that the best negro workers are the "raw natives," fresh from the bush, uncontaminated by previous experience of wage-earning. Nevertheless, no one effectively contends that they ought to be preserved from the deterioration which we bring, since no one effectively doubts that it is good to increase the world's production at no matter what cost.

The belief in the importance of production has a fanatical irrationality and ruthlessness. So long as something is produced, what it is that is produced seems to be thought a matter of no account. Our whole economic system encourages this view, since fear of unemployment makes any kind of work a boon to wage-earners. The mania for increasing production has turned men's thoughts away from much more important problems, and has prevented the world from getting the benefits it might have got out of the increased productivity of labor.

When we are fed and clothed and housed, further material goods are needed only for ostentation.With modern methods, a certain proportion of the population, without working long hours, could do all the work that is really necessary in the way of producing commodities. The time which is now spent in producing luxuries could be spent partly in enjoyment and country holidays, partly in better education, partly in work that is not manual or subserving manual work. We could, if we wished, have far more science and art, more diffused knowledge and mental cultivation, more leisure for wage-earners, and more capacity for intelligent pleasures. At present not only wages, but almost all earned incomes, can only be obtained by working much longer hours than men ought to work. A man who earns £800 a year by hard work could not, as a rule, earn £400 a year by half as much work. Often he could not earn anything if he were not willing to work practically all day and every day. Because of the excessive belief in the value of production, it is thought right and proper for men to work long hours, and the good that might result from shorter hours is not realized. And all the cruelties of the industrial system, not only in Europe but even more in the tropics, arouse only an occasional feeble protest from a few philanthropists. This is because, owing to the distortion produced by our present economic methods, men's conscious desires, in such matters, cover only a very small part, and that not the most important part, of the real needs affected by industrial work. If this is to be remedied, it can only be by a different economic system, in which the relation of activity to needs will be less concealed and more direct.

The purpose of maximizing production will not be achieved in the long run if our present industrial system continues. Our present system is wasteful of human material, partly through damage to the health and efficiency of industrial workers, especially when women and children are employed, partly through the fact that the best workers tend to have small families and that the more civilized races are in danger of gradual extinction. Every great city is a center of race-deterioration. For the case of London this has been argued with a wealth of statistical detail by Sir H. Llewelyn Smith; and it cannot easily be doubted that it is equally true in other cases. The same is true of material resources: the minerals, the virgin forests, and the newly developed wheatfields of the world are being exhausted with a reckless prodigality which entails almost a certainty of hardship for future generations.

Socialists see the remedy in State ownership of land and capital, combined with a more just system of distribution. It cannot be denied that our present system of distribution is indefensible from every point of view, including the point of view of justice. Our system of distribution is regulated by law, and is capable of being changed in many respects which familiarity makes us regard as natural and inevitable. We may distinguish four chief sources of recognized legal rights to private property: (1) a man's right to what he has made himself; (2) the right to interest on capital which has been lent; (3) the ownership of land; (4) inheritance. These form a crescendo of respectability: capital is more respectable than labor, land is more respectable than capital, and any form of wealth is more respectable when it is inherited than when it has been acquired by our own exertions.

A man's right to the produce of his own labor has never, in fact, had more than a very limited recognition from the law. The early socialists, especially the English forerunners of Marx, used to insist upon this right as the basis of a just system of distribution, but in the complication of modern industrial processes it is impossible to say what a man has produced. What proportion of the goods carried by a railway should belong to the goods porters concerned in their journey? When a surgeon saves a man's life by an operation, what proportion of the commodities which the man subsequently produces can the surgeon justly claim? Such problems are insoluble. And there is no special justice, even if they were soluble, in allowing to each man what he himself produces. Some men are stronger, healthier, cleverer, than others, but there is no reason for increasing these natural injustices by the artificial injustices of the law. The principle recommends itself partly as a way of abolishing the very rich, partly as a way of stimulating people to work hard. But the first of these objects can be better obtained in other ways, and the second ceases to be obviously desirable as soon as we cease to worship money.

Interest arises naturally in any community in which private property is unrestricted and theft is punished, because some of the most economical processes of production are slow, and those who have the skill to perform them may not have the means of living while they are being completed. But the power of lending money gives such great wealth and influence to private capitalists that unless strictly controlled it is not compatible with any real freedom for the rest of the population. Its effects at present, both in the industrial world and in international politics, are so bad that it seems imperatively necessary to devise some means of curbing its power.

Private property in land has no justification except historically through power of the sword. In the beginning of feudal times, certain men had enough military strength to be able to force those whom they disliked not to live in a certain area. Those whom they chose to leave on the land became their serfs, and were forced to work for them in return for the gracious permission to stay. In order to establish law in place of private force, it was necessary, in the main, to leave undisturbed the rights which had been acquired by the sword. The land became the property of those who had conquered it, and the serfs were allowed to give rent instead of service. There is no justification for private property in land, except the historical necessity to conciliate turbulent robbers who would not otherwise have obeyed the law. This necessity arose in Europe many centuries ago, but in Africa the whole process is often quite recent. It is by this process, slightly disguised, that the Kimberley diamond mines and the Rand gold mines were acquired in spite of prior native rights. It is a singular example of human inertia that men should have continued until now to endure the tyranny and extortion which a small minority are able to inflict by their possession of the land. No good to the community, of any sort or kind, results from the private ownership of land. If men were reasonable, they would decree that it should cease to-morrow, with no compensation beyond a moderate life income to the present holders.

The mere abolition of rent would not remove injustice, since it would confer a capricious advantage upon the occupiers of the best sites and the most fertile land. It is necessary that there should be rent, but it should be paid to the State or to some body which performs public services; or, if the total rental were more than is required for such purposes, it might be paid into a common fund and divided equally among the population. Such a method would be just, and would not only help to relieve poverty, but would prevent wasteful employment of land and the tyranny of local magnates. Much that appears as the power of capital is really the power of the landowner—for example, the power of railway companies and mine-owners. The evil and injustice of the present system are glaring, but men's patience of preventable evils to which they are accustomed is so great that it is impossible to guess when they will put an end to this strange absurdity.

Inheritance, which is the source of the greater part of the unearned income in the world, is regarded by most men as a natural right. Sometimes, as in England, the right is inherent in the owner of property, who may dispose of it in any way that seems good to him. Sometimes, as in France, his right is limited by the right of his family to inherit at least a portion of what he has to leave. But neither the right to dispose of property by will nor the right of children to inherit from parents has any basis outside the instincts of possession and family pride.

There may be reasons for allowing a man whose work is exceptionally fruitful—for instance, an inventor—to enjoy a larger income than is enjoyed by the average citizen, but there can be no good reason for allowing this privilege to descend to his children and grandchildren and so on for ever. The effect is to produce an idle and exceptionally fortunate class, who are influential through their money, and opposed to reform for fear it should be directed against themselves. Their whole habit of thought becomes timid, since they dread being forced to acknowledge that their position is indefensible; yet snobbery and the wish to secure their favor leads almost the whole middle-class to ape their manners and adopt their opinions. In this way they become a poison infecting the outlook of almost all educated people.

It is sometimes said that without the incentive of inheritance men would not work so well. The great captains of industry, we are assured, are actuated by the desire to found a family, and would not devote their lives to unremitting toil without the hope of gratifying this desire. I do not believe that any large proportion of really useful work is done from this motive. Ordinary work is done for the sake of a living, and the very best work is done for the interest of the work itself. Even the captains of industry, who are thought (perhaps by themselves as well as by others) to be aiming at founding a family, are probably more actuated by love of power and by the adventurous pleasure of great enterprises. And if there were some slight diminution in the amount of work done, it would be well worth while in order to get rid of the idle rich, with the oppression, feebleness, and corruption which they inevitably introduce.

The present system of distribution is not based upon any principle. Starting from a system imposed by conquest, the arrangements made by the conquerors for their own benefit were stereotyped by the law, and have never been fundamentally reconstructed. On what principles ought the reconstruction to be based?

Socialism, which is the most widely advocated scheme of reconstruction, aims chiefly at *justice*: the present inequalities of wealth are unjust, and socialism would abolish them. It is not essential to socialism that all men should have the same income, but it is essential that inequalities should be justified, in each case, by inequality of need or of service performed. There can be no disputing that the present system is grossly unjust, and that almost all that is unjust in it is harmful. But I do not think justice alone is a sufficient principle upon which to base an economic reconstruction. Justice would be secured if all were equally unhappy, as well as if all were equally happy. Justice, by itself, when once realized, contains no source of new life. The old type of Marxian revolutionary socialist never dwelt, in imagination, upon the life of communities after the establishment of the millennium. He imagined that, like the Prince and Princess in a fairy story, they would live happily ever after. But that is not a condition possible to human nature. Desire, activity, purpose, are essential to a tolerable life, and a millennium, though it may be a joy in prospect, would be intolerable if it were actually achieved.

The more modern socialists, it is true, have lost most of the religious fervor which characterized the pioneers, and view socialism as a tendency rather than a definite goal. But they still retain the view that what is of most political importance to a man is his income, and that the principal aim of a democratic politician ought to be to increase the wages of labor. I believe this involves too passive a conception of what constitutes happiness. It is true that, in the industrial world, large sections of the population are too poor to have any possibility of a good life; but it is not true that a good life will come of itself with a diminution of poverty. Very few of the well-to-do classes have a good life at present, and perhaps socialism would only substitute the evils which now afflict the more prosperous in place of the evils resulting from destitution.

In the existing labor movement, although it is one of the most vital sources of change, there are certain tendencies against which reformers ought to be on their guard. The labor movement is in essence a movement in favor of justice, based upon the belief that the sacrifice of the many to the few is not necessary now, whatever may have been the case in the past. When labor was less productive and education was less widespread, an aristocratic civilization may have been the only one possible: it may have been necessary that the many should contribute to the life of the few, if the few were to transmit and increase the world's possessions in art and thought and civilized existence. But this necessity is past or rapidly passing, and there is no longer any valid objection to the claims of justice. The labor movement is morally irresistible, and is not now seriously opposed except by prejudice and simple self-assertion. All living thought is on its side; what is against it is traditional and dead. But although it itself is living, it is not by any means certain that it will make for life.

Labor is led by current political thought in certain directions which would become repressive and dangerous if they were to remain strong after labor had triumphed. The aspirations of the labor movement are, on the whole, opposed by the great majority of the educated classes, who feel a menace, not only or chiefly to their personal comfort, but to the civilized life in which they have their part, which they profoundly believe to be important to the world. Owing to the opposition of the educated classes, labor, when it is revolutionary and vigorous, tends to despise all that the educated classes represent. When it is more respectful, as its leaders tend to be in England, the subtle and almost unconscious influence of educated men is apt to sap revolutionary ardor, producing doubt and uncertainty instead of the swift, simple assurance by which victory might have been won. The very sympathy which the best men in the well-to-do classes extend to labor, their very readiness to admit the justice of its claims,may have the effect of softening the opposition of labor leaders to the *status quo*, and of opening their minds to the suggestion that no fundamental change is possible. Since these influences affect leaders much more than the rank and file, they tend to produce in the rank and file a distrust of leaders, and a desire to seek out new leaders who will be less ready to concede the claims of the more fortunate classes. The result may be in the end a labor movement as hostile to the life of the mind as some terrified property-owners believe it to be at present.

The claims of justice, narrowly interpreted, may reinforce this tendency. It may be thought unjust that some men should have larger incomes or shorter hours of work than other men. But efficiency in mental work, including the work of education, certainly requires more comfort and longer periods of rest than are required for efficiency in physical work, if only because mental work is not physiologically wholesome. If this is not recognized, the life of the mind may suffer through short-sightedness even more than through deliberate hostility.

Education suffers at present, and may long continue to suffer, through the desire of parents that their children should earn money as soon as possible. Every one knows that the half-time system, for example, is bad; but the power of organized labor keeps it in existence. It is clear that the cure for this evil, as for those that are concerned with the population question, is to relieve parents of the expense of their children's education, and at the same time to take away their right to appropriate their children's earnings.

The way to prevent any dangerous opposition of labor to the life of the mind is not to oppose the labor movement, which is too strong to be opposed with justice. The right way is, to show by actual practice that thought is useful to labor, that without thought its positive aims cannot be achieved, and that there are men in the world of thought who are willing to devote their energies to helping labor in its struggle. Such men, if they are wise and sincere, can prevent labor from becoming destructive of what is living in the intellectual world.

Another danger in the aims of organized labor is the danger of conservatism as to methods of production. Improvements of machinery or organization bring great advantages to employers, but involve temporary and sometimes permanent loss to the wage-earners. For this reason, and also from mere instinctive dislike of any change of habits, strong labor organizations are often obstacles to technical progress. The ultimate basis of all social progress must be increased technical efficiency, a greater result from a given amount of labor. If labor were to offer an effective opposition to this kind of progress, it would in the long run paralyze all other progress. The way to overcome the opposition of labor is not by hostility or moral homilies, but by giving to labor the direct interest in economical processes which now belongs to the employers. Here, as elsewhere, the unprogressive part of a movement which is essentially progressive is to be eliminated, not by decrying the whole movement but by giving it a wider sweep, making it more progressive, and leading it to demand an even greater change in the structure of society than any that it had contemplated in its inception.

The most important purpose that political institutions can achieve is to keep alive in individuals creativeness, vigor, vitality, and the joy of life. These things existed, for example, in Elizabethan England in a way in which they do not exist now. They stimulated adventure, poetry, music, fine architecture, and set going the whole movement out of which England's greatness has sprung in every direction in which England has been great. These things coexisted with injustice, but outweighed it, and made a national life more admirable than any that is likely to exist under socialism.

What is wanted in order to keep men full of vitality is opportunity, not security. Security is merely a refuge from fear; opportunity is the source of hope. The chief test of an economic system is not whether it makes men prosperous, or whether it secures distributive justice (though these are both very desirable), but whether it leaves men's instinctive growth unimpeded. To achieve this purpose, there are two main conditions which it should fulfil: it should not cramp men's private affections, and it should give the greatest possible outlet to the impulse of creation. There is in most men, until it becomes atrophied by disuse, an instinct of constructiveness, a wish to make something. The men who achieve most are, as a rule, those in whom this instinct is strongest: such men become artists, men of science, statesmen, empire-builders, or captains of industry, according to the accidents of temperament and opportunity. The most beneficent and the most harmful careers are inspired by this impulse. Without it, the world would sink to the level of Tibet: it would subsist, as it is always prone to do, on the wisdom of its ancestors, and each generation would sink more deeply into a lifeless traditionalism.

But it is not only the remarkable men who have the instinct of constructiveness, though it is they who have it most strongly. It is almost universal in boys, and in men it usually survives in a greater or less degree, according to the greater or less outlet which it is able to find. Work inspired by this instinct is satisfying, even when it is irksome and difficult, because every effort is as natural as the effort of a dog pursuing a hare. The chief defect of the present capitalistic system is that work done for wages very seldom affords any outlet for the creative impulse. The man who works for wages has no choice as to what he shall make: the whole creativeness of the processes concentrate in the employer who orders the work to be done. For this reason the work becomes a merely external means to a certain result, the earning of wages. Employers grow indignant about the trade union rules for limitation of output, but they have no right to be indignant,since they do not permit the men whom they employ to have any share in the purpose for which the work is undertaken. And so the process of production, which should form one instinctive cycle, becomes divided into separate purposes, which can no longer provide any satisfaction of instinct for those who do the work.

This result is due to our industrial system, but it would not be avoided by socialism. In a socialist community, the State would be the employer, and the individual workman would have almost as little control over his work as he has at present. Such control as he could exercise would be indirect, through political channels, and would be too slight and roundabout to afford any appreciable satisfaction. It is to be feared that instead of an increase of self-direction, there would only be an increase of mutual interference.

The total abolition of private capitalistic enterprise, which is demanded by Marxian socialism, seems scarcely necessary. Most men who construct sweeping systems of reform, like most of those who defend the *status quo*, do not allow enough for the importance of exceptions and the undesirability of rigid system. Provided the sphere of capitalism is restricted, and a large proportion of the population are rescued from its dominion, there is no reason to wish it wholly abolished. As a competitor and a rival, it might serve a useful purpose in preventing more democratic enterprises from sinking into sloth and technical conservatism. But it is of the very highest importance that capitalism should become the exception rather than the rule, and that the bulk of the world's industry should be conducted on a more democratic system.

Much of what is to be said against militarism in the State is also to be said against capitalism in the economic sphere. Economic organizations, in the pursuit of efficiency, grow larger and larger, and there is no possibility of reversing this process. The causes of their growth are technical, and large organizations must be accepted as an essential part of civilized society. But there is no reason why their government should be centralized and monarchical. The present economic system, by robbing most men of initiative, is one of the causes of the universal weariness which devitalizes urban and industrial populations, making them perpetually seek excitement, and leading them to welcome even the outbreak of war as a relief from the dreary monotony of their daily lives.

If the vigor of the nation is to be preserved, if we are to retain any capacity for new ideas, if we are not to sink into a Chinese condition of stereotyped immobility, the monarchical organization of industry must be swept away. All large businesses must become democratic and federal in their government. The whole wage-earning system is an abomination, not only because of the social injustice which it causes and perpetuates, but also because it separates the man who does the work from the purpose for which the work is done. The whole of the controlling purpose is concentrated in the capitalist; the purpose of the wage-earner is not the produce, but the wages. The purpose of the capitalist is to secure the maximum of work for the minimum of wages; the purpose of the wage-earner is to secure the maximum of wages for the minimum of work. A system involving this essential conflict of interests cannot be expected to work smoothly or successfully, or to produce a community with any pride in efficiency.

Two movements exist, one already well advanced, the other in its infancy, which seem capable, between them, of effecting most of what is needed. The two movements I mean are the coöperative movement and syndicalism. The coöperative movement is capable of replacing the wage system over a very wide field, but it is not easy to see how it could be applied to such things as railways. It is just in these cases that the principles of syndicalism are most easily applicable.

If organization is not to crush individuality, membership of an organization ought to be voluntary, not compulsory, and ought always to carry with it a voice in the management. This is not the case with economic organizations, which give no opportunity for the pride and pleasure that men find in an activity of their own choice, provided it is not utterly monotonous.

It must be admitted, however, that much of the mechanical work which is necessary in industry is probably not capable of being made interesting in itself. But it will seem less tedious than it does at present if those who do it have a voice in the management of their industry. And men who desire leisure for other occupations might be given the opportunity of doing uninteresting work during a few hours of the day for a low wage; this would give an opening to all who wished for some activity not immediately profitable to themselves. When everything that is possible has been done to make work interesting, the residue will have to be made endurable, as almost all work is at present, by the inducement of rewards outside the hours of labor. But if these rewards are to be satisfactory, it is essential that the uninteresting work should not necessarily absorb a man's whole energies, and that opportunities should exist for more or less continuous activities during the remaining hours. Such a system might be an immeasurable boon to artists, men of letters, and others who produce for their own satisfaction works which the public does not value soon enough to secure a living for the producers; and apart from such rather rare cases, it might provide an opportunity for young men and women with intellectual ambitions to continue their education after they have left school, or to prepare themselves for careers which require an exceptionally long training.

The evils of the present system result from the separation between the several interests of consumer, producer, and capitalist. No one of these three has the same interests as the community or as either of the other two. The cooperative system amalgamates the interests of consumer and capitalist; syndicalism would amalgamate the interests of producer and capitalist. Neither amalgamates all three, or makes the interests of those who direct industry quite identical with those of the community. Neither, therefore, would wholly prevent industrial strife, or obviate the need of the State as arbitrator. But either would be better than the present system, and probably a mixture of both would cure most of the evils of industrialism as it exists now. It is surprising that, while men and women have struggled to achieve political democracy, so little has been done to introduce democracy in industry. I believe incalculable benefits might result from industrial democracy, either on the coöperative model or with recognition of a trade or industry as a unit for purposes of government, with some kind of Home Rule such as syndicalism aims at securing. There is no reason why all governmental units should be geographical: this system was necessary in the past because of the slowness of means of communication, but it is not necessary now. By some such system many men might come to feel again a pride in their work, and to find again that outlet for the creative impulse which is now denied to all but a fortunate few. Such a system requires the abolition of the land-owner and the restriction of the capitalist, but does not entail equality of earnings. And unlike socialism, it is not a static or final system: it is hardly more than a framework for energy and initiative. It is only by some such method, I believe, that the free growth of the individual can be reconciled with the huge technical organizations which have been rendered necessary by industrialism.

V
EDUCATION

NO political theory is adequate unless it is applicable to children as well as to men and women. Theorists are mostly childless, or, if they have children, they are carefully screened from the disturbances which would be caused by youthful turmoil. Some of them have written books on education, but without, as a rule, having any actual children present to their minds while they wrote. Those educational theorists who have had a knowledge of children, such as the inventors of Kindergarten and the Montessori system, have not always had enough realization of the ultimate goal of education to be able to deal successfully with advanced instruction. I have not the knowledge either of children or of education which would enable me to supply whatever defects there may be in the writings of others. But some questions, concerning education as a political institution, are involved in any hope of social reconstruction, and are not usually considered by writers on educational theory. It is these questions that I wish to discuss.

The power of education in forming character and opinion is very great and very generally recognized. The genuine beliefs, though not usually the professed precepts, of parents and teachers are almost unconsciously acquired by most children; and even if they depart from these beliefs in later life, something of them remains deeply implanted, ready to emerge in a time of stress or crisis. Education is, as a rule, the strongest force on the side of what exists and against fundamental change: threatened institutions, while they are still powerful, possess themselves of the educational machine, and instil a respect for their own excellence into the malleable minds of the young. Reformers retort by trying to oust their opponents from their position of vantage. The children themselves are not considered by either party; they are merely so much material, to be recruited into one army or the other. If the children themselves were considered, education would not aim at making them belong to this party or that, but at enabling them to choose intelligently between the parties; it would aim at making them able to think, not at making them think what their teachers think. Education as a political weapon could not exist if we respected the rights of children. If we respected the rights of children, we should educate them so as to give them the knowledge and the mental habits required for forming independent opinions; but education as a political institution endeavors to form habits and to circumscribe knowledge in such a way as to make one set of opinions inevitable.

The two principles of *justice* and *liberty*, which cover a very great deal of the social reconstruction required, are not by themselves sufficient where education is concerned. Justice, in the literal sense of equal rights, is obviously not wholly possible as regards children. And as for liberty, it is, to begin with, essentially negative: it condemns all avoidable interference with freedom, without giving a positive principle of construction. But education is essentially constructive, and requires some positive conception of what constitutes a good life. And although liberty is to be respected in education as much as is compatible with instruction,and although a very great deal more liberty than is customary can be allowed without loss to instruction, yet it is clear that some departure from complete liberty is unavoidable if children are to be taught anything, except in the case of unusually intelligent children who are kept isolated from more normal companions. This is one reason for the great responsibility which rests upon teachers: the children must, necessarily, be more or less at the mercy of their elders, and cannot make themselves the guardians of their own interests. Authority in education is to some extent unavoidable, and those who educate have to find a way of exercising authority in accordance with the *spirit* of liberty.

Where authority is unavoidable, what is needed is *reverence*. A man who is to educate really well, and is to make the young grow and develop into their full stature, must be filled through and through with the spirit of reverence. It is reverence towards others that is lacking in those who advocate machine-made cast-iron systems: militarism, capitalism, Fabian scientific organization, and all the other prisons into which reformers and reactionaries try to force the human spirit. In education,with its codes of rules emanating from a Government office, its large classes and fixed curriculum and overworked teachers, its determination to produce a dead level of glib mediocrity, the lack of reverence for the child is all but universal. Reverence requires imagination and vital warmth; it requires most imagination in respect of those who have least actual achievement or power. The child is weak and superficially foolish, the teacher is strong, and in an every-day sense wiser than the child. The teacher without reverence, or the bureaucrat without reverence, easily despises the child for these outward inferiorities. He thinks it is his duty to "mold" the child: in imagination he is the potter with the clay. And so he gives to the child some unnatural shape, which hardens with age, producing strains and spiritual dissatisfactions, out of which grow cruelty and envy, and the belief that others must be compelled to undergo the same distortions.

Tho man who has reverence will not think it his duty to "mold" the young. He feels in all that lives, but especially in human beings, and most of all in children, something sacred, indefinable, unlimited, something individual and strangely precious, the growing principle of life, an embodied fragment of the dumb striving of the world. In the presence of a child he feels an unaccountable humility—a humility not easily defensible on any rational ground, and yet somehow nearer to wisdom than the easy self-confidence of many parents and teachers. The outward helplessness of the child and the appeal of dependence make him conscious of the responsibility of a trust. His imagination shows him what the child may become, for good or evil, how its impulses may be developed or thwarted, how its hopes must be dimmed and the life in it grow less living, how its trust will be bruised and its quick desires replaced by brooding will. All this gives him a longing to help the child in its own battle; he would equip and strengthen it, not for some outside end proposed by the State or by any other impersonal authority, but for the ends which the child's own spirit is obscurely seeking. The man who feels this can wield the authority of an educator without infringing the principle of liberty.

It is not in a spirit of reverence that education is conducted by States and Churches and the great institutions that are subservient to them. What is considered in education is hardly ever the boy or girl, the young man or young woman, but almost always, in some form, the maintenance of the existing order. When the individual is considered, it is almost exclusively with a view to worldly success—making money or achieving a good position. To be ordinary, and to acquire the art of getting on, is the ideal which is set before the youthful mind, except by a few rare teachers who have enough energy of belief to break through the system within which they are expected to work. Almost all education has a political motive: it aims at strengthening some group, national or religious or even social, in the competition with other groups. It is this motive, in the main, which determines the subjects taught, the knowledge offered and the knowledge withheld, and also decides what mental habits the pupils are expected to acquire. Hardly anything is done to foster the inward growth of mind and spirit; in fact, those who have had most education are very often atrophied in their mental and spiritual life, devoid of impulse, and possessing only certain mechanical aptitudes which take the place of living thought.

Some of the things which education achieves at present must continue to be achieved by education in any civilized country. All children must continue to be taught how to read and write, and some must continue to acquire the knowledge needed for such professions as medicine or law or engineering. The higher education required for the sciences and the arts is necessary for those to whom it is suited. Except in history and religion and kindred matters, the actual instruction is only inadequate, not positively harmful. The instruction might be given in a more liberal spirit, with more attempt to show its ultimate uses; and of course much of it is traditional and dead. But in the main it is necessary, and would have to form a part of any educational system.

It is in history and religion and other controversial subjects that the actual instruction is positively harmful. These subjects touch the interests by which schools are maintained; and the interests maintain the schools in order that certain views on these subjects may be instilled. History, in every country, is so taught as to magnify that country: children learn to believe that their own country has always been in the right and almost always victorious, that it has produced almost all the great men, and that it is in all respects superior to all other countries.Since these beliefs are flattering, they are easily absorbed, and hardly ever dislodged from instinct by later knowledge.

To take a simple and almost trivial example: the facts about the battle of Waterloo are known in great detail and with minute accuracy; but the facts as taught in elementary schools will be widely different in England, France, and Germany. The ordinary English boy imagines that the Prussians played hardly any part; the ordinary German boy imagines that Wellington was practically defeated when the day was retrieved by Blücher's gallantry. If the facts were taught accurately in both countries, national pride would not be fostered to the same extent, neither nation would feel so certain of victory in the event of war, and the willingness to fight would be diminished. It is this result which has to be prevented. Every State wishes to promote national pride, and is conscious that this cannot be done by unbiased history. The defenseless children are taught by distortions and suppressions and suggestions. The false ideas as to the history of the world which are taught in the various countries are of a kind which encourages strife and serves to keep alive a bigoted nationalism. If good relations between States were desired, one of the first steps ought to be to submit all teaching of history to an international commission, which should produce neutral textbooks free from the patriotic bias which is now demanded everywhere.

Exactly the same thing applies to religion. Elementary schools are practically always in the hands either of some religious body or of a State which has a certain attitude towards religion. A religious body exists through the fact that its members all have certain definite beliefs on subjects as to which the truth is not ascertainable. Schools conducted by religious bodies have to prevent the young, who are often inquiring by nature, from discovering that these definite beliefs are opposed by others which are no more unreasonable, and that many of the men best qualified to judge think that there is no good evidence in favor of any definite belief. When the State is militantly secular, as in France, State schools become as dogmatic as those that are in the hands of the Churches (I understand that the word "God" must not be mentioned in a French elementary school). The result in all these cases is the same: free inquiry is checked, and on the most important matter in the world the child is met with dogma or with stony silence.

It is not only in elementary education that these evils exist. In more advanced education they take subtler forms, and there is more attempt to conceal them, but they are still present. Eton and Oxford set a certain stamp upon a man's mind, just as a Jesuit College does. It can hardly be said that Eton and Oxford have a *conscious* purpose, but they have a purpose which is none the less strong and effective for not being formulated. In almost all who have been through them they produce a worship of "good form," which is as destructive to life and thought as the medieval Church. "Good form" is quite compatible with a superficial open-mindedness, a readiness to hear all sides, and a certain urbanity towards opponents. But it is not compatible with fundamental open-mindedness, or with any inward readiness to give weight to the other side. Its essence is the assumption that what is most important is a certain kind of behavior, a behavior which minimizes friction between equals and delicately impresses inferiors with a conviction of their own crudity. As a political weapon for preserving the privileges of the rich in a snobbish democracy it is unsurpassable. As a means of producing an agreeable social *milieu* for those who have money with no strong beliefs or unusual desires it has some merit. In every other respect it is abominable.

The evils of "good form" arise from two sources: its perfect assurance of its own rightness, and its belief that correct manners are more to be desired than intellect, or artistic creation, or vital energy, or any of the other sources of progress in the world. Perfect assurance, by itself, is enough to destroy all mental progress in those who have it. And when it is combined with contempt for the angularities and awkwardnesses that are almost invariably associated with great mental power, it becomes a source of destruction to all who come in contact with it. "Good form" is itself dead and incapable of growth; and by its attitude to those who are without it it spreads its own death to many who might otherwise have life. The harm which it has done to well-to-do Englishmen, and to men whose abilities have led the well-to-do to notice them, is incalculable.

The prevention of free inquiry is unavoidable so long as the purpose of education is to produce belief rather than thought, to compel the young to hold positive opinions on doubtful matters rather than to let them see the doubtfulness and be encouraged to independence of mind. Education ought to foster the wish for truth, not the conviction that some particular creed is the truth. But it is creeds that hold men together in fighting organizations: Churches, States, political parties. It is intensity of belief in a creed that produces efficiency in fighting: victory comes to those who feel the strongest certainty about matters on which doubt is the only rational attitude. To produce this intensity of belief and this efficiency in fighting, the child's nature is warped, and its free outlook is cramped, by cultivating inhibitions as a check to the growth of new ideas. In those whose minds are not very active the result is the omnipotence of prejudice; while the few whose thought cannot be wholly killed become cynical, intellectually hopeless, destructively critical, able to make all that is living seem foolish, unable themselves to supply the creative impulses which they destroy in others.

The success in fighting which is achieved by suppressing freedom of thought is brief and very worthless. In the long run mental vigor is as essential to success as it is to a good life. The conception of education as a form of drill, a means of producing unanimity through slavishness, is very common, and is defended chiefly on the ground that it leads to victory. Those who enjoy parallels from ancient history will point to the victory of Sparta over Athens to enforce their moral. But it is Athens that has had power over men's thoughts and imaginations, not Sparta: any one of us, if we could be born again into some past epoch, would rather be born an Athenian than a Spartan. And in the modern world so much intellect is required in practical affairs that even the external victory is more likely to be won by intelligence than by docility. Education in credulity leads by quick stages to mental decay; it is only by keeping alive the spirit of free inquiry that the indispensable minimum of progress can be achieved.

Certain mental habits are commonly instilled by those who are engaged in educating: obedience and discipline, ruthlessness in the struggle for worldly success, contempt towards opposing groups, and an unquestioning credulity, a passive acceptance of the teacher's wisdom. All these habits are against life. Instead of obedience and discipline, we ought to aim at preserving independence and impulse. Instead of ruthlessness, education should try to develop justice in thought. Instead of contempt, it ought to instil reverence, and the attempt at understanding; towards the opinions of others it ought to produce, not necessarily acquiescence, but only such opposition as is combined with imaginative apprehension and a clear realization of the grounds for opposition. Instead of credulity, the object should be to stimulate constructive doubt, the love of mental adventure, the sense of worlds to conquer by enterprise and boldness in thought.Contentment with the *status quo*, and subordination of the individual pupil to political aims, owing to the indifference to the things of the mind, are the immediate causes of these evils; but beneath these causes there is one more fundamental, the fact that education is treated as a means of acquiring power over the pupil, not as a means of nourishing his own growth. It is in this that lack of reverence shows itself; and it is only by more reverence that a fundamental reform can be effected.

Obedience and discipline are supposed to be indispensable if order is to be kept in a class, and if any instruction is to be given. To some extent this is true; but the extent is much less than it is thought to be by those who regard obedience and discipline as in themselves desirable. Obedience, the yielding of one's will to outside direction, is the counterpart of authority. Both may be necessary in certain cases. Refractory children, lunatics, and criminals may require authority, and may need to be forced to obey. But in so far as this is necessary it is a misfortune: what is to be desired is the free choice of ends with which it is not necessary to interfere. And educational reformers have shown that this is far more possible than our fathers would ever have believed.

What makes obedience seem necessary in schools is the large classes and overworked teachers demanded by a false economy. Those who have no experience of teaching are incapable of imagining the expense of spirit entailed by any really living instruction. They think that teachers can reasonably be expected to work as many hours as bank clerks. Intense fatigue and irritable nerves are the result, and an absolute necessity of performing the day's task mechanically. But the task cannot be performed mechanically except by exacting obedience.

If we took education seriously, and thought it as important to keep alive the minds of children as to secure victory in war, we should conduct education quite differently: we should make sure of achieving the end, even if the expense were a hundredfold greater than it is. To many men and women a small amount of teaching is a delight, and can be done with a fresh zest and life which keeps most pupils interested without any need of discipline. The few who do not become interested might be separated from the rest, and given a different kind of instruction. A teacher ought to have only as much teaching as can be done, on most days, with actual pleasure in the work, and with an awareness of the pupil's mental needs. The result would be a relation of friendliness instead of hostility between teacher and pupil, a realization on the part of most pupils that education serves to develop their own lives and is not merely an outside imposition, interfering with play and demanding many hours of sitting still. All that is necessary to this end is a (greater expenditure of money), to secure teachers with more leisure and with a natural love of teaching.

Discipline, as it exists in schools, is very largely an evil. There is a kind of discipline which is necessary to almost all achievement, and which perhaps is not sufficiently valued by those who react against the purely external discipline of traditional methods. The desirable kind of discipline is the kind that comes from within, which consists in the power of pursuing a distant object steadily, foregoing and suffering many things on the way. This involves the subordination of impulse to will, the power of a directing action by large creative desires even at moments when they are not vividly alive. Without this, no serious ambition, good or bad, can be realized, no consistent purpose can dominate. This kind of discipline is very necessary, but can only result from strong desires for ends not immediately attainable, and can only be produced by education if education fosters such desires, which it seldom does at present. Such discipline springs from one's own will, not from outside authority. It is not this kind which is sought in most schools, and it is not this kind which seems to me an evil.

Although elementary education encourages the undesirable discipline that consists in passive obedience, and although hardly any existing education encourages the moral discipline of consistent self-direction, there is a certain kind of purely mental discipline which is produced by the traditional higher education. The kind I mean is that which enables a man to concentrate his thoughts at will upon any matter that he has occasion to consider, regardless of preoccupations or boredom or intellectual difficulty. This quality, though it has no important intrinsic excellence, greatly enhances the efficiency of the mind as an instrument. It is this that enables a lawyer to master the scientific details of a patent case which he forgets as soon as judgment has been given, or a civil servant to deal quickly with many different administrative questions in succession. It is this that enables men to forget private cares during business hours. In a complicated world it is a very necessary faculty for those whose work requires mental concentration.

Success in producing mental discipline is the chief merit of traditional higher education. I doubt whether it can be achieved except by compelling or persuading active attention to a prescribed task. It is for this reason chiefly that I do not believe methods such as Madame Montessori's applicable when the age of childhood has been passed. The essence of her method consists in giving a choice of occupations, any one of which is interesting to most children, and all of which are instructive. The child's attention is wholly spontaneous, as in play; it enjoys acquiring knowledge in this way, and does not acquire any knowledge which it does not desire. I am convinced that this is the best method of education with young children: the actual results make it almost impossible to think otherwise. But it is difficult to see how this method can lead to control of attention by the will. Many things which must be thought about are uninteresting, and even those that are interesting at first often become very wearisome before they have been considered as long as is necessary. The power of giving prolonged attention is very important, and it is hardly to be widely acquired except as a habit induced originally by outside pressure. Some few boys, it is true, have sufficiently strong intellectual desires to be willing to undergo all that is necessary by their own initiative and free will; but for all others an external inducement is required in order to make them learn any subject thoroughly. There is among educational reformers a certain fear of demanding great efforts, and in the world at large a growing unwillingness to be bored. Both these tendencies have their good sides, but both also have their dangers. The mental discipline which is jeopardized can be preserved by mere advice without external compulsion whenever a boy's intellectual interest and ambition can be sufficiently stimulated. A good teacher ought to be able to do this for any boy who is capable of much mental achievement; and for many of the others the present purely bookish education is probably not the best. In this way, so long as the importance of mental discipline is realized, it can probably be attained, whenever it is attainable, by appealing to the pupil's consciousness of his own needs. So long as teachers are not expected to succeed by this method, it is easy for them to slip into a slothful dullness, and blame their pupils when the fault is really their own.

Ruthlessness in the economic struggle will almost unavoidably be taught in schools so long as the economic structure of society remains unchanged. This must be particularly the case in middle-class schools, which depend for their numbers upon the good opinion of parents, and secure the good opinion of parents by advertising the successes of pupils. This is one of many ways in which the competitive organization of the State is harmful. Spontaneous and disinterested desire for knowledge is not at all uncommon in the young, and might be easily aroused in many in whom it remains latent. But it is remorselessly checked by teachers who think only of examinations, diplomas, and degrees. For the abler boys there is no time for thought, no time for the indulgence of intellectual taste, from the moment of first going to school until the moment of leaving the university. From first to last there is nothing but one long drudgery of examination tips and textbook facts. The most intelligent, at the end, are disgusted with learning, longing only to forget it and to escape into a life of action. Yet there, as before, the economic machine holds them prisoners, and all their spontaneous desires are bruised and thwarted.

The examination system, and the fact that instruction is treated mainly as training for a livelihood, leads the young to regard knowledge, from a purely utilitarian point of view, as the road to money, not as the gateway to wisdom. This would not matter so much if it affected only those who have no genuine intellectual interests. But unfortunately it affects most those whose intellectual interests are strongest, since it is upon them that the pressure of examinations falls with most severity. To them most, but to all in some degree, education appears as a means of acquiring superiority over others; it is infected through and through with ruthlessness and glorification of social inequality. Any free, disinterested consideration shows that, whatever inequalities might remain in a Utopia, the actual inequalities are almost all contrary to justice. But our educational system tends to conceal this from all except the failures, since those who succeed are on the way to profit by the inequalities, with every encouragement from the men who have directed their education.

Passive acceptance of the teacher's wisdom is easy to most boys and girls. It involves no effort of independent thought, and seems rational because the teacher knows more than his pupils; it is moreover the way to win the favor of the teacher unless he is a very exceptional man. Yet the habit of passive acceptance is a disastrous one in later life. It causes men to seek a leader, and to accept as a leader whoever is established in that position. It makes the power of Churches, Governments, party caucuses, and all the other organizations by which plain men are misled into supporting old systems which are harmful to the nation and to themselves. It is possible that there would not be much independence of thought even if education did everything to promote it; but there would certainly be more than there is at present. If the object were to make pupils think, rather than to make them accept certain conclusions,education would be conducted quite differently: there would be less rapidity of instruction and more discussion, more occasions when pupils were encouraged to express themselves, more attempt to make education concern itself with matters in which the pupils felt some interest.

Above all, there would be an endeavor to rouse and stimulate the love of mental adventure. The world in which we live is various and astonishing: some of the things that seem plainest grow more and more difficult the more they are considered; other things, which might have been thought quite impossible to discover, have nevertheless been laid bare by genius and industry. The powers of thought, the vast regions which it can master, the much more vast regions which it can only dimly suggest to imagination, give to those whose minds have traveled beyond the daily round an amazing richness of material, an escape from the triviality and wearisomeness of familiar routine, by which the whole of life is filled with interest, and the prison walls of the commonplace are broken down. The same love of adventure which takes men to the South Pole, the same passion for a conclusive trial of strength which leads some men to welcome war, can find in creative thought an outlet which is neither wasteful nor cruel, but increases the dignity of man by incarnating in life some of that shining splendor which the human spirit is bringing down out of the unknown. To give this joy, in a greater or less measure, to all who are capable of it, is the supreme end for which the education of the mind is to be valued.

It will be said that the joy of mental adventure must be rare, that there are few who can appreciate it, and that ordinary education can take no account of so aristocratic a good. I do not believe this. The joy of mental adventure is far commoner in the young than in grown men and women. Among children it is very common, and grows naturally out of the period of make-believe and fancy. It is rare in later life because everything is done to kill it during education. Men fear thought as they fear nothing else on earth—more than ruin, more even than death. Thought is subversive and revolutionary, destructive and terrible; thought is merciless to privilege, established institutions, and comfortable habits; thought is anarchic and lawless, indifferent to authority, careless of the well-tried wisdom of the ages. Thought looks into the pit of hell and is not afraid. It sees man, a feeble speck, surrounded by unfathomable depths of silence; yet it bears itself proudly, as unmoved as if it were lord of the universe. Thought is great and swift and free, the light of the world, and the chief glory of man.

But if thought is to become the possession of many, not the privilege of the few, we must have done with fear. It is fear that holds men back—fear lest their cherished beliefs should prove delusions, fear lest the institutions by which they live should prove harmful, fear lest they themselves should prove less worthy of respect than they have supposed themselves to be. "Should the working man think freely about property? Then what will become of us, the rich? Should young men and young women think freely about sex? Then what will become of morality? Should soldiers think freely about war? Then what will become of military discipline? Away with thought! Back into the shades of prejudice, lest property, morals, and war should be endangered! Better men should be stupid, slothful, and oppressive than that their thoughts should be free. For if their thoughts were free they might not think as we do. And at all costs this disaster must be averted." So the opponents of thought argue in the unconscious depths of their souls. And so they act in their churches, their schools, and their universities.

No institution inspired by fear can further life. Hope, not fear, is the creative principle in human affairs. All that has made man great has sprung from the attempt to secure what is good, not from the struggle to avert what was thought evil. It is because modern education is so seldom inspired by a great hope that it so seldom achieves a great result. The wish to preserve the past rather than the hope of creating the future dominates the minds of those who control the teaching of the young. Education should not aim at a passive awareness of dead facts, but at an activity directed towards the world that our efforts are to create. It should be inspired, not by a regretful hankering after the extinct beauties of Greece and the Renaissance, but by a shining vision of the society that is to be, of the triumphs that thought will achieve in the time to come, and of the ever-widening horizon of man's survey over the universe. Those who are taught in this spirit will be filled with life and hope and joy, able to bear their part in bringing to mankind a future less somber than the past, with faith in the glory that human effort can create.

VI
MARRIAGE AND THE POPULATION QUESTION

THE influence of the Christian religion on daily life has decayed very rapidly throughout Europe during the last hundred years. Not only has the proportion of nominal believers declined, but even among those who believe the intensity and dogmatism of belief is enormously diminished. But there is one social institution which is still profoundly affected by the Christian tradition—I mean the institution of marriage. The law and public opinion as regards marriage are dominated even now to a very great extent by the teachings of the Church, which continue to influence in this way the lives of men, women, and children in their most intimate concerns.

It is marriage as a political institution that I wish to consider, not marriage as a matter for the private morality of each individual. Marriage is regulated by law, and is regarded as a matter in which the community has a right to interfere. It is only the action of the community in regard to marriage that I am concerned to discuss: whether the present action furthers the life of the community, and if not, in what ways it ought to be changed.

There are two questions to be asked in regard to any marriage system: first, how it affects the development and character of the men and women concerned; secondly, what is its influence on the propagation and education of children. These two questions are entirely distinct, and a system may well be desirable from one of these two points of view when it is very undesirable from the other. I propose first to describe the present English law and public opinion and practice in regard to the relations of the sexes, then to consider their effects as regards children, and finally to consider how these effects, which are bad, could be obviated by a system which would also have a better influence on the character and development of men and women.

The law in England is based upon the expectation that the great majority of marriages will be lifelong. A marriage can only be dissolved if either the wife or the husband, but not both, can be proved to have committed adultery. In case the husband is the "guilty party," he must also be guilty of cruelty or desertion. Even when these conditions are fulfilled, in practice only the well-to-do can be divorced, because the expense is very great.A marriage cannot be dissolved for insanity or crime, or for cruelty, however abominable, or for desertion, or for adultery by both parties; and it cannot be dissolved for any cause whatever if both husband and wife have agreed that they wish it dissolved. In all these cases the law regards the man and woman as bound together for life. A special official, the

King's Proctor, is employed to prevent divorce when there is collusion and when both parties have committed adultery.

This interesting system embodies the opinions held by the Church of England some fifty years ago, and by most Nonconformists then and now. It rests upon the assumption that adultery is sin, and that when this sin has been committed by one party to the marriage, the other is entitled to revenge if he is rich. But when both have committed the same sin, or when the one who has not committed it feels no righteous anger, the right to revenge does not exist. As soon as this point of view is understood, the law, which at first seems somewhat strange, is seen to be perfectly consistent. It rests, broadly speaking, upon four propositions: (1) that sexual intercourse outside marriage is sin; (2) that resentment of adultery by the "innocent"party is a righteous horror of wrong-doing; (3) that his resentment, but nothing else, may be rightly regarded as making a common life impossible; (4) that the poor have no right to fine feelings. The Church of England, under the influence of the High Church, has ceased to believe the third of these propositions, but it still believes the first and second, and does nothing actively to show that it disbelieves the fourth.

The penalty for infringing the marriage law is partly financial, but depends mainly upon public opinion. A rather small section of the public genuinely believes that sexual relations outside marriage are wicked; those who believe this are naturally kept in ignorance of the conduct of friends who feel otherwise, and are able to go through life not knowing how others live or what others think. This small section of the public regards as depraved not only actions, but opinions, which are contrary to its principles. It is able to control the professions of politicians through its influence on elections, and the votes of the House of Lords through the presence of the Bishops. By these means it governs legislation, and makes any change in the marriage law almost impossible. It is able, also, to secure in most cases that a man who openly infringes the marriage law shall be dismissed from his employment or ruined by the defection of his customers or clients. A doctor or lawyer, or a tradesman in a country town, cannot make a living, nor can a politician be in Parliament, if he is publicly known to be "immoral." Whatever a man's own conduct may be, he is not likely to defend publicly those who have been branded, lest some of the odium should fall on him. Yet so long as a man has not been branded, few men will object to him, whatever they may know privately of his behavior in these respects.

Owing to the nature of the penalty, it falls very unequally upon different professions. An actor or journalist usually escapes all punishment. An urban workingman can almost always do as he likes. A man of private means, unless he wishes to take part in public life, need not suffer at all if he has chosen his friends suitably. Women, who formerly suffered more than men, now suffer less, since there are large circles in which no social penalty is inflicted, and a very rapidly increasing number of women who do not believe the conventional code. But for the majority of men outside the working classes the penalty is still sufficiently severe to be prohibitive.

The result of this state of things is a widespread but very flimsy hypocrisy, which allows many infractions of the code, and forbids only those which must become public. A man may not live openly with a woman who is not his wife, an unmarried woman may not have a child, and neither man nor woman may get into the divorce court. Subject to these restrictions, there is in practice very great freedom. It is this practical freedom which makes the state of the law seem tolerable to those who do not accept the principles upon which it is based. What has to be sacrificed to propitiate the holders of strict views is not pleasure, but only children and a common life and truth and honesty. It cannot be supposed that this is the result desired by those who maintain the code, but equally it cannot be denied that this is the result which they do in fact achieve. Extra-matrimonial relations which do not lead to children and are accompanied by a certain amount of deceit remain unpunished, but severe penalties fall on those which are honest or lead to children.

Within marriage, the expense of children leads to continually greater limitation of families. The limitation is greatest among those who have most sense of parental responsibility and most wish to educate their children well, since it is to them that the expense of children is most severe. But although the economic motive for limiting families has hitherto probably been the strongest, it is being continually reinforced by another. Women are acquiring freedom—not merely outward and formal freedom, but inward freedom, enabling them to think and feel genuinely, not according to received maxims. To the men who have prated confidently of women's natural instincts, the result would be surprising if they were aware of it. Very large numbers of women, when they are sufficiently free to think for themselves, do not desire to have children, or at most desire one child in order not to miss the experience which a child brings. There are women who are intelligent and active-minded who resent the slavery to the body which is involved in having children. There are ambitious women, who desire a career which leaves no time for children. There are women who love pleasure and gaiety, and women who love the admiration of men; such women will at least postpone child-bearing until their youth is past. All these classes of women are rapidly becoming more numerous, and it may be safely assumed that their numbers will continue to increase for many years to come.

It is too soon to judge with any confidence as to the effects of women's freedom upon private life and upon the life of the nation. But I think it is not too soon to see that it will be profoundly different from the effect expected by the pioneers of the women's movement. Men have invented, and women in the past have often accepted, a theory that women are the guardians of the race, that their life centers in motherhood, that all their instincts and desires are directed, consciously or unconsciously, to this end. Tolstoy's Natacha illustrates this theory: she is charming, gay, liable to passion, until she is married; then she becomes merely a virtuous mother, without any mental life. This result has Tolstoy's entire approval. It must be admitted that it is very desirable from the point of view of the nation, whatever we may think of it in relation to private life. It must also be admitted that it is probably common among women who are physically vigorous and not highly civilized. But in countries like France and England it is becoming increasingly rare. More and more women find motherhood unsatisfying, not what their needs demand. And more and more there comes to be a conflict between their personal development and the future of the community. It is difficult to know what ought to be done to mitigate this conflict, but I think it is worth while to see what are likely to be its effects if it is not mitigated.

Owing to the combination of economic prudence with the increasing freedom of women, there is at present a selective birth-rate of a very singular kind.In France the population is practically stationary, and in England it is rapidly becoming so; this means that some sections are dwindling while others are increasing. Unless some change occurs, the sections that are dwindling will practically become extinct, and the population will be almost wholly replenished from the sections that are now increasing.The sections that are dwindling include the whole middle-class and the skilled artisans. The sections that are increasing are the very poor, the shiftless and drunken, the feeble-minded—feeble-minded women, especially, are apt to be very prolific. There is an increase in those sections of the population which still actively believe the Catholic religion, such as the Irish and the Bretons, because the Catholic religion forbids limitation of families. Within the classes that are dwindling, it is the best elements that are dwindling most rapidly. Working-class boys of exceptional ability rise, by means of scholarships, into the professional class; they naturally desire to marry into the class to which they belong by education, not into the class from which they spring; but as they have no money beyond what they earn, they cannot marry young, or afford a large family. The result is that in each generation the best elements are extracted from the working classes and artificially sterilized, at least in comparison with those who are left. In the professional classes the young women who have initiative,energy, or intelligence are as a rule not inclined to marry young, or to have more than one or two children when they do marry. Marriage has been in the past the only obvious means of livelihood for women; pressure from parents and fear of becoming an old maid combined to force many women to marry in spite of a complete absence of inclination for the duties of a wife. But now a young woman of ordinary intelligence can easily earn her own living, and can acquire freedom and experience without the permanent ties of a husband and a family of children. The result is that if she marries she marries late.

For these reasons, if an average sample of children were taken out of the population of England, and their parents were examined, it would be found that prudence, energy, intellect, and enlightenment were less common among the parents than in the population in general; while shiftlessness, feeble-mindedness, stupidity, and superstition were more common than in the population in general. It would be found that those who are prudent or energetic or intelligent or enlightened actually fail to reproduce their own numbers; that is to say, they do not on the average have as many as two children each who survive infancy. On the other hand, those who have the opposite qualities have, on the average, more than two children each, and more than reproduce their own numbers.

It is impossible to estimate the effect which this will have upon the character of the population without a much greater knowledge of heredity than exists at present. But so long as children continue to live with their parents, parental example and early education must have a great influence in developing their character, even if we leave heredity entirely out of account. Whatever may be thought of genius, there can be no doubt that intelligence, whether through heredity or through education, tends to run in families, and that the decay of the families in which it is common must lower the mental standard of the population. It seems unquestionable that if our economic system and our moral standards remain unchanged, there will be, in the next two or three generations, a rapid change for the worse in the character of the population in all civilized countries, and an actual diminution of numbers in the most civilized.

The diminution of numbers, in all likelihood,will rectify itself in time through the elimination of those characteristics which at present lead to a small birth-rate. Men and women who can still believe the Catholic faith will have a biological advantage; gradually a race will grow up which will be impervious to all the assaults of reason, and will believe imperturbably that limitation of families leads to hell-fire. Women who have mental interests, who care about art or literature or politics, who desire a career or who value their liberty, will gradually grow rarer, and be more and more replaced by a placid maternal type which has no interests outside the home and no dislike of the burden of motherhood. This result, which ages of masculine domination have vainly striven to achieve, is likely to be the final outcome of women's emancipation and of their attempt to enter upon a wider sphere than that to which the jealousy of men confined them in the past.

Perhaps, if the facts could be ascertained, it would be found that something of the same kind occurred in the Roman Empire. The decay of energy and intelligence during the second, third, and fourth centuries of our era has always remained more or less mysterious. But there is reason to think that then, as now, the best elements of the population in each generation failed to reproduce themselves, and that the least vigorous were, as a rule, those to whom the continuance of the race was due. One might be tempted to suppose that civilization, when it has reached a certain height, becomes unstable, and tends to decay through some inherent weakness, some failure to adapt the life of instinct to the intense mental life of a period of high culture. But such vague theories have always something glib and superstitious which makes them worthless as scientific explanations or as guides to action. It is not by a literary formula, but by detailed and complex thought, that a true solution is to be found.

Let us first be clear as to what we desire. There is no importance in an increasing population; on the contrary, if the population of Europe were stationary, it would be much easier to promote economic reform and to avoid war. What is regrettable *at present* is not the decline of the birth-rate in itself, but the fact that the decline is greatest in the best elements of the population. There is reason, however, to fear in the future three bad results: first, an absolute decline in the numbers of English, French, and Germans; secondly, as a consequence of this decline, their subjugation by less civilized races and the extinction of their tradition; thirdly, a revival of their numbers on a much lower plane of civilization, after generations of selection of those who have neither intelligence nor foresight. If this result is to be avoided, the present unfortunate selectiveness of the birth-rate must be somehow stopped.

The problem is one which applies to the whole of Western civilization. There is no difficulty in discovering a theoretical solution, but there is great difficulty in persuading men to adopt a solution in practice, because the effects to be feared are not immediate and the subject is one upon which people are not in the habit of using their reason. If a rational solution is ever adopted, the cause will probably be international rivalry. It is obvious that if one State—say Germany—adopted a rational means of dealing with the matter, it would acquire an enormous advantage over other States unless they did likewise. After the war, it is possible that population questions will attract more attention than they did before, and it is likely that they will be studied from the point of view of international rivalry. This motive, unlike reason and humanity, is perhaps strong enough to overcome men's objections to a scientific treatment of the birth-rate.

In the past, at most periods and in most societies, the instincts of men and women led of themselves to a more than sufficient birth-rate; Malthus's statement of the population question had been true enough up to the time when he wrote. It is still true of barbarous and semi-civilized races, and of the worst elements among civilized races. But it has become false as regards the more civilized half of the population in Western Europe and America. Among them, instinct no longer suffices to keep numbers even stationary.

We may sum up the reasons for this in order of importance, as follows:—

1. The expense of children is very great if parents are conscientious.

2. An increasing number of women desire to have no children, or only one or two, in order not to be hampered in their own careers.

3. Owing to the excess of women, a large number of women remain unmarried. These women, though not debarred in practice from relations with men, are debarred by the code from having children. In this class are to be found an enormous and increasing number of women who earn their own living as typists, in shops, or otherwise. The war has opened many employments to women from which they were formerly excluded, and this change is probably only in part temporary.

If the sterilizing of the best parts of the population is to be arrested, the first and most pressing necessity is the removal of the economic motives for limiting families. The expense of children ought to be borne wholly by the community. Their food and clothing and education ought to be provided, not only to the very poor as a matter of charity, but to all classes as a matter of public interest. In addition to this, a woman who is capable of earning money, and who abandons wage-earning for motherhood, ought to receive from the State as nearly as possible what she would have received if she had not had children. The only condition attached to State maintenance of the mother and the children should be that both parents are physically and mentally sound in all ways likely to affect the children. Those who are not sound should not be debarred from having children, but should continue, as at present, to bear the expense of children themselves.

It ought to be recognized that the law is only concerned with marriage through the question of children, and should be indifferent to what is called "morality," which is based upon custom and texts of the Bible, not upon any real consideration of the needs of the community. The excess women, who at present are in every way discouraged from having children, ought no longer to be discouraged. If the State is to undertake the expense of children, it has the right, on eugenic grounds, to know who the father is, and to demand a certain stability in a union. But there is no reason to demand or expect a lifelong stability, or to exact any ground for divorce beyond mutual consent. This would make it possible for the women who must at present remain unmarried to have children if they wished it. In this way an enormous and unnecessary waste would be prevented, and a great deal of needless unhappiness would be avoided.

There is no necessity to begin such a system all at once. It might be begun tentatively with certain exceptionally desirable sections of the community. It might then be extended gradually, with the experience of its working which had been derived from the first experiment. If the birth-rate were very much increased,the eugenic conditions exacted might be made more strict.

There are of course various practical difficulties in the way of such a scheme: the opposition of the Church and the upholders of traditional morality, the fear of weakening parental responsibility, and the expense. All these, however, might be overcome. But there remains one difficulty which it seems impossible to overcome completely in England, and that is, that the whole conception is anti-democratic, since it regards some men as better than others, and would demand that the State should bestow a better education upon the children of some men than upon the children of others. This is contrary to all the principles of progressive politics in England. For this reason it can hardly be expected that any such method of dealing with the population question will ever be adopted in its entirety in this country. Something of the sort may well be done in Germany, and if so, it will assure German hegemony as no merely military victory could do. But among ourselves we can only hope to see it adopted in some partial, piecemeal fashion, and probably only after a change in the economic structure of society which will remove most of the artificial inequalities that progressive parties are rightly trying to diminish.

So far we have been considering the question of the reproduction of the race, rather than the effect of sex relations in fostering or hindering the development of men and women. From the point of view of the race, what seems needed is a complete removal of the economic burdens due to children from all parents who are not physically or mentally unfit, and as much freedom in the law as is compatible with public knowledge of paternity. Exactly the same changes seem called for when the question is considered from the point of view of the men and women concerned.

In regard to marriage, as with all the other traditional bonds between human beings, a very extraordinary change is taking place, wholly inevitable, wholly necessary as a stage in the development of a new life, but by no means wholly satisfactory until it is completed. All the traditional bonds were based on *authority*—of the king, the feudal baron, the priest, the father, the husband. All these bonds, just because they were based on authority, are dissolving or already dissolved, and the creation of other bonds to take their place is as yet very incomplete.For this reason human relations have at present an unusual triviality, and do less than they did formerly to break down the hard walls of the Ego.

The ideal of marriage in the past depended upon the authority of the husband, which was admitted as a right by the wife. The husband was free, the wife was a willing slave. In all matters which concerned husband and wife jointly, it was taken for granted that the husband's fiat should be final. The wife was expected to be faithful, while the husband, except in very religious societies, was only expected to throw a decent veil over his infidelities. Families could not be limited except by continence, and a wife had no recognized right to demand continence, however she might suffer from frequent children.

So long as the husband's right to authority was unquestioningly believed by both men and women, this system was fairly satisfactory, and afforded to both a certain instinctive fulfilment which is rarely achieved among educated people now. Only one will, the husband's, had to be taken into account, and there was no need of the difficult adjustments required when common decisions have to be reached by two equal wills.The wife's desires were not treated seriously enough to enable them to thwart the husband's needs, and the wife herself, unless she was exceptionally selfish, did not seek self-development, or see in marriage anything but an opportunity for duties. Since she did not seek or expect much happiness, she suffered less, when happiness was not attained, than a woman does now: her suffering contained no element of indignation or surprise, and did not readily turn into bitterness and sense of injury.

The saintly, self-sacrificing woman whom our ancestors praised had her place in a certain organic conception of society, the conception of the ordered hierarchy of authorities which dominated the Middle Ages. She belongs to the same order of ideas as the faithful servant, the loyal subject, and the orthodox son of the Church. This whole order of ideas has vanished from the civilized world, and it is to be hoped that it has vanished for ever, in spite of the fact that the society which it produced was vital and in some ways full of nobility. The old order has been destroyed by the new ideals of justice and liberty, beginning with religion, passing on to politics, and reaching at last the private relations of marriage and the family. When once the question has been asked, "Why should a woman submit to a man?" when once the answers derived from tradition and the Bible have ceased to satisfy, there is no longer any possibility of maintaining the old subordination. To every man who has the power of thinking impersonally and freely, it is obvious, as soon as the question is asked, that the rights of women are precisely the same as the rights of men. Whatever dangers and difficulties, whatever temporary chaos, may be incurred in the transition to equality, the claims of reason are so insistent and so clear that no opposition to them can hope to be long successful.

Mutual liberty, which is now demanded, is making the old form of marriage impossible. But a new form, which shall be an equally good vehicle for instinct, and an equal help to spiritual growth, has not yet been developed. For the present, women who are conscious of liberty as something to be preserved are also conscious of the difficulty of preserving it. The wish for mastery is an ingredient in most men's sexual passions, especially in those which are strong and serious. It survives in many men whose theories are entirely opposed to despotism. The result is a fight for liberty on the one side and for life on the other. Women feel that they must protect their individuality; men feel, often very dumbly, that the repression of instinct which is demanded of them is incompatible with vigor and initiative. The clash of these opposing moods makes all real mingling of personalities impossible; the man and woman remain hard, separate units, continually asking themselves whether anything of value to themselves is resulting from the union. The effect is that relations tend to become trivial and temporary, a pleasure rather than the satisfaction of a profound need, an excitement, not an attainment. The fundamental loneliness into which we are born remains untouched, and the hunger for inner companionship remains unappeased.

No cheap and easy solution of this trouble is possible. It is a trouble which affects most the most civilized men and women, and is an outcome of the increasing sense of individuality which springs inevitably from mental progress. I doubt if there is any radical cure except in some form of religion, so firmly and sincerely believed as to dominate even the life of instinct. The individual is not the end and aim of his own being: outside the individual, there is the community, the future of mankind, the immensity of the universe in which all our hopes and fears are a mere pin-point. A man and woman with reverence for the spirit of life in each other, with an equal sense of their own unimportance beside the whole life of man, may become comrades without interference with liberty, and may achieve the union of instinct without doing violence to the life of mind and spirit. As religion dominated the old form of marriage, so religion must dominate the new. But it must be a new religion, based upon liberty, justice, and love, not upon authority and law and hell-fire.

A bad effect upon the relations of men and women has been produced by the romantic movement, through directing attention to what ought to be an incidental good, not the purpose for which relations exist. Love is what gives intrinsic value to a marriage, and, like art and thought, it is one of the supreme things which make human life worth preserving. But though there is no good marriage without love, the best marriages have a purpose which goes beyond love. The love of two people for each other is too circumscribed, too separate from the community, to be by itself the main purpose of a good life. It is not in itself a sufficient source of activities, it is not sufficiently prospective, to make an existence in which ultimate satisfaction can be found. It brings its great moments, and then its times which are less great, which are unsatisfying because they are less great. It becomes, sooner or later, retrospective, a tomb of dead joys, not a well-spring of new life. This evil is inseparable from any purpose which is to be achieved in a single supreme emotion. The only adequate purposes are those which stretch out into the future, which can never be fully achieved, but are always growing, and infinite with the infinity of human endeavor. And it is only when love is linked to some infinite purpose of this kind that it can have the seriousness and depth of which it is capable.

For the great majority of men and women seriousness in sex relations is most likely to be achieved through children. Children are to most people rather a need than a desire: instinct is as a rule only consciously directed towards what used to lead to children. The desire for children is apt to develop in middle life, when the adventure of one's own existence is past, when the friendships of youth seem less important than they once did, when the prospect of a lonely old age begins to terrify, and the feeling of having no share in the future becomes oppressive. Then those who, while they were young, have had no sense that children would be a fulfilment of their needs, begin to regret their former contempt for the normal, and to envy acquaintances whom before they had thought humdrum. But owing to economic causes it is often impossible for the young, and especially for the best of the young, to have children without sacrificing things of vital importance to their own lives. And so youth passes, and the need is felt too late.

Needs without corresponding desires have grown increasingly common as life has grown more different from that primitive existence from which our instincts are derived, and to which, rather than to that of the present day, they are still very largely adapted. An unsatisfied need produces, in the end, as much pain and as much distortion of character as if it had been associated with a conscious desire. For this reason, as well as for the sake of the race, it is important to remove the present economic inducements to childlessness. There is no necessity whatever to urge parenthood upon those who feel disinclined to it, but there is necessity not to place obstacles in the way of those who have no such disinclination.

In speaking of the importance of preserving seriousness in the relations of men and women, I do not mean to suggest that relations which are not serious are always harmful. Traditional morality has erred by laying stress on what ought not to happen, rather than on what ought to happen. What is important is that men and women should find, sooner or later, the best relation of which their natures are capable. It is not always possible to know in advance what will be the best, or to be sure of not missing the best if everything that can be doubted is rejected. Among primitive races, a man wants a female, a woman wants a male, and there is no such differentiation as makes one a much more suitable companion than another. But with the increasing complexity of disposition that civilized life brings, it becomes more and more difficult to find the man or woman who will bring happiness, and more and more necessary to make it not too difficult to acknowledge a mistake.

The present marriage law is an inheritance from a simpler age, and is supported, in the main, by unreasoning fears and by contempt for all that is delicate and difficult in the life of the mind. Owing to the law, large numbers of men and women are condemned, so far as their ostensible relations are concerned, to the society of an utterly uncongenial companion, with all the embittering consciousness that escape is practically impossible. In these circumstances, happier relations with others are often sought, but they have to be clandestine, without a common life, and without children. Apart from the great evil of being clandestine, such relations have some almost inevitable drawbacks. They are liable to emphasize sex unduly, to be exciting and disturbing; and it is hardly possible that they should bring a real satisfaction of instinct. It is the combination of love, children, and a common life that makes the best relation between a man and a woman. The law at present confines children and a common life within the bonds of monogamy, but it cannot confine love. By forcing many to separate love from children and a common life, the law cramps their lives, prevents them from reaching the full measure of their possible development, and inflicts a wholly unnecessary torture upon those who are not content to become frivolous.

To sum up: The present state of the law, of public opinion, and of our economic system is tending to degrade the quality of the race, by making the worst half of the population the parents of more than half of the next generation. At the same time, women's claim to liberty is making the old form of marriage a hindrance to the development of both men and women. A new system is required, if the European nations are not to degenerate, and if the relations of men and women are to have the strong happiness and organic seriousness which belonged to the best marriages in the past. The new system must be based upon the fact that to produce children is a service to the State, and ought not to expose parents to heavy pecuniary penalties. It will have to recognize that neither the law nor public opinion should concern itself with the private relations of men and women, except where children are concerned. It ought to remove the inducements to make relations clandestine and childless. It ought to admit that, although lifelong monogamy is best when it is successful, the increasing complexity of our needs makes it increasingly often a failure for which divorce is the best preventive. Here, as elsewhere, liberty is the basis of political wisdom.And when liberty has been won, what remains to be desired must be left to the conscience and religion of individual men and women.

VII
RELIGION AND THE CHURCHES

ALMOST all the changes which the world has undergone since the end of the Middle Ages are due to the discovery and diffusion of new knowledge. This was the primary cause of the Renaissance, the Reformation, and the industrial revolution. It was also, very directly, the cause of the decay of dogmatic religion. The study of classical texts and early Church history, Copernican astronomy and physics, Darwinian biology and comparative anthropology, have each in turn battered down some part of the edifice of Catholic dogma, until, for almost all thinking and instructed people, the most that seems defensible is some inner spirit, some vague hope, and some not very definite feeling of moral obligation. This result might perhaps have remained limited to the educated minority but for the fact that the Churches have almost everywhere opposed political progress with the same bitterness with which they have opposed progress in thought. Political conservatism has brought the Churches into conflict with whatever was vigorous in the working classes, and has spread free thought in wide circles which might otherwise have remained orthodox for centuries. The decay of dogmatic religion is, for good or evil, one of the most important facts in the modern world. Its effects have hardly yet begun to show themselves: what they will be it is impossible to say, but they will certainly be profound and far-reaching.

Religion is partly personal, partly social: to the Protestant primarily personal, to the Catholic primarily social. It is only when the two elements are intimately blended that religion becomes a powerful force in moulding society. The Catholic Church, as it existed from the time of Constantine to the time of the Reformation, represented a blending which would have seemed incredible if it had not been actually achieved, the blending of Christ and Cæsar, of the morality of humble submission with the pride of Imperial Rome. Those who loved the one could find it in the Thebaid; those who loved the other could admire it in the pomp of metropolitan archbishops. In St. Francis and Innocent III the same two sides of the Church are still represented. But since the Reformation personal religion has been increasingly outside the Catholic Church, while the religion which has remained Catholic has been increasingly a matter of institutions and politics and historic continuity. This division has weakened the force of religion: religious bodies have not been strengthened by the enthusiasm and single-mindedness of the men in whom personal religion is strong, and these men have not found their teaching diffused and made permanent by the power of ecclesiastical institutions.

The Catholic Church achieved, during the Middle Ages, the most organic society and the most harmonious inner synthesis of instinct, mind, and spirit, that the

Western world has ever known. St. Francis, Thomas Aquinas, and Dante represent its summit as regards individual development. The cathedrals, the mendicant Orders, and the triumph of the Papacy over the Empire represent its supreme political success. But the perfection which had been achieved was a narrow perfection: instinct, mind, and spirit all suffered from curtailment in order to fit into the pattern; laymen found themselves subject to the Church in ways which they resented, and the Church used its power for rapacity and oppression. The perfect synthesis was an enemy to new growth, and after the time of Dante all that was living in the world had first to fight for its right to live against the representatives of the old order. This fight is even now not ended. Only when it is quite ended, both in the external world of politics and in the internal world of men's own thoughts, will it be possible for a new organic society and a new inner synthesis to take the place which the Church held for a thousand years.

The clerical profession suffers from two causes, one of which it shares with some other professions, while the other is peculiar to itself. The cause peculiar to it is the convention that clergymen are more virtuous than other men. Any average selection of mankind, set apart and told that it excels the rest in virtue, must tend to sink below the average. This is an ancient commonplace in regard to princes and those who used to be called "the great." But it is no less true as regards those of the clergy who are not genuinely and by nature as much better than the average as they are conventionally supposed to be. The other source of harm to the clerical profession is endowments. Property which is only available for those who will support an established institution has a tendency to warp men's judgments as to the excellence of the institution. The tendency is aggravated when the property is associated with social consideration and opportunities for petty power. It is at its worst when the institution is tied by law to an ancient creed, almost impossible to change, and yet quite out of touch with the unfettered thought of the present day. All these causes combine to damage the moral force of the Church.

It is not so much that the creed of the Church is the wrong one. What is amiss is the mere existence of a creed. As soon as income, position, and power are dependent upon acceptance of no matter what creed, intellectual honesty is imperiled. Men will tell themselves that a formal assent is justified by the good which it will enable them to do. They fail to realize that, in those whose mental life has any vigor, loss of complete intellectual integrity puts an end to the power of doing good, by producing gradually in all directions an inability to see truth simply. The strictness of party discipline has introduced the same evil in politics; there, because the evil is comparatively new, it is visible to many who think it unimportant as regards the Church. But the evil is greater as regards the Church, because religion is of more importance than politics, and because it is more necessary that the exponents of religion should be wholly free from taint.

The evils we have been considering seem inseparable from the existence of a professional priesthood. If religion is not to be harmful in a world of rapid change, it must, like the Society of Friends, be carried on by men who have other occupations during the week, who do their religious work from enthusiasm, without receiving any payment. And such men, because they know the everyday world, are not likely to fall into a remote morality which no one regards as applicable to common life. Being free, they will not be bound to reach certain conclusions decided in advance, but will be able to consider moral and religious questions genuinely, without bias. Except in a quite stationary society, no religious life can be living or a real support to the spirit unless it is freed from the incubus of a professional priesthood.

It is largely for these reasons that so little of what is valuable in morals and religion comes nowadays from the men who are eminent in the religious world. It is true that among professed believers there are many who are wholly sincere, who feel still the inspiration which Christianity brought before it had been weakened by the progress of knowledge. These sincere believers are valuable to the world because they keep alive the conviction that the life of the spirit is what is of most importance to men and women. Some of them, in all the countries now at war, have had the courage to preach peace and love in the name of Christ, and have done what lay in their power to mitigate the bitterness of hatred. All praise is due to these men, and without them the world would be even worse than it is.

But it is not through even the most sincere and courageous believers in the traditional religion that a new spirit can come into the world. It is not through them that religion can be brought back to those who have lost it because their minds were active, not because their spirit was dead. Believers in the traditional religion necessarily look to the past for inspiration rather than to the future. They seek wisdom in the teaching of Christ, which, admirable as it is, remains quite inadequate for many of the social and spiritual issues of modern life. Art and intellect and all the problems of government are ignored in the Gospels. Those who, like Tolstoy, endeavor seriously to take the Gospels as a guide to life are compelled to regard the ignorant peasant as the best type of man, and to brush aside political questions by an extreme and impracticable anarchism.

If a religious view of life and the world is ever to reconquer the thoughts and feelings of free-minded men and women, much that we are accustomed to associate with religion will have to be discarded. The first and greatest change that is required is to establish a morality of initiative, not a morality of submission, a morality of hope rather than fear, of things to be done rather than of things to be left undone. It is not the whole duty of man to slip through the world so as to escape the wrath of God. The world is *our* world, and it rests with us to make it a heaven or a hell. The power is ours, and the kingdom and the glory would be ours also if we had courage and insight to create them. The religious life that we must seek will not be one of occasional solemnity and superstitious prohibitions, it will not be sad or ascetic, it will concern itself little with rules of conduct. It will be inspired by a vision of what human life may be, and will be happy with the joy of creation, living in a large free world of initiative and hope. It will love mankind, not for what they are to the outward eye, but for what imagination shows that they have it in them to become. It will not readily condemn, but it will give praise to positive achievement rather than negative sinlessness, to the joy of life, the quick affection, the creative insight, by which the world may grow young and beautiful and filled with vigor.

"Religion" is a word which has many meanings and a long history. In origin, it was concerned with certain rites, inherited from a remote past, performed originally for some reason long since forgotten, and associated from time to time with various myths to account for their supposed importance. Much of this lingers still. A religious man is one who goes to church, a communicant, one who "practises," as Catholics say. How he behaves otherwise, or how he feels concerning life and man's place in the world, does not bear upon the question whether he is "religious" in this simple but historically correct sense. Many men and women are religious in this sense without having in their natures anything that deserves to be called religion in the sense in which I mean the word.The mere familiarity of the Church service has made them impervious to it; they are unconscious of all the history and human experience by which the liturgy has been enriched, and unmoved by the glibly repeated words of the Gospel, which condemn almost all the activities of those who fancy themselves disciples of Christ. This fate must overtake any habitual rite: it is impossible that it should continue to produce much effect after it has been performed so often as to grow mechanical.

The activities of men may be roughly derived from three sources, not in actual fact sharply separate one from another, but sufficiently distinguishable to deserve different names. The three sources I mean are instinct, mind, and spirit, and of these three it is the life of the spirit that makes religion.

The life of instinct includes all that man shares with the lower animals, all that is concerned with self-preservation and reproduction and the desires and impulses derivative from these. It includes vanity and love of possessions, love of family, and even much of what makes love of country. It includes all the impulses that are essentially concerned with the biological success of oneself or one's group—for among gregarious animals the life of instinct includes the group. The impulses which it includes may not in fact make for success, and may often in fact militate against it, but are nevertheless those of which success is the *raison d'être*, those which express the animal nature of man and his position among a world of competitors.

The life of the mind is the life of pursuit of knowledge, from mere childish curiosity up to the greatest efforts of thought. Curiosity exists in animals, and serves an obvious biological purpose; but it is only in men that it passes beyond the investigation of particular objects which may be edible or poisonous, friendly or hostile. Curiosity is the primary impulse out of which the whole edifice of scientific knowledge has grown. Knowledge has been found so useful that most actual acquisition of it is no longer prompted by curiosity; innumerable other motives now contribute to foster the intellectual life. Nevertheless, direct love of knowledge and dislike of error still play a very large part, especially with those who are most successful in learning. No man acquires much knowledge unless the acquisition is in itself delightful to him, apart from any consciousness of the use to which the knowledge may be put. The impulse to acquire knowledge and the activities which center round it constitute what I mean by the life of the mind. The life of the mind consists of thought which is wholly or partially impersonal, in the sense that it concerns itself with objects on their own account, and not merely on account of their bearing upon our instinctive life.

The life of the spirit centers round impersonal feeling, as the life of the mind centers round impersonal thought. In this sense, all art belongs to the life of the spirit, though its greatness is derived from its being also intimately bound up with the life of instinct. Art starts from instinct and rises into the region of the spirit; religion starts from the spirit and endeavors to dominate and inform the life of instinct. It is possible to feel the same interest in the joys and sorrows of others as in our own, to love and hate independently of all relation to ourselves, to care about the destiny of man and the development of the universe without a thought that we are personally involved. Reverence and worship, the sense of an obligation to mankind, the feeling of imperativeness and acting under orders which traditional religion has interpreted as Divine inspiration, all belong to the life of the spirit. And deeper than all these lies the sense of a mystery half revealed, of a hidden wisdom and glory, of a transfiguring vision in which common things lose their solid importance and become a thin veil behind which the ultimate truth of the world is dimly seen. It is such feelings that are the source of religion, and if they were to die most of what is best would vanish out of life.

Instinct, mind, and spirit are all essential to a full life; each has its own excellence and its own corruption. Each can attain a spurious excellence at the expense of the others; each has a tendency to encroach upon the others; but in the life which is to be sought all three will be developed in coördination, and intimately blended in a single harmonious whole. Among uncivilized men instinct is supreme, and mind and spirit hardly exist. Among educated men at the present day mind is developed, as a rule, at the expense of both instinct and spirit, producing a curious inhumanity and lifelessness, a paucity of both personal and impersonal desires, which leads to cynicism and intellectual destructiveness. Among ascetics and most of those who would be called saints, the life of the spirit has been developed at the expense of instinct and mind, producing an outlook which is impossible to those who have a healthy animal life and to those who have a love of active thought. It is not in any of these one-sided developments that we can find wisdom or a philosophy which will bring new life to the civilized world.

Among civilized men and women at the present day it is rare to find instinct, mind, and spirit in harmony. Very few have achieved a practical philosophy which gives its due place to each; as a rule, instinct is at war with either mind or spirit, and mind and spirit are at war with each other. This strife compels men and women to direct much of their energy inwards, instead of being able to expend it all in objective activities. When a man achieves a precarious inward peace by the defeat of a part of his nature, his vital force is impaired, and his growth is no longer quite healthy. If men are to remain whole, it is very necessary that they should achieve a reconciliation of instinct, mind, and spirit.

Instinct is the source of vitality, the bond that unites the life of the individual with the life of the race, the basis of all profound sense of union with others, and the means by which the collective life nourishes the life of the separate units. But instinct by itself leaves us powerless to control the forces of Nature, either in ourselves or in our physical environment, and keeps us in bondage to the same unthinking impulse by which the trees grow. Mind can liberate us from this bondage, by the power of impersonal thought, which enables us to judge critically the purely biological purposes towards which instinct more or less blindly tends. But mind, in its dealings with instinct, is *merely* critical: so far as instinct is concerned, the unchecked activity of the mind is apt to be destructive and to generate cynicism. Spirit is an antidote to the cynicism of mind: it universalizes the emotions that spring from instinct, and by universalizing them makes them impervious to mental criticism. And when thought is informed by spirit it loses its cruel, destructive quality; it no longer promotes the death of instinct, but only its purification from insistence and ruthlessness and its emancipation from the prison walls of accidental circumstance. It is instinct that gives force, mind that gives the means of directing force to desired ends, and spirit that suggests impersonal uses for force of a kind that thought cannot discredit by criticism. This is an outline of the parts that instinct, mind, and spirit would play in a harmonious life.

Instinct, mind, and spirit are each a help to the others when their development is free and unvitiated; but when corruption comes into any one of the three, not only does that one fail, but the others also become poisoned. All three must grow together. And if they are to grow to their full stature in any one man or woman, that man or woman must not be isolated, but must be one of a society where growth is not thwarted and made crooked.

The life of instinct, when it is unchecked by mind or spirit, consists of instinctive cycles, which begin with impulses to more or less definite acts, and pass on to satisfaction of needs through the consequences of these impulsive acts. Impulse and desire are not directed towards the whole cycle, but only towards its initiation: the rest is left to natural causes. We desire to eat, but we do not desire to be nourished unless we are valetudinarians. Yet without the nourishment eating is a mere momentary pleasure, not part of the general impulse to life. Men desire sexual intercourse, but they do not as a rule desire children strongly or often. Yet without the hope of children and its occasional realization, sexual intercourse remains for most people an isolated and separate pleasure, not uniting their personal life with the life of mankind, not continuous with the central purposes by which they live, and not capable of bringing that profound sense of fulfilment which comes from completion by children. Most men, unless the impulse is atrophied through disuse, feel a desire to create something, great or small according to their capacities. Some few are able to satisfy this desire: some happy men can create an Empire, a science, a poem, or a picture. The men of science, who have less difficulty than any others in finding an outlet for creativeness, are the happiest of intelligent men in the modern world, since their creative activity affords full satisfaction to mind and spirit as well as to the instinct of creation.In them a beginning is to be seen of the new way of life which is to be sought; in their happiness we may perhaps find the germ of a future happiness for all mankind. The rest, with few exceptions, are thwarted in their creative impulses. They cannot build their own house or make their own garden, or direct their own labor to producing what their free choice would lead them to produce. In this way the instinct of creation, which should lead on to the life of mind and spirit, is checked and turned aside. Too often it is turned to destruction, as the only effective action which remains possible. Out of its defeat grows envy, and out of envy grows the impulse to destroy the creativeness of more fortunate men. This is one of the greatest sources of corruption in the life of instinct.

The life of instinct is important, not only on its own account, or because of the direct usefulness of the actions which it inspires, but also because, if it is unsatisfactory, the individual life becomes detached and separated from the general life of man. All really profound sense of unity with others depends upon instinct, upon coöperation or agreement in some instinctive purpose. This is most obvious in the relations of men and women and parents and children. But it is true also in wider relations. It is true of large assemblies swayed by a strong common emotion, and even of a whole nation in times of stress. It is part of what makes the value of religion as a social institution. Where this feeling is wholly absent, other human beings seem distant and aloof. Where it is actively thwarted, other human beings become objects of instinctive hostility. The aloofness or the instinctive hostility may be masked by religious love, which can be given to all men regardless of their relation to ourselves. But religious love does not bridge the gulf that parts man from man: it looks across the gulf, it views others with compassion or impersonal sympathy, but it does not live with the same life with which they live. Instinct alone can do this, but only when it is fruitful and sane and direct. To this end it is necessary that instinctive cycles should be fairly often completed, not interrupted in the middle of their course. At present they are constantly interrupted, partly by purposes which conflict with them for economic or other reasons, partly by the pursuit of pleasure, which picks out the most agreeable part of the cycle and avoids the rest. In this way instinct is robbed of its importance and seriousness; it becomes incapable of bringing any real fulfilment, its demands grow more and more excessive, and life becomes no longer a whole with a single movement, but a series of detached moments, some of them pleasurable, most of them full of weariness and discouragement.

The life of the mind, although supremely excellent in itself, cannot bring health into the life of instinct, except when it results in a not too difficult outlet for the instinct of creation. In other cases it is, as a rule, too widely separated from instinct, too detached, too destitute of inward growth, to afford either a vehicle for instinct or a means of subtilizing and refining it. Thought is in its essence impersonal and detached, instinct is in its essence personal and tied to particular circumstances: between the two, unless both reach a high level, there is a war which is not easily appeased. This is the fundamental reason for vitalism, futurism, pragmatism, and the various other philosophies which advertise themselves as vigorous and virile. All these represent the attempt to find a mode of thought which shall not be hostile to instinct. The attempt, in itself, is deserving of praise, but the solution offered is far too facile. What is proposed amounts to a subordination of thought to instinct, a refusal to allow thought to achieve its own ideal. Thought which does not rise above what is personal is not thought in any true sense: it is merely a more or less intelligent use of instinct. It is thought and spirit that raise man above the level of the brutes. By discarding them we may lose the proper excellence of men, but cannot acquire the excellence of animals. Thought must achieve its full growth before a reconciliation with instinct is attempted.

When refined thought and unrefined instinct coexist, as they do in many intellectual men, the result is a complete disbelief in any important good to be achieved by the help of instinct. According to their disposition, some such men will as far as possible discard instinct and become ascetic, while others will accept it as a necessity, leaving it degraded and separated from all that is really important in their lives. Either of these courses prevents instinct from remaining vital, or from being a bond with others; either produces a sense of physical solitude, a gulf across which the minds and spirits of others may speak, but not their instincts. To very many men, the instinct of patriotism, when the war broke out, was the first instinct that had bridged the gulf, the first that had made them feel a really profound unity with others.This instinct, just because, in its intense form, it was new and unfamiliar, had remained uninfected by thought, not paralyzed or devitalized by doubt and cold detachment. The sense of unity which it brought is capable of being brought by the instinctive life of more normal times, if thought and spirit are not hostile to it. And so long as this sense of unity is absent, instinct and spirit cannot be in harmony, nor can the life of the community have vigor and the seeds of new growth.

The life of the mind, because of its detachment, tends to separate a man inwardly from other men, so long as it is not balanced by the life of the spirit. For this reason, mind without spirit can render instinct corrupt or atrophied, but cannot add any excellence to the life of instinct. On this ground, some men are hostile to thought. But no good purpose is served by trying to prevent the growth of thought, which has its own insistence, and if checked in the directions in which it tends naturally, will turn into other directions where it is more harmful. And thought is in itself god-like: if the opposition between thought and instinct were irreconcilable, it would be thought that ought to conquer. But the opposition is not irreconciliable:all that is necessary is that both thought and instinct should be informed by the life of the spirit.

In order that human life should have vigor, it is necessary for the instinctive impulses to be strong and direct; but in order that human life should be good, these impulses must be dominated and controlled by desires less personal and ruthless, less liable to lead to conflict than those that are inspired by instinct alone. Something impersonal and universal is needed over and above what springs out of the principle of individual growth. It is this that is given by the life of the spirit.

Patriotism affords an example of the kind of control which is needed. Patriotism is compounded out of a number of instinctive feelings and impulses: love of home, love of those whose ways and outlook resemble our own, the impulse to coöperation in a group, the sense of pride in the achievements of one's group. All these impulses and desires, like everything belonging to the life of instinct, are personal, in the sense that the feelings and actions which they inspire towards others are determined by the relation of those others to ourselves, not by what those others are intrinsically. All these impulses and desires unite to produce a love of man's own country which is more deeply implanted in the fiber of his being, and more closely united to his vital force, than any love not rooted in instinct. But if spirit does not enter in to generalize love of country, the exclusiveness of instinctive love makes it a source of hatred of other countries. What spirit can effect is to make us realize that other countries equally are worthy of love, that the vital warmth which makes us love our own country reveals to us that it deserves to be loved, and that only the poverty of our nature prevents us from loving all countries as we love our own. In this way instinctive love can be extended in imagination, and a sense of the value of all mankind can grow up, which is more living and intense than any that is possible to those whose instinctive love is weak. Mind can only show us that it is irrational to love our own country best; it can weaken patriotism, but cannot strengthen the love of all mankind. Spirit alone can do this, by extending and universalizing the love that is born of instinct. And in doing this it checks and purifies whatever is insistent or ruthless or oppressively personal in the life of instinct.

The same extension through spirit is necessary with other instinctive loves, if they are not to be enfeebled or corrupted by thought. The love of husband and wife is capable of being a very good thing, and when men and women are sufficiently primitive nothing but instinct and good fortune is needed to make it reach a certain limited perfection. But as thought begins to assert its right to criticize instinct the old simplicity becomes impossible. The love of husband and wife, as unchecked instinct leaves it, is too narrow and personal to stand against the shafts of satire, until it is enriched by the life of the spirit. The romantic view of marriage, which our fathers and mothers professed to believe, will not survive an imaginative peregrination down a street of suburban villas, each containing its couple, each couple having congratulated themselves as they first crossed the threshold, that here they could love in peace, without interruption from others, without contact with the cold outside world. The separateness and stuffiness, the fine names for cowardices and timid vanities, that are shut within the four walls of thousands upon thousands of little villas, present themselves coldly and mercilessly to those in whom mind is dominant at the expense of spirit.

Nothing is good in the life of a human being except the very best that his nature can achieve. As men advance, things which have been good cease to be good, merely because something better is possible. So it is with the life of instinct: for those whose mental life is strong, much that was really good while mind remained less developed has now become bad merely through the greater degree of truth in their outlook on the world. The instinctive man in love feels that his emotion is unique, that the lady of his heart has perfections such as no other woman ever equaled. The man who has acquired the power of impersonal thought realizes, when he is in love, that he is one of so many millions of men who are in love at this moment, that not more than one of all the millions can be right in thinking his love supreme, and that it is not likely that that one is oneself. He perceives that the state of being in love in those whose instinct is unaffected by thought or spirit, is a state of illusion, serving the ends of Nature and making a man a slave to the life of the species, not a willing minister to the impersonal ends which he sees to be good. Thought rejects this slavery; for no end that Nature may have in view will thought abdicate, or forgo its right to think truly. "Better the world should perish than that I or any other human being should believe a lie"—this is the religion of thought, in whose scorching flames the dross of the world is being burnt away. It is a good religion, and its work of destruction must be completed. But it is not all that man has need of. New growth must come after the destruction, and new growth can come only through the spirit.

Both patriotism and the love of man and woman, when they are merely instinctive, have the same defects: their exclusions, their enclosing walls, their indifference or hostility to the outside world. It is through this that thought is led to satire, that comedy has infected what men used to consider their holiest feelings. The satire and the comedy are justified, but not the death of instinct which they may produce if they remain in supreme command. They are justified, not as the last word of wisdom but as the gateway of pain through which men pass to a new life, where instinct is purified and yet nourished by the deeper desires and insight of spirit.

The man who has the life of the spirit within him views the love of man and woman, both in himself and in others, quite differently from the man who is exclusively dominated by mind. He sees, in his moments of insight, that in all human beings there is something deserving of love, something mysterious, something appealing, a cry out of the night, a groping journey, and a possible victory. When his instinct loves, he welcomes its help in seeing and feeling the value of the human being whom he loves. Instinct becomes a reinforcement to spiritual insight. What instinct tells him spiritual insight confirms, however much the mind may be aware of littlenesses, limitations, and enclosing walls that prevent the spirit from shining forth. His spirit divines in all men what his instinct shows him in the object of his love.

The love of parents for children has need of the same transformation. The purely instinctive love, unchecked by thought, uninformed by spirit, is exclusive, ruthless, and unjust. No benefit to others is felt, by the purely instinctive parent, to be worth an injury to one's own children. Honor and conventional morality place certain important practical limitations on the vicarious selfishness of parents, since a civilized community exacts a certain minimum before it will give respect. But within the limits allowed by public opinion, parental affection,when it is merely instinctive, will seek the advantage of children without regard to others. Mind can weaken the impulse to injustice, and diminish the force of instinctive love, but it cannot keep the whole force of instinctive love and turn it to more universal ends. Spirit can do this. It can leave the instinctive love of children undimmed, and extend the poignant devotion of a parent, in imagination, to the whole world. And parental love itself will prompt the parent who has the life of the spirit to give to his children the sense of justice, the readiness for service, the reverence, the will that controls self-seeking, which he feels to be a greater good than any personal success.

The life of the spirit has suffered in recent times by its association with traditional religion, by its apparent hostility to the life of the mind, and by the fact that it has seemed to center in renunciation. The life of the spirit demands readiness for renunciation when the occasion arises, but is in its essence as positive and as capable of enriching individual existence as mind and instinct are. It brings with it the joy of vision, of the mystery and profundity of the world, of the contemplation of life, and above all the joy of universal love. It liberates those who have it from the prison-house of insistent personal passion and mundane cares. It gives freedom and breadth and beauty to men's thoughts and feelings, and to all their relations with others. It brings the solution of doubts, the end of the feeling that all is vanity. It restores harmony between mind and instinct, and leads the separated unit back into his place in the life of mankind. For those who have once entered the world of thought, it is only through spirit that happiness and peace can return.

VIII
WHAT WE CAN DO

WHAT can we do for the world while we live?

Many men and women would wish to serve mankind, but they are perplexed and their power seems infinitesimal. Despair seizes them; those who have the strongest passion suffer most from the sense of impotence, and are most liable to spiritual ruin through lack of hope.

So long as we think only of the immediate future, it seems that what we can do is not much. It is probably impossible for us to bring the war to an end. We cannot destroy the excessive power of the State or of private property. We cannot, here and now, bring new life into education. In such matters, though we may see the evil, we cannot quickly cure it by any of the ordinary methods of politics. We must recognize that the world is ruled in a wrong spirit, and that a change of spirit will not come from one day to the next. Our expectations must not be for to-morrow, but for the time when what is thought now by a few shall have become the common thought of many. If we have courage and patience, we can think the thoughts and feel the hopes by which, sooner or later, men will be inspired, and weariness and discouragement will be turned into energy and ardor. For this reason, the first thing we have to do is to be clear in our own minds as to the kind of life we think good and the kind of change that we desire in the world.

The ultimate power of those whose thought is vital is far greater than it seems to men who suffer from the irrationality of contemporary politics. Religious toleration was once the solitary speculation of a few bold philosophers. Democracy, as a theory, arose among a handful of men in Cromwell's army; by them, after the Restoration, it was carried to America, where it came to fruition in the War of Independence. From America, Lafayette and the other Frenchmen who fought by the side of Washington brought the theory of democracy to France, where it united itself with the teaching of Rousseau and inspired the Revolution. Socialism, whatever we may think of its merits,is a great and growing power, which is transforming economic and political life; and socialism owes its origin to a very small number of isolated theorists. The movement against the subjection of women, which has become irresistible and is not far from complete triumph, began in the same way with a few impracticable idealists—Mary Wollstonecraft, Shelley, John Stuart Mill. The power of thought, in the long run, is greater than any other human power. Those who have the ability to think and the imagination to think in accordance with men's needs, are likely to achieve the good they aim at sooner or later, though probably not while they are still alive.

But those who wish to gain the world by thought must be content to lose it as a support in the present. Most men go through life without much questioning, accepting the beliefs and practices which they find current, feeling that the world will be their ally if they do not put themselves in opposition to it. New thought about the world is incompatible with this comfortable acquiescence; it requires a certain intellectual detachment, a certain solitary energy, a power of inwardly dominating the world and the outlook that the world engenders. Without some willingness to be lonely new thought cannot be achieved. And it will not be achieved to any purpose if the loneliness is accompanied by aloofness, so that the wish for union with others dies, or if intellectual detachment leads to contempt. It is because the state of mind required is subtle and difficult, because it is hard to be intellectually detached yet not aloof, that fruitful thought on human affairs is not common, and that most theorists are either conventional or sterile. The right kind of thought is rare and difficult, but it is not impotent. It is not the fear of impotence that need turn us aside from thought if we have the wish to bring new hope into the world.

In seeking a political theory which is to be useful at any given moment, what is wanted is not the invention of a Utopia, but the discovery of the best direction of movement. The direction which is good at one time may be superficially very different from that which is good at another time. Useful thought is that which indicates the right direction for the present time. But in judging what is the right direction there are two general principles which are always applicable.

1. The growth and vitality of individuals and communities is to be promoted as far as possible.

2. The growth of one individual or one community is to be as little as possible at the expense of another.

The second of these principles, as applied by an individual in his dealings with others, is the principle of *reverence*, that the life of another has the same importance which we feel in our own life. As applied impersonally in politics, it is the principle of *liberty*, or rather it includes the principle of liberty as a part. Liberty in itself is a negative principle; it tells us not to interfere, but does not give any basis for construction. It shows that many political and social institutions are bad and ought to be swept away, but it does not show what ought to be put in their place. For this reason a further principle is required, if our political theory is not to be purely destructive.

The combination of our two principles is not in practice an easy matter. Much of the vital energy of the world runs into channels which are oppressive. The Germans have shown themselves extraordinarily full of vital energy, but unfortunately in a form which seems incompatible with the vitality of their neighbors. Europe in general has more vital energy than Africa, but it has used its energy to drain Africa, through industrialism, of even such life as the negroes possessed. The vitality of southeastern Europe is being drained to supply cheap labor for the enterprise of American millionaires. The vitality of men has been in the past a hindrance to the development of women, and it is possible that in the near future women may become a similar hindrance to men. For such reasons the principle of reverence, though not in itself sufficient, is of very great importance, and is able to indicate many of the political changes that the world requires.

In order that both principles may be capable of being satisfied, what is needed is a unifying or integration, first of our individual lives, then of the life of the community and of the world, without sacrifice of individuality. The life of an individual, the life of a community, and even the life of mankind, ought to be, not a number of separate fragments but in some sense a whole. When this is the case, the growth of the individual is fostered, and is not incompatible with the growth of other individuals. In this way the two principles are brought into harmony.

What integrates an individual life is a consistent creative purpose or unconscious direction. Instinct alone will not suffice to give unity to the life of a civilized man or woman: there must be some dominant object, an ambition, a desire for scientific or artistic creation, a religious principle, or strong and lasting affections. Unity of life is very difficult for a man or woman who has suffered a certain kind of defeat, the kind by which what should have been the dominant impulse is checked and made abortive. Most professions inflict this kind of defeat upon a man at the very outset. If a man becomes a journalist, he probably has to write for a newspaper whose politics he dislikes; this kills his pride in work and his sense of independence. Most medical men find it very hard to succeed without humbug, by which whatever scientific conscience they may have had is destroyed. Politicians are obliged, not only to swallow the party program but to pretend to be saints, in order to conciliate religious supporters; hardly any man can enter Parliament without hypocrisy. In no profession is there any respect for the native pride without which a man cannot remain whole; the world ruthlessly crushes it out, because it implies independence, and men desire to enslave others more than they desire to be free themselves. Inward freedom is infinitely precious, and a society which will preserve it is immeasurably to be desired.

The principle of growth in a man is not crushed necessarily by preventing him from doing some definite thing, but it is often crushed by persuading him to do something else. The things that crush growth are those that produce a sense of impotence in the directions in which the vital impulse wishes to be effective. The worst things are those to which the will assents. Often, chiefly from failure of self-knowledge, a man's will is on a lower level than his impulse: his impulse is towards some kind of creation, while his will is towards a conventional career, with a sufficient income and the respect of his contemporaries. The stereotyped illustration is the artist who produces shoddy work to please the public. But something of the artist's definiteness of impulse exists in very many men who are not artists. Because the impulse is deep and dumb, because what is called common sense is often against it, because a young man can only follow it if he is willing to set up his own obscure feelings against the wisdom and prudent maxims of elders and friends, it happens in ninety-nine cases out of a hundred that the creative impulse, out of which a free and vigorous life might have sprung, is checked and thwarted at the very outset: the young man consents to become a tool, not an independent workman; a mere means to the fulfilment of others, not the artificer of what his own nature feels to be good. In the moment when he makes this act of consent something dies within him. He can never again become a whole man, never again have the undamaged self-respect, the upright pride, which might have kept him happy in his soul in spite of all outward troubles and difficulties—except, indeed, through conversion and a fundamental change in his way of life.

Outward prohibitions, to which the will gives no assent, are far less harmful than the subtler inducements which seduce the will. A serious disappointment in love may cause the most poignant pain, but to a vigorous man it will not do the same inward damage as is done by marrying for money. The achievement of this or that special desire is not what is essential: what is essential is the direction, the *kind* of effectiveness which is sought. When the fundamental impulse is opposed by will, it is made to feel helpless: it has no longer enough hope to be powerful as a motive. Outward compulsion does not do the same damage unless it produces the same sense of impotence; and it will not produce the same sense of impotence if the impulse is strong and courageous. Some thwarting of special desires is unavoidable even in the best imaginable community, since some men's desires, unchecked, lead to the oppression or destruction of others. In a good community Napoleon could not have been allowed the profession of his choice, but he might have found happiness as a pioneer in Western America. He could not have found happiness as a City clerk, and no tolerable organization of society would compel him to become a City clerk.

The integration of an individual life requires that it should embody whatever creative impulse a man may possess, and that his education should have been such as to elicit and fortify this impulse. The integration of a community requires that the different creative impulses of different men and women should work together towards some common life, some common purpose, not necessarily conscious, in which all the members of the community find a help to their individual fulfilment. Most of the activities that spring from vital impulses consist of two parts: one creative, which furthers one's own life and that of others with the same kind of impulse or circumstances, and one possessive, which hinders the life of some group with a different kind of impulse or circumstances.

For this reason, much of what is in itself most vital may nevertheless work against life, as, for example, seventeenth-century Puritanism did in England, or as nationalism does throughout Europe at the present day. Vitality easily leads to strife or oppression, and so to loss of vitality. War, at its outset, integrates the life of a nation, but it disintegrates the life of the world, and in the long run the life of a nation too, when it is as severe as the present war.

The war has made it clear that it is impossible to produce a secure integration of the life of a single community while the relations between civilized countries are governed by aggressiveness and suspicion. For this reason any really powerful movement of reform will have to be international. A merely national movement is sure to fail through fear of danger from without. Those who desire a better world, or even a radical improvement in their own country, will have to cooperate with those who have similar desires in other countries, and to devote much of their energy to overcoming that blind hostility which the war has intensified. It is not in partial integrations, such as patriotism alone can produce, that any ultimate hope is to be found. The problem is, in national and international questions as in the individual life, to keep what is creative in vital impulses, and at the same time to turn into other channels the part which is at present destructive.

Men's impulses and desires may be divided into those that are creative and those that are possessive. Some of our activities are directed to creating what would not otherwise exist, others are directed towards acquiring or retaining what exists already. The typical creative impulse is that of the artist; the typical possessive impulse is that of property. The best life is that in which creative impulses play the largest part and possessive impulses the smallest. The best institutions are those which produce the greatest possible creativeness and the least possessiveness compatible with self-preservation. Possessiveness may be defensive or aggressive: in the criminal law it is defensive, and in criminals it is aggressive. It may perhaps be admitted that the criminal law is less abominable than the criminal, and that defensive possessiveness is unavoidable so long as aggressive possessiveness exists. But not even the most purely defensive forms of possessiveness are in themselves admirable; indeed, as soon as they are strong they become hostile to the creative impulses. "Take no thought, saying, What shall we eat? or What shall we drink, or Wherewithal shall we be clothed?" Whoever has known a strong creative impulse has known the value of this precept in its exact and literal sense: it is preoccupation with possessions, more than anything else, that prevents men from living freely and nobly. The State and Property are the great embodiments of possessiveness; it is for this reason that they are against life, and that they issue in war. Possession means taking or keeping some good thing which another is prevented from enjoying; creation means putting into the world a good thing which otherwise no one would be able to enjoy. Since the material goods of the world must be divided among the population, and since some men are by nature brigands, there must be defensive possession, which will be regulated, in a good community, by some principle of impersonal justice. But all this is only the preface to a good life or good political institutions, in which creation will altogether outweigh possession, and distributive justice will exist as an uninteresting matter of course.

The supreme principle, both in politics and in private life, should be *to promote all that is creative, and so to diminish the impulses and desires that center round possession*. The State at present is very largely an embodiment of possessive impulses: internally, it protects the rich against the poor; externally, it uses force for the exploitation of inferior races, and for competition with other States. Our whole economic system is concerned exclusively with possession; yet the production of goods is a form of creation, and except in so far as it is irredeemably mechanical and monotonous, it might afford a vehicle for creative impulses. A great deal might be achieved towards this end by forming the producers of a certain kind of commodity into an autonomous democracy, subject to State control as regards the price of their commodity but not as to the manner of its production.

Education, marriage, and religion are essentially creative, yet all three have been vitiated by the intrusion of possessive motives. Education is usually treated as a means of prolonging the *status quo* by instilling prejudices, rather than of creating free thought and a noble outlook by the example of generous feeling and the stimulus of mental adventure. In marriage, love, which is creative, is kept in chains by jealousy, which is possessive. Religion, which should set free the creative vision of the spirit, is usually more concerned to repress the life of instinct and to combat the subversiveness of thought. In all these ways the fear that grows out of precarious possession has replaced the hope inspired by creative force. The wish to plunder others is recognized, in theory, to be bad; but the fear of being plundered is little better. Yet these two motives between them dominate nine-tenths of politics and private life.

The creative impulses in different men are essentially harmonious, since what one man creates cannot be a hindrance to what another is wishing to create. It is the possessive impulses that involve conflict. Although, morally and politically, the creative and possessive impulses are opposites, yet psychologically either passes easily into the other, according to the accidents of circumstance and opportunity. The genesis of impulses and the causes which make them change ought to be studied; education and social institutions ought to be made such as to strengthen the impulses which harmonize in different men, and to weaken those that involve conflict. I have no doubt that what might be accomplished in this way is almost unlimited.

It is rather through impulse than through will that individual lives and the life of the community can derive the strength unity of a single direction. Will is of two kinds, of which one is directed outward and the other inward. The first, which is directed outward, is called into play by external obstacles, either the opposition of others or the technical difficulties of an undertaking. This kind of will is an expression of strong impulse or desire, whenever instant success is impossible; it exists in all whose life is vigorous, and only decays when their vital force is enfeebled. It is necessary to success in any difficult enterprise, and without it great achievement is very rare. But the will which is directed inward is only necessary in so far as there is an inner conflict of impulses or desires; a perfectly harmonious nature would have no occasion for inward will. Such perfect harmony is of course a scarcely realizable ideal: in all men impulses arise which are incompatible with their central purpose, and which must be checked if their life as a whole is not to be a failure. But this will happen least with those whose central impulses are strongest; and it will happen less often in a society which aims at freedom than in a society like ours, which is full of artificial incompatibilities created by antiquated institutions and a tyrannous public opinion. The power to exert inward will when the occasion arises must always be needed by those who wish their lives to embody some central purpose, but with better institutions the occasions when inward will is necessary might be made fewer and less important. This result is very much to be desired, because when will checks impulses which are only accidentally harmful, it diverts a force which might be spent on overcoming outward obstacles, and if the impulses checked are strong and serious, it actually diminishes the vital force available. A life full of inhibitions is likely not to remain a very vigorous life but to become listless and without zest. Impulse tends to die when it is constantly held in check, and if it does not die, it is apt to work underground, and issue in some form much worse than that in which it has been checked. For these reasons the necessity for using inward will ought to be avoided as much as possible, and consistency of action ought to spring rather from consistency of impulse than from control of impulse by will.

The unifying of life ought not to demand the suppression of the casual desires that make amusement and play; on the contrary, everything ought to be done to make it easy to combine the main purposes of life with all kinds of pleasure that are not in their nature harmful. Such things as habitual drunkenness, drugs, cruel sports, or pleasure in inflicting pain are essentially harmful, but most of the amusements that civilized men naturally enjoy are either not harmful at all or only accidentally harmful through some effect which might be avoided in a better society. What is needed is, not asceticism or a drab Puritanism, but capacity for strong impulses and desires directed towards large creative ends. When such impulses and desires are vigorous, they bring with them, of themselves, what is needed to make a good life.

But although amusement and adventure ought to have their share, it is impossible to create a good life if they are what is mainly desired. Subjectivism, the habit of directing thought and desire to our own states of mind rather than to something objective, inevitably makes life fragmentary and unprogressive. The man to whom amusement is the end of life tends to lose interest gradually in the things out of which he has been in the habit of obtaining amusement, since he does not value these things on their own account, but on account of the feelings which they arouse in him. When they are no longer amusing, boredom drives him to seek some new stimulus, which fails him in its turn. Amusement consists in a series of moments without any essential continuity; a purpose which unifies life is one which requires some prolonged activity, and is like building a monument rather than a child's castle in the sand.

Subjectivism has other forms beside the mere pursuit of amusement. Many men, when they are in love, are more interested in their own emotion than in the object of their love; such love does not lead to any essential union, but leaves fundamental separateness undiminished.As soon as the emotion grows less vivid the experience has served its purpose, and there seems no motive for prolonging it. In another way, the same evil of subjectivism was fostered by Protestant religion and morality, since they directed attention to sin and the state of the soul rather than to the outer world and our relations with it. None of these forms of subjectivism can prevent a man's life from being fragmentary and isolated. Only a life which springs out of dominant impulses directed to objective ends can be a satisfactory whole, or be intimately united with the lives of others.

The pursuit of pleasure and the pursuit of virtue alike suffer from subjectivism: Epicureanism and Stoicism are infected with the same taint. Marcus Aurelius, enacting good laws in order that he might be virtuous, is not an attractive figure. Subjectivism is a natural outcome of a life in which there is much more thought than action: while outer things are being remembered or desired, not actually experienced, they seem to become mere ideas. What they are in themselves becomes less interesting to us than the effects which they produce in our own minds. Such a result tends to be brought about by increasing civilization, because increasing civilization continually diminishes the need for vivid action and enhances the opportunities for thought. But thought will not have this bad result if it is active thought, directed towards achieving some purpose; it is only passive thought that leads to subjectivism. What is needed is to keep thought in intimate union with impulses and desires, making it always itself an activity with an objective purpose. Otherwise, thought and impulse become enemies, to the great detriment of both.

In order to make the lives of average men and women less fragmentary and separate, and to give greater opportunity for carrying out creative impulses, it is not enough to know the goal we wish to reach, or to proclaim the excellence of what we desire to achieve. It is necessary to understand the effect of institutions and beliefs upon the life of impulse, and to discover ways of improving this effect by a change in institutions. And when this intellectual work has been done, our thought will still remain barren unless we can bring it into relation with some powerful political force. The only powerful political force from which any help is to be expected in bringing about such changes as seem needed is Labor. The changes required are very largely such as Labor may be expected to welcome, especially during the time of hardship after the war. When the war is over, labor discontent is sure to be very prevalent throughout Europe, and to constitute a political force by means of which a great and sweeping reconstruction may be effected.

The civilized world has need of fundamental change if it is to be saved from decay—change both in its economic structure and in its philosophy of life. Those of us who feel the need of change must not sit still in dull despair: we can, if we choose, profoundly influence the future. We can discover and preach the kind of change that is required—the kind that preserves what is positive in the vital beliefs of our time, and, by eliminating what is negative and inessential, produces a synthesis to which all that is not purely reactionary can give allegiance. As soon as it has become clear what *kind* of change is required, it will be possible to work out its parts in more detail. But until the war is ended there is little use in detail, since we do not know what kind of world the war will leave. The only thing that seems indubitable is that much new thought will be required in the new world produced by the war. Traditional views will give little help. It is clear that men's most important actions are not guided by the sort of motives that are emphasized in traditional political philosophies. The impulses by which the war has been produced and sustained come out of a deeper region than that of most political argument. And the opposition to the war on the part of those few who have opposed it comes from the same deep region. A political theory, if it is to hold in times of stress, must take account of the impulses that underlie explicit thought: it must appeal to them, and it must discover how to make them fruitful rather than destructive.

Economic systems have a great influence in promoting or destroying life. Except slavery, the present industrial system is the most destructive of life that has ever existed. Machinery and large-scale production are ineradicable, and must survive in any better system which is to replace the one under which we live. Industrial federal democracy is probably the best direction for reform to take.

Philosophies of life, when they are widely believed, also have a very great influence on the vitality of a community. The most widely accepted philosophy of life at present is that what matters most to a man's happiness is his income. This philosophy, apart from other demerits, is harmful because it leads men to aim at a result rather than an activity, an enjoyment of material goods in which men are not differentiated, rather than a creative impulse which embodies each man's individuality. More refined philosophies, such as are instilled by higher education, are too apt to fix attention on the past rather than the future, and on correct behavior rather than effective action. It is not in such philosophies that men will find the energy to bear lightly the weight of tradition and of ever-accumulating knowledge.

The world has need of a philosophy, or a religion, which will promote life. But in order to promote life it is necessary to value something other than mere life. Life devoted only to life is animal without any real human value, incapable of preserving men permanently from weariness and the feeling that all is vanity. If life is to be fully human it must serve some end which seems, in some sense, outside human life, some end which is impersonal and above mankind, such as God or truth or beauty. Those who best promote life do not have life for their purpose. They aim rather at what seems like a gradual incarnation, a bringing into our human existence of something eternal, something that appears to imagination to live in a heaven remote from strife and failure and the devouring jaws of Time. Contact with this eternal world—even if it be only a world of our imagining—brings a strength and a fundamental peace which cannot be wholly destroyed by the struggles and apparent failures of our temporal life. It is this happy contemplation of what is eternal that Spinoza calls the intellectual love of God. To those who have once known it, it is the key of wisdom.

What we have to do practically is different for each one of us, according to our capacities and opportunities. But if we have the life of the spirit within us, what we must do and what we must avoid will become apparent to us.

By contact with what is eternal, by devoting our life to bringing something of the Divine into this troubled world, we can make our own lives creative even now, even in the midst of the cruelty and strife and hatred that surround us on every hand. To make the individual life creative is far harder in a community based on possession than it would be in such a community as human effort may be able to build up in the future. Those who are to begin the regeneration of the world must face loneliness, opposition, poverty, obloquy. They must be able to live by truth and love, with a rational unconquerable hope; they must be honest and wise, fearless, and guided by a consistent purpose. A body of men and women so inspired will conquer—first the difficulties and perplexities of their individual lives, then, in time, though perhaps only in a long time, the outer world. Wisdom and hope are what the world needs; and though it fights against them, it gives its respect to them in the end.

When the Goths sacked Rome, St. Augustine wrote the "City of God," putting a spiritual hope in place of the material reality that had been destroyed. Throughout the centuries that followed St. Augustine's hope lived and gave life, while Rome sank to a village of hovels. For us, too, it is necessary to create a new hope, to build up by our thought a better world than the one which is hurling itself into ruin. Because the times are bad, more is required of us than would be required in normal times. Only a supreme fire of thought and spirit can save future generations from the death that has befallen the generation which we knew and loved.

It has been my good fortune to come in contact as a teacher with young men of many different nations—young men in whom hope was alive, in whom the creative energy existed that would have realized in the world some part at least of the imagined beauty by which they lived. They have been swept into the war, some on one side, some on the other. Some are still fighting, some are maimed for life, some are dead; of those who survive it is to be feared that many will have lost the life of the spirit, that hope will have died, that energy will be spent, and that the years to come will be only a weary journey towards the grave. Of all this tragedy, not a few of those who teach seem to have no feeling: with ruthless logic, they prove that these young men have been sacrificed unavoidably for some coldly abstract end; undisturbed themselves, they lapse quickly into comfort after any momentary assault of feeling. In such men the life of the spirit is dead. If it were living, it would go out to meet the spirit in the young, with a love as poignant as the love of father or mother. It would be unaware of the bounds of self; their tragedy would be its own. Something would cry out: "No, this is not right; this is not good; this is not a holy cause, in which the brightness of youth is destroyed and dimmed. It is we, the old, who have sinned; we have sent these young men to the battlefield for our evil passions, our spiritual death, our failure to live generously out of the warmth of the heart and out of the living vision of the spirit. Let us come out of this death, for it is we who are dead, not the young men who have died through our fear of life. Their very ghosts have more life than we: they hold us up for ever to the shame and obloquy of all the ages to come. Out of their ghosts must come life, and it is we whom they must vivify."

POLITICAL IDEALS

Chapter I

Political Ideals

In dark days, men need a clear faith and a well-grounded hope; and as the outcome of these, the calm courage which takes no account of hardships by the way. The times through which we are passing have afforded to many of us a confirmation of our faith. We see that the things we had thought evil are really evil, and we know more definitely than we ever did before the directions in which men must move if a better world is to arise on the ruins of the one which is now hurling itself into destruction. We see that men's political dealings with one another are based on wholly wrong ideals, and can only be saved by quite different ideals from continuing to be a source of suffering, devastation, and sin.

Political ideals must be based upon ideals for the individual life. The aim of politics should be to make the lives of individuals as good as possible. There is nothing for the politician to consider outside or above the various men, women, and children who compose the world. The problem of politics is to adjust the relations of human beings in such a way that each severally may have as much of good in his existence as possible. And this problem requires that we should first consider what it is that we think good in the individual life.

To begin with, we do not want all men to be alike. We do not want to lay down a pattern or type to which men of all sorts are to be made by some means or another to approximate. This is the ideal of the impatient administrator. A bad teacher will aim at imposing his opinion, and turning out a set of pupils all of whom will give the same definite answer on a doubtful point. Mr. Bernard Shaw is said to hold that *Troilus and Cressida* is the best of Shakespeare's plays. Although I disagree with this opinion, I should welcome it in a pupil as a sign of individuality; but most teachers would not tolerate such a heterodox view. Not only teachers, but all commonplace persons in authority, desire in their subordinates that kind of uniformity which makes their actions easily predictable and never inconvenient. The result is that they crush initiative and individuality when they can, and when they cannot, they quarrel with it.

It is not one ideal for all men, but a separate ideal for each separate man, that has to be realized if possible. Every man has it in his being to develop into something good or bad: there is a best possible for him, and a worst possible. His circumstances will determine whether his capacities for good are developed or crushed, and whether his bad impulses are strengthened or gradually diverted into better channels.

But although we cannot set up in any detail an ideal of character which is to be universally applicable—although we cannot say, for instance, that all men ought to be industrious, or self-sacrificing, or fond of music—there are some broad principles which can be used to guide our estimates as to what is possible or desirable.

We may distinguish two sorts of goods, and two corresponding sorts of impulses. There are goods in regard to which individual possession is possible, and there are goods in which all can share alike. The food and clothing of one man is not the food and clothing of another; if the supply is insufficient, what one man has is obtained at the expense of some other man. This applies to material goods generally, and therefore to the greater part of the present economic life of the world. On the other hand, mental and spiritual goods do not belong to one man to the exclusion of another. If one man knows a science, that does not prevent others from knowing it; on the contrary, it helps them to acquire the knowledge. If one man is a great artist or poet, that does not prevent others from painting pictures or writing poems, but helps to create the atmosphere in which such things are possible. If one man is full of good-will toward others, that does not mean that there is less good-will to be shared among the rest; the more good-will one man has, the more he is likely to create among others. In such matters there is no *possession*, because there is not a definite amount to be shared; any increase anywhere tends to produce an increase everywhere.

There are two kinds of impulses, corresponding to the two kinds of goods. There are *possessive* impulses, which aim at acquiring or retaining private goods that cannot be shared; these center in the impulse of property. And there are *creative* or constructive impulses, which aim at bringing into the world or making available for use the kind of goods in which there is no privacy and no possession.

The best life is the one in which the creative impulses play the largest part and the possessive impulses the smallest. This is no new discovery. The Gospel says: "Take no thought, saying, What shall we eat? or What shall we drink? or, Wherewithal shall we be clothed?" The thought we give to these things is taken away from matters of more importance. And what is worse, the habit of mind engendered by thinking of these things is a bad one; it leads to competition, envy, domination, cruelty, and almost all the moral evils that infest the world. In particular, it leads to the predatory use of force. Material possessions can be taken by force and enjoyed by the robber. Spiritual possessions cannot be taken in this way. You may kill an artist or a thinker, but you cannot acquire his art or his thought. You may put a man to death because he loves his fellow-men, but you will not by so doing acquire the love which made his happiness. Force is impotent in such matters; it is only as regards material goods that it is effective. For this reason the men who believe in force are the men whose thoughts and desires are preoccupied with material goods.

The possessive impulses, when they are strong, infect activities which ought to be purely creative. A man who has made some valuable discovery may be filled with jealousy of a rival discoverer. If one man has found a cure for cancer and another has found a cure for consumption, one of them may be delighted if the other man's discovery turns out a mistake, instead of regretting the suffering of patients which would otherwise have been avoided. In such cases, instead of desiring knowledge for its own sake, or for the sake of its usefulness, a man is desiring it as a means to reputation. Every creative impulse is shadowed by a possessive impulse; even the aspirant to saintliness may be jealous of the more successful saint. Most affection is accompanied by some tinge of jealousy, which is a possessive impulse intruding into the creative region. Worst of all, in this direction, is the sheer envy of those who have missed everything worth having in life, and who are instinctively bent on preventing others from enjoying what they have not had. There is often much of this in the attitude of the old toward the young.

There is in human beings, as in plants and animals, a certain natural impulse of growth, and this is just as true of mental as of physical development. Physical development is helped by air and nourishment and exercise, and may be hindered by the sort of treatment which made Chinese women's feet small. In just the same way mental development may be helped or hindered by outside influences. The outside influences that help are those that merely provide encouragement or mental food or opportunities for exercising mental faculties. The influences that hinder are those that interfere with growth by applying any kind of force, whether discipline or authority or fear or the tyranny of public opinion or the necessity of engaging in some totally incongenial occupation. Worst of all influences are those that thwart or twist a man's fundamental impulse, which is what shows itself as conscience in the moral sphere; such influences are likely to do a man an inward danger from which he will never recover.

Those who realize the harm that can be done to others by any use of force against them, and the worthlessness of the goods that can be acquired by force, will be very full of respect for the liberty of others; they will not try to bind them or fetter them; they will be slow to judge and swift to sympathize; they will treat every human being with a kind of tenderness, because the principle of good in him is at once fragile and infinitely precious. They will not condemn those who are unlike themselves; they will know and feel that individuality brings differences and uniformity means death. They will wish each human being to be as much a living thing and as little a mechanical product as it is possible to be; they will cherish in each one just those things which the harsh usage of a ruthless world would destroy. In one word, all their dealings with others will be inspired by a deep impulse of *reverence*.

What we shall desire for individuals is now clear: strong creative impulses, overpowering and absorbing the instinct of possession; reverence for others; respect for the fundamental creative impulse in ourselves. A certain kind of self-respect or native pride is necessary to a good life; a man must not have a sense of utter inward defeat if he is to remain whole, but must feel the courage and the hope and the will to live by the best that is in him, whatever outward or inward obstacles it may encounter. So far as it lies in a man's own power, his life will realize its best possibilities if it has three things: creative rather than possessive impulses, reverence for others, and respect for the fundamental impulse in himself.

Political and social institutions are to be judged by the good or harm that they do to individuals. Do they encourage creativeness rather than possessiveness? Do they embody or promote a spirit of reverence between human beings? Do they preserve self-respect?

In all these ways the institutions under which we live are very far indeed from what they ought to be.

Institutions, and especially economic systems, have a profound influence in molding the characters of men and women. They may encourage adventure and hope, or timidity and the pursuit of safety. They may open men's minds to great possibilities, or close them against everything but the risk of obscure misfortune. They may make a man's happiness depend upon what he adds to the general possessions of the world, or upon what he can secure for himself of the private goods in which others cannot share. Modern capitalism forces the wrong decision of these alternatives upon all who are not heroic or exceptionally fortunate.

Men's impulses are molded, partly by their native disposition, partly by opportunity and environment, especially early environment. Direct preaching can do very little to change impulses, though it can lead people to restrain the direct expression of them, often with the result that the impulses go underground and come to the surface again in some contorted form. When we have discovered what kinds of impulse we desire, we must not rest content with preaching, or with trying to produce the outward manifestation without the inner spring; we must try rather to alter institutions in the way that will, of itself, modify the life of impulse in the desired direction.

At present our institutions rest upon two things: property and power. Both of these are very unjustly distributed; both, in the actual world, are of great importance to the happiness of the individual. Both are possessive goods; yet without them many of the goods in which all might share are hard to acquire as things are now.

Without property, as things are, a man has no freedom, and no security for the necessities of a tolerable life; without power, he has no opportunity for initiative. If men are to have free play for their creative impulses, they must be liberated from sordid cares by a certain measure of security, and they must have a sufficient share of power to be able to exercise initiative as regards the course and conditions of their lives.

Few men can succeed in being creative rather than possessive in a world which is wholly built on competition, where the great majority would fall into utter destitution if they became careless as to the acquisition of material goods, where honor and power and respect are given to wealth rather than to wisdom, where the law embodies and consecrates the injustice of those who have toward those who have not. In such an environment even those whom nature has endowed with great creative gifts become infected with the poison of competition. Men combine in groups to attain more strength in the scramble for material goods, and loyalty to the group spreads a halo of quasi-idealism round the central impulse of greed. Trade-unions and the Labor party are no more exempt from this vice than other parties and other sections of society; though they are largely inspired by the hope of a radically better world. They are too often led astray by the immediate object of securing for themselves a large share of material goods. That this desire is in accordance with justice, it is impossible to deny; but something larger and more constructive is needed as a political ideal, if the victors of to-morrow are not to become the oppressors of the day after. The inspiration and outcome of a reforming movement ought to be freedom and a generous spirit, not niggling restrictions and regulations.

The present economic system concentrates initiative in the hands of a small number of very rich men. Those who are not capitalists have, almost always, very little choice as to their activities when once they have selected a trade or profession; they are not part of the power that moves the mechanism, but only a passive portion of the machinery. Despite political democracy, there is still an extraordinary degree of difference in the power of self-direction belonging to a capitalist and to a man who has to earn his living. Economic affairs touch men's lives, at most times, much more intimately than political questions. At present the man who has no capital usually has to sell himself to some large organization, such as a railway company, for example. He has no voice in its management, and no liberty in politics except what his trade-union can secure for him. If he happens to desire a form of liberty which is not thought important by his trade-union, he is powerless; he must submit or starve.

Exactly the same thing happens to professional men. Probably a majority of journalists are engaged in writing for newspapers whose politics they disagree with; only a man of wealth can own a large newspaper, and only an accident can enable the point of view or the interests of those who are not wealthy to find expression in a newspaper. A large part of the best brains of the country are in the civil service, where the condition of their employment is silence about the evils which cannot be concealed from them. A Nonconformist minister loses his livelihood if his views displease his congregation; a member of Parliament loses his seat if he is not sufficiently supple or sufficiently stupid to follow or share all the turns and twists of public opinion. In every walk of life, independence of mind is punished by failure, more and more as economic organizations grow larger and more rigid. Is it surprising that men become increasingly docile, increasingly ready to submit to dictation and to forego the right of thinking for themselves? Yet along such lines civilization can only sink into a Byzantine immobility.

Fear of destitution is not a motive out of which a free creative life can grow, yet it is the chief motive which inspires the daily work of most wage-earners. The hope of possessing more wealth and power than any man ought to have, which is the corresponding motive of the rich, is quite as bad in its effects; it compels men to close their minds against justice, and to prevent themselves from thinking honestly on social questions while in the depths of their hearts they uneasily feel that their pleasures are bought by the miseries of others. The injustices of destitution and wealth alike ought to be rendered impossible. Then a great fear would be removed from the lives of the many, and hope would have to take on a better form in the lives of the few.

But security and liberty are only the negative conditions for good political institutions. When they have been won, we need also the positive condition: encouragement of creative energy. Security alone might produce a smug and stationary society; it demands creativeness as its counterpart, in order to keep alive the adventure and interest of life, and the movement toward perpetually new and better things. There can be no final goal for human institutions; the best are those that most encourage progress toward others still better. Without effort and change, human life cannot remain good. It is not a finished Utopia that we ought to desire, but a world where imagination and hope are alive and active.

It is a sad evidence of the weariness mankind has suffered from excessive toil that his heavens have usually been places where nothing ever happened or changed. Fatigue produces the illusion that only rest is needed for happiness; but when men have rested for a time, boredom drives them to renewed activity. For this reason, a happy life must be one in which there is activity. If it is also to be a useful life, the activity ought to be as far as possible creative, not merely predatory or defensive. But creative activity requires imagination and originality, which are apt to be subversive of the *status quo*. At present, those who have power dread a disturbance of the *status quo*, lest their unjust privileges should be taken away. In combination with the instinct for conventionality,[1] which man shares with the other gregarious animals, those who profit by the existing order have established a system which punishes originality and starves imagination from the moment of first going to school down to the time of death and burial. The whole spirit in which education is conducted needs to be changed, in order that children may be encouraged to think and feel for themselves, not to acquiesce passively in the thoughts and feelings of others. It is not rewards after the event that will produce initiative, but a certain mental atmosphere. There have been times when such an atmosphere existed: the great days of Greece, and Elizabethan England, may serve as examples. But in our own day the tyranny of vast machine-like organizations, governed from above by men who know and care little for the lives of those whom they control, is killing individuality and freedom of mind, and forcing men more and more to conform to a uniform pattern.

[1] In England this is called "a sense of humor."

Vast organizations are an inevitable element in modern life, and it is useless to aim at their abolition, as has been done by some reformers, for instance, William Morris. It is true that they make the preservation of individuality more difficult, but what is needed is a way of combining them with the greatest possible scope for individual initiative.

One very important step toward this end would be to render democratic the government of every organization. At present, our legislative institutions are more or less democratic, except for the important fact that women are excluded. But our administration is still purely bureaucratic, and our economic organizations are monarchical or oligarchic. Every limited liability company is run by a small number of self-appointed or coöpted directors. There can be no real freedom or democracy until the men who do the work in a business also control its management.

Another measure which would do much to increase liberty would be an increase of self-government for subordinate groups, whether geographical or economic or defined by some common belief, like religious sects. A modern state is so vast and its machinery is so little understood that even when a man has a vote he does not feel himself any effective part of the force which determines its policy. Except in matters where he can act in conjunction with an exceptionally powerful group, he feels himself almost impotent, and the government remains a remote impersonal circumstance, which must be simply endured, like the weather. By a share in the control of smaller bodies, a man might regain some of that sense of personal opportunity and responsibility which belonged to the citizen of a city-state in ancient Greece or medieval Italy.

When any group of men has a strong corporate consciousness—such as belongs, for example, to a nation or a trade or a religious body—liberty demands that it should be free to decide for itself all matters which are of great importance to the outside world. This is the basis of the universal claim for national independence. But nations are by no means the only groups which ought to have self-government for their internal concerns. And nations, like other groups, ought not to have complete liberty of action in matters which are of equal concern to foreign nations. Liberty demands self-government, but not the right to interfere with others. The greatest degree of liberty is not secured by anarchy. The reconciliation of liberty with government is a difficult problem, but it is one which any political theory must face.

The essence of government is the use of force in accordance with law to secure certain ends which the holders of power consider desirable. The coercion of an individual or a group by force is always in itself more or less harmful. But if there were no government, the result would not be an absence of force in men's relations to each other; it would merely be the exercise of force by those who had strong predatory instincts, necessitating either slavery or a perpetual readiness to repel force with force on the part of those whose instincts were less violent. This is the state of affairs at present in international relations, owing to the fact that no international government exists. The results of anarchy between states should suffice to persuade us that anarchism has no solution to offer for the evils of the world.

There is probably one purpose, and only one, for which the use of force by a government is beneficent, and that is to diminish the total amount of force used m the world. It is clear, for example, that the legal prohibition of murder diminishes the total amount of violence in the world. And no one would maintain that parents should have unlimited freedom to ill-treat their children. So long as some men wish to do violence to others, there cannot be complete liberty, for either the wish to do violence must be restrained, or the victims must be left to suffer. For this reason, although individuals and societies should have the utmost freedom as regards their own affairs, they ought not to have complete freedom as regards their dealings with others. To give freedom to the strong to oppress the weak is not the way to secure the greatest possible amount of freedom in the world. This is the basis of the socialist revolt against the kind of freedom which used to be advocated by *laissez-faire* economists.

Democracy is a device—the best so far invented—for diminishing as much as possible the interference of governments with liberty. If a nation is divided into two sections which cannot both have their way, democracy theoretically insures that the majority shall have their way. But democracy is not at all an adequate device unless it is accompanied by a very great amount of devolution. Love of uniformity, or the mere pleasure of interfering, or dislike of differing tastes and temperaments, may often lead a majority to control a minority in matters which do not really concern the majority. We should none of us like to have the internal affairs of Great Britain settled by a parliament of the world, if ever such a body came into existence. Nevertheless, there are matters which such a body could settle much better than any existing instrument of government.

The theory of the legitimate use of force in human affairs, where a government exists, seems clear. Force should only be used against those who attempt to use force against others, or against those who will not respect the law in cases where a common decision is necessary and a minority are opposed to the action of the majority. These seem legitimate occasions for the use of force; and they should be legitimate occasions in international affairs, if an international government existed. The problem of the legitimate occasions for the use of force in the absence of a government is a different one, with which we are not at present concerned.

Although a government must have the power to use force, and may on occasion use it legitimately, the aim of the reformers to have such institutions as will diminish the need for actual coercion will be found to have this effect. Most of us abstain, for instance, from theft, not because it is illegal, but because we feel no desire to steal. The more men learn to live creatively rather than possessively, the less their wishes will lead them to thwart others or to attempt violent interference with their liberty. Most of the conflicts of interests, which lead individuals or organizations into disputes, are purely imaginary, and would be seen to be so if men aimed more at the goods in which all can share, and less at those private possessions that are the source of strife. In proportion as men live creatively, they cease to wish to interfere with others by force. Very many matters in which, at present, common action is thought indispensable, might well be left to individual decision. It used to be thought absolutely necessary that all the inhabitants of a country should have the same religion, but we now know that there is no such necessity. In like manner it will be found, as men grow more tolerant in their instincts, that many uniformities now insisted upon are useless and even harmful.

Good political institutions would weaken the impulse toward force and domination in two ways: first, by increasing the opportunities for the creative impulses, and by shaping education so as to strengthen these impulses; secondly, by diminishing the outlets for the possessive instincts. The diffusion of power, both in the political and the economic sphere, instead of its concentration in the hands of officials and captains of industry, would greatly diminish the opportunities for acquiring the habit of command, out of which the desire for exercising tyranny is apt to spring. Autonomy, both for districts and for organizations, would leave fewer occasions when governments were called upon to make decisions as to other people's concerns. And the abolition of capitalism and the wage system would remove the chief incentive to fear and greed, those correlative passions by which all free life is choked and gagged.

Few men seem to realize how many of the evils from which we suffer are wholly unnecessary, and that they could be abolished by a united effort within a few years. If a majority in every civilized country so desired, we could, within twenty years, abolish all abject poverty, quite half the illness in the world, the whole economic slavery which binds down nine tenths of our population; we could fill the world with beauty and joy, and secure the reign of universal peace. It is only because men are apathetic that this is not achieved, only because imagination is sluggish, and what always has been is regarded as what always must be. With good-will, generosity, intelligence, these things could be brought about.

Chapter II

Capitalism and the Wage System

I

The world is full of preventible evils which most men would be glad to see prevented.

Nevertheless, these evils persist, and nothing effective is done toward abolishing them.

This paradox produces astonishment in inexperienced reformers, and too often produces disillusionment in those who have come to know the difficulty of changing human institutions.

War is recognized as an evil by an immense majority in every civilized country; but this recognition does not prevent war.

The unjust distribution of wealth must be obviously an evil to those who are not prosperous, and they are nine tenths of the population. Nevertheless it continues unabated.

The tyranny of the holders of power is a source of needless suffering and misfortune to very large sections of mankind; but power remains in few hands, and tends, if anything, to grow more concentrated.

I wish first to study the evils of our present institutions, and the causes of the very limited success of reformers in the past, and then to suggest reasons for the hope of a more lasting and permanent success in the near future.

The war has come as a challenge to all who desire a better world. The system which cannot save mankind from such an appalling disaster is at fault somewhere, and cannot be amended in any lasting way unless the danger of great wars in the future can be made very small.

But war is only the final flower of an evil tree. Even in times of peace, most men live lives of monotonous labor, most women are condemned to a drudgery which almost kills the possibility of happiness before youth is past, most children are allowed to grow up in ignorance of all that would enlarge their thoughts or stimulate their imagination. The few who are more fortunate are rendered illiberal by their unjust privileges, and oppressive through fear of the awakening indignation of the masses. From the highest to the lowest, almost all men are absorbed in the economic struggle: the struggle to acquire what is their due or to retain what is not their due. Material possessions, in fact or in desire, dominate our outlook, usually to the exclusion of all generous and creative impulses.

Possessiveness—the passion to have and to hold—is the ultimate source of war, and the foundation of all the ills from which the political world is suffering. Only by diminishing the strength of this passion and its hold upon our daily lives can new institutions bring permanent benefit to mankind.

Institutions which will diminish the sway of greed are possible, but only through a complete reconstruction of our whole economic system. Capitalism and the wage system must be abolished; they are twin monsters which are eating up the life of the world. In place of them we need a system which will hold in cheek men's predatory impulses, and will diminish the economic injustice that allows some to be rich in idleness while others are poor in spite of unremitting labor; but above all we need a system which will destroy the tyranny of the employer, by making men at the same time secure against destitution and able to find scope for individual initiative in the control of the industry by which they live. A better system can do all these things, and can be established by the democracy whenever it grows weary of enduring evils which there is no reason to endure.

We may distinguish four purposes at which an economic system may aim: first, it may aim at the greatest possible production of goods and at facilitating technical progress; second, it may aim at securing distributive justice; third, it may aim at giving security against destitution; and, fourth, it may aim at liberating creative impulses and diminishing possessive impulses.

Of these four purposes the last is the most important. Security is chiefly important as a means to it. State socialism, though it might give material security and more justice than we have at present, would probably fail to liberate creative impulses or produce a progressive society.

Our present system fails in all four purposes. It is chiefly defended on the ground that it achieves the first of the four purposes, namely, the greatest possible production of material goods, but it only does this in a very short-sighted way, by methods which are wasteful in the long run both of human material and of natural resources.

Capitalistic enterprise involves a ruthless belief in the importance of increasing material production to the utmost possible extent now and in the immediate future. In obedience to this belief, new portions of the earth's surface are continually brought under the sway of industrialism. Vast tracts of Africa become recruiting grounds for the labor required in the gold and diamond mines of the Rand, Rhodesia, and Kimberley; for this purpose, the population is demoralized, taxed, driven into revolt, and exposed to the contamination of European vice and disease. Healthy and vigorous races from Southern Europe are tempted to America, where sweating and slum life reduce their vitality if they do not actually cause their death. What damage is done to our own urban populations by the conditions under which they live, we all know. And what is true of the human riches of the world is no less true of the physical resources. The mines, forests, and wheat-fields of the world are all being exploited at a rate which must practically exhaust them at no distant date. On the side of material production, the world is living too fast; in a kind of delirium, almost all the energy of the world has rushed into the immediate production of something, no matter what, and no matter at what cost. And yet our present system is defended on the ground that it safeguards progress!

It cannot be said that our present economic system is any more successful in regard to the other three objects which ought to be aimed at. Among the many obvious evils of capitalism and the wage system, none are more glaring than that they encourage predatory instincts, that they allow economic injustice, and that they give great scope to the tyranny of the employer.

As to predatory instincts, we may say, broadly speaking, that in a state of nature there would be two ways of acquiring riches—one by production, the other by robbery. Under our existing system, although what is recognized as robbery is forbidden, there are nevertheless many ways of becoming rich without contributing anything to the wealth of the community. Ownership of land or capital, whether acquired or inherited, gives a legal right to a permanent income. Although most people have to produce in order to live, a privileged minority are able to live in luxury without producing anything at all. As these are the men who are not only the most fortunate but also the most respected, there is a general desire to enter their ranks, and a widespread unwillingness to face the fact that there is no justification whatever for incomes derived in this way. And apart from the passive enjoyment of rent or interest, the methods of acquiring wealth are very largely predatory. It is not, as a rule, by means of useful inventions, or of any other action which increases the general wealth of the community, that men amass fortunes; it is much more often by skill in exploiting or circumventing others. Nor is it only among the rich that our present régime promotes a narrowly acquisitive spirit. The constant risk of destitution compels most men to fill a great part of their time and thought with the economic struggle. There is a theory that this increases the total output of wealth by the community. But for reasons to which I shall return later, I believe this theory to be wholly mistaken.

Economic injustice is perhaps the most obvious evil of our present system. It would be utterly absurd to maintain that the men who inherit great wealth deserve better of the community than those who have to work for their living. I am not prepared to maintain that economic justice requires an exactly equal income for everybody. Some kinds of work require a larger income for efficiency than others do; but there is economic injustice as soon as a man has more than his share, unless it is because his efficiency in his work requires it, or as a reward for some definite service. But this point is so obvious that it needs no elaboration.

The modern growth of monopolies in the shape of trusts, cartels, federations of employers and so on has greatly increased the power of the capitalist to levy toll on the community. This tendency will not cease of itself, but only through definite action on the part of those who do not profit by the capitalist régime. Unfortunately the distinction between the proletariat and the capitalist is not so sharp as it was in the minds of socialist theorizers. Trade-unions have funds in various securities; friendly societies are large capitalists; and many individuals eke out their wages by invested savings. All this increases the difficulty of any clear-cut radical change in our economic system. But it does not diminish the desirability of such a change.

Such a system as that suggested by the French syndicalists, in which each trade would be self-governing and completely independent, without the control of any central authority, would not secure economic justice. Some trades are in a much stronger bargaining position than others. Coal and transport, for example, could paralyze the national life, and could levy blackmail by threatening to do so. On the other hand, such people as school teachers, for example, could rouse very little terror by the threat of a strike and would be in a very weak bargaining position. Justice can never be secured by any system of unrestrained force exercised by interested parties in their own interests. For this reason the abolition of the state, which the syndicalists seem to desire, would be a measure not compatible with economic justice.

The tyranny of the employer, which at present robs the greater part of most men's lives of all liberty and all initiative, is unavoidable so long as the employer retains the right of dismissal with consequent loss of pay. This right is supposed to be essential in order that men may have an incentive to work thoroughly. But as men grow more civilized, incentives based on hope become increasingly preferable to those that are based on fear. It would be far better that men should be rewarded for working well than that they should be punished for working badly. This system is already in operation in the civil service, where a man is only dismissed for some exceptional degree of vice or virtue, such as murder or illegal abstention from it. Sufficient pay to ensure a livelihood ought to be given to every person who is willing to work, independently of the question whether the particular work at which he is skilled is wanted at the moment or not. If it is not wanted, some new trade which is wanted ought to be taught at the public expense. Why, for example, should a hansom-cab driver be allowed to suffer on account of the introduction of taxies? He has not committed any crime, and the fact that his work is no longer wanted is due to causes entirely outside his control. Instead of being allowed to starve, he ought to be given instruction in motor driving or in whatever other trade may seem most suitable. At present, owing to the fact that all industrial changes tend to cause hardships to some section of wage-earners, there is a tendency to technical conservatism on the part of labor, a dislike of innovations, new processes, and new methods. But such changes, if they are in the permanent interest of the community, ought to be carried out without allowing them to bring unmerited loss to those sections of the community whose labor is no longer wanted in the old form. The instinctive conservatism of mankind is sure to make all processes of production change more slowly than they should. It is a pity to add to this by the avoidable conservatism which is forced upon organized labor at present through the unjust workings of a change.

It will be said that men will not work well if the fear of dismissal does not spur them on. I think it is only a small percentage of whom this would be true at present. And those of whom it would be true might easily become industrious if they were given more congenial work or a wiser training. The residue who cannot be coaxed into industry by any such methods are probably to be regarded as pathological cases, requiring medical rather than penal treatment. And against this residue must be set the very much larger number who are now ruined in health or in morale by the terrible uncertainty of their livelihood and the great irregularity of their employment. To very many, security would bring a quite new possibility of physical and moral health.

The most dangerous aspect of the tyranny of the employer is the power which it gives him of interfering with men's activities outside their working hours. A man may be dismissed because the employer dislikes his religion or his politics, or chooses to think his private life immoral. He may be dismissed because he tries to produce a spirit of independence among his fellow employees. He may fail completely to find employment merely on the ground that he is better educated than most and therefore more dangerous. Such cases actually occur at present. This evil would not be remedied, but rather intensified, under state socialism, because, where the State is the only employer, there is no refuge from its prejudices such as may now accidentally arise through the differing opinions of different men. The State would be able to enforce any system of beliefs it happened to like, and it is almost certain that it would do so. Freedom of thought would be penalized, and all independence of spirit would die out.

Any rigid system would involve this evil. It is very necessary that there should be diversity and lack of complete systematization. Minorities must be able to live and develop their opinions freely. If this is not secured, the instinct of persecution and conformity will force all men into one mold and make all vital progress impossible.

For these reasons, no one ought to be allowed to suffer destitution so long as he or she is *willing* to work. And no kind of inquiry ought to be made into opinion or private life. It is only on this basis that it is possible to build up an economic system not founded upon tyranny and terror.

II

The power of the economic reformer is limited by the technical productivity of labor. So long as it was necessary to the bare subsistence of the human race that most men should work very long hours for a pittance, so long no civilization was possible except an aristocratic one; if there were to be men with sufficient leisure for any mental life, there had to be others who were sacrificed for the good of the few. But the time when such a system was necessary has passed away with the progress of machinery. It would be possible now, if we had a wise economic system, for all who have mental needs to find satisfaction for them. By a few hours a day of manual work, a man can produce as much as is necessary for his own subsistence; and if he is willing to forgo luxuries, that is all that the community has a right to demand of him. It ought to be open to all who so desire to do short hours of work for little pay, and devote their leisure to whatever pursuit happens to attract them. No doubt the great majority of those who chose this course would spend their time in mere amusement, as most of the rich do at present. But it could not be said, in such a society, that they were parasites upon the labor of others. And there would be a minority who would give their hours of nominal idleness to science or art or literature, or some other pursuit out of which fundamental progress may come. In all such matters, organization and system can only do harm. The one thing that can be done is to provide opportunity, without repining at the waste that results from most men failing to make good use of the opportunity.

But except in cases of unusual laziness or eccentric ambition, most men would elect to do a full day's work for a full day's pay. For these, who would form the immense majority, the important thing is that ordinary work should, as far as possible, afford interest and independence and scope for initiative. These things are more important than income, as soon as a certain minimum has been reached. They can be secured by gild socialism, by industrial self-government subject to state control as regards the relations of a trade to the rest of the community. So far as I know, they cannot be secured in any other way.

Guild socialism, as advocated by Mr. Orage and the "New Age," is associated with a polemic against "political" action, and in favor of direct economic action by trade-unions. It shares this with syndicalism, from which most of what is new in it is derived. But I see no reason for this attitude; political and economic action seem to me equally necessary, each in its own time and place. I think there is danger in the attempt to use the machinery of the present capitalist state for socialistic purposes. But there is need of political action to transform the machinery of the state, side by side with the transformation which we hope to see in economic institutions. In this country, neither transformation is likely to be brought about by a sudden revolution; we must expect each to come step by step, if at all, and I doubt if either could or should advance very far without the other.

The economic system we should ultimately wish to see would be one in which the state would be the sole recipient of economic rent, while private capitalistic enterprises should be replaced by self-governing combinations of those who actually do the work. It ought to be optional whether a man does a whole day's work for a whole day's pay, or half a day's work for half a day's pay, except in cases where such an arrangement would cause practical inconvenience. A man's pay should not cease through the accident of his work being no longer needed, but should continue so long as he is willing to work, a new trade being taught him at the public expense, if necessary. Unwillingness to work should be treated medically or educationally, when it could not be overcome by a change to some more congenial occupation.

The workers in a given industry should all be combined in one autonomous unit, and their work should not be subject to any outside control. The state should fix the price at which they produce, but should leave the industry self-governing in all other respects. In fixing prices, the state should, as far as possible, allow each industry to profit by any improvements which it might introduce into its own processes, but should endeavor to prevent undeserved loss or gain through changes in external economic conditions. In this way there would be every incentive to progress, with the least possible danger of unmerited destitution. And although large economic organizations will continue, as they are bound to do, there will be a diffusion of power which will take away the sense of individual impotence from which men and women suffer at present.

III

Some men, though they may admit that such a system would be desirable, will argue that it is impossible to bring it about, and that therefore we must concentrate on more immediate objects.

I think it must be conceded that a political party ought to have proximate aims, measures which it hopes to carry in the next session or the next parliament, as well as a more distant goal. Marxian socialism, as it existed in Germany, seemed to me to suffer in this way: although the party was numerically powerful, it was politically weak, because it had no minor measures to demand while waiting for the revolution. And when, at last, German socialism was captured by those who desired a less impracticable policy, the modification which occurred was of exactly the wrong kind: acquiescence in bad policies, such as militarism and imperialism, rather than advocacy of partial reforms which, however inadequate, would still have been steps in the right direction.

A similar defect was inherent in the policy of French syndicalism as it existed before the war. Everything was to wait for the general strike; after adequate preparation, one day the whole proletariat would unanimously refuse to work, the property owners would acknowledge their defeat, and agree to abandon all their privileges rather than starve. This is a dramatic conception; but love of drama is a great enemy of true vision. Men cannot be trained, except under very rare circumstances, to do something suddenly which is very different from what they have been doing before. If the general strike were to succeed, the victors, despite their anarchism, would be compelled at once to form an administration, to create a new police force to prevent looting and wanton destruction, to establish a provisional government issuing dictatorial orders to the various sections of revolutionaries. Now the syndicalists are opposed in principle to all political action; they would feel that they were departing from their theory in taking the necessary practical steps, and they would be without the required training because of their previous abstention from politics. For these reasons it is likely that, even after a syndicalist revolution, actual power would fall into the hands of men who were not really syndicalists.

Another objection to a program which is to be realized suddenly at some remote date by a revolution or a general strike is that enthusiasm flags when there is nothing to do meanwhile, and no partial success to lessen the weariness of waiting. The only sort of movement which can succeed by such methods is one where the sentiment and the program are both very simple, as is the case in rebellions of oppressed nations. But the line of demarcation between capitalist and wage-earner is not sharp, like the line between Turk and Armenian, or between an Englishman and a native of India. Those who have advocated the social revolution have been mistaken in their political methods, chiefly because they have not realized how many people there are in the community whose sympathies and interests lie half on the side of capital, half on the side of labor. These people make a clear-cut revolutionary policy very difficult.

For these reasons, those who aim at an economic reconstruction which is not likely to be completed to-morrow must, if they are to have any hope of success, be able to approach their goal by degrees, through measures which are of some use in themselves, even if they should not ultimately lead to the desired end. There must be activities which train men for those that they are ultimately to carry out, and there must be possible achievements in the near future, not only a vague hope of a distant paradise.

But although I believe that all this is true, I believe no less firmly that really vital and radical reform requires some vision beyond the immediate future, some realization of what human beings might make of human life if they chose. Without some such hope, men will not have the energy and enthusiasm necessary to overcome opposition, or the steadfastness to persist when their aims are for the moment unpopular. Every man who has really sincere desire for any great amelioration in the conditions of life has first to face ridicule, then persecution, then cajolery and attempts at subtle corruption. We know from painful experience how few pass unscathed through these three ordeals. The last especially, when the reformer is shown all the kingdoms of the earth, is difficult, indeed almost impossible, except for those who have made their ultimate goal vivid to themselves by clear and definite thought.

Economic systems are concerned essentially with the production and distribution of material goods. Our present system is wasteful on the production side, and unjust on the side of distribution. It involves a life of slavery to economic forces for the great majority of the community, and for the minority a degree of power over the lives of others which no man ought to have. In a good community the production of the necessaries of existence would be a mere preliminary to the important and interesting part of life, except for those who find a pleasure in some part of the work of producing necessaries. It is not in the least necessary that economic needs should dominate man as they do at present. This is rendered necessary at present, partly by the inequalities of wealth, partly by the fact that things of real value, such as a good education, are difficult to acquire, except for the well-to-do.

Private ownership of land and capital is not defensible on grounds of justice, or on the ground that it is an economical way of producing what the community needs. But the chief objections to it are that it stunts the lives of men and women, that it enshrines a ruthless possessiveness in all the respect which is given to success, that it leads men to fill the greater part of their time and thought with the acquisition of purely material goods, and that it affords a terrible obstacle to the advancement of civilization and creative energy.

The approach to a system free from these evils need not be sudden; it is perfectly possible to proceed step by step towards economic freedom and industrial self-government. It is not true that there is any outward difficulty in creating the kind of institutions that we have been considering. If organized labor wishes to create them, nothing could stand in its way. The difficulty involved is merely the difficulty of inspiring men with hope, of giving them enough imagination to see that the evils from which they suffer are unnecessary, and enough thought to understand how the evils are to be cured. This is a difficulty which can be overcome by time and energy. But it will not be overcome if the leaders of organized labor have no breadth of outlook, no vision, no hopes beyond some slight superficial improvement within the framework of the existing system. Revolutionary action may be unnecessary, but revolutionary thought is indispensable, and, as the outcome of thought, a rational and constructive hope.

Chapter III: Pitfalls in Socialism

I

In its early days, socialism was a revolutionary movement of which the object was the liberation of the wage-earning classes and the establishment of freedom and justice. The passage from capitalism to the new régime was to be sudden and violent: capitalists were to be expropriated without compensation, and their power was not to be replaced by any new authority.

Gradually a change came over the spirit of socialism. In France, socialists became members of the government, and made and unmade parliamentary majorities. In Germany, social democracy grew so strong that it became impossible for it to resist the temptation to barter away some of its intransigeance in return for government recognition of its claims. In England, the Fabians taught the advantage of reform as against revolution, and of conciliatory bargaining as against irreconcilable antagonism.

The method of gradual reform has many merits as compared to the method of revolution, and I have no wish to preach revolution. But gradual reform has certain dangers, to wit, the ownership or control of businesses hitherto in private hands, and by encouraging legislative interference for the benefit of various sections of the wage-earning classes. I think it is at least doubtful whether such measures do anything at all to contribute toward the ideals which inspired the early socialists and still inspire the great majority of those who advocate some form of socialism.

Let us take as an illustration such a measure as state purchase of railways. This is a typical object of state socialism, thoroughly practicable, already achieved in many countries, and clearly the sort of step that must be taken in any piecemeal approach to complete collectivism. Yet I see no reason to believe that any real advance toward democracy, freedom, or economic justice is achieved when a state takes over the railways after full compensation to the shareholders.

Economic justice demands a diminution, if not a total abolition, of the proportion of the national income which goes to the recipients of rent and interest. But when the holders of railway shares are given government stock to replace their shares, they are given the prospect of an income in perpetuity equal to what they might reasonably expect to have derived from their shares. Unless there is reason to expect a great increase in the earnings of railways, the whole operation does nothing to alter the distribution of wealth. This could only be effected if the present owners were expropriated, or paid less than the market value, or given a mere life-interest as compensation. When full value is given, economic justice is not advanced in any degree.

There is equally little advance toward freedom. The men employed on the railway have no more voice than they had before in the management of the railway, or in the wages and conditions of work. Instead of having to fight the directors, with the possibility of an appeal to the government, they now have to fight the government directly; and experience does not lead to the view that a government department has any special tenderness toward the claims of labor. If they strike, they have to contend against the whole organized power of the state, which they can only do successfully if they happen to have a strong public opinion on their side. In view of the influence which the state can always exercise on the press, public opinion is likely to be biased against them, particularly when a nominally progressive government is in power. There will no longer be the possibility of divergences between the policies of different railways. Railway men in England derived advantages for many years from the comparatively liberal policy of the North Eastern Railway, which they were able to use as an argument for a similar policy elsewhere. Such possibilities are excluded by the dead uniformity of state administration.

And there is no real advance toward democracy. The administration of the railways will be in the hands of officials whose bias and associations separate them from labor, and who will develop an autocratic temper through the habit of power. The democratic machinery by which these officials are nominally controlled is cumbrous and remote, and can only be brought into operation on first-class issues which rouse the interest of the whole nation. Even then it is very likely that the superior education of the officials and the government, combined with the advantages of their position, will enable them to mislead the public as to the issues, and alienate the general sympathy even from the most excellent cause.

I do not deny that these evils exist at present; I say only that they will not be remedied by such measures as the nationalization of railways in the present economic and political environment. A greater upheaval, and a greater change in men's habits of mind, is necessary for any really vital progress.

II

State socialism, even in a nation which possesses the form of political democracy, is not a truly democratic system. The way in which it fails to be democratic may be made plain by an analogy from the political sphere. Every democrat recognizes that the Irish ought to have self-government for Irish affairs, and ought not to be told that they have no grievance because they share in the

Parliament of the United Kingdom. It is essential to democracy that any group of citizens whose interests or desires separate them at all widely from the rest of the community should be free to decide their internal affairs for themselves. And what is true of national or local groups is equally true of economic groups, such as miners or railway men. The national machinery of general elections is by no means sufficient to secure for groups of this kind the freedom which they ought to have.

The power of officials, which is a great and growing danger in the modern state, arises from the fact that the majority of the voters, who constitute the only ultimate popular control over officials, are as a rule not interested in any one particular question, and are therefore not likely to interfere effectively against an official who is thwarting the wishes of the minority who are interested. The official is nominally subject to indirect popular control, but not to the control of those who are directly affected by his action. The bulk of the public will either never hear about the matter in dispute, or, if they do hear, will form a hasty opinion based upon inadequate information, which is far more likely to come from the side of the officials than from the section of the community which is affected by the question at issue. In an important political issue, some degree of knowledge is likely to be diffused in time; but in other matters there is little hope that this will happen.

It may be said that the power of officials is much less dangerous than the power of capitalists, because officials have no economic interests that are opposed to those of wage-earners. But this argument involves far too simple a theory of political human nature—a theory which orthodox socialism adopted from the classical political economy, and has tended to retain in spite of growing evidence of its falsity. Economic self-interest, and even economic class-interest, is by no means the only important political motive. Officials, whose salary is generally quite unaffected by their decisions on particular questions, are likely, if they are of average honesty, to decide according to their view of the public interest; but their view will none the less have a bias which will often lead them wrong. It is important to understand this bias before entrusting our destinies too unreservedly to government departments.

The first thing to observe is that, in any very large organization, and above all in a great state, officials and legislators are usually very remote from those whom they govern, and not imaginatively acquainted with the conditions of life to which their decisions will be applied. This makes them ignorant of much that they ought to know, even when they are industrious and willing to learn whatever can be taught by statistics and blue-books. The one thing they understand intimately is the office routine and the administrative rules. The result is an undue anxiety to secure a uniform system. I have heard of a French minister of education taking out his watch, and remarking, "At this moment all the children of such and such an age in France are learning so and so." This is the ideal of the administrator, an ideal utterly fatal to free growth, initiative, experiment, or any far reaching innovation. Laziness is not one of the motives recognized in textbooks on political theory, because all ordinary knowledge of human nature is considered unworthy of the dignity of these works; yet we all know that laziness is an immensely powerful motive with all but a small minority of mankind.

Unfortunately, in this case laziness is reinforced by love of power, which leads energetic officials to create the systems which lazy officials like to administer. The energetic official inevitably dislikes anything that he does not control. His official sanction must be obtained before anything can be done. Whatever he finds in existence he wishes to alter in some way, so as to have the satisfaction of feeling his power and making it felt. If he is conscientious, he will think out some perfectly uniform and rigid scheme which he believes to be the best possible, and he will then impose this scheme ruthlessly, whatever promising growths he may have to lop down for the sake of symmetry. The result inevitably has something of the deadly dullness of a new rectangular town, as compared with the beauty and richness of an ancient city which has lived and grown with the separate lives and individualities of many generations. What has grown is always more living than what has been decreed; but the energetic official will always prefer the tidiness of what he has decreed to the apparent disorder of spontaneous growth.

The mere possession of power tends to produce a love of power, which is a very dangerous motive, because the only sure proof of power consists in preventing others from doing what they wish to do. The essential theory of democracy is the diffusion of power among the whole people, so that the evils produced by one man's possession of great power shall be obviated. But the diffusion of power through democracy is only effective when the voters take an interest in the question involved. When the question does not interest them, they do not attempt to control the administration, and all actual power passes into the hands of officials.

For this reason, the true ends of democracy are not achieved by state socialism or by any system which places great power in the hands of men subject to no popular control except that which is more or less indirectly exercised through parliament.

Any fresh survey of men's political actions shows that, in those who have enough energy to be politically effective, love of power is a stronger motive than economic self-interest. Love of power actuates the great millionaires, who have far more money than they can spend, but continue to amass wealth merely in order to control more and more of the world's finance. Love of power is obviously the ruling motive of many politicians. It is also the chief cause of wars, which are admittedly almost always a bad speculation from the mere point of view of wealth. For this reason, a new economic system which merely attacks economic motives and does not interfere with the concentration of power is not likely to effect any very great improvement in the world. This is one of the chief reasons for regarding state socialism with suspicion.

III

The problem of the distribution of power is a more difficult one than the problem of the distribution of wealth. The machinery of representative government has concentrated on *ultimate* power as the only important matter, and has ignored immediate executive power. Almost nothing has been done to democratize administration. Government officials, in virtue of their income, security, and social position, are likely to be on the side of the rich, who have been their daily associates ever since the time of school and college. And whether or not they are on the side of the rich, they are not likely, for the reasons we have been considering, to be genuinely in favor of progress. What applies to government officials applies also to members of Parliament, with the sole difference that they have had to recommend themselves to a constituency. This, however, only adds hypocrisy to the other qualities of a ruling caste. Whoever has stood in the lobby of the House of Commons watching members emerge with wandering eye and hypothetical smile, until the constituent is espied, his arm taken, "my dear fellow" whispered in his ear, and his steps guided toward the inner precincts—whoever, observing this, has realized that these are the arts by which men become and remain legislators, can hardly fail to feel that democracy as it exists is not an absolutely perfect instrument of government. It is a painful fact that the ordinary voter, at any rate in England, is quite blind to insincerity. The man who does not care about any definite political measures can generally be won by corruption or flattery, open or concealed; the man who is set on securing reforms will generally prefer an ambitious windbag to a man who desires the public good without possessing a ready tongue. And the ambitious windbag, as soon as he has become a power by the enthusiasm he has aroused, will sell his influence to the governing clique, sometimes openly, sometimes by the more subtle method of intentionally failing at a crisis. This is part of the normal working of democracy as embodied in representative institutions. Yet a cure must be found if democracy is not to remain a farce.

One of the sources of evil in modern large democracies is the fact that most of the electorate have no direct or vital interest in most of the questions that arise. Should Welsh children be allowed the use of the Welsh language in schools? Should gipsies be compelled to abandon their nomadic life at the bidding of the education authorities? Should miners have an eight-hour day? Should Christian

Scientists be compelled to call in doctors in case of serious illness? These are matters of passionate interest to certain sections of the community, but of very little interest to the great majority. If they are decided according to the wishes of the numerical majority, the intense desires of a minority will be overborne by the very slight and uninformed whims of the indifferent remainder. If the minority are geographically concentrated, so that they can decide elections in a certain number of constituencies, like the Welsh and the miners, they have a good chance of getting their way, by the wholly beneficent process which its enemies describe as log-rolling. But if they are scattered and politically feeble, like the gipsies and the Christian Scientists, they stand a very poor chance against the prejudices of the majority. Even when they are geographically concentrated, like the Irish, they may fail to obtain their wishes, because they arouse some hostility or some instinct of domination in the majority. Such a state of affairs is the negation of all democratic principles.

The tyranny of the majority is a very real danger. It is a mistake to suppose that the majority is necessarily right. On every new question the majority is always wrong at first. In matters where the state must act as a whole, such as tariffs, for example, decision by majorities is probably the best method that can be devised. But there are a great many questions in which there is no need of a uniform decision. Religion is recognized as one of these. Education ought to be one, provided a certain minimum standard is attained. Military service clearly ought to be one. Wherever divergent action by different groups is possible without anarchy, it ought to be permitted. In such cases it will be found by those who consider past history that, whenever any new fundamental issue arises, the majority are in the wrong, because they are guided by prejudice and habit. Progress comes through the gradual effect of a minority in converting opinion and altering custom. At one time—not so very long ago—it was considered monstrous wickedness to maintain that old women ought not to be burnt as witches. If those who held this opinion had been forcibly suppressed, we should still be steeped in medieval superstition. For such reasons, it is of the utmost importance that the majority should refrain from imposing its will as regards matters in which uniformity is not absolutely necessary.

IV

The cure for the evils and dangers which we have been considering is a very great extension of devolution and federal government. Wherever there is a national consciousness, as in Wales and Ireland, the area in which it exists ought to be allowed to decide all purely local affairs without external interference. But there are many matters which ought to be left to the management, not of local groups, but of trade groups, or of organizations embodying some set of opinions. In the

East, men are subject to different laws according to the religion they profess. Something of this kind is necessary if any semblance of liberty is to exist where there is great divergence in beliefs.

Some matters are essentially geographical; for instance, gas and water, roads, tariffs, armies and navies. These must be decided by an authority representing an area. How large the area ought to be, depends upon accidents of topography and sentiment, and also upon the nature of the matter involved. Gas and water require a small area, roads a somewhat larger one, while the only satisfactory area for an army or a navy is the whole planet, since no smaller area will prevent war.

But the proper unit in most economic questions, and also in most questions that are intimately concerned with personal opinions, is not geographical at all. The internal management of railways ought not to be in the hands of the geographical state, for reasons which we have already considered. Still less ought it to be in the hands of a set of irresponsible capitalists. The only truly democratic system would be one which left the internal management of railways in the hands of the men who work on them. These men should elect the general manager, and a parliament of directors if necessary. All questions of wages, conditions of labor, running of trains, and acquisition of material, should be in the hands of a body responsible only to those actually engaged in the work of the railway.

The same arguments apply to other large trades: mining, iron and steel, cotton, and so on. British trade-unionism, it seems to me, has erred in conceiving labor and capital as both permanent forces, which were to be brought to some equality of strength by the organization of labor. This seems to me too modest an ideal. The ideal which I should wish to substitute involves the conquest of democracy and self-government in the economic sphere as in the political sphere, and the total abolition of the power now wielded by the capitalist. The man who works on a railway ought to have a voice in the government of the railway, just as much as the man who works in a state has a right to a voice in the management of his state. The concentration of business initiative in the hands of the employers is a great evil, and robs the employees of their legitimate share of interest in the larger problems of their trade.

French syndicalists were the first to advocate the system of trade autonomy as a better solution than state socialism. But in their view the trades were to be independent, almost like sovereign states at present. Such a system would not promote peace, any more than it does at present in international relations. In the affairs of any body of men, we may broadly distinguish what may be called questions of home politics from questions of foreign politics. Every group sufficiently well-marked to constitute a political entity ought to be autonomous in regard to internal matters, but not in regard to those that directly affect the outside world. If two groups are both entirely free as regards their relations to each other, there is no way of averting the danger of an open or covert appeal to force. The relations of a group of men to the outside world ought, whenever possible, to be controlled by a neutral authority. It is here that the state is necessary for adjusting the relations between different trades. The men who make some commodity should be entirely free as regards hours of labor, distribution of the total earnings of the trade, and all questions of business management. But they should not be free as regards the price of what they produce, since price is a matter concerning their relations to the rest of the community. If there were nominal freedom in regard to price, there would be a danger of a constant tug-of-war, in which those trades which were most immediately necessary to the existence of the community could always obtain an unfair advantage. Force is no more admirable in the economic sphere than in dealings between states. In order to secure the maximum of freedom with the minimum of force, the universal principle is: *Autonomy within each politically important group, and a neutral authority for deciding questions involving relations between groups*. The neutral authority should, of course, rest on a democratic basis, but should, if possible, represent a constituency wider than that of the groups concerned. In international affairs the only adequate authority would be one representing all civilized nations.

In order to prevent undue extension of the power of such authorities, it is desirable and necessary that the various autonomous groups should be very jealous of their liberties, and very ready to resist by political means any encroachments upon their independence. State socialism does not tolerate such groups, each with their own officials responsible to the group. Consequently it abandons the internal affairs of a group to the control of men not responsible to that group or specially aware of its needs. This opens the door to tyranny and to the destruction of initiative. These dangers are avoided by a system which allows any group of men to combine for any given purpose, provided it is not predatory, and to claim from the central authority such self-government as is necessary to the carrying out of the purpose. Churches of various denominations afford an instance. Their autonomy was won by centuries of warfare and persecution. It is to be hoped that a less terrible struggle will be required to achieve the same result in the economic sphere. But whatever the obstacles, I believe the importance of liberty is as great in the one case as it has been admitted to be in the other.

Chapter IV: Individual Liberty and Public Control

I

Society cannot exist without law and order, and cannot advance except through the initiative of vigorous innovators. Yet law and order are always hostile to innovations, and innovators are almost always, to some extent, anarchists. Those whose minds are dominated by fear of a relapse towards barbarism will emphasize the importance of law and order, while those who are inspired by the hope of an advance towards civilization will usually be more conscious of the need of individual initiative. Both temperaments are necessary, and wisdom lies in allowing each to operate freely where it is beneficent. But those who are on the side of law and order, since they are reinforced by custom and the instinct for upholding the *status quo*, have no need of a reasoned defense. It is the innovators who have difficulty in being allowed to exist and work. Each generation believes that this difficulty is a thing of the past, but each generation is only tolerant of *past* innovations. Those of its own day are met with the same persecution as though the principle of toleration had never been heard of.

"In early society," says Westermarck, "customs are not only moral rules, but the only moral rules ever thought of. The savage strictly complies with the Hegelian command that no man must have a private conscience. The following statement, which refers to the Tinnevelly Shanars, may be quoted as a typical example: 'Solitary individuals amongst them rarely adopt any new opinions, or any new course of procedure. They follow the multitude to do evil, and they follow the multitude to do good. They think in herds.'"

Those among ourselves who have never thought a thought or done a deed in the slightest degree different from the thoughts and deeds of our neighbors will congratulate themselves on the difference between us and the savage. But those who have ever attempted any real innovation cannot help feeling that the people they know are not so very unlike the Tinnevelly Shanars.

Under the influence of socialism, even progressive opinion, in recent years, has been hostile to individual liberty. Liberty is associated, in the minds of reformers, with *laissez-faire*, the Manchester School, and the exploitation of women and children which resulted from what was euphemistically called "free competition." All these things were evil, and required state interference; in fact, there is need of an immense increase of state action in regard to cognate evils which still exist. In everything that concerns the economic life of the community, as regards both distribution and conditions of production, what is required is more public control, not less—how much more, I do not profess to know.

Another direction in which there is urgent need of the substitution of law and order for anarchy is international relations. At present, each sovereign state has complete individual freedom, subject only to the sanction of war. This individual freedom will have to be curtailed in regard to external relations if wars are ever to cease.

But when we pass outside the sphere of material possessions, we find that the arguments in favor of public control almost entirely disappear.

Religion, to begin with, is recognized as a matter in which the state ought not to interfere. Whether a man is Christian, Mahometan, or Jew is a question of no public concern, so long as he obeys the laws; and the laws ought to be such as men of all religions can obey. Yet even here there are limits. No civilized state would tolerate a religion demanding human sacrifice. The English in India put an end to suttee, in spite of a fixed principle of non-interference with native religious customs. Perhaps they were wrong to prevent suttee, yet almost every European would have done the same. We cannot *effectively* doubt that such practices ought to be stopped, however we may theorize in favor of religious liberty.

In such cases, the interference with liberty is imposed from without by a higher civilization. But the more common case, and the more interesting, is when an independent state interferes on behalf of custom against individuals who are feeling their way toward more civilized beliefs and institutions.

"In New South Wales," says Westermarck, "the first-born of every lubra used to be eaten by the tribe 'as part of a religious ceremony.' In the realm of Khai-muh, in China, according to a native account, it was customary to kill and devour the eldest son alive. Among certain tribes in British Columbia the first child is often sacrificed to the sun. The Indians of Florida, according to Le Moyne de Morgues, sacrificed the first-born son to the chief....'"

There are pages and pages of such instances.

There is nothing analogous to these practices among ourselves. When the first-born in Florida was told that his king and country needed him, this was a mere mistake, and with us mistakes of this kind do not occur. But it is interesting to inquire how these superstitions died out, in such cases, for example, as that of Khai-muh, where foreign compulsion is improbable. We may surmise that some parents, under the selfish influence of parental affection, were led to doubt whether the sun would really be angry if the eldest child were allowed to live. Such rationalism would be regarded as very dangerous, since it was calculated to damage the harvest. For generations the opinion would be cherished in secret by a handful of cranks, who would not be able to act upon it. At last, by concealment or flight, a few parents would save their children from the sacrifice. Such parents would be regarded as lacking all public spirit, and as willing to endanger the community for their private pleasure. But gradually it would appear that the state remained intact, and the crops were no worse than in former years. Then, by a fiction, a child would be deemed to have been sacrificed if it was solemnly dedicated to agriculture or some other work of national importance chosen by the chief. It would be many generations before the child would be allowed to choose its own occupation after it had grown old enough to know its own tastes and capacities. And during all those generations, children would be reminded that only an act of grace had allowed them to live at all, and would exist under the shadow of a purely imaginary duty to the state.

The position of those parents who first disbelieved in the utility of infant sacrifice illustrates all the difficulties which arise in connection with the adjustment of individual freedom to public control. The authorities, believing the sacrifice necessary for the good of the community, were bound to insist upon it; the parents, believing it useless, were equally bound to do everything in their power toward saving the child. How ought both parties to act in such a case?

The duty of the skeptical parent is plain: to save the child by any possible means, to preach the uselessness of the sacrifice in season and out of season, and to endure patiently whatever penalty the law may indict for evasion. But the duty of the authorities is far less clear. So long as they remain firmly persuaded that the universal sacrifice of the first-born is indispensable, they are bound to persecute those who seek to undermine this belief. But they will, if they are conscientious, very carefully examine the arguments of opponents, and be willing in advance to admit that these arguments *may* be sound. They will carefully search their own hearts to see whether hatred of children or pleasure in cruelty has anything to do with their belief. They will remember that in the past history of Khai-muh there are innumerable instances of beliefs, now known to be false, on account of which those who disagreed with the prevalent view were put to death. Finally they will reflect that, though errors which are traditional are often wide-spread, new beliefs seldom win acceptance unless they are nearer to the truth than what they replace; and they will conclude that a new belief is probably either an advance, or so unlikely to become common as to be innocuous. All these considerations will make them hesitate before they resort to punishment.

II

The study of past times and uncivilized races makes it clear beyond question that the customary beliefs of tribes or nations are almost invariably false. It is difficult to divest ourselves completely of the customary beliefs of our own age and nation, but it is not very difficult to achieve a certain degree of doubt in regard to them. The Inquisitor who burnt men at the stake was acting with true humanity if all his beliefs were correct; but if they were in error at any point, he was inflicting a wholly unnecessary cruelty. A good working maxim in such matters is this: Do not trust customary beliefs so far as to perform actions which must be disastrous unless the beliefs in question are wholly true. The world would be utterly bad, in the opinion of the average Englishman, unless he could say "Britannia rules the waves"; in the opinion of the average German, unless he could say "Deutschland über alles." For the sake of these beliefs, they are willing to destroy European civilization. If the beliefs should happen to be false, their action is regrettable.

One fact which emerges from these considerations is that no obstacle should be placed in the way of thought and its expression, nor yet in the way of statements of fact. This was formerly common ground among liberal thinkers, though it was never quite realized in the practice of civilized countries. But it has recently become, throughout Europe, a dangerous paradox, on account of which men suffer imprisonment or starvation. For this reason it has again become worth stating. The grounds for it are so evident that I should be ashamed to repeat them if they were not universally ignored. But in the actual world it is very necessary to repeat them.

To attain complete truth is not given to mortals, but to advance toward it by successive steps is not impossible. On any matter of general interest, there is usually, in any given community at any given time, a received opinion, which is accepted as a matter of course by all who give no special thought to the matter. Any questioning of the received opinion rouses hostility, for a number of reasons.

The most important of these is the instinct of conventionality, which exists in all gregarious animals and often leads them to put to death any markedly peculiar member of the herd.

The next most important is the feeling of insecurity aroused by doubt as to the beliefs by which we are in the habit of regulating our lives. Whoever has tried to explain the philosophy of Berkeley to a plain man will have seen in its unadulterated form the anger aroused by this feeling. What the plain man derives from Berkeley's philosophy at a first hearing is an uncomfortable suspicion that nothing is solid, so that it is rash to sit on a chair or to expect the floor to sustain us. Because this suspicion is uncomfortable, it is irritating, except to those who regard the whole argument as merely nonsense. And in a more or less analogous way any questioning of what has been taken for granted destroys the feeling of standing on solid ground, and produces a condition of bewildered fear.

A third reason which makes men dislike novel opinions is that vested interests are bound up with old beliefs. The long fight of the church against science, from Giordano Bruno to Darwin, is attributable to this motive among others. The horror of socialism which existed in the remote past was entirely attributable to this cause. But it would be a mistake to assume, as is done by those who seek economic motives everywhere, that vested interests are the principal source of anger against novelties in thought. If this were the case, intellectual progress would be much more rapid than it is.

The instinct of conventionality, horror of uncertainty, and vested interests, all militate against the acceptance of a new idea. And it is even harder to think of a new idea than to get it accepted; most people might spend a lifetime in reflection without ever making a genuinely original discovery.

In view of all these obstacles, it is not likely that any society at any time will suffer from a plethora of heretical opinions. Least of all is this likely in a modern civilized society, where the conditions of life are in constant rapid change, and demand, for successful adaptation, an equally rapid change in intellectual outlook. There should be an attempt, therefore, to encourage, rather than discourage, the expression of new beliefs and the dissemination of knowledge tending to support them. But the very opposite is, in fact, the case. From childhood upward, everything is done to make the minds of men and women conventional and sterile. And if, by misadventure, some spark of imagination remains, its unfortunate possessor is considered unsound and dangerous, worthy only of contempt in time of peace and of prison or a traitor's death in time of war. Yet such men are known to have been in the past the chief benefactors of mankind, and are the very men who receive most honor as soon as they are safely dead.

The whole realm of thought and opinion is utterly unsuited to public control; it ought to be as free, and as spontaneous as is possible to those who know what others have believed. The state is justified in insisting that children shall be educated, but it is not justified in forcing their education to proceed on a uniform plan and to be directed to the production of a dead level of glib uniformity. Education, and the life of the mind generally, is a matter in which individual initiative is the chief thing needed; the function of the state should begin and end with insistence on some kind of education, and, if possible, a kind which promotes mental individualism, not a kind which happens to conform to the prejudices of government officials.

III

Questions of practical morals raise more difficult problems than questions of mere opinion. The thugs honestly believe it their duty to commit murders, but the government does not acquiesce. The conscientious objectors honestly hold the opposite opinion, and again the government does not acquiesce. Killing is a state prerogative; it is equally criminal to do it unbidden and not to do it when bidden. The same applies to theft, unless it is on a large scale or by one who is already rich. Thugs and thieves are men who use force in their dealings with their neighbors, and we may lay it down broadly that the private use of force should be prohibited except in rare cases, however conscientious may be its motive. But this principle will not justify compelling men to use force at the bidding of the state, when they do not believe it justified by the occasion. The punishment of conscientious objectors seems clearly a violation of individual liberty within its legitimate sphere.

It is generally assumed without question that the state has a right to punish certain kinds of sexual irregularity. No one doubts that the Mormons sincerely believed polygamy to be a desirable practice, yet the United States required them to abandon its legal recognition, and probably any other Christian country would have done likewise. Nevertheless, I do not think this prohibition was wise. Polygamy is legally permitted in many parts of the world, but is not much practised except by chiefs and potentates. If, as Europeans generally believe, it is an undesirable custom, it is probable that the Mormons would have soon abandoned it, except perhaps for a few men of exceptional position. If, on the other hand, it had proved a successful experiment, the world would have acquired a piece of knowledge which it is now unable to possess. I think in all such cases the law should only intervene when there is some injury inflicted without the consent of the injured person.

It is obvious that men and women would not tolerate having their wives or husbands selected by the state, whatever eugenists might have to say in favor of such a plan. In this it seems clear that ordinary public opinion is in the right, not because people choose wisely, but because any choice of their own is better than a forced marriage. What applies to marriage ought also to apply to the choice of a trade or profession; although some men have no marked preferences, most men greatly prefer some occupations to others, and are far more likely to be useful citizens if they follow their preferences than if they are thwarted by a public authority.

The case of the man who has an intense conviction that he ought to do a certain kind of work is peculiar, and perhaps not very common; but it is important because it includes some very important individuals. Joan of Arc and Florence Nightingale defied convention in obedience to a feeling of this sort; reformers and agitators in unpopular causes, such as Mazzini, have belonged to this class; so have many men of science. In cases of this kind the individual conviction deserves the greatest respect, even if there seems no obvious justification for it. Obedience to the impulse is very unlikely to do much harm, and may well do great good. The practical difficulty is to distinguish such impulses from desires which produce similar manifestations. Many young people wish to be authors without having an impulse to write any particular book, or wish to be painters without having an impulse to create any particular picture. But a little experience will usually show the difference between a genuine and a spurious impulse; and there is less harm in indulging the spurious impulse for a time than in thwarting the impulse which is genuine. Nevertheless, the plain man almost always has a tendency to thwart the genuine impulse, because it seems anarchic and unreasonable, and is seldom able to give a good account of itself in advance.

What is markedly true of some notable personalities is true, in a lesser degree, of almost every individual who has much vigor or force of life; there is an impulse towards activity of some kind, as a rule not very definite in youth, but growing gradually more sharply outlined under the influence of education and opportunity. The direct impulse toward a kind of activity for its own sake must be distinguished from the desire for the expected effects of the activity. A young man may desire the rewards of great achievement without having any spontaneous impulse toward the activities which lead to achievement. But those who actually achieve much, although they may desire the rewards, have also something in their nature which inclines them to choose a certain kind of work as the road which they must travel if their ambition is to be satisfied. This artist's impulse, as it may be called, is a thing of infinite value to the individual, and often to the world; to respect it in oneself and in others makes up nine tenths of the good life. In most human beings it is rather frail, rather easily destroyed or disturbed; parents and teachers are too often hostile to it, and our economic system crushes out its last remnants in young men and young women. The result is that human beings cease to be individual, or to retain the native pride that is their birthright; they become machine-made, tame, convenient for the bureaucrat and the drill-sergeant, capable of being tabulated in statistics without anything being omitted. This is the fundamental evil resulting from lack of liberty; and it is an evil which is being continually intensified as population grows more dense and the machinery of organization grows more efficient.

The things that men desire are many and various: admiration, affection, power, security, ease, outlets for energy, are among the commonest of motives. But such abstractions do not touch what makes the difference between one man and another. Whenever I go to the zoölogical gardens, I am struck by the fact that all the movements of a stork have some common quality, differing from the movements of a parrot or an ostrich. It is impossible to put in words what the common quality is, and yet we feel that each thing an animal does is the sort of thing we might expect that animal to do. This indefinable quality constitutes the individuality of the animal, and gives rise to the pleasure we feel in watching the animal's actions. In a human being, provided he has not been crushed by an economic or governmental machine, there is the same kind of individuality, a something distinctive without which no man or woman can achieve much of importance, or retain the full dignity which is native to human beings. It is this distinctive individuality that is loved by the artist, whether painter or writer. The artist himself, and the man who is creative in no matter what direction, has more of it than the average man. Any society which crushes this quality, whether intentionally or by accident, must soon become utterly lifeless and traditional, without hope of progress and without any purpose in its being. To preserve and strengthen the impulse that makes individuality should be the foremost object of all political institutions.

IV

We now arrive at certain general principles in regard to individual liberty and public control.

The greater part of human impulses may be divided into two classes, those which are possessive and those which are constructive or creative. Social institutions are the garments or embodiments of impulses, and may be classified roughly according to the impulses which they embody. Property is the direct expression of possessiveness; science and art are among the most direct expressions of creativeness. Possessiveness is either defensive or aggressive; it seeks either to retain against a robber, or to acquire from a present holder. In either case an attitude of hostility toward others is of its essence. It would be a mistake to suppose that defensive possessiveness is always justifiable, while the aggressive kind is always blameworthy; where there is great injustice in the *status quo*, the exact opposite may be the case, and ordinarily neither is justifiable.

State interference with the actions of individuals is necessitated by possessiveness. Some goods can be acquired or retained by force, while others cannot. A wife can be acquired by force, as the Romans acquired the Sabine women; but a wife's affection cannot be acquired in this way. There is no record that the Romans desired the affection of the Sabine women; and those in whom possessive impulses are strong tend to care chiefly for the goods that force can secure. All material goods belong to this class. Liberty in regard to such goods, if it were unrestricted, would make the strong rich and the weak poor. In a capitalistic society, owing to the partial restraints imposed by law, it makes cunning men rich and honest men poor, because the force of the state is put at men's disposal, not according to any just or rational principle, but according to a set of traditional maxims of which the explanation is purely historical.

In all that concerns possession and the use of force, unrestrained liberty involves anarchy and injustice. Freedom to kill, freedom to rob, freedom to defraud, no longer belong to individuals, though they still belong to great states, and are exercised by them in the name of patriotism. Neither individuals nor states ought to be free to exert force on their own initiative, except in such sudden emergencies as will subsequently be admitted in justification by a court of law. The reason for this is that the exertion of force by one individual against another is always an evil on both sides, and can only be tolerated when it is compensated by some overwhelming resultant good. In order to minimize the amount of force actually exerted in the world, it is necessary that there should be a public authority, a repository of practically irresistible force, whose function should be primarily to repress the private use of force. A use of force is *private* when it is exerted by one of the interested parties, or by his friends or accomplices, not by a public neutral authority according to some rule which is intended to be in the public interest.

The régime of private property under which we live does much too little to restrain the private use of force. When a man owns a piece of land, for example, he may use force against trespassers, though they must not use force against him. It is clear that some restriction of the liberty of trespass is necessary for the cultivation of the land. But if such powers are to be given to an individual, the state ought to satisfy itself that he occupies no more land than he is warranted in occupying in the public interest, and that the share of the produce of the land that comes to him is no more than a just reward for his labors. Probably the only way in which such ends can be achieved is by state ownership of land. The possessors of land and capital are able at present, by economic pressure, to use force against those who have no possessions. This force is sanctioned by law, while force exercised by the poor against the rich is illegal. Such a state of things is unjust, and does not diminish the use of private force as much as it might be diminished.

The whole realm of the possessive impulses, and of the use of force to which they give rise, stands in need of control by a public neutral authority, in the interests of liberty no less than of justice. Within a nation, this public authority will naturally be the state; in relations between nations, if the present anarchy is to cease, it will have to be some international parliament.

But the motive underlying the public control of men's possessive impulses should always be the increase of liberty, both by the prevention of private tyranny and by the liberation of creative impulses. If public control is not to do more harm than good, it must be so exercised as to leave the utmost freedom of private initiative in all those ways that do not involve the private use of force. In this respect all governments have always failed egregiously, and there is no evidence that they are improving.

The creative impulses, unlike those that are possessive, are directed to ends in which one man's gain is not another man's loss. The man who makes a scientific discovery or writes a poem is enriching others at the same time as himself. Any increase in knowledge or good-will is a gain to all who are affected by it, not only to the actual possessor. Those who feel the joy of life are a happiness to others as well as to themselves. Force cannot create such things, though it can destroy them; no principle of distributive justice applies to them, since the gain of each is the gain of all. For these reasons, the creative part of a man's activity ought to be as free as possible from all public control, in order that it may remain spontaneous and full of vigor. The only function of the state in regard to this part of the individual life should be to do everything possible toward providing outlets and opportunities.

In every life a part is governed by the community, and a part by private initiative. The part governed by private initiative is greatest in the most important individuals, such as men of genius and creative thinkers. This part ought only to be restricted when it is predatory; otherwise, everything ought to be done to make it as great and as vigorous as possible. The object of education ought not to be to make all men think alike, but to make each think in the way which is the fullest expression of his own personality. In the choice of a means of livelihood all young men and young women ought, as far as possible, to be able to choose what is attractive to them; if no money-making occupation is attractive, they ought to be free to do little work for little pay, and spend their leisure as they choose. Any kind of censure on freedom of thought or on the dissemination of knowledge is, of course, to be condemned utterly.

Huge organizations, both political and economic, are one of the distinguishing characteristics of the modern world. These organizations have immense power, and often use their power to discourage originality in thought and action. They ought, on the contrary, to give the freest scope that is possible without producing anarchy or violent conflict. They ought not to take cognizance of any part of a man's life except what is concerned with the legitimate objects of public control, namely, possessions and the use of force. And they ought, by devolution, to leave as large a share of control as possible in the hands of individuals and small groups. If this is not done, the men at the head of these vast organizations will infallibly become tyrannous through the habit of excessive power, and will in time interfere in ways that crush out individual initiative.

The problem which faces the modern world is the combination of individual initiative with the increase in the scope and size of organizations. Unless it is solved, individuals will grow less and less full of life and vigor, and more and more passively submissive to conditions imposed upon them. A society composed of such individuals cannot be progressive or add much to the world's stock of mental and spiritual possessions. Only personal liberty and the encouragement of initiative can secure these things. Those who resist authority when it encroaches upon the legitimate sphere of the individual are performing a service to society, however little society may value it. In regard to the past, this is universally acknowledged; but it is no less true in regard to the present and the future.

Chapter V: National Independence and Internationalism

In the relations between states, as in the relations of groups within a single state, what is to be desired is independence for each as regards internal affairs, and law rather than private force as regards external affairs. But as regards groups within a state, it is internal independence that must be emphasized, since that is what is lacking; subjection to law has been secured, on the whole, since the end of the Middle Ages. In the relations between states, on the contrary, it is law and a central government that are lacking, since independence exists for external as for internal affairs. The stage we have reached in the affairs of Europe corresponds to the stage reached in our internal affairs during the Wars of the Roses, when turbulent barons frustrated the attempt to make them keep the king's peace. Thus, although the goal is the same in the two cases, the steps to be taken in order to achieve it are quite different.

There can be no good international system until the boundaries of states coincide as nearly as possible with the boundaries of nations.

But it is not easy to say what we mean by a nation. Are the Irish a nation? Home Rulers say yes, Unionists say no. Are the Ulstermen a nation? Unionists say yes, Home Rulers say no. In all such cases it is a party question whether we are to call a group a nation or not. A German will tell you that the Russian Poles are a nation, but as for the Prussian Poles, they, of course, are part of Prussia. Professors can always be hired to prove, by arguments of race or language or history, that a group about which there is a dispute is, or is not, a nation, as may be desired by those whom the professors serve. If we are to avoid all these controversies, we must first of all endeavor to find some definition of a nation.

A nation is not to be defined by affinities of language or a common historical origin, though these things often help to produce a nation. Switzerland is a nation, despite diversities of race, religion, and language. England and Scotland now form one nation, though they did not do so at the time of the Civil War. This is shown by Cromwell's saying, in the height of the conflict, that he would rather be subject to the domain of the royalists than to that of the Scotch. Great Britain was one state before it was one nation; on the other hand, Germany was one nation before it was one state.

What constitutes a nation is a sentiment and an instinct, a sentiment of similarity and an instinct of belonging to the same group or herd. The instinct is an extension of the instinct which constitutes a flock of sheep, or any other group of gregarious animals. The sentiment which goes with this is like a milder and more extended form of family feeling. When we return to England after being on the Continent, we feel something friendly in the familiar ways, and it is easy to believe that Englishmen on the whole are virtuous, while many foreigners are full of designing wickedness.

Such feelings make it easy to organize a nation into a state. It is not difficult, as a rule, to acquiesce in the orders of a national government. We feel that it is our government, and that its decrees are more or less the same as those which we should have given if we ourselves had been the governors. There is an instinctive and usually unconscious sense of a common purpose animating the members of a nation. This becomes especially vivid when there is war or a danger of war. Any one who, at such a time, stands out against the orders of his government feels an inner conflict quite different from any that he would feel in standing out against the orders of a foreign government in whose power he might happen to find himself. If he stands out, he does so with some more or less conscious hope that his government may in time come to think as he does; whereas, in standing out against a foreign government, no such hope is necessary. This group instinct, however it may have arisen, is what constitutes a nation, and what makes it important that the boundaries of nations should also be the boundaries of states.

National sentiment is a fact, and should be taken account of by institutions. When it is ignored, it is intensified and becomes a source of strife. It can only be rendered harmless by being given free play, so long as it is not predatory. But it is not, in itself, a good or admirable feeling. There is nothing rational and nothing desirable in a limitation of sympathy which confines it to a fragment of the human race. Diversities of manners and customs and traditions are, on the whole, a good thing, since they enable different nations to produce different types of excellence. But in national feeling there is always latent or explicit an element of hostility to foreigners. National feeling, as we know it, could not exist in a nation which was wholly free from external pressure of a hostile kind.

And group feeling produces a limited and often harmful kind of morality. Men come to identify the good with what serves the interests of their own group, and the bad with what works against those interests, even if it should happen to be in the interests of mankind as a whole. This group morality is very much in evidence during war, and is taken for granted in men's ordinary thought. Although almost all Englishmen consider the defeat of Germany desirable for the good of the world, yet nevertheless most of them honor a German for fighting for his country, because it has not occurred to them that his actions ought to be guided by a morality higher than that of the group.

A man does right, as a rule, to have his thoughts more occupied with the interests of his own nation than with those of others, because his actions are more likely to affect his own nation. But in time of war, and in all matters which are of equal concern to other nations and to his own, a man ought to take account of the universal welfare, and not allow his survey to be limited by the interest, or supposed interest, of his own group or nation.

So long as national feeling exists, it is very important that each nation should be self-governing as regards its internal affairs. Government can only be carried on by force and tyranny if its subjects view it with hostile eyes, and they will so view it if they feel that it belongs to an alien nation. This principle meets with difficulties in cases where men of different nations live side by side in the same area, as happens in some parts of the Balkans. There are also difficulties in regard to places which, for some geographical reason, are of great international importance, such as the Suez Canal and the Panama Canal. In such cases the purely local desires of the inhabitants may have to give way before larger interests. But in general, at any rate as applied to civilized communities, the principle that the boundaries of nations ought to coincide with the boundaries of states has very few exceptions.

This principle, however, does not decide how the relations between states are to be regulated, or how a conflict of interests between rival states is to be decided. At present, every great state claims absolute sovereignty, not only in regard to its internal affairs but also in regard to its external actions. This claim to absolute sovereignty leads it into conflict with similar claims on the part of other great states. Such conflicts at present can only be decided by war or diplomacy, and diplomacy is in essence nothing but the threat of war. There is no more justification for the claim to absolute sovereignty on the part of a state than there would be for a similar claim on the part of an individual. The claim to absolute sovereignty is, in effect, a claim that all external affairs are to be regulated purely by force, and that when two nations or groups of nations are interested in a question, the decision shall depend solely upon which of them is, or is believed to be, the stronger. This is nothing but primitive anarchy, "the war of all against all," which Hobbes asserted to be the original state of mankind.

There cannot be secure peace in the world, or any decision of international questions according to international law, until states are willing to part with their absolute sovereignty as regards their external relations, and to leave the decision in such matters to some international instrument of government. An international government will have to be legislative as well as judicial. It is not enough that there should be a Hague tribunal, deciding matters according to some already existing system of international law; it is necessary also that there should be a body capable of enacting international law, and this body will have to have the power of transferring territory from one state to another, when it is persuaded that adequate grounds exist for such a transference. Friends of peace will make a mistake if they unduly glorify the *status quo*. Some nations grow, while others dwindle; the population of an area may change its character by emigration and immigration. There is no good reason why states should resent changes in their boundaries under such conditions, and if no international authority has power to make changes of this kind, the temptations to war will sometimes become irresistible.

The international authority ought to possess an army and navy, and these ought to be the only army and navy in existence. The only legitimate use of force is to diminish the total amount of force exercised in the world. So long as men are free to indulge their predatory instincts, some men or groups of men will take advantage of this freedom for oppression and robbery. Just as the police are necessary to prevent the use of force by private citizens, so an international police will be necessary to prevent the lawless use of force by separate states.

But I think it is reasonable to hope that if ever an international government, possessed of the only army and navy in the world, came into existence, the need of force to enact obedience to its decisions would be very temporary. In a short time the benefits resulting from the substitution of law for anarchy would become so obvious that the international government would acquire an unquestioned authority, and no state would dream of rebelling against its decisions. As soon as this stage had been reached, the international army and navy would become unnecessary.

We have still a very long road to travel before we arrive at the establishment of an international authority, but it is not very difficult to foresee the steps by which this result will be gradually reached. There is likely to be a continual increase in the practice of submitting disputes to arbitration, and in the realization that the supposed conflicts of interest between different states are mainly illusory. Even where there is a real conflict of interest, it must in time become obvious that neither of the states concerned would suffer as much by giving way as by fighting. With the progress of inventions, war, when it does occur, is bound to become increasingly destructive. The civilized races of the world are faced with the alternative of coöperation or mutual destruction. The present war is making this alternative daily more evident. And it is difficult to believe that, when the enmities which it has generated have had time to cool, civilized men will deliberately choose to destroy civilization, rather than acquiesce in the abolition of war.

The matters in which the interests of nations are supposed to clash are mainly three: tariffs, which are a delusion; the exploitation of inferior races, which is a crime; pride of power and dominion, which is a schoolboy folly.

The economic argument against tariffs is familiar, and I shall not repeat it. The only reason why it fails to carry conviction is the enmity between nations. Nobody proposes to set up a tariff between England and Scotland, or between Lancashire and Yorkshire. Yet the arguments by which tariffs between nations are supported might be used just as well to defend tariffs between counties.

Universal free trade would indubitably be of economic benefit to mankind, and would be adopted to-morrow if it were not for the hatred and suspicion which nations feel one toward another. From the point of view of preserving the peace of the world, free trade between the different civilized states is not so important as the open door in their dependencies. The desire for exclusive markets is one of the most potent causes of war.

Exploiting what are called "inferior races" has become one of the main objects of European statecraft. It is not only, or primarily, trade that is desired, but opportunities for investment; finance is more concerned in the matter than industry. Rival diplomatists are very often the servants, conscious or unconscious, of rival groups of financiers. The financiers, though themselves of no particular nation, understand the art of appealing to national prejudice, and of inducing the taxpayer to incur expenditure of which they reap the benefit. The evils which they produce at home, and the devastation that they spread among the races whom they exploit, are part of the price which the world has to pay for its acquiescence in the capitalist régime.

But neither tariffs nor financiers would be able to cause serious trouble, if it were not for the sentiment of national pride. National pride might be on the whole beneficent, if it took the direction of emulation in the things that are important to civilization. If we prided ourselves upon our poets, our men of science, or the justice and humanity of our social system, we might find in national pride a stimulus to useful endeavors. But such matters play a very small part. National pride, as it exists now, is almost exclusively concerned with power and dominion, with the extent of territory that a nation owns, and with its capacity for enforcing its will against the opposition of other nations. In this it is reinforced by group morality. To nine citizens out of ten it seems self-evident, whenever the will of their own nation clashes with that of another, that their own nation must be in the right. Even if it were not in the right on the particular issue, yet it stands in general for so much nobler ideals than those represented by the other nation to the dispute, that any increase in its power is bound to be for the good of mankind. Since all nations equally believe this of themselves, all are equally ready to insist upon the victory of their own side in any dispute in which they believe that they have a good hope of victory. While this temper persists, the hope of international coöperation must remain dim.

If men could divest themselves of the sentiment of rivalry and hostility between different nations, they would perceive that the matters in which the interests of different nations coincide immeasurably outweigh those in which they clash; they would perceive, to begin with, that trade is not to be compared to warfare; that the man who sells you goods is not doing you an injury. No one considers that the butcher and the baker are his enemies because they drain him of money. Yet as soon as goods come from a foreign country, we are asked to believe that we suffer a terrible injury in purchasing them. No one remembers that it is by means of goods exported that we purchase them. But in the country to which we export, it is the goods we send which are thought dangerous, and the goods we buy are forgotten. The whole conception of trade, which has been forced upon us by manufacturers who dreaded foreign competition, by trusts which desired to secure monopolies, and by economists poisoned by the virus of nationalism, is totally and absolutely false. Trade results simply from division of labor. A man cannot himself make all the goods of which he has need, and therefore he must exchange his produce with that of other people. What applies to the individual, applies in exactly the same way to the nation. There is no reason to desire that a nation should itself produce all the goods of which it has need; it is better that it should specialize upon those goods which it can produce to most advantage, and should exchange its surplus with the surplus of other goods produced by other countries. There is no use in sending goods out of the country except in order to get other goods in return. A butcher who is always willing to part with his meat but not willing to take bread from the baker, or boots from the bootmaker, or clothes from the tailor, would soon find himself in a sorry plight. Yet he would be no more foolish than the protectionist who desires that we should send goods abroad without receiving payment in the shape of goods imported from abroad.

The wage system has made people believe that what a man needs is work. This, of course, is absurd. What he needs is the goods produced by work, and the less work involved in making a given amount of goods, the better. But owing to our economic system, every economy in methods of production enables employers to dismiss some of their employees, and to cause destitution, where a better system would produce only an increase of wages or a diminution in the hours of work without any corresponding diminution of wages.

Our economic system is topsyturvy. It makes the interest of the individual conflict with the interest of the community in a thousand ways in which no such conflict ought to exist. Under a better system the benefits of free trade and the evils of tariffs would be obvious to all.

Apart from trade, the interests of nations coincide in all that makes what we call civilization. Inventions and discoveries bring benefit to all. The progress of science is a matter of equal concern to the whole civilized world. Whether a man of science is an Englishman, a Frenchman, or a German is a matter of no real importance. His discoveries are open to all, and nothing but intelligence is required in order to profit by them. The whole world of art and literature and learning is international; what is done in one country is not done for that country, but for mankind. If we ask ourselves what are the things that raise mankind above the brutes, what are the things that make us think the human race more valuable than any species of animals, we shall find that none of them are things in which any one nation can have exclusive property, but all are things in which the whole world can share. Those who have any care for these things, those who wish to see mankind fruitful in the work which men alone can do, will take little account of national boundaries, and have little care to what state a man happens to owe allegiance.

The importance of international coöperation outside the sphere of politics has been brought home to me by my own experience. Until lately I was engaged in teaching a new science which few men in the world were able to teach. My own work in this science was based chiefly upon the work of a German and an Italian. My pupils came from all over the civilized world: France, Germany, Austria, Russia, Greece, Japan, China, India, and America. None of us was conscious of any sense of national divisions. We felt ourselves an outpost of civilization, building a new road into the virgin forest of the unknown. All coöperated in the common task, and in the interest of such a work the political enmities of nations seemed trivial, temporary, and futile.

But it is not only in the somewhat rarefied atmosphere of abstruse science that international coöperation is vital to the progress of civilization. All our economic problems, all the questions of securing the rights of labor, all the hopes of freedom at home and humanity abroad, rest upon the creation of international good-will.

So long as hatred, suspicion, and fear dominate the feelings of men toward each other, so long we cannot hope to escape from the tyranny of violence and brute force. Men must learn to be conscious of the common interests of mankind in which all are at one, rather than of those supposed interests in which the nations are divided. It is not necessary, or even desirable, to obliterate the differences of manners and custom and tradition between different nations. These differences enable each nation to make its own distinctive contribution to the sum total of the world's civilization.

What is to be desired is not cosmopolitanism, not the absence of all national characteristics that one associates with couriers, *wagon-lit* attendants, and others, who have had everything distinctive obliterated by multiple and trivial contacts with men of every civilized country. Such cosmopolitanism is the result of loss, not gain. The international spirit which we should wish to see produced will be something added to love of country, not something taken away. Just as patriotism does not prevent a man from feeling family affection, so the international spirit ought not to prevent a man from feeling affection for his own country. But it will somewhat alter the character of that affection. The things which he will desire for his own country will no longer be things which can only be acquired at the expense of others, but rather those things in which the excellence of any one country is to the advantage of all the world. He will wish his own country to be great in the arts of peace, to be eminent in thought and science, to be magnanimous and just and generous. He will wish it to help mankind on the way toward that better world of liberty and international concord which must be realized if any happiness is to be left to man. He will not desire for his country the passing triumphs of a narrow possessiveness, but rather the enduring triumph of having helped to embody in human affairs something of that spirit of brotherhood which Christ taught and which the Christian churches have forgotten. He will see that this spirit embodies not only the highest morality, but also the truest wisdom, and the only road by which the nations, torn and bleeding with the wounds which scientific madness has inflicted, can emerge into a life where growth is possible and joy is not banished at the frenzied call of unreal and fictitious duties. Deeds inspired by hate are not duties, whatever pain and self-sacrifice they may involve. Life and hope for the world are to be found only in the deeds of love.

MYSTICISM AND LOGIC AND OTHER ESSAYS

I

MYSTICISM AND LOGIC

Metaphysics, or the attempt to conceive the world as a whole by means of thought, has been developed, from the first, by the union and conflict of two very different human impulses, the one urging men towards mysticism, the other urging them towards science. Some men have achieved greatness through one of these impulses alone, others through the other alone: in Hume, for example, the scientific impulse reigns quite unchecked, while in Blake a strong hostility to science co-exists with profound mystic insight. But the greatest men who have been philosophers have felt the need both of science and of mysticism: the attempt to harmonise the two was what made their life, and what always must, for all its arduous uncertainty, make philosophy, to some minds, a greater thing than either science or religion.

Before attempting an explicit characterisation of the scientific and the mystical impulses, I will illustrate them by examples from two philosophers whose greatness lies in the very intimate blending which they achieved. The two philosophers I mean are Heraclitus and Plato.

Heraclitus, as everyone knows, was a believer in universal flux: time builds and destroys all things. From the few fragments that remain, it is not easy to discover how he arrived at his opinions, but there are some sayings that strongly suggest scientific observation as the source.

"The things that can be seen, heard, and learned," he says, "are what I prize the most." This is the language of the empiricist, to whom observation is the sole guarantee of truth. "The sun is new every day," is another fragment; and this opinion, in spite of its paradoxical character, is obviously inspired by scientific reflection, and no doubt seemed to him to obviate the difficulty of understanding how the sun can work its way underground from west to east during the night. Actual observation must also have suggested to him his central doctrine, that Fire is the one permanent substance, of which all visible things are passing phases. In combustion we see things change utterly, while their flame and heat rise up into the air and vanish.

"This world, which is the same for all," he says, "no one of gods or men has made; but it was ever, is now, and ever shall be, an ever-living Fire, with measures kindling, and measures going out."

"The transformations of Fire are, first of all, sea; and half of the sea is earth, half whirlwind."

This theory, though no longer one which science can accept, is nevertheless scientific in spirit. Science, too, might have inspired the famous saying to which Plato alludes: "You cannot step twice into the same rivers; for fresh waters are ever flowing in upon you." But we find also another statement among the extant fragments: "We step and do not step into the same rivers; we are and are not."

The comparison of this statement, which is mystical, with the one quoted by Plato, which is scientific, shows how intimately the two tendencies are blended in the system of Heraclitus. Mysticism is, in essence, little more than a certain intensity and depth of feeling in regard to what is believed about the universe; and this kind of feeling leads Heraclitus, on the basis of his science, to strangely poignant sayings concerning life and the world, such as:

"Time is a child playing draughts, the kingly power is a child's."

It is poetic imagination, not science, which presents Time as despotic lord of the world, with all the irresponsible frivolity of a child. It is mysticism, too, which leads Heraclitus to assert the identity of opposites: "Good and ill are one," he says; and again: "To God all things are fair and good and right, but men hold some things wrong and some right."

Much of mysticism underlies the ethics of Heraclitus. It is true that a scientific determinism alone might have inspired the statement: "Man's character is his fate"; but only a mystic would have said:

"Every beast is driven to the pasture with blows"; and again:

"It is hard to fight with one's heart's desire. Whatever it wishes to get, it purchases at the cost of soul"; and again:

"Wisdom is one thing. It is to know the thought by which all things are steered through all things."

Examples might be multiplied, but those that have been given are enough to show the character of the man: the facts of science, as they appeared to him, fed the flame in his soul, and in its light he saw into the depths of the world by the reflection of his own dancing swiftly penetrating fire. In such a nature we see the true union of the mystic and the man of science—the highest eminence, as I think, that it is possible to achieve in the world of thought.

In Plato, the same twofold impulse exists, though the mystic impulse is distinctly the stronger of the two, and secures ultimate victory whenever the conflict is sharp. His description of the cave is the classical statement of belief in a knowledge and reality truer and more real than that of the senses:

"Imagine a number of men living in an underground cavernous chamber, with an entrance open to the light, extending along the entire length of the cavern, in which they have been confined, from their childhood, with their legs and necks so shackled that they are obliged to sit still and look straight forwards, because their chains render it impossible for them to turn their heads round: and imagine a bright fire burning some way off, above and behind them, and an elevated roadway passing between the fire and the prisoners, with a low wall built along it, like the screens which conjurors put up in front of their audience, and above which they exhibit their wonders.

I have it, he replied.

Also figure to yourself a number of persons walking behind this wall, and carrying with them statues of men, and images of other animals, wrought in wood and stone and all kinds of materials, together with various other articles, which overtop the wall; and, as you might expect, let some of the passers-by be talking, and others silent.

You are describing a strange scene, and strange prisoners.

They resemble us, I replied.

Now consider what would happen if the course of nature brought them a release from their fetters, and a remedy for their foolishness, in the following manner. Let us suppose that one of them has been released, and compelled suddenly to stand up, and turn his neck round and walk with open eyes towards the light; and let us suppose that he goes through all these actions with pain, and that the dazzling splendour renders him incapable of discerning those objects of which he used formerly to see the shadows. What answer should you expect him to make, if some one were to tell him that in those days he was watching foolish phantoms, but that now he is somewhat nearer to reality, and is turned towards things more real, and sees more correctly; above all, if he were to point out to him the several objects that are passing by, and question him, and compel him to answer what they are? Should you not expect him to be puzzled, and to regard his old visions as truer than the objects now forced upon his notice?

Yes, much truer....

Hence, I suppose, habit will be necessary to enable him to perceive objects in that upper world. At first he will be most successful in distinguishing shadows; then he will discern the reflections of men and other things in water, and afterwards the realities; and after this he will raise his eyes to encounter the light of the moon and stars, finding it less difficult to study the heavenly bodies and the heaven itself by night, than the sun and the sun's light by day.

Doubtless.

Last of all, I imagine, he will be able to observe and contemplate the nature of the sun, not as it *appears* in water or on alien ground, but as it is in itself in its own territory.

Of course.

His next step will be to draw the conclusion, that the sun is the author of the seasons and the years, and the guardian of all things in the visible world, and in a manner the cause of all those things which he and his companions used to see.

Obviously, this will be his next step....

Now this imaginary case, my dear Glancon, you must apply in all its parts to our former statements, by comparing the region which the eye reveals to the prison house, and the light of the fire therein to the power of the sun: and if, by the upward ascent and the contemplation of the upper world, you understand the mounting of the soul into the intellectual region, you will hit the tendency of my own surmises, since you desire to be told what they are; though, indeed, God only knows whether they are correct. But, be that as it may, the view which I take of the subject is to the following effect. In the world of knowledge, the essential Form of Good is the limit of our enquiries, and can barely be perceived; but, when perceived, we cannot help concluding that it is in every case the source of all that is bright and beautiful,—in the visible world giving birth to light and its master, and in the intellectual world dispensing, immediately and with full authority, truth and reason;—and that whosoever would act wisely, either in private or in public, must set this Form of Good before his eyes."

But in this passage, as throughout most of Plato's teaching, there is an identification of the good with the truly real, which became embodied in the philosophical tradition, and is still largely operative in our own day. In thus allowing a legislative function to the good, Plato produced a divorce between philosophy and science, from which, in my opinion, both have suffered ever since and are still suffering. The man of science, whatever his hopes may be, must lay them aside while he studies nature; and the philosopher, if he is to achieve truth, must do the same. Ethical considerations can only legitimately appear when the truth has been ascertained: they can and should appear as determining our feeling towards the truth, and our manner of ordering our lives in view of the truth, but not as themselves dictating what the truth is to be.

There are passages in Plato—among those which illustrate the scientific side of his mind—where he seems clearly aware of this. The most noteworthy is the one in which Socrates, as a young man, is explaining the theory of ideas to Parmenides.

After Socrates has explained that there is an idea of the good, but not of such things as hair and mud and dirt, Parmenides advises him "not to despise even the meanest things," and this advice shows the genuine scientific temper. It is with this impartial temper that the mystic's apparent insight into a higher reality and a hidden good has to be combined if philosophy is to realise its greatest possibilities. And it is failure in this respect that has made so much of idealistic philosophy thin, lifeless, and insubstantial. It is only in marriage with the world that our ideals can bear fruit: divorced from it, they remain barren. But marriage with the world is not to be achieved by an ideal which shrinks from fact, or demands in advance that the world shall conform to its desires.

Parmenides himself is the source of a peculiarly interesting strain of mysticism which pervades Plato's thought—the mysticism which may be called "logical" because it is embodied in theories on logic. This form of mysticism, which appears, so far as the West is concerned, to have originated with Parmenides, dominates the reasonings of all the great mystical metaphysicians from his day to that of Hegel and his modern disciples. Reality, he says, is uncreated, indestructible, unchanging, indivisible; it is "immovable in the bonds of mighty chains, without beginning and without end; since coming into being and passing away have been driven afar, and true belief has cast them away." The fundamental principle of his inquiry is stated in a sentence which would not be out of place in Hegel: "Thou canst not know what is not—that is impossible—nor utter it; for it is the same thing that can be thought and that can be." And again: "It needs must be that what can be thought and spoken of is; for it is possible for it to be, and it is not possible for what is nothing to be." The impossibility of change follows from this principle; for what is past can be spoken of, and therefore, by the principle, still is.

Mystical philosophy, in all ages and in all parts of the world, is characterised by certain beliefs which are illustrated by the doctrines we have been considering.

There is, first, the belief in insight as against discursive analytic knowledge: the belief in a way of wisdom, sudden, penetrating, coercive, which is contrasted with the slow and fallible study of outward appearance by a science relying wholly upon the senses. All who are capable of absorption in an inward passion must have experienced at times the strange feeling of unreality in common objects, the loss of contact with daily things, in which the solidity of the outer world is lost, and the soul seems, in utter loneliness, to bring forth, out of its own depths, the mad dance of fantastic phantoms which have hitherto appeared as independently real and living. This is the negative side of the mystic's initiation: the doubt concerning common knowledge, preparing the way for the reception of what seems a higher wisdom. Many men to whom this negative experience is familiar do not pass beyond it, but for the mystic it is merely the gateway to an ampler world.

The mystic insight begins with the sense of a mystery unveiled, of a hidden wisdom now suddenly become certain beyond the possibility of a doubt. The sense of certainty and revelation comes earlier than any definite belief. The definite beliefs at which mystics arrive are the result of reflection upon the inarticulate experience gained in the moment of insight. Often, beliefs which have no real connection with this moment become subsequently attracted into the central nucleus; thus in addition to the convictions which all mystics share, we find, in many of them, other convictions of a more local and temporary character, which no doubt become amalgamated with what was essentially mystical in virtue of their subjective certainty. We may ignore such inessential accretions, and confine ourselves to the beliefs which all mystics share.

The first and most direct outcome of the moment of illumination is belief in the possibility of a way of knowledge which may be called revelation or insight or intuition, as contrasted with sense, reason, and analysis, which are regarded as blind guides leading to the morass of illusion. Closely connected with this belief is the conception of a Reality behind the world of appearance and utterly different from it. This Reality is regarded with an admiration often amounting to worship; it is felt to be always and everywhere close at hand, thinly veiled by the shows of sense, ready, for the receptive mind, to shine in its glory even through the apparent folly and wickedness of Man. The poet, the artist, and the lover are seekers after that glory: the haunting beauty that they pursue is the faint reflection of its sun. But the mystic lives in the full light of the vision: what others dimly seek he knows, with a knowledge beside which all other knowledge is ignorance.

The second characteristic of mysticism is its belief in unity, and its refusal to admit opposition or division anywhere. We found Heraclitus saying "good and ill are one"; and again he says, "the way up and the way down is one and the same." The same attitude appears in the simultaneous assertion of contradictory propositions, such as: "We step and do not step into the same rivers; we are and are not." The assertion of Parmenides, that reality is one and indivisible, comes from the same impulse towards unity. In Plato, this impulse is less prominent, being held in check by his theory of ideas; but it reappears, so far as his logic permits, in the doctrine of the primacy of the Good.

A third mark of almost all mystical metaphysics is the denial of the reality of Time. This is an outcome of the denial of division; if all is one, the distinction of past and future must be illusory. We have seen this doctrine prominent in Parmenides; and among moderns it is fundamental in the systems of Spinoza and Hegel.

The last of the doctrines of mysticism which we have to consider is its belief that all evil is mere appearance, an illusion produced by the divisions and oppositions of the analytic intellect. Mysticism does not maintain that such things as cruelty, for example, are good, but it denies that they are real: they belong to that lower world of phantoms from which we are to be liberated by the insight of the vision. Sometimes—for example in Hegel, and at least verbally in Spinoza—not only evil, but good also, is regarded as illusory, though nevertheless the emotional attitude towards what is held to be Reality is such as would naturally be associated with the belief that Reality is good. What is, in all cases, ethically characteristic of mysticism is absence of indignation or protest, acceptance with joy, disbelief in the ultimate truth of the division into two hostile camps, the good and the bad. This attitude is a direct outcome of the nature of the mystical experience: with its sense of unity is associated a feeling of infinite peace. Indeed it may be suspected that the feeling of peace produces, as feelings do in dreams, the whole system of associated beliefs which make up the body of mystic doctrine. But this is a difficult question, and one on which it cannot be hoped that mankind will reach agreement.

Four questions thus arise in considering the truth or falsehood of mysticism, namely:

I. Are there two ways of knowing, which may be called respectively reason and intuition? And if so, is either to be preferred to the other?

II. Is all plurality and division illusory?

III. Is time unreal?

IV. What kind of reality belongs to good and evil?

On all four of these questions, while fully developed mysticism seems to me mistaken, I yet believe that, by sufficient restraint, there is an element of wisdom to be learned from the mystical way of feeling, which does not seem to be attainable in any other manner. If this is the truth, mysticism is to be commended as an attitude towards life, not as a creed about the world. The meta-physical creed, I shall maintain, is a mistaken outcome of the emotion, although this emotion, as colouring and informing all other thoughts and feelings, is the inspirer of whatever is best in Man. Even the cautious and patient investigation of truth by science, which seems the very antithesis of the mystic's swift certainty, may be fostered and nourished by that very spirit of reverence in which mysticism lives and moves.

I. REASON AND INTUITION

Of the reality or unreality of the mystic's world I know nothing. I have no wish to deny it, nor even to declare that the insight which reveals it is not a genuine insight. What I do wish to maintain—and it is here that the scientific attitude becomes imperative—is that insight, untested and unsupported, is an insufficient guarantee of truth, in spite of the fact that much of the most important truth is first suggested by its means. It is common to speak of an opposition between instinct and reason; in the eighteenth century, the opposition was drawn in favour of reason, but under the influence of Rousseau and the romantic movement instinct was given the preference, first by those who rebelled against artificial forms of government and thought, and then, as the purely rationalistic defence of traditional theology became increasingly difficult, by all who felt in science a menace to creeds which they associated with a spiritual outlook on life and the world. Bergson, under the name of "intuition," has raised instinct to the position of sole arbiter of metaphysical truth. But in fact the opposition of instinct and reason is mainly illusory. Instinct, intuition, or insight is what first leads to the beliefs which subsequent reason confirms or confutes; but the confirmation, where it is possible, consists, in the last analysis, of agreement with other beliefs no less instinctive. Reason is a harmonising, controlling force rather than a creative one. Even in the most purely logical realm, it is insight that first arrives at what is new.

Where instinct and reason do sometimes conflict is in regard to single beliefs, held instinctively, and held with such determination that no degree of inconsistency with other beliefs leads to their abandonment. Instinct, like all human faculties, is liable to error. Those in whom reason is weak are often unwilling to admit this as regards themselves, though all admit it in regard to others. Where instinct is least liable to error is in practical matters as to which right judgment is a help to survival: friendship and hostility in others, for instance, are often felt with extraordinary discrimination through very careful disguises. But even in such matters a wrong impression may be given by reserve or flattery; and in matters less directly practical, such as philosophy deals with, very strong instinctive beliefs are sometimes wholly mistaken, as we may come to know through their perceived inconsistency with other equally strong beliefs. It is such considerations that necessitate the harmonising mediation of reason, which tests our beliefs by their mutual compatibility, and examines, in doubtful cases, the possible sources of error on the one side and on the other. In this there is no opposition to instinct as a whole, but only to blind reliance upon some one interesting aspect of instinct to the exclusion of other more commonplace but not less trustworthy aspects. It is such one-sidedness, not instinct itself, that reason aims at correcting.

These more or less trite maxims may be illustrated by application to Bergson's advocacy of "intuition" as against "intellect." There are, he says, "two profoundly different ways of knowing a thing. The first implies that we move round the object: the second that we enter into it. The first depends on the point of view at which we are placed and on the symbols by which we express ourselves. The second neither depends on a point of view nor relies on any symbol. The first kind of knowledge may be said to stop at the *relative*; the second, in those cases where it is possible, to attain the *absolute*." The second of these, which is intuition, is, he says, "the kind of *intellectual sympathy* by which one places oneself within an object in order to coincide with what is unique in it and therefore inexpressible". In illustration, he mentions self-knowledge: "there is one reality, at least, which we all seize from within, by intuition and not by simple analysis. It is our own personality in its flowing through time—our self which endures" The rest of Bergson's philosophy consists in reporting, through the imperfect medium of words, the knowledge gained by intuition, and the consequent complete condemnation of all the pretended knowledge derived from science and common sense.

This procedure, since it takes sides in a conflict of instinctive beliefs, stands in need of justification by proving the greater trustworthiness of the beliefs on one side than of those on the other. Bergson attempts this justification in two ways, first by explaining that intellect is a purely practical faculty to secure biological success, secondly by mentioning remarkable feats of instinct in animals and by pointing out characteristics of the world which, though intuition can apprehend them, are baffling to intellect as he interprets it.

Of Bergson's theory that intellect is a purely practical faculty, developed in the struggle for survival, and not a source of true beliefs, we may say, first, that it is only through intellect that we know of the struggle for survival and of the biological ancestry of man: if the intellect is misleading, the whole of this merely inferred history is presumably untrue. If, on the other hand, we agree with him in thinking that evolution took place as Darwin believed, then it is not only intellect, but all our faculties, that have been developed under the stress of practical utility. Intuition is seen at its best where it is directly useful, for example in regard to other people's characters and dispositions. Bergson apparently holds that capacity for this kind of knowledge is less explicable by the struggle for existence than, for example, capacity for pure mathematics. Yet the savage deceived by false friendship is likely to pay for his mistake with his life; whereas even in the most civilised societies men are not put to death for mathematical incompetence. All the most striking of his instances of intuition in animals have a very direct survival value. The fact is, of course, that both intuition and intellect have been developed because they are useful, and that, speaking broadly, they are useful when they give truth and become harmful when they give falsehood. Intellect, in civilised man, like artistic capacity, has occasionally been developed beyond the point where it is useful to the individual; intuition, on the other hand, seems on the whole to diminish as civilisation increases. It is greater, as a rule, in children than in adults, in the uneducated than in the educated.Probably in dogs it exceeds anything to be found in human beings. But those who see in these facts a recommendation of intuition ought to return to running wild in the woods, dyeing themselves with woad and living on hips and haws.

Let us next examine whether intuition possesses any such infallibility as Bergson claims for it. The best instance of it, according to him, is our acquaintance with ourselves; yet self-knowledge is proverbially rare and difficult. Most men, for example, have in their nature meannesses, vanities, and envies of which they are quite unconscious, though even their best friends can perceive them without any difficulty. It is true that intuition has a convincingness which is lacking to intellect: while it is present, it is almost impossible to doubt its truth. But if it should appear, on examination, to be at least as fallible as intellect, its greater subjective certainty becomes a demerit, making it only the more irresistibly deceptive. Apart from self-knowledge, one of the most notable examples of intuition is the knowledge people believe themselves to possess of those with whom they are in love: the wall between different personalities seems to become transparent, and people think they see into another soul as into their own. Yet deception in such cases is constantly practised with success; and even where there is no intentional deception, experience gradually proves, as a rule, that the supposed insight was illusory, and that the slower more groping methods of the intellect are in the long run more reliable.

Bergson maintains that intellect can only deal with things in so far as they resemble what has been experienced in the past, while intuition has the power of apprehending the uniqueness and novelty that always belong to each fresh moment. That there is something unique and new at every moment, is certainly true; it is also true that this cannot be fully expressed by means of intellectual concepts. Only direct acquaintance can give knowledge of what is unique and new. But direct acquaintance of this kind is given fully in sensation, and does not require, so far as I can see, any special faculty of intuition for its apprehension. It is neither intellect nor intuition, but sensation, that supplies new data; but when the data are new in any remarkable manner, intellect is much more capable of dealing with them than intuition would be. The hen with a brood of ducklings no doubt has intuition which seems to place her inside them, and not merely to know them analytically; but when the ducklings take to the water, the whole apparent intuition is seen to be illusory, and the hen is left helpless on the shore. Intuition, in fact, is an aspect and development of instinct, and, like all instinct, is admirable in those customary surroundings which have moulded the habits of the animal in question, but totally incompetent as soon as the surroundings are changed in a way which demands some non-habitual mode of action.

The theoretical understanding of the world, which is the aim of philosophy, is not a matter of great practical importance to animals, or to savages, or even to most civilised men. It is hardly to be supposed, therefore, that the rapid, rough and ready methods of instinct or intuition will find in this field a favourable ground for their application. It is the older kinds of activity, which bring out our kinship with remote generations of animal and semi-human ancestors, that show intuition at its best. In such matters as self-preservation and love, intuition will act sometimes (though not always) with a swiftness and precision which are astonishing to the critical intellect. But philosophy is not one of the pursuits which illustrate our affinity with the past: it is a highly refined, highly civilised pursuit, demanding, for its success, a certain liberation from the life of instinct, and even, at times, a certain aloofness from all mundane hopes and fears. It is not in philosophy, therefore, that we can hope to see intuition at its best. On the contrary, since the true objects of philosophy, and the habit of thought demanded for their apprehension, are strange, unusual, and remote, it is here, more almost than anywhere else, that intellect proves superior to intuition, and that quick unanalysed convictions are least deserving of uncritical acceptance.

In advocating the scientific restraint and balance, as against the self-assertion of a confident reliance upon intuition, we are only urging, in the sphere of knowledge, that largeness of contemplation, that impersonal disinterestedness, and that freedom from practical preoccupations which have been inculcated by all the great religions of the world. Thus our conclusion, however it may conflict with the explicit beliefs of many mystics, is, in essence, not contrary to the spirit which inspires those beliefs, but rather the outcome of this very spirit as applied in the realm of thought.

II. UNITY AND PLURALITY

One of the most convincing aspects of the mystic illumination is the apparent revelation of the oneness of all things, giving rise to pantheism in religion and to monism in philosophy. An elaborate logic, beginning with Parmenides, and culminating in Hegel and his followers, has been gradually developed, to prove that the universe is one indivisible Whole, and that what seem to be its parts, if considered as substantial and self-existing, are mere illusion. The conception of a Reality quite other than the world of appearance, a reality one, indivisible, and unchanging, was introduced into Western philosophy by Parmenides, not, nominally at least, for mystical or religious reasons, but on the basis of a logical argument as to the impossibility of not-being, and most subsequent metaphysical systems are the outcome of this fundamental idea.

The logic used in defence of mysticism seems to be faulty as logic, and open to technical criticisms, which I have explained elsewhere. I shall not here repeat these criticisms, since they are lengthy and difficult, but shall instead attempt an analysis of the state of mind from which mystical logic has arisen.

Belief in a reality quite different from what appears to the senses arises with irresistible force in certain moods, which are the source of most mysticism, and of most metaphysics. While such a mood is dominant, the need of logic is not felt, and accordingly the more thoroughgoing mystics do not employ logic, but appeal directly to the immediate deliverance of their insight. But such fully developed mysticism is rare in the West. When the intensity of emotional conviction subsides, a man who is in the habit of reasoning will search for logical grounds in favour of the belief which he finds in himself. But since the belief already exists, he will be very hospitable to any ground that suggests itself. The paradoxes apparently proved by his logic are really the paradoxes of mysticism, and are the goal which he feels his logic must reach if it is to be in accordance with insight. The resulting logic has rendered most philosophers incapable of giving any account of the world of science and daily life. If they had been anxious to give such an account, they would probably have discovered the errors of their logic; but most of them were less anxious to understand the world of science and daily life than to convict it of unreality in the interests of a super-sensible "real" world.

It is in this way that logic has been pursued by those of the great philosophers who were mystics. But since they usually took for granted the supposed insight of the mystic emotion, their logical doctrines were presented with a certain dryness, and were believed by their disciples to be quite independent of the sudden illumination from which they sprang. Nevertheless their origin clung to them, and they remained—to borrow a useful word from Mr. Santayana—"malicious" in regard to the world of science and common sense. It is only so that we can account for the complacency with which philosophers have accepted the inconsistency of their doctrines with all the common and scientific facts which seem best established and most worthy of belief.

The logic of mysticism shows, as is natural, the defects which are inherent in anything malicious. The impulse to logic, not felt while the mystic mood is dominant, reasserts itself as the mood fades, but with a desire to retain the vanishing insight, or at least to prove that it *was* insight, and that what seems to contradict it is illusion. The logic which thus arises is not quite disinterested or candid, and is inspired by a certain hatred of the daily world to which it is to be applied. Such an attitude naturally does not tend to the best results. Everyone knows that to read an author simply in order to refute him is not the way to understand him; and to read the book of Nature with a conviction that it is all illusion is just as unlikely to lead to understanding. If our logic is to find the common world intelligible, it must not be hostile, but must be inspired by a genuine acceptance such as is not usually to be found among metaphysicians.

III. TIME

The unreality of time is a cardinal doctrine of many metaphysical systems, often nominally based, as already by Parmenides, upon logical arguments, but originally derived, at any rate in the founders of new systems, from the certainty which is born in the moment of mystic insight. As a Persian Sufi poet says:

"Past and future are what veil God from our sight.Burn up both of them with fire! How longWilt thou be partitioned by these segments as a reed?"

The belief that what is ultimately real must be immutable is a very common one: it gave rise to the metaphysical notion of substance, and finds, even now, a wholly illegitimate satisfaction in such scientific doctrines as the conservation of energy and mass.

It is difficult to disentangle the truth and the error in this view. The arguments for the contention that time is unreal and that the world of sense is illusory must, I think, be regarded as fallacious. Nevertheless there is some sense—easier to feel than to state—in which time is an unimportant and superficial characteristic of reality. Past and future must be acknowledged to be as real as the present, and a certain emancipation from slavery to time is essential to philosophic thought. The importance of time is rather practical than theoretical, rather in relation to our desires than in relation to truth. A truer image of the world, I think, is obtained by picturing things as entering into the stream of time from an eternal world outside, than from a view which regards time as the devouring tyrant of all that is. Both in thought and in feeling, even though time be real, to realise the unimportance of time is the gate of wisdom.

That this is the case may be seen at once by asking ourselves why our feelings towards the past are so different from our feelings towards the future. The reason for this difference is wholly practical: our wishes can affect the future but not the past, the future is to some extent subject to our power, while the past is unalterably fixed. But every future will some day be past: if we see the past truly now, it must, when it was still future, have been just what we now see it to be, and what is now future must be just what we shall see it to be when it has become past. The felt difference of quality between past and future, therefore, is not an intrinsic difference, but only a difference in relation to us: to impartial contemplation, it ceases to exist. And impartiality of contemplation is, in the intellectual sphere, that very same virtue of disinterestedness which, in the sphere of action, appears as justice and unselfishness. Whoever wishes to see the world truly, to rise in thought above the tyranny of practical desires, must learn to overcome the difference of attitude towards past and future, and to survey the whole stream of time in one comprehensive vision.

The kind of way in which, as it seems to me, time ought not to enter into our theoretic philosophical thought, may be illustrated by the philosophy which has become associated with the idea of evolution, and which is exemplified by Nietzsche, pragmatism, and Bergson. This philosophy, on the basis of the development which has led from the lowest forms of life up to man, sees in *progress* the fundamental law of the universe, and thus admits the difference between *earlier* and *later* into the very citadel of its contemplative outlook. With its past and future history of the world, conjectural as it is, I do not wish to quarrel. But I think that, in the intoxication of a quick success, much that is required for a true understanding of the universe has been forgotten. Something of Hellenism, something, too, of Oriental resignation, must be combined with its hurrying Western self-assertion before it can emerge from the ardour of youth into the mature wisdom of manhood. In spite of its appeals to science, the true scientific philosophy, I think, is something more arduous and more aloof, appealing to less mundane hopes, and requiring a severer discipline for its successful practice.

Darwin's *Origin of Species* persuaded the world that the difference between different species of animals and plants is not the fixed immutable difference that it appears to be. The doctrine of natural kinds, which had rendered classification easy and definite, which was enshrined in the Aristotelian tradition, and protected by its supposed necessity for orthodox dogma, was suddenly swept away for ever out of the biological world. The difference between man and the lower animals, which to our human conceit appears enormous, was shown to be a gradual achievement, involving intermediate being who could not with certainty be placed either within or without the human family. The sun and the planets had already been shown by Laplace to be very probably derived from a primitive more or less undifferentiated nebula. Thus the old fixed landmarks became wavering and indistinct, and all sharp outlines were blurred. Things and species lost their boundaries, and none could say where they began or where they ended.

But if human conceit was staggered for a moment by its kinship with the ape, it soon found a way to reassert itself, and that way is the "philosophy" of evolution.A process which led from the am[oe]ba to Man appeared to the philosophers to be obviously a progress—though whether the am[oe]ba would agree with this opinion is not known. Hence the cycle of changes which science had shown to be the probable history of the past was welcomed as revealing a law of development towards good in the universe—an evolution or unfolding of an idea slowly embodying itself in the actual. But such a view, though it might satisfy Spencer and those whom we may call Hegelian evolutionists, could not be accepted as adequate by the more whole-hearted votaries of change. An ideal to which the world continuously approaches is, to these minds, too dead and static to be inspiring. Not only the aspiration, but the ideal too, must change and develop with the course of evolution: there must be no fixed goal, but a continual fashioning of fresh needs by the impulse which is life and which alone gives unity to the process.

Life, in this philosophy, is a continuous stream, in which all divisions are artificial and unreal. Separate things, beginnings and endings, are mere convenient fictions: there is only smooth unbroken transition. The beliefs of to-day may count as true to-day, if they carry us along the stream; but to-morrow they will be false, and must be replaced by new beliefs to meet the new situation. All our thinking consists of convenient fictions, imaginary congealings of the stream: reality flows on in spite of all our fictions, and though it can be lived, it cannot be conceived in thought. Somehow, without explicit statement, the assurance is slipped in that the future, though we cannot foresee it, will be better than the past or the present: the reader is like the child which expects a sweet because it has been told to open its mouth and shut its eyes. Logic, mathematics,physics disappear in this philosophy, because they are too "static"; what is real is no impulse and movement towards a goal which, like the rainbow, recedes as we advance, and makes every place different when it reaches it from what it appeared to be at a distance.

I do not propose to enter upon a technical examination of this philosophy. I wish only to maintain that the motives and interests which inspire it are so exclusively practical, and the problems with which it deals are so special, that it can hardly be regarded as touching any of the questions that, to my mind, constitute genuine philosophy.

The predominant interest of evolutionism is in the question of human destiny, or at least of the destiny of Life. It is more interested in morality and happiness than in knowledge for its own sake. It must be admitted that the same may be said of many other philosophies, and that a desire for the kind of knowledge which philosophy can give is very rare. But if philosophy is to attain truth, it is necessary first and foremost that philosophers should acquire the disinterested intellectual curiosity which characterises the genuine man of science. Knowledge concerning the future—which is the kind of knowledge that must be sought if we are to know about human destiny—is possible within certain narrow limits. It is impossible to say how much the limits may be enlarged with the progress of science. But what is evident is that any proposition about the future belongs by its subject-matter to some particular science, and is to be ascertained, if at all, by the methods of that science. Philosophy is not a short cut to the same kind of results as those of the other sciences: if it is to be a genuine study, it must have a province of its own, and aim at results which the other sciences can neither prove nor disprove.

Evolutionism, in basing itself upon the notion of *progress*, which is change from the worse to the better, allows the notion of time, as it seems to me, to become its tyrant rather than its servant, and thereby loses that impartiality of contemplation which is the source of all that is best in philosophic thought and feeling. Metaphysicians, as we saw, have frequently denied altogether the reality of time. I do not wish to do this; I wish only to preserve the mental outlook which inspired the denial, the attitude which, in thought, regards the past as having the same reality as the present and the same importance as the future. "In so far," says Spinoza,"as the mind conceives a thing according to the dictate of reason, it will be equally affected whether the idea is that of a future, past, or present thing." It is this "conceiving according to the dictate of reason" that I find lacking in the philosophy which is based on evolution.

IV. GOOD AND EVIL

Mysticism maintains that all evil is illusory, and sometimes maintains the same view as regards good, but more often holds that all Reality is good. Both views are to be found in Heraclitus: "Good and ill are one," he says, but again, "To God all things are fair and good and right, but men hold some things wrong and some right." A similar twofold position is to be found in Spinoza, but he uses the word "perfection" when he means to speak of the good that is not merely human. "By reality and perfection I mean the same thing," he says; but elsewhere we find the definition: "By *good* I shall mean that which we certainly know to be useful to us."Thus perfection belongs to Reality in its own nature, but goodness is relative to ourselves and our needs, and disappears in an impartial survey. Some such distinction, I think, is necessary in order to understand the ethical outlook of mysticism: there is a lower mundane kind of good and evil, which divides the world of appearance into what seem to be conflicting parts; but there is also a higher, mystical kind of good, which belongs to Reality and is not opposed by any correlative kind of evil.

It is difficult to give a logically tenable account of this position without recognising that good and evil are subjective, that what is good is merely that towards which we have one kind of feeling, and what is evil is merely that towards which we have another kind of feeling. In our active life, where we have to exercise choice, and to prefer this to that of two possible acts, it is necessary to have a distinction of good and evil, or at least of better and worse. But this distinction, like everything pertaining to action, belongs to what mysticism regards as the world of illusion, if only because it is essentially concerned with time. In our contemplative life, where action is not called for, it is possible to be impartial, and to overcome the ethical dualism which action requires. So long as we remain *merely* impartial, we may be content to say that both the good and the evil of action are illusions. But if, as we must do if we have the mystic vision, we find the whole world worthy of love and worship, if we see

"The earth, and every common sight....Apparell'd in celestial light,"

we shall say that there is a higher good than that of action, and that this higher good belongs to the whole world as it is in reality. In this way the twofold attitude and the apparent vacillation of mysticism are explained and justified.

The possibility of this universal love and joy in all that exists is of supreme importance for the conduct and happiness of life, and gives inestimable value to the mystic emotion, apart from any creeds which may be built upon it. But if we are not to be led into false beliefs, it is necessary to realise exactly *what* the mystic emotion reveals. It reveals a possibility of human nature—a possibility of a nobler, happier, freer life than any that can be otherwise achieved. But it does not reveal anything about the non-human, or about the nature of the universe in general. Good and bad, and even the higher good that mysticism finds everywhere, are the reflections of our own emotions on other things, not part of the substance of things as they are in themselves. And therefore, an impartial contemplation, freed from all pre-occupation with Self, will not judge things good or bad, although it is very easily combined with that feeling of universal love which leads the mystic to say that the whole world is good.

The philosophy of evolution, through the notion of progress, is bound up with the ethical dualism of the worse and the better, and is thus shut out, not only from the kind of survey which discards good and evil altogether from its view, but also from the mystical belief in the goodness of everything. In this way the distinction of good and evil, like time, becomes a tyrant in this philosophy, and introduces into thought the restless selectiveness of action. Good and evil, like time, are, it would seem, not general or fundamental in the world of thought, but late and highly specialised members of the intellectual hierarchy.

Although, as we saw, mysticism can be interpreted so as to agree with the view that good and evil are not intellectually fundamental, it must be admitted that here we are no longer in verbal agreement with most of the great philosophers and religious teachers of the past. I believe, however, that the elimination of ethical considerations from philosophy is both scientifically necessary and—though this may seem a paradox—an ethical advance. Both these contentions must be briefly defended.

The hope of satisfaction to our more human desires—the hope of demonstrating that the world has this or that desirable ethical characteristic—is not one which, so far as I can see, a scientific philosophy can do anything whatever to satisfy. The difference between a good world and a bad one is a difference in the particular characteristics of the particular things that exist in these worlds: it is not a sufficiently abstract difference to come within the province of philosophy. Love and hate, for example, are ethical opposites, but to philosophy they are closely analogous attitudes towards objects. The general form and structure of those attitudes towards objects which constitute mental phenomena is a problem for philosophy, but the difference between love and hate is not a difference of form or structure, and therefore belongs rather to the special science of psychology than to philosophy. Thus the ethical interests which have often inspired philosophers must remain in the background: some kind of ethical interest may inspire the whole study, but none must obtrude in the detail or be expected in the special results which are sought.

If this view seems at first sight disappointing, we may remind ourselves that a similar change has been found necessary in all the other sciences. The physicist or chemist is not now required to prove the ethical importance of his ions or atoms; the biologist is not expected to prove the utility of the plants or animals which he dissects. In pre-scientific ages this was not the case. Astronomy, for example, was studied because men believed in astrology: it was thought that the movements of the planets had the most direct and important bearing upon the lives of human beings. Presumably, when this belief decayed and the disinterested study of astronomy began, many who had found astrology absorbingly interesting decided that astronomy had too little human interest to be worthy of study. Physics, as it appears in Plato's Timæus for example, is full of ethical notions: it is an essential part of its purpose to show that the earth is worthy of admiration. The modern physicist, on the contrary, though he has no wish to deny that the earth is admirable, is not concerned, as physicist, with its ethical attributes: he is merely concerned to find out facts, not to consider whether they are good or bad. In psychology, the scientific attitude is even more recent and more difficult than in the physical sciences: it is natural to consider that human nature is either good or bad, and to suppose that the difference between good and bad, so all-important in practice, must be important in theory also. It is only during the last century that an ethically neutral psychology has grown up; and here too, ethical neutrality has been essential to scientific success.

In philosophy, hitherto, ethical neutrality has been seldom sought and hardly ever achieved. Men have remembered their wishes, and have judged philosophies in relation to their wishes. Driven from the particular sciences, the belief that the notions of good and evil must afford a key to the understanding of the world has sought a refuge in philosophy. But even from this last refuge, if philosophy is not to remain a set of pleasing dreams, this belief must be driven forth. It is a commonplace that happiness is not best achieved by those who seek it directly; and it would seem that the same is true of the good. In thought, at any rate, those who forget good and evil and seek only to know the facts are more likely to achieve good than those who view the world through the distorting medium of their own desires.

We are thus brought back to our seeming paradox, that a philosophy which does not seek to impose upon the world its own conceptions of good and evil is not only more likely to achieve truth, but is also the outcome of a higher ethical standpoint than one which, like evolutionism and most traditional systems, is perpetually appraising the universe and seeking to find in it an embodiment of present ideals. In religion, and in every deeply serious view of the world and of human destiny, there is an element of submission, a realisation of the limits of human power, which is somewhat lacking in the modern world, with its quick material successes and its insolent belief in the boundless possibilities of progress. "He that loveth his life shall lose it"; and there is danger lest, through a too confident love of life, life itself should lose much of what gives it its highest worth. The submission which religion inculcates in action is essentially the same in spirit as that which science teaches in thought; and the ethical neutrality by which its victories have been achieved is the outcome of that submission.

The good which it concerns us to remember is the good which it lies in our power to create—the good in our own lives and in our attitude towards the world. Insistence on belief in an external realisation of the good is a form of self-assertion, which, while it cannot secure the external good which it desires, can seriously impair the inward good which lies within our power, and destroy that reverence towards fact which constitutes both what is valuable in humility and what is fruitful in the scientific temper.

Human beings cannot, of course, wholly transcend human nature; something subjective, if only the interest that determines the direction of our attention, must remain in all our thought. But scientific philosophy comes nearer to objectivity than any other human pursuit, and gives us, therefore, the closest constant and the most intimate relation with the outer world that it is possible to achieve. To the primitive mind, everything is either friendly or hostile; but experience has shown that friendliness and hostility are not the conceptions by which the world is to be understood. Scientific philosophy thus represents, though as yet only in a nascent condition, a higher form of thought than any pre-scientific belief or imagination, and, like every approach to self-transcendence, it brings with it a rich reward in increase of scope and breadth and comprehension. Evolutionism, in spite of its appeals to particular scientific facts, fails to be a truly scientific philosophy because of its slavery to time, its ethical preoccupations, and its predominant interest in our mundane concerns and destiny. A truly scientific philosophy will be more humble, more piecemeal, more arduous, offering less glitter of outward mirage to flatter fallacious hopes, but more indifferent to fate, and more capable of accepting the world without the tyrannous imposition of our human and temporary demands.

II

THE PLACE OF SCIENCE IN A LIBERAL EDUCATION

I

Science, to the ordinary reader of newspapers, is represented by a varying selection of sensational triumphs, such as wireless telegraphy and aeroplanes, radio-activity and the marvels of modern alchemy. It is not of this aspect of science that I wish to speak. Science, in this aspect, consists of detached up-to-date fragments, interesting only until they are replaced by something newer and more up-to-date, displaying nothing of the systems of patiently constructed knowledge out of which, almost as a casual incident, have come the practically useful results which interest the man in the street. The increased command over the forces of nature which is derived from science is undoubtedly an amply sufficient reason for encouraging scientific research, but this reason has been so often urged and is so easily appreciated that other reasons, to my mind quite as important, are apt to be overlooked. It is with these other reasons, especially with the intrinsic value of a scientific habit of mind in forming our outlook on the world, that I shall be concerned in what follows.

The instance of wireless telegraphy will serve to illustrate the difference between the two points of view. Almost all the serious intellectual labour required for the possibility of this invention is due to three men—Faraday, Maxwell, and Hertz. In alternating layers of experiment and theory these three men built up the modern theory of electromagnetism, and demonstrated the identity of light with electromagnetic waves. The system which they discovered is one of profound intellectual interest, bringing together and unifying an endless variety of apparently detached phenomena, and displaying a cumulative mental power which cannot but afford delight to every generous spirit. The mechanical details which remained to be adjusted in order to utilise their discoveries for a practical system of telegraphy demanded, no doubt, very considerable ingenuity, but had not that broad sweep and that universality which could give them intrinsic interest as an object of disinterested contemplation.

From the point of view of training the mind, of giving that well-informed, impersonal outlook which constitutes culture in the good sense of this much-misused word, it seems to be generally held indisputable that a literary education is superior to one based on science. Even the warmest advocates of science are apt to rest their claims on the contention that culture ought to be sacrificed to utility. Those men of science who respect culture, when they associate with men learned in the classics, are apt to admit, not merely politely, but sincerely, a certain inferiority on their side, compensated doubtless by the services which science renders to humanity, but none the less real. And so long as this attitude exists among men of science, it tends to verify itself: the intrinsically valuable aspects of science tend to be sacrificed to the merely useful, and little attempt is made to preserve that leisurely, systematic survey by which the finer quality of mind is formed and nourished.

But even if there be, in present fact, any such inferiority as is supposed in the educational value of science, this is, I believe, not the fault of science itself, but the fault of the spirit in which science is taught. If its full possibilities were realised by those who teach it, I believe that its capacity of producing those habits of mind which constitute the highest mental excellence would be at least as great as that of literature, and more particularly of Greek and Latin literature. In saying this I have no wish whatever to disparage a classical education. I have not myself enjoyed its benefits, and my knowledge of Greek and Latin authors is derived almost wholly from translations. But I am firmly persuaded that the Greeks fully deserve all the admiration that is bestowed upon them, and that it is a very great and serious loss to be unacquainted with their writings. It is not by attacking them, but by drawing attention to neglected excellences in science, that I wish to conduct my argument.

One defect, however, does seem inherent in a purely classical education—namely, a too exclusive emphasis on the past. By the study of what is absolutely ended and can never be renewed, a habit of criticism towards the present and the future is engendered. The qualities in which the present excels are qualities to which the study of the past does not direct attention, and to which, therefore, the student of Greek civilisation may easily become blind. In what is new and growing there is apt to be something crude, insolent, even a little vulgar, which is shocking to the man of sensitive taste; quivering from the rough contact, he retires to the trim gardens of a polished past, forgetting that they were reclaimed from the wilderness by men as rough and earth-soiled as those from whom he shrinks in his own day. The habit of being unable to recognise merit until it is dead is too apt to be the result of a purely bookish life, and a culture based wholly on the past will seldom be able to pierce through everyday surroundings to the essential splendour of contemporary things, or to the hope of still greater splendour in the future.

"My eyes saw not the men of old;And now their age away has rolled.I weep—to think I shall not seeThe heroes of posterity."

So says the Chinese poet; but such impartiality is rare in the more pugnacious atmosphere of the West, where the champions of past and future fight a never-ending battle, instead of combining to seek out the merits of both.

This consideration, which militates not only against the exclusive study of the classics, but against every form of culture which has become static, traditional, and academic, leads inevitably to the fundamental question: What is the true end of education? But before attempting to answer this question it will be well to define the sense in which we are to use the word "education." For this purpose, I shall distinguish the sense in which I mean to use it from two others, both perfectly legitimate, the one broader and the other narrower than the sense in which I mean to use the word.

In the broader sense, education will include not only what we learn through instruction, but all that we learn through personal experience—the formation of character through the education of life. Of this aspect of education, vitally important as it is, I will say nothing, since its consideration would introduce topics quite foreign to the question with which we are concerned.

In the narrower sense, education may be confined to instruction, the imparting of definite information on various subjects, because such information, in and for itself, is useful in daily life. Elementary education—reading, writing, and arithmetic—is almost wholly of this kind. But instruction, necessary as it is, does not *per se* constitute education in the sense in which I wish to consider it.

Education, in the sense in which I mean it, may be defined as *the formation, by means of instruction, of certain mental habits and a certain outlook on life and the world.* It remains to ask ourselves, what mental habits, and what sort of outlook, can be hoped for as the result of instruction? When we have answered this question, we can attempt to decide what science has to contribute to the formation of the habits and outlook which we desire.

Our whole life is built about a certain number—not a very small number—of primary instincts and impulses. Only what is in some way connected with these instincts and impulses appears to us desirable or important; there is no faculty, whether "reason" or "virtue" or whatever it may be called, that can take our active life and our hopes and fears outside the region controlled by these first movers of all desire. Each of them is like a queen-bee, aided by a hive of workers gathering honey; but when the queen is gone the workers languish and die, and the cells remain empty of their expected sweetness. So with each primary impulse in civilised man: it is surrounded and protected by a busy swarm of attendant derivative desires, which store up in its service whatever honey the surrounding world affords. But if the queen-impulse dies, the death-dealing influence, though retarded a little by habit, spreads slowly through all the subsidiary impulses, and a whole tract of life becomes inexplicably colourless. What was formerly full of zest, and so obviously worth doing that it raised no questions, has now grown dreary and purposeless: with a sense of disillusion we inquire the meaning of life, and decide, perhaps, that all is vanity. The search for an outside meaning that can *compel* an inner response must always be disappointed: all "meaning" must be at bottom related to our primary desires, and when they are extinct no miracle can restore to the world the value which they reflected upon it.

The purpose of education, therefore, cannot be to create any primary impulse which is lacking in the uneducated; the purpose can only be to enlarge the scope of those that human nature provides, by increasing the number and variety of attendant thoughts, and by showing where the most permanent satisfaction is to be found. Under the impulse of a Calvinistic horror of the "natural man," this obvious truth has been too often misconceived in the training of the young; "nature" has been falsely regarded as excluding all that is best in what is natural, and the endeavour to teach virtue has led to the production of stunted and contorted hypocrites instead of full-grown human beings. From such mistakes in education a better psychology or a kinder heart is beginning to preserve the present generation; we need, therefore, waste no more words on the theory that the purpose of education is to thwart or eradicate nature.

But although nature must supply the initial force of desire, nature is not, in the civilised man, the spasmodic, fragmentary, and yet violent set of impulses that it is in the savage. Each impulse has its constitutional ministry of thought and knowledge and reflection, through which possible conflicts of impulses are foreseen, and temporary impulses are controlled by the unifying impulse which may be called wisdom. In this way education destroys the crudity of instinct, and increases through knowledge the wealth and variety of the individual's contacts with the outside world, making him no longer an isolated fighting unit, but a citizen of the universe, embracing distant countries, remote regions of space, and vast stretches of past and future within the circle of his interests. It is this simultaneous softening in the insistence of desire and enlargement of its scope that is the chief moral end of education.

Closely connected with this moral end is the more purely intellectual aim of education, the endeavour to make us see and imagine the world in an objective manner, as far as possible as it is in itself, and not merely through the distorting medium of personal desire. The complete attainment of such an objective view is no doubt an ideal, indefinitely approachable, but not actually and fully realisable. Education, considered as a process of forming our mental habits and our outlook on the world, is to be judged successful in proportion as its outcome approximates to this ideal; in proportion, that is to say, as it gives us a true view of our place in society, of the relation of the whole human society to its non-human environment, and of the nature of the non-human world as it is in itself apart from our desires and interests. If this standard is admitted, we can return to the consideration of science, inquiring how far science contributes to such an aim, and whether it is in any respect superior to its rivals in educational practice.

II

Two opposite and at first sight conflicting merits belong to science as against literature and art. The one, which is not inherently necessary, but is certainly true at the present day, is hopefulness as to the future of human achievement, and in particular as to the useful work that may be accomplished by any intelligent student. This merit and the cheerful outlook which it engenders prevent what might otherwise be the depressing effect of another aspect of science, to my mind also a merit, and perhaps its greatest merit—I mean the irrelevance of human passions and of the whole subjective apparatus where scientific truth is concerned. Each of these reasons for preferring the study of science requires some amplification. Let us begin with the first.

In the study of literature or art our attention is perpetually riveted upon the past: the men of Greece or of the Renaissance did better than any men do now; the triumphs of former ages, so far from facilitating fresh triumphs in our own age, actually increase the difficulty of fresh triumphs by rendering originality harder of attainment; not only is artistic achievement not cumulative, but it seems even to depend upon a certain freshness and *naïveté* of impulse and vision which civilisation tends to destroy. Hence comes, to those who have been nourished on the literary and artistic productions of former ages, a certain peevishness and undue fastidiousness towards the present, from which there seems no escape except into the deliberate vandalism which ignores tradition and in the search after originality achieves only the eccentric. But in such vandalism there is none of the simplicity and spontaneity out of which great art springs: theory is still the canker in its core, and insincerity destroys the advantages of a merely pretended ignorance.

The despair thus arising from an education which suggests no pre-eminent mental activity except that of artistic creation is wholly absent from an education which gives the knowledge of scientific method. The discovery of scientific method, except in pure mathematics, is a thing of yesterday; speaking broadly, we may say that it dates from Galileo. Yet already it has transformed the world, and its success proceeds with ever-accelerating velocity. In science men have discovered an activity of the very highest value in which they are no longer, as in art, dependent for progress upon the appearance of continually greater genius, for in science the successors stand upon the shoulders of their predecessors; where one man of supreme genius has invented a method, a thousand lesser men can apply it. No transcendent ability is required in order to make useful discoveries in science; the edifice of science needs its masons, bricklayers, and common labourers as well as its foremen, master-builders, and architects. In art nothing worth doing can be done without genius; in science even a very moderate capacity can contribute to a supreme achievement.

In science the man of real genius is the man who invents a new method. The notable discoveries are often made by his successors, who can apply the method with fresh vigour, unimpaired by the previous labour of perfecting it; but the mental calibre of the thought required for their work, however brilliant, is not so great as that required by the first inventor of the method. There are in science immense numbers of different methods, appropriate to different classes of problems; but over and above them all, there is something not easily definable, which may be called *the* method of science. It was formerly customary to identify this with the inductive method, and to associate it with the name of Bacon. But the true inductive method was not discovered by Bacon, and the true method of science is something which includes deduction as much as induction, logic and mathematics as much as botany and geology. I shall not attempt the difficult task of stating what the scientific method is, but I will try to indicate the temper of mind out of which the scientific method grows, which is the second of the two merits that were mentioned above as belonging to a scientific education.

The kernel of the scientific outlook is a thing so simple, so obvious, so seemingly trivial, that the mention of it may almost excite derision. The kernel of the scientific outlook is the refusal to regard our own desires, tastes, and interests as affording a key to the understanding of the world. Stated thus baldly, this may seem no more than a trite truism. But to remember it consistently in matters arousing our passionate partisanship is by no means easy, especially where the available evidence is uncertain and inconclusive. A few illustrations will make this clear.

Aristotle, I understand, considered that the stars must move in circles because the circle is the most perfect curve. In the absence of evidence to the contrary, he allowed himself to decide a question of fact by an appeal to æsthetico-moral considerations. In such a case it is at once obvious to us that this appeal was unjustifiable. We know now how to ascertain as a fact the way in which the heavenly bodies move, and we know that they do not move in circles, or even in accurate ellipses, or in any other kind of simply describable curve. This may be painful to a certain hankering after simplicity of pattern in the universe, but we know that in astronomy such feelings are irrelevant. Easy as this knowledge seems now, we owe it to the courage and insight of the first inventors of scientific method, and more especially of Galileo.

We may take as another illustration Malthus's doctrine of population. This illustration is all the better for the fact that his actual doctrine is now known to be largely erroneous. It is not his conclusions that are valuable, but the temper and method of his inquiry. As everyone knows, it was to him that Darwin owed an essential part of his theory of natural selection, and this was only possible because Malthus's outlook was truly scientific. His great merit lies in considering man not as the object of praise or blame, but as a part of nature, a thing with a certain characteristic behaviour from which certain consequences must follow. If the behaviour is not quite what Malthus supposed, if the consequences are not quite what he inferred, that may falsify his conclusions, but does not impair the value of his method. The objections which were made when his doctrine was new—that it was horrible and depressing, that people ought not to act as he said they did, and so on—were all such as implied an unscientific attitude of mind; as against all of them, his calm determination to treat man as a natural phenomenon marks an important advance over the reformers of the eighteenth century and the Revolution.

Under the influence of Darwinism the scientific attitude towards man has now become fairly common, and is to some people quite natural, though to most it is still a difficult and artificial intellectual contortion. There is however, one study which is as yet almost wholly untouched by the scientific spirit—I mean the study of philosophy. Philosophers and the public imagine that the scientific spirit must pervade pages that bristle with allusions to ions, germ-plasms, and the eyes of shell-fish. But as the devil can quote Scripture, so the philosopher can quote science. The scientific spirit is not an affair of quotation, of externally acquired information, any more than manners are an affair of the etiquette-book. The scientific attitude of mind involves a sweeping away of all other desires in the interests of the desire to know—it involves suppression of hopes and fears, loves and hates, and the whole subjective emotional life, until we become subdued to the material, able to see it frankly, without preconceptions, without bias, without any wish except to see it as it is, and without any belief that what it is must be determined by some relation, positive or negative, to what we should like it to be, or to what we can easily imagine it to be.

Now in philosophy this attitude of mind has not as yet been achieved. A certain self-absorption, not personal, but human, has marked almost all attempts to conceive the universe as a whole. Mind, or some aspect of it—thought or will or sentience—has been regarded as the pattern after which the universe is to be conceived, for no better reason, at bottom, than that such a universe would not seem strange, and would give us the cosy feeling that every place is like home. To conceive the universe as essentially progressive or essentially deteriorating, for example, is to give to our hopes and fears a cosmic importance which *may*, of course, be justified, but which we have as yet no reason to suppose justified. Until we have learnt to think of it in ethically neutral terms, we have not arrived at a scientific attitude in philosophy; and until we have arrived at such an attitude, it is hardly to be hoped that philosophy will achieve any solid results.

I have spoken so far largely of the negative aspect of the scientific spirit, but it is from the positive aspect that its value is derived. The instinct of constructiveness, which is one of the chief incentives to artistic creation, can find in scientific systems a satisfaction more massive than any epic poem. Disinterested curiosity, which is the source of almost all intellectual effort, finds with astonished delight that science can unveil secrets which might well have seemed for ever undiscoverable. The desire for a larger life and wider interests, for an escape from private circumstances, and even from the whole recurring human cycle of birth and death, is fulfilled by the impersonal cosmic outlook of science as by nothing else. To all these must be added, as contributing to the happiness of the man of science, the admiration of splendid achievement, and the consciousness of inestimable utility to the human race. A life devoted to science is therefore a happy life, and its happiness is derived from the very best sources that are open to dwellers on this troubled and passionate planet.

III

A FREE MAN'S WORSHIP

To Dr. Faustus in his study Mephistopheles told the history of the Creation, saying:

"The endless praises of the choirs of angels had begun to grow wearisome; for, after all, did he not deserve their praise? Had he not given them endless joy? Would it not be more amusing to obtain undeserved praise, to be worshipped by beings whom he tortured? He smiled inwardly, and resolved that the great drama should be performed.

"For countless ages the hot nebula whirled aimlessly through space. At length it began to take shape, the central mass threw off planets, the planets cooled, boiling seas and burning mountains heaved and tossed, from black masses of cloud hot sheets of rain deluged the barely solid crust. And now the first germ of life grew in the depths of the ocean, and developed rapidly in the fructifying warmth into vast forest trees, huge ferns springing from the damp mould, sea monsters breeding, fighting, devouring, and passing away. And from the monsters, as the play unfolded itself, Man was born, with the power of thought, the knowledge of good and evil, and the cruel thirst for worship. And Man saw that all is passing in this mad, monstrous world, that all is struggling to snatch, at any cost, a few brief moments of life before Death's inexorable decree. And Man said: 'There is a hidden purpose, could we but fathom it, and the purpose is good; for we must reverence something, and in the visible world there is nothing worthy of reverence.' And Man stood aside from the struggle, resolving that God intended harmony to come out of chaos by human efforts. And when he followed the instincts which God had transmitted to him from his ancestry of beasts of prey, he called it Sin, and asked God to forgive him. But he doubted whether he could be justly forgiven, until he invented a divine Plan by which God's wrath was to have been appeased. And seeing the present was bad, he made it yet worse, that thereby the future might be better. And he gave God thanks for the strength that enabled him to forgo even the joys that were possible. And God smiled; and when he saw that Man had become perfect in renunciation and worship, he sent another sun through the sky, which crashed into Man's sun; and all returned again to nebula.

"'Yes,' he murmured, 'it was a good play; I will have it performed again.'"

Such, in outline, but even more purposeless, more void of meaning, is the world which Science presents for our belief. Amid such a world, if anywhere, our ideals henceforward must find a home. That Man is the product of causes which had no prevision of the end they were achieving; that his origin, his growth, his hopes and fears, his loves and his beliefs, are but the outcome of accidental collocations of atoms; that no fire, no heroism, no intensity of thought and feeling, can preserve an individual life beyond the grave; that all the labours of the ages, all the devotion, all the inspiration, all the noonday brightness of human genius, are destined to extinction in the vast death of the solar system, and that the whole temple of Man's achievement must inevitably be buried beneath the débris of a universe in ruins—all these things, if not quite beyond dispute, are yet so nearly certain, that no philosophy which rejects them can hope to stand. Only within the scaffolding of these truths, only on the firm foundation of unyielding despair, can the soul's habitation henceforth be safely built.

How, in such an alien and inhuman world, can so powerless a creature as Man preserve his aspirations untarnished? A strange mystery it is that Nature, omnipotent but blind, in the revolutions of her secular hurryings through the abysses of space, has brought forth at last a child, subject still to her power, but gifted with sight, with knowledge of good and evil, with the capacity of judging all the works of his unthinking Mother. In spite of Death, the mark and seal of the parental control, Man is yet free, during his brief years, to examine, to criticise, to know, and in imagination to create. To him alone, in the world with which he is acquainted, this freedom belongs; and in this lies his superiority to the resistless forces that control his outward life.

The savage, like ourselves, feels the oppression of his impotence before the powers of Nature; but having in himself nothing that he respects more than Power, he is willing to prostrate himself before his gods, without inquiring whether they are worthy of his worship. Pathetic and very terrible is the long history of cruelty and torture, of degradation and human sacrifice, endured in the hope of placating the jealous gods: surely, the trembling believer thinks, when what is most precious has been freely given, their lust for blood must be appeased, and more will not be required. The religion of Moloch—as such creeds may be generically called—is in essence the cringing submission of the slave, who dare not, even in his heart, allow the thought that his master deserves no adulation. Since the independence of ideals is not yet acknowledged, Power may be freely worshipped, and receive an unlimited respect, despite its wanton infliction of pain.

But gradually, as morality grows bolder, the claim of the ideal world begins to be felt; and worship, if it is not to cease, must be given to gods of another kind than those created by the savage. Some, though they feel the demands of the ideal, will still consciously reject them, still urging that naked Power is worthy of worship. Such is the attitude inculcated in God's answer to Job out of the whirlwind: the divine power and knowledge are paraded, but of the divine goodness there is no hint. Such also is the attitude of those who, in our own day, base their morality upon the struggle for survival, maintaining that the survivors are necessarily the fittest. But others, not content with an answer so repugnant to the moral sense, will adopt the position which we have become accustomed to regard as specially religious, maintaining that, in some hidden manner, the world of fact is really harmonious with the world of ideals. Thus, Man creates God, all-powerful and all-good, the mystic unity of what is and what should be.

But the world of fact, after all, is not good; and, in submitting our judgment to it, there is an element of slavishness from which our thoughts must be purged. For in all things it is well to exalt the dignity of Man, by freeing him as far as possible from the tyranny of non-human Power. When we have realised that Power is largely bad, that man, with his knowledge of good and evil, is but a helpless atom in a world which has no such knowledge, the choice is again presented to us: Shall we worship Force, or shall we worship Goodness? Shall our God exist and be evil, or shall he be recognised as the creation of our own conscience?

The answer to this question is very momentous, and affects profoundly our whole morality. The worship of Force, to which Carlyle and Nietzsche and the creed of Militarism have accustomed us, is the result of failure to maintain our own ideals against a hostile universe: it is itself a prostrate submission to evil, a sacrifice of our best to Moloch. If strength indeed is to be respected, let us respect rather the strength of those who refuse that false "recognition of facts" which fails to recognise that facts are often bad. Let us admit that, in the world we know, there are many things that would be better otherwise, and that the ideals to which we do and must adhere are not realised in the realm of matter. Let us preserve our respect for truth, for beauty, for the ideal of perfection which life does not permit us to attain, though none of these things meet with the approval of the unconscious universe. If Power is bad, as it seems to be, let us reject it from our hearts. In this lies Man's true freedom: in determination to worship only the God created by our own love of the good, to respect only the heaven which inspires the insight of our best moments. In action, in desire, we must submit perpetually to the tyranny of outside forces; but in thought, in aspiration, we are free, free from our fellow-men, free from the petty planet on which our bodies impotently crawl, free even, while we live, from the tyranny of death. Let us learn, then, that energy of faith which enables us to live constantly in the vision of the good; and let us descend, in action, into the world of fact, with that vision always before us.

When first the opposition of fact and ideal grows fully visible, a spirit of fiery revolt, of fierce hatred of the gods, seems necessary to the assertion of freedom. To defy with Promethean constancy a hostile universe, to keep its evil always in view, always actively hated, to refuse no pain that the malice of Power can invent, appears to be the duty of all who will not bow before the inevitable. But indignation is still a bondage, for it compels our thoughts to be occupied with an evil world; and in the fierceness of desire from which rebellion springs there is a kind of self-assertion which it is necessary for the wise to overcome. Indignation is a submission of our thoughts, but not of our desires; the Stoic freedom in which wisdom consists is found in the submission of our desires, but not of our thoughts. From the submission of our desires springs the virtue of resignation; from the freedom of our thoughts springs the whole world of art and philosophy, and the vision of beauty by which, at last, we half reconquer the reluctant world. But the vision of beauty is possible only to unfettered contemplation, to thoughts not weighted by the load of eager wishes; and thus Freedom comes only to those who no longer ask of life that it shall yield them any of those personal goods that are subject to the mutations of Time.

Although the necessity of renunciation is evidence of the existence of evil, yet Christianity, in preaching it, has shown a wisdom exceeding that of the Promethean philosophy of rebellion. It must be admitted that, of the things we desire, some, though they prove impossible, are yet real goods; others, however, as ardently longed for, do not form part of a fully purified ideal. The belief that what must be renounced is bad, though sometimes false, is far less often false than untamed passion supposes; and the creed of religion, by providing a reason for proving that it is never false, has been the means of purifying our hopes by the discovery of many austere truths.

But there is in resignation a further good element: even real goods, when they are unattainable, ought not to be fretfully desired. To every man comes, sooner or later, the great renunciation. For the young, there is nothing unattainable; a good thing desired with the whole force of a passionate will, and yet impossible, is to them not credible. Yet, by death, by illness, by poverty, or by the voice of duty, we must learn, each one of us, that the world was not made for us, and that, however beautiful may be the things we crave, Fate may nevertheless forbid them. It is the part of courage, when misfortune comes, to bear without repining the ruin of our hopes, to turn away our thoughts from vain regrets. This degree of submission to Power is not only just and right: it is the very gate of wisdom.

But passive renunciation is not the whole of wisdom; for not by renunciation alone can we build a temple for the worship of our own ideals. Haunting foreshadowings of the temple appear in the realm of imagination, in music, in architecture, in the untroubled kingdom of reason, and in the golden sunset magic of lyrics, where beauty shines and glows, remote from the touch of sorrow, remote from the fear of change, remote from the failures and disenchantments of the world of fact. In the contemplation of these things the vision of heaven will shape itself in our hearts, giving at once a touchstone to judge the world about us, and an inspiration by which to fashion to our needs whatever is not incapable of serving as a stone in the sacred temple.

Except for those rare spirits that are born without sin, there is a cavern of darkness to be traversed before that temple can be entered. The gate of the cavern is despair, and its floor is paved with the gravestones of abandoned hopes. There Self must die; there the eagerness, the greed of untamed desire must be slain, for only so can the soul be freed from the empire of Fate. But out of the cavern the Gate of Renunciation leads again to the daylight of wisdom, by whose radiance a new insight, a new joy, a new tenderness, shine forth to gladden the pilgrim's heart.

When, without the bitterness of impotent rebellion, we have learnt both to resign ourselves to the outward rule of Fate and to recognise that the non-human world is unworthy of our worship, it becomes possible at last so to transform and refashion the unconscious universe, so to transmute it in the crucible of imagination, that a new image of shining gold replaces the old idol of clay. In all the multiform facts of the world—in the visual shapes of trees and mountains and clouds, in the events of the life of man, even in the very omnipotence of Death—the insight of creative idealism can find the reflection of a beauty which its own thoughts first made. In this way mind asserts its subtle mastery over the thoughtless forces of Nature. The more evil the material with which it deals, the more thwarting to untrained desire, the greater is its achievement in inducing the reluctant rock to yield up its hidden treasures, the prouder its victory in compelling the opposing forces to swell the pageant of its triumph. Of all the arts, Tragedy is the proudest, the most triumphant; for it builds its shining citadel in the very centre of the enemy's country, on the very summit of his highest mountain; from its impregnable watchtowers, his camps and arsenals, his columns and forts, are all revealed; within its walls the free life continues, while the legions of Death and Pain and Despair, and all the servile captains of tyrant Fate, afford the burghers of that dauntless city new spectacles of beauty. Happy those sacred ramparts, thrice happy the dwellers on that all-seeing eminence. Honour to those brave warriors who, through countless ages of warfare, have preserved for us the priceless heritage of liberty, and have kept undefiled by sacrilegious invaders the home of the unsubdued.

But the beauty of Tragedy does but make visible a quality which, in more or less obvious shapes, is present always and everywhere in life. In the spectacle of Death, in the endurance of intolerable pain, and in the irrevocableness of a vanished past, there is a sacredness, an overpowering awe, a feeling of the vastness, the depth, the inexhaustible mystery of existence, in which, as by some strange marriage of pain, the sufferer is bound to the world by bonds of sorrow. In these moments of insight, we lose all eagerness of temporary desire, all struggling and striving for petty ends, all care for the little trivial things that, to a superficial view, make up the common life of day by day; we see, surrounding the narrow raft illumined by the flickering light of human comradeship, the dark ocean on whose rolling waves we toss for a brief hour; from the great night without, a chill blast breaks in upon our refuge; all the loneliness of humanity amid hostile forces is concentrated upon the individual soul, which must struggle alone, with what of courage it can command, against the whole weight of a universe that cares nothing for its hopes and fears. Victory, in this struggle with the powers of darkness, is the true baptism into the glorious company of heroes, the true initiation into the overmastering beauty of human existence. From that awful encounter of the soul with the outer world, renunciation, wisdom, and charity are born; and with their birth a new life begins. To take into the inmost shrine of the soul the irresistible forces whose puppets we seem to be—Death and change, the irrevocableness of the past, and the powerlessness of man before the blind hurry of the universe from vanity to vanity—to feel these things and know them is to conquer them.

This is the reason why the Past has such magical power. The beauty of its motionless and silent pictures is like the enchanted purity of late autumn, when the leaves, though one breath would make them fall, still glow against the sky in golden glory. The Past does not change or strive; like Duncan, after life's fitful fever it sleeps well; what was eager and grasping, what was petty and transitory, has faded away, the things that were beautiful and eternal shine out of it like stars in the night. Its beauty, to a soul not worthy of it, is unendurable; but to a soul which has conquered Fate it is the key of religion.

The life of Man, viewed outwardly, is but a small thing in comparison with the forces of Nature. The slave is doomed to worship Time and Fate and Death, because they are greater than anything he finds in himself, and because all his thoughts are of things which they devour. But, great as they are, to think of them greatly, to feel their passionless splendour, is greater still. And such thought makes us free men; we no longer bow before the inevitable in Oriental subjection, but we absorb it, and make it a part of ourselves. To abandon the struggle for private happiness, to expel all eagerness of temporary desire, to burn with passion for eternal things—this is emancipation, and this is the free man's worship. And this liberation is effected by a contemplation of Fate; for Fate itself is subdued by the mind which leaves nothing to be purged by the purifying fire of Time.

United with his fellow-men by the strongest of all ties, the tie of a common doom, the free man finds that a new vision is with him always, shedding over every daily task the light of love. The life of Man is a long march through the night, surrounded by invisible foes, tortured by weariness and pain, towards a goal that few can hope to reach, and where none may tarry long. One by one, as they march, our comrades vanish from our sight, seized by the silent orders of omnipotent Death. Very brief is the time in which we can help them, in which their happiness or misery is decided. Be it ours to shed sunshine on their path, to lighten their sorrows by the balm of sympathy, to give them the pure joy of a never-tiring affection, to strengthen failing courage, to instil faith in hours of despair. Let us not weigh in grudging scales their merits and demerits, but let us think only of their need—of the sorrows, the difficulties, perhaps the blindnesses, that make the misery of their lives; let us remember that they are fellow-sufferers in the same darkness, actors in the same tragedy with ourselves. And so, when their day is over, when their good and their evil have become eternal by the immortality of the past, be it ours to feel that, where they suffered, where they failed, no deed of ours was the cause; but wherever a spark of the divine fire kindled in their hearts, we were ready with encouragement, with sympathy, with brave words in which high courage glowed.

Brief and powerless is Man's life; on him and all his race the slow, sure doom falls pitiless and dark. Blind to good and evil, reckless of destruction, omnipotent matter rolls on its relentless way; for Man, condemned to-day to lose his dearest, to-morrow himself to pass through the gate of darkness, it remains only to cherish, ere yet the blow falls, the lofty thoughts that ennoble his little day; disdaining the coward terrors of the slave of Fate, to worship at the shrine that his own hands have built; undismayed by the empire of chance, to preserve a mind free from the wanton tyranny that rules his outward life; proudly defiant of the irresistible forces that tolerate, for a moment, his knowledge and his condemnation, to sustain alone, a weary but unyielding Atlas, the world that his own ideals have fashioned despite the trampling march of unconscious power.

IV

THE STUDY OF MATHEMATICS

In regard to every form of human activity it is necessary that the question should be asked from time to time, What is its purpose and ideal? In what way does it contribute to the beauty of human existence? As respects those pursuits which contribute only remotely, by providing the mechanism of life, it is well to be reminded that not the mere fact of living is to be desired, but the art of living in the contemplation of great things. Still more in regard to those avocations which have no end outside themselves, which are to be justified, if at all, as actually adding to the sum of the world's permanent possessions, it is necessary to keep alive a knowledge of their aims, a clear prefiguring vision of the temple in which creative imagination is to be embodied.

The fulfilment of this need, in what concerns the studies forming the material upon which custom has decided to train the youthful mind, is indeed sadly remote—so remote as to make the mere statement of such a claim appear preposterous. Great men, fully alive to the beauty of the contemplations to whose service their lives are devoted, desiring that others may share in their joys, persuade mankind to impart to the successive generations the mechanical knowledge without which it is impossible to cross the threshold. Dry pedants possess themselves of the privilege of instilling this knowledge: they forget that it is to serve but as a key to open the doors of the temple; though they spend their lives on the steps leading up to those sacred doors, they turn their backs upon the temple so resolutely that its very existence is forgotten, and the eager youth, who would press forward to be initiated to its domes and arches, is bidden to turn back and count the steps.

Mathematics, perhaps more even than the study of Greece and Rome, has suffered from this oblivion of its due place in civilisation. Although tradition has decreed that the great bulk of educated men shall know at least the elements of the subject, the reasons for which the tradition arose are forgotten, buried beneath a great rubbish-heap of pedantries and trivialities. To those who inquire as to the purpose of mathematics, the usual answer will be that it facilitates the making of machines, the travelling from place to place, and the victory over foreign nations, whether in war or commerce. If it be objected that these ends—all of which are of doubtful value—are not furthered by the merely elementary study imposed upon those who do not become expert mathematicians, the reply, it is true, will probably be that mathematics trains the reasoning faculties. Yet the very men who make this reply are, for the most part, unwilling to abandon the teaching of definite fallacies, known to be such, and instinctively rejected by the unsophisticated mind of every intelligent learner. And the reasoning faculty itself is generally conceived, by those who urge its cultivation, as merely a means for the avoidance of pitfalls and a help in the discovery of rules for the guidance of practical life. All these are undeniably important achievements to the credit of mathematics; yet it is none of these that entitles mathematics to a place in every liberal education. Plato, we know, regarded the contemplation of mathematical truths as worthy of the Deity; and Plato realised, more perhaps than any other single man, what those elements are in human life which merit a place in heaven. There is in mathematics, he says, "something which is *necessary* and cannot be set aside ... and, if I mistake not, of divine necessity; for as to the human necessities of which the Many talk in this connection, nothing can be more ridiculous than such an application of the words. *Cleinias.* And what are these necessities of knowledge, Stranger, which are divine and not human? *Athenian.* Those things without some use or knowledge of which a man cannot become a God to the world, nor a spirit, nor yet a hero, nor able earnestly to think and care for man".Such was Plato's judgment of mathematics; but the mathematicians do not read Plato, while those who read him know no mathematics, and regard his opinion upon this question as merely a curious aberration.

Mathematics, rightly viewed, possesses not only truth, but supreme beauty—a beauty cold and austere, like that of sculpture, without appeal to any part of our weaker nature, without the gorgeous trappings of painting or music, yet sublimely pure, and capable of a stern perfection such as only the greatest art can show. The true spirit of delight, the exaltation, the sense of being more than man, which is the touchstone of the highest excellence, is to be found in mathematics as surely as in poetry. What is best in mathematics deserves not merely to be learnt as a task, but to be assimilated as a part of daily thought, and brought again and again before the mind with ever-renewed encouragement. Real life is, to most men, a long second-best, a perpetual compromise between the ideal and the possible; but the world of pure reason knows no compromise, no practical limitations, no barrier to the creative activity embodying in splendid edifices the passionate aspiration after the perfect from which all great work springs. Remote from human passions, remote even from the pitiful facts of nature, the generations have gradually created an ordered cosmos, where pure thought can dwell as in its natural home, and where one, at least, of our nobler impulses can escape from the dreary exile of the actual world.

So little, however, have mathematicians aimed at beauty, that hardly anything in their work has had this conscious purpose. Much, owing to irrepressible instincts, which were better than avowed beliefs, has been moulded by an unconscious taste; but much also has been spoilt by false notions of what was fitting. The characteristic excellence of mathematics is only to be found where the reasoning is rigidly logical: the rules of logic are to mathematics what those of structure are to architecture. In the most beautiful work, a chain of argument is presented in which every link is important on its own account, in which there is an air of ease and lucidity throughout, and the premises achieve more than would have been thought possible, by means which appear natural and inevitable. Literature embodies what is general in particular circumstances whose universal significance shines through their individual dress; but mathematics endeavours to present whatever is most general in its purity, without any irrelevant trappings.

How should the teaching of mathematics be conducted so as to communicate to the learner as much as possible of this high ideal? Here experience must, in a great measure, be our guide; but some maxims may result from our consideration of the ultimate purpose to be achieved.

One of the chief ends served by mathematics, when rightly taught, is to awaken the learner's belief in reason, his confidence in the truth of what has been demonstrated, and in the value of demonstration. This purpose is not served by existing instruction; but it is easy to see ways in which it might be served. At present, in what concerns arithmetic, the boy or girl is given a set of rules, which present themselves as neither true nor false, but as merely the will of the teacher, the way in which, for some unfathomable reason, the teacher prefers to have the game played. To some degree, in a study of such definite practical utility, this is no doubt unavoidable; but as soon as possible, the reasons of rules should be set forth by whatever means most readily appeal to the childish mind. In geometry, instead of the tedious apparatus of fallacious proofs for obvious truisms which constitutes the beginning of Euclid, the learner should be allowed at first to assume the truth of everything obvious, and should be instructed in the demonstrations of theorems which are at once startling and easily verifiable by actual drawing, such as those in which it is shown that three or more lines meet in a point. In this way belief is generated; it is seen that reasoning may lead to startling conclusions, which nevertheless the facts will verify; and thus the instinctive distrust of whatever is abstract or rational is gradually overcome. Where theorems are difficult, they should be first taught as exercises in geometrical drawing, until the figure has become thoroughly familiar; it will then be an agreeable advance to be taught the logical connections of the various lines or circles that occur. It is desirable also that the figure illustrating a theorem should be drawn in all possible cases and shapes, that so the abstract relations with which geometry is concerned may of themselves emerge as the residue of similarity amid such great apparent diversity. In this way the abstract demonstrations should form but a small part of the instruction, and should be given when, by familiarity with concrete illustrations, they have come to be felt as the natural embodiment of visible fact. In this early stage proofs should not be given with pedantic fullness; definitely fallacious methods, such as that of superposition, should be rigidly excluded from the first, but where, without such methods, the proof would be very difficult, the result should be rendered acceptable by arguments and illustrations which are explicitly contrasted with demonstrations.

In the beginning of algebra, even the most intelligent child finds, as a rule, very great difficulty. The use of letters is a mystery, which seems to have no purpose except mystification. It is almost impossible, at first, not to think that every letter stands for some particular number, if only the teacher would reveal *what* number it stands for. The fact is, that in algebra the mind is first taught to consider general truths, truths which are not asserted to hold only of this or that particular thing, but of any one of a whole group of things. It is in the power of understanding and discovering such truths that the mastery of the intellect over the whole world of things actual and possible resides; and ability to deal with the general as such is one of the gifts that a mathematical education should bestow. But how little, as a rule, is the teacher of algebra able to explain the chasm which divides it from arithmetic, and how little is the learner assisted in his groping efforts at comprehension! Usually the method that has been adopted in arithmetic is continued: rules are set forth, with no adequate explanation of their grounds; the pupil learns to use the rules blindly,and presently, when he is able to obtain the answer that the teacher desires, he feels that he has mastered the difficulties of the subject. But of inner comprehension of the processes employed he has probably acquired almost nothing.

When algebra has been learnt, all goes smoothly until we reach those studies in which the notion of infinity is employed—the infinitesimal calculus and the whole of higher mathematics. The solution of the difficulties which formerly surrounded the mathematical infinite is probably the greatest achievement of which our own age has to boast. Since the beginnings of Greek thought these difficulties have been known; in every age the finest intellects have vainly endeavoured to answer the apparently unanswerable questions that had been asked by Zeno the Eleatic. At last Georg Cantor has found the answer, and has conquered for the intellect a new and vast province which had been given over to Chaos and old Night. It was assumed as self-evident, until Cantor and Dedekind established the opposite, that if, from any collection of things, some were taken away, the number of things left must always be less than the original number of things. This assumption, as a matter of fact, holds only of finite collections; and the rejection of it, where the infinite is concerned, has been shown to remove all the difficulties that had hitherto baffled human reason in this matter, and to render possible the creation of an exact science of the infinite. This stupendous fact ought to produce a revolution in the higher teaching of mathematics; it has itself added immeasurably to the educational value of the subject, and it has at last given the means of treating with logical precision many studies which, until lately, were wrapped in fallacy and obscurity. By those who were educated on the old lines, the new work is considered to be appallingly difficult, abstruse, and obscure; and it must be confessed that the discoverer, as is so often the case, has hardly himself emerged from the mists which the light of his intellect is dispelling. But inherently, the new doctrine of the infinite, to all candid and inquiring minds, has facilitated the mastery of higher mathematics; for hitherto, it has been necessary to learn, by a long process of sophistication, to give assent to arguments which, on first acquaintance, were rightly judged to be confused and erroneous. So far from producing a fearless belief in reason, a bold rejection of whatever failed to fulfil the strictest requirements of logic, a mathematical training, during the past two centuries, encouraged the belief that many things, which a rigid inquiry would reject as fallacious, must yet be accepted because they work in what the mathematician calls "practice." By this means, a timid, compromising spirit, or else a sacerdotal belief in mysteries not intelligible to the profane, has been bred where reason alone should have ruled. All this it is now time to sweep away; let those who wish to penetrate into the arcana of mathematics be taught at once the true theory in all its logical purity, and in the concatenation established by the very essence of the entities concerned.

If we are considering mathematics as an end in itself, and not as a technical training for engineers, it is very desirable to preserve the purity and strictness of its reasoning. Accordingly, those who have attained a sufficient familiarity with its easier portions should be led backward from propositions to which they have assented as self-evident to more and more fundamental principles from which what had previously appeared as premises can be deduced. They should be taught—what the theory of infinity very aptly illustrates—that many propositions seem self-evident to the untrained mind which, nevertheless, a nearer scrutiny shows to be false. By this means they will be led to a sceptical inquiry into first principles, an examination of the foundations upon which the whole edifice of reasoning is built, or, to take perhaps a more fitting metaphor, the great trunk from which the spreading branches spring. At this stage, it is well to study afresh the elementary portions of mathematics, asking no longer merely whether a given proposition is true, but also how it grows out of the central principles of logic. Questions of this nature can now be answered with a precision and certainty which were formerly quite impossible; and in the chains of reasoning that the answer requires the unity of all mathematical studies at last unfolds itself.

In the great majority of mathematical text-books there is a total lack of unity in method and of systematic development of a central theme. Propositions of very diverse kinds are proved by whatever means are thought most easily intelligible, and much space is devoted to mere curiosities which in no way contribute to the main argument. But in the greatest works, unity and inevitability are felt as in the unfolding of a drama; in the premisses a subject is proposed for consideration, and in every subsequent step some definite advance is made towards mastery of its nature. The love of system, of interconnection, which is perhaps the inmost essence of the intellectual impulse, can find free play in mathematics as nowhere else. The learner who feels this impulse must not be repelled by an array of meaningless examples or distracted by amusing oddities, but must be encouraged to dwell upon central principles, to become familiar with the structure of the various subjects which are put before him, to travel easily over the steps of the more important deductions. In this way a good tone of mind is cultivated, and selective attention is taught to dwell by preference upon what is weighty and essential.

When the separate studies into which mathematics is divided have each been viewed as a logical whole, as a natural growth from the propositions which constitute their principles, the learner will be able to understand the fundamental science which unifies and systematises the whole of deductive reasoning. This is symbolic logic—a study which, though it owes its inception to Aristotle, is yet, in its wider developments, a product, almost wholly, of the nineteenth century, and is indeed, in the present day, still growing with great rapidity. The true method of discovery in symbolic logic, and probably also the best method for introducing the study to a learner acquainted with other parts of mathematics, is the analysis of actual examples of deductive reasoning, with a view to the discovery of the principles employed. These principles, for the most part, are so embedded in our ratiocinative instincts, that they are employed quite unconsciously, and can be dragged to light only by much patient effort. But when at last they have been found, they are seen to be few in number, and to be the sole source of everything in pure mathematics. The discovery that all mathematics follows inevitably from a small collection of fundamental laws is one which immeasurably enhances the intellectual beauty of the whole; to those who have been oppressed by the fragmentary and incomplete nature of most existing chains of deduction this discovery comes with all the overwhelming force of a revelation; like a palace emerging from the autumn mist as the traveller ascends an Italian hill-side, the stately storeys of the mathematical edifice appear in their due order and proportion, with a new perfection in every part.

Until symbolic logic had acquired its present development, the principles upon which mathematics depends were always supposed to be philosophical, and discoverable only by the uncertain, unprogressive methods hitherto employed by philosophers. So long as this was thought, mathematics seemed to be not autonomous, but dependent upon a study which had quite other methods than its own. Moreover, since the nature of the postulates from which arithmetic, analysis, and geometry are to be deduced was wrapped in all the traditional obscurities of metaphysical discussion, the edifice built upon such dubious foundations began to be viewed as no better than a castle in the air. In this respect, the discovery that the true principles are as much a part of mathematics as any of their consequences has very greatly increased the intellectual satisfaction to be obtained. This satisfaction ought not to be refused to learners capable of enjoying it, for it is of a kind to increase our respect for human powers and our knowledge of the beauties belonging to the abstract world.

Philosophers have commonly held that the laws of logic, which underlie mathematics, are laws of thought, laws regulating the operations of our minds. By this opinion the true dignity of reason is very greatly lowered: it ceases to be an investigation into the very heart and immutable essence of all things actual and possible, becoming, instead, an inquiry into something more or less human and subject to our limitations. The contemplation of what is non-human, the discovery that our minds are capable of dealing with material not created by them, above all, the realisation that beauty belongs to the outer world as to the inner, are the chief means of overcoming the terrible sense of impotence, of weakness, of exile amid hostile powers, which is too apt to result from acknowledging the all-but omnipotence of alien forces. To reconcile us, by the exhibition of its awful beauty, to the reign of Fate—which is merely the literary personification of these forces—is the task of tragedy. But mathematics takes us still further from what is human, into the region of absolute necessity, to which not only the actual world, but every possible world, must conform; and even here it builds a habitation, or rather finds a habitation eternally standing, where our ideals are fully satisfied and our best hopes are not thwarted. It is only when we thoroughly understand the entire independence of ourselves, which belongs to this world that reason finds, that we can adequately realise the profound importance of its beauty.

Not only is mathematics independent of us and our thoughts, but in another sense we and the whole universe of existing things are independent of mathematics. The apprehension of this purely ideal character is indispensable, if we are to understand rightly the place of mathematics as one among the arts. It was formerly supposed that pure reason could decide, in some respects, as to the nature of the actual world: geometry, at least, was thought to deal with the space in which we live. But we now know that pure mathematics can never pronounce upon questions of actual existence: the world of reason, in a sense, controls the world of fact, but it is not at any point creative of fact, and in the application of its results to the world in time and space, its certainty and precision are lost among approximations and working hypotheses. The objects considered by mathematicians have, in the past, been mainly of a kind suggested by phenomena; but from such restrictions the abstract imagination should be wholly free. A reciprocal liberty must thus be accorded: reason cannot dictate to the world of facts, but the facts cannot restrict reason's privilege of dealing with whatever objects its love of beauty may cause to seem worthy of consideration. Here, as elsewhere, we build up our own ideals out of the fragments to be found in the world; and in the end it is hard to say whether the result is a creation or a discovery.

It is very desirable, in instruction, not merely to persuade the student of the accuracy of important theorems, but to persuade him in the way which itself has, of all possible ways, the most beauty. The true interest of a demonstration is not, as traditional modes of exposition suggest, concentrated wholly in the result; where this does occur, it must be viewed as a defect, to be remedied, if possible, by so generalising the steps of the proof that each becomes important in and for itself. An argument which serves only to prove a conclusion is like a story subordinated to some moral which it is meant to teach: for æsthetic perfection no part of the whole should be merely a means. A certain practical spirit, a desire for rapid progress, for conquest of new realms, is responsible for the undue emphasis upon results which prevails in mathematical instruction. The better way is to propose some theme for consideration—in geometry, a figure having important properties; in analysis, a function of which the study is illuminating, and so on. Whenever proofs depend upon some only of the marks by which we define the object to be studied, these marks should be isolated and investigated on their own account. For it is a defect, in an argument, to employ more premisses than the conclusion demands: what mathematicians call elegance results from employing only the essential principles in virtue of which the thesis is true. It is a merit in Euclid that he advances as far as he is able to go without employing the axiom of parallels—not, as is often said, because this axiom is inherently objectionable, but because, in mathematics, every new axiom diminishes the generality of the resulting theorems, and the greatest possible generality is before all things to be sought.

Of the effects of mathematics outside its own sphere more has been written than on the subject of its own proper ideal. The effect upon philosophy has, in the past, been most notable, but most varied; in the seventeenth century, idealism and rationalism, in the eighteenth, materialism and sensationalism, seemed equally its offspring. Of the effect which it is likely to have in the future it would be very rash to say much; but in one respect a good result appears probable. Against that kind of scepticism which abandons the pursuit of ideals because the road is arduous and the goal not certainly attainable, mathematics, within its own sphere, is a complete answer. Too often it is said that there is no absolute truth, but only opinion and private judgment; that each of us is conditioned, in his view of the world, by his own peculiarities, his own taste and bias; that there is no external kingdom of truth to which, by patience and discipline, we may at last obtain admittance, but only truth for me, for you, for every separate person. By this habit of mind one of the chief ends of human effort is denied, and the supreme virtue of candour, of fearless acknowledgment of what is, disappears from our moral vision. Of such scepticism mathematics is a perpetual reproof; for its edifice of truths stands unshakable and inexpungable to all the weapons of doubting cynicism.

The effects of mathematics upon practical life, though they should not be regarded as the motive of our studies, may be used to answer a doubt to which the solitary student must always be liable. In a world so full of evil and suffering, retirement into the cloister of contemplation, to the enjoyment of delights which, however noble, must always be for the few only, cannot but appear as a somewhat selfish refusal to share the burden imposed upon others by accidents in which justice plays no part. Have any of us the right, we ask, to withdraw from present evils, to leave our fellow-men unaided, while we live a life which, though arduous and austere, is yet plainly good in its own nature? When these questions arise, the true answer is, no doubt, that some must keep alive the sacred fire, some must preserve, in every generation, the haunting vision which shadows forth the goal of so much striving. But when, as must sometimes occur, this answer seems too cold, when we are almost maddened by the spectacle of sorrows to which we bring no help, then we may reflect that indirectly the mathematician often does more for human happiness than any of his more practically active contemporaries. The history of science abundantly proves that a body of abstract propositions—even if, as in the case of conic sections, it remains two thousand years without effect upon daily life—may yet, at any moment, be used to cause a revolution in the habitual thoughts and occupations of every citizen. The use of steam and electricity—to take striking instances—is rendered possible only by mathematics. In the results of abstract thought the world possesses a capital of which the employment in enriching the common round has no hitherto discoverable limits. Nor does experience give any means of deciding what parts of mathematics will be found useful. Utility, therefore, can be only a consolation in moments of discouragement, not a guide in directing our studies.

For the health of the moral life, for ennobling the tone of an age or a nation, the austerer virtues have a strange power, exceeding the power of those not informed and purified by thought. Of these austerer virtues the love of truth is the chief, and in mathematics, more than elsewhere, the love of truth may find encouragement for waning faith. Every great study is not only an end in itself, but also a means of creating and sustaining a lofty habit of mind; and this purpose should be kept always in view throughout the teaching and learning of mathematics.

V

MATHEMATICS AND THE METAPHYSICIANS

The nineteenth century, which prided itself upon the invention of steam and evolution, might have derived a more legitimate title to fame from the discovery of pure mathematics. This science, like most others, was baptised long before it was born; and thus we find writers before the nineteenth century alluding to what they called pure mathematics. But if they had been asked what this subject was, they would only have been able to say that it consisted of Arithmetic, Algebra, Geometry, and so on. As to what these studies had in common, and as to what distinguished them from applied mathematics, our ancestors were completely in the dark.

Pure mathematics was discovered by Boole, in a work which he called the *Laws of Thought* (1854). This work abounds in asseverations that it is not mathematical, the fact being that Boole was too modest to suppose his book the first ever written on mathematics. He was also mistaken in supposing that he was dealing with the laws of thought: the question how people actually think was quite irrelevant to him, and if his book had really contained the laws of thought, it was curious that no one should ever have thought in such a way before. His book was in fact concerned with formal logic, and this is the same thing as mathematics.

Pure mathematics consists entirely of assertions to the effect that, if such and such a proposition is true of *anything*, then such and such another proposition is true of that thing. It is essential not to discuss whether the first proposition is really true, and not to mention what the anything is, of which it is supposed to be true. Both these points would belong to applied mathematics. We start, in pure mathematics, from certain rules of inference, by which we can infer that *if* one proposition is true, then so is some other proposition. These rules of inference constitute the major part of the principles of formal logic. We then take any hypothesis that seems amusing, and deduce its consequences. *If* our hypothesis is about *anything*, and not about some one or more particular things, then our deductions constitute mathematics. Thus mathematics may be defined as the subject in which we never know what we are talking about, nor whether what we are saying is true. People who have been puzzled by the beginnings of mathematics will, I hope, find comfort in this definition, and will probably agree that it is accurate.

As one of the chief triumphs of modern mathematics consists in having discovered what mathematics really is, a few more words on this subject may not be amiss. It is common to start any branch of mathematics—for instance,

Geometry—with a certain number of primitive ideas, supposed incapable of definition, and a certain number of primitive propositions or axioms, supposed incapable of proof. Now the fact is that, though there are indefinables and indemonstrables in every branch of applied mathematics, there are none in pure mathematics except such as belong to general logic. Logic, broadly speaking, is distinguished by the fact that its propositions can be put into a form in which they apply to anything whatever. All pure mathematics—Arithmetic, Analysis, and Geometry—is built up by combinations of the primitive ideas of logic, and its propositions are deduced from the general axioms of logic, such as the syllogism and the other rules of inference. And this is no longer a dream or an aspiration. On the contrary, over the greater and more difficult part of the domain of mathematics, it has been already accomplished; in the few remaining cases, there is no special difficulty, and it is now being rapidly achieved. Philosophers have disputed for ages whether such deduction was possible; mathematicians have sat down and made the deduction. For the philosophers there is now nothing left but graceful acknowledgments.

The subject of formal logic, which has thus at last shown itself to be identical with mathematics, was, as every one knows, invented by Aristotle, and formed the chief study (other than theology) of the Middle Ages. But Aristotle never got beyond the syllogism, which is a very small part of the subject, and the schoolmen never got beyond Aristotle. If any proof were required of our superiority to the mediæval doctors, it might be found in this. Throughout the Middle Ages, almost all the best intellects devoted themselves to formal logic, whereas in the nineteenth century only an infinitesimal proportion of the world's thought went into this subject. Nevertheless, in each decade since 1850 more has been done to advance the subject than in the whole period from Aristotle to Leibniz. People have discovered how to make reasoning symbolic, as it is in Algebra, so that deductions are effected by mathematical rules. They have discovered many rules besides the syllogism, and a new branch of logic, called the Logic of Relatives,has been invented to deal with topics that wholly surpassed the powers of the old logic, though they form the chief contents of mathematics.

It is not easy for the lay mind to realise the importance of symbolism in discussing the foundations of mathematics, and the explanation may perhaps seem strangely paradoxical. The fact is that symbolism is useful because it makes things difficult. (This is not true of the advanced parts of mathematics, but only of the beginnings.) What we wish to know is, what can be deduced from what. Now, in the beginnings, everything is self-evident; and it is very hard to see whether one self-evident proposition follows from another or not. Obviousness is always the enemy to correctness. Hence we invent some new and difficult symbolism, in which nothing seems obvious. Then we set up certain rules for operating on the symbols, and the whole thing becomes mechanical. In this way we find out what must be taken as premiss and what can be demonstrated or defined. For instance, the whole of Arithmetic and Algebra has been shown to require three indefinable notions and five indemonstrable propositions. But without a symbolism it would have been very hard to find this out. It is so obvious that two and two are four, that we can hardly make ourselves sufficiently sceptical to doubt whether it can be proved. And the same holds in other cases where self-evident things are to be proved.

But the proof of self-evident propositions may seem, to the uninitiated, a somewhat frivolous occupation. To this we might reply that it is often by no means self-evident that one obvious proposition follows from another obvious proposition; so that we are really discovering new truths when we prove what is evident by a method which is not evident. But a more interesting retort is, that since people have tried to prove obvious propositions, they have found that many of them are false.Self-evidence is often a mere will-o'-the-wisp, which is sure to lead us astray if we take it as our guide. For instance, nothing is plainer than that a whole always has more terms than a part, or that a number is increased by adding one to it. But these propositions are now known to be usually false. Most numbers are infinite, and if a number is infinite you may add ones to it as long as you like without disturbing it in the least. One of the merits of a proof is that it instils a certain doubt as to the result proved; and when what is obvious can be proved in some cases, but not in others, it becomes possible to suppose that in these other cases it is false.

The great master of the art of formal reasoning, among the men of our own day, is an Italian, Professor Peano, of the University of Turin.He has reduced the greater part of mathematics (and he or his followers will, in time, have reduced the whole) to strict symbolic form, in which there are no words at all. In the ordinary mathematical books, there are no doubt fewer words than most readers would wish. Still, little phrases occur, such as *therefore, let us assume, consider*, or *hence it follows*. All these, however, are a concession, and are swept away by Professor Peano. For instance, if we wish to learn the whole of Arithmetic, Algebra, the Calculus, and indeed all that is usually called pure mathematics (except Geometry), we must start with a dictionary of three words. One symbol stands for *zero*, another for *number*, and a third for *next after*. What these ideas mean, it is necessary to know if you wish to become an arithmetician. But after symbols have been invented for these three ideas, not another word is required in the whole development. All future symbols are symbolically explained by means of these three. Even these three can be explained by means of the notions of *relation* and *class*; but this requires the Logic of Relations, which Professor Peano has never taken up. It must be admitted that what a mathematician has to know to begin with is not much. There are at most a dozen notions out of which all the notions in all pure mathematics (including Geometry) are compounded.

Professor Peano, who is assisted by a very able school of young Italian disciples, has shown how this may be done; and although the method which he has invented is capable of being carried a good deal further than he has carried it, the honour of the pioneer must belong to him.

Two hundred years ago, Leibniz foresaw the science which Peano has perfected, and endeavoured to create it. He was prevented from succeeding by respect for the authority of Aristotle, whom he could not believe guilty of definite, formal fallacies; but the subject which he desired to create now exists, in spite of the patronising contempt with which his schemes have been treated by all superior persons. From this "Universal Characteristic," as he called it, he hoped for a solution of all problems, and an end to all disputes. "If controversies were to arise," he says, "there would be no more need of disputation between two philosophers than between two accountants. For it would suffice to take their pens in their hands, to sit down to their desks, and to say to each other (with a friend as witness, if they liked), 'Let us calculate.'" This optimism has now appeared to be somewhat excessive; there still are problems whose solution is doubtful, and disputes which calculation cannot decide. But over an enormous field of what was formerly controversial, Leibniz's dream has become sober fact. In the whole philosophy of mathematics, which used to be at least as full of doubt as any other part of philosophy, order and certainty have replaced the confusion and hesitation which formerly reigned. Philosophers, of course, have not yet discovered this fact, and continue to write on such subjects in the old way. But mathematicians, at least in Italy, have now the power of treating the principles of mathematics in an exact and masterly manner, by means of which the certainty of mathematics extends also to mathematical philosophy. Hence many of the topics which used to be placed among the great mysteries—for example, the natures of infinity, of continuity, of space, time and motion—are now no longer in any degree open to doubt or discussion. Those who wish to know the nature of these things need only read the works of such men as Peano or Georg Cantor; they will there find exact and indubitable expositions of all these quondam mysteries.

In this capricious world, nothing is more capricious than posthumous fame. One of the most notable examples of posterity's lack of judgment is the Eleatic Zeno. This man, who may be regarded as the founder of the philosophy of infinity, appears in Plato's Parmenides in the privileged position of instructor to Socrates. He invented four arguments, all immeasurably subtle and profound, to prove that motion is impossible, that Achilles can never overtake the tortoise, and that an arrow in flight is really at rest. After being refuted by Aristotle, and by every subsequent philosopher from that day to our own, these arguments were reinstated, and made the basis of a mathematical renaissance, by a German professor, who probably never dreamed of any connection between himself and

Zeno. Weierstrass,by strictly banishing from mathematics the use of infinitesimals, has at last shown that we live in an unchanging world, and that the arrow in its flight is truly at rest. Zeno's only error lay in inferring (if he did infer) that, because there is no such thing as a state of change, therefore the world is in the same state at any one time as at any other. This is a consequence which by no means follows; and in this respect, the German mathematician is more constructive than the ingenious Greek. Weierstrass has been able, by embodying his views in mathematics, where familiarity with truth eliminates the vulgar prejudices of common sense, to invest Zeno's paradoxes with the respectable air of platitudes; and if the result is less delightful to the lover of reason than Zeno's bold defiance, it is at any rate more calculated to appease the mass of academic mankind.

Zeno was concerned, as a matter of fact, with three problems, each presented by motion, but each more abstract than motion, and capable of a purely arithmetical treatment. These are the problems of the infinitesimal, the infinite, and continuity. To state clearly the difficulties involved, was to accomplish perhaps the hardest part of the philosopher's task. This was done by Zeno. From him to our own day, the finest intellects of each generation in turn attacked the problems, but achieved, broadly speaking, nothing. In our own time, however, three men—Weierstrass, Dedekind, and Cantor—have not merely advanced the three problems, but have completely solved them. The solutions, for those acquainted with mathematics, are so clear as to leave no longer the slightest doubt or difficulty. This achievement is probably the greatest of which our age has to boast; and I know of no age (except perhaps the golden age of Greece) which has a more convincing proof to offer of the transcendent genius of its great men. Of the three problems, that of the infinitesimal was solved by Weierstrass; the solution of the other two was begun by Dedekind, and definitively accomplished by Cantor.

The infinitesimal played formerly a great part in mathematics. It was introduced by the Greeks, who regarded a circle as differing infinitesimally from a polygon with a very large number of very small equal sides. It gradually grew in importance, until, when Leibniz invented the Infinitesimal Calculus, it seemed to become the fundamental notion of all higher mathematics. Carlyle tells, in his *Frederick the Great*, how Leibniz used to discourse to Queen Sophia Charlotte of Prussia concerning the infinitely little, and how she would reply that on that subject she needed no instruction—the behaviour of courtiers had made her thoroughly familiar with it. But philosophers and mathematicians—who for the most part had less acquaintance with courts—continued to discuss this topic, though without making any advance. The Calculus required continuity, and continuity was supposed to require the infinitely little; but nobody could discover what the infinitely little might be. It was plainly not quite zero, because a sufficiently large number of infinitesimals, added together, were seen to make up a finite whole. But nobody could point out any fraction which was not zero, and yet not finite. Thus there was a deadlock. But at last Weierstrass discovered that the infinitesimal was not needed at all, and that everything could be accomplished without it. Thus there was no longer any need to suppose that there was such a thing. Nowadays, therefore, mathematicians are more dignified than Leibniz: instead of talking about the infinitely small, they talk about the infinitely great—a subject which, however appropriate to monarchs, seems, unfortunately, to interest them even less than the infinitely little interested the monarchs to whom Leibniz discoursed.

The banishment of the infinitesimal has all sorts of odd consequences, to which one has to become gradually accustomed. For example, there is no such thing as the next moment. The interval between one moment and the next would have to be infinitesimal, since, if we take two moments with a finite interval between them, there are always other moments in the interval. Thus if there are to be no infinitesimals, no two moments are quite consecutive, but there are always other moments between any two. Hence there must be an infinite number of moments between any two; because if there were a finite number one would be nearest the first of the two moments, and therefore next to it. This might be thought to be a difficulty; but, as a matter of fact, it is here that the philosophy of the infinite comes in, and makes all straight.

The same sort of thing happens in space. If any piece of matter be cut in two, and then each part be halved, and so on, the bits will become smaller and smaller, and can theoretically be made as small as we please. However small they may be, they can still be cut up and made smaller still. But they will always have *some* finite size, however small they may be. We never reach the infinitesimal in this way, and no finite number of divisions will bring us to points. Nevertheless there *are* points, only these are not to be reached by successive divisions. Here again, the philosophy of the infinite shows us how this is possible, and why points are not infinitesimal lengths.

As regards motion and change, we get similarly curious results. People used to think that when a thing changes, it must be in a state of change, and that when a thing moves, it is in a state of motion. This is now known to be a mistake. When a body moves, all that can be said is that it is in one place at one time and in another at another. We must not say that it will be in a neighbouring place at the next instant, since there is no next instant. Philosophers often tell us that when a body is in motion, it changes its position within the instant. To this view Zeno long ago made the fatal retort that every body always is where it is; but a retort so simple and brief was not of the kind to which philosophers are accustomed to give weight, and they have continued down to our own day to repeat the same phrases which roused the Eleatic's destructive ardour. It was only recently that it became possible to explain motion in detail in accordance with Zeno's platitude, and in opposition to the philosopher's paradox. We may now at last indulge the comfortable belief that a body in motion is just as truly where it is as a body at rest. Motion consists merely in the fact that bodies are sometimes in one place and sometimes in another, and that they are at intermediate places at intermediate times. Only those who have waded through the quagmire of philosophic speculation on this subject can realise what a liberation from antique prejudices is involved in this simple and straightforward commonplace.

The philosophy of the infinitesimal, as we have just seen, is mainly negative. People used to believe in it, and now they have found out their mistake. The philosophy of the infinite, on the other hand, is wholly positive. It was formerly supposed that infinite numbers, and the mathematical infinite generally, were self-contradictory. But as it was obvious that there were infinities—for example, the number of numbers—the contradictions of infinity seemed unavoidable, and philosophy seemed to have wandered into a "cul-de-sac." This difficulty led to Kant's antinomies, and hence, more or less indirectly, to much of Hegel's dialectic method. Almost all current philosophy is upset by the fact (of which very few philosophers are as yet aware) that all the ancient and respectable contradictions in the notion of the infinite have been once for all disposed of. The method by which this has been done is most interesting and instructive. In the first place, though people had talked glibly about infinity ever since the beginnings of Greek thought, nobody had ever thought of asking, What is infinity? If any philosopher had been asked for a definition of infinity, he might have produced some unintelligible rigmarole, but he would certainly not have been able to give a definition that had any meaning at all. Twenty years ago, roughly speaking, Dedekind and Cantor asked this question, and, what is more remarkable, they answered it. They found, that is to say, a perfectly precise definition of an infinite number or an infinite collection of things. This was the first and perhaps the greatest step. It then remained to examine the supposed contradictions in this notion. Here Cantor proceeded in the only proper way. He took pairs of contradictory propositions, in which both sides of the contradiction would be usually regarded as demonstrable, and he strictly examined the supposed proofs. He found that all proofs adverse to infinity involved a certain principle, at first sight obviously true, but destructive, in its consequences, of almost all mathematics. The proofs favourable to infinity, on the other hand, involved no principle that had evil consequences. It thus appeared that common sense had allowed itself to be taken in by a specious maxim, and that, when once this maxim was rejected, all went well.

The maxim in question is, that if one collection is part of another, the one which is a part has fewer terms than the one of which it is a part. This maxim is true of finite numbers. For example, Englishmen are only some among Europeans, and there are fewer Englishmen than Europeans. But when we come to infinite numbers, this is no longer true. This breakdown of the maxim gives us the precise definition of infinity. A collection of terms is infinite when it contains as parts other collections which have just as many terms as it has. If you can take away some of the terms of a collection, without diminishing the number of terms, then there are an infinite number of terms in the collection. For example, there are just as many even numbers as there are numbers altogether, since every number can be doubled. This may be seen by putting odd and even numbers together in one row, and even numbers alone in a row below:—

1, 2, 3, 4, 5, *ad infinitum.*
2, 4, 6, 8, 10, *ad infinitum.*

There are obviously just as many numbers in the row below as in the row above, because there is one below for each one above. This property, which was formerly thought to be a contradiction, is now transformed into a harmless definition of infinity, and shows, in the above case, that the number of finite numbers is infinite.

But the uninitiated may wonder how it is possible to deal with a number which cannot be counted. It is impossible to count up *all* the numbers, one by one, because, however many we may count, there are always more to follow. The fact is that counting is a very vulgar and elementary way of finding out how many terms there are in a collection. And in any case, counting gives us what mathematicians call the *ordinal* number of our terms; that is to say, it arranges our terms in an order or series, and its result tells us what type of series results from this arrangement. In other words, it is impossible to count things without counting some first and others afterwards, so that counting always has to do with order. Now when there are only a finite number of terms, we can count them in any order we like; but when there are an infinite number, what corresponds to counting will give us quite different results according to the way in which we carry out the operation. Thus the ordinal number, which results from what, in a general sense may be called counting, depends not only upon how many terms we have, but also (where the number of terms is infinite) upon the way in which the terms are arranged.

The fundamental infinite numbers are not ordinal, but are what is called *cardinal*. They are not obtained by putting our terms in order and counting them, but by a different method, which tells us, to begin with, whether two collections have the same number of terms, or, if not, which is the greater.It does not tell us, in the way in which counting does, *what* number of terms a collection has; but if we define a number as the number of terms in such and such a collection, then this method enables us to discover whether some other collection that may be mentioned has more or fewer terms. An illustration will show how this is done.

If there existed some country in which, for one reason or another, it was impossible to take a census, but in which it was known that every man had a wife and every woman a husband, then (provided polygamy was not a national institution) we should know, without counting, that there were exactly as many men as there were women in that country, neither more nor less. This method can be applied generally. If there is some relation which, like marriage, connects the things in one collection each with one of the things in another collection, and vice versa, then the two collections have the same number of terms. This was the way in which we found that there are as many even numbers as there are numbers. Every number can be doubled, and every even number can be halved, and each process gives just one number corresponding to the one that is doubled or halved. And in this way we can find any number of collections each of which has just as many terms as there are finite numbers. If every term of a collection can be hooked on to a number, and all the finite numbers are used once, and only once, in the process, then our collection must have just as many terms as there are finite numbers. This is the general method by which the numbers of infinite collections are defined.

But it must not be supposed that all infinite numbers are equal. On the contrary, there are infinitely more infinite numbers than finite ones. There are more ways of arranging the finite numbers in different types of series than there are finite numbers. There are probably more points in space and more moments in time than there are finite numbers. There are exactly as many fractions as whole numbers, although there are an infinite number of fractions between any two whole numbers. But there are more irrational numbers than there are whole numbers or fractions. There are probably exactly as many points in space as there are irrational numbers, and exactly as many points on a line a millionth of an inch long as in the whole of infinite space. There is a greatest of all infinite numbers, which is the number of things altogether, of every sort and kind. It is obvious that there cannot be a greater number than this, because,if everything has been taken, there is nothing left to add. Cantor has a proof that there is no greatest number, and if this proof were valid, the contradictions of infinity would reappear in a sublimated form. But in this one point, the master has been guilty of a very subtle fallacy, which I hope to explain in some future work.

We can now understand why Zeno believed that Achilles cannot overtake the tortoise and why as a matter of fact he can overtake it. We shall see that all the people who disagreed with Zeno had no right to do so, because they all accepted premises from which his conclusion followed. The argument is this: Let Achilles and the tortoise start along a road at the same time, the tortoise (as is only fair) being allowed a handicap. Let Achilles go twice as fast as the tortoise, or ten times or a hundred times as fast. Then he will never reach the tortoise. For at every moment the tortoise is somewhere and Achilles is somewhere; and neither is ever twice in the same place while the race is going on. Thus the tortoise goes to just as many places as Achilles does, because each is in one place at one moment, and in another at any other moment. But if Achilles were to catch up with the tortoise, the places where the tortoise would have been would be only part of the places where Achilles would have been. Here, we must suppose, Zeno appealed to the maxim that the whole has more terms than the part.Thus if Achilles were to overtake the tortoise, he would have been in more places than the tortoise; but we saw that he must, in any period, be in exactly as many places as the tortoise. Hence we infer that he can never catch the tortoise. This argument is strictly correct, if we allow the axiom that the whole has more terms than the part. As the conclusion is absurd, the axiom must be rejected, and then all goes well. But there is no good word to be said for the philosophers of the past two thousand years and more, who have all allowed the axiom and denied the conclusion.

The retention of this axiom leads to absolute contradictions, while its rejection leads only to oddities. Some of these oddities, it must be confessed, are very odd. One of them, which I call the paradox of Tristram Shandy, is the converse of the Achilles, and shows that the tortoise, if you give him time, will go just as far as Achilles. Tristram Shandy, as we know, employed two years in chronicling the first two days of his life, and lamented that, at this rate, material would accumulate faster than he could deal with it, so that, as years went by, he would be farther and farther from the end of his history. Now I maintain that, if he had lived for ever, and had not wearied of his task, then, even if his life had continued as event fully as it began, no part of his biography would have remained unwritten. For consider: the hundredth day will be described in the hundredth year, the thousandth in the thousandth year, and so on. Whatever day we may choose as so far on that he cannot hope to reach it, that day will be described in the corresponding year. Thus any day that may be mentioned will be written up sooner or later, and therefore no part of the biography will remain permanently unwritten. This paradoxical but perfectly true proposition depends upon the fact that the number of days in all time is no greater than the number of years.

Thus on the subject of infinity it is impossible to avoid conclusions which at first sight appear paradoxical, and this is the reason why so many philosophers have supposed that there were inherent contradictions in the infinite. But a little practice enables one to grasp the true principles of Cantor's doctrine, and to acquire new and better instincts as to the true and the false. The oddities then become no odder than the people at the antipodes, who used to be thought impossible because they would find it so inconvenient to stand on their heads.

The solution of the problems concerning infinity has enabled Cantor to solve also the problems of continuity. Of this, as of infinity, he has given a perfectly precise definition, and has shown that there are no contradictions in the notion so defined. But this subject is so technical that it is impossible to give any account of it here.

The notion of continuity depends upon that of *order*, since continuity is merely a particular type of order. Mathematics has, in modern times, brought order into greater and greater prominence. In former days, it was supposed (and philosophers are still apt to suppose) that quantity was the fundamental notion of mathematics. But nowadays, quantity is banished altogether, except from one little corner of Geometry, while order more and more reigns supreme. The investigation of different kinds of series and their relations is now a very large part of mathematics, and it has been found that this investigation can be conducted without any reference to quantity, and, for the most part, without any reference to number. All types of series are capable of formal definition, and their properties can be deduced from the principles of symbolic logic by means of the Algebra of Relatives.The notion of a limit, which is fundamental in the greater part of higher mathematics, used to be defined by means of quantity, as a term to which the terms of some series approximate as nearly as we please. But nowadays the limit is defined quite differently, and the series which it limits may not approximate to it at all. This improvement also is due to Cantor, and it is one which has revolutionised mathematics. Only order is now relevant to limits. Thus, for instance, the smallest of the infinite integers is the limit of the finite integers, though all finite integers are at an infinite distance from it. The study of different types of series is a general subject of which the study of ordinal numbers (mentioned above) is a special and very interesting branch. But the unavoidable technicalities of this subject render it impossible to explain to any but professed mathematicians.

Geometry, like Arithmetic, has been subsumed, in recent times, under the general study of order. It was formerly supposed that Geometry was the study of the nature of the space in which we live, and accordingly it was urged, by those who held that what exists can only be known empirically, that Geometry should really be regarded as belonging to applied mathematics. But it has gradually appeared, by the increase of non-Euclidean systems, that Geometry throws no more light upon the nature of space than Arithmetic throws upon the population of the United States. Geometry is a whole collection of deductive sciences based on a corresponding collection of sets of axioms. One set of axioms is Euclid's; other equally good sets of axioms lead to other results. Whether Euclid's axioms are true, is a question as to which the pure mathematician is indifferent; and, what is more, it is a question which it is theoretically impossible to answer with certainty in the affirmative. It might possibly be shown, by very careful measurements, that Euclid's axioms are false; but no measurements could ever assure us (owing to the errors of observation) that they are exactly true. Thus the geometer leaves to the man of science to decide, as best he may, what axioms are most nearly true in the actual world. The geometer takes any set of axioms that seem interesting, and deduces their consequences. What defines Geometry, in this sense, is that the axioms must give rise to a series of more than one dimension. And it is thus that Geometry becomes a department in the study of order.

In Geometry, as in other parts of mathematics, Peano and his disciples have done work of the very greatest merit as regards principles. Formerly, it was held by philosophers and mathematicians alike that the proofs in Geometry depended on the figure; nowadays, this is known to be false. In the best books there are no figures at all. The reasoning proceeds by the strict rules of formal logic from a set of axioms laid down to begin with. If a figure is used, all sorts of things seem obviously to follow, which no formal reasoning can prove from the explicit axioms, and which, as a matter of fact, are only accepted because they are obvious. By banishing the figure, it becomes possible to discover *all* the axioms that are needed; and in this way all sorts of possibilities, which would have otherwise remained undetected, are brought to light.

One great advance, from the point of view of correctness, has been made by introducing points as they are required, and not starting, as was formerly done, by assuming the whole of space. This method is due partly to Peano, partly to another Italian named Fano. To those unaccustomed to it, it has an air of somewhat wilful pedantry. In this way, we begin with the following axioms: (1) There is a class of entities called *points*. (2) There is at least one point. (3) If *a* be a point, there is at least one other point besides *a*. Then we bring in the straight line joining two points, and begin again with (4), namely, on the straight line joining *a* and *b*, there is at least one other point besides *a* and *b*. (5) There is at least one point not on the line *ab*. And so we go on, till we have the means of obtaining as many points as we require. But the word *space*, as Peano humorously remarks, is one for which Geometry has no use at all.

The rigid methods employed by modern geometers have deposed Euclid from his pinnacle of correctness. It was thought, until recent times, that, as Sir Henry Savile remarked in 1621, there were only two blemishes in Euclid, the theory of parallels and the theory of proportion. It is now known that these are almost the only points in which Euclid is free from blemish. Countless errors are involved in his first eight propositions. That is to say, not only is it doubtful whether his axioms are true, which is a comparatively trivial matter, but it is certain that his propositions do not follow from the axioms which he enunciates. A vastly greater number of axioms, which Euclid unconsciously employs, are required for the proof of his propositions. Even in the first proposition of all, where he constructs an equilateral triangle on a given base, he uses two circles which are assumed to intersect. But no explicit axiom assures us that they do so, and in some kinds of spaces they do not always intersect. It is quite doubtful whether our space belongs to one of these kinds or not. Thus Euclid fails entirely to prove his point in the very first proposition. As he is certainly not an easy author, and is terribly long-winded, he has no longer any but an historical interest. Under these circumstances, it is nothing less than a scandal that he should still be taught to boys in England.A book should have either intelligibility or correctness; to combine the two is impossible, but to lack both is to be unworthy of such a place as Euclid has occupied in education.

The most remarkable result of modern methods in mathematics is the importance of symbolic logic and of rigid formalism. Mathematicians, under the influence of Weierstrass, have shown in modern times a care for accuracy, and an aversion to slipshod reasoning, such as had not been known among them previously since the time of the Greeks. The great inventions of the seventeenth century—Analytical Geometry and the Infinitesimal Calculus—were so fruitful in new results that mathematicians had neither time nor inclination to examine their foundations. Philosophers, who should have taken up the task, had too little mathematical ability to invent the new branches of mathematics which have now been found necessary for any adequate discussion. Thus mathematicians were only awakened from their "dogmatic slumbers" when Weierstrass and his followers showed that many of their most cherished propositions are in general false. Macaulay, contrasting the certainty of mathematics with the uncertainty of philosophy, asks who ever heard of a reaction against Taylor's theorem? If he had lived now, he himself might have heard of such a reaction, for this is precisely one of the theorems which modern investigations have overthrown. Such rude shocks to mathematical faith have produced that love of formalism which appears, to those who are ignorant of its motive, to be mere outrageous pedantry.

The proof that all pure mathematics, including Geometry, is nothing but formal logic, is a fatal blow to the Kantian philosophy. Kant, rightly perceiving that Euclid's propositions could not be deduced from Euclid's axioms without the help of the figures, invented a theory of knowledge to account for this fact; and it accounted so successfully that, when the fact is shown to be a mere defect in Euclid, and not a result of the nature of geometrical reasoning, Kant's theory also has to be abandoned. The whole doctrine of *a priori* intuitions, by which Kant explained the possibility of pure mathematics, is wholly inapplicable to mathematics in its present form. The Aristotelian doctrines of the schoolmen come nearer in spirit to the doctrines which modern mathematics inspire; but the schoolmen were hampered by the fact that their formal logic was very defective, and that the philosophical logic based upon the syllogism showed a corresponding narrowness. What is now required is to give the greatest possible development to mathematical logic, to allow to the full the importance of relations, and then to found upon this secure basis a new philosophical logic, which may hope to borrow some of the exactitude and certainty of its mathematical foundation. If this can be successfully accomplished, there is every reason to hope that the near future will be as great an epoch in pure philosophy as the immediate past has been in the principles of mathematics. Great triumphs inspire great hopes; and pure thought may achieve, within our generation, such results as will place our time, in this respect, on a level with the greatest age of Greece.

VI

ON SCIENTIFIC METHOD IN PHILOSOPHY

When we try to ascertain the motives which have led men to the investigation of philosophical questions, we find that, broadly speaking, they can be divided into two groups, often antagonistic, and leading to very divergent systems. These two groups of motives are, on the one hand, those derived from religion and ethics, and, on the other hand, those derived from science. Plato, Spinoza, and Hegel may be taken as typical of the philosophers whose interests are mainly religious and ethical, while Leibniz, Locke, and Hume may be taken as representatives of the scientific wing. In Aristotle, Descartes, Berkeley, and Kant we find both groups of motives strongly present.

Herbert Spencer, in whose honour we are assembled to-day, would naturally be classed among scientific philosophers: it was mainly from science that he drew his data, his formulation of problems, and his conception of method. But his strong religious sense is obvious in much of his writing, and his ethical pre-occupations are what make him value the conception of evolution—that conception in which, as a whole generation has believed, science and morals are to be united in fruitful and indissoluble marriage.

It is my belief that the ethical and religious motives in spite of the splendidly imaginative systems to which they have given rise, have been on the whole a hindrance to the progress of philosophy, and ought now to be consciously thrust aside by those who wish to discover philosophical truth. Science, originally, was entangled in similar motives, and was thereby hindered in its advances. It is, I maintain, from science, rather than from ethics and religion, that philosophy should draw its inspiration.

But there are two different ways in which a philosophy may seek to base itself upon science. It may emphasise the most general *results* of science, and seek to give even greater generality and unity to these results. Or it may study the *methods* of science, and seek to apply these methods, with the necessary adaptations, to its own peculiar province. Much philosophy inspired by science has gone astray through preoccupation with the *results* momentarily supposed to have been achieved. It is not results, but *methods* that can be transferred with profit from the sphere of the special sciences to the sphere of philosophy. What I wish to bring to your notice is the possibility and importance of applying to philosophical problems certain broad principles of method which have been found successful in the study of scientific questions.

The opposition between a philosophy guided by scientific method and a philosophy dominated by religious and ethical ideas may be illustrated by two notions which are very prevalent in the works of philosophers, namely the notion of *the universe*, and the notion of *good and evil*. A philosopher is expected to tell us something about the nature of the universe as a whole, and to give grounds for either optimism or pessimism. Both these expectations seem to me mistaken. I believe the conception of "the universe" to be, as its etymology indicates, a mere relic of pre-Copernican astronomy: and I believe the question of optimism and pessimism to be one which the philosopher will regard as outside his scope, except, possibly, to the extent of maintaining that it is insoluble.

In the days before Copernicus, the conception of the "universe" was defensible on scientific grounds: the diurnal revolution of the heavenly bodies bound them together as all parts of one system, of which the earth was the centre. Round this apparent scientific fact, many human desires rallied: the wish to believe Man important in the scheme of things, the theoretical desire for a comprehensive understanding of the Whole, the hope that the course of nature might be guided by some sympathy with our wishes. In this way, an ethically inspired system of metaphysics grew up, whose anthropocentrism was apparently warranted by the geocentrism of astronomy. When Copernicus swept away the astronomical basis of this system of thought, it had grown so familiar, and had associated itself so intimately with men's aspirations, that it survived with scarcely diminished force—survived even Kant's "Copernican revolution," and is still now the unconscious premiss of most metaphysical systems.

The oneness of the world is an almost undiscussed postulate of most metaphysics. "Reality is not merely one and self-consistent, but is a system of reciprocally determinate parts"—such a statement would pass almost unnoticed as a mere truism. Yet I believe that it embodies a failure to effect thoroughly the "Copernican revolution," and that the apparent oneness of the world is merely the oneness of what is seen by a single spectator or apprehended by a single mind. The Critical Philosophy, although it intended to emphasise the subjective element in many apparent characteristics of the world, yet, by regarding the world in itself as unknowable, so concentrated attention upon the subjective representation that its subjectivity was soon forgotten. Having recognised the categories as the work of the mind, it was paralysed by its own recognition, and abandoned in despair the attempt to undo the work of subjective falsification. In part, no doubt, its despair was well founded, but not, I think, in any absolute or ultimate sense. Still less was it a ground for rejoicing, or for supposing that the nescience to which it ought to have given rise could be legitimately exchanged for a metaphysical dogmatism.

I

As regards our present question, namely, the question of the unity of the world, the right method, as I think, has been indicated by William James."Let us now turn our backs upon ineffable or unintelligible ways of accounting for the world's oneness, and inquire whether, instead of being a principle, the 'oneness' affirmed may not merely be a name like 'substance' descriptive of the fact that certain *specific and verifiable connections* are found among the parts of the experiential flux.... We can easily conceive of things that shall have no connection whatever with each other. We may assume them to inhabit different times and spaces, as the dreams of different persons do even now. They may be so unlike and incommensurable, and so inert towards one another, as never to jostle or interfere. Even now there may actually be whole universes so disparate from ours that we who know ours have no means of perceiving that they exist. We conceive their diversity, however; and by that fact the whole lot of them form what is known in logic as 'a universe of discourse.' To form a universe of discourse argues, as this example shows, no further kind of connexion. The importance attached by certain monistic writers to the fact that any chaos may become a universe by merely being named, is to me incomprehensible." We are thus left with two kinds of unity in the experienced world; the one what we may call the epistemological unity, due merely to the fact that my experienced world is what *one* experience selects from the sum total of existence: the other that tentative and partial unity exhibited in the prevalence of scientific laws in those portions of the world which science has hitherto mastered. Now a generalisation based upon either of these kinds of unity would be fallacious. That the things which we experience have the common property of being experienced by us is a truism from which obviously nothing of importance can be deducible: it is clearly fallacious to draw from the fact that whatever we experience is experienced the conclusion that therefore everything must be experienced. The generalisation of the second kind of unity, namely, that derived from scientific laws, would be equally fallacious, though the fallacy is a trifle less elementary. In order to explain it let us consider for a moment what is called the reign of law. People often speak as though it were a remarkable fact that the physical world is subject to invariable laws. In fact, however, it is not easy to see how such a world could fail to obey general laws. Taking any arbitrary set of points in space, there is a function of the time corresponding to these points, i.e. expressing the motion of a particle which traverses these points: this function may be regarded as a general law to which the behaviour of such a particle is subject. Taking all such functions for all the particles in the universe, there will be theoretically some one formula embracing them all, and this formula may be regarded as the single and supreme law of the spatio-temporal world. Thus what is surprising in physics is not the existence of general laws, but their extreme simplicity. It is not the uniformity of nature that should surprise us, for, by sufficient analytic ingenuity, any conceivable course of nature might be shown to exhibit uniformity. What should surprise us is the fact that the uniformity is simple enough for us to be able to discover it. But it is just this characteristic of simplicity in the laws of nature hitherto discovered which it would be fallacious to generalise, for it is obvious that simplicity has been a part cause of their discovery, and can, therefore, give no ground for the supposition that other undiscovered laws are equally simple.

The fallacies to which these two kinds of unity have given rise suggest a caution as regards all use in philosophy of general *results* that science is supposed to have achieved. In the first place, in generalising these results beyond past experience, it is necessary to examine very carefully whether there is not some reason making it more probable that these results should hold of all that has been experienced than that they should hold of things universally. The sum total of what is experienced by mankind is a selection from the sum total of what exists, and any general character exhibited by this selection may be due to the manner of selecting rather than to the general character of that from which experience selects. In the second place, the most general results of science are the least certain and the most liable to be upset by subsequent research. In utilizing these results as the basis of a philosophy, we sacrifice the most valuable and remarkable characteristic of scientific method,namely, that, although almost everything in science is found sooner or later to require some correction, yet this correction is almost always such as to leave untouched, or only slightly modified, the greater part of the results which have been deduced from the premiss subsequently discovered to be faulty. The prudent man of science acquires a certain instinct as to the kind of uses which may be made of present scientific beliefs without incurring the danger of complete and utter refutation from the modifications likely to be introduced by subsequent discoveries. Unfortunately the use of scientific generalisations of a sweeping kind as the basis of philosophy is just that kind of use which an instinct of scientific caution would avoid, since, as a rule, it would only lead to true results if the generalisation upon which it is based stood in *no* need of correction.

We may illustrate these general considerations by means of two examples, namely, the conservation of energy and the principle of evolution.

(1) Let us begin with the conservation of energy, or, as Herbert Spencer used to call it, the persistence of force. He says:

"Before taking a first step in the rational interpretation of Evolution, it is needful to recognise, not only the facts that Matter is indestructible and Motion continuous, but also the fact that Force persists. An attempt to assign the *causes* of Evolution would manifestly be absurd if that agency to which the metamorphosis in general and in detail is due, could either come into existence or cease to exist. The succession of phenomena would in such case be altogether arbitrary, and deductive Science impossible."

This paragraph illustrates the kind of way in which the philosopher is tempted to give an air of absoluteness and necessity to empirical generalisations, of which only the approximate truth in the regions hitherto investigated can be guaranteed by the unaided methods of science. It is very often said that the persistence of something or other is a necessary presupposition of all scientific investigation, and this presupposition is then thought to be exemplified in some quantity which physics declares to be constant. There are here, as it seems to me, three distinct errors. First, the detailed scientific investigation of nature does not *presuppose* any such general laws as its results are found to verify. Apart from particular observations, science need presuppose nothing except the general principles of logic, and these principles are not laws of nature, for they are merely hypothetical, and apply not only to the actual world but to whatever is *possible*. The second error consists in the identification of a constant quantity with a persistent entity. Energy is a certain function of a physical system, but is not a thing or substance persisting throughout the changes of the system. The same is true of mass, in spite of the fact that mass has often been defined as *quantity of matter*. The whole conception of quantity, involving, as it does, numerical measurement based largely upon conventions, is far more artificial, far more an embodiment of mathematical convenience, than is commonly believed by those who philosophise on physics. Thus even if (which I cannot for a moment admit) the persistence of some entity were among the necessary postulates of science, it would be a sheer error to infer from this the constancy of any physical quantity, or the *a priori* necessity of any such constancy which may be empirically discovered. In the third place, it has become more and more evident with the progress of physics that large generalisations, such as the conservation of energy or mass, are far from certain and are very likely only approximate. Mass, which used to be regarded as the most indubitable of physical quantities, is now generally believed to vary according to velocity, and to be, in fact, a vector quantity which at a given moment is different in different directions. The detailed conclusions deduced from the supposed constancy of mass for such motions as used to be studied in physics will remain very nearly exact, and therefore over the field of the older investigations very little modification of the older results is required. But as soon as such a principle as the conservation of mass or of energy is erected into a universal *a priori* law, the slightest failure in absolute exactness is fatal, and the whole philosophic structure raised upon this foundation is necessarily ruined. The prudent philosopher, therefore, though he may with advantage study the methods of physics, will be very chary of basing anything upon what happen at the moment to be the most general results apparently obtained by those methods.

(2) The philosophy of evolution, which was to be our second example, illustrates the same tendency to hasty generalisation, and also another sort, namely, the undue preoccupation with ethical notions. There are two kinds of evolutionist philosophy, of which both Hegel and Spencer represent the older and less radical kind, while Pragmatism and Bergson represent the more modern and revolutionary variety. But both these sorts of evolutionism have in common the emphasis on *progress*, that is, upon a continual change from the worse to the better, or from the simpler to the more complex. It would be unfair to attribute to Hegel any scientific motive or foundation, but all the other evolutionists, including Hegel's modern disciples, have derived their impetus very largely from the history of biological development. To a philosophy which derives a law of universal progress from this history there are two objections. First, that this history itself is concerned with a very small selection of facts confined to an infinitesimal fragment of space and time, and even on scientific grounds probably not an average sample of events in the world at large. For we know that decay as well as growth is a normal occurrence in the world. An extra-terrestrial philosopher, who had watched a single youth up to the age of twenty-one and had never come across any other human being, might conclude that it is the nature of human beings to grow continually taller and wiser in an indefinite progress towards perfection; and this generalisation would be just as well founded as the generalisation which evolutionists base upon the previous history of this planet. Apart, however, from this scientific objection to evolutionism, there is another, derived from the undue admixture of ethical notions in the very idea of progress from which evolutionism derives its charm. Organic life, we are told, has developed gradually from the protozoon to the philosopher, and this development, we are assured, is indubitably an advance. Unfortunately it is the philosopher, not the protozoon, who gives us this assurance, and we can have no security that the impartial outsider would agree with the philosopher's self-complacent assumption. This point has been illustrated by the philosopher Chuang Tzŭ in the following instructive anecdote:

"The Grand Augur, in his ceremonial robes, approached the shambles and thus addressed the pigs: 'How can you object to die? I shall fatten you for three months. I shall discipline myself for ten days and fast for three. I shall strew fine grass, and place you bodily upon a carved sacrificial dish. Does not this satisfy you?'

Then, speaking from the pigs' point of view, he continued: 'It is better, perhaps, after all, to live on bran and escape the shambles....'

'But then,' added he, speaking from his own point of view, 'to enjoy honour when alive one would readily die on a war-shield or in the headsman's basket.'

So he rejected the pigs' point of view and adopted his own point of view. In what sense, then, was he different from the pigs?"

I much fear that the evolutionists too often resemble the Grand Augur and the pigs.

The ethical element which has been prominent in many of the most famous systems of philosophy is, in my opinion, one of the most serious obstacles to the victory of scientific method in the investigation of philosophical questions. Human ethical notions, as Chuang Tzŭ perceived, are essentially anthropocentric, and involve, when used in metaphysics, an attempt, however veiled, to legislate for the universe on the basis of the present desires of men. In this way they interfere with that receptivity to fact which is the essence of the scientific attitude towards the world. To regard ethical notions as a key to the understanding of the world is essentially pre-Copernican. It is to make man, with the hopes and ideals which he happens to have at the present moment, the centre of the universe and the interpreter of its supposed aims and purposes. Ethical metaphysics is fundamentally an attempt, however disguised, to give legislative force to our own wishes. This may, of course, be questioned, but I think that it is confirmed by a consideration of the way in which ethical notions arise. Ethics is essentially a product of the gregarious instinct, that is to say, of the instinct to co-operate with those who are to form our own group against those who belong to other groups. Those who belong to our own group are good; those who belong to hostile groups are wicked. The ends which are pursued by our own group are desirable ends, the ends pursued by hostile groups are nefarious. The subjectivity of this situation is not apparent to the gregarious animal, which feels that the general principles of justice are on the side of its own herd. When the animal has arrived at the dignity of the metaphysician, it invents ethics as the embodiment of its belief in the justice of its own herd. So the Grand Augur invokes ethics as the justification of Augurs in their conflicts with pigs. But, it may be said, this view of ethics takes no account of such truly ethical notions as that of self-sacrifice. This, however, would be a mistake. The success of gregarious animals in the struggle for existence depends upon co-operation within the herd, and co-operation requires sacrifice, to some extent, of what would otherwise be the interest of the individual. Hence arises a conflict of desires and instincts, since both self-preservation and the preservation of the herd are biological ends to the individual. Ethics is in origin the art of recommending to others the sacrifices required for co-operation with oneself. Hence, by reflexion, it comes, through the operation of social justice, to recommend sacrifices by oneself, but all ethics, however refined, remains more or less subjective. Even vegetarians do not hesitate, for example, to save the life of a man in a fever, although in doing so they destroy the lives of many millions of microbes. The view of the world taken by the philosophy derived from ethical notions is thus never impartial and therefore never fully scientific. As compared with science, it fails to achieve the imaginative liberation from self which is necessary to such understanding of the world as man can hope to achieve, and the philosophy which it inspires is always more or less parochial, more or less infected with the prejudices of a time and a place.

I do not deny the importance or value, within its own sphere, of the kind of philosophy which is inspired by ethical notions. The ethical work of Spinoza, for example, appears to me of the very highest significance, but what is valuable in such work is not any metaphysical theory as to the nature of the world to which it may give rise, nor indeed anything which can be proved or disproved by argument. What is valuable is the indication of some new way of feeling towards life and the world, some way of feeling by which our own existence can acquire more of the characteristics which we must deeply desire. The value of such work, however immeasurable it is, belongs with practice and not with theory. Such theoretic importance as it may possess is only in relation to human nature, not in relation to the world at large. The scientific philosophy, therefore, which aims only at understanding the world and not directly at any other improvement of human life, cannot take account of ethical notions without being turned aside from that submission to fact which is the essence of the scientific temper.

II

If the notion of the universe and the notion of good and evil are extruded from scientific philosophy, it may be asked what specific problems remain for the philosopher as opposed to the man of science? It would be difficult to give a precise answer to this question, but certain characteristics may be noted as distinguishing the province of philosophy from that of the special sciences.

In the first place a philosophical proposition must be general. It must not deal specially with things on the surface of the earth, or with the solar system, or with any other portion of space and time. It is this need of generality which has led to the belief that philosophy deals with the universe as a whole. I do not believe that this belief is justified, but I do believe that a philosophical proposition must be applicable to everything that exists or may exist. It might be supposed that this admission would be scarcely distinguishable from the view which I wish to reject. This, however, would be an error, and an important one. The traditional view would make the universe itself the subject of various predicates which could not be applied to any particular thing in the universe, and the ascription of such peculiar predicates to the universe would be the special business of philosophy. I maintain, on the contrary, that there are no propositions of which the "universe" is the subject; in other words, that there is no such thing as the "universe." What I do maintain is that there are general propositions which may be asserted of each individual thing, such as the propositions of logic. This does not involve that all the things there are form a whole which could be regarded as another thing and be made the subject of predicates. It involves only the assertion that there are properties which belong to each separate thing, not that there are properties belonging to the whole of things collectively. The philosophy which I wish to advocate may be called logical atomism or absolute pluralism, because, while maintaining that there are many things, it denies that there is a whole composed of those things. We shall see, therefore, that philosophical propositions, instead of being concerned with the whole of things collectively, are concerned with all things distributively; and not only must they be concerned with all things, but they must be concerned with such properties of all things as do not depend upon the accidental nature of the things that there happen to be, but are true of any possible world, independently of such facts as can only be discovered by our senses.

This brings us to a second characteristic of philosophical propositions, namely, that they must be *a priori*. A philosophical proposition must be such as can be neither proved nor disproved by empirical evidence. Too often we find in philosophical books arguments based upon the course of history, or the convolutions of the brain, or the eyes of shell-fish. Special and accidental facts of this kind are irrelevant to philosophy, which must make only such assertions as would be equally true however the actual world were constituted.

We may sum up these two characteristics of philosophical propositions by saying that *philosophy is the science of the possible*. But this statement unexplained is liable to be misleading, since it may be thought that the possible is something other than the general, whereas in fact the two are indistinguishable.

Philosophy, if what has been said is correct, becomes indistinguishable from logic as that word has now come to be used. The study of logic consists, broadly speaking, of two not very sharply distinguished portions. On the one hand it is concerned with those general statements which can be made concerning everything without mentioning any one thing or predicate or relation, such for example as "if x is a member of the class α and every member of α is a member of β, then x is a member of the class β, whatever x, α, and β may be." On the other hand, it is concerned with the analysis and enumeration of logical *forms*, i.e. with the kinds of propositions that may occur, with the various types of facts, and with the classification of the constituents of facts. In this way logic provides an inventory of possibilities, a repertory of abstractly tenable hypotheses.

It might be thought that such a study would be too vague and too general to be of any very great importance, and that, if its problems became at any point sufficiently definite, they would be merged in the problems of some special science. It appears, however, that this is not the case. In some problems, for example, the analysis of space and time, the nature of perception, or the theory of judgment, the discovery of the logical form of the facts involved is the hardest part of the work and the part whose performance has been most lacking hitherto. It is chiefly for want of the right logical hypothesis that such problems have hitherto been treated in such an unsatisfactory manner, and have given rise to those contradictions or antinomies in which the enemies of reason among philosophers have at all times delighted.

By concentrating attention upon the investigation of logical forms, it becomes possible at last for philosophy to deal with its problems piecemeal, and to obtain, as the sciences do, such partial and probably not wholly correct results as subsequent investigation can utilise even while it supplements and improves them. Most philosophies hitherto have been constructed all in one block, in such a way that, if they were not wholly correct, they were wholly incorrect, and could not be used as a basis for further investigations. It is chiefly owing to this fact that philosophy, unlike science, has hitherto been unprogressive, because each original philosopher has had to begin the work again from the beginning, without being able to accept anything definite from the work of his predecessors. A scientific philosophy such as I wish to recommend will be piecemeal and tentative like other sciences; above all, it will be able to invent hypotheses which, even if they are not wholly true, will yet remain fruitful after the necessary corrections have been made. This possibility of successive approximations to the truth is, more than anything else, the source of the triumphs of science, and to transfer this possibility to philosophy is to ensure a progress in method whose importance it would be almost impossible to exaggerate.

The essence of philosophy as thus conceived is analysis, not synthesis. To build up systems of the world, like Heine's German professor who knit together fragments of life and made an intelligible system out of them, is not, I believe, any more feasible than the discovery of the philosopher's stone. What is feasible is the understanding of general forms, and the division of traditional problems into a number of separate and less baffling questions. "Divide and conquer" is the maxim of success here as elsewhere.

Let us illustrate these somewhat general maxims by examining their application to the philosophy of space, for it is only in application that the meaning or importance of a method can be understood. Suppose we are confronted with the problem of space as presented in Kant's Transcendental Æsthetic, and suppose we wish to discover what are the elements of the problem and what hope there is of obtaining a solution of them. It will soon appear that three entirely distinct problems, belonging to different studies, and requiring different methods for their solution, have been confusedly combined in the supposed single problem with which Kant is concerned. There is a problem of logic, a problem of physics, and a problem of theory of knowledge. Of these three, the problem of logic can be solved exactly and perfectly; the problem of physics can probably be solved with as great a degree of certainty and as great an approach to exactness as can be hoped in an empirical region; the problem of theory of knowledge, however, remains very obscure and very difficult to deal with. Let us see how these three problems arise.

(1) The logical problem has arisen through the suggestions of non-Euclidean geometry. Given a body of geometrical propositions, it is not difficult to find a minimum statement of the axioms from which this body of propositions can be deduced. It is also not difficult, by dropping or altering some of these axioms, to obtain a more general or a different geometry, having, from the point of view of pure mathematics, the same logical coherence and the same title to respect as the more familiar Euclidean geometry. The Euclidean geometry itself is true perhaps of actual space (though this is doubtful), but certainly of an infinite number of purely arithmetical systems, each of which, from the point of view of abstract logic, has an equal and indefeasible right to be called a Euclidean space. Thus space as an object of logical or mathematical study loses its uniqueness; not only are there many kinds of spaces, but there are an infinity of examples of each kind, though it is difficult to find any kind of which the space of physics may be an example, and it is impossible to find any kind of which the space of physics is certainly an example. As an illustration of one possible logical system of geometry we may consider all relations of three terms which are analogous in certain formal respects to the relation "between" as it appears to be in actual

space. A space is then defined by means of one such three-term relation. The points of the space are all the terms which have this relation to something or other, and their order in the space in question is determined by this relation. The points of one space are necessarily also points of other spaces, since there are necessarily other three-term relations having those same points for their field. The space in fact is not determined by the class of its points, but by the ordering three-term relation. When enough abstract logical properties of such relations have been enumerated to determine the resulting kind of geometry, say, for example, Euclidean geometry, it becomes unnecessary for the pure geometer in his abstract capacity to distinguish between the various relations which have all these properties. He considers the whole class of such relations, not any single one among them. Thus in studying a given kind of geometry the pure mathematician is studying a certain class of relations defined by means of certain abstract logical properties which take the place of what used to be called axioms. The nature of geometrical *reasoning* therefore is purely deductive and purely logical; if any special epistemological peculiarities are to be found in geometry, it must not be in the reasoning, but in our knowledge concerning the axioms in some given space.

(2) The physical problem of space is both more interesting and more difficult than the logical problem.The physical problem may be stated as follows: to find in the physical world, or to construct from physical materials, a space of one of the kinds enumerated by the logical treatment of geometry. This problem derives its difficulty from the attempt to accommodate to the roughness and vagueness of the real world some system possessing the logical clearness and exactitude of pure mathematics. That this can be done with a certain degree of approximation is fairly evident If I see three people *A*, *B*, and *C* sitting in a row, I become aware of the fact which may be expressed by saying that *B* is between *A* and *C* rather than that *A* is between *B* and *C*, or *C* is between *A* and *B*. This relation of "between" which is thus perceived to hold has some of the abstract logical properties of those three-term relations which, we saw, give rise to a geometry, but its properties fail to be exact, and are not, as empirically given, amenable to the kind of treatment at which geometry aims. In abstract geometry we deal with points, straight lines, and planes; but the three people *A*, *B*, and *C* whom I see sitting in a row are not exactly points, nor is the row exactly a straight line. Nevertheless physics, which formally assumes a space containing points, straight lines, and planes, is found empirically to give results applicable to the sensible world. It must therefore be possible to find an interpretation of the points, straight lines, and planes of physics in terms of physical data, or at any rate in terms of data together with such hypothetical additions as seem least open to question. Since all data suffer from a lack of mathematical precision through being of a certain size and somewhat vague in outline, it is plain that if such a notion as that of a point is to find any application to empirical material, the point must be

neither a datum nor a hypothetical addition to data, but a *construction* by means of data with their hypothetical additions. It is obvious that any hypothetical filling out of data is less dubious and unsatisfactory when the additions are closely analogous to data than when they are of a radically different sort. To assume, for example, that objects which we see continue, after we have turned away our eyes, to be more or less analogous to what they were while we were looking, is a less violent assumption than to assume that such objects are composed of an infinite number of mathematical points. Hence in the physical study of the geometry of physical space, points must not be assumed *ab initio* as they are in the logical treatment of geometry, but must be constructed as systems composed of data and hypothetical analogues of data. We are thus led naturally to define a physical point as a certain class of those objects which are the ultimate constituents of the physical world. It will be the class of all those objects which, as one would naturally say, *contain* the point. To secure a definition giving this result, without previously assuming that physical objects are composed of points, is an agreeable problem in mathematical logic. The solution of this problem and the perception of its importance are due to my friend Dr. Whitehead. The oddity of regarding a point as a class of physical entities wears off with familiarity, and ought in any case not to be felt by those who maintain, as practically every one does, that points are mathematical fictions. The word "fiction" is used glibly in such connexions by many men who seem not to feel the necessity of explaining how it can come about that a fiction can be so useful in the study of the actual world as the points of mathematical physics have been found to be. By our definition, which regards a point as a class of physical objects, it is explained both how the use of points can lead to important physical results, and how we can nevertheless avoid the assumption that points are themselves entities in the physical world.

Many of the mathematically convenient properties of abstract logical spaces cannot be either known to belong or known not to belong to the space of physics. Such are all the properties connected with continuity. For to know that actual space has these properties would require an infinite exactness of sense-perception. If actual space is continuous, there are nevertheless many possible non-continuous spaces which will be empirically indistinguishable from it; and, conversely, actual space may be non-continuous and yet empirically indistinguishable from a possible continuous space. Continuity, therefore, though obtainable in the *a priori* region of arithmetic, is not with certainty obtainable in the space or time of the physical world: whether these are continuous or not would seem to be a question not only unanswered but for ever unanswerable. From the point of view of philosophy, however, the discovery that a question is unanswerable is as complete an answer as any that could possibly be obtained. And from the point of view of physics, where no empirical means of distinction can be found, there can be no empirical objection to the mathematically simplest assumption, which is that of continuity.

The subject of the physical theory of space is a very large one, hitherto little explored. It is associated with a similar theory of time, and both have been forced upon the attention of philosophically minded physicists by the discussions which have raged concerning the theory of relativity.

(3) The problem with which Kant is concerned in the Transcendental Æsthetic is primarily the epistemological problem: "How do we come to have knowledge of geometry *a priori*?" By the distinction between the logical and physical problems of geometry, the bearing and scope of this question are greatly altered. Our knowledge of pure geometry is *a priori* but is wholly logical. Our knowledge of physical geometry is synthetic, but is not *a priori*. Our knowledge of pure geometry is hypothetical, and does not enable us to assert, for example, that the axiom of parallels is true in the physical world. Our knowledge of physical geometry, while it does enable us to assert that this axiom is approximately verified, does not, owing to the inevitable inexactitude of observation, enable us to assert that it is verified *exactly*. Thus, with the separation which we have made between pure geometry and the geometry of physics, the Kantian problem collapses. To the question, "How is synthetic *a priori* knowledge possible?" we can now reply, at any rate so far as geometry is concerned, "It is not possible," if "synthetic" means "not deducible from logic alone." Our knowledge of geometry, like the rest of our knowledge, is derived partly from logic, partly from sense, and the peculiar position which in Kant's day geometry appeared to occupy is seen now to be a delusion. There are still some philosophers, it is true, who maintain that our knowledge that the axiom of parallels, for example, is true of actual space, is not to be accounted for empirically, but is as Kant maintained derived from an *a priori* intuition. This position is not logically refutable, but I think it loses all plausibility as soon as we realise how complicated and derivative is the notion of physical space. As we have seen, the application of geometry to the physical world in no way demands that there should really be points and straight lines among physical entities. The principle of economy, therefore, demands that we should abstain from assuming the existence of points and straight lines. As soon, however, as we accept the view that points and straight lines are complicated constructions by means of classes of physical entities, the hypothesis that we have an *a priori* intuition enabling us to know what happens to straight lines when they are produced indefinitely becomes extremely strained and harsh; nor do I think that such an hypothesis would ever have arisen in the mind of a philosopher who had grasped the nature of physical space. Kant, under the influence of Newton, adopted, though with some vacillation, the hypothesis of absolute space, and this hypothesis, though logically unobjectionable, is removed by Occam's razor, since absolute space is an unnecessary entity in the explanation of the physical world. Although, therefore, we cannot refute the Kantian theory of an *a priori* intuition, we can remove its grounds one by one through an analysis of the problem. Thus, here as in many other philosophical

questions, the analytic method, while not capable of arriving at a demonstrative result, is nevertheless capable of showing that all the positive grounds in favour of a certain theory are fallacious and that a less unnatural theory is capable of accounting for the facts.

Another question by which the capacity of the analytic method can be shown is the question of realism. Both those who advocate and those who combat realism seem to me to be far from clear as to the nature of the problem which they are discussing. If we ask: "Are our objects of perception *real* and are they *independent* of the percipient?" it must be supposed that we attach some meaning to the words "real" and "independent," and yet, if either side in the controversy of realism is asked to define these two words, their answer is pretty sure to embody confusions such as logical analysis will reveal.

Let us begin with the word "real." There certainly are objects of perception, and therefore, if the question whether these objects are real is to be a substantial question, there must be in the world two sorts of objects, namely, the real and the unreal, and yet the unreal is supposed to be essentially what there is not. The question what properties must belong to an object in order to make it real is one to which an adequate answer is seldom if ever forthcoming. There is of course the Hegelian answer, that the real is the self-consistent and that nothing is self-consistent except the Whole; but this answer, true or false, is not relevant in our present discussion, which moves on a lower plane and is concerned with the status of objects of perception among other objects of equal fragmentariness. Objects of perception are contrasted, in the discussions concerning realism, rather with psychical states on the one hand and matter on the other hand than with the all-inclusive whole of things. The question we have therefore to consider is the question as to what can be meant by assigning "reality" to some but not all of the entities that make up the world. Two elements, I think, make up what is felt rather than thought when the word "reality" is used in this sense. A thing is real if it persists at times when it is not perceived; or again, a thing is real when it is correlated with other things in a way which experience has led us to expect. It will be seen that reality in either of these senses is by no means necessary to a thing, and that in fact there might be a whole world in which nothing was real in either of these senses. It might turn out that the objects of perception failed of reality in one or both of these respects, without its being in any way deducible that they are not parts of the external world with which physics deals. Similar remarks will apply to the word "independent." Most of the associations of this word are bound up with ideas as to causation which it is not now possible to maintain. *A* is independent of *B* when *B* is not an indispensable part of the *cause* of *A*. But when it is recognised that causation is nothing more than correlation, and that there are correlations of simultaneity as well as of succession, it becomes evident that there is no uniqueness in a series of casual antecedents of a given event, but that, at any point where there is a correlation of simultaneity, we can pass from one line of antecedents to another in order to obtain a new series of causal antecedents. It will be necessary to specify the causal law according to which the antecedents are to be considered. I received a letter the other day from a correspondent who had been puzzled by various philosophical questions. After enumerating them he says: "These questions led me from Bonn to Strassburg, where I found Professor Simmel." Now, it would be absurd to deny that these questions caused his body to move from Bonn to Strassburg, and yet it must be supposed that a set of purely mechanical antecedents could also be found which would account for this transfer of matter from one place to another. Owing to this plurality of causal series antecedent to a given event, the notion of *the* cause becomes indefinite, and the question of independence becomes correspondingly ambiguous. Thus, instead of asking simply whether *A* is independent of *B*, we ought to ask whether there is a series determined by such and such causal laws leading from *B* to *A*. This point is important in connexion with the particular question of objects of perception. It may be that no objects quite like those which we perceive ever exist unperceived;in this case there will be a causal law according to which objects of perception are not independent of being perceived. But even if this be the case, it may nevertheless also happen that there are purely physical causal laws determining the occurrence of objects which are perceived by means of other objects which perhaps are not perceived. In that case, in regard to such causal laws objects of perception will be independent of being perceived. Thus the question whether objects of perception are independent of being perceived is, as it stands, indeterminate, and the answer will be yes or no according to the method adopted of making it determinate. I believe that this confusion has borne a very large part in prolonging the controversies on this subject, which might well have seemed capable of remaining for ever undecided. The view which I should wish to advocate is that objects of perception do not persist unchanged at times when they are not perceived, although probably objects more or less resembling them do exist at such times; that objects of perception are part, and the only empirically knowable part, of the actual subject-matter of physics, and are themselves properly to be called physical; that purely physical laws exist determining the character and duration of objects of perception without any reference to the fact that they are perceived; and that in the establishment of such laws the propositions of physics do not presuppose any propositions of psychology or even the existence of mind. I do not know whether realists would recognise such a view as realism. All that I should claim for it is, that it avoids difficulties which seem to me to beset both realism and idealism as hitherto advocated, and that it avoids the appeal which they have made to ideas which logical analysis shows to be ambiguous. A further defence and elaboration of the positions which I advocate, but for which time is lacking now, will be found indicated in my book on *Our Knowledge of the External World*.

The adoption of scientific method in philosophy, if I am not mistaken, compels us to abandon the hope of solving many of the more ambitious and humanly interesting problems of traditional philosophy. Some of these it relegates, though with little expectation of a successful solution, to special sciences, others it shows to be such as our capacities are essentially incapable of solving. But there remain a large number of the recognised problems of philosophy in regard to which the method advocated gives all those advantages of division into distinct questions, of tentative, partial, and progressive advance, and of appeal to principles with which, independently of temperament, all competent students must agree. The failure of philosophy hitherto has been due in the main to haste and ambition: patience and modesty, here as in other sciences, will open the road to solid and durable progress.

VII

THE ULTIMATE CONSTITUENTS OF MATTER

I wish to discuss in this article no less a question than the ancient metaphysical query, "What is matter?" The question, "What is matter?" in so far as it concerns philosophy, is, I think, already capable of an answer which in principle will be as complete as an answer can hope to be; that is to say, we can separate the problem into an essentially soluble and an essentially insoluble portion, and we can now see how to solve the essentially soluble portion, at least as regards its main outlines. It is these outlines which I wish to suggest in the present article. My main position, which is realistic, is, I hope and believe, not remote from that of Professor Alexander, by whose writings on this subject I have profited greatly. It is also in close accord with that of Dr. Nunn.

Common sense is accustomed to the division of the world into mind and matter. It is supposed by all who have never studied philosophy that the distinction between mind and matter is perfectly clear and easy, that the two do not at any point overlap, and that only a fool or a philosopher could be in doubt as to whether any given entity is mental or material. This simple faith survives in Descartes and in a somewhat modified form in Spinoza, but with Leibniz it begins to disappear, and from his day to our own almost every philosopher of note has criticised and rejected the dualism of common sense. It is my intention in this article to defend this dualism; but before defending it we must spend a few moments on the reasons which have prompted its rejection.

Our knowledge of the material world is obtained by means of the senses, of sight and touch and so on. At first it is supposed that things are just as they seem, but two opposite sophistications soon destroy this naïve belief. On the one hand the physicists cut up matter into molecules, atoms, corpuscles, and as many more such subdivisions as their future needs may make them postulate, and the units at which they arrive are uncommonly different from the visible, tangible objects of daily life. A unit of matter tends more and more to be something like an electromagnetic field filling all space, though having its greatest intensity in a small region. Matter consisting of such elements is as remote from daily life as any metaphysical theory. It differs from the theories of metaphysicians only in the fact that its practical efficacy proves that it contains some measure of truth and induces business men to invest money on the strength of it; but, in spite of its connection with the money market, it remains a metaphysical theory none the less.

The second kind of sophistication to which the world of common sense has been subjected is derived from the psychologists and physiologists. The physiologists point out that what we see depends upon the eye, that what we hear depends upon the ear, and that all our senses are liable to be affected by anything which affects the brain, like alcohol or hasheesh. Psychologists point out how much of what we think we see is supplied by association or unconscious inference, how much is mental interpretation, and how doubtful is the residuum which can be regarded as crude datum. From these facts it is argued by the psychologists that the notion of a datum passively received by the mind is a delusion, and it is argued by the physiologists that even if a pure datum of sense could be obtained by the analysis of experience, still this datum could not belong, as common sense supposes, to the outer world, since its whole nature is conditioned by our nerves and sense organs, changing as they change in ways which it is thought impossible to connect with any change in the matter supposed to be perceived. This physiologist's argument is exposed to the rejoinder, more specious than solid, that our knowledge of the existence of the sense organs and nerves is obtained by that very process which the physiologist has been engaged in discrediting, since the existence of the nerves and sense organs is only known through the evidence of the senses themselves. This argument may prove that some reinterpretation of the results of physiology is necessary before they can acquire metaphysical validity. But it does not upset the physiological argument in so far as this constitutes merely a *reductio ad absurdum* of naïve realism.

These various lines of argument prove, I think, that some part of the beliefs of common sense must be abandoned. They prove that, if we take these beliefs as a whole, we are forced into conclusions which are in part self-contradictory; but such arguments cannot of themselves decide what portion of our common-sense beliefs is in need of correction. Common sense believes that what we see is physical, outside the mind, and continuing to exist if we shut our eyes or turn them in another direction. I believe that common sense is right in regarding what we see as physical and (in one of several possible senses) outside the mind, but is probably wrong in supposing that it continues to exist when we are no longer looking at it. It seems to me that the whole discussion of matter has been obscured by two errors which support each other. The first of these is the error that what we see, or perceive through any of our other senses, is subjective: the second is the belief that what is physical must be persistent. Whatever physics may regard as the ultimate constituents of matter, it always supposes these constituents to be indestructible. Since the immediate data of sense are not indestructible but in a state of perpetual flux, it is argued that these data themselves cannot be among the ultimate constituents of matter. I believe this to be a sheer mistake. The persistent particles of mathematical physics I regard as logical constructions, symbolic fictions enabling us to express compendiously very complicated assemblages of facts; and, on the other hand, I believe that the actual data in sensation, the immediate objects of sight or touch or hearing, are extra-mental, purely physical, and among the ultimate constituents of matter.

My meaning in regard to the impermanence of physical entities may perhaps be made clearer by the use of Bergson's favourite illustration of the cinematograph. When I first read Bergson's statement that the mathematician conceives the world after the analogy of a cinematograph, I had never seen a cinematograph, and my first visit to one was determined by the desire to verify Bergson's statement, which I found to be completely true, at least so far as I am concerned. When, in a picture palace, we see a man rolling down hill, or running away from the police, or falling into a river, or doing any of those other things to which men in such places are addicted, we know that there is not really only one man moving, but a succession of films, each with a different momentary man. The illusion of persistence arises only through the approach to continuity in the series of momentary men. Now what I wish to suggest is that in this respect the cinema is a better metaphysician than common sense, physics, or philosophy. The real man too, I believe, however the police may swear to his identity, is really a series of momentary men, each different one from the other, and bound together, not by a numerical identity, but by continuity and certain intrinsic causal laws. And what applies to men applies equally to tables and chairs, the sun, moon and stars. Each of these is to be regarded, not as one single persistent entity, but as a series of entities succeeding each other in time, each lasting for a very brief period, though probably not for a mere mathematical instant. In saying this I am only urging the same kind of division in time as we are accustomed to acknowledge in the case of space. A body which fills a cubic foot will be admitted to consist of many smaller bodies, each occupying only a very tiny volume; similarly a thing which persists for an hour is to be regarded as composed of many things of less duration. A true theory of matter requires a division of things into time-corpuscles as well as into space-corpuscles.

The world may be conceived as consisting of a multitude of entities arranged in a certain pattern. The entities which are arranged I shall call "particulars." The arrangement or pattern results from relations among particulars. Classes or series of particulars, collected together on account of some property which makes it convenient to be able to speak of them as wholes, are what I call logical constructions or symbolic fictions. The particulars are to be conceived, not on the analogy of bricks in a building, but rather on the analogy of notes in a symphony. The ultimate constituents of a symphony (apart from relations) are the notes, each of which lasts only for a very short time. We may collect together all the notes played by one instrument: these may be regarded as the analogues of the successive particulars which common sense would regard as successive states of one "thing." But the "thing" ought to be regarded as no more "real" or "substantial" than, for example, the rôle of the trombone. As soon as "things" are conceived in this manner it will be found that the difficulties in the way of regarding immediate objects of sense as physical have largely disappeared.

When people ask, "Is the object of sense mental or physical?" they seldom have any clear idea either what is meant by "mental" or "physical," or what criteria are to be applied for deciding whether a given entity belongs to one class or the other. I do not know how to give a sharp definition of the word "mental," but something may be done by enumerating occurrences which are indubitably mental: believing, doubting, wishing, willing, being pleased or pained, are certainly mental occurrences; so are what we may call experiences, seeing, hearing, smelling, perceiving generally. But it does not follow from this that what is seen, what is heard, what is smelt, what is perceived, must be mental. When I see a flash of lightning, my seeing of it is mental, but what I see, although it is not quite the same as what anybody else sees at the same moment, and although it seems very unlike what the physicist would describe as a flash of lightning, is not mental. I maintain, in fact, that if the physicist could describe truly and fully all that occurs in the physical world when there is a flash of lightning, it would contain as a constituent what I see, and also what is seen by anybody else who would commonly be said to see the same flash. What I mean may perhaps be made plainer by saying that if my body could remain in exactly the same state in which it is, although my mind had ceased to exist, precisely that object which I now see when I see the flash would exist, although of course I should not see it, since my seeing is mental. The principal reasons which have led people to reject this view have, I think, been two: first, that they did not adequately distinguish between my seeing and what I see; secondly, that the causal dependence of what I see upon my body has made people suppose that what I see cannot be "outside" me. The first of these reasons need not detain us, since the confusion only needs to be pointed out in order to be obviated; but the second requires some discussion, since it can only be answered by removing current misconceptions, on the one hand as to the nature of space, and on the other, as to the meaning of causal dependence.

When people ask whether colours, for example, or other secondary qualities are inside or outside the mind, they seem to suppose that their meaning must be clear, and that it ought to be possible to say yes or no without any further discussion of the terms involved. In fact, however, such terms as "inside" or "outside" are very ambiguous. What is meant by asking whether this or that is "in" the mind? The mind is not like a bag or a pie; it does not occupy a certain region in space, or, if (in a sense) it does, what is in that region is presumably part of the brain, which would not be said to be in the mind. When people say that sensible qualities are in the mind, they do not mean "spatially contained in" in the sense in which the blackbirds were in the pie. We might regard the mind as an assemblage of particulars, namely, what would be called "states of mind," which would belong together in virtue of some specific common quality. The common quality of all states of mind would be the quality designated by the word "mental"; and besides this we should have to suppose that each separate person's states of mind have some common characteristic distinguishing them from the states of mind of other people. Ignoring this latter point, let us ask ourselves whether the quality designated by the word "mental" does, as a matter of observation, actually belong to objects of sense, such as colours or noises. I think any candid person must reply that, however difficult it may be to know what we mean by "mental," it is not difficult to see that colours and noises are not mental in the sense of having that intrinsic peculiarity which belongs to beliefs and wishes and volitions, but not to the physical world. Berkeley advances on this subject a plausible argument which seems to me to rest upon an ambiguity in the word "pain." He argues that the realist supposes the heat which he feels in approaching a fire to be something outside his mind, but that as he approaches nearer and nearer to the fire the sensation of heat passes imperceptibly into pain, and that no one could regard pain as something outside the mind. In reply to this argument, it should be observed in the first place that the heat of which we are immediately aware is not in the fire but in our own body. It is only by inference that the fire is judged to be the cause of the heat which we feel in our body. In the second place (and this is the more important point), when we speak of pain we may mean one of two things: we may mean the object of the sensation or other experience which has the quality of being painful,or we may mean the quality of painfulness itself. When a man says he has a pain in his great toe, what he means is that he has a sensation associated with his great toe and having the quality of painfulness. The sensation itself, like every sensation, consists in experiencing a sensible object, and the experiencing has that quality of painfulness which only mental occurrences can have, but which may belong to thoughts or desires, as well as to sensations. But in common language we speak of the sensible object experienced in a painful sensation as a pain, and it is this way of speaking which causes the confusion upon which the plausibility of Berkeley's argument depends. It would be absurd to attribute the quality of painfulness to anything non-mental, and hence it comes to be thought that what we call a pain in the toe must be mental. In fact, however, it is not the sensible object in such a case which is painful, but the sensation, that is to say, the experience of the sensible object. As the heat which we experience from the fire grows greater, the experience passes gradually from being pleasant to being painful, but neither the pleasure nor the pain is a quality of the object experienced as opposed to the experience, and it is therefore a fallacy to argue that this object must be mental on the ground that painfulness can only be attributed to what is mental.

If, then, when we say that something is in the mind we mean that it has a certain recognisable intrinsic characteristic such as belongs to thoughts and desires, it must be maintained on grounds of immediate inspection that objects of sense are not in any mind.

A different meaning of "in the mind" is, however, to be inferred from the arguments advanced by those who regard sensible objects as being in the mind. The arguments used are, in the main, such as would prove the causal dependence of objects of sense upon the percipient. Now the notion of causal dependence is very obscure and difficult, much more so in fact than is generally realised by philosophers. I shall return to this point in a moment. For the present, however, accepting the notion of causal dependence without criticism, I wish to urge that the dependence in question is rather upon our bodies than upon our minds. The visual appearance of an object is altered if we shut one eye, or squint, or look previously at something dazzling; but all these are bodily acts, and the alterations which they effect are to be explained by physiology and optics, not by psychology.They are in fact of exactly the same kind as the alterations effected by spectacles or a microscope. They belong therefore to the theory of the physical world, and can have no bearing upon the question whether what we see is causally dependent upon the mind. What they do tend to prove, and what I for my part have no wish to deny, is that what we see is causally dependent upon our body and is not, as crude common sense would suppose, something which would exist equally if our eyes and nerves and brain were absent, any more than the visual appearance presented by an object seen through a microscope would remain if the microscope were removed. So long as it is supposed that the physical world is composed of stable and more or less permanent constituents, the fact that what we see is changed by changes in our body appears to afford reason for regarding what we see as not an ultimate constituent of matter. But if it is recognised that the ultimate constituents of matter are as circumscribed in duration as in spatial extent, the whole of this difficulty vanishes.

There remains, however, another difficulty, connected with space. When we look at the sun we wish to know something about the sun itself, which is ninety-three million miles away; but what we see is dependent upon our eyes, and it is difficult to suppose that our eyes can affect what happens at a distance of ninety-three million miles. Physics tells us that certain electromagnetic waves start from the sun, and reach our eyes after about eight minutes. They there produce disturbances in the rods and cones, thence in the optic nerve, thence in the brain. At the end of this purely physical series, by some odd miracle, comes the experience which we call "seeing the sun," and it is such experiences which form the whole and sole reason for our belief in the optic nerve, the rods and cones, the ninety-three million miles, the electromagnetic waves, and the sun itself. It is this curious oppositeness of direction between the order of causation as affirmed by physics, and the order of evidence as revealed by theory of knowledge, that causes the most serious perplexities in regard to the nature of physical reality. Anything that invalidates our seeing, as a source of knowledge concerning physical reality, invalidates also the whole of physics and physiology. And yet, starting from a common-sense acceptance of our seeing, physics has been led step by step to the construction of the causal chain in which our seeing is the last link, and the immediate object which we see cannot be regarded as that initial cause which we believe to be ninety-three million miles away, and which we are inclined to regard as the "real" sun.

I have stated this difficulty as forcibly as I can, because I believe that it can only be answered by a radical analysis and reconstruction of all the conceptions upon whose employment it depends.

Space, time, matter and cause, are the chief of these conceptions. Let us begin with the conception of cause.

Causal dependence, as I observed a moment ago, is a conception which it is very dangerous to accept at its face value. There exists a notion that in regard to any event there is something which may be called *the* cause of that event—some one definite occurrence, without which the event would have been impossible and with which it becomes necessary. An event is supposed to be dependent upon its cause in some way which in it is not dependent upon other things. Thus men will urge that the mind is dependent upon the brain, or, with equal plausibility, that the brain is dependent upon the mind. It seems not improbable that if we had sufficient knowledge we could infer the state of a man's mind from the state of his brain, or the state of his brain from the state of his mind. So long as the usual conception of causal dependence is retained, this state of affairs can be used by the materialist to urge that the state of our brain causes our thoughts, and by the idealist to urge that our thoughts cause the state of our brain. Either contention is equally valid or equally invalid. The fact seems to be that there are many correlations of the sort which may be called causal, and that, for example, either a physical or a mental event can be predicted, theoretically, either from a sufficient number of physical antecedents or from a sufficient number of mental antecedents. To speak of *the* cause of an event is therefore misleading. Any set of antecedents from which the event can theoretically be inferred by means of correlations might be called a cause of the event. But to speak of *the* cause is to imply a uniqueness which does not exist.

The relevance of this to the experience which we call "seeing the sun" is obvious. The fact that there exists a chain of antecedents which makes our seeing dependent upon the eyes and nerves and brain does not even tend to show that there is not another chain of antecedents in which the eyes and nerves and brain as physical things are ignored. If we are to escape from the dilemma which seemed to arise out of the physiological causation of what we see when we say we see the sun, we must find, at least in theory, a way of stating causal laws for the physical world, in which the units are not material things, such as the eyes and nerves and brain, but momentary particulars of the same sort as our momentary visual object when we look at the sun. The sun itself and the eyes and nerves and brain must be regarded as assemblages of momentary particulars. Instead of supposing, as we naturally do when we start from an uncritical acceptance of the apparent dicta of physics, that *matter* is what is "really real" in the physical world, and that the immediate objects of sense are mere phantasms, we must regard matter as a logical construction, of which the constituents will be just such evanescent particulars as may, when an observer happens to be present, become data of sense to that observer. What physics regards as the sun of eight minutes ago will be a whole assemblage of particulars, existing at different times, spreading out from a centre with the velocity of light, and containing among their number all those visual data which are seen by people who are now looking at the sun. Thus the sun of eight minutes ago is a class of particulars, and what I see when I now look at the sun is one member of this class. The various particulars constituting this class will be correlated with each other by a certain continuity and certain intrinsic laws of variation as we pass outwards from the centre, together with certain modifications correlated extrinsically with other particulars which are not members of this class. It is these extrinsic modifications which represent the sort of facts that, in our former account, appeared as the influence of the eyes and nerves in modifying the appearance of the sun.

The *prima facie* difficulties in the way of this view are chiefly derived from an unduly conventional theory of space. It might seem at first sight as if we had packed the world much fuller than it could possibly hold. At every place between us and the sun, we said, there is to be a particular which is to be a member of the sun as it was a few minutes ago. There will also, of course, have to be a particular which is a member of any planet or fixed star that may happen to be visible from that place. At the place where I am, there will be particulars which will be members severally of all the "things" I am now said to be perceiving. Thus throughout the world, everywhere, there will be an enormous number of particulars coexisting in the same place. But these troubles result from contenting ourselves too readily with the merely three-dimensional space to which schoolmasters have accustomed us. The space of the real world is a space of six dimensions, and as soon as we realise this we see that there is plenty of room for all the particulars for which we want to find positions. In order to realise this we have only to return for a moment from the polished space of physics to the rough and untidy space of our immediate sensible experience. The space of one man's sensible objects is a three-dimensional space. It does not appear probable that two men ever both perceive at the same time any one sensible object; when they are said to see the same thing or hear the same noise, there will always be some difference, however slight, between the actual shapes seen or the actual sounds heard. If this is so, and if, as is generally assumed, position in space is purely relative, it follows that the space of one man's objects and the space of another man's objects have no place in common, that they are in fact different spaces, and not merely different parts of one space. I mean by this that such immediate spatial relations as are perceived to hold between the different parts of the sensible space perceived by one man, do not hold between parts of sensible spaces perceived by different men. There are therefore a multitude of three-dimensional spaces in the world: there are all those perceived by observers, and presumably also those which are not perceived, merely because no observer is suitably situated for perceiving them.

But although these spaces do not have to one another the same kind of spatial relations as obtain between the parts of one of them, it is nevertheless possible to arrange these spaces themselves in a three-dimensional order. This is done by means of the correlated particulars which we regard as members (or aspects) of one physical thing. When a number of people are said to see the same object, those who would be said to be near to the object see a particular occupying a larger part of their field of vision than is occupied by the corresponding particular seen by people who would be said to be farther from the thing. By means of such considerations it is possible, in ways which need not now be further specified, to arrange all the different spaces in a three-dimensional series. Since each of the spaces is itself three-dimensional, the whole world of particulars is thus arranged in a six-dimensional space, that is to say, six co-ordinates will be required to assign completely the position of any given particular, namely, three to assign its position in its own space and three more to assign the position of its space among the other spaces.

There are two ways of classifying particulars: we may take together all those that belong to a given "perspective," or all those that are, as common sense would say, different "aspects" of the same "thing." For example, if I am (as is said) seeing the sun, what I see belongs to two assemblages: (1) the assemblage of all my present objects of sense, which is what I call a "perspective";(2) the assemblage of all the different particulars which would be called aspects of the sun of eight minutes ago—this assemblage is what I define as *being* the sun of eight minutes ago. Thus "perspectives" and "things" are merely two different ways of classifying particulars. It is to be observed that there is no *a priori* necessity for particulars to be susceptible of this double classification. There may be what might be called "wild" particulars, not having the usual relations by which the classification is effected; perhaps dreams and hallucinations are composed of particulars which are "wild" in this sense.

The exact definition of what is meant by a perspective is not quite easy. So long as we confine ourselves to visible objects or to objects of touch we might define the perspective of a given particular as "all particulars which have a simple (direct) spatial relation to the given particular." Between two patches of colour which I see now, there is a direct spatial relation which I equally see. But between patches of colour seen by different men there is only an indirect constructed spatial relation by means of the placing of "things" in physical space (which is the same as the space composed of perspectives). Those particulars which have direct spatial relations to a given particular will belong to the same perspective. But if, for example, the sounds which I hear are to belong to the same perspective with the patches of colour which I see, there must be particulars which have no direct spatial relation and yet belong to the same perspective. We cannot define a perspective as all the data of one percipient at one time, because we wish to allow the possibility of perspectives which are not perceived by any one. There will be need, therefore, in defining a perspective, of some principle derived neither from psychology nor from space.

Such a principle may be obtained from the consideration of *time*. The one all-embracing time, like the one all-embracing space, is a construction; there is no *direct* time-relation between particulars belonging to my perspective and particulars belonging to another man's. On the other hand, any two particulars of which I am aware are either simultaneous or successive, and their simultaneity or successiveness is sometimes itself a datum to me. We may therefore define the perspective to which a given particular belongs as "all particulars simultaneous with the given particular," where "simultaneous" is to be understood as a direct simple relation, not the derivative constructed relation of physics. It may be observed that the introduction of "local time" suggested by the principle of relativity has effected, for purely scientific reasons, much the same multiplication of times as we have just been advocating.

The sum-total of all the particulars that are (directly) either simultaneous with or before or after a given particular may be defined as the "biography" to which that particular belongs. It will be observed that, just as a perspective need not be actually perceived by any one, so a biography need not be actually lived by any one. Those biographies that are lived by no one are called "official."

The definition of a "thing" is effected by means of continuity and of correlations which have a certain differential independence of other "things." That is to say, given a particular in one perspective, there will usually in a neighbouring perspective be a very similar particular, differing from the given particular, to the first order of small quantities, according to a law involving only the difference of position of the two perspectives in perspective space, and not any of the other "things" in the universe. It is this continuity and differential independence in the law of change as we pass from one perspective to another that defines the class of particulars which is to be called "one thing."

Broadly speaking, we may say that the physicist finds it convenient to classify particulars into "things," while the psychologist finds it convenient to classify them into "perspectives" and "biographies," since one perspective *may* constitute the momentary data of one percipient, and one biography *may* constitute the whole of the data of one percipient throughout his life.

We may now sum up our discussion. Our object has been to discover as far as possible the nature of the ultimate constituents of the physical world. When I speak of the "physical world," I mean, to begin with, the world dealt with by physics. It is obvious that physics is an empirical science, giving us a certain amount of knowledge and based upon evidence obtained through the senses. But partly through the development of physics itself, partly through arguments derived from physiology, psychology or metaphysics, it has come to be thought that the immediate data of sense could not themselves form part of the ultimate constituents of the physical world, but were in some sense "mental," "in the mind," or "subjective." The grounds for this view, in so far as they depend upon physics, can only be adequately dealt with by rather elaborate constructions depending upon symbolic logic, showing that out of such materials as are provided by the senses it is possible to construct classes and series having the properties which physics assigns to matter. Since this argument is difficult and technical, I have not embarked upon it in this article. But in so far as the view that sense-data are "mental" rests upon physiology, psychology, or metaphysics, I have tried to show that it rests upon confusions and prejudices—prejudices in favour of permanence in the ultimate constituents of matter, and confusions derived from unduly simple notions as to space, from the causal correlation of sense-data with sense-organs, and from failure to distinguish between sense-data and sensations. If what we have said on these subjects is valid, the existence of sense-data is logically independent of the existence of mind, and is causally dependent upon the *body* of the percipient rather than upon his mind. The causal dependence upon the body of the percipient, we found, is a more complicated matter than it appears to be, and, like all causal dependence, is apt to give rise to erroneous beliefs through misconceptions as to the nature of causal correlation. If we have been right in our contentions, sense-data are merely those among the ultimate constituents of the physical world, of which we happen to be immediately aware; they themselves are purely physical, and all that is mental in connection with them is our awareness of them, which is irrelevant to their nature and to their place in physics.

Unduly simple notions as to space have been a great stumbling-block to realists. When two men look at the same table, it is supposed that what the one sees and what the other sees are in the same place. Since the shape and colour are not quite the same for the two men, this raises a difficulty, hastily solved, or rather covered up, by declaring what each sees to be purely "subjective"—though it would puzzle those who use this glib word to say what they mean by it. The truth seems to be that space—and time also—is much more complicated than it would appear to be from the finished structure of physics, and that the one all-embracing three-dimensional space is a logical construction, obtained by means of correlations from a crude space of six dimensions. The particulars occupying this six-dimensional space, classified in one way, form "things," from which with certain further manipulations we can obtain what physics can regard as matter; classified in another way, they form "perspectives" and "biographies," which may, if a suitable percipient happens to exist, form respectively the sense-data of a momentary or of a total experience. It is only when physical "things" have been dissected into series of classes of particulars, as we have done, that the conflict between the point of view of physics and the point of view of psychology can be overcome. This conflict, if what has been said is not mistaken, flows from different methods of classification, and vanishes as soon as its source is discovered.

In favour of the theory which I have briefly outlined, I do not claim that it is *certainly* true. Apart from the likelihood of mistakes, much of it is avowedly hypothetical. What I do claim for the theory is that it *may* be true, and that this is more than can be said for any other theory except the closely analogous theory of Leibniz. The difficulties besetting realism, the confusions obstructing any philosophical account of physics, the dilemma resulting from discrediting sense-data, which yet remain the sole source of our knowledge of the outer world—all these are avoided by the theory which I advocate. This does not prove the theory to be true, since probably many other theories might be invented which would have the same merits. But it does prove that the theory has a better chance of being true than any of its present competitors, and it suggests that what can be known with certainty is likely to be discoverable by taking our theory as a starting-point, and gradually freeing it from all such assumptions as seem irrelevant, unnecessary, or unfounded. On these grounds, I recommend it to attention as a hypothesis and a basis for further work, though not as itself a finished or adequate solution of the problem with which it deals.

VIII

THE RELATION OF SENSE-DATA TO PHYSICS

I. THE PROBLEM STATED

Physics is said to be an empirical science, based upon observation and experiment.

It is supposed to be verifiable, i.e. capable of calculating beforehand results subsequently confirmed by observation and experiment.

What can we learn by observation and experiment?

Nothing, so far as physics is concerned, except immediate data of sense: certain patches of colour, sounds, tastes, smells, etc., with certain spatio-temporal relations.

The supposed contents of the physical world are *prima facie* very different from these: molecules have no colour, atoms make no noise, electrons have no taste, and corpuscles do not even smell.

If such objects are to be verified, it must be solely through their relation to sense-data: they must have some kind of correlation with sense-data, and must be verifiable through their correlation *alone*.

But how is the correlation itself ascertained? A correlation can only be ascertained empirically by the correlated objects being constantly *found* together. But in our case, only one term of the correlation, namely, the sensible term, is ever *found*: the other term seems essentially incapable of being found. Therefore, it would seem, the correlation with objects of sense, by which physics was to be verified, is itself utterly and for ever unverifiable.

There are two ways of avoiding this result.

(1) We may say that we know some principle *a priori*, without the need of empirical verification, e.g. that our sense-data have *causes* other than themselves, and that something can be known about these causes by inference from their effects. This way has been often adopted by philosophers. It may be necessary to adopt this way to some extent, but in so far as it is adopted physics ceases to be empirical or based upon experiment and observation alone. Therefore, this way is to be avoided as much as possible.

(2) We may succeed in actually defining the objects of physics as functions of sense-data. Just in so far as physics leads to expectations, this *must* be possible, since we can only *expect* what can be experienced. And in so far as the physical state of affairs is inferred from sense-data, it must be capable of expression as a

function of sense-data. The problem of accomplishing this expression leads to much interesting logico-mathematical work.

In physics as commonly set forth, sense-data appear as functions of physical objects: when such-and-such waves impinge upon the eye, we see such-and-such colours, and so on. But the waves are in fact inferred from the colours, not vice versa. Physics cannot be regarded as validly based upon empirical data until the waves have been expressed as functions of the colours and other sense-data.

Thus if physics is to be verifiable we are faced with the following problem: Physics exhibits sense-data as functions of physical objects, but verification is only possible if physical objects can be exhibited as functions of sense-data. We have therefore to solve the equations giving sense-data in terms of physical objects, so as to make them instead give physical objects in terms of sense-data.

II. CHARACTERISTICS OF SENSE-DATA

When I speak of a "sense-datum," I do not mean the whole of what is given in sense at one time. I mean rather such a part of the whole as might be singled out by attention: particular patches of colour, particular noises, and so on. There is some difficulty in deciding what is to be considered *one* sense-datum: often attention causes divisions to appear where, so far as can be discovered, there were no divisions before. An observed complex fact, such as that this patch of red is to the left of that patch of blue, is also to be regarded as a datum from our present point of view: epistemologically, it does not differ greatly from a simple sense-datum as regards its function in giving knowledge. Its *logical* structure is very different, however, from that of sense: *sense* gives acquaintance with particulars, and is thus a two-term relation in which the object can be *named* but not *asserted*, and is inherently incapable of truth or falsehood, whereas the observation of a complex fact, which may be suitably called perception, is not a two-term relation, but involves the propositional form on the object-side, and gives knowledge of a truth, not mere acquaintance with a particular. This logical difference, important as it is, is not very relevant to our present problem; and it will be convenient to regard data of perception as included among sense-data for the purposes of this paper. It is to be observed that the particulars which are constituents of a datum of perception are always sense-data in the strict sense.

Concerning sense-data, we know that they are there while they are data, and this is the epistemological basis of all our knowledge of external particulars. (The meaning of the word "external" of course raises problems which will concern us later.) We do not know, except by means of more or less precarious inferences, whether the objects which are at one time sense-data continue to exist at times when they are not data. Sense-data at the times when they are data are all that we directly and primitively know of the external world; hence in epistemology the fact that they are *data* is all-important. But the fact that they are all that we directly know gives, of course, no presumption that they are all that there is. If we could construct an impersonal metaphysic, independent of the accidents of our knowledge and ignorance, the privileged position of the actual data would probably disappear, and they would probably appear as a rather haphazard selection from a mass of objects more or less like them. In saying this, I assume only that it is probable that there are particulars with which we are not acquainted. Thus the special importance of sense-data is in relation to epistemology, not to metaphysics. In this respect, physics is to be reckoned as metaphysics: it is impersonal, and nominally pays no special attention to sense-data. It is only when we ask how physics can be *known* that the importance of sense-data re-emerges.

III. SENSIBILIA

I shall give the name *sensibilia* to those objects which have the same metaphysical and physical status as sense-data, without necessarily being data to any mind. Thus the relation of a *sensibile* to a sense-datum is like that of a man to a husband: a man becomes a husband by entering into the relation of marriage, and similarly a *sensibile* becomes a sense-datum by entering into the relation of acquaintance. It is important to have both terms; for we wish to discuss whether an object which is at one time a sense-datum can still exist at a time when it is not a sense-datum. We cannot ask "Can sense-data exist without being given?" for that is like asking "Can husbands exist without being married?" We must ask "Can *sensibilia* exist without being given?" and also "Can a particular *sensibile* be at one time a sense-datum, and at another not?" Unless we have the word *sensibile* as well as the word "sense-datum," such questions are apt to entangle us in trivial logical puzzles.

It will be seen that all sense-data are *sensibilia*. It is a metaphysical question whether all *sensibilia* are sense-data, and an epistemological question whether there exist means of inferring *sensibilia* which are not data from those that are.

A few preliminary remarks, to be amplified as we proceed, will serve to elucidate the use which I propose to make of *sensibilia*.

I regard sense-data as not mental, and as being, in fact, part of the actual subject-matter of physics. There are arguments, shortly to be examined, for their subjectivity, but these arguments seem to me only to prove *physiological* subjectivity, i.e. causal dependence on the sense-organs, nerves, and brain. The appearance which a thing presents to us is causally dependent upon these, in exactly the same way as it is dependent upon intervening fog or smoke or coloured glass. Both dependences are contained in the statement that the appearance which a piece of matter presents when viewed from a given place is a function not only of the piece of matter, but also of the intervening medium. (The terms used in this statement—"matter," "view from a given place," "appearance," "intervening medium"—will all be defined in the course of the present paper.) We have not the means of ascertaining how things appear from places not surrounded by brain and nerves and sense-organs, because we cannot leave the body; but continuity makes it not unreasonable to suppose that they present *some* appearance at such places. Any such appearance would be included among *sensibilia*. If—*per impossibile*—there were a complete human body with no mind inside it, all those *sensibilia* would exist, in relation to that body, which would be sense-data if there were a mind in the body. What the mind adds to *sensibilia*, in fact, is *merely* awareness: everything else is physical or physiological.

IV. SENSE-DATA ARE PHYSICAL

Before discussing this question it will be well to define the sense in which the terms "mental" and "physical" are to be used. The word "physical," in all preliminary discussions, is to be understood as meaning "what is dealt with by physics." Physics, it is plain, tells us something about some of the constituents of the actual world; what these constituents are may be doubtful, but it is they that are to be called physical, whatever their nature may prove to be.

The definition of the term "mental" is more difficult, and can only be satisfactorily given after many difficult controversies have been discussed and decided. For present purposes therefore I must content myself with assuming a dogmatic answer to these controversies. I shall call a particular "mental" when it is aware of something, and I shall call a fact "mental" when it contains a mental particular as a constituent.

It will be seen that the mental and the physical are not necessarily mutually exclusive, although I know of no reason to suppose that they overlap.

The doubt as to the correctness of our definition of the "mental" is of little importance in our present discussion. For what I am concerned to maintain is that sense-data are physical, and this being granted it is a matter of indifference in our present inquiry whether or not they are also mental. Although I do not hold, with Mach and James and the "new realists," that the difference between the mental and the physical is *merely* one of arrangement, yet what I have to say in the present paper is compatible with their doctrine and might have been reached from their standpoint.

In discussions on sense-data, two questions are commonly confused, namely:

(1) Do sensible objects persist when we are not sensible of them? in other words, do *sensibilia* which are data at a certain time sometimes continue to exist at times when they are not data? And (2) are sense-data mental or physical?

I propose to assert that sense-data are physical, while yet maintaining that they probably never persist unchanged after ceasing to be data. The view that they do not persist is often thought, quite erroneously in my opinion, to imply that they are mental; and this has, I believe, been a potent source of confusion in regard to our present problem. If there were, as some have held, a *logical impossibility* in sense-data persisting after ceasing to be data, that certainly would tend to show that they were mental; but if, as I contend, their non-persistence is merely a probable inference from empirically ascertained causal laws, then it carries no such implication with it, and we are quite free to treat them as part of the subject-matter of physics.

Logically a sense-datum is an object, a particular of which the subject is aware. It does not contain the subject as a part, as for example beliefs and volitions do. The existence of the sense-datum is therefore not logically dependent upon that of the subject; for the only way, so far as I know, in which the existence of *A* can be *logically* dependent upon the existence of *B* is when *B* is part of *A*. There is therefore no *a priori* reason why a particular which is a sense-datum should not persist after it has ceased to be a datum, nor why other similar particulars should not exist without ever being data. The view that sense-data are mental is derived, no doubt, in part from their physiological subjectivity, but in part also from a failure to distinguish between sense-data and "sensations." By a sensation I mean the fact consisting in the subject's awareness of the sense-datum. Thus a sensation is a complex of which the subject is a constituent and which therefore is mental. The sense-datum, on the other hand, stands over against the subject as that external object of which in sensation the subject is aware. It is true that the sense-datum is in many cases in the subject's body, but the subject's body is as distinct from the subject as tables and chairs are, and is in fact merely a part of the material world. So soon, therefore, as sense-data are clearly distinguished from sensations, and as their subjectivity is recognised to be physiological not psychical, the chief obstacles in the way of regarding them as physical are removed.

V. "SENSIBILIA" AND "THINGS"

But if "sensibilia" are to be recognised as the ultimate constituents of the physical world, a long and difficult journey is to be performed before we can arrive either at the "thing" of common sense or at the "matter" of physics. The supposed impossibility of combining the different sense-data which are regarded as appearances of the same "thing" to different people has made it seem as though these "sensibilia" must be regarded as mere subjective phantasms. A given table will present to one man a rectangular appearance, while to another it appears to have two acute angles and two obtuse angles; to one man it appears brown, while to another, towards whom it reflects the light, it appears white and shiny. It is said, not wholly without plausibility, that these different shapes and different colours cannot co-exist simultaneously in the same place, and cannot therefore both be constituents of the physical world. This argument I must confess appeared to me until recently to be irrefutable. The contrary opinion has, however, been ably maintained by Dr. T.P. Nunn in an article entitled: "Are Secondary Qualities Independent of Perception?"The supposed impossibility derives its apparent force from the phrase: "*in the same place*," and it is precisely in this phrase that its weakness lies. The conception of space is too often treated in philosophy—even by those who on reflection would not defend such treatment—as though it were as given, simple, and unambiguous as Kant, in his psychological innocence, supposed. It is the unperceived ambiguity of the word "place" which, as we shall shortly see, has caused the difficulties to realists and given an undeserved advantage to their opponents. Two "places" of different kinds are involved in every sense-datum, namely the place *at* which it appears and the place *from* which it appears. These belong to different spaces, although, as we shall see, it is possible, with certain limitations, to establish a correlation between them.What we call the different appearances of the same thing to different observers are each in a space private to the observer concerned. No place in the private world of one observer is identical with a place in the private world of another observer. There is therefore no question of combining the different appearances in the one place; and the fact that they cannot all exist in one place affords accordingly no ground whatever for questioning their physical reality. The "thing" of common sense may in fact be identified with the whole class of its appearances—where, however, we must include among appearances not only those which are actual sense-data, but also those "sensibilia," if any, which, on grounds of continuity and resemblance, are to be regarded as belonging to the same system of appearances, although there happen to be no observers to whom they are data.

An example may make this clearer. Suppose there are a number of people in a room, all seeing, as they say, the same tables and chairs, walls and pictures. No two of these people have exactly the same sense-data, yet there is sufficient similarity among their data to enable them to group together certain of these data as appearances of one "thing" to the several spectators, and others as appearances of another "thing." Besides the appearances which a given thing in the room presents to the actual spectators, there are, we may suppose, other appearances which it would present to other possible spectators. If a man were to sit down between two others, the appearance which the room would present to him would be intermediate between the appearances which it presents to the two others: and although this appearance would not exist as it is without the sense organs, nerves and brain, of the newly arrived spectator, still it is not unnatural to suppose that, from the position which he now occupies, *some* appearance of the room existed before his arrival. This supposition, however, need merely be noticed and not insisted upon.

Since the "thing" cannot, without indefensible partiality, be identified with any single one of its appearances, it came to be thought of as something distinct from all of them and underlying them. But by the principle of Occam's razor, if the class of appearances will fulfil the purposes for the sake of which the thing was invented by the prehistoric metaphysicians to whom common sense is due, economy demands that we should identify the thing with the class of its appearances. It is not necessary to *deny* a substance or substratum underlying these appearances; it is merely expedient to abstain from asserting this unnecessary entity. Our procedure here is precisely analogous to that which has swept away from the philosophy of mathematics the useless menagerie of metaphysical monsters with which it used to be infested.

VI. CONSTRUCTIONS VERSUS INFERENCES

Before proceeding to analyse and explain the ambiguities of the word "place," a few general remarks on method are desirable. The supreme maxim in scientific philosophising is this:

Wherever possible, logical constructions are to be substituted for inferred entities.

Some examples of the substitution of construction for inference in the realm of mathematical philosophy may serve to elucidate the uses of this maxim. Take first the case of irrationals. In old days, irrationals were inferred as the supposed limits of series of rationals which had no rational limit; but the objection to this procedure was that it left the existence of irrationals merely optative, and for this reason the stricter methods of the present day no longer tolerate such a definition. We now define an irrational number as a certain class of ratios, thus constructing it logically by means of ratios, instead of arriving at it by a doubtful inference from them. Take again the case of cardinal numbers. Two equally numerous collections appear to have something in common: this something is supposed to be their cardinal number. But so long as the cardinal number is inferred from the collections, not constructed in terms of them, its existence must remain in doubt, unless in virtue of a metaphysical postulate *ad hoc*. By defining the cardinal number of a given collection as the class of all equally numerous collections, we avoid the necessity of this metaphysical postulate, and thereby remove a needless element of doubt from the philosophy of arithmetic. A similar method, as I have shown elsewhere, can be applied to classes themselves, which need not be supposed to have any metaphysical reality, but can be regarded as symbolically constructed fictions.

The method by which the construction proceeds is closely analogous in these and all similar cases. Given a set of propositions nominally dealing with the supposed inferred entities, we observe the properties which are required of the supposed entities in order to make these propositions true. By dint of a little logical ingenuity, we then construct some logical function of less hypothetical entities which has the requisite properties. This constructed function we substitute for the supposed inferred entities, and thereby obtain a new and less doubtful interpretation of the body of propositions in question. This method, so fruitful in the philosophy of mathematics, will be found equally applicable in the philosophy of physics, where, I do not doubt, it would have been applied long ago but for the fact that all who have studied this subject hitherto have been completely ignorant of mathematical logic. I myself cannot claim originality in the application of this method to physics, since I owe the suggestion and the stimulus for its application entirely to my friend and collaborator Dr. Whitehead, who is engaged in applying it to the more mathematical portions of the region intermediate between sense-data and the points, instants and particles of physics.

A complete application of the method which substitutes constructions for inferences would exhibit matter wholly in terms of sense-data, and even, we may add, of the sense-data of a single person, since the sense-data of others cannot be known without some element of inference. This, however, must remain for the present an ideal, to be approached as nearly as possible, but to be reached, if at all, only after a long preliminary labour of which as yet we can only see the very beginning. The inferences which are unavoidable can, however, be subjected to certain guiding principles. In the first place they should always be made perfectly explicit, and should be formulated in the most general manner possible. In the second place the inferred entities should, whenever this can be done, be similar to those whose existence is given, rather than, like the Kantian *Ding an sich*, something wholly remote from the data which nominally support the inference. The inferred entities which I shall allow myself are of two kinds: (*a*) the sense-data of other people, in favour of which there is the evidence of testimony, resting ultimately upon the analogical argument in favour of minds other than my own; (*b*) the "sensibilia" which would appear from places where there happen to be no minds, and which I suppose to be real although they are no one's data. Of these two classes of inferred entities, the first will probably be allowed to pass unchallenged. It would give me the greatest satisfaction to be able to dispense with it, and thus establish physics upon a solipsistic basis; but those—and I fear they are the majority—in whom the human affections are stronger than the desire for logical economy, will, no doubt, not share my desire to render solipsism scientifically satisfactory. The second class of inferred entities raises much more serious questions. It may be thought monstrous to maintain that a thing can present any appearance at all in a place where no sense organs and nervous structure exist through which it could appear. I do not myself feel the monstrosity; nevertheless I should regard these supposed appearances only in the light of a hypothetical scaffolding, to be used while the edifice of physics is being raised, though possibly capable of being removed as soon as the edifice is completed. These "sensibilia" which are not data to anyone are therefore to be taken rather as an illustrative hypothesis and as an aid in preliminary statement than as a dogmatic part of the philosophy of physics in its final form.

VII. PRIVATE SPACE AND THE SPACE OF PERSPECTIVES

We have now to explain the ambiguity in the word "place," and how it comes that two places of different sorts are associated with every sense-datum, namely the place *at* which it is and the place *from* which it is perceived. The theory to be advocated is closely analogous to Leibniz's monadology, from which it differs chiefly in being less smooth and tidy.

The first fact to notice is that, so far as can be discovered, no sensibile is ever a datum to two people at once. The things seen by two different people are often closely similar, so similar that the same *words* can be used to denote them, without which communication with others concerning sensible objects would be impossible. But, in spite of this similarity, it would seem that some difference always arises from difference in the point of view. Thus each person, so far as his sense-data are concerned, lives in a private world. This private world contains its own space, or rather spaces, for it would seem that only experience teaches us to correlate the space of sight with the space of touch and with the various other spaces of other senses. This multiplicity of private spaces, however, though interesting to the psychologist, is of no great importance in regard to our present problem, since a merely solipsistic experience enables us to correlate them into the one private space which embraces all our own sense-data. The place *at* which a sense-datum is, is a place in private space. This place therefore is different from any place in the private space of another percipient. For if we assume, as logical economy demands, that all position is relative, a place is only definable by the things in or around it, and therefore the same place cannot occur in two private worlds which have no common constituent. The question, therefore, of combining what we call different appearances of the same thing in the same place does not arise, and the fact that a given object appears to different spectators to have different shapes and colours affords no argument against the physical reality of all these shapes and colours.

In addition to the private spaces belonging to the private worlds of different percipients, there is, however, another space, in which one whole private world counts as a point, or at least as a spatial unit. This might be described as the space of points of view, since each private world may be regarded as the appearance which the universe presents from a certain point of view. I prefer, however, to speak of it as the space of *perspectives*, in order to obviate the suggestion that a private world is only real when someone views it. And for the same reason, when I wish to speak of a private world without assuming a percipient, I shall call it a "perspective."

We have now to explain how the different perspectives are ordered in one space. This is effected by means of the correlated "sensibilia" which are regarded as the appearances, in different perspectives, of one and the same thing. By moving, and by testimony, we discover that two different perspectives, though they cannot both contain the same "sensibilia," may nevertheless contain very similar ones; and the spatial order of a certain group of "sensibilia" in a private space of one perspective is found to be identical with, or very similar to, the spatial order of the correlated "sensibilia" in the private space of another perspective. In this way one "sensibile" in one perspective is correlated with one "sensibile" in another. Such correlated "sensibilia" will be called "appearances of one thing." In Leibniz's monadology, since each monad mirrored the whole universe, there was in each perspective a "sensibile" which was an appearance of each thing. In our system of perspectives, we make no such assumption of completeness. A given thing will have appearances in some perspectives, but presumably not in certain others. The "thing" being defined as the class of its appearances, if κ is the class of perspectives in which a certain thing θ appears, then θ is a member of the multiplicative class of κ, κ being a class of mutually exclusive classes of "sensibilia." And similarly a perspective is a member of the multiplicative class of the things which appear in it.

The arrangement of perspectives in a space is effected by means of the differences between the appearances of a given thing in the various perspectives. Suppose, say, that a certain penny appears in a number of different perspectives; in some it looks larger and in some smaller, in some it looks circular, in others it presents the appearance of an ellipse of varying eccentricity. We may collect together all those perspectives in which the appearance of the penny is circular. These we will place on one straight line, ordering them in a series by the variations in the apparent size of the penny. Those perspectives in which the penny appears as a straight line of a certain thickness will similarly be placed upon a plane (though in this case there will be many different perspectives in which the penny is of the same size; when one arrangement is completed these will form a circle concentric with the penny), and ordered as before by the apparent size of the penny. By such means, all those perspectives in which the penny presents a visual appearance can be arranged in a three-dimensional spatial order. Experience shows that the same spatial order of perspectives would have resulted if, instead of the penny, we had chosen any other thing which appeared in all the perspectives in question, or any other method of utilising the differences between the appearances of the same things in different perspectives. It is this empirical fact which has made it possible to construct the one all-embracing space of physics.

The space whose construction has just been explained, and whose elements are whole perspectives, will be called "perspective-space."

VIII. THE PLACING OF "THINGS" AND "SENSIBILIA" IN PERSPECTIVE SPACE

The world which we have so far constructed is a world of six dimensions, since it is a three-dimensional series of perspectives, each of which is itself three-dimensional. We have now to explain the correlation between the perspective space and the various private spaces contained within the various perspectives severally. It is by means of this correlation that the one three-dimensional space of physics is constructed; and it is because of the unconscious performance of this correlation that the distinction between perspective space and the percipient's private space has been blurred, with disastrous results for the philosophy of physics. Let us revert to our penny: the perspectives in which the penny appears larger are regarded as being nearer to the penny than those in which it appears smaller, but as far as experience goes the apparent size of the penny will not grow beyond a certain limit, namely, that where (as we say) the penny is so near the eye that if it were any nearer it could not be seen. By touch we may prolong the series until the penny touches the eye, but no further. If we have been travelling along a line of perspectives in the previously defined sense, we may, however, by imagining the penny removed, prolong the line of perspectives by means, say, of another penny; and the same may be done with any other line of perspectives defined by means of the penny. All these lines meet in a certain place, that is, in a certain perspective. This perspective will be defined as "the place where the penny is."

It is now evident in what sense two places in constructed physical space are associated with a given "sensibile." There is first the place which is the perspective of which the "sensibile" is a member. This is the place *from* which the "sensibile" appears. Secondly there is the place where the thing is of which the "sensibile" is a member, in other words an appearance; this is the place *at* which the "sensibile" appears. The "sensibile" which is a member of one perspective is correlated with another perspective, namely, that which is the place where the thing is of which the "sensibile" is an appearance. To the psychologist the "place from which" is the more interesting, and the "sensibile" accordingly appears to him subjective and where the percipient is. To the physicist the "place at which" is the more interesting, and the "sensibile" accordingly appears to him physical and external. The causes, limits and partial justification of each of these two apparently incompatible views are evident from the above duplicity of places associated with a given "sensibile."

We have seen that we can assign to a physical thing a place in the perspective space. In this way different parts of our body acquire positions in perspective space, and therefore there is a meaning (whether true or false need not much concern us) in saying that the perspective to which our sense-data belong is inside our head. Since our mind is correlated with the perspective to which our sense-data belong, we may regard this perspective as being the position of our mind in perspective space. If, therefore, this perspective is, in the above defined sense, inside our head, there is a good meaning for the statement that the mind is in the head. We can now say of the various appearances of a given thing that some of them are nearer to the thing than others; those are nearer which belong to perspectives that are nearer to "the place where the thing is." We can thus find a meaning, true or false, for the statement that more is to be learnt about a thing by examining it close to than by viewing it from a distance. We can also find a meaning for the phrase "the things which intervene between the subject and a thing of which an appearance is a datum to him." One reason often alleged for the subjectivity of sense-data is that the appearance of a thing may change when we find it hard to suppose that the thing itself has changed—for example, when the change is due to our shutting our eyes, or to our screwing them up so as to make the thing look double. If the thing is defined as the class of its appearances (which is the definition adopted above), there is of course necessarily *some* change in the thing whenever any one of its appearances changes. Nevertheless there is a very important distinction between two different ways in which the appearances may change. If after looking at a thing I shut my eyes, the appearance of my eyes changes in every perspective in which there is such an appearance, whereas most of the appearances of the thing will remain unchanged. We may say, as a matter of definition, that a thing changes when, however near to the thing an appearance of it may be, there are changes in appearances as near as, or still nearer to, the thing. On the other hand we shall say that the change is in some other thing if all appearances of the thing which are at not more than a certain distance from the thing remain unchanged, while only comparatively distant appearances of the thing are altered. From this consideration we are naturally led to the consideration of *matter*, which must be our next topic.

IX. THE DEFINITION OF MATTER

We defined the "physical thing" as the class of its appearances, but this can hardly be taken as a definition of matter. We want to be able to express the fact that the appearance of a thing in a given perspective is causally affected by the matter between the thing and the perspective. We have found a meaning for "between a thing and a perspective." But we want matter to be something other than the whole class of appearances of a thing, in order to state the influence of matter on appearances.

We commonly assume that the information we get about a thing is more accurate when the thing is nearer. Far off, we see it is a man; then we see it is Jones; then we see he is smiling. Complete accuracy would only be attainable as a limit: if the appearances of Jones as we approach him tend towards a limit, that limit may be taken to be what Jones really is. It is obvious that from the point of view of physics the appearances of a thing close to "count" more than the appearances far off. We may therefore set up the following tentative definition:

The *matter* of a given thing is the limit of its appearances as their distance from the thing diminishes.

It seems probable that there is something in this definition, but it is not quite satisfactory, because empirically there is no such limit to be obtained from sense-data. The definition will have to be eked out by constructions and definitions. But probably it suggests the right direction in which to look.

We are now in a position to understand in outline the reverse journey from matter to sense-data which is performed by physics. The appearance of a thing in a given perspective is a function of the matter composing the thing and of the intervening matter. The appearance of a thing is altered by intervening smoke or mist, by blue spectacles or by alterations in the sense-organs or nerves of the percipient (which also must be reckoned as part of the intervening medium). The nearer we approach to the thing, the less its appearance is affected by the intervening matter. As we travel further and further from the thing, its appearances diverge more and more from their initial character; and the causal laws of their divergence are to be stated in terms of the matter which lies between them and the thing. Since the appearances at very small distances are less affected by causes other than the thing itself, we come to think that the limit towards which these appearances tend as the distance diminishes is what the thing "really is," as opposed to what it merely seems to be. This, together with its necessity for the statement of causal laws, seems to be the source of the entirely erroneous feeling that matter is more "real" than sense-data.

Consider for example the infinite divisibility of matter. In looking at a given thing and approaching it, one sense-datum will become several, and each of these will again divide. Thus *one* appearance may represent *many* things, and to this process there seems no end. Hence in the limit, when we approach indefinitely near to the thing there will be an indefinite number of units of matter corresponding to what, at a finite distance, is only one appearance. This is how infinite divisibility arises.

The whole causal efficacy of a thing resides in its matter. This is in some sense an empirical fact, but it would be hard to state it precisely, because "causal efficacy" is difficult to define.

What can be known empirically about the matter of a thing is only approximate, because we cannot get to know the appearances of the thing from very small distances, and cannot accurately infer the limit of these appearances. But it *is* inferred *approximately* by means of the appearances we can observe. It then turns out that these appearances can be exhibited by physics as a function of the matter in our immediate neighbourhood; e.g. the visual appearance of a distant object is a function of the light-waves that reach the eyes. This leads to confusions of thought, but offers no real difficulty.

One appearance, of a visible object for example, is not sufficient to determine its other simultaneous appearances, although it goes a certain distance towards determining them. The determination of the hidden structure of a thing, so far as it is possible at all, can only be effected by means of elaborate dynamical inferences.

X. TIME

It seems that the one all-embracing time is a construction, like the one all-embracing space. Physics itself has become conscious of this fact through the discussions connected with relativity.

Between two perspectives which both belong to one person's experience, there will be a direct time-relation of before and after. This suggests a way of dividing history in the same sort of way as it is divided by different experiences, but without introducing experience or anything mental: we may define a "biography" as everything that is (directly) earlier or later than, or simultaneous with, a given "sensibile." This will give a series of perspectives, which *might* all form parts of one person's experience, though it is not necessary that all or any of them should actually do so. By this means, the history of the world is divided into a number of mutually exclusive biographies.

We have now to correlate the times in the different biographies. The natural thing would be to say that the appearances of a given (momentary) thing in two different perspectives belonging to different biographies are to be taken as simultaneous; but this is not convenient. Suppose *A* shouts to *B*, and *B* replies as soon as he hears *A's* shout. Then between *A's* hearing of his own shout and his hearing of *B's* there is an interval; thus if we made *A's* and *B's* hearing of the same shout exactly simultaneous with each other, we should have events exactly simultaneous with a given event but not with each other. To obviate this, we assume a "velocity of sound." That is, we assume that the time when *B* hears *A's* shout is half-way between the time when *A* hears his own shout and the time when he hears *B's*. In this way the correlation is effected.

What has been said about sound applies of course equally to light. The general principle is that the appearances, in different perspectives, which are to be grouped together as constituting what a certain thing is at a certain moment, are not to be all regarded as being at that moment. On the contrary they spread outward from the thing with various velocities according to the nature of the appearances. Since no *direct* means exist of correlating the time in one biography with the time in another, this temporal grouping of the appearances belonging to a given thing at a given moment is in part conventional. Its motive is partly to secure the verification of such maxims as that events which are exactly simultaneous with the same event are exactly simultaneous with one another, partly to secure convenience in the formulation of causal laws.

XI. THE PERSISTENCE OF THINGS AND MATTER

Apart from any of the fluctuating hypotheses of physics, three main problems arise in connecting the world of physics with the world of sense, namely:

1. the construction of a single space;
2. the construction of a single time;
3. the construction of permanent things or matter.

We have already considered the first and second of these problems; it remains to consider the third.

We have seen how correlated appearances in different perspectives are combined to form one "thing" at one moment in the all-embracing time of physics. We have now to consider how appearances at different times are combined as belonging to one "thing," and how we arrive at the persistent "matter" of physics. The assumption of permanent substance, which technically underlies the procedure of physics, cannot of course be regarded as metaphysically legitimate: just as the one thing simultaneously seen by many people is a construction, so the one thing seen at different times by the same or different people must be a construction, being in fact nothing but a certain grouping of certain "sensibilia."

We have seen that the momentary state of a "thing" is an assemblage of "sensibilia," in different perspectives, not all simultaneous in the one constructed time, but spreading out from "the place where the thing is" with velocities depending upon the nature of the "sensibilia." The time *at* which the "thing" is in this state is the lower limit of the times at which these appearances occur. We have now to consider what leads us to speak of another set of appearances as belonging to the same "thing" at a different time.

For this purpose, we may, at least to begin with, confine ourselves within a single biography. If we can always say when two "sensibilia" in a given biography are appearances of one thing, then, since we have seen how to connect "sensibilia" in different biographies as appearances of the same momentary state of a thing, we shall have all that is necessary for the complete construction of the history of a thing.

It is to be observed, to begin with, that the identity of a thing for common sense is not always correlated with the identity of matter for physics. A human body is one persisting thing for common sense, but for physics its matter is constantly changing. We may say, broadly, that the common-sense conception is based upon continuity in appearances at the ordinary distances of sense-data, while the physical conception is based upon the continuity of appearances at very small distances from the thing. It is probable that the common-sense conception is not capable of complete precision. Let us therefore concentrate our attention upon the conception of the persistence of matter in physics.

The first characteristic of two appearances of the same piece of matter at different times is *continuity*. The two appearances must be connected by a series of intermediaries, which, if time and space form compact series, must themselves form a compact series. The colour of the leaves is different in autumn from what it is in summer; but we believe that the change occurs gradually, and that, if the colours are different at two given times, there are intermediate times at which the colours are intermediate between those at the given times.

But there are two considerations that are important as regards continuity.

First, it is largely hypothetical. We do not observe any one thing continuously, and it is merely a hypothesis to assume that, while we are not observing it, it passes through conditions intermediate between those in which it is perceived. During uninterrupted observation, it is true, continuity is nearly verified; but even here, when motions are very rapid, as in the case of explosions, the continuity is not actually capable of direct verification. Thus we can only say that the sense-data are found to *permit* a hypothetical complement of "sensibilia" such as will preserve continuity, and that therefore there *may* be such a complement. Since, however, we have already made such use of hypothetical "sensibilia," we will let this point pass, and admit such "sensibilia" as are required to preserve continuity.

Secondly, continuity is not a sufficient criterion of material identity. It is true that in many cases, such as rocks, mountains, tables, chairs, etc., where the appearances change slowly, continuity is sufficient, but in other cases, such as the parts of an approximately homogeneous fluid, it fails us utterly. We can travel by sensibly continuous gradations from any one drop of the sea at any one time to any other drop at any other time. We infer the motions of sea-water from the effects of the current, but they cannot be inferred from direct sensible observation together with the assumption of continuity.

The characteristic required in addition to continuity is conformity with the laws of dynamics. Starting from what common sense regards as persistent things, and making only such modifications as from time to time seem reasonable, we arrive at assemblages of "sensibilia" which are found to obey certain simple laws, namely those of dynamics. By regarding "sensibilia" at different times as belonging to the same piece of matter, we are able to define *motion*, which presupposes the assumption or construction of something persisting throughout the time of the motion. The motions which are regarded as occurring, during a period in which all the "sensibilia" and the times of their appearance are given, will be different according to the manner in which we combine "sensibilia" at different times as belonging to the same piece of matter. Thus even when the whole history of the world is given in every particular, the question what motions take place is still to a certain extent arbitrary even after the assumption of continuity. Experience shows that it is possible to determine motions in such a way as to satisfy the laws of dynamics, and that this determination, roughly and on the whole, is fairly in agreement with the common-sense opinions about persistent things. This determination, therefore, is adopted, and leads to a criterion by which we can determine, sometimes practically, sometimes only theoretically, whether two appearances at different times are to be regarded as belonging to the same piece of matter. The persistence of all matter throughout all time can, I imagine, be secured by definition.

To recommend this conclusion, we must consider what it is that is proved by the empirical success of physics. What is proved is that its hypotheses, though unverifiable where they go beyond sense-data, are at no point in contradiction with sense-data, but, on the contrary, are ideally such as to render all sense-data calculable when a sufficient collection of "sensibilia" is given. Now physics has found it empirically possible to collect sense-data into series, each series being regarded as belonging to one "thing," and behaving, with regard to the laws of physics, in a way in which series not belonging to one thing would in general not behave. If it is to be unambiguous whether two appearances belong to the same thing or not, there must be only one way of grouping appearances so that the resulting things obey the laws of physics. It would be very difficult to prove that this is the case, but for our present purposes we may let this point pass, and assume that there is only one way. Thus we may lay down the following definition: *Physical things are those series of appearances whose matter obeys the laws of physics*. That such series exist is an empirical fact, which constitutes the verifiability of physics.

XII. ILLUSIONS, HALLUCINATIONS, AND DREAMS

It remains to ask how, in our system, we are to find a place for sense-data which apparently fail to have the usual connection with the world of physics. Such sense-data are of various kinds, requiring somewhat different treatment. But all are of the sort that would be called "unreal," and therefore, before embarking upon the discussion, certain logical remarks must be made upon the conceptions of reality and unreality.

Mr. A. Wolf says:

"The conception of mind as a system of transparent activities is, I think, also untenable because of its failure to account for the very possibility of dreams and hallucinations. It seems impossible to realise how a bare, transparent activity can be directed to what is not there, to apprehend what is not given."

This statement is one which, probably, most people would endorse. But it is open to two objections. First it is difficult to see how an activity, however un-"transparent," can be directed towards a nothing: a term of a relation cannot be a mere nonentity. Secondly, no reason is given, and I am convinced that none can be given, for the assertion that dream-objects are not "there" and not "given." Let us take the second point first.

(1) The belief that dream-objects are not given comes, I think, from failure to distinguish, as regards waking life, between the sense-datum and the corresponding "thing." In dreams, there is no such corresponding "thing" as the dreamer supposes; if, therefore, the "thing" were given in waking life, as e.g. Meinong maintains,then there would be a difference in respect of givenness between dreams and waking life. But if, as we have maintained, what is given is never the thing, but merely one of the "sensibilia" which compose the thing, then what we apprehend in a dream is just as much given as what we apprehend in waking life.

Exactly the same argument applies as to the dream-objects being "there." They have their position in the private space of the perspective of the dreamer; where they fail is in their correlation with other private spaces and therefore with perspective space. But in the only sense in which "there" can be a datum, they are "there" just as truly as any of the sense-data of waking life.

(2) The conception of "illusion" or "unreality," and the correlative conception of "reality," are generally used in a way which embodies profound logical confusions. Words that go in pairs, such as "real" and "unreal," "existent" and "non-existent," "valid" and "invalid," etc., are all derived from the one fundamental pair, "true" and "false." Now "true" and "false" are applicable only—except in derivative significations—to *propositions*. Thus wherever the above pairs can be significantly applied, we must be dealing either with

propositions or with such incomplete phrases as only acquire meaning when put into a context which, with them, forms a proposition. Thus such pairs of words can be applied to *descriptions*,but not to proper names: in other words, they have no application whatever to data, but only to entities or non-entities described in terms of data.

Let us illustrate by the terms "existence" and "non-existence." Given any datum *x*, it is meaningless either to assert or to deny that *x* "exists." We might be tempted to say: "Of course *x* exists, for otherwise it could not be a datum." But such a statement is really meaningless, although it is significant and true to say "My present sense-datum exists," and it may also be true that "*x* is my present sense-datum." The inference from these two propositions to "*x* exists" is one which seems irresistible to people unaccustomed to logic; yet the apparent proposition inferred is not merely false, but strictly meaningless. To say "My present sense-datum exists" is to say (roughly): "There is an object of which 'my present sense-datum' is a description." But we cannot say: "There is an object of which '*x*' is a description," because '*x*' is (in the case we are supposing) a name, not a description. Dr. Whitehead and I have explained this point fully elsewhere (*loc. cit.*) with the help of symbols, without which it is hard to understand; I shall not therefore here repeat the demonstration of the above propositions, but shall proceed with their application to our present problem.

The fact that "existence" is only applicable to descriptions is concealed by the use of what are grammatically proper names in a way which really transforms them into descriptions. It is, for example, a legitimate question whether Homer existed; but here "Homer" means "the author of the Homeric poems," and is a description. Similarly we may ask whether God exists; but then "God" means "the Supreme Being" or "the *ens realissimum*" or whatever other description we may prefer. If "God" were a proper name, God would have to be a datum; and then no question could arise as to His existence. The distinction between existence and other predicates, which Kant obscurely felt, is brought to light by the theory of descriptions, and is seen to remove "existence" altogether from the fundamental notions of metaphysics.

What has been said about "existence" applies equally to "reality," which may, in fact, be taken as synonymous with "existence." Concerning the immediate objects in illusions, hallucinations, and dreams, it is meaningless to ask whether they "exist" or are "real." There they are, and that ends the matter. But we may legitimately inquire as to the existence or reality of "things" or other "sensibilia" inferred from such objects. It is the unreality of these "things" and other "sensibilia," together with a failure to notice that they are not data, which has led to the view that the objects of dreams are unreal.

We may now apply these considerations in detail to the stock arguments against realism, though what is to be said will be mainly a repetition of what others have said before.

(1) We have first the variety of normal appearances, supposed to be incompatible. This is the case of the different shapes and colours which a given thing presents to different spectators. Locke's water which seems both hot and cold belongs to this class of cases. Our system of different perspectives fully accounts for these cases, and shows that they afford no argument against realism.

(2) We have cases where the correlation between different senses is unusual. The bent stick in water belongs here. People say it looks bent but is straight: this only means that it is straight to the touch, though bent to sight. There is no "illusion," but only a false inference, if we think that the stick would feel bent to the touch. The stick would look just as bent in a photograph, and, as Mr. Gladstone used to say, "the photograph cannot lie."The case of seeing double also belongs here, though in this case the cause of the unusual correlation is physiological, and would therefore not operate in a photograph. It is a mistake to ask whether the "thing" is duplicated when we see it double. The "thing" is a whole system of "sensibilia," and it is only those visual "sensibilia" which are data to the percipient that are duplicated. The phenomenon has a purely physiological explanation; indeed, in view of our having two eyes, it is in less need of explanation than the single visual sense-datum which we normally obtain from the things on which we focus.

(3) We come now to cases like dreams, which may, at the moment of dreaming, contain nothing to arouse suspicion, but are condemned on the ground of their supposed incompatibility with earlier and later data. Of course it often happens that dream-objects fail to behave in the accustomed manner: heavy objects fly, solid objects melt, babies turn into pigs or undergo even greater changes. But none of these unusual occurrences *need* happen in a dream, and it is not on account of such occurrences that dream-objects are called "unreal." It is their lack of continuity with the dreamer's past and future that makes him, when he wakes, condemn them; and it is their lack of correlation with other private worlds that makes others condemn them. Omitting the latter ground, our reason for condemning them is that the "things" which we infer from them cannot be combined according to the laws of physics with the "things" inferred from waking sense-data. This might be used to condemn the "things" inferred from the data of dreams. Dream-data are no doubt appearances of "things," but not of such "things" as the dreamer supposes. I have no wish to combat psychological theories of dreams, such as those of the psycho-analysts. But there certainly are cases where (whatever psychological causes may contribute) the presence of physical causes also is very evident. For instance, a door banging may produce a dream of a naval engagement, with images of battleships and sea and smoke. The

whole dream will be an appearance of the door banging, but owing to the peculiar condition of the body (especially the brain) during sleep, this appearance is not that expected to be produced by a door banging, and thus the dreamer is led to entertain false beliefs. But his sense-data are still physical, and are such as a completed physics would include and calculate.

(4) The last class of illusions are those which cannot be discovered within one person's experience, except through the discovery of discrepancies with the experiences of others. Dreams might conceivably belong to this class, if they were jointed sufficiently neatly into waking life; but the chief instances are recurrent sensory hallucinations of the kind that lead to insanity. What makes the patient, in such cases, become what others call insane is the fact that, within his own experience, there is nothing to show that the hallucinatory sense-data do not have the usual kind of connection with "sensibilia" in other perspectives. Of course he may learn this through testimony, but he probably finds it simpler to suppose that the testimony is untrue and that he is being wilfully deceived. There is, so far as I can see, no theoretical criterion by which the patient can decide, in such a case, between the two equally satisfactory hypotheses of his madness and of his friends' mendacity.

From the above instances it would appear that abnormal sense-data, of the kind which we regard as deceptive, have intrinsically just the same status as any others, but differ as regards their correlations or causal connections with other "sensibilia" and with "things." Since the usual correlations and connections become part of our unreflective expectations, and even seem, except to the psychologist, to form part of our data, it comes to be thought, mistakenly, that in such cases the data are unreal, whereas they are merely the causes of false inferences. The fact that correlations and connections of unusual kinds occur adds to the difficulty of inferring things from sense and of expressing physics in terms of sense-data. But the unusualness would seem to be always physically or physiologically explicable, and therefore raises only a complication, not a philosophical objection.

I conclude, therefore, that no valid objection exists to the view which regards sense-data as part of the actual substance of the physical world, and that, on the other hand, this view is the only one which accounts for the empirical verifiability of physics. In the present paper, I have given only a rough preliminary sketch. In particular, the part played by *time* in the construction of the physical world is, I think, more fundamental than would appear from the above account. I should hope that, with further elaboration, the part played by unperceived "sensibilia" could be indefinitely diminished, probably by invoking the history of a "thing" to eke out the inferences derivable from its momentary appearance.

IX

ON THE NOTION OF CAUSE

In the following paper I wish, first, to maintain that the word "cause" is so inextricably bound up with misleading associations as to make its complete extrusion from the philosophical vocabulary desirable; secondly, to inquire what principle, if any, is employed in science in place of the supposed "law of causality" which philosophers imagine to be employed; thirdly, to exhibit certain confusions, especially in regard to teleology and determinism, which appear to me to be connected with erroneous notions as to causality.

All philosophers, of every school, imagine that causation is one of the fundamental axioms or postulates of science, yet, oddly enough, in advanced sciences such as gravitational astronomy, the word "cause" never occurs. Dr. James Ward, in his *Naturalism and Agnosticism*, makes this a ground of complaint against physics: the business of those who wish to ascertain the ultimate truth about the world, he apparently thinks, should be the discovery of causes, yet physics never even seeks them. To me it seems that philosophy ought not to assume such legislative functions, and that the reason why physics has ceased to look for causes is that, in fact, there are no such things. The law of causality, I believe, like much that passes muster among philosophers, is a relic of a bygone age, surviving, like the monarchy, only because it is erroneously supposed to do no harm. In order to find out what philosophers commonly understand by "cause," I consulted Baldwin's *Dictionary*, and was rewarded beyond my expectations, for I found the following three mutually incompatible definitions:—

"CAUSALITY. (1) The necessary connection of events in the time-series....

"CAUSE (notion of). Whatever may be included in the thought or perception of a process as taking place in consequence of another process....

"CAUSE AND EFFECT. (1) Cause and effect ... are correlative terms denoting any two distinguishable things, phases, or aspects of reality, which are so related to each other that whenever the first ceases to exist the second comes into existence immediately after, and whenever the second comes into existence the first has ceased to exist immediately before."

Let us consider these three definitions in turn. The first, obviously, is unintelligible without a definition of "necessary." Under this head, Baldwin's *Dictionary* gives the following:—

"NECESSARY. That is necessary which not only is true, but would be true under all circumstances. Something more than brute compulsion is, therefore, involved in the conception; there is a general law under which the thing takes place."

The notion of cause is so intimately connected with that of necessity that it will be no digression to linger over the above definition, with a view to discovering, if possible, *some* meaning of which it is capable; for, as it stands, it is very far from having any definite signification.

The first point to notice is that, if any meaning is to be given to the phrase "would be true under all circumstances," the subject of it must be a propositional function, not a proposition. A proposition is simply true or false, and that ends the matter: there can be no question of "circumstances." "Charles I's head was cut off" is just as true in summer as in winter, on Sundays as on Mondays. Thus, when it is worth saying that something "would be true under all circumstances," the something in question must be a propositional function, i.e., an expression containing a variable, and becoming a proposition when a value is assigned to the variable; the varying "circumstances" alluded to are then the different values of which the variable is capable. Thus if "necessary" means "what is true under all circumstances," then "if x is a man, x is mortal" is necessary, because it is true for any possible value of x. Thus, we should be led to the following definition: —

"NECESSARY is a predicate of a propositional function, meaning that it is true for all possible values of its argument or arguments."

Unfortunately, however, the definition in Baldwin's *Dictionary* says that what is necessary is not only "true under all circumstances" but is also "true." Now these two are incompatible. Only propositions can be "true," and only propositional functions can be "true under all circumstances." Hence the definition as it stands is nonsense. What is meant seems to be this: "A proposition is necessary when it is a value of a propositional function which is true under all circumstances, i.e. for all values of its argument or arguments." But if we adopt this definition, the same proposition will be necessary or contingent according as we choose one or other of its terms as the argument to our propositional function. For example, "if Socrates is a man, Socrates is mortal," is necessary if Socrates is chosen as argument, but not if *man* or *mortal* is chosen. Again, "if Socrates is a man, Plato is mortal," will be necessary if either Socrates or *man* is chosen as argument, but not if Plato or *mortal* is chosen. However, this difficulty can be overcome by specifying the constituent which is to be regarded as argument, and we thus arrive at the following definition:

"A proposition is *necessary* with respect to a given constituent if it remains true when that constituent is altered in any way compatible with the proposition remaining significant."

We may now apply this definition to the definition of causality quoted above. It is obvious that the argument must be the time at which the earlier event occurs. Thus an instance of causality will be such as: "If the event e_1 occurs at the time t_1, it will be followed by the event e_2." This proposition is intended to be necessary with respect to t_1, i.e. to remain true however t_1 may be varied. Causality, as a universal law, will then be the following: "Given any event e_1, there is an event e_2 such that, whenever e_1 occurs, e_2 occurs later." But before this can be considered precise, we must specify how much later e_2 is to occur. Thus the principle becomes:—

"Given any event e_1, there is an event e_2 and a time-interval τ such that, whenever e_1 occurs, e_2 follows after an interval τ."

I am not concerned as yet to consider whether this law is true or false. For the present, I am merely concerned to discover what the law of causality is supposed to be. I pass, therefore, to the other definitions quoted above.

The second definition need not detain us long, for two reasons. First, because it is psychological: not the "thought or perception" of a process, but the process itself, must be what concerns us in considering causality. Secondly, because it is circular: in speaking of a process as "taking place in consequence of" another process, it introduces the very notion of because which was to be defined.

The third definition is by far the most precise; indeed as regards clearness it leaves nothing to be desired. But a great difficulty is caused by the temporal contiguity of cause and effect which the definition asserts. No two instants are contiguous, since the time-series is compact; hence either the cause or the effect or both must, if the definition is correct, endure for a finite time; indeed, by the wording of the definition it is plain that both are assumed to endure for a finite time. But then we are faced with a dilemma: if the cause is a process involving change within itself, we shall require (if causality is universal) causal relations between its earlier and later parts; moreover, it would seem that only the later parts can be relevant to the effect, since the earlier parts are not contiguous to the effect, and therefore (by the definition) cannot influence the effect. Thus we shall be led to diminish the duration of the cause without limit, and however much we may diminish it, there will still remain an earlier part which might be altered without altering the effect, so that the true cause, as defined, will not have been reached, for it will be observed that the definition excludes plurality of causes. If, on the other hand, the cause is purely static, involving no change within itself, then, in the first place, no such cause is to be found in nature, and in the second place, it seems strange—too strange to be accepted, in spite of bare logical possibility—that the cause, after existing placidly for some time, should suddenly explode into the effect, when it might just as well have done so at any earlier time, or have gone on unchanged without producing its effect. This dilemma, therefore, is fatal to the view that cause and effect can be contiguous in time; if there are causes and effects, they must be separated by a finite time-interval τ, as was assumed in the above interpretation of the first definition.

What is essentially the same statement of the law of causality as the one elicited above from the first of Baldwin's definitions is given by other philosophers. Thus John Stuart Mill says:—

"The Law of Causation, the recognition of which is the main pillar of inductive science, is but the familiar truth, that invariability of succession is found by observation to obtain between every fact in nature and some other fact which has preceded it."

And Bergson, who has rightly perceived that the law as stated by philosophers is worthless, nevertheless continues to suppose that it is used in science. Thus he says:—

"Now, it is argued, this law [the law of causality] means that every phenomenon is determined by its conditions, or, in other words, that the same causes produce the same effects."

And again:—

"We perceive physical phenomena, and these phenomena obey laws. This means: (1) That phenomena *a, b, c, d*, previously perceived, can occur again in the same shape; (2) that a certain phenomenon P, which appeared after the conditions *a, b, c, d*, and after these conditions only, will not fail to recur as soon as the same conditions are again present."

A great part of Bergson's attack on science rests on the assumption that it employs this principle. In fact, it employs no such principle, but philosophers—even Bergson—are too apt to take their views on science from each other, not from science. As to what the principle is, there is a fair consensus among philosophers of different schools. There are, however, a number of difficulties which at once arise. I omit the question of plurality of causes for the present, since other graver questions have to be considered. Two of these, which are forced on our attention by the above statement of the law, are the following:—

(1) What is meant by an "event"?

(2) How long may the time-interval be between cause and effect?

(1) An "event," in the statement of the law, is obviously intended to be something that is likely to recur since otherwise the law becomes trivial. It follows that an "event" is not a particular, but some universal of which there may be many instances. It follows also that an "event" must be something short of the whole state of the universe, since it is highly improbable that this will recur. What is meant by an "event" is something like striking a match, or dropping a penny into the slot of an automatic machine. If such an event is to recur, it must not be defined too narrowly: we must not state with what degree of force the match is to be struck, nor what is to be the temperature of the penny. For if such

considerations were relevant, our "event" would occur at most once, and the law would cease to give information. An "event," then, is a universal defined sufficiently widely to admit of many particular occurrences in time being instances of it.

(2) The next question concerns the time-interval. Philosophers, no doubt, think of cause and effect as contiguous in time, but this, for reasons already given, is impossible. Hence, since there are no infinitesimal time-intervals, there must be some finite lapse of time τ between cause and effect. This, however, at once raises insuperable difficulties. However short we make the interval τ, something may happen during this interval which prevents the expected result. I put my penny in the slot, but before I can draw out my ticket there is an earthquake which upsets the machine and my calculations. In order to be sure of the expected effect, we must know that there is nothing in the environment to interfere with it. But this means that the supposed cause is not, by itself, adequate to insure the effect. And as soon as we include the environment, the probability of repetition is diminished, until at last, when the whole environment is included, the probability of repetition becomes almost *nil.*

In spite of these difficulties, it must, of course, be admitted that many fairly dependable regularities of sequence occur in daily life. It is these regularities that have suggested the supposed law of causality; where they are found to fail, it is thought that a better formulation could have been found which would have never failed. I am far from denying that there may be such sequences which in fact never do fail. It may be that there will never be an exception to the rule that when a stone of more than a certain mass, moving with more than a certain velocity, comes in contact with a pane of glass of less than a certain thickness, the glass breaks. I also do not deny that the observation of such regularities, even when they are not without exceptions, is useful in the infancy of a science: the observation that unsupported bodies in air usually fall was a stage on the way to the law of gravitation. What I deny is that science assumes the existence of invariable uniformities of sequence of this kind, or that it aims at discovering them. All such uniformities, as we saw, depend upon a certain vagueness in the definition of the "events." That bodies fall is a vague qualitative statement; science wishes to know how fast they fall. This depends upon the shape of the bodies and the density of the air. It is true that there is more nearly uniformity when they fall in a vacuum; so far as Galileo could observe, the uniformity is then complete. But later it appeared that even there the latitude made a difference, and the altitude. Theoretically, the position of the sun and moon must make a difference. In short, every advance in a science takes us farther away from the crude uniformities which are first observed, into greater differentiation of antecedent and consequent, and into a continually wider circle of antecedents recognised as relevant.

The principle "same cause, same effect," which philosophers imagine to be vital to science, is therefore utterly otiose. As soon as the antecedents have been given sufficiently fully to enable the consequent to be calculated with some exactitude, the antecedents have become so complicated that it is very unlikely they will ever recur. Hence, if this were the principle involved, science would remain utterly sterile.

The importance of these considerations lies partly in the fact that they lead to a more correct account of scientific procedure, partly in the fact that they remove the analogy with human volition which makes the conception of cause such a fruitful source of fallacies. The latter point will become clearer by the help of some illustrations. For this purpose, I shall consider a few maxims which have played a great part in the history of philosophy.

(1) "Cause and effect must more or less resemble each other." This principle was prominent in the philosophy of occasionalism, and is still by no means extinct. It is still often thought, for example, that mind could not have grown up in a universe which previously contained nothing mental, and one ground for this belief is that matter is too dissimilar from mind to have been able to cause it. Or, more particularly, what are termed the nobler parts of our nature are supposed to be inexplicable, unless the universe always contained something at least equally noble which could cause them. All such views seem to depend upon assuming some unduly simplified law of causality; for, in any legitimate sense of "cause" and "effect," science seems to show that they are usually very widely dissimilar, the "cause" being, in fact, two states of the whole universe, and the "effect" some particular event.

(2) "Cause is analogous to volition, since there must be an intelligible *nexus* between cause and effect." This maxim is, I think, often unconsciously in the imaginations of philosophers who would reject it when explicitly stated. It is probably operative in the view we have just been considering, that mind could not have resulted from a purely material world. I do not profess to know what is meant by "intelligible"; it seems to mean "familiar to imagination." Nothing is less "intelligible," in any other sense, than the connection between an act of will and its fulfilment. But obviously the sort of nexus desired between cause and effect is such as could only hold between the "events" which the supposed law of causality contemplates; the laws which replace causality in such a science as physics leave no room for any two events between which a nexus could be sought.

(3) "The cause *compels* the effect in some sense in which the effect does not compel the cause." This belief seems largely operative in the dislike of determinism; but, as a matter of fact, it is connected with our second maxim, and falls as soon as that is abandoned. We may define "compulsion" as follows: "Any

set of circumstances is said to compel A when A desires to do something which the circumstances prevent, or to abstain from something which the circumstances cause." This presupposes that some meaning has been found for the word "cause"—a point to which I shall return later. What I want to make clear at present is that compulsion is a very complex notion, involving thwarted desire. So long as a person does what he wishes to do, there is no compulsion, however much his wishes may be calculable by the help of earlier events. And where desire does not come in, there can be no question of compulsion. Hence it is, in general, misleading to regard the cause as compelling the effect.

A vaguer form of the same maxim substitutes the word "determine" for the word "compel"; we are told that the cause *determines* the effect in a sense in which the effect does not *determine* the cause. It is not quite clear what is meant by "determining"; the only precise sense, so far as I know, is that of a function or one-many relations. If we admit plurality of causes, but not of effects, that is, if we suppose that, given the cause, the effect must be such and such, but, given the effect, the cause may have been one of many alternatives, then we may say that the cause determines the effect, but not the effect the cause. Plurality of causes, however, results only from conceiving the effect vaguely and narrowly and the cause precisely and widely. Many antecedents may "cause" a man's death, because his death is vague and narrow. But if we adopt the opposite course, taking as the "cause" the drinking of a dose of arsenic, and as the "effect" the whole state of the world five minutes later, we shall have plurality of effects instead of plurality of causes. Thus, the supposed lack of symmetry between "cause" and "effect" is illusory.

(4) "A cause cannot operate when it has ceased to exist, because what has ceased to exist is nothing." This is a common maxim, and a still more common unexpressed prejudice. It has, I fancy, a good deal to do with the attractiveness of Bergson's "*durée*": since the past has effects now, it must still exist in some sense. The mistake in this maxim consists in the supposition that causes "operate" at all. A volition "operates" when what it wills takes place; but nothing can operate except a volition. The belief that causes "operate" results from assimilating them, consciously or unconsciously, to volitions. We have already seen that, if there are causes at all, they must be separated by a finite interval of time from their effects, and thus cause their effects after they have ceased to exist.

It may be objected to the above definition of a volition "operating" that it only operates when it "causes" what it wills, not when it merely happens to be followed by what it wills. This certainly represents the usual view of what is meant by a volition "operating," but as it involves the very view of causation which we are engaged in combating, it is not open to us as a definition. We may say that a volition "operates" when there is some law in virtue of which a similar volition in rather similar circumstances will usually be followed by what it wills.

But this is a vague conception, and introduces ideas which we have not yet considered. What is chiefly important to notice is that the usual notion of "operating" is not open to us if we reject, as I contend that we should, the usual notion of causation.

(5) "A cause cannot operate except where it is." This maxim is very widespread; it was urged against Newton, and has remained a source of prejudice against "action at a distance." In philosophy it has led to a denial of transient action, and thence to monism or Leibnizian monadism. Like the analogous maxim concerning temporal contiguity, it rests upon the assumption that causes "operate," i.e. that they are in some obscure way analogous to volitions. And, as in the case of temporal contiguity, the inferences drawn from this maxim are wholly groundless.

I return now to the question, What law or laws can be found to take the place of the supposed law of causality?

First, without passing beyond such uniformities of sequence as are contemplated by the traditional law, we may admit that, if any such sequence has been observed in a great many cases, and has never been found to fail, there is an inductive probability that it will be found to hold in future cases. If stones have hitherto been found to break windows, it is probable that they will continue to do so. This, of course, assumes the inductive principle, of which the truth may reasonably be questioned; but as this principle is not our present concern, I shall in this discussion treat it as indubitable. We may then say, in the case of any such frequently observed sequence, that the earlier event is the *cause* and the later event the *effect*.

Several considerations, however, make such special sequences very different from the traditional relation of cause and effect. In the first place, the sequence, in any hitherto unobserved instance, is no more than probable, whereas the relation of cause and effect was supposed to be necessary. I do not mean by this merely that we are not sure of having discovered a true case of cause and effect; I mean that, even when we have a case of cause and effect in our present sense, all that is meant is that on grounds of observation, it is probable that when one occurs the other will also occur. Thus in our present sense, A may be the cause of B even if there actually are cases where B does not follow A. Striking a match will be the cause of its igniting, in spite of the fact that some matches are damp and fail to ignite.

In the second place, it will not be assumed that *every* event has some antecedent which is its cause in this sense; we shall only believe in causal sequences where we find them, without any presumption that they always are to be found.

In the third place, *any* case of sufficiently frequent sequence will be causal in our present sense; for example, we shall not refuse to say that night is the cause of day. Our repugnance to saying this arises from the ease with which we can imagine the sequence to fail, but owing to the fact that cause and effect must be separated by a finite interval of time, *any* such sequence *might* fail through the interposition of other circumstances in the interval. Mill, discussing this instance of night and day, says:—

"It is necessary to our using the word cause, that we should believe not only that the antecedent always *has* been followed by the consequent, but that as long as the present constitution of things endures, it always *will* be so."

In this sense, we shall have to give up the hope of finding causal laws such as Mill contemplated; any causal sequence which we have observed may at any moment be falsified without a falsification of any laws of the kind that the more advanced sciences aim at establishing.

In the fourth place, such laws of probable sequence, though useful in daily life and in the infancy of a science, tend to be displaced by quite different laws as soon as a science is successful. The law of gravitation will illustrate what occurs in any advanced science. In the motions of mutually gravitating bodies, there is nothing that can be called a cause, and nothing that can be called an effect; there is merely a formula. Certain differential equations can be found, which hold at every instant for every particle of the system, and which, given the configuration and velocities at one instant, or the configurations at two instants, render the configuration at any other earlier or later instant theoretically calculable. That is to say, the configuration at any instant is a function of that instant and the configurations at two given instants. This statement holds throughout physics, and not only in the special case of gravitation. But there is nothing that could be properly called "cause" and nothing that could be properly called "effect" in such a system.

No doubt the reason why the old "law of causality" has so long continued to pervade the books of philosophers is simply that the idea of a function is unfamiliar to most of them, and therefore they seek an unduly simplified statement. There is no question of repetitions of the "same" cause producing the "same" effect; it is not in any sameness of causes and effects that the constancy of scientific law consists, but in sameness of relations. And even "sameness of relations" is too simple a phrase; "sameness of differential equations" is the only correct phrase. It is impossible to state this accurately in non-mathematical language; the nearest approach would be as follows: "There is a constant relation between the state of the universe at any instant and the rate of change in the rate at which any part of the universe is changing at that instant, and this relation is many-one, i.e. such that the rate of change in the rate of change is determinate when the state of the universe is given." If the "law of causality" is to be something actually discoverable in the practice of science, the above proposition has a better right to the name than any "law of causality" to be found in the books of philosophers.

In regard to the above principle, several observations must be made—

(1) No one can pretend that the above principle is *a priori* or self-evident or a "necessity of thought." Nor is it, in any sense, a premiss of science: it is an empirical generalisation from a number of laws which are themselves empirical generalisations.

(2) The law makes no difference between past and future: the future "determines" the past in exactly the same sense in which the past "determines" the future. The word "determine," here, has a purely logical significance: a certain number of variables "determine" another variable if that other variable is a function of them.

(3) The law will not be empirically verifiable unless the course of events within some sufficiently small volume will be approximately the same in any two states of the universe which only differ in regard to what is at a considerable distance from the small volume in question. For example, motions of planets in the solar system must be approximately the same however the fixed stars may be distributed, provided that all the fixed stars are very much farther from the sun than the planets are. If gravitation varied directly as the distance, so that the most remote stars made the most difference to the motions of the planets, the world might be just as regular and just as much subject to mathematical laws as it is at present, but we could never discover the fact.

(4) Although the old "law of causality" is not assumed by science, something which we may call the "uniformity of nature" is assumed, or rather is accepted on inductive grounds. The uniformity of nature does not assert the trivial principle "same cause, same effect," but the principle of the permanence of laws. That is to say, when a law exhibiting, e.g. an acceleration as a function of the configuration has been found to hold throughout the observable past, it is expected that it will continue to hold in the future, or that, if it does not itself hold, there is some other law, agreeing with the supposed law as regards the past, which will hold for the future. The ground of this principle is simply the inductive ground that it has been found to be true in very many instances; hence the principle cannot be considered certain, but only probable to a degree which cannot be accurately estimated.

The uniformity of nature, in the above sense, although it is assumed in the practice of science, must not, in its generality, be regarded as a kind of major premiss, without which all scientific reasoning would be in error. The assumption that *all* laws of nature are permanent has, of course, less probability than the assumption that this or that particular law is permanent; and the assumption that a particular law is permanent for all time has less probability than the assumption that it will be valid up to such and such a date. Science, in any given case, will assume what the case requires, but no more. In constructing the *Nautical Almanac* for 1915 it will assume that the law of gravitation will remain true up to the end of that year; but it will make no assumption as to 1916 until it comes to the next volume of the almanac. This procedure is, of course, dictated by the fact that the uniformity of nature is not known *a priori*, but is an empirical generalisation, like "all men are mortal." In all such cases, it is better to argue immediately from the given particular instances to the new instance, than to argue by way of a major premiss; the conclusion is only probable in either case, but acquires a higher probability by the former method than by the latter.

In all science we have to distinguish two sorts of laws: first, those that are empirically verifiable but probably only approximate; secondly, those that are not verifiable, but may be exact. The law of gravitation, for example, in its applications to the solar system, is only empirically verifiable when it is assumed that matter outside the solar system may be ignored for such purposes; we believe this to be only approximately true, but we cannot empirically verify the law of universal gravitation which we believe to be exact. This point is very important in connection with what we may call "relatively isolated systems." These may be defined as follows:—

A system relatively isolated during a given period is one which, within some assignable margin of error, will behave in the same way throughout that period, however the rest of the universe may be constituted.

A system may be called "practically isolated" during a given period if, although there *might* be states of the rest of the universe which would produce more than the assigned margin of error, there is reason to believe that such states do not in fact occur.

Strictly speaking, we ought to specify the respect in which the system is relatively isolated. For example, the earth is relatively isolated as regards falling bodies, but not as regards tides; it is *practically* isolated as regards economic phenomena, although, if Jevons' sunspot theory of commercial crises had been true, it would not have been even practically isolated in this respect.

It will be observed that we cannot prove in advance that a system is isolated. This will be inferred from the observed fact that approximate uniformities can be stated for this system alone. If the complete laws for the whole universe were known, the isolation of a system could be deduced from them; assuming, for example, the law of universal gravitation, the practical isolation of the solar system in this respect can be deduced by the help of the fact that there is very little matter in its neighbourhood. But it should be observed that isolated systems are only important as providing a possibility of *discovering* scientific laws; they have no theoretical importance in the finished structure of a science.

The case where one event A is said to "cause" another event B, which philosophers take as fundamental, is really only the most simplified instance of a practically isolated system. It may happen that, as a result of general scientific laws, whenever A occurs throughout a certain period, it is followed by B; in that case, A and B form a system which is practically isolated throughout that period. It is, however, to be regarded as a piece of good fortune if this occurs; it will always be due to special circumstances, and would not have been true if the rest of the universe had been different though subject to the same laws.

The essential function which causality has been supposed to perform is the possibility of inferring the future from the past, or, more generally, events at any time from events at certain assigned times. Any system in which such inference is possible may be called a "deterministic" system. We may define a deterministic system as follows:—

A system is said to be "deterministic" when, given certain data, $e_1, e_2, ..., e_n$, at times $t_1, t_2, ..., t_n$ respectively, concerning this system, if E_t is the state of the system at any time t, there is a functional relation of the form

$$E_t = f(e_1, t_1, e_2, t_2, ..., e_n, t_n, t). \quad \text{(A)}$$

The system will be "deterministic throughout a given period" if t, in the above formula, may be any time within that period, though outside that period the formula may be no longer true. If the universe, as a whole, is such a system, determinism is true of the universe; if not, not. A system which is part of a deterministic system I shall call "determined"; one which is not part of any such system I shall call "capricious."

The events $e_1, e_2, ..., e_n$ I shall call "determinants" of the system. It is to be observed that a system which has one set of determinants will in general have many. In the case of the motions of the planets, for example, the configurations of the solar system at any two given times will be determinants.

We may take another illustration from the hypothesis of psycho-physical parallelism. Let us assume, for the purposes of this illustration, that to a given state of brain a given state of mind always corresponds, and vice versa, i.e. that there is a one-one relation between them, so that each is a function of the other. We may also assume, what is practically certain, that to a given state of a certain brain a given state of the whole material universe corresponds, since it is highly improbable that a given brain is ever twice in exactly the same state. Hence there will be a one-one relation between the state of a given person's mind and the state of the whole material universe. It follows that, if n states of the material universe are determinants of the material universe, then n states of a given man's mind are determinants of the whole material and mental universe—assuming, that is to say, that psycho-physical parallelism is true.

The above illustration is important in connection with a certain confusion which seems to have beset those who have philosophised on the relation of mind and matter. It is often thought that, if the state of the mind is determinate when the state of the brain is given, and if the material world forms a deterministic system, then mind is "subject" to matter in some sense in which matter is not "subject" to mind. But if the state of the brain is also determinate when the state of the mind is given, it must be exactly as true to regard matter as subject to mind as it would be to regard mind as subject to matter. We could, theoretically, work out the history of mind without ever mentioning matter, and then, at the end, deduce that matter must meanwhile have gone through the corresponding history. It is true that if the relation of brain to mind were many-one, not one-one, there would be a one-sided dependence of mind on brain, while conversely, if the relation were one-many, as Bergson supposes, there would be a one-aided dependence of brain on mind. But the dependence involved is, in any case, only logical; it does not mean that we shall be compelled to do things we desire not to do, which is what people instinctively imagine it to mean.

As another illustration we may take the case of mechanism and teleology. A system may be defined as "mechanical" when it has a set of determinants that are purely material, such as the positions of certain pieces of matter at certain times. It is an open question whether the world of mind and matter, as we know it, is a mechanical system or not; let us suppose, for the sake of argument, that it is a mechanical system. This supposition—so I contend—throws no light whatever on the question whether the universe is or is not a "teleological" system. It is difficult to define accurately what is meant by a "teleological" system, but the argument is not much affected by the particular definition we adopt. Broadly, a teleological system is one in which purposes are realised, i.e. in which certain desires—those that are deeper or nobler or more fundamental or more universal or what not—are followed by their realisation. Now the fact—if it be a fact—that the universe is mechanical has no bearing whatever on the question whether it is teleological in the above sense. There might be a mechanical system in which all wishes were realised, and there might be one in which all wishes were thwarted. The question whether, or how far, our actual world is teleological, cannot, therefore, be settled by proving that it is mechanical, and the desire that it should be teleological is no ground for wishing it to be not mechanical.

There is, in all these questions, a very great difficulty in avoiding confusion between what we can infer and what is in fact determined. Let us consider, for a moment, the various senses in which the future may be "determined." There is one sense—and a very important one—in which it is determined quite independently of scientific laws, namely, the sense that it will be what it will be. We all regard the past as determined simply by the fact that it has happened; but for the accident that memory works backward and not forward, we should regard the future as equally determined by the fact that it will happen. "But," we are told, "you cannot alter the past, while you can to some extent alter the future." This view seems to me to rest upon just those errors in regard to causation which it has been my object to remove. You cannot make the past other than it was—true, but this is a mere application of the law of contradiction. If you already know what the past was, obviously it is useless to wish it different. But also you cannot make the future other than it will be; this again is an application of the law of contradiction. And if you happen to know the future—e.g. in the case of a forthcoming eclipse—it is just as useless to wish it different as to wish the past different. "But," it will be rejoined, "our wishes can *cause* the future, sometimes, to be different from what it would be if they did not exist, and they can have no such effect upon the past." This, again, is a mere tautology. An effect being *defined* as something subsequent to its cause, obviously we can have no *effect* upon the past. But that does not mean that the past would not have been different if our present wishes had been different. Obviously, our present wishes are conditioned by the past, and therefore could not have been different unless the past had been different; therefore, if our present wishes were different, the past would be different. Of course, the past cannot be different from what it was, but no more can our present wishes be different from what they are; this again is merely the law of contradiction. The facts seem to be merely (1) that wishing generally depends upon ignorance, and is therefore commoner in regard to the future than in regard to the past; (2) that where a wish concerns the future, it and its realisation very often form a "practically independent system," i.e. many wishes regarding the future are realised. But there seems no doubt that the main difference in our feelings arises from the accidental fact that the past but not the future can be known by memory.

Although the sense of "determined" in which the future is determined by the mere fact that it will be what it will be is sufficient (at least so it seems to me) to refute some opponents of determinism, notably M. Bergson and the pragmatists, yet it is not what most people have in mind when they speak of the future as determined. What they have in mind is a formula by means of which the future can be exhibited, and at least theoretically calculated, as a function of the past. But at this point we meet with a great difficulty, which besets what has been said above about deterministic systems, as well as what is said by others.

If formulæ of any degree of complexity, however great, are admitted, it would seem that any system, whose state at a given moment is a function of certain measurable quantities, must be a deterministic system. Let us consider, in illustration, a single material particle, whose co-ordinates at time t are x_t, y_t, z_t. Then, however, the particle moves, there must be, theoretically, functions f_1, f_2, f_3, such that

$$x_t = f_t(t), \quad y_t = f_2(t), \quad z_t = f_3(t).$$

It follows that, theoretically, the whole state of the material universe at time t must be capable of being exhibited as a function of t. Hence our universe will be deterministic in the sense defined above. But if this be true, no information is conveyed about the universe in stating that it is deterministic. It is true that the formulæ involved may be of strictly infinite complexity, and therefore not practically capable of being written down or apprehended. But except from the point of view of our knowledge, this might seem to be a detail: in itself, if the above considerations are sound, the material universe *must* be deterministic, *must* be subject to laws.

This, however, is plainly not what was intended. The difference between this view and the view intended may be seen as follows. Given some formula which fits the facts hitherto—say the law of gravitation—there will be an infinite number of other formulæ, not empirically distinguishable from it in the past, but diverging from it more and more in the future. Hence, even assuming that there are persistent laws, we shall have no reason for assuming that the law of the inverse square will hold in future; it may be some other hitherto indistinguishable law that will hold. We cannot say that *every* law which has held hitherto must hold in the future, because past facts which obey one law will also obey others, hitherto indistinguishable but diverging in future. Hence there must, at every moment, be laws hitherto unbroken which are now broken for the first time. What science does, in fact, is to select the *simplest* formula that will fit the facts. But this, quite obviously, is merely a methodological precept, not a law of Nature. If the simplest formula ceases, after a time, to be applicable, the simplest formula that remains applicable is selected, and science has no sense that an axiom has been falsified. We are thus left with the brute fact that, in many departments of science, quite simple laws have hitherto been found to hold. This fact cannot be regarded as having any *a priori* ground, nor can it be used to support inductively the opinion that the same laws will continue; for at every moment laws hitherto true are being falsified, though in the advanced sciences these laws are less simple than those that have remained true. Moreover it would be fallacious to argue inductively from the state of the advanced sciences to the future state of the others, for it may well be that the advanced sciences are advanced simply because, hitherto, their subject-matter has obeyed simple and easily ascertainable laws, while the subject-matter of other sciences has not done so.

The difficulty we have been considering seems to be met partly, if not wholly, by the principle that the *time* must not enter explicitly into our formulæ. All mechanical laws exhibit acceleration as a function of configuration, not of configuration and time jointly; and this principle of the irrelevance of the time may be extended to all scientific laws. In fact we might interpret the "uniformity of nature" as meaning just this, that no scientific law involves the time as an argument, unless, of course, it is given in an integrated form, in which case *lapse* of time, though not absolute time, may appear in our formulæ. Whether this consideration suffices to overcome our difficulty completely, I do not know; but in any case it does much to diminish it.

It will serve to illustrate what has been said if we apply it to the question of free will.

(1) Determinism in regard to the will is the doctrine that our volitions belong to some deterministic system, i.e. are "determined" in the sense defined above. Whether this doctrine is true or false, is a mere question of fact; no *a priori* considerations (if our previous discussions have been correct) can exist on either side. On the one hand, there is no *a priori* category of causality, but merely certain observed uniformities. As a matter of fact, there are observed uniformities in regard to volitions; thus there is some empirical evidence that volitions are determined. But it would be very rash to maintain that the evidence is overwhelming, and it is quite possible that some volitions, as well as some other things, are not determined, except in the sense in which we found that everything must be determined.

(2) But, on the other hand, the subjective sense of freedom, sometimes alleged against determinism, has no bearing on the question whatever. The view that it has a bearing rests upon the belief that causes compel their effects, or that nature enforces obedience to its laws as governments do. These are mere anthropomorphic superstitions, due to assimilation of causes with volitions and of natural laws with human edicts. We feel that our will is not compelled, but that only means that it is not other than we choose it to be. It is one of the demerits of the traditional theory of causality that it has created an artificial opposition between determinism and the freedom of which we are introspectively conscious.

(3) Besides the general question whether volitions are determined, there is the further question whether they are *mechanically* determined, i.e. whether they are part of what was above defined as a mechanical system. This is the question whether they form part of a system with purely material determinants, i.e. whether there are laws which, given certain material data, make all volitions functions of those data. Here again, there is empirical evidence up to a point, but it is not conclusive in regard to all volitions. It is important to observe, however that even if volitions are part of a mechanical system, this by no means implies any supremacy of matter over mind. It may well be that the same system which is susceptible of material determinants is also susceptible of mental determinants; thus a mechanical system may be determined by sets of volitions, as well as by sets of material facts. It would seem, therefore, that the reasons which make people dislike the view that volitions are mechanically determined are fallacious.

(4) The notion of *necessity*, which is often associated with determinism, is a confused notion not legitimately deducible from determinism. Three meanings are commonly confounded when necessity is spoken of:—

(α) An *action* is necessary when it will be performed however much the agent may wish to do otherwise. Determinism does not imply that actions are necessary in this sense.

(β) A *propositional function* is necessary when all its values are true. This sense is not relevant to our present discussion.

(γ) A *proposition* is necessary with respect to a given constituent when it is the value, with that constituent as argument, of a necessary propositional function, in other words, when it remains true however that constituent may be varied. In this sense, in a deterministic system, the connection of a volition with its determinants is necessary, if the time at which the determinants occur be taken as the constituent to be varied, the time-interval between the determinants and the volition being kept constant. But this sense of necessity is purely logical, and has no emotional importance.

We may now sum up our discussion of causality. We found first that the law of causality, as usually stated by philosophers, is false, and is not employed in science. We then considered the nature of scientific laws, and found that, instead of stating that one event A is always followed by another event B, they stated functional relations between certain events at certain times, which we called determinants, and other events at earlier or later times or at the same time. We were unable to find any *a priori* category involved: the existence of scientific laws appeared as a purely empirical fact, not necessarily universal, except in a trivial and scientifically useless form. We found that a system with one set of determinants may very likely have other sets of a quite different kind, that, for example, a mechanically determined system may also be teleologically or volitionally determined. Finally we considered the problem of free will: here we found that the reasons for supposing volitions to be determined are strong but not conclusive, and we decided that even if volitions are mechanically determined, that is no reason for denying freedom in the sense revealed by introspection, or for supposing that mechanical events are not determined by volitions. The problem of free will *versus* determinism is therefore, if we were right, mainly illusory, but in part not yet capable of being decisively solved.

X

KNOWLEDGE BY ACQUAINTANCE AND KNOWLEDGE BY DESCRIPTION

The object of the following paper is to consider what it is that we know in cases where we know propositions about "the so-and-so" without knowing who or what the so-and-so is. For example, I know that the candidate who gets most votes will be elected, though I do not know who is the candidate who will get most votes. The problem I wish to consider is: What do we know in these cases, where the subject is merely described? I have considered this problem elsewhere from a purely logical point of view; but in what follows I wish to consider the question in relation to theory of knowledge as well as in relation to logic, and in view of the above-mentioned logical discussions, I shall in this paper make the logical portion as brief as possible.

In order to make clear the antithesis between "acquaintance" and "description," I shall first of all try to explain what I mean by "acquaintance." I say that I am *acquainted* with an object when I have a direct cognitive relation to that object, i.e. when I am directly aware of the object itself. When I speak of a cognitive relation here, I do not mean the sort of relation which constitutes judgment, but the sort which constitutes presentation. In fact, I think the relation of subject and object which I call acquaintance is simply the converse of the relation of object and subject which constitutes presentation. That is, to say that S has acquaintance with O is essentially the same thing as to say that O is presented to S. But the associations and natural extensions of the word *acquaintance* are different from those of the word *presentation*. To begin with, as in most cognitive words, it is natural to say that I am acquainted with an object even at moments when it is not actually before my mind, provided it has been before my mind, and will be again whenever occasion arises. This is the same sense in which I am said to know that 2+2=4 even when I am thinking of something else. In the second place, the word *acquaintance* is designed to emphasise, more than the word *presentation*, the relational character of the fact with which we are concerned. There is, to my mind, a danger that, in speaking of presentation, we may so emphasise the object as to lose sight of the subject. The result of this is either to lead to the view that there is no subject, whence we arrive at materialism; or to lead to the view that what is presented is part of the subject, whence we arrive at idealism, and should arrive at solipsism but for the most desperate contortions. Now I wish to preserve the dualism of subject and object in my terminology, because this dualism seems to me a fundamental fact concerning cognition. Hence, I prefer the word *acquaintance* because it emphasises the need of a subject which is acquainted.

When we ask what are the kinds of objects with which we are acquainted, the first and most obvious example is *sense-data*. When I see a colour or hear a noise, I have direct acquaintance with the colour or the noise. The sense-datum with which I am acquainted in these cases is generally, if not always, complex. This is particularly obvious in the case of sight. I do not mean, of course, merely that the supposed physical object is complex, but that the direct sensible object is complex and contains parts with spatial relations. Whether it is possible to be aware of a complex without being aware of its constituents is not an easy question, but on the whole, it would seem that there is no reason why it should not be possible. This question arises in an acute form in connection with self-consciousness, which we must now briefly consider.

In introspection, we seem to be immediately aware of varying complexes, consisting of objects in various cognitive and conative relations to ourselves. When I see the sun, it often happens that I am aware of my seeing the sun, in addition to being aware of the sun; and when I desire food, it often happens that I am aware of my desire for food. But it is hard to discover any state of mind in which I am aware of myself alone, as opposed to a complex of which I am a constituent. The question of the nature of self-consciousness is too large and too slightly connected with our subject, to be argued at length here. It is difficult, but probably not impossible, to account for plain facts if we assume that we do not have acquaintance with ourselves. It is plain that we are not only *acquainted* with the complex "Self-acquainted-with-A," but we also *know* the proposition "I am acquainted with A." Now here the complex has been analysed, and if "I" does not stand for something which is a direct object of acquaintance, we shall have to suppose that "I" is something known by description. If we wished to maintain the view that there is no acquaintance with Self, we might argue as follows: We are acquainted with *acquaintance*, and we know that it is a relation. Also we are acquainted with a complex in which we perceive that acquaintance is the relating relation. Hence, we know that this complex must have a constituent which is that which is acquainted, i.e. must have a subject-term as well as an object-term. This subject-term we define as "I." Thus "I" means "the subject-term in awareness of which *I* am aware." But as a definition this cannot be regarded as a happy effort. It would seem necessary, therefore, either to suppose that I am acquainted with myself, and that "I," therefore, requires no definition, being merely the proper name of a certain object, or to find some other analysis of self-consciousness. Thus self-consciousness cannot be regarded as throwing light on the question whether we can know a complex without knowing its constituents. This question, however, is not important for our present purposes, and I shall therefore not discuss it further.

The awarenesses we have considered so far have all been awarenesses of particular existents, and might all in a large sense be called sense-data. For, from the point of view of theory of knowledge, introspective knowledge is exactly on a level with knowledge derived from sight or hearing. But, in addition to awareness of the above kind of objects, which may be called awareness of *particulars*; we have also (though not quite in the same sense) what may be called awareness of *universals*. Awareness of universals is called *conceiving*, and a universal of which we are aware is called a *concept*. Not only are we aware of particular yellows, but if we have seen a sufficient number of yellows and have sufficient intelligence, we are aware of the universal *yellow*; this universal is the subject in such judgments as "yellow differs from blue" or "yellow resembles blue less than green does." And the universal yellow is the predicate in such judgments as "this is yellow," where "this" is a particular sense-datum. And universal relations, too,are objects of awarenesses; up and down, before and after, resemblance, desire, awareness itself, and so on, would seem to be all of them objects of which we can be aware.

In regard to relations, it might be urged that we are never aware of the universal relation itself, but only of complexes in which it is a constituent. For example, it may be said that we do not know directly such a relation as *before*, though we understand such a proposition as "this is before that," and may be directly aware of such a complex as "this being before that." This view, however, is difficult to reconcile with the fact that we often know propositions in which the relation is the subject, or in which the relata are not definite given objects, but "anything." For example, we know that if one thing is before another, and the other before a third, then the first is before the third; and here the things concerned are not definite things, but "anything." It is hard to see how we could know such a fact about "before" unless we were acquainted with "before," and not merely with actual particular cases of one given object being before another given object. And more directly: A judgment such as "this is before that," where this judgment is derived from awareness of a complex, constitutes an analysis, and we should not understand the analysis if we were not acquainted with the meaning of the terms employed. Thus, we must suppose that we are acquainted with the meaning of "before," and not merely with instances of it.

There are thus at least two sorts of objects of which we are aware, namely, particulars and universals. Among particulars I include all existents, and all complexes of which one or more constituents are existents, such as this-before-that, this-above-that, the-yellowness-of-this. Among universals I include all objects of which no particular is a constituent. Thus, the disjunction "universal-particular" includes all objects. We might also call it the disjunction "abstract-concrete." It is not quite parallel with the opposition "concept-percept," because things remembered or imagined belong with particulars, but can hardly be called percepts. (On the other hand, universals with which we are acquainted may be identified with concepts.)

It will be seen that among the objects with which we are acquainted are not included physical objects (as opposed to sense-data), nor other people's minds. These things are known to us by what I call "knowledge by description," which we must now consider.

By a "description" I mean any phrase of the form "a so-and-so" or "the so-and-so." A phrase of the form "a so-and-so" I shall call an "ambiguous" description; a phrase of the form "the so-and-so" (in the singular) I shall call a "definite" description. Thus "a man" is an ambiguous description, and "the man with the iron mask" is a definite description. There are various problems connected with ambiguous descriptions, but I pass them by, since they do not directly concern the matter I wish to discuss. What I wish to discuss is the nature of our knowledge concerning objects in cases where we know that there is an object answering to a definite description, though we are not *acquainted* with any such object. This is a matter which is concerned exclusively with *definite* descriptions. I shall, therefore, in the sequel, speak simply of "descriptions" when I mean "definite descriptions." Thus a description will mean any phrase of the form "the so-and-so" in the singular.

I shall say that an object is "known by description" when we know that it is "*the* so-and-so," i.e. when we know that there is one object, and no more, having a certain property; and it will generally be implied that we do not have knowledge of the same object by acquaintance. We know that the man with the iron mask existed, and many propositions are known about him; but we do not know who he was. We know that the candidate who gets most votes will be elected, and in this case we are very likely also acquainted (in the only sense in which one can be acquainted with some one else) with the man who is, in fact, the candidate who will get most votes, but we do not know which of the candidates he is, i.e. we do not know any proposition of the form "A is the candidate who will get most votes" where A is one of the candidates by name. We shall say that we have "*merely* descriptive knowledge" of the so-and-so when, although we know that the so-and-so exists, and although we may possibly be acquainted with the object which is, in fact, the so-and-so, yet we do not know any proposition "*a* is the so-and-so," where *a* is something with which we are acquainted.

When we say "the so-and-so exists," we mean that there is just one object which is the so-and-so. The proposition "*a* is the so-and-so" means that *a* has the property so-and-so, and nothing else has. "Sir Joseph Larmor is the Unionist candidate" means "Sir Joseph Larmor is a Unionist candidate, and no one else is." "The Unionist candidate exists" means "some one is a Unionist candidate, and no one else is." Thus, when we are acquainted with an object which we know to be the so-and-so, we know that the so-and-so exists but we may know that the so-and-so exists when we are not acquainted with any object which we know to be the so-and-so, and even when we are not acquainted with any object which, in fact, is the so-and-so.

Common words, even proper names, are usually really descriptions. That is to say, the thought in the mind of a person using a proper name correctly can generally only be expressed explicitly if we replace the proper name by a description. Moreover, the description required to express the thought will vary for different people, or for the same person at different times. The only thing constant (so long as the name is rightly used) is the object to which the name applies. But so long as this remains constant, the particular description involved usually makes no difference to the truth or falsehood of the proposition in which the name appears.

Let us take some illustrations. Suppose some statement made about Bismarck. Assuming that there is such a thing as direct acquaintance with oneself, Bismarck himself might have used his name directly to designate the particular person with whom he was acquainted. In this case, if he made a judgment about himself, he himself might be a constituent of the judgment. Here the proper name has the direct use which it always wishes to have, as simply standing for a certain object, and not for a description of the object. But if a person who knew Bismarck made a judgment about him, the case is different. What this person was acquainted with were certain sense-data which he connected (rightly, we will suppose) with Bismarck's body. His body as a physical object, and still more his mind, were only known as the body and the mind connected with these sense-data. That is, they were known by description. It is, of course, very much a matter of chance which characteristics of a man's appearance will come into a friend's mind when he thinks of him; thus the description actually in the friend's mind is accidental. The essential point is that he knows that the various descriptions all apply to the same entity, in spite of not being acquainted with the entity in question.

When we, who did not know Bismarck, make a judgment about him, the description in our minds will probably be some more or less vague mass of historical knowledge—far more, in most cases, than is required to identify him. But, for the sake of illustration, let us assume that we think of him as "the first Chancellor of the German Empire." Here all the words are abstract except "German." The word "German" will again have different meanings for different people. To some it will recall travels in Germany, to some the look of Germany on the map, and so on. But if we are to obtain a description which we know to be applicable, we shall be compelled, at some point, to bring in a reference to a particular with which we are acquainted. Such reference is involved in any mention of past, present, and future (as opposed to definite dates), or of here and there, or of what others have told us. Thus it would seem that, in some way or other, a description known to be applicable to a particular must involve some reference to a particular with which we are acquainted, if our knowledge about the thing described is not to be merely what follows logically from the description. For example, "the most long-lived of men" is a description which must apply to some man, but we can make no judgments concerning this man which involve knowledge about him beyond what the description gives. If, however, we say, "the first Chancellor of the German Empire was an astute diplomatist," we can only be assured of the truth of our judgment in virtue of something with which we are acquainted—usually a testimony heard or read. Considered psychologically, apart from the information we convey to others, apart from the fact about the actual Bismarck, which gives importance to our judgment, the thought we really have contains the one or more particulars involved, and otherwise consists wholly of concepts. All names of places—London, England, Europe, the earth, the Solar System—similarly involve, when used, descriptions which start from some one or more particulars with which we are acquainted. I suspect that even the Universe, as considered by metaphysics, involves such a connection with particulars. In logic, on the contrary, where we are concerned not merely with what does exist, but with whatever might or could exist or be, no reference to actual particulars is involved.

It would seem that, when we make a statement about something only known by description, we often *intend* to make our statement, not in the form involving the description, but about the actual thing described. That is to say, when we say anything about Bismarck, we should like, if we could, to make the judgment which Bismarck alone can make, namely, the judgment of which he himself is a constituent. In this we are necessarily defeated, since the actual Bismarck is unknown to us. But we know that there is an object B called Bismarck, and that B was an astute diplomatist. We can thus *describe* the proposition we should like to affirm, namely, "B was an astute diplomatist," where B is the object which was Bismarck. What enables us to communicate in spite of the varying descriptions we employ is that we know there is a true proposition concerning the actual Bismarck, and that, however we may vary the description (so long as the description is correct), the proposition described is still the same. This proposition, which is described and is known to be true, is what interests us; but we are not acquainted with the proposition itself, and do not know *it*, though we know it is true.

It will be seen that there are various stages in the removal from acquaintance with particulars: there is Bismarck to people who knew him, Bismarck to those who only know of him through history, the man with the iron mask, the longest-lived of men. These are progressively further removed from acquaintance with particulars, and there is a similar hierarchy in the region of universals. Many universals, like many particulars, are only known to us by description. But here, as in the case of particulars, knowledge concerning what is known by description is ultimately reducible to knowledge concerning what is known by acquaintance.

The fundamental epistemological principle in the analysis of propositions containing descriptions is this: *Every proposition which we can understand must be composed wholly of constituents with which we are acquainted.* From what has been said already, it will be plain why I advocate this principle, and how I propose to meet the case of propositions which at first sight contravene it. Let us begin with the reasons for supposing the principle true.

The chief reason for supposing the principle true is that it seems scarcely possible to believe that we can make a judgment or entertain a supposition without knowing what it is that we are judging or supposing about. If we make a judgment about (say) Julius Cæsar, it is plain that the actual person who was Julius Cæsar is not a constituent of the judgment. But before going further, it may be well to explain what I mean when I say that this or that is a constituent of a judgment, or of a proposition which we understand. To begin with judgments: a judgment, as an occurrence, I take to be a relation of a mind to several entities, namely, the entities which compose what is judged. If, e.g. I judge that A loves B, the judgment as an event consists in the existence, at a certain moment, of a specific four-term relation, called *judging*, between me and A and love and B. That is to say, at the time when I judge, there is a certain complex whose terms are myself and A and love and B, and whose relating relation is *judging*. My reasons for this view have been set forth elsewhere,and I shall not repeat them here. Assuming this view of judgment, the constituents of the judgment are simply the constituents of the complex which is the judgment. Thus, in the above case, the constituents are myself and A and love and B and judging. But myself and judging are constituents shared by all my judgments; thus the *distinctive* constituents of the particular judgment in question are A and love and B. Coming now to what is meant by "understanding a proposition," I should say that there is another relation possible between me and A and love and B, which is called my *supposing* that A loves B.When we can *suppose* that A loves B, we "understand the proposition" *A loves B*. Thus we often understand a proposition in cases where we have not enough knowledge to make a judgment. Supposing, like judging, is a many-term relation, of which a mind is one term. The other terms of the relation are called the constituents of the proposition supposed. Thus the principle which I enunciated may be re-stated as follows: *Whenever a relation of supposing or judging occurs, the terms to which the supposing or judging mind is related by the relation of supposing or judging must be terms with which the mind in question is acquainted.* This is merely to say that we cannot make a judgment or a supposition without knowing what it is that we are making our judgment or supposition about. It seems to me that the truth of this principle is evident as soon as the principle is understood; I shall, therefore, in what follows, assume the principle, and use it as a guide in analysing judgments that contain descriptions.

Returning now to Julius Cæsar, I assume that it will be admitted that he himself is not a constituent of any judgment which I can make. But at this point it is necessary to examine the view that judgments are composed of something called "ideas," and that it is the "idea" of Julius Cæsar that is a constituent of my judgment. I believe the plausibility of this view rests upon a failure to form a right theory of descriptions. We may mean by my "idea" of Julius Cæsar the things that I know about him, e.g. that he conquered Gaul, was assassinated on the Ides of March, and is a plague to schoolboys. Now I am admitting, and indeed contending, that in order to discover what is actually in my mind when I judge about Julius Cæsar, we must substitute for the proper name a description made up of some of the things I know about him. (A description which will often serve to express my thought is "the man whose name was *Julius Cæsar*." For whatever else I may have forgotten about him, it is plain that when I mention him I have not forgotten that that was his name.) But although I think the theory that judgments consist of ideas may have been suggested in some such way, yet I think the theory itself is fundamentally mistaken. The view seems to be that there is some mental existent which may be called the "idea" of something outside the mind of the person who has the idea, and that, since judgment is a mental event, its constituents must be constituents of the mind of the person judging. But in this view ideas become a veil between us and outside things—we never really, in knowledge, attain to the things we are supposed to be knowing about, but only to the ideas of those things. The relation of mind, idea, and object, on this view, is utterly obscure, and, so far as I can see, nothing discoverable by inspection warrants the intrusion of the idea between the mind and the object. I suspect that the view is fostered by the dislike of relations, and that it is felt the mind could not know objects unless there were something "in" the mind which could be called the state of knowing the object. Such a view, however, leads at once to a vicious endless regress, since the relation of idea to object will have to be explained by supposing that the idea itself has an idea of the object, and so on *ad infinitum*. I therefore see no reason to believe that, when we are acquainted with an object, there is in us something which can be called the "idea" of the object. On the contrary, I hold that acquaintance is wholly a relation, not demanding any such constituent of the mind as is supposed by advocates of "ideas." This is, of course, a large question, and one which would take us far from our subject if it were adequately discussed. I therefore content myself with the above indications, and with the corollary that, in judging, the actual objects concerning which we judge, rather than any supposed purely mental entities, are constituents of the complex which is the judgment.

When, therefore, I say that we must substitute for "Julius Cæsar" some description of Julius Cæsar, in order to discover the meaning of a judgment nominally about him, I am not saying that we must substitute an idea. Suppose our description is "the man whose name was *Julius Cæsar*." Let our judgment be

"Julius Cæsar was assassinated." Then it becomes "the man whose name was *Julius Cæsar* was assassinated." Here *Julius Cæsar* is a noise or shape with which we are acquainted, and all the other constituents of the judgment (neglecting the tense in "was") are *concepts* with which we are acquainted. Thus our judgment is wholly reduced to constituents with which we are acquainted, but Julius Cæsar himself has ceased to be a constituent of our judgment. This, however, requires a proviso, to be further explained shortly, namely that "the man whose name was *Julius Cæsar*" must not, as a whole, be a constituent of our judgment, that is to say, this phrase must not, as a whole, have a meaning which enters into the judgment. Any right analysis of the judgment, therefore, must break up this phrase, and not treat it as a subordinate complex which is part of the judgment. The judgment "the man whose name was *Julius Cæsar* was assassinated" may be interpreted as meaning "one and only one man was called *Julius Cæsar*, and that one was assassinated." Here it is plain that there is no constituent corresponding to the phrase "the man whose name was *Julius Cæsar*." Thus there is no reason to regard this phrase as expressing a constituent of the judgment, and we have seen that this phrase must be broken up if we are to be acquainted with all the constituents of the judgment. This conclusion, which we have reached from considerations concerned with the theory of knowledge, is also forced upon us by logical considerations, which must now be briefly reviewed.

It is common to distinguish two aspects, *meaning* and *denotation*, such phrases as "the author of Waverley." The meaning will be a certain complex, consisting (at least) of authorship and Waverley with some relation; the denotation will be Scott. Similarly "featherless bipeds" will have a complex meaning, containing as constituents the presence of two feet and the absence of feathers, while its denotation will be the class of men. Thus when we say "Scott is the author of Waverley" or "men are the same as featherless bipeds," we are asserting an identity of denotation, and this assertion is worth making because of the diversity of meaning.I believe that the duality of meaning and denotation, though capable of a true interpretation, is misleading if taken as fundamental. The denotation, I believe, is not a constituent of the proposition, except in the case of proper names, i.e. of words which do not assign a property to an object, but merely and solely name it. And I should hold further that, in this sense, there are only two words which are strictly proper names of particulars, namely, "I" and "this."

One reason for not believing the denotation to be a constituent of the proposition is that we may know the proposition even when we are not acquainted with the denotation. The proposition "the author of Waverley is a novelist" was known to people who did not know that "the author of Waverley" denoted Scott. This reason has been already sufficiently emphasised.

A second reason is that propositions concerning "the so-and-so" are possible even when "the so-and-so" has no denotation. Take, e.g. "the golden mountain does not exist" or "the round square is self-contradictory."If we are to preserve the duality of meaning and denotation, we have to say, with Meinong, that there are such objects as the golden mountain and the round square, although these objects do not have being. We even have to admit that the existent round square is existent, but does not exist. Meinong does not regard this as a contradiction, but I fail to see that it is not one. Indeed, it seems to me evident that the judgment "there is no such object as the round square" does not presuppose that there is such an object. If this is admitted, however, we are led to the conclusion that, by parity of form, no judgment concerning "the so-and-so" actually involves the so-and-so as a constituent.

Miss Jones contends that there is no difficulty in admitting contradictory predicates concerning such an object as "the present King of France," on the ground that this object is in itself contradictory. Now it might, of course, be argued that this object, unlike the round square, is not self-contradictory, but merely non-existent. This, however, would not go to the root of the matter. The real objection to such an argument is that the law of contradiction ought not to be stated in the traditional form "A is not both B and not B," but in the form "no proposition is both true and false." The traditional form only applies to certain propositions, namely, to those which attribute a predicate to a subject. When the law is stated of propositions, instead of being stated concerning subjects and predicates, it is at once evident that propositions about the present King of France or the round square can form no exception, but are just as incapable of being both true and false as other propositions. Miss Jones argues that "Scott is the author of Waverley" asserts identity of denotation between *Scott* and *the author of Waverley*. But there is some difficulty in choosing among alternative meanings of this contention. In the first place, it should be observed that *the author of Waverley* is not a *mere* name, like *Scott*. *Scott* is merely a noise or shape conventionally used to designate a certain person; it gives us no information about that person, and has nothing that can be called meaning as opposed to denotation. (I neglect the fact, considered above, that even proper names, as a rule, really stand for descriptions.) But *the author of Waverley* is not merely conventionally a name for Scott; the element of mere convention belongs here to the separate words, *the* and *author* and *of* and *Waverley*. Given what these words stand for, *the author of Waverley* is no longer arbitrary. When it is said that Scott is the author of Waverley, we are not stating that these are two *names* for one man, as we should be if we said "Scott is Sir Walter." A man's name is what he is called, but however much Scott had been called the author of Waverley, that would not have made him be the author; it was necessary for him actually to write Waverley, which was a fact having nothing to do with names.

If, then, we are asserting identity of denotation, we must not mean by *denotation* the mere relation of a name to the thing named. In fact, it would be nearer to the truth to say that the *meaning* of "Scott" is the *denotation* of "the author of Waverley." The relation of "Scott" to Scott is that "Scott" means Scott, just as the relation of "author" to the concept which is so called is that "author" means this concept. Thus if we distinguish meaning and denotation in "the author of Waverley," we shall have to say that "Scott" has meaning but not denotation. Also when we say "Scott is the author of Waverley," the *meaning* of "the author of Waverley" is relevant to our assertion. For if the denotation alone were relevant, any other phrase with the same denotation would give the same proposition. Thus "Scott is the author of Marmion" would be the same proposition as "Scott is the author of Waverley." But this is plainly not the case, since from the first we learn that Scott wrote Marmion and from the second we learn that he wrote Waverley, but the first tells us nothing about Waverley and the second nothing about Marmion. Hence the meaning of "the author of Waverley," as opposed to the denotation, is certainly relevant to "Scott is the author of Waverley."

We have thus agreed that "the author of Waverley" is not a mere name, and that its meaning is relevant in propositions in which it occurs. Thus if we are to say, as Miss Jones does, that "Scott is the author of Waverley" asserts an identity of denotation, we must regard the denotation of "the author of Waverley" as the denotation of what is *meant* by "the author of Waverley." Let us call the meaning of "the author of Waverley" M. Thus M is what "the author of Waverley" means. Then we are to suppose that "Scott is the author of Waverley" means "Scott is the denotation of M." But here we are explaining our proposition by another of the same form, and thus we have made no progress towards a real explanation. "The denotation of M," like "the author of Waverley," has both meaning and denotation, on the theory we are examining. If we call its meaning M', our proposition becomes "Scott is the denotation of M'." But this leads at once to an endless regress. Thus the attempt to regard our proposition as asserting identity of denotation breaks down, and it becomes imperative to find some other analysis. When this analysis has been completed, we shall be able to reinterpret the phrase "identity of denotation," which remains obscure so long as it is taken as fundamental.

The first point to observe is that, in any proposition about "the author of Waverley," provided Scott is not explicitly mentioned, the denotation itself, i.e. Scott, does not occur, but only the concept of denotation, which will be represented by a variable. Suppose we say "the author of Waverley was the author of Marmion," we are certainly not saying that both were Scott—we may have forgotten that there was such a person as Scott. We are saying that there is some man who was the author of Waverley and the author of Marmion. That is to say, there is some one who wrote Waverley and Marmion, and no one else wrote them. Thus the identity is that of a variable, i.e. of an indefinite subject, "some one." This is why we can understand propositions about "the author of Waverley," without knowing who he was. When we say "the author of Waverley was a poet," we mean "one and only one man wrote Waverley, and he was a poet"; when we say "the author of Waverley was Scott" we mean "one and only one man wrote Waverley, and he was Scott." Here the identity is between a variable, i.e. an indeterminate subject ("he"), and Scott; "the author of Waverley" has been analysed away, and no longer appears as a constituent of the proposition.

The reason why it is imperative to analyse away the phrase "the author of Waverley" may be stated as follows. It is plain that when we say "the author of Waverley is the author of Marmion," the *is* expresses identity. We have seen also that the common *denotation*, namely Scott, is not a constituent of this proposition, while the *meanings* (if any) of "the author of Waverley" and "the author of Marmion" are not identical. We have seen also that, in any sense in which the meaning of a word is a constituent of a proposition in whose verbal expression the word occurs, "Scott" means the actual man Scott, in the same sense (so far as concerns our present discussion) in which "author" means a certain universal. Thus, if "the author of Waverley" were a subordinate complex in the above proposition, its *meaning* would have to be what was said to be identical with the *meaning* of "the author of Marmion." This is plainly not the case; and the only escape is to say that "the author of Waverley" does not, by itself, have a meaning, though phrases of which it is part do have a meaning. That is, in a right analysis of the above proposition, "the author of Waverley" must disappear. This is effected when the above proposition is analysed as meaning: "Some one wrote Waverley and no one else did, and that some one also wrote Marmion and no one else did." This may be more simply expressed by saying that the propositional function "x wrote Waverley and Marmion, and no one else did" is capable of truth, i.e. some value of x makes it true, but no other value does. Thus the true subject of our judgment is a propositional function, i.e. a complex containing an undetermined constituent, and becoming a proposition as soon as this constituent is determined.

We may now define the denotation of a phrase. If we know that the proposition "a is the so-and-so" is true, i.e. that a is so-and-so and nothing else is, we call a the denotation of the phrase "the so-and-so." A very great many of the propositions we naturally make about "the so-and-so" will remain true or remain false if we substitute a for "the so-and-so," where a is the denotation of "the so-and-so." Such propositions will also remain true or remain false if we substitute for "the so-and-so" any other phrase having the same denotation. Hence, as practical men, we become interested in the denotation more than in the description, since the denotation decides as to the truth or falsehood of so many statements in which the description occurs. Moreover, as we saw earlier in considering the relations of description and acquaintance, we often wish to reach the denotation, and are only hindered by lack of acquaintance: in such cases the description is merely the means we employ to get as near as possible to the denotation. Hence it naturally comes to be supposed that the denotation is part of the proposition in which the description occurs. But we have seen, both on logical and on epistemological grounds, that this is an error. The actual object (if any) which is the denotation is not (unless it is explicitly mentioned) a constituent of propositions in which descriptions occur; and this is the reason why, in order to understand such propositions, we need acquaintance with the constituents of the description, but do not need acquaintance with its denotation. The first result of analysis, when applied to propositions whose grammatical subject is "the so-and-so," is to substitute a variable as subject; i.e. we obtain a proposition of the form: "There is *something* which alone is so-and-so, and that *something* is such-and-such." The further analysis of propositions concerning "the so-and-so" is thus merged in the problem of the nature of the variable, i.e. of the meanings of *some*, *any*, and *all*. This is a difficult problem, concerning which I do not intend to say anything at present.

To sum up our whole discussion. We began by distinguishing two sorts of knowledge of objects, namely, knowledge by *acquaintance* and knowledge by *description*. Of these it is only the former that brings the object itself before the mind. We have acquaintance with sense-data, with many universals, and possibly with ourselves, but not with physical objects or other minds. We have *descriptive* knowledge of an object when we know that it is *the* object having some property or properties with which we are acquainted; that is to say, when we know that the property or properties in question belong to one object and no more, we are said to have knowledge of that one object by description, whether or not we are acquainted with the object. Our knowledge of physical objects and of other minds is only knowledge by description, the descriptions involved being usually such as involve sense-data. All propositions intelligible to us, whether or not they primarily concern things only known to us by description, are composed wholly of constituents with which we are acquainted, for a constituent with which we are not acquainted is unintelligible to us. A judgment, we found, is not composed of mental constituents called "ideas," but consists of an occurrence whose constituents are a mind and certain objects, particulars or universals. (One at least must be a universal.) When a judgment is rightly analysed, the objects which are constituents of it must all be objects with which the mind which is a constituent of it is acquainted. This conclusion forces us to analyse descriptive phrases occurring in propositions, and to say that the objects denoted by such phrases are not constituents of judgments in which such phrases occur (unless these objects are explicitly mentioned). This leads us to the view (recommended also on purely logical grounds) that when we say "the author of Marmion was the author of Waverley," Scott himself is not a constituent of our judgment, and that the judgment cannot be explained by saying that it affirms identity of denotation with diversity of meaning. It also, plainly, does not assert identity of meaning. Such judgments, therefore, can only be analysed by breaking up the descriptive phrases, introducing a variable, and making propositional functions the ultimate subjects. In fact, "the so-and-so is such-and-such" will mean that "*x* is so-and-so and nothing else is, and *x* is such-and-such" is capable of truth. The analysis of such judgments involves many fresh problems, but the discussion of these problems is not undertaken in the present paper.

THE PRACTICE AND THEORY OF BOLSHEVISM

PREFACE

The Russian Revolution is one of the great heroic events of the world's history. It is natural to compare it to the French Revolution, but it is in fact something of even more importance. It does more to change daily life and the structure of society: it also does more to change men's beliefs. The difference is exemplified by the difference between Marx and Rousseau: the latter sentimental and soft, appealing to emotion, obliterating sharp outlines; the former systematic like Hegel, full of hard intellectual content, appealing to historic necessity and the technical development of industry, suggesting a view of human beings as puppets in the grip of omnipotent material forces. Bolshevism combines the characteristics of the French Revolution with those of the rise of Islam; and the result is something radically new, which can only be understood by a patient and passionate effort of imagination.

Before entering upon any detail, I wish to state, as clearly and unambiguously as I can, my own attitude towards this new thing.

By far the most important aspect of the Russian Revolution is as an attempt to realize Communism. I believe that Communism is necessary to the world, and I believe that the heroism of Russia has fired men's hopes in a way which was essential to the realization of Communism in the future. Regarded as a splendid attempt, without which ultimate success would have been very improbable, Bolshevism deserves the gratitude and admiration of all the progressive part of mankind.

But the method by which Moscow aims at establishing Communism is a pioneer method, rough and dangerous, too heroic to count the cost of the opposition it arouses. I do not believe that by this method a stable or desirable form of Communism can be established. Three issues seem to me possible from the present situation. The first is the ultimate defeat of Bolshevism by the forces of capitalism. The second is the victory of the Bolshevists accompanied by a complete loss of their ideals and a régime of Napoleonic imperialism. The third is a prolonged world-war, in which civilization will go under, and all its manifestations (including Communism) will be forgotten.

It is because I do not believe that the methods of the Third International can lead to the desired goal that I have thought it worth while to point out what seem to me undesirable features in the present state of Russia. I think there are lessons to be learnt which must be learnt if the world is ever to achieve what is desired by those in the West who have sympathy with the original aims of the Bolsheviks. I do not think these lessons can be learnt except by facing frankly and fully whatever elements of failure there are in Russia. I think these elements of failure are less attributable to faults of detail than to an impatient philosophy, which aims at creating a new world without sufficient preparation in the opinions and feelings of ordinary men and women.

But although I do not believe that Communism can be realized immediately by the spread of Bolshevism, I do believe that, if Bolshevism falls, it will have contributed a legend and a heroic attempt without which ultimate success might never have come. A fundamental economic reconstruction, bringing with it very far-reaching changes in ways of thinking and feeling, in philosophy and art and private relations, seems absolutely necessary if industrialism is to become the servant of man instead of his master. In all this, I am at one with the Bolsheviks; politically, I criticize them only when their methods seem to involve a departure from their own ideals.

There is, however, another aspect of Bolshevism from which I differ more fundamentally. Bolshevism is not merely a political doctrine; it is also a religion, with elaborate dogmas and inspired scriptures. When Lenin wishes to prove some proposition, he does so, if possible, by quoting texts from Marx and Engels. A full-fledged Communist is not merely a man who believes that land and capital should be held in common, and their produce distributed as nearly equally as possible. He is a man who entertains a number of elaborate and dogmatic beliefs—such as philosophic materialism, for example—which may be true, but are not, to a scientific temper, capable of being known to be true with any certainty. This habit, of militant certainty about objectively doubtful matters, is one from which, since the Renaissance, the world has been gradually emerging, into that temper of constructive and fruitful scepticism which constitutes the scientific outlook. I believe the scientific outlook to be immeasurably important to the human race. If a more just economic system were only attainable by closing men's minds against free inquiry, and plunging them back into the intellectual prison of the middle ages, I should consider the price too high. It cannot be denied that, over any short period of time, dogmatic belief is a help in fighting. If all Communists become religious fanatics, while supporters of capitalism retain a sceptical temper, it may be assumed that the Communists will win, while in the contrary case the capitalists would win. It seems evident, from the attitude of the capitalist world to Soviet Russia, of the Entente to the Central Empires, and of England to Ireland and India, that there is no depth of cruelty, perfidy or brutality from which the present holders of power will shrink when they feel themselves threatened. If, in order to oust them, nothing short of religious fanaticism will serve, it is they who are the prime sources of the resultant evil. And it is permissible to hope that, when they have been dispossessed, fanaticism will fade, as other fanaticisms have faded in the past.

The present holders of power are evil men, and the present manner of life is doomed. To make the transition with a minimum of bloodshed, with a maximum of preservation of whatever has value in our existing civilization, is a difficult problem. It is this problem which has chiefly occupied my mind in writing the following pages. I wish I could think that its solution would be facilitated by some slight degree of moderation and humane feeling on the part of those who enjoy unjust privileges in the world as it is.

The present work is the outcome of a visit to Russia, supplemented by much reading and discussion both before and after. I have thought it best to record what I saw separately from theoretical considerations, and I have endeavoured to state my impressions without any bias for or against the Bolsheviks. I received at their hands the greatest kindness and courtesy, and I owe them a debt of gratitude for the perfect freedom which they allowed me in my investigations. I am conscious that I was too short a time in Russia to be able to form really reliable judgments; however, I share this drawback with most other westerners who have written on Russia since the October Revolution. I feel that Bolshevism is a matter of such importance that it is necessary, for almost every political question, to define one's attitude in regard to it; and I have hopes that I may help others to define their attitude, even if only by way of opposition to what I have written.

I have received invaluable assistance from my secretary, Miss D.W. Black, who was in Russia shortly after I had left. The chapter on Art and Education is written by her throughout. Neither is responsible for the other's opinions.

BERTRAND RUSSELL

September, 1920.

PART I

THE PRESENT CONDITION OF RUSSIA

I

WHAT IS HOPED FROM BOLSHEVISM

To understand Bolshevism it is not sufficient to know facts; it is necessary also to enter with sympathy or imagination into a new spirit. The chief thing that the Bolsheviks have done is to create a hope, or at any rate to make strong and widespread a hope which was formerly confined to a few. This aspect of the movement is as easy to grasp at a distance as it is in Russia—perhaps even easier, because in Russia present circumstances tend to obscure the view of the distant future. But the actual situation in Russia can only be understood superficially if we forget the hope which is the motive power of the whole. One might as well describe the Thebaid without mentioning that the hermits expected eternal bliss as the reward of their sacrifices here on earth.

I cannot share the hopes of the Bolsheviks any more than those of the Egyptian anchorites; I regard both as tragic delusions, destined to bring upon the world centuries of darkness and futile violence. The principles of the Sermon on the Mount are admirable, but their effect upon average human nature was very different from what was intended. Those who followed Christ did not learn to love their enemies or to turn the other cheek. They learned instead to use the Inquisition and the stake, to subject the human intellect to the yoke of an ignorant and intolerant priesthood, to degrade art and extinguish science for a thousand years. These were the inevitable results, not of the teaching, but of fanatical belief in the teaching. The hopes which inspire Communism are, in the main, as admirable as those instilled by the Sermon on the Mount, but they are held as fanatically, and are likely to do as much harm. Cruelty lurks in our instincts, and fanaticism is a camouflage for cruelty. Fanatics are seldom genuinely humane, and those who sincerely dread cruelty will be slow to adopt a fanatical creed. I do not know whether Bolshevism can be prevented from acquiring universal power. But even if it cannot, I am persuaded that those who stand out against it, not from love of ancient injustice, but in the name of the free spirit of Man, will be the bearers of the seeds of progress, from which, when the world's gestation is accomplished, new life will be born.

The war has left throughout Europe a mood of disillusionment and despair which calls aloud for a new religion, as the only force capable of giving men the energy to live vigorously. Bolshevism has supplied the new religion. It promises glorious things: an end of the injustice of rich and poor, an end of economic slavery, an end of war. It promises an end of the disunion of classes which poisons political life and threatens our industrial system with destruction. It promises an end to commercialism, that subtle falsehood that leads men to appraise everything by its money value, and to determine money value often merely by the caprices of idle plutocrats. It promises a world where all men and women shall be kept sane by work, and where all work shall be of value to the community, not only to a few wealthy vampires. It is to sweep away listlessness and pessimism and weariness and all the complicated miseries of those whose circumstances allow idleness and whose energies are not sufficient to force activity. In place of palaces and hovels, futile vice and useless misery, there is to be wholesome work, enough but not too much, all of it useful, performed by men and women who have no time for pessimism and no occasion for despair.

The existing capitalist system is doomed. Its injustice is so glaring that only ignorance and tradition could lead wage-earners to tolerate it. As ignorance diminishes, tradition becomes weakened, and the war destroyed the hold upon men's minds of everything merely traditional. It may be that, through the influence of America, the capitalist system will linger for another fifty years; but it will grow continually weaker, and can never recover the position of easy dominance which it held in the nineteenth century. To attempt to bolster it up is a useless diversion of energies which might be expended upon building something new. Whether the new thing will be Bolshevism or something else, I do not know; whether it will be better or worse than capitalism, I do not know. But that a radically new order of society will emerge, I feel no doubt. And I also feel no doubt that the new order will be either some form of Socialism or a reversion to barbarism and petty war such as occurred during the barbarian invasion. If Bolshevism remains the only vigorous and effective competitor of capitalism, I believe that no form of Socialism will be realized, but only chaos and destruction. This belief, for which I shall give reasons later, is one of the grounds upon which I oppose Bolshevism. But to oppose it from the point of view of a supporter of capitalism would be, to my mind, utterly futile and against the movement of history in the present age.

The effect of Bolshevism as a revolutionary hope is greater outside Russia than within the Soviet Republic. Grim realities have done much to kill hope among those who are subject to the dictatorship of Moscow. Yet even within Russia, the Communist party, in whose hands all political power is concentrated, still lives by hope, though the pressure of events has made the hope severe and stern and somewhat remote. It is this hope that leads to concentration upon the rising generation. Russian Communists often avow that there is little hope for those who are already adult, and that happiness can only come to the children who have grown up under the new régime and been moulded from the first to the group-mentality that Communism requires. It is only after the lapse of a generation that they hope to create a Russia that shall realize their vision.

In the Western World, the hope inspired by Bolshevism is more immediate, less shot through with tragedy. Western Socialists who have visited Russia have seen fit to suppress the harsher features of the present régime, and have disseminated a belief among their followers that the millennium would be quickly realized there if there were no war and no blockade. Even those Socialists who are not Bolsheviks for their own country have mostly done very little to help men in appraising the merits or demerits of Bolshevik methods. By this lack of courage they have exposed Western Socialism to the danger of becoming Bolshevik through ignorance of the price that has to be paid and of the uncertainty as to whether the desired goal will be reached in the end. I believe that the West is capable of adopting less painful and more certain methods of reaching Socialism than those that have seemed necessary in Russia. And I believe that while some forms of Socialism are immeasurably better than capitalism, others are even worse. Among those that are worse I reckon the form which is being achieved in Russia, not only in itself, but as a more insuperable barrier to further progress.

In judging of Bolshevism from what is to be seen in Russia at present, it is necessary to disentangle various factors which contribute to a single result. To begin with, Russia is one of the nations that were defeated in the war; this has produced a set of circumstances resembling those found in Germany and Austria. The food problem, for example, appears to be essentially similar in all three countries. In order to arrive at what is specifically Bolshevik,we must first eliminate what is merely characteristic of a country which has suffered military disaster. Next we come to factors which are Russian, which Russian Communists share with other Russians, but not with other Communists. There is, for example, a great deal of disorder and chaos and waste, which shocks Westerners (especially Germans) even when they are in close political sympathy with the Bolsheviks. My own belief is that, although, with the exception of a few very able men, the Russian Government is less efficient in organization than the Germans or the Americans would be in similar circumstances, yet it represents what is most efficient in Russia, and does more to prevent chaos than any possible alternative government would do. Again, the intolerance and lack of liberty which has been inherited from the Tsarist régime is probably to be regarded as Russian rather than Communist. If a Communist Party were to acquire power in England, it would probably be met by a less irresponsible opposition, and would be able to show itself far more tolerant than any government can hope to be in Russia if it is to escape assassination. This, however, is a matter of degree. A great part of the despotism which characterizes the Bolsheviks belongs to the essence of their social philosophy, and would have to be reproduced, even if in a milder form, wherever that philosophy became dominant.

It is customary among the apologists of Bolshevism in the West to excuse its harshness on the ground that it has been produced by the necessity of fighting the Entente and its mercenaries. Undoubtedly it is true that this necessity has produced many of the worst elements in the present state of affairs. Undoubtedly, also, the Entente has incurred a heavy load of guilt by its peevish and futile opposition. But the expectation of such opposition was always part of Bolshevik theory. A general hostility to the first Communist State was both foreseen and provoked by the doctrine of the class war. Those who adopt the Bolshevik standpoint must reckon with the embittered hostility of capitalist States; it is not worth while to adopt Bolshevik methods unless they can lead to good in spite of this hostility. To say that capitalists are wicked and we have no responsibility for their acts is unscientific; it is, in particular, contrary to the Marxian doctrine of economic determinism. The evils produced in Russia by the enmity of the Entente are therefore to be reckoned as essential in the Bolshevik method of transition to Communism, not as specially Russian. I am not sure that we cannot even go a step further.The exhaustion and misery caused by unsuccessful war were necessary to the success of the Bolsheviks; a prosperous population will not embark by such methods upon a fundamental economic reconstruction. One can imagine England becoming Bolshevik after an unsuccessful war involving the loss of India—no improbable contingency in the next few years. But at present the average wage-earner in England will not risk what he has for the doubtful gain of a revolution. A condition of widespread misery may, therefore, be taken as indispensable to the inauguration of Communism, unless, indeed, it were possible to establish Communism more or less peacefully, by methods which would not, even temporarily, destroy the economic life of the country. If the hopes which inspired Communism at the start, and which still inspire its Western advocates, are ever to be realized, the problem of minimizing violence in the transition must be faced. Unfortunately, violence is in itself delightful to most really vigorous revolutionaries, and they feel no interest in the problem of avoiding it as far as possible. Hatred of enemies is easier and more intense than love of friends. But from men who are more anxious to injure opponents than to benefit the world at large no great good is to be expected.

II

GENERAL CHARACTERISTICS

I entered Soviet Russia on May 11th and recrossed the frontier on June 16th. The Russian authorities only admitted me on the express condition that I should travel with the British Labour Delegation, a condition with which I was naturally very willing to comply, and which that Delegation kindly allowed me to fulfil. We were conveyed from the frontier to Petrograd, as well as on subsequent journeys, in a special *train de luxe*; covered with mottoes about the Social Revolution and the Proletariat of all countries; we were received everywhere by regiments of soldiers, with the Internationale being played on the regimental band while civilians stood bare-headed and soldiers at the salute; congratulatory orations were made by local leaders and answered by prominent Communists who accompanied us; the entrances to the carriages were guarded by magnificent Bashkir cavalry-men in resplendent uniforms; in short, everything was done to make us feel like the Prince of Wales. Innumerable functions were arranged for us: banquets, public meetings, military reviews, etc.

The assumption was that we had come to testify to the solidarity of British Labour with Russian Communism, and on that assumption the utmost possible use was made of us for Bolshevik propaganda. We, on the other hand, desired to ascertain what we could of Russian conditions and Russian methods of government, which was impossible in the atmosphere of a royal progress. Hence arose an amicable contest, degenerating at times into a game of hide and seek: while they assured us how splendid the banquet or parade was going to be, we tried to explain how much we should prefer a quiet walk in the streets. I, not being a member of the Delegation, felt less obligation than my companions did to attend at propaganda meetings where one knew the speeches by heart beforehand. In this way, I was able, by the help of neutral interpreters, mostly English or American, to have many conversations with casual people whom I met in the streets or on village greens, and to find out how the whole system appears to the ordinary non-political man and woman. The first five days we spent in Petrograd, the next eleven in Moscow. During this time we were living in daily contact with important men in the Government, so that we learned the official point of view without difficulty. I saw also what I could of the intellectuals in both places. We were all allowed complete freedom to see politicians of opposition parties, and we naturally made full use of this freedom. We saw Mensheviks, Social Revolutionaries of different groups, and Anarchists; we saw them without the presence of any Bolsheviks, and they spoke freely after they had overcome their initial fears. I had an hour's talk with Lenin, virtually *tête-à-tête*; I met Trotsky, though only in company; I spent a night in the country with Kamenev; and I saw a great deal of other men who, though less known outside Russia, are of considerable importance in the Government.

At the end of our time in Moscow we all felt a desire to see something of the country, and to get in touch with the peasants, since they form about 85 per cent, of the population. The Government showed the greatest kindness in meeting our wishes, and it was decided that we should travel down the Volga from Nijni Novgorod to Saratov, stopping at many places, large and small, and talking freely with the inhabitants. I found this part of the time extraordinarily instructive. I learned to know more than I should have thought possible of the life and outlook of peasants, village schoolmasters, small Jew traders, and all kinds of people. Unfortunately, my friend, Clifford Allen, fell ill, and my time was much taken up with him. This had, however, one good result, namely, that I was able to go on with the boat to Astrakhan, as he was too ill to be moved off it. This not only gave me further knowledge of the country, but made me acquainted with Sverdlov, Acting Minister of Transport, who was travelling on the boat to organize the movement of oil from Baku up the Volga, and who was one of the ablest as well as kindest people whom I met in Russia.

One of the first things that I discovered after passing the Red Flag which marks the frontier of Soviet Russia, amid a desolate region of marsh, pine wood, and barbed wire entanglements, was the profound difference between the theories of actual Bolsheviks and the version of those theories current among advanced Socialists in this country. Friends of Russia here think of the dictatorship of the proletariat as merely a new form of representative government, in which only working men and women have votes, and the constituencies are partly occupational, not geographical. They think that "proletariat" means "proletariat," but "dictatorship" does not quite mean "dictatorship." This is the opposite of the truth. When a Russian Communist speaks of dictatorship, he means the word literally, but when he speaks of the proletariat, he means the word in a Pickwickian sense. He means the "class-conscious" part of the proletariat, *i.e.*, the Communist Party.He includes people by no means proletarian (such as Lenin and Tchicherin) who have the right opinions, and he excludes such wage-earners as have not the right opinions, whom he classifies as lackeys of the *bourgeoisie*. The Communist who sincerely believes the party creed is convinced that private property is the root of all evil; he is so certain of this that he shrinks from no measures, however harsh, which seem necessary for constructing and preserving the Communist State. He spares himself as little as he spares others. He works sixteen hours a day, and foregoes his Saturday half-holiday. He volunteers for any difficult or dangerous work which needs to be done, such as clearing away piles of infected corpses left by Kolchak or Denikin. In spite of his position of power and his control of supplies, he lives an austere life. He is not pursuing personal ends, but aiming at the creation of a new social order. The same motives, however, which make him austere make him also ruthless. Marx has taught that Communism is fatally predestined to come about; this fits in with the Oriental traits in the Russian character, and produces a state of mind not unlike that of the early successors of Mahomet. Opposition is crushed without mercy, and without shrinking from the methods of the Tsarist police, many of whom are still employed at their old work. Since all evils are due to private property, the evils of the Bolshevik régime while it has to fight private property will automatically cease as soon as it has succeeded.

These views are the familiar consequences of fanatical belief. To an English mind they reinforce the conviction upon which English life has been based ever since 1688, that kindliness and tolerance are worth all the creeds in the world—a view which, it is true, we do not apply to other nations or to subject races.

In a very novel society it is natural to seek for historical parallels. The baser side of the present Russian Government is most nearly paralleled by the Directoire in France, but on its better side it is closely analogous to the rule of Cromwell. The sincere Communists (and all the older members of the party have proved their sincerity by years of persecution) are not unlike the Puritan soldiers in their stern politico-moral purpose. Cromwell's dealings with Parliament are not unlike Lenin's with the Constituent Assembly. Both, starting from a combination of democracy and religious faith, were driven to sacrifice democracy to religion enforced by military dictatorship. Both tried to compel their countries to live at a higher level of morality and effort than the population found tolerable. Life in modern Russia, as in Puritan England, is in many ways contrary to instinct. And if the Bolsheviks ultimately fall, it will be for the reason for which the Puritans fell: because there comes a point at which men feel that amusement and ease are worth more than all other goods put together.

Far closer than any actual historical parallel is the parallel of Plato's Republic. The Communist Party corresponds to the guardians; the soldiers have about the same status in both; there is in Russia an attempt to deal with family life more or less as Plato suggested. I suppose it may be assumed that every teacher of Plato throughout the world abhors Bolshevism, and that every Bolshevik regards Plato as an antiquated *bourgeois*. Nevertheless, the parallel is extraordinarily exact between Plato's Republic and the régime which the better Bolsheviks are endeavouring to create.

Bolshevism is internally aristocratic and externally militant. The Communists in many ways resemble the British public-school type: they have all the good and bad traits of an aristocracy which is young and vital. They are courageous, energetic, capable of command, always ready to serve the State; on the other hand, they are dictatorial, lacking in ordinary consideration for the plebs. They are practically the sole possessors of power, and they enjoy innumerable advantages in consequence. Most of them, though far from luxurious, have better food than other people. Only people of some political importance can obtain motor-cars or telephones. Permits for railway journeys, for making purchases at the Soviet stores (where prices are about one-fiftieth of what they are in the market), for going to the theatre, and so on, are, of course, easier to obtain for the friends of those in power than for ordinary mortals. In a thousand ways, the Communists have a life which is happier than that of the rest of the community. Above all, they are less exposed to the unwelcome attentions of the police and the extraordinary commission.

The Communist theory of international affairs is exceedingly simple. The revolution foretold by Marx, which is to abolish capitalism throughout the world, happened to begin in Russia, though Marxian theory would seem to demand that it should begin in America. In countries where the revolution has not yet broken out, the sole duty of a Communist is to hasten its advent. Agreements with capitalist States can only be make-shifts, and can never amount on either side to a sincere peace. No real good can come to any country without a bloody revolution: English Labour men may fancy that a peaceful evolution is possible, but they will find their mistake. Lenin told me that he hopes to see a Labour Government in England, and would wish his supporters to work for it, but solely in order that the futility of Parliamentarism may be conclusively demonstrated to the British working man. Nothing will do any real good except the arming of the proletariat and the disarming of the *bourgeoisie*. Those who preach anything else are social traitors or deluded fools.

For my part, after weighing this theory carefully, and after admitting the whole of its indictment of *bourgeois* capitalism, I find myself definitely and strongly opposed to it. The Third International is an organization which exists to promote the class-war and to hasten the advent of revolution everywhere. My objection is not that capitalism is less bad than the Bolsheviks believe, but that Socialism is less good, not in its best form, but in the only form which is likely to be brought about by war.The evils of war, especially of civil war, are certain and very great; the gains to be achieved by victory are problematical. In the course of a desperate struggle, the heritage of civilization is likely to be lost, while hatred, suspicion, and cruelty become normal in the relations of human beings. In order to succeed in war, a concentration of power is necessary, and from concentration of power the very same evils flow as from the capitalist concentration of wealth. For these reasons chiefly, I cannot support any movement which aims at world revolution. The damage to civilization done by revolution in one country may be repaired by the influence of another in which there has been no revolution; but in a universal cataclysm civilization might go under for a thousand years. But while I cannot advocate world revolution, I cannot escape from the conclusion that the Governments of the leading capitalist countries are doing everything to bring it about. Abuse of our power against Germany, Russia, and India (to say nothing of any other countries) may well bring about our downfall, and produce those very evils which the enemies of Bolshevism most dread.

The true Communist is thoroughly international. Lenin, for example, so far as I could judge, is not more concerned with the interests of Russia than with those of other countries; Russia is, at the moment, the protagonist of the social revolution, and, as such, valuable to the world, but Lenin would sacrifice Russia rather than the revolution, if the alternative should ever arise. This is the orthodox attitude, and is no doubt genuine in many of the leaders. But nationalism is natural and instinctive; through pride in the revolution, it grows again even in the breasts of Communists. Through the Polish war, the Bolsheviks have acquired the support of national feeling, and their position in the country has been immensely strengthened.

The only time I saw Trotsky was at the Opera in Moscow. The British Labour Delegation were occupying what had been the Tsar's box. After speaking with us in the ante-chamber, he stepped to the front of the box and stood with folded arms while the house cheered itself hoarse. Then he spoke a few sentences, short and sharp, with military precision, winding up by calling for "three cheers for our brave fellows at the front," to which the audience responded as a London audience would have responded in the autumn of 1914. Trotsky and the Red Army undoubtedly now have behind them a great body of nationalist sentiment. The reconquest of Asiatic Russia has even revived what is essentially an imperialist way of feeling, though this would be indignantly repudiated by many of those in whom I seemed to detect it. Experience of power is inevitably altering Communist theories, and men who control a vast governmental machine can hardly have quite the same outlook on life as they had when they were hunted fugitives. If the Bolsheviks remain in power, it is much to be feared that their Communism will fade, and that they will increasingly resemble any other Asiatic Government—for example, our own Government in India.

III

LENIN, TROTSKY AND GORKY

The first question I asked him was as to how far he recognized the peculiarity of English economic and political conditions? I was anxious to know whether advocacy of violent revolution is an indispensable condition of joining the Third International, although I did not put this question directly because others were asking it officially. His answer was unsatisfactory to me. He admitted that there is little chance of revolution in England now, and that the working man is not yet disgusted with Parliamentary government. But he hopes that this result may be brought about by a Labour Ministry. He thinks that, if Mr. Henderson, for instance, were to become Prime Minister, nothing of importance would be done; organized Labour would then, so he hopes and believes, turn to revolution. On this ground, he wishes his supporters in this country to do everything in their power to secure a Labour majority in Parliament; he does not advocate abstention from Parliamentary contests, but participation with a view to making Parliament obviously contemptible. The reasons which make attempts at violent revolution seem to most of us both improbable and undesirable in this country carry no weight with him, and seem to him mere *bourgeois* prejudices. When I suggested that whatever is possible in England can be achieved without bloodshed, he waved aside the suggestion as fantastic. I got little impression of knowledge or psychological imagination as regards Great Britain. Indeed the whole tendency of Marxianism is against psychological imagination, since it attributes everything in politics to purely material causes.

I asked him next whether he thought it possible to establish Communism firmly and fully in a country containing such a large majority of peasants. He admitted that it was difficult, and laughed over the exchange the peasant is compelled to make, of food for paper; the worthlessness of Russian paper struck him as comic. But he said—what is no doubt true—that things will right themselves when there are goods to offer to the peasant. For this he looks partly to electrification in industry, which, he says, is a technical necessity in Russia, but will take ten years to complete. He spoke with enthusiasm, as they all do, of the great scheme for generating electrical power by means of peat. Of course, he looks to the raising of the blockade as the only radical cure; but he was not very hopeful of this being achieved thoroughly or permanently except through revolutions in other countries. Peace between Bolshevik Russia and capitalist countries, he said, must always be insecure; the Entente might be led by weariness and mutual dissensions to conclude peace, but he felt convinced that the peace would be of brief duration. I found in him, as in almost all leading Communists, much less eagerness than existed in our delegation for peace and the raising of the blockade.

He believes that nothing of real value can be achieved except through world revolution and the abolition of capitalism; I felt that he regarded the resumption of trade with capitalist countries as a mere palliative of doubtful value.

He described the division between rich and poor peasants, and the Government propaganda among the latter against the former, leading to acts of violence which he seemed to find amusing. He spoke as though the dictatorship over the peasant would have to continue a long time, because of the peasant's desire for free trade. He said he knew from statistics (what I can well believe) that the peasants have had more to eat these last two years than they ever had before, "and yet they are against us," he added a little wistfully. I asked him what to reply to critics who say that in the country he has merely created peasant proprietorship, not Communism; he replied that that is not quite the truth, but he did not say what the truth is.

The last question I asked him was whether resumption of trade with capitalist countries, if it took place, would not create centres of capitalist influence, and make the preservation of Communism more difficult? It had seemed to me that the more ardent Communists might well dread commercial intercourse with the outer world, as leading to an infiltration of heresy, and making the rigidity of the present system almost impossible. I wished to know whether he had such a feeling. He admitted that trade would create difficulties, but said they would be less than those of the war. He said that two years ago neither he nor his colleagues thought they could survive against the hostility of the world. He attributes their survival to the jealousies and divergent interests of the different capitalist nations; also to the power of Bolshevik propaganda. He said the Germans had laughed when the Bolsheviks proposed to combat guns with leaflets, but that the event had proved the leaflets quite as powerful. I do not think he recognizes that the Labour and Socialist parties have had any part in the matter. He does not seem to know that the attitude of British Labour has done a great deal to make a first-class war against Russia impossible, since it has confined the Government to what could be done in a hole-and-corner way, and denied without a too blatant mendacity.

He thoroughly enjoys the attacks of Lord Northcliffe, to whom he wishes to send a medal for Bolshevik propaganda. Accusations of spoliation, he remarked, may shock the *bourgeois*, but have an opposite effect upon the proletarian.

I think if I had met him without knowing who he was, I should not have guessed that he was a great man; he struck me as too opinionated and narrowly orthodox. His strength comes, I imagine, from his honesty, courage, and unwavering faith—religious faith in the Marxian gospel, which takes the place of the Christian martyr's hopes of Paradise, except that it is less egotistical. He has as little love of liberty as the Christians who suffered under Diocletian, and retaliated when they acquired power. Perhaps love of liberty is incompatible with whole-hearted belief in a panacea for all human ills. If so, I cannot but rejoice in the sceptical temper of the Western world. I went to Russia a Communist; but contact with those who have no doubts has intensified a thousandfold my own doubts, not as to Communism in itself, but as to the wisdom of holding a creed so firmly that for its sake men are willing to inflict widespread misery.

Trotsky, whom the Communists do not by any means regard as Lenin's equal, made more impression upon me from the point of view of intelligence and personality, though not of character. I saw too little of him, however, to have more than a very superficial impression. He has bright eyes, military bearing, lightning intelligence and magnetic personality. He is very good-looking, with admirable wavy hair; one feels he would be irresistible to women. I felt in him a vein of gay good humour, so long as he was not crossed in any way. I thought, perhaps wrongly, that his vanity was even greater than his love of power—the sort of vanity that one associates with an artist or actor. The comparison with Napoleon was forced upon one. But I had no means of estimating the strength of his Communist conviction, which may be very sincere and profound.

An extraordinary contrast to both these men was Gorky, with whom I had a brief interview in Petrograd. He was in bed, apparently very ill and obviously heart-broken. He begged me, in anything I might say about Russia, always to emphasize what Russia has suffered. He supports the Government—as I should do, if I were a Russian—not because he thinks it faultless, but because the possible alternatives are worse. One felt in him a love of the Russian people which makes their present martyrdom almost unbearable, and prevents the fanatical faith by which the pure Marxians are upheld. I felt him the most lovable, and to me the most sympathetic, of all the Russians I saw. I wished for more knowledge of his outlook, but he spoke with difficulty and was constantly interrupted by terrible fits of coughing, so that I could not stay. All the intellectuals whom I met—a class who have suffered terribly—expressed their gratitude to him for what he has done on their behalf. The materialistic conception of history is all very well, but some care for the higher things of civilization is a relief. The Bolsheviks are sometimes said to have done great things for art, but I could not discover that they had done more than preserve something of what existed before. When I questioned one of them on the subject, he grew impatient, and said: "We haven't time for a new art, any more than for a new religion." Unavoidably, although the Government favours art as much as it can, the atmosphere is one in which art cannot flourish, because art is anarchic and resistant to organization. Gorky has done all that one man could to preserve the intellectual and artistic life of Russia. I feared that he was dying, and that, perhaps, it was dying too. But he recovered, and I hope it will recover also.

IV

ART AND EDUCATION

It has often been said that, whatever the inadequacy of Bolshevik organization in other fields, in art and in education at least they have made great progress.

To take first of all art: it is true that they began by recognizing, as perhaps no other revolutionary government would, the importance and spontaneity of the artistic impulse, and therefore while they controlled or destroyed the counter-revolutionary in all other social activities, they allowed the artist, whatever his political creed, complete freedom to continue his work. Moreover, as regards clothing and rations they treated him especially well. This, and the care devoted to the upkeep of churches, public monuments, and museums, are well-known facts, to which there has already been ample testimony.

The preservation of the old artistic community practically intact was the more remarkable in view of the pronounced sympathy of most of them with the old régime. The theory, however, was that art and politics belonged to two separate realms; but great honour would of course be the portion of those artists who would be inspired by the revolution.

Three years' experience, however, have proved the falsity of this doctrine and led to a divorce between art and popular feeling which a sensitive observer cannot fail to remark. It is glaringly apparent in the hitherto most vital of all Russian arts, the theatre. The artists have continued to perform the old classics in tragedy or comedy, and the old-style operette. The theatre programmes have remained the same for the last two years, and, but for the higher standard of artistic performance, might belong to the theatres of Paris or London. As one sits in the theatre, one is so acutely conscious of the discrepancy between the daily life of the audience and that depicted in the play that the latter seems utterly dead and meaningless. To some of the more fiery Communists it appears that a mistake has been made. They complain that *bourgeois* art is being preserved long after its time, they accuse the artists of showing contempt for their public, of being as untouched by the revolutionary mood as an elderly *bourgeoise* bewailing the loss of her personal comfort; they would like to see only the revolutionary mood embodied in art, and to achieve this would make a clean sweep, enforcing the writing and performance of nothing but revolutionary plays and the painting of revolutionary pictures. Nor can it be argued that they are wrong as to the facts: it is plain that the preservation of the old artistic tradition has served very little purpose; but on the other hand it is equally plain that an artist cannot be drilled like a military recruit. There is, fortunately, no sign that these tactics will be directly adopted, but in an indirect fashion they are already being applied. An artist is not to blame if his temperament leads him to draw cartoons of leading Bolsheviks, or satirize the various comical aspects—and they are many—of the Soviet régime. To force such a man, however, to turn his talent only against Denikin, Yudenitch and Kolchak, or the leaders of the Entente, is momentarily good for Communism, but it is discouraging to the artist, and may prove in the long run bad for art, and possibly for Communism also. It is plain from the religious nature of Communism in Russia, that such controlling of the impulse to artistic creation is inevitable, and that propaganda art alone can flourish in such an atmosphere. For example, no poetry or literature that is not orthodox will reach the printing press. It is so easy to make the excuse of lack of paper and the urgent need for manifestoes. Thus there may well come to be a repetition of the attitude of the mediæval Church to the sagas and legends of the people, except that, in this case, it is the folk tales which will be preserved, and the more sensitive and civilized products banned. The only poet who seems to be much spoken of at present in Russia is one who writes rough popular songs. There are revolutionary odes, but one may hazard a guess that they resemble our patriotic war poetry.

I said that this state of affairs may in the long run be bad for art, but the contrary may equally well prove to be the truth. It is of course discouraging and paralysing to the old-style artist, and it is death to the old individual art which depended on subtlety and oddity of temperament, and arose very largely from the complicated psychology of the idle. There it stands, this old art, the purest monument to the nullity of the art-for-art's-sake doctrine, like a rich exotic plant of exquisite beauty, still apparently in its glory, till one perceives that the roots are cut, and that leaf by leaf it is gradually fading away.

But, unlike the Puritans in this respect, the Bolsheviks have not sought to dig up the roots, and there are signs that the paralysis is merely temporary. Moreover, individual art is not the only form, and in particular the plastic arts have shown that they can live by mass action, and flourish under an intolerant faith. Communist artists of the future may erect public buildings surpassing in beauty the mediæval churches, they may paint frescoes, organize pageants, make Homeric songs about their heroes. Communist art will begin, and is beginning now, in the propaganda pictures, and stories such as those designed for peasants and children. There is, for instance, a kind of Rake's Progress or "How she became a Communist," in which the Entente leaders make a sorry and grotesque appearance. Lenin and Trotsky already figure in woodcuts as Moses and Aaron, deliverers of their people, while the mother and child who illustrate the statistics of the maternity exhibition have the grace and beauty of mediæval madonnas. Russia is only now emerging from the middle ages, and the Church tradition in painting is passing with incredible smoothness into the service of Communist doctrine. These pictures have, too, an oriental flavour: there are brown Madonnas in the Russian churches, and such an one illustrates the statistics of infant mortality in India, while the Russian mother, broad-footed, in gay petticoat and kerchief, sits in a starry meadow suckling her baby from a very ample white breast. I think that this movement towards the Church tradition may be unconscious and instinctive, and would perhaps be deplored by many Communists, for whom grandiose bad Rodin statuary and the crudity of cubism better express what they mean by revolution. But this revolution is Russian and not French, and its art, if all goes well, should inevitably bear the popular Russian stamp. It is would-be primitive and popular art that is vulgar. Such at least is the reflection engendered by an inspection of Russian peasant work as compared with the spirit of *Children's Tales*.

The Russian peasant's artistic impulse is no legend. Besides the carving and embroidery which speak eloquently to peasant skill, one observes many instances in daily life. He will climb down, when his slowly-moving train stops by the wayside, to gather branches and flowers with which he will decorate the railway carriage both inside and out, he will work willingly at any task which has beauty for its object, and was all too prone under the old régime to waste his time and his employer's material in fashioning small metal or wooden objects with his hands.

If the *bourgeois* tradition then will not serve, there is a popular tradition which is still live and passionate and which may perhaps persist. Unhappily it has a formidable enemy in the organization and development of industry, which is far more dangerous to art than Communist doctrine. Indeed, industry in its early stages seems everywhere doomed to be the enemy of beauty and instinctive life. One might hope that this would not prove to be so in Russia, the first Socialist State, as yet unindustrial, able to draw on the industrial experience of the whole world, were it not that one discovers with a certain misgiving in the Bolshevik leaders the rasping arid temperament of those to whom the industrial machine is an end in itself, and, in addition, reflects that these industrially minded men have as yet no practical experience, nor do there exist men of goodwill to help them. It does not seem reasonable to hope that Russia can pass through the period of industrialization without a good deal of mismanagement, involving waste resulting in too long hours, child labour and other evils with which the West is all too familiar. What the Bolsheviks would not therefore willingly do to art, the Juggernaut which they are bent on setting in motion may accomplish for them.

The next generation in Russia will have to consist of practical hard-working men, the old-style artists will die off and successors will not readily arise. A State which is struggling with economic difficulties is bound to be slow to admit an artistic vocation, since this involves exemption from practical work. Moreover the majority of minds always turn instinctively to the real need of the moment. A man therefore who is adapted by talent and temperament to becoming an opera singer, will under the pressure of Communist enthusiasm and Government encouragement turn his attention to economics. (I am here quoting an actual instance.) The whole Russian people at this stage in their development strike one as being forced by the logic of their situation to make a similar choice.

It may be all to the good that there should be fewer professional artists, since some of the finest work has been done by men and groups of men to whom artistic expression was only a pastime. They were not hampered by the solemnity and reverence for art which too often destroy the spontaneity of the professional. Indeed a revival of this attitude to art is one of the good results which may be hoped for from a Communist revolution in a more advanced industrial community. There the problem of education will be to stimulate the creative impulses towards art and science so that men may know how to employ their leisure hours. Work in the factory can never be made to provide an adequate outlet. The only hope, if men are to remain human beings under industrialism, is to reduce hours to the minimum. But this is only possible when production and organization are highly efficient, which will not be the case for a long time in Russia. Hence not only does it appear that the number of artists will grow less, but that the number of people undamaged in their artistic impulses and on that account able to create or appreciate as amateurs is likely to be deplorably small. It is in this damaging effect of industry on human instinct that the immediate danger to art in Russia lies.

The effect of industry on the crafts is quite obvious. A craftsman who is accustomed to work with his hands, following the tradition developed by his ancestors, is useless when brought face to face with a machine. And the man who can handle the machine will only be concerned with quantity and utility in the first instance. Only gradually do the claims of beauty come to be recognized. Compare the modern motor car with the first of its species, or even, since the same law seems to operate in nature, the prehistoric animal with its modern descendant. The same relation exists between them as between man and the ape, or the horse and the hipparion. The movement of life seems to be towards ever greater delicacy and complexity, and man carries it forward in the articles that he makes and the society that he develops. Industry is a new tool, difficult to handle, but it will produce just as beautiful objects as did the mediæval builder and craftsman, though not until it has been in being for a long time and belongs to tradition.

One may expect, therefore, that while the crafts in Russia will lose in artistic value, the drama, sculpture and painting and all those arts which have nothing to do with the machine and depend entirely upon mental and spiritual inspiration will receive an impetus from the Communist faith. Whether the flowering period will be long or short depends partly on the political situation, but chiefly on the rapidity of industrial development. It may be that the machine will ultimately conquer the Communist faith and grind out the human impulses, and Russia become during this transition period as inartistic and soulless as was America until quite recent years. One would like to hope that mechanical progress will be swift and social idealism sufficiently strong to retain control. But the practical difficulties are almost insuperable.

Such signs of the progress of art as it is possible to notice at this early stage would seem to bear out the above argument. For instance, an attempt is being made to foster the continuation of peasant embroidery, carving, &c., in the towns. It is done by people who have evidently lost the tradition already. They are taught to copy the models which are placed in the Peasant Museum, but there is no comparison between the live little wooden lady who smiles beneath the glass case, and the soulless staring-eyed creature who is offered for sale, nor between the quite ordinary carved fowl one may buy and the amusing life-like figure one may merely gaze at.

But when one comes to art directly inspired by Communism it is a different story. Apart from the propaganda pictures already referred to, there are propaganda plays performed by the Red Army in its spare moments, and there are the mass pageant plays performed on State occasions. I had the good fortune to witness one of each kind.

The play was called *Zarevo* (The Dawn), and was performed on a Saturday night on a small stage in a small hall in an entirely amateur fashion. It represented Russian life just before the revolution. It was intense and tragic and passionately acted. Dramatic talent is not rare in Russia. Almost the only comic relief was provided by the Tsarist police, who made one appearance towards the end, got up like comic military characters in a musical comedy—just as, in mediæval miracle plays, the comic character was Satan. The play's intention was to show a typical Russian working-class family. There were the old father, constantly drunk on vodka, alternately maudlin and scolding; the old mother; two sons, the one a Communist and the other an Anarchist; the wife of the Communist, who did dressmaking; her sister, a prostitute; and a young girl of *bourgeois* family, also a Communist, involved in a plot with the Communist son, who was of course the hero of the play.

The first act revealed the stern and heroic Communist maintaining his views despite the reproaches of father and mother and the nagging of his wife. It showed also the Anarchist brother (as might be expected from the Bolshevik hostility to Anarchism) as an unruly, lazy, ne'er-do-well, with a passionate love for Sonia, the young *bourgeoise*, which was likely to become dangerous if not returned. She, on the other hand, obviously preferred the Communist. It was clear that he returned her love, but it was not quite clear that he would wish the relation to be anything more than platonic comradeship in the service of their common ideal. An unsuccessful strike, bringing want and danger from the police, together with increasing jealousy on the part of the Anarchist, led up to the tragic dénouement. I was not quite definite as to how this was brought about. All violent action was performed off the stage, and this made the plot at times difficult to follow. But it seemed that the Anarchist in a jealous rage forged a letter from his brother to bring Sonia to a rendezvous, and there murdered her, at the same time betraying his brother to the police. When the latter came to effect his arrest, and accuse him also, as the most likely person, of the murder, the Anarchist was seized with remorse and confessed. Both were therefore led away together. Once the plot is sketched, the play calls for no comment. It had not great merit, though it is unwise to hazard a judgment on a play whose dialogue was not fully interpreted, but it was certainly real, and the link between audience and performers was established as it never seemed to be in the professional theatre. After the performance, the floor was cleared for dancing, and the audience were in a mood of thorough enjoyment.

The pageant of the "World Commune," which was performed at the opening of the Third International Congress in Petrograd, was a still more important and significant phenomenon. I do not suppose that anything of the kind has been staged since the days of the mediæval mystery plays. It was, in fact, a mystery play designed by the High Priests of the Communist faith to instruct the people. It was played on the steps of an immense white building that was once the Stock Exchange, a building with a classical colonnade on three sides of it, with a vast flight of steps in front, that did not extend the whole width of the building but left at each side a platform that was level with the floor of the colonnade. In front of this building a wide road ran from a bridge over one arm of the river to a bridge over the other, so that the stretches of water and sky on either side seemed to the eye of imagination like the painted wings of a gigantic stage. Two battered red columns of fantastic design, that were once light towers to guide ships, stood on either side midway between the extremities of the building and the water, but on the opposite side of the road. These two towers were beflagged and illuminated and carried the limelight, and between and behind them was gathered a densely packed audience of forty or fifty thousand people. The play began at sundown, while the sky was still red away to the right and the palaces on the far bank to the left still aglow with the setting sun, and it continued under the magic of the darkening sky. At first the beauty and grandeur of the setting drew the attention away from the performers, but gradually one became aware that on the platform before the columns kings and queens and courtiers in sumptuous conventional robes, and attended by soldiers, were conversing in dumb show with one another. A few climbed the steps of a small wooden platform that was set up in the middle, and one indicated by a lifted hand that here should be built a monument to the power of capitalism over the earth. All gave signs of delight. Sentimental music was heard, and the gay company fell to waltzing away the hours. Meanwhile, from below on the road level, there streamed out of the darkness on either side of the building and up the half-lit steps, their fetters ringing in harmony with the music, the enslaved and toiling masses coming in response to command to build the monument for their masters. It is impossible to describe the exquisite beauty of the slow movement of those dark figures aslant the broad flight of steps; individual expressions were of course indistinguishable, and yet the movement and attitude of the groups conveyed pathos and patient endurance as well as any individual speech or gesture in the ordinary theatre. Some groups carried hammer and anvil, and others staggered under enormous blocks of stone. Love for the ballet has perhaps made the Russians understand the art of moving groups of actors in unison. As I watched these processions climbing the steps in apparently careless and spontaneous fashion, and yet producing so graceful a result, I remembered the mad leap of the archers down the stage in *Prince Igor*, which is also apparently careless and spontaneous and full of wild and irregular beauty, yet never varies a hair-breadth from one performance to the next.

For a time the workers toiled in the shadow in their earthly world, and dancing continued in the lighted paradise of the rulers above, until presently, in sign that the monument was complete, a large yellow disc was hoisted amid acclamation above the highest platform between the columns. But at the same moment a banner was uplifted amongst the people, and a small figure was seen gesticulating. Angry fists were shaken and the banner and speaker disappeared, only to reappear almost immediately in another part of the dense crowd. Again hostility, until finally among the French workers away up on the right, the first Communist manifesto found favour. Rallying around their banner the *communards* ran shouting down the steps, gathering supporters as they came. Above, all is confusion, kings and queens scuttling in unroyal fashion with flying velvet robes to safe citadels right and left, while the army prepares to defend the main citadel of capitalism with its golden disc of power. The *communards* scale the steps to the fortress which they finally capture, haul down the disc and set their banner in its place. The merry music of the *Carmagnole* is heard, and the victors are seen expressing their delight by dancing first on one foot and then on the other, like marionettes. Below, the masses dance with them in a frenzy of joy. But a pompous procession of Prussian legions is seen approaching, and, amid shrieks and wails of despair, the people are driven back, and their leaders set in a row and shot. Thereafter came one of the most moving scenes in the drama. Several dark-clad women appeared carrying a black pall supported on sticks, which they set in front of the bodies of the leaders so that it stood out, an irregular pointed black shape against the white columns behind. But for this melancholy monument the stage was now empty. Thick clouds of black smoke arose from braziers on either side and obscured the steps and the platform. Through the smoke came the distant sound of Chopin's *Marche Funèbre*, and as the air became clearer white figures could be dimly seen moving around the black pall in a solemn dance of mourning. Behind them the columns shone ghostly and unreal against the glimmering mauve rays of an uncertain and watery dawn.

The second part of the pageant opened in July 1914. Once again the rulers were feasting and the workers at toil, but the scene was enlivened by the presence of the leaders of the Second International, a group of decrepit professorial old men, who waddled in in solemn procession carrying tomes full of international learning. They sat in a row between the rulers and the people, deep in study, spectacles on nose. The call to war was the signal for a dramatic appeal from the workers to these leaders, who refused to accept the Red Flag, but weakly received patriotic flags from their respective governments. Jaurès, elevated to be the symbol of protest, towered above the people, crying in a loud voice, but fell back immediately as the assassin's shot rang out. Then the people divided into their national groups and the war began. It was at this point that "God Save the King" was played as the English soldiers marched out, in a comic manner which made one think of it as "*Gawd* save the King." Other national anthems were burlesqued in a similar fashion, but none quite so successfully. A ridiculous effigy of the Tsar with a knout in his hand now occupied the symbolic position and dominated the scene. The incidents of the war which affected Russia were then played. Spectacular cavalry charges on the road, marching soldiers, batteries of artillery, a pathetic procession of cripples and nurses, and other scenes too numerous to describe, made up that part of the pageant devoted to the war.

Then came the Russian Revolution in all its stages. Cars dashed by full of armed men, red flags appeared everywhere, the people stormed the citadel and hauled down the effigy of the Tsar. The Kerensky Government assumed control and drove them forth to war again, but soon they returned to the charge, destroyed the Provisional Government, and hoisted all the emblems of the Russian Soviet Republic. The Entente leaders, however, were seen preparing their troops for battle, and the pageant went on to show the formation of the Red Army under its emblem the Red Star. White figures with golden trumpets appeared foretelling victory for the proletariat. The last scene, the World Commune, is described in the words of the abstract, taken from a Russian newspaper, as follows:—

Cannon shots announce the breaking of the blockade against Soviet Russia, and the victory of the World Proletariat. The Red Army returns from the front, and passes in triumphant review before the leaders of the Revolution. At their feet lie the crowns of kings and the gold of the bankers. Ships draped with flags are seen carrying workers from the west. The workers of the whole world, with the emblems of labour, gather for the celebration of the World Commune. In the heavens luminous inscriptions in different languages appear, greeting the Congress: "Long live the Third International! Workers of the world, unite! Triumph to the sounds of the hymn of the World Commune, the 'International'."

Even so glowing an account, however, hardly does it justice. It had the pomp and majesty of the Day of Judgment itself. Rockets climbed the skies and peppered them with a thousand stars, fireworks blazed on all sides, garlanded and beflagged ships moved up and down the river, chariots bearing the emblems of prosperity, grapes and corn, travelled slowly along the road. The Eastern peoples came carrying gifts and emblems. The actors, massed upon the steps, waved triumphant hands, trumpets sounded, and the song of the International from ten thousand throats rose like a mighty wave engulfing the whole.

Though the end of this drama may have erred on the side of the grandiose, this may perhaps be forgiven the organizers in view of the occasion for which they prepared it. Nothing, however, could detract from the beauty and dramatic power of the opening and of many of the scenes. Moreover, the effects obtained by movement in the mass were almost intoxicating. The first entrance of the masses gave a sense of dumb and patient force that was moving in the extreme, and the frenzied delight of the dancing crowd at the victory of the French *communards* stirred one to ecstasy. The pageant lasted for five hours or more, and was as exhausting emotionally as the Passion Play is said to be. I had the vision of a great period of Communist art, more especially of such open-air spectacles, which should have the grandeur and scope and eternal meaning of the plays of ancient Greece, the mediæval mysteries, or the Shakespearean theatre. In building, writing, acting, even in painting, work would be done, as it once was, by groups, not by one hand or mind, and evolution would proceed slowly until once again the individual emerged from the mass.

In considering Education under the Bolshevik régime, the same two factors which I have already dealt with in discussing art, namely industrial development and the communist doctrine, must be taken into account. Industrial development is in reality one of the tenets of Communism, but as it is one which in Russia is likely to endanger the doctrine as a whole I have thought it better to consider it as a separate item.

As in the matter of art, so in education, those who have given unqualified praise seem to have taken the short and superficial view. It is hardly necessary to launch into descriptions of the crèches, country homes or palaces for children, where Montessori methods prevail, where the pupils cultivate their little gardens, model in plasticine, draw and sing and act, and dance their Eurythmic dances barefoot on floors once sacred to the tread of the nobility. I saw a reception and distributing house in Petrograd with which no fault could be found from the point of view of scientific organization. The children were bright-eyed and merry, and the rooms airy and clean. I saw, too, a performance by school children in Moscow which included some quite wonderful Eurythmic dancing, in particular an interpretation of Grieg's *Tanz in der Halle des Bergkönigs* by the Dalcroze method, but with a colour and warmth which were Russian, and in odd contrast to the mathematical precision associated with most Dalcroze performances.

But in spite of the obvious merit of such institutions as exist, misgivings would arise. To begin with, it must be remembered that it is necessary first to admit that children should be delivered up almost entirely to the State. Nominally, the mother still comes to see her child in these schools, but in actual fact, the drafting of children to the country must intervene, and the whole temper of the authorities seemed to be directed towards breaking the link between mother and child. To some this will seem an advantage, and it is a point which admits of lengthy discussion, but as it belongs rather to the question of women and the family under Communism, I can do no more than mention it here.

Then, again, it must be remembered that the tactics of the Bolsheviks towards such schools as existed under the old régime in provincial towns and villages, have not been the same as their tactics towards the theatres. The greater number of these schools are closed, in part, it would seem, from lack of personnel, and in part from fear of counter-revolutionary propaganda. The result is that, though those schools which they have created are good and organized on modern lines, on the whole there would seem to be less diffusion of child education than before. In this, as in most other departments, the Bolsheviks show themselves loath to attempt anything which cannot be done on a large scale and impregnated with Communist doctrine. It goes without saying that Communist doctrine is taught in schools, as Christianity has been taught hitherto, moreover the Communist teachers show bitter hostility to other teachers who do not accept the doctrine. At the children's entertainment alluded to above, the dances and poems performed had nearly all some close relation to Communism, and a teacher addressed the children for something like an hour and a half on the duties of Communists and the errors of Anarchism.

This teaching of Communism, however necessary it may appear for the building of the Communist state of the future, does seem to me to be an evil in that it is done emotionally and fanatically, with an appeal to hate and militant ardour rather than to constructive reason. It binds the free intellect and destroys initiative. An industrial state needs not only obedient and patient workers and artists, it needs also men and women with initiative in scientific research. It is idle to provide channels for scientific research later if it is to be choked at the source. That source is an enquiring and free intellect]unhampered by iron dogma. Beneficial to artistic and emotional development therefore, the teaching of Communism as a faith may well be most pernicious to the scientific and intellectual side of education, and will lead direct to the pragmatist view of knowledge and scientific research which the Church and the capitalist already find it so convenient to adopt.

But to come to the chief and most practical question, the relation of education to industry. Sooner or later education in Russia must become subordinate to the needs of industrial development. That the Bolsheviks already realize this is proved by the articles of Lunacharsky which recently appeared in *Le Phare* (Geneva). It was the spectre of industry that haunted me throughout the consideration of education as in the consideration of art, and what I have said above of its dangers to the latter seems to me also to apply here. Montessori schools belong, in my view, to that stage in industrial development when education is directed as much towards leisure occupations as towards preparation for professional life. Possibly the fine flower of useless scientific enquiry belongs to this stage also. Nobody in Russia is likely to have much leisure for a good many years to come, if the Bolshevik programme of industrial development is efficiently carried out. And there seemed to me to be something pathetic and almost cruel in this varied and agreeable education of the child, when one reflected on the long hours of grinding toil to which he was soon to be subject in workshop or factory. For I repeat that I do not believe industrial work in the early days of industry can be made tolerable to the worker. Once again I experienced the dread of seeing the ideals of the Russian revolutionaries go down before the logic of necessity. They are beginning to pride themselves on being hard, practical men, and it seems quite reasonable to fear that they should come to regard this full and humane development of the child as a mere luxury and ultimately neglect it. Worse still, the few of these schools which already exist may perhaps become exclusive to the Communists and their children, or that company of Samurai which is to leaven and govern the mass of the people. If so, they will soon come to resemble our public schools, in that they will prepare, in an artificial play atmosphere, men who will pass straight to the position of leaders, while the portion of the proletariat who serve under them will be reading and writing, just so much technical training as is necessary, and Communist doctrine.

This is a nightmare hypothesis, but the difficulties of the practical problem seem to warrant its entertainment. The number of people in Russia who can even read and write is extremely small, the need to get them employed industrially as rapidly as possible is very great, hence the system of education which develops out of this situation cannot be very ambitious or enlightened. Further it will have to continue over a sufficiently long period of time to allow of the risk of its becoming stable and traditional. In adult education already the pupil comes for a short period, learns Communism, reading and writing—there is hardly time to give him much more—and returns to leaven the army or his native village. In achieving this the Bolsheviks are already doing a very important and valuable work, but they cannot hope for a long while to become the model of public instruction which they have hitherto been represented to be. And the conditions of their becoming so ultimately are adherence to their ideals through a very long period of stress, and a lessening of fanaticism in their Communist teaching, conditions which, unhappily, seem to be mutually incompatible.

The whole of the argument set out in this chapter may be summed up in the statement of one fact which the mere idealist is prone to overlook, namely that Russia is a country at a stage in economic development not much more advanced than America in the pioneer days. The old civilization was aristocratic and exotic; it could not survive in the modern world. It is true that it produced great men, but its foundations were rotten. The new civilization may, for the moment, be less productive of individual works of genius, but it has a new solidity and gives promise of a new unity. It may be that I have taken too hopeful a view and that the future evolution of Russia will have as little connection with the life and tradition of its present population as modern America with the life of the Red Indian tribes. The fact that there exists in Russia a population at a far higher stage of culture, which will be industrially educated, not exterminated, militates against this hypothesis, but the need for education may make progress slower than it was in the United States.

One would not have looked for the millennium of Communism, nor even for valuable art and educational experiment in the America of early railroading and farming days. Nor must one look for such things from Russia yet. It may be that during the next hundred years there, economic evolution will obscure Communist ideals, until finally, in a country that has reached the stage of present-day America, the battle will be fought out again to a victorious and stable issue. Unless, indeed, the Marxian scripture prove to be not infallible, and faith and heroic devotion show themselves capable of triumphing over economic necessity.

V

COMMUNISM AND THE SOVIET CONSTITUTION

Before I went to Russia I imagined that I was going to see an interesting experiment in a new form of representative government. I did see an interesting experiment, but not in representative government. Everyone who is interested in Bolshevism knows the series of elections, from the village meeting to the All-Russian Soviet, by which the people's commissaries are supposed to derive their power. We were told that, by the recall, the occupational constituencies, and so on, a new and far more perfect machinery had been devised for ascertaining and registering the popular will. One of the things we hoped to study was the question whether the Soviet system is really superior to Parliamentarism in this respect.

We were not able to make any such study, because the Soviet system is moribund.No conceivable system of free election would give majorities to the Communists, either in town or country. Various methods are therefore adopted for giving the victory to Government candidates. In the first place, the voting is by show of hands, so that all who vote against the Government are marked men. In the second place, no candidate who is not a Communist can have any printing done, the printing works being all in the hands of the State. In the third place, he cannot address any meetings, because the halls all belong to the State. The whole of the press is, of course, official; no independent daily is permitted. In spite of all these obstacles, the Mensheviks have succeeded in winning about seats out of 1,500 on the Moscow Soviet, by being known in certain large factories where the electoral campaign could be conducted by word of mouth. They won, in fact, every seat that they contested.

But although the Moscow Soviet is nominally sovereign in Moscow, it is really only a body of electors who choose the executive committee of forty, out of which, in turn, is chosen the Presidium, consisting of nine men who have all the power. The Moscow Soviet, as a whole, meets rarely; the Executive Committee is supposed to meet once a week, but did not meet while we were in Moscow. The Presidium, on the contrary, meets daily. Of course, it is easy for the Government to exercise pressure over the election of the executive committee, and again over the election of the Presidium. It must be remembered that effective protest is impossible, owing to the absolutely complete suppression of free speech and free Press. The result is that the Presidium of the Moscow Soviet consists only of orthodox Communists.

Kamenev, the President of the Moscow Soviet, informed us that the recall is very frequently employed; he said that in Moscow there are, on an average, thirty recalls a month. I asked him what were the principal reasons for the recall, and he mentioned four: drinking, going to the front (and being, therefore, incapable of performing the duties), change of politics on the part of the electors, and failure to make a report to the electors once a fortnight, which all members of the Soviet are expected to do. It is evident that the recall affords opportunities for governmental pressure, but I had no chance of finding out whether it is used for this purpose.

In country districts the method employed is somewhat different. It is impossible to secure that the village Soviet shall consist of Communists, because, as a rule, at any rate in the villages I saw, there are no Communists. But when I asked in the villages how they were represented on the Volost (the next larger area) or the Gubernia, I was met always with the reply that they were not represented at all. I could not verify this, and it is probably an overstatement, but all concurred in the assertion that if they elected a non-Communist representative he could not obtain a pass on the railway and, therefore, could not attend the Volost or Gubernia Soviet. I saw a meeting of the Gubernia Soviet of Saratov. The representation is so arranged that the town workers have an enormous preponderance over the surrounding peasants; but even allowing for this, the proportion of peasants seemed astonishingly small for the centre of a very important agricultural area.

The All-Russian Soviet, which is constitutionally the supreme body, to which the People's Commissaries are responsible, meets seldom, and has become increasingly formal. Its sole function at present, so far as I could discover, is to ratify, without discussion, previous decisions of the Communist Party on matters (especially concerning foreign policy) upon which the constitution requires its decision.

All real power is in the hands of the Communist Party, who number about 600,000 in a population of about 120 millions. I never came across a Communist by chance: the people whom I met in the streets or in the villages, when I could get into conversation with them, almost invariably said they were of no party. The only other answer I ever had was from some of the peasants, who openly stated that they were Tsarists. It must be said that the peasants' reasons for disliking the Bolsheviks are very inadequate. It is said—and all I saw confirmed the assertion—that the peasants are better off than they ever were before. I saw no one—man, woman, or child—who looked underfed in the villages. The big landowners are dispossessed, and the peasants have profited. But the towns and the army still need nourishing, and the Government has nothing to give the peasants in return for food except paper, which the peasants resent having to take. It is a singular fact that Tsarist roubles are worth ten times as much as Soviet roubles, and are much commoner in the country. Although they are illegal, pocket-books full of them are openly displayed in the market places. I do not think it should be inferred that the peasants expect a Tsarist restoration: they are merely actuated by custom and dislike of novelty. They have never heard of the blockade; consequently they cannot understand why the Government is unable to give them the clothes and agricultural implements that they need. Having got their land, and being ignorant of affairs outside their own neighbourhood, they wish their own village to be independent, and would resent the demands of any Government whatever.

Within the Communist Party there are, of course, as always in a bureaucracy, different factions, though hitherto the external pressure has prevented disunion. It seemed to me that the personnel of the bureaucracy could be divided into three classes. There are first the old revolutionists, tested by years of persecution. These men have most of the highest posts. Prison and exile have made them tough and fanatical and rather out of touch with their own country. They are honest men, with a profound belief that Communism will regenerate the world. They think themselves utterly free from sentiment, but, in fact, they are sentimental about Communism and about the régime that they are creating; they cannot face the fact that what they are creating is not complete Communism, and that Communism is anathema to the peasant, who wants his own land and nothing else. They are pitiless in punishing corruption or drunkenness when they find either among officials; but they have built up a system in which the temptations to petty corruption are tremendous, and their own materialistic theory should persuade them that under such a system corruption must be rampant.

The second class in the bureaucracy, among whom are to be found most of the men occupying political posts just below the top, consists of *arrivistes*, who are enthusiastic Bolsheviks because of the material success of Bolshevism. With them must be reckoned the army of policemen, spies, and secret agents, largely inherited from the Tsarist times, who make their profit out of the fact that no one can live except by breaking the law. This aspect of Bolshevism is exemplified by the Extraordinary Commission, a body practically independent of the Government, possessing its own regiments, who are better fed than the Red Army. This body has the power of imprisoning any man or woman without trial on such charges as speculation or counter-revolutionary activity. It has shot thousands without proper trial, and though now it has nominally lost the power of inflicting the death penalty, it is by no means certain that it has altogether lost it in fact. It has spies everywhere, and ordinary mortals live in terror of it.

The third class in the bureaucracy consists of men who are not ardent Communists, who have rallied to the Government since it has proved itself stable, and who work for it either out of patriotism or because they enjoy the opportunity of developing their ideas freely without the obstacle of traditional institutions. Among this class are to be found men of the type of the successful business man, men with the same sort of ability as is found in the American self-made Trust magnate, but working for success and power, not for money. There is no doubt that the Bolsheviks are successfully solving the problem of enlisting this kind of ability in the public service, without permitting it to amass wealth as it does in capitalist communities. This is perhaps their greatest success so far, outside the domain of war. It makes it possible to suppose that, if Russia is allowed to have peace, an amazing industrial development may take place, making Russia a rival of the United States. The Bolsheviks are industrialists in all their aims; they love everything in modern industry except the excessive rewards of the capitalists. And the harsh discipline to which they are subjecting the workers is calculated, if anything can, to give them the habits of industry and honesty which have hitherto been lacking, and the lack of which alone prevents Russia from being one of the foremost industrial countries.

VI

THE FAILURE OF RUSSIAN INDUSTRY

At first sight it is surprising that Russian industry should have collapsed as badly as it has done, and still more surprising that the efforts of the Communists have not been more successful in reviving it. As I believe that the continued efficiency of industry is the main condition for success in the transition to a Communist State, I shall endeavour to analyse the causes of the collapse, with a view to the discovery of ways by which it can be avoided elsewhere.

Of the fact of the collapse there can be no doubt. The Ninth Congress of the Communist Party (March-April, 1920) speaks of "the incredible catastrophes of public economy," and in connection with transport, which is one of the vital elements of the problem, it acknowledges "the terrible collapse of the transport and the railway system," and urges the introduction of "measures which cannot be delayed and which are to obviate the complete paralysis of the railway system and, together with this, the ruin of the Soviet Republic." Almost all those who have visited Russia would confirm this view of the gravity of the situation. In the factories, in great works like those of Putilov and Sornovo, very little except war work is being done; machinery stands idle and plant is becoming unusable. One sees hardly any new manufactured articles in Russia, beyond a certain very inadequate quantity of clothes and boots—always excepting what is needed for the army. And the difficulty of obtaining food is conclusive evidence of the absence of goods such as are needed by the peasants.

How has this state of affairs arisen? And why does it continue?

A great deal of disorganization occurred before the first revolution and under Kerensky. Russian industry was partly dependent on Poland; the war was conducted by methods of reckless extravagance, especially as regards rolling-stock; under Kerensky there was a tendency to universal holiday, under the impression that freedom had removed the necessity for work. But when all this is admitted to the full, it remains true that the state of industry under the Bolsheviks is much worse than even under Kerensky.

The first and most obvious reason for this is that Russia was quite unusually dependent upon foreign assistance. Not only did the machinery in the factories and the locomotives on the railways come from abroad, but the organizing and technical brains in industry were mainly foreign. When the Entente became hostile to Russia, the foreigners in Russian industry either left the country or assisted counter-revolution. Even those who were in fact loyal naturally became suspect, and could not well be employed in responsible posts, any more than

Germans could in England during the war. The native Russians who had technical or business skill were little better; they almost all practised sabotage in the first period of the Bolshevik régime. One hears amusing stories of common sailors frantically struggling with complicated accounts, because no competent accountant would work for the Bolsheviks.

But those days passed. When the Government was seen to be stable, a great many of those who had formerly sabotaged it became willing to accept posts under it, and are now in fact so employed, often at quite exceptional salaries. Their importance is thoroughly realized. One resolution at the above-mentioned Congress says (I quote verbally the unedited document which was given to us in Moscow):

Being of opinion that without a scientific organization of industry, even the widest application of compulsory labour service, as the great labour heroism of the working class, will not only fail to secure the establishment of a powerful socialist production, but will also fail to assist the country to free itself from the clutches of poverty—the Congress considers it imperative to register all able specialists of the various departments of public economy and widely to utilize them for the purpose of industrial organization.

The Congress considers the elucidation for the wide masses of the workers of the tremendous character of the economic problems of the country to be one of the chief problems of industrial and general political agitation and propaganda; and of equal importance to this, technical education, and administrative and scientific technical experience. The Congress makes it obligatory on all the members of the party mercilessly to fight that particular obnoxious form, the ignorant conceit which deems the working class capable of solving all problems without the assistance *in the most responsible cases* of specialists of the bourgeois school, the management. Demagogic elements who speculate on this kind of prejudice in the more backward section of our working classes, can have no place in the ranks of the party of Scientific Socialism.

But Russia alone is unable to supply the amount of skill required, and is very deficient in technical instructors, as well as in skilled workmen. One was told, over and over again, that the first step in improvement would be the obtaining of spare parts for locomotives. It seems strange that these could not be manufactured in Russia. To some extent they can be, and we were shown locomotives which had been repaired on Communist Saturdays. But in the main the machinery for making spare parts is lacking and the skill required for its manufacture does not exist. Thus dependence on the outside world persists, and the blockade continues to do its deadly work of spreading hunger, demoralization and despair.

The food question is intimately bound up with the question of industry. There is a vicious circle, for not only does the absence of manufactured goods cause a food shortage in the towns, but the food shortage, in turn, diminishes the strength of the workers and makes them less able to produce goods. I cannot but think that there has been some mismanagement as regards the food question. For example, in Petrograd many workers have allotments and often work in them for eight hours after an eight hours' day in their regular employment. But the food produced in the allotments is taken for general consumption, not left to each individual producer. This is in accordance with Communist theory, but of course greatly diminishes the incentive to work, and increases the red tape and administrative machinery.

Lack of fuel has been another very grave source of trouble. Before the war coal came mostly from Poland and the Donetz Basin. Poland is lost to Russia, and the Donetz Basin was in the hands of Denikin, who so destroyed the mines before retreating that they are still not in working order. The result is a practically complete absence of coal. Oil, which is equally important in Russia, was also lacking until the recent recovery of Baku. All that I saw on the Volga made me believe that real efficiency has been shown in reorganizing the transport of oil, and doubtless this will do something to revive industry. But the oil used to be worked very largely by Englishmen, and English machinery is much needed for refining it. In the meantime, Russia has had to depend upon wood, which involves immense labour. Most of the houses are not warmed in winter, so that people live in a temperature below freezing-point. Another consequence of lack of fuel was the bursting of water-pipes, so that people in Petrograd, for the most part, have to go down to the Neva to fetch their water—a considerable addition to the labour of an already overworked day.

I find it difficult to believe that, if greater efficiency had existed in the Government, the food and fuel difficulties could not have been considerably alleviated. In spite of the needs of the army, there are still many horses in Russia; I saw troops of thousands of horses on the Volga, which apparently belonged to Kalmuk tribes. By the help of carts and sledges, it ought to be possible, without more labour than is warranted by the importance of the problem, to bring food and timber into Moscow and Petrograd. It must be remembered that both cities are surrounded by forests, and Moscow at least is surrounded by good agricultural land. The Government has devoted all its best energies hitherto to the two tasks of war and propaganda, while industry and the food problem have been left to a lesser degree of energy and intelligence. It is no doubt probable that, if peace is secured, the economic problems will receive more attention than hitherto. But the Russian character seems less adapted to steady work of an unexciting nature than to heroic efforts on great occasions; it has immense passive endurance, but not much active tenacity. Whether, with the menace of foreign invasion removed, enough day-by-day detailed energy would exist for the reorganization of industry, is a doubtful question, as to which only time can decide.

This leads to the conclusion—which I think is adopted by most of the leading men in Russia—that it will be very difficult indeed to save the revolution without outside economic assistance. Outside assistance from capitalist countries is dangerous to the principles of Communism, as well as precarious from the likelihood of fresh causes of quarrel. But the need of help is urgent, and if the policy of promoting revolution elsewhere were to succeed, it would probably render the nations concerned temporarily incapable of supplying Russian needs. It is, therefore, necessary for Russia to accept the risks and uncertainties involved in attempting to make peace with the Entente and to trade with America. By continuing war, Russia can do infinite damage to us, especially in Asia, but cannot hope, for many years, to achieve any degree of internal prosperity. The situation, therefore, is one in which, even from the narrowest point of view, peace is to the interest of both parties.

It is difficult for an outsider with only superficial knowledge to judge of the efforts which have been made to reorganize industry without outside help. These efforts have chiefly taken the form of industrial conscription. Workers in towns seek to escape to the country, in order to have enough to eat; but this is illegal and severely punished. The same Communist Report from which I have already quoted speaks on this subject as follows:

Labour Desertion.—Owing to the fact that a considerable part of the workers either in search of better food conditions or often for the purposes of speculation, voluntarily leave their places of employment or change from place to place, which inevitably harms production and deteriorates the general position of the working class, the Congress considers one of the most urgent problems of Soviet Government and of the Trade Union organization to be established as the firm, systematic and insistent struggle with labour desertion, The way to fight this is to publish a list of desertion fines, the creation of a labour Detachment of Deserters under fine, and, finally, internment in concentration camps.

It is hoped to extend the system to the peasantry:

The defeat of the White Armies and the problems of peaceful construction in connection with the incredible catastrophes of public economy demand an extraordinary effort of all the powers of the proletariat and the drafting into the process of public labour of the wide masses of the peasantry.

On the vital subject of transport, in a passage of which I have already quoted a fragment, the Communist Party declares:

For the most immediate future transport remains the centre of the attention and the efforts of the Soviet Government. The improvement of transport is the indispensable basis upon which even the most moderate success in all other spheres of production and first of all in the provision question can be gained.

The chief difficulty with regard to the improvement of transport is the weakness of the Transport Trade Union, which is due in the first case to the heterogeneity of the personnel of the railways, amongst whom there are still a number of those who belong to the period of disorganization, and, secondly, to the fact that the most class-conscious and best elements of the railway proletariat were at the various fronts of the civil war.

Considering wide Trade Union assistance to the railway workers to be one of the principal tasks of the Party, and as the only condition under which transport can be raised to its height, the Congress at the same time recognizes the inflexible necessity of employing exclusive and extraordinary measures (martial law, and so forth). Such necessity is the result of the terrible collapse of the transport and the railroad system and is to introduce measures which cannot be delayed and which are to obviate the complete paralysis of the railway system and, together with this, the ruin of the Soviet Republic.

The general attitude to the militarization of labour is stated in the Resolution with which this section of the Proceedings begins:

The ninth Congress approves of the decision of the Central Committee of the Russian Communist Party on the mobilization of the industrial proletariat, compulsory labour service, militarization of production and the application of military detachments to economic needs.

In connection with the above, the Congress decrees that the Party organization should in every way assist the Trade Unions and the Labour Sections in registering all skilled workers with a view of employing them in the various branches of production with the same consistency and strictness as was done, and is being carried out at the present time, in relation to the commanding staff for army needs.

Every skilled worker is to return to his particular trade Exceptions, i.e. the retention of the skilled worker in any other branch of Soviet service, is allowed only with the sanction of the corresponding central and local authorities.

It is, of course, evident that in these measures the Bolsheviks have been compelled to travel a long way from the ideals which originally inspired the revolution. But the situation is so desperate that they could not be blamed if their measures were successful. In a shipwreck all hands must turn to, and it would be ridiculous to prate of individual liberty. The most distressing feature of the situation is that these stern laws seem to have produced so little effect. Perhaps in the course of years Russia might become self-supporting without help from the outside world, but the suffering meantime would be terrible. The early hopes of the revolution would fade more and more. Every failure of industry, every tyrannous regulation brought about by the desperate situation, is used by the Entente as a justification of its policy. If a man is deprived of food and drink, he will grow weak, lose his reason, and finally die. This is not usually considered a good reason for inflicting death by starvation. But where nations are concerned, the weakness and struggles are regarded as morally culpable, and are held to justify further punishment. So at least it has been in the case of Russia. Nothing produced a doubt in our governing minds as to the rightness of our policy except the strength of the Red Army and the fear of revolution in Asia. Is it surprising that professions of humanitarian feeling on the part of English people are somewhat coldly received in Soviet Russia?

VII

DAILY LIFE IN MOSCOW

Daily life in Moscow, so far as I could discover, has neither the horrors depicted by the Northcliffe Press nor the delights imagined by the more ardent of our younger Socialists.

On the one hand, there is no disorder, very little crime, not much insecurity for those who keep clear of politics. Everybody works hard; the educated people have, by this time, mostly found their way into Government offices or teaching or some other administrative profession in which their education is useful. The theatres, the opera and the ballet continue as before, and are quite admirable; some of the seats are paid for, others are given free to members of trade unions. There is, of course, no drunkenness, or at any rate so little that none of us ever saw a sign of it. There is very little prostitution, infinitely less than in any other capital. Women are safer from molestation than anywhere else in the world. The whole impression is one of virtuous, well-ordered activity.

On the other hand, life is very hard for all except men in good posts. It is hard, first of all, owing to the food shortage. This is familiar to all who have interested themselves in Russia, and it is unnecessary to dwell upon it. What is less realized is that most people work much longer hours than in this country. The eight-hour day was introduced with a flourish of trumpets; then, owing to the pressure of the war, it was extended to ten hours in certain trades. But no provision exists against extra work at other jobs, and very many people do extra work, because the official rates do not afford a living wage. This is not the fault of the Government, at any rate as regards the major part; it is due chiefly to war and blockade. When the day's work is over, a great deal of time has to be spent in fetching food and water and other necessaries of life. The sight of the workers going to and fro, shabbily clad, with the inevitable bundle in one hand and tin can in the other, through streets almost entirely empty of traffic, produces the effect of life in some vast village, rather than in an important capital city.

Holidays, such as are common throughout all but the very poorest class in this country, are very difficult in Russia. A train journey requires a permit, which is only granted on good reasons being shown; with the present shortage of transport, this regulation is quite unavoidable. Railway queues are a common feature in Moscow; it often takes several days to get a permit. Then, when it has been obtained, it may take several more days to get a seat in a train. The ordinary trains are inconceivably crowded, far more so, though that seems impossible, than London trains at the busiest hour. On the shorter journeys, passengers are even known to ride on the roof and buffers, or cling like flies to the sides of the waggons. People in Moscow travel to the country whenever they can afford the time and get a permit, because in the country there is enough to eat. They go to stay with relations—most people in Moscow, in all classes, but especially among manual workers, have relations in the country. One cannot, of course, go to an hotel as one would in other countries. Hotels have been taken over by the State, and the rooms in them (when they are still used) are allocated by the police to people whose business is recognized as important by the authorities. Casual travel is therefore impossible even on a holiday.

Journeys have vexations in addition to the slowness and overcrowding of the trains. Police search the travellers for evidences of "speculation," especially for food. The police play, altogether, a much greater part in daily life than they do in other countries—much greater than they did, for example, in Prussia twenty-five years ago, when there was a vigorous campaign against Socialism. Everybody breaks the law almost daily, and no one knows which among his acquaintances is a spy of the Extraordinary Commission. Even in the prisons, among prisoners, there are spies, who are allowed certain privileges but not their liberty.

Newspapers are not taken in, except by very few people, but they are stuck up in public places, where passers-by occasionally glance at them.There is very little to read; owing to paper shortage, books are rare, and money to buy them is still rarer. One does not see people reading, as one does here in the Underground for example. There is practically no social life, partly because of the food shortage, partly because, when anybody is arrested, the police are apt to arrest everybody whom they find in his company, or who comes to visit him. And once arrested, a man or woman, however innocent, may remain for months in prison without trial. While we were in Moscow, forty social revolutionaries and Anarchists were hunger-striking to enforce their demand to be tried and to be allowed visits. I was told that on the eighth day of the strike the Government consented to try them, and that few could be proved guilty of any crime; but I had no means of verifying this.

Industrial conscription is, of course, rigidly enforced. Every man and woman has to work, and slacking is severely punished, by prison or a penal settlement. Strikes are illegal, though they sometimes occur. By proclaiming itself the friend of the proletarian, the Government has been enabled to establish an iron discipline, beyond the wildest dreams of the most autocratic American magnate. And by the same professions the Government has led Socialists from other countries to abstain from reporting unpleasant features in what they have seen.

The Tolstoyans, of whom I saw the leaders, are obliged by their creed to resist every form of conscription, though some have found ways of compromising. The law concerning conscientious objectors to military service is practically the same as ours, and its working depends upon the temper of the tribunal before which a man comes. Some conscientious objectors have been shot; on the other hand, some have obtained absolute exemption.

Life in Moscow, as compared to life in London, is drab, monotonous, and depressed. I am not, of course, comparing life there with that of the rich here, but with that of the average working-class family. When it is realized that the highest wages are about fifteen shillings a month, this is not surprising. I do not think that life could, under any system, be very cheerful in a country so exhausted by war as Russia, so I am not saying this as a criticism of the Bolsheviks. But I do think there might be less police interference, less vexatious regulation, and more freedom for spontaneous impulses towards harmless enjoyments.

Religion is still very strong. I went into many churches, where I saw obviously famished priests in gorgeous vestments, and a congregation enormously devout. Generally, more than half the congregation were men, and among the men many were soldiers. This applies to the towns as well as to the country. In Moscow I constantly saw people in the streets crossing themselves.

There is a theory that the Moscow working man feels himself free from capitalist domination, and therefore bears hardships gladly. This is no doubt true of the minority who are active Communists, but I do not think it has any truth for the others. The average working man, to judge by a rather hasty impression, feels himself the slave of the Government, and has no sense whatever of having been liberated from a tyranny.

I recognize to the full the reasons for the bad state of affairs, in the past history of Russia and the recent policy of the Entente. But I have thought it better to record impressions frankly, trusting the readers to remember that the Bolsheviks have only a very limited share of responsibility for the evils from which Russia is suffering.

VIII

TOWN AND COUNTRY

The problem of inducing the peasants to feed the towns is one which Russia shares with Central Europe, and from what one hears Russia has been less unsuccessful than some other countries in dealing with this problem. For the Soviet Government, the problem is mainly concentrated in Moscow and Petrograd; the other towns are not very large, and are mostly in the centre of rich agricultural districts. It is true that in the North even the rural population normally depends upon food from more southerly districts; but the northern population is small. It is commonly said that the problem of feeding Moscow and Petrograd is a transport problem, but I think this is only partially true. There is, of course, a grave deficiency of rolling-stock, especially of locomotives in good repair. But Moscow is surrounded by very good land. In the course of a day's motoring in the neighbourhood, I saw enough cows to supply milk to the whole child population of Moscow, although what I had come to see was children's sanatoria, not farms. All kinds of food can be bought in the market at high prices. I travelled over a considerable extent of Russian railways, and saw a fair number of goods trains. For all these reasons, I feel convinced that the share of the transport problem in the food difficulties has been exaggerated. Of course, transport plays a larger part in the shortage in Petrograd than in Moscow, because food comes mainly from south of Moscow. In Petrograd, most of the people one sees in the streets show obvious signs of under-feeding. In Moscow, the visible signs are much less frequent, but there is no doubt that under-feeding, though not actual starvation, is nearly universal.

The Government supplies rations to everyone who works in the towns at a very low fixed price. The official theory is that the Government has a monopoly of the food and that the rations are sufficient to sustain life. The fact is that the rations are not sufficient, and that they are only a portion of the food supply of Moscow. Moreover, people complain, I do not know how truly, that the rations are delivered irregularly; some say, about every other day. Under these circumstances, almost everybody, rich or poor, buys food in the market, where it costs about fifty times the fixed Government price. A pound of butter costs about a month's wages. In order to be able to afford extra food, people adopt various expedients. Some do additional work, at extra rates, after their official day's work is over. For, though there is supposed to be by law an eight-hours day, extended to ten in certain vital industries, the wage paid for it is not a living wage, and there is nothing to prevent a man from undertaking other work in his spare time.

But the usual resource is what is called "speculation," i.e., buying and selling. Some person formerly rich sells clothes or furniture or jewellery in return for food; the buyer sells again at an enhanced price, and so on through perhaps twenty hands, until a final purchaser is found in some well-to-do peasant or *nouveau riche* speculator. Again, most people have relations in the country, whom they visit from time to time, bringing back with them great bags of flour. It is illegal for private persons to bring food into Moscow, and the trains are searched; but, by corruption or cunning, experienced people can elude the search. The food market is illegal, and is raided occasionally; but as a rule it is winked at. Thus, the attempt to suppress private commerce has resulted in an amount of unprofessional buying and selling which far exceeds what happens in capitalist countries. It takes up a great deal of time that might be more profitably employed; and, being illegal, it places practically the whole population of Moscow at the mercy of the police. Moreover, it depends largely upon the stores of goods belonging to those who were formerly rich, and when these are expended, the whole system must collapse, unless industry has meanwhile been re-established on a sound basis.

It is clear that the state of affairs is unsatisfactory, but, from the Government's point of view, it is not easy to see what ought to be done. The urban and industrial population is mainly concerned in carrying on the work of government and supplying munitions to the army. These are very necessary tasks, the cost of which ought to be defrayed out of taxation. A moderate tax in kind on the peasants would easily feed Moscow and Petrograd. But the peasants take no interest in war or government. Russia is so vast that invasion of one part does not touch another part; and the peasants are too ignorant to have any national consciousness, such as one takes for granted in England or France or Germany. The peasants will not willingly part with a portion of their produce merely for purposes of national defence, but only for the goods they need—clothes, agricultural implements, &c.—which the Government, owing to the war and the blockade, is not in a position to supply.

When the food shortage was at its worst, the Government antagonized the peasants by forced requisitions, carried out with great harshness by the Red Army. This method has been modified, but the peasants still part unwillingly with their food, as is natural in view of the uselessness of paper and the enormously higher prices offered by private buyers.

The food problem is the main cause of popular opposition to the Bolsheviks, yet I cannot see how any popular policy could have been adopted. The Bolsheviks are disliked by the peasants because they take so much food; they are disliked in the towns because they take so little. What the peasants want is what is called free trade, i.e., de-control of agricultural produce. If this policy were adopted, the towns would be faced by utter starvation, not merely by hunger and hardship. It is an entire misconception to suppose that the peasants cherish any hostility to the Entente. The *Daily News* of July 13th, in an otherwise excellent leading article, speaks of "the growing hatred of the Russian peasant, who is neither a Communist nor a Bolshevik, for the Allies generally and this country in particular." The typical Russian peasant has never heard of the Allies or of this country; he does not know that there is a blockade; all he knows is that he used to have six cows but the Government reduced him to one for the sake of poorer peasants, and that it takes his corn (except what is needed for his own family) at a very low price. The reasons for these actions do not interest him, since his horizon is bounded by his own village. To a remarkable extent, each village is an independent unit. So long as the Government obtains the food and soldiers that it requires, it does not interfere, and leaves untouched the old village communism, which is extraordinarily unlike Bolshevism and entirely dependent upon a very primitive stage of culture.

The Government represents the interests of the urban and industrial population, and is, as it were, encamped amid a peasant nation, with whom its relations are rather diplomatic and military than governmental in the ordinary sense. The economic situation, as in Central Europe, is favourable to the country and unfavourable to the towns. If Russia were governed democratically, according to the will of the majority, the inhabitants of Moscow and Petrograd would die of starvation. As it is, Moscow and Petrograd just manage to live, by having the whole civil and military power of the State devoted to their needs. Russia affords the curious spectacle of a vast and powerful Empire, prosperous at the periphery, but faced with dire want at the centre. Those who have least prosperity have most power; and it is only through their excess of power that they are enabled to live at all. The situation is due at bottom to two facts: that almost the whole industrial energies of the population have had to be devoted to war, and that the peasants do not appreciate the importance of the war or the fact of the blockade.

It is futile to blame the Bolsheviks for an unpleasant and difficult situation which it has been impossible for them to avoid. Their problem is only soluble in one of two ways: by the cessation of the war and the blockade, which would enable them to supply the peasants with the goods they need in exchange for food; or by the gradual development of an independent Russian industry. This latter method would be slow, and would involve terrible hardships, but some of the ablest men in the Government believe it to be possible if peace cannot be achieved. If we force this method upon Russia by the refusal of peace and trade, we shall forfeit the only inducement we can hold out for friendly relations; we shall render the Soviet State unassailable and completely free to pursue the policy of promoting revolution everywhere. But the industrial problem is a large subject, which has been already discussed in Chapter

IX

INTERNATIONAL POLICY

In the course of these chapters, I have had occasion to mention disagreeable features of the Bolshevik régime. But it must always be remembered that these are chiefly due to the fact that the industrial life of Russia has been paralysed except as ministering to the wants of the Army, and that the Government has had to wage a bitter and doubtful civil and external war, involving the constant menace of domestic enemies. Harshness, espionage, and a curtailment of liberty result unavoidably from these difficulties. I have no doubt whatever that the sole cure for the evils from which Russia is suffering is peace and trade. Peace and trade would put an end to the hostility of the peasants, and would at once enable the Government to depend upon popularity rather than force. The character of the Government would alter rapidly under such conditions. Industrial conscription, which is now rigidly enforced, would become unnecessary. Those who desire a more liberal spirit would be able to make their voices heard without the feeling that they were assisting reaction and the national enemies. The food difficulties would cease, and with them the need for an autocratic system in the towns.

It must not be assumed, as is common with opponents of Bolshevism, that any other Government could easily be established in Russia. I think everyone who has been in Russia recently is convinced that the existing Government is stable. It may undergo internal developments, and might easily, but for Lenin, become a Bonapartist military autocracy. But this would be a change from within—not perhaps a very great change—and would probably do little to alter the economic system. From what I saw of the Russian character and of the opposition parties, I became persuaded that Russia is not ready for any form of democracy, and needs a strong Government. The Bolsheviks represent themselves as the Allies of Western advanced Socialism, and from this point of view they are open to grave criticism. For their international programme there is, to my mind, nothing to be said. But as a national Government, stripped of their camouflage, regarded as the successors of Peter the Great, they are performing a necessary though unamiable task. They are introducing, as far as they can, American efficiency among a lazy and undisciplined population. They are preparing to develop the natural resources of their country by the methods of State Socialism, for which, in Russia, there is much to be said. In the Army they are abolishing illiteracy, and if they had peace, they would do great things for education everywhere.

But if we continue to refuse peace and trade, I do not think the Bolsheviks will go under. Russia will endure great hardships, in the years to come as before. But the Russians are inured to misery as no Western nation is; they can live and work under conditions which we should find intolerable. The Government will be driven more and more, from mere self-preservation, into a policy of imperialism. The Entente has been doing everything to expose Germany to a Russian invasion of arms and leaflets, by allowing Poland to engage in war and compelling Germany to disarm. All Asia lies open to Bolshevik ambitions. Almost the whole of the former Russian Empire in Asia is quite firmly in their grasp. Trains are running at a reasonable speed to Turkestan, and I saw cotton from there being loaded on to Volga steamers. In Persia and Turkey, revolts are taking place, with Bolshevik support. It is only a question of a few years before India will be in touch with the Red Army. If we continue to antagonize the Bolsheviks, I do not see what force exists that can prevent them from acquiring the whole of Asia within ten years.

The Russian Government is not yet definitely imperialistic in spirit, and would still prefer peace to conquest. The country is weary of war and denuded of goods. But if the Western Powers insist upon war, another spirit, which is already beginning to show itself, will become dominant. Conquest will be the only alternative to submission. Asiatic conquest will not be difficult. But for us, from the imperialist standpoint, it will mean utter ruin. And for the Continent it will mean revolutions, civil wars, economic cataclysms. The policy of crushing Bolshevism by force was always foolish and criminal; it has now become impossible and fraught with disaster. Our own Government, it would seem, have begun to realize the dangers, but apparently, they do not realize them sufficiently to enforce their view against opposition.

In the Theses presented to the Second Congress of the Third International (July 1920), there is a very interesting article by Lenin called "First Sketch of the Theses on National and Colonial Questions" (*Theses*, pp. 40-47). The following passages seemed to me particularly illuminating: —

The present world-situation in politics places on the order of the day the dictatorship of the proletariat; and all the events of world politics are inevitably concentrated round one centre of gravity: the struggle of the international bourgeoisie against the Soviet Republic, which inevitably groups round it, on the one hand the Sovietist movements of the advanced working men of all countries, on the other hand all the national movements of emancipation of colonies and oppressed nations which have been convinced by a bitter experience that there is no salvation for them except in the victory of the Soviet Government over world-imperialism.

We cannot therefore any longer confine ourselves to recognizing and proclaiming the union of the workers of all countries. It is henceforth necessary to pursue the realization of the strictest union of all the national and colonial movements of emancipation with Soviet Russia, by giving to this union forms corresponding to the degree of evolution of the proletarian movement among the proletariat of each country, or of the democratic-bourgeois movement of emancipation among the workers and peasants of backward countries or backward nationalities.

The federal principle appears to us as a transitory form towards the complete unity of the workers of all countries.

This is the formula for co-operation with Sinn Fein or with Egyptian and Indian nationalism. It is further defined later. In regard to backward countries, Lenin says, we must have in view: —

The necessity of the co-operation of all Communists in the democratic-bourgeois movement of emancipation in those countries.

Again:

"The Communist International must conclude temporary alliances with the bourgeois democracy of backward countries, but must never fuse with it." The class-conscious proletariat must "show itself particularly circumspect towards the survivals of national sentiment in countries long oppressed," and must "consent to certain useful concessions."

The Asiatic policy of the Russian Government was adopted as a move against the British Empire, and as a method of inducing the British Government to make peace. It plays a larger part in the schemes of the leading Bolsheviks than is realized by the Labour Party in this country. Its method is not, for the present, to preach Communism, since the Persians and Hindoos are considered scarcely ripe for the doctrines of Marx. It is nationalist movements that are supported by money and agitators from Moscow. The method of quasi-independent states under Bolshevik protection is well understood. It is obvious that this policy affords opportunities for imperialism, under the cover of propaganda, and there is no doubt that some among the Bolsheviks are fascinated by its imperialist aspect. The importance officially attached to the Eastern policy is illustrated by the fact that it was the subject of the concluding portion of Lenin's speech to the recent Congress of the Third International (July 1920).

Bolshevism, like everything Russian, is partly Asiatic in character. One may distinguish two distinct trends, developing into two distinct policies. On the one side are the practical men, who wish to develop Russia industrially, to secure the gains of the Revolution nationally, to trade with the West, and gradually settle down into a more or less ordinary State. These men have on their side the fact of the economic exhaustion of Russia, the danger of ultimate revolt against Bolshevism if life continues to be as painful as it is at present, and the natural sentiment of humanity that wishes to relieve the sufferings of the people; also the fact that, if revolutions elsewhere produce a similar collapse of industry, they will make it impossible for Russia to receive the outside help which is urgently needed. In the early days, when the Government was weak, they had unchallenged control of policy, but success has made their position less secure.

On the other side there is a blend of two quite different aims: first, the desire to promote revolution in the Western nations, which is in line with Communist theory, and is also thought to be the only way of obtaining a really secure peace; secondly, the desire for Asiatic dominion, which is probably accompanied in the minds of some with dreams of sapphires and rubies and golden thrones and all the glories of their forefather Solomon. This desire produces an unwillingness to abandon the Eastern policy, although it is realized that, until it is abandoned, peace with capitalist England is impossible. I do not know whether there are some to whom the thought occurs that if England were to embark on revolution we should become willing to abandon India to the Russians. But I am certain that the converse thought occurs, namely that, if India could be taken from us, the blow to imperialist feeling might lead us to revolution. In either case, the two policies, of revolution in the West and conquest (disguised as liberation of oppressed peoples) in the East, work in together, and dovetail into a strongly coherent whole.

Bolshevism as a social phenomenon is to be reckoned as a religion, not as an ordinary political movement. The important and effective mental attitudes to the world may be broadly divided into the religious and the scientific. The scientific attitude is tentative and piecemeal, believing what it finds evidence for, and no more. Since Galileo, the scientific attitude has proved itself increasingly capable of ascertaining important facts and laws, which are acknowledged by all competent people regardless of temperament or self-interest or political pressure. Almost all the progress in the world from the earliest times is attributable to science and the scientific temper; almost all the major ills are attributable to religion.

By a religion I mean a set of beliefs held as dogmas, dominating the conduct of life, going beyond or contrary to evidence, and inculcated by methods which are emotional or authoritarian, not intellectual. By this definition, Bolshevism is a religion: that its dogmas go beyond or contrary to evidence, I shall try to prove in what follows. Those who accept Bolshevism become impervious to scientific evidence, and commit intellectual suicide. Even if all the doctrines of Bolshevism were true, this would still be the case, since no unbiased examination of them is tolerated. One who believes, as I do, that the free intellect is the chief engine of human progress, cannot but be fundamentally opposed to Bolshevism, as much as to the Church of Rome.

Among religions, Bolshevism is to be reckoned with Mohammedanism rather than with Christianity and Buddhism. Christianity and Buddhism are primarily personal religions, with mystical doctrines and a love of contemplation. Mohammedanism and Bolshevism are practical, social, unspiritual, concerned to win the empire of this world. Their founders would not have resisted the third of the temptations in the wilderness. What Mohammedanism did for the Arabs, Bolshevism may do for the Russians. As Ali went down before the politicians who only rallied to the Prophet after his success, so the genuine Communists may go down before those who are now rallying to the ranks of the Bolsheviks.]If so, Asiatic empire with all its pomps and splendours may well be the next stage of development, and Communism may seem, in historical retrospect, as small a part of Bolshevism as abstinence from alcohol is of Mohammedanism. It is true that, as a world force, whether for revolution or for empire, Bolshevism must sooner or later be brought by success into a desperate conflict with America; and America is more solid and strong, as yet, than anything that Mohammed's followers had to face. But the doctrines of Communism are almost certain, in the long run, to make progress among American wage-earners, and the opposition of America is therefore not likely to be eternal. Bolshevism may go under in Russia, but even if it does it will spring up again elsewhere, since it is ideally suited to an industrial population in distress. What is evil in it is mainly due to the fact that it has its origin in distress; the problem is to disentangle the good from the evil, and induce the adoption of the good in countries not goaded into ferocity by despair.

Russia is a backward country, not yet ready for the methods of equal co-operation which the West is seeking to substitute for arbitrary power in politics and industry. In Russia, the methods of the Bolsheviks are probably more or less unavoidable; at any rate, I am not prepared to criticize them in their broad lines. But they are not the methods appropriate to more advanced countries, and our Socialists will be unnecessarily retrograde if they allow the prestige of the Bolsheviks to lead them into slavish imitation. It will be a far less excusable error in our reactionaries if, by their unteachableness, they compel the adoption of violent methods. We have a heritage of civilization and mutual tolerance which is important to ourselves and to the world. Life in Russia has always been fierce and cruel, to a far greater degree than with us, and out of the war has come a danger that this fierceness and cruelty may become universal. I have hopes that in England this may be avoided through the moderation of both sides. But it is essential to a happy issue that melodrama should no longer determine our views of the Bolsheviks: they are neither angels to be worshipped nor devils to be exterminated, but merely bold and able men attempting with great skill an almost impossible task.

PART II

I

THE MATERIALISTIC THEORY OF HISTORY

The materialistic conception of history, as it is called, is due to Marx, and underlies the whole Communist philosophy. I do not mean, of course, that a man could not be a Communist without accepting it, but that in fact it is accepted by the Communist Party, and that it profoundly influences their views as to politics and tactics. The name does not convey at all accurately what is meant by the theory. It means that all the mass-phenomena of history are determined by economic motives. This view has no essential connection with materialism in the philosophic sense. Materialism in the philosophic sense may be defined as the theory that all apparently mental occurrences either are really physical, or at any rate have purely physical causes. Materialism in this sense also was preached by Marx, and is accepted by all orthodox Marxians. The arguments for and against it are long and complicated, and need not concern us, since, in fact, its truth or falsehood has little or no bearing on politics.

In particular, philosophic materialism does not prove that economic causes are fundamental in politics. The view of Buckle, for example, according to which climate is one of the decisive factors, is equally compatible with materialism. So is the Freudian view, which traces everything to sex. There are innumerable ways of viewing history which are materialistic in the philosophic sense without being economic or falling within the Marxian formula. Thus the "materialistic conception of history" may be false even if materialism in the philosophic sense should be true.

On the other hand, economic causes might be at the bottom of all political events even if philosophic materialism were false. Economic causes operate through men's desire for possessions, and would be supreme if this desire were supreme, even if desire could not, from a philosophic point of view, be explained in materialistic terms.

There is, therefore, no logical connection either way between philosophic materialism and what is called the "materialistic conception of history."

It is of some moment to realize such facts as this, because otherwise political theories are both supported and opposed for quite irrelevant reasons, and arguments of theoretical philosophy are employed to determine questions which depend upon concrete facts of human nature. This mixture damages both philosophy and politics, and is therefore important to avoid.

For another reason, also, the attempt to base a political theory upon a philosophical doctrine is undesirable. The philosophical doctrine of materialism, if true at all, is true everywhere and always; we cannot expect exceptions to it, say, in Buddhism or in the Hussite movement. And so it comes about that people whose politics are supposed to be a consequence of their metaphysics grow absolute and sweeping, unable to admit that a general theory of history is likely, at best, to be only true on the whole and in the main. The dogmatic character of Marxian Communism finds support in the supposed philosophic basis of the doctrine; it has the fixed certainty of Catholic theology, not the changing fluidity and sceptical practicality of modern science.

Treated as a practical approximation, not as an exact metaphysical law, the materialistic conception of history has a very large measure of truth. Take, as an instance of its truth, the influence of industrialism upon ideas. It is industrialism, rather than the arguments of Darwinians and Biblical critics, that has led to the decay of religious belief in the urban working class. At the same time, industrialism has revived religious belief among the rich. In the eighteenth-century French aristocrats mostly became free-thinkers; now their descendants are mostly Catholics, because it has become necessary for all the forces of reaction to unite against the revolutionary proletariat. Take, again, the emancipation of women. Plato, Mary Wolstonecraft, and John Stuart Mill produced admirable arguments, but influenced only a few impotent idealists. The war came, leading to the employment of women in industry on a large scale, and instantly the arguments in favour of votes for women were seen to be irresistible. More than that, traditional sexual morality collapsed, because its whole basis was the economic dependence of women upon their fathers and husbands. Changes in such a matter as sexual morality bring with them profound alterations in the thoughts and feelings of ordinary men and women; they modify law, literature, art, and all kinds of institutions that seem remote from economics.

Such facts as these justify Marxians in speaking, as they do, of "bourgeois ideology," meaning that kind of morality which has been imposed upon the world by the possessors of capital. Contentment with one's lot may be taken as typical of the virtues preached by the rich to the poor. They honestly believe it is a virtue—at any rate they did formerly. The more religious among the poor also believed it, partly from the influence of authority, partly from an impulse to submission, what MacDougall calls "negative self-feeling," which is commoner than some people think. Similarly men preached the virtue of female chastity, and women usually accepted their teaching; both really believed the doctrine, but its persistence was only possible through the economic power of men. This led erring women to punishment here on earth, which made further punishment hereafter seem probable. When the economic penalty ceased, the conviction of sinfulness gradually decayed. In such changes we see the collapse of "bourgeois ideology."

But in spite of the fundamental importance of economic facts in determining the politics and beliefs of an age or nation, I do not think that non-economic factors can be neglected without risks of errors which may be fatal in practice.

The most obvious non-economic factor, and the one the neglect of which has led Socialists most astray, is nationalism. Of course a nation, once formed, has economic interests which largely determine its politics; but it is not, as a rule, economic motives that decide what group of human beings shall form a nation. Trieste, before the war, considered itself Italian, although its whole prosperity as a port depended upon its belonging to Austria. No economic motive can account for the opposition between Ulster and the rest of Ireland. In Eastern Europe, the Balkanization produced by self-determination has been obviously disastrous from an economic point of view, and was demanded for reasons which were in essence sentimental. Throughout the war wage-earners, with only a few exceptions, allowed themselves to be governed by nationalist feeling, and ignored the traditional Communist exhortation: "Workers of the world, unite." According to Marxian orthodoxy, they were misled by cunning capitalists, who made their profit out of the slaughter. But to any one capable of observing psychological facts, it is obvious that this is largely a myth. Immense numbers of capitalists were ruined by the war; those who were young were just as liable to be killed as the proletarians were. No doubt commercial rivalry between England and Germany had a great deal to do with causing the war; but rivalry is a different thing from profit-seeking. Probably by combination English and German capitalists could have made more than they did out of rivalry, but the rivalry was instinctive, and its economic form was accidental. The capitalists were in the grip of nationalist instinct as much as their proletarian "dupes." In both classes some have gained by the war; but the universal will to war was not produced by the hope of gain. It was produced by a different set of instincts, and one which Marxian psychology fails to recognize adequately.

The Marxian assumes that a man's "herd," from the point of view of herd-instinct, is his class, and that he will combine with those whose economic class-interest is the same as his. This is only very partially true in fact. Religion has been the most decisive factor in determining a man's herd throughout long periods of the world's history. Even now a Catholic working man will vote for a Catholic capitalist rather than for an unbelieving Socialist. In America the divisions in local elections are mainly on religious lines. This is no doubt convenient for the capitalists, and tends to make them religious men; but the capitalists alone could not produce the result. The result is produced by the fact that many working men prefer the advancement of their creed to the improvement of their livelihood.

However deplorable such a state of mind may be, it is not necessarily due to capitalist lies.

All politics are governed by human desires. The materialist theory of history, in the last analysis, requires the assumption that every politically conscious person is governed by one single desire—the desire to increase his own share of commodities; and, further, that his method of achieving this desire will usually be to seek to increase the share of his class, not only his own individual share. But this assumption is very far from the truth. Men desire power, they desire satisfactions for their pride and their self-respect. They desire victory over rivals so profoundly that they will invent a rivalry for the unconscious purpose of making a victory possible. All these motives cut across the pure economic motive in ways that are practically important.

There is need of a treatment of political motives by the methods of psycho-analysis. In politics, as in private life, men invent myths to rationalize their conduct. If a man thinks that the only reasonable motive in politics is economic self-advancement, he will persuade himself that the things he wishes to do will make him rich. When he wants to fight the Germans, he tells himself that their competition is ruining his trade. If, on the other hand, he is an "idealist," who holds that his politics should aim at the advancement of the human race, he will tell himself that the crimes of the Germans demand their humiliation. The Marxian sees through this latter camouflage, but not through the former. To desire one's own economic advancement is comparatively reasonable; to Marx, who inherited eighteenth-century rationalist psychology from the British orthodox economists, self-enrichment seemed the natural aim of a man's political actions. But modern psychology has dived much deeper into the ocean of insanity upon which the little barque of human reason insecurely floats. The intellectual optimism of a bygone age is no longer possible to the modern student of human nature. Yet it lingers in Marxism, making Marxians rigid and Procrustean in their treatment of the life of instinct. Of this rigidity the materialistic conception of history is a prominent instance.

In the next chapter I shall attempt to outline a political psychology which seems to me more nearly true than that of Marx.

II

DECIDING FORCES IN POLITICS

The larger events in the political life of the world are determined by the interaction of material conditions and human passions. The operation of the passions on the material conditions is modified by intelligence. The passions themselves may be modified by alien intelligence guided by alien passions. So far, such modification has been wholly unscientific, but it may in time become as precise as engineering.

The classification of the passions which is most convenient in political theory is somewhat different from that which would be adopted in psychology.

We may begin with desires for the necessaries of life: food, drink, sex, and (in cold climates) clothing and housing. When these are threatened, there is no limit to the activity and violence that men will display.

Planted upon these primitive desires are a number of secondary desires. Love of property, of which the fundamental political importance is obvious, may be derived historically and psychologically from the hoarding instinct. Love of the good opinion of others (which we may call vanity) is a desire which man shares with many animals; it is perhaps derivable from courtship, but has great survival value, among gregarious animals, in regard to others besides possible mates. Rivalry and love of power are perhaps developments of jealousy; they are akin, but not identical.

These four passions—acquisitiveness, vanity, rivalry, and love of power—are, after the basic instincts, the prime movers of almost all that happens in politics. Their operation is intensified and regularized by herd instinct. But herd instinct, by its very nature, cannot be a prime mover, since it merely causes the herd to act in unison, without determining what the united action is to be. Among men, as among other gregarious animals, the united action, in any given circumstances, is determined partly by the common passions of the herd, partly by imitation of leaders. The art of politics consists in causing the latter to prevail over the former.

Of the four passions we have enumerated, only one, namely acquisitiveness, is concerned at all directly with men's relations to their material conditions. The other three—vanity, rivalry, and love of power—are concerned with social relations. I think this is the source of what is erroneous in the Marxian interpretation of history, which tacitly assumes that acquisitiveness is the source of all political actions. It is clear that many men willingly forego wealth for the sake of power and glory, and that nations habitually sacrifice riches to rivalry with other nations. The desire for some form of superiority is common to almost all energetic men. No social system which attempts to thwart it can be stable, since the lazy majority will never be a match for the energetic minority.

What is called "virtue" is an offshoot of vanity: it is the habit of acting in a manner which others praise.

The operation of material conditions may be illustrated by the statement (Myers's *Dawn of History*) that four of the greatest movements of conquest have been due to drought in Arabia, causing the nomads of that country to migrate into regions already inhabited. The last of these four movements was the rise of Islam. In these four cases, the primal need of food and drink was enough to set events in motion; but as this need could only be satisfied by conquest, the four secondary passions must have very soon come into play. In the conquests of modern industrialism, the secondary passions have been almost wholly dominant, since those who directed them had no need to fear hunger or thirst. It is the potency of vanity and love of power that gives hope for the industrial future of Soviet Russia, since it enables the Communist State to enlist in its service men whose abilities might give them vast wealth in a capitalistic society.

Intelligence modifies profoundly the operation of material conditions. When America was first discovered, men only desired gold and silver; consequently the portions first settled were not those that are now most profitable. The Bessemer process created the German iron and steel industry; inventions requiring oil have created a demand for that commodity which is one of the chief influences in international politics.

The intelligence which has this profound effect on politics is not political, but scientific and technical: it is the kind of intelligence which discovers how to make nature minister to human passions. Tungsten had no value until it was found to be useful in the manufacture of shells and electric light, but now people will, if necessary, kill each other in order to acquire tungsten. Scientific intelligence is the cause of this change.

The progress or retrogression of the world depends,broadly speaking, upon the balance between acquisitiveness and rivalry. The former makes for progress, the latter for retrogression. When intelligence provides improved methods of production, these may be employed to increase the general share of goods, or to set apart more of the labour power of the community for the business of killing its rivals. Until 1914, acquisitiveness had prevailed, on the whole, since the fall of Napoleon; the past six years have seen a prevalence of the instinct of rivalry. Scientific intelligence makes it possible to indulge this instinct more fully than is possible for primitive peoples, since it sets free more men from the labour of producing necessaries. It is possible that scientific intelligence may, in time, reach the point when it will enable rivalry to exterminate the human race. This is the most hopeful method of bringing about an end of war.

For those who do not like this method, there is another: the study of scientific psychology and physiology. The physiological causes of emotions have begun to be known, through the studies of such men as Cannon (*Bodily Changes in Pain, Hunger, Fear and Rage*). In time, it may become possible, by physiological means, to alter the whole emotional nature of a population. It will then depend upon the passions of the rulers how this power is used. Success will come to the State which discovers how to promote pugnacity to the extent required for external war, but not to the extent which would lead to domestic dissensions. There is no method by which it can be insured that rulers shall desire the good of mankind, and therefore there is no reason to suppose that the power to modify men's emotional nature would cause progress.

If men desired to diminish rivalry, there is an obvious method. Habits of power intensify the passion of rivalry; therefore a State in which power is concentrated will, other things being equal, be more bellicose than one in which power is diffused. For those who dislike wars, this is an additional argument against all forms of dictatorship. But dislike of war is far less common than we used to suppose; and those who like war can use the same argument to support dictatorship.

III

BOLSHEVIK CRITICISM OF DEMOCRACY

The Bolshevik argument against Parliamentary democracy as a method of achieving Socialism is a powerful one. My answer to it lies rather in pointing out what I believe to be fallacies in the Bolshevik method, from which I conclude that no swift method exists of establishing any desirable form of Socialism. But let us first see what the Bolshevik argument is.

In the first place, it assumes that those to whom it is addressed are absolutely certain that Communism is desirable, so certain that they are willing, if necessary, to force it upon an unwilling population at the point of the bayonet. It then proceeds to argue that, while capitalism retains its hold over propaganda and its means of corruption, Parliamentary methods are very unlikely to give a majority for Communism in the House of Commons, or to lead to effective action by such a majority even if it existed. Communists point out how the people are deceived, and how their chosen leaders have again and again betrayed them. From this they argue that the destruction of capitalism must be sudden and catastrophic; that it must be the work of a minority; and that it cannot be effected constitutionally or without violence. It is therefore, in their view, the duty of the Communist party in a capitalist country to prepare for armed conflict, and to take all possible measure for disarming the bourgeoisie and arming that part of the proletariat which is willing to support the Communists.

There is an air of realism and disillusionment about this position, which makes it attractive to those idealists who wish to think themselves cynics. But I think there are various points in which it fails to be as realistic as it pretends.

In the first place, it makes much of the treachery of Labour leaders in constitutional movements, but does not consider the possibility of the treachery of Communist leaders in a revolution. To this the Marxian would reply that in constitutional movements men are bought, directly or indirectly, by the money of the capitalists, but that revolutionary Communism would leave the capitalists no money with which to attempt corruption. This has been achieved in Russia, and could be achieved elsewhere. But selling oneself to the capitalists is not the only possible form of treachery. It is also possible, having acquired power, to use it for one's own ends instead of for the people. This is what I believe to be likely to happen in Russia: the establishment of a bureaucratic aristocracy, concentrating authority in its own hands, and creating a régime just as oppressive and cruel as that of capitalism. Marxians never sufficiently recognize that love of power is quite as strong a motive, and quite as great a source of injustice, as love of money; yet this must be obvious to any unbiased student of politics. It is also obvious that the method of violent revolution leading to a minority dictatorship is one peculiarly calculated to create habits of despotism which would survive the crisis by which they were generated. Communist politicians are likely to become just like the politicians of other parties: a few will be honest, but the great majority will merely cultivate the art of telling a plausible tale with a view to tricking the people into entrusting them with power. The only possible way by which politicians as a class can be improved is the political and psychological education of the people, so that they may learn to detect a humbug. In England men have reached the point of suspecting a good speaker, but if a man speaks badly they think he must be honest. Unfortunately, virtue is not so widely diffused as this theory would imply.

In the second place, it is assumed by the Communist argument that, although capitalist propaganda can prevent the majority from becoming Communists, yet capitalist laws and police forces cannot prevent the Communists, while still a minority, from acquiring a supremacy of military power. It is thought that secret propaganda can undermine the army and navy, although it is admittedly impossible to get the majority to vote at elections for the programme of the Bolsheviks. This view is based upon Russian experience, where the army and navy had suffered defeat and had been brutally ill used by incompetent Tsarist authorities. The argument has no application to more efficient and successful States. Among the Germans, even in defeat, it was the civilian population that began the revolution.

There is a further assumption in the Bolshevik argument which seems to me quite unwarrantable. It is assumed that the capitalist governments will have learned nothing from the experience of Russia. Before the Russian Revolution, governments had not studied Bolshevik theory. And defeat in war created a revolutionary mood throughout Central and Eastern Europe. But now the holders of power are on their guard. There seems no reason whatever to suppose that they will supinely permit a preponderance of armed force to pass into the hands of those who wish to overthrow them, while, according to the Bolshevik theory, they are still sufficiently popular to be supported by a majority at the polls. Is it not as clear as noonday that in a democratic country it is more difficult for the proletariat to destroy the Government by arms than to defeat it in a general election? Seeing the immense advantages of a Government in dealing with rebels, it seems clear that rebellion could have little hope of success unless a very large majority supported it. Of course, if the army and navy were specially revolutionary, they might effect an unpopular revolution; but this situation, though something like it occurred in Russia, is hardly to be expected in the Western nations. This whole Bolshevik theory of revolution by a minority is one which might just conceivably have succeeded as a secret plot, but becomes impossible as soon as it is openly avowed and advocated.

But perhaps it will be said that I am caricaturing the Bolshevik doctrine of revolution. It is urged by advocates of this doctrine, quite truly, that all political events are brought about by minorities, since the majority are indifferent to politics. But there is a difference between a minority in which the indifferent acquiesce, and a minority so hated as to startle the indifferent into belated action. To make the Bolshevik doctrine reasonable, it is necessary to suppose that they believe the majority can be induced to acquiesce, at least temporarily, in the revolution made by the class-conscious minority. This, again, is based upon Russian experience: desire for peace and land led to a widespread support of the Bolsheviks in November 1917 on the part of people who have subsequently shown no love for Communism.

I think we come here to an essential part of Bolshevik philosophy. In the moment of revolution, Communists are to have some popular cry by which they win more support than mere Communism could win. Having thus acquired the State machine, they are to use it for their own ends. But this, again, is a method which can only be practised successfully so long as it is not avowed. It is to some extent habitual in politics. The Unionists in 1900 won a majority on the Boer War, and used it to endow brewers and Church schools. The Liberals in 1906 won a majority on Chinese labour, and used it to cement the secret alliance with France and to make an alliance with Tsarist Russia. President Wilson, in 1916, won his majority on neutrality, and used it to come into the war. This method is part of the stock-in-trade of democracy. But its success depends upon repudiating it until the moment comes to practise it. Those who, like the Bolsheviks, have the honesty to proclaim in advance their intention of using power for other ends than those for which it was given them, are not likely to have a chance of carrying out their designs.

What seems to me to emerge from these considerations is this: That in a democratic and politically educated country, armed revolution in favour of Communism would have no chance of succeeding unless it were supported by a larger majority than would be required for the election of a Communist Government by constitutional methods. It is possible that, if such a Government came into existence, and proceeded to carry out its programme, it would be met by armed resistance on the part of capital, including a large proportion of the officers in the army and navy. But in subduing this resistance it would have the support of that great body of opinion which believes in legality and upholds the constitution. Moreover, having, by hypothesis, converted a majority of the nation, a Communist Government could be sure of loyal help from immense numbers of workers, and would not be forced, as the Bolsheviks are in Russia, to suspect treachery everywhere. Under these circumstances, I believe that the resistance of the capitalists could be quelled without much difficulty, and would receive little support from moderate people. Whereas, in a minority revolt of

Communists against a capitalist Government, all moderate opinion would be on the side of capitalism.

The contention that capitalist propaganda is what prevents the adoption of Communism by wage-earners is only very partially true. Capitalist propaganda has never been able to prevent the Irish from voting against the English, though it has been applied to this object with great vigour. It has proved itself powerless, over and over again, in opposing nationalist movements which had almost no moneyed support. It has been unable to cope with religious feeling. And those industrial populations which would most obviously benefit by Socialism have, in the main, adopted it, in spite of the opposition of employers. The plain truth is that Socialism does not arouse the same passionate interest in the average citizen as is roused by nationality and used to be roused by religion. It is not unlikely that things may change in this respect: we may be approaching a period of economic civil wars comparable to that of the religious civil wars that followed the Reformation. In such a period, nationalism is submerged by party: British and German Socialists, or British and German capitalists, will feel more kinship with each other than with compatriots of the opposite political camp. But when that day comes, there will be no difficulty, in highly industrial countries, in securing Socialist majorities; if Socialism is not then carried without bloodshed, it will be due to the unconstitutional action of the rich, not to the need of revolutionary violence on the part of the advocates of the proletariat. Whether such a state of opinion grows up or not depends mainly upon the stubbornness or conciliatoriness of the possessing classes, and, conversely, upon the moderation or violence of those who desire fundamental economic change. The majority which Bolsheviks regard as unattainable is chiefly prevented by the ruthlessness of their own tactics.

Apart from all arguments of detail, there are two broad objections to violent revolution in a democratic community. The first is that, when once the principle of respecting majorities as expressed at the ballot-box is abandoned, there is no reason to suppose that victory will be secured by the particular minority to which one happens to belong.There are many minorities besides Communists: religious minorities, teetotal minorities, militarist minorities, capitalist minorities. Any one of these could adopt the method of obtaining power advocated by the Bolsheviks, and any one would be just as likely to succeed as they are. What restrains these minorities, more or less, at present, is respect for the law and the constitution. Bolsheviks tacitly assume that every other party will preserve this respect while they themselves, unhindered, prepare the revolution. But if their philosophy of violence becomes popular, there is not the slightest reason to suppose that they will be its beneficiaries. They believe that Communism is for the good of the majority; they ought to believe that they can persuade the majority on this question, and to have the patience to set about the task of winning by propaganda.

The second argument of principle against the method of minority violence is that abandonment of law, when it becomes widespread, lets loose the wild beast, and gives a free rein to the primitive lusts and egoisms which civilization in some degree curbs. Every student of mediæval thought must have been struck by the extraordinarily high value placed upon law in that period. The reason was that, in countries infested by robber barons, law was the first requisite of progress. We, in the modern world, take it for granted that most people will be law-abiding, and we hardly realize what centuries of effort have gone to making such an assumption possible. We forget how many of the good things that we unquestionably expect would disappear out of life if murder, rape, and robbery with violence became common. And we forget even more how very easily this might happen. The universal class-war foreshadowed by the Third International, following upon the loosening of restraints produced by the late war, and combined with a deliberate inculcation of disrespect for law and constitutional government, might, and I believe would, produce a state of affairs in which it would be habitual to murder men for a crust of bread, and in which women would only be safe while armed men protected them. The civilized nations have accepted democratic government as a method of settling internal disputes without violence. Democratic government may have all the faults attributed to it, but it has the one great merit that people are, on the whole, willing to accept it as a substitute for civil war in political disputes. Whoever sets to work to weaken this acceptance, whether in Ulster or in Moscow, is taking a fearful responsibility. Civilization is not so stable that it cannot be broken up; and a condition of lawless violence is not one out of which any good thing is likely to emerge. For this reason, if for no other, revolutionary violence in a democracy is infinitely dangerous.

IV

REVOLUTION AND DICTATORSHIP

The Bolsheviks have a very definite programme for achieving Communism—a programme which has been set forth by Lenin repeatedly, and quite recently in the reply of the Third International to the questionnaire submitted by the Independent Labour Party.

Capitalists, we are assured, will stick at nothing in defence of their privileges. It is the nature of man, in so far as he is politically conscious, to fight for the interests of his class so long as classes exist. When the conflict is not pushed to extremes, methods of conciliation and political deception may be preferable to actual physical warfare; but as soon as the proletariat make a really vital attack upon the capitalists, they will be met by guns and bayonets. This being certain and inevitable, it is as well to be prepared for it, and to conduct propaganda accordingly. Those who pretend that pacific methods can lead to the realization of Communism are false friends to the wage-earners; intentionally or unintentionally, they are covert allies of the bourgeoisie.

There must, then, according to Bolshevik theory, be armed conflict sooner or later, if the injustices of the present economic system are ever to be remedied. Not only do they assume armed conflict: they have a fairly definite conception of the way in which it is to be conducted. This conception has been carried out in Russia, and is to be carried out, before very long, in every civilized country. The Communists, who represent the class-conscious wage-earners, wait for some propitious moment when events have caused a mood of revolutionary discontent with the existing Government. They then put themselves at the head of the discontent, carry through a successful revolution, and in so doing acquire the arms, the railways, the State treasure, and all the other resources upon which the power of modern Governments is built. They then confine political power to Communists, however small a minority they may be of the whole nation. They set to work to increase their number by propaganda and the control of education. And meanwhile, they introduce Communism into every department of economic life as quickly as possible.

Ultimately, after a longer or shorter period, according to circumstances, the nation will be converted to Communism, the relics of capitalist institutions will have been obliterated, and it will be possible to restore freedom. But the political conflicts to which we are accustomed will not reappear. All the burning political questions of our time, according to the Communists, are questions of class conflict, and will disappear when the division of classes disappears. Accordingly the State will no longer be required, since the State is essentially an engine of power designed to give the victory to one side in the class conflict. Ordinary States are designed to give the victory to the capitalists; the proletarian State (Soviet Russia) is designed to give the victory to the wage-earners. As soon as the community contains only wage-earners, the State will cease to have any functions. And so, through a period of dictatorship, we shall finally arrive at a condition very similar to that aimed at by Anarchist Communism.

Three questions arise in regard to this method of reaching Utopia. First, would the ultimate state foreshadowed by the Bolsheviks be desirable in itself? Secondly, would the conflict involved in achieving it by the Bolshevik method be so bitter and prolonged that its evils would outweigh the ultimate good? Thirdly, is this method likely to lead, in the end, to the state which the Bolsheviks desire, or will it fail at some point and arrive at a quite different result? If we are to be Bolsheviks, we must answer all these questions in a sense favourable to their programme.

As regards the first question, I have no hesitation in answering it in a manner favourable to Communism. It is clear that the present inequalities of wealth are unjust. In part, they may be defended as affording an incentive to useful industry, but I do not think this defence will carry us very far. However, I have argued this question before in my book on *Roads to Freedom*, and I will not spend time upon it now. On this matter, I concede the Bolshevik case. It is the other two questions that I wish to discuss.

Our second question was: Is the ultimate good aimed at by the Bolsheviks sufficiently great to be worth the price that, according to their own theory, will have to be paid for achieving it?

If anything human were absolutely certain, we might answer this question affirmatively with some confidence. The benefits of Communism, if it were once achieved, might be expected to be lasting; we might legitimately hope that further change would be towards something still better, not towards a revival of ancient evils. But if we admit, as we must do, that the outcome of the Communist revolution is in some degree uncertain, it becomes necessary to count the cost; for a great part of the cost is all but certain.

Since the revolution of October, 1917, the Soviet Government has been at war with almost all the world, and has had at the same time to face civil war at home. This is not to be regarded as accidental, or as a misfortune which could not be foreseen. According to Marxian theory, what has happened was bound to happen. Indeed, Russia has been wonderfully fortunate in not having to face an even more desperate situation. First and foremost, the world was exhausted by the war, and in no mood for military adventures. Next, the Tsarist régime was the worst in Europe, and therefore rallied less support than would be secured by any other capitalist Government. Again, Russia is vast and agricultural, making it capable of resisting both invasion and blockade better than Great Britain or France or Germany. The only other country that could have resisted with equal success is the United States, which is at present very far removed from a proletarian revolution, and likely long to remain the chief bulwark of the capitalist system. It is evident that Great Britain, attempting a similar revolution, would be forced by starvation to yield within a few months, provided America led a policy of blockade. The same is true, though in a less degree, of continental countries. Therefore, unless and until an international Communist revolution becomes possible, we must expect that any other nation following Russia's example will have to pay an even higher price than Russia has had to pay.

Now the price that Russia is having to pay is very great. The almost universal poverty might be thought to be a small evil in comparison with the ultimate gain, but it brings with it other evils of which the magnitude would be acknowledged even by those who have never known poverty and therefore make light of it. Hunger brings an absorption in the question of food, which, to most people, makes life almost purely animal. The general shortage makes people fierce, and reacts upon the political atmosphere. The necessity of inculcating Communism produces a hot-house condition, where every breath of fresh air must be excluded: people are to be taught to think in a certain way, and all free intelligence becomes taboo. The country comes to resemble an immensely magnified Jesuit College. Every kind of liberty is banned as being "*bourgeois*"; but it remains a fact that intelligence languishes where thought is not free.

All this, however, according to the leaders of the Third International, is only a small beginning of the struggle, which must become world-wide before it achieves victory. In their reply to the Independent Labour Party they say:

It is probable that upon the throwing off of the chains of the capitalist Governments, the revolutionary proletariat of Europe will meet the resistance of Anglo-Saxon capital in the persons of British and American capitalists who will attempt to blockade it. It is then possible that the revolutionary proletariat of Europe will rise in union with the peoples of the East and commence a revolutionary struggle, the scene of which will be the entire world, to deal a final blow to British and American capitalism (*The Times*, July 30, 1920).

The war here prophesied, if it ever takes place, will be one compared to which the late war will come to seem a mere affair of outposts. Those who realize the destructiveness of the late war, the devastation and impoverishment, the lowering of the level of civilization throughout vast areas, the general increase of hatred and savagery, the letting loose of bestial instincts which had been curbed during peace—those who realize all this will hesitate to incur inconceivably greater horrors, even if they believe firmly that Communism in itself is much to be desired. An economic system cannot be considered apart from the population which is to carry it out; and the population resulting from such a world-war as Moscow calmly contemplates would be savage, bloodthirsty and ruthless to an extent that must make any system a mere engine of oppression and cruelty.

This brings us to our third question: Is the system which Communists regard as their goal likely to result from the adoption of their methods? This is really the most vital question of the three.

Advocacy of Communism by those who believe in Bolshevik methods rests upon the assumption that there is no slavery except economic slavery, and that when all goods are held in common there must be perfect liberty. I fear this is a delusion.

There must be administration, there must be officials who control distribution. These men, in a Communist State, are the repositories of power. So long as they control the army, they are able, as in Russia at this moment, to wield despotic power even if they are a small minority. The fact that there is Communism—to a certain extent—does not mean that there is liberty. If the Communism were more complete, it would not necessarily mean more freedom; there would still be certain officials in control of the food supply, and these officials could govern as they pleased so long as they retained the support of the soldiers. This is not mere theory: it is the patent lesson of the present condition of Russia. The Bolshevik theory is that a small minority are to seize power, and are to hold it until Communism is accepted practically universally, which, they admit, may take a long time. But power is sweet, and few men surrender it voluntarily. It is especially sweet to those who have the habit of it, and the habit becomes most ingrained in those who have governed by bayonets, without popular support. Is it not almost inevitable that men placed as the Bolsheviks are placed in Russia, and as they maintain that the Communists must place themselves wherever the social revolution succeeds, will be loath to relinquish their monopoly of power, and will find reasons for remaining until some new revolution ousts them? Would it not be fatally easy for them, without altering economic structure, to decree large salaries for high Government officials, and so reintroduce the old inequalities of wealth? What motive would they have for not doing so? What motive is possible except idealism, love of mankind, non-economic motives of the sort that Bolsheviks decry? The system created by violence and the forcible rule of a minority must necessarily allow of tyranny and exploitation; and if human nature is what Marxians assert it to be, why should the rulers neglect such opportunities of selfish advantage?

It is sheer nonsense to pretend that the rulers of a great empire such as Soviet Russia, when they have become accustomed to power, retain the proletarian psychology, and feel that their class-interest is the same as that of the ordinary working man. This is not the case in fact in Russia now, however the truth may be concealed by fine phrases. The Government has a class-consciousness and a class-interest quite distinct from those of the genuine proletarian, who is not to be confounded with the paper proletarian of the Marxian schema. In a capitalist state, the Government and the capitalists on the whole hang together, and form one class; in Soviet Russia, the Government has absorbed the capitalist mentality together with the governmental, and the fusion has given increased strength to the upper class. But I see no reason whatever to expect equality or freedom to result from such a system, except reasons derived from a false psychology and a mistaken analysis of the sources of political power.

I am compelled to reject Bolshevism for two reasons: First, because the price mankind must pay to achieve Communism by Bolshevik methods is too terrible; and secondly because, even after paying the price, I do not believe the result would be what the Bolsheviks profess to desire.

But if their methods are rejected, how are we ever to arrive at a better economic system? This is not an easy question, and I shall treat it in a separate chapter.

V

MECHANISM AND THE INDIVIDUAL

Is it possible to effect a fundamental reform of the existing economic system by any other method than that of Bolshevism? The difficulty of answering this question is what chiefly attracts idealists to the dictatorship of the proletariat. If, as I have argued, the method of violent revolution and Communist rule is not likely to have the results which idealists desire, we are reduced to despair unless we can see hope in other methods. The Bolshevik arguments against all other methods are powerful. I confess that, when the spectacle of present-day Russia forced me to disbelieve in Bolshevik methods, I was at first unable to see any way of curing the essential evils of capitalism. My first impulse was to abandon political thinking as a bad job, and to conclude that the strong and ruthless must always exploit the weaker and kindlier sections of the population. But this is not an attitude that can be long maintained by any vigorous and temperamentally hopeful person. Of course, if it were the truth, one would have to acquiesce. Some people believe that by living on sour milk one can achieve immortality. Such optimists are answered by a mere refutation; it is not necessary to go on and point out some other way of escaping death. Similarly an argument that Bolshevism will not lead to the millennium would remain valid even if it could be shown that the millennium cannot be reached by any other road. But the truth in social questions is not quite like truth in physiology or physics, since it depends upon men's beliefs. Optimism tends to verify itself by making people impatient of avoidable evils; while despair, on the other hand, makes the world as bad as it believes it to be. It is therefore imperative for those who do not believe in Bolshevism to put some other hope in its place.

I think there are two things that must be admitted: first, that many of the worst evils of capitalism might survive under Communism; secondly, that the cure for these evils cannot be sudden, since it requires changes in the average mentality.

What are the chief evils of the present system? I do not think that mere inequality of wealth, in itself, is a very grave evil. If everybody had enough, the fact that some have more than enough would be unimportant. With a very moderate improvement in methods of production, it would be easy to ensure that everybody should have enough, even under capitalism, if wars and preparations for wars were abolished. The problem of poverty is by no means insoluble within the existing system, except when account is taken of psychological factors and the uneven distribution of power.

The graver evils of the capitalist system all arise from its uneven distribution of power. The possessors of capital wield an influence quite out of proportion to their numbers or their services to the community. They control almost the whole of education and the press; they decide what the average man shall know or not know; the cinema has given them a new method of propaganda, by which they enlist the support of those who are too frivolous even for illustrated papers. Very little of the intelligence of the world is really free: most of it is, directly or indirectly, in the pay of business enterprises or wealthy philanthropists. To satisfy capitalist interests, men are compelled to work much harder and more monotonously than they ought to work, and their education is scamped. Wherever, as in barbarous or semi-civilized countries, labour is too weak or too disorganized to protect itself, appalling cruelties are practised for private profit. Economic and political organizations become more and more vast, leaving less and less room for individual development and initiative. It is this sacrifice of the individual to the machine that is the fundamental evil of the modern world.

To cure this evil is not easy, because efficiency is promoted, at any given moment, though not in the long run, by sacrificing the individual to the smooth working of a vast organization, whether military or industrial. In war and in commercial competition, it is necessary to control individual impulses, to treat men as so many "bayonets" or "sabres" or "hands," not as a society of separate people with separate tastes and capacities. Some sacrifice of individual impulses is, of course, essential to the existence of an ordered community, and this degree of sacrifice is, as a rule, not regretable even from the individual's point of view. But what is demanded in a highly militarized or industrialized nation goes far beyond this very moderate degree. A society which is to allow much freedom to the individual must be strong enough to be not anxious about home defence, moderate enough to refrain from difficult external conquests, and rich enough to value leisure and a civilized existence more than an increase of consumable commodities.

But where the material conditions for such a state of affairs exist, the psychological conditions are not likely to exist unless power is very widely diffused throughout the community. Where power is concentrated in a few, it will happen, unless those few are very exceptional people, that they will value tangible achievements in the way of increase in trade or empire more than the slow and less obvious improvements that would result from better education combined with more leisure. The joys of victory are especially great to the holders of power, while the evils of a mechanical organization fall almost exclusively upon the less influential. For these reasons, I do not believe that any community in which power is much concentrated will long refrain from conflicts of the kind involving a sacrifice of what is most valuable in the individual. In Russia at this moment, the sacrifice of the individual is largely inevitable, because of the severity of the economic and military struggle. But I did not feel, in the Bolsheviks, any consciousness of the magnitude of this misfortune, or any realization of the importance of the individual as against the State. Nor do I believe that men who do realize this are likely to succeed, or to come to the top, in times when everything has to be done against personal liberty. The Bolshevik theory requires that every country, sooner or later, should go through what Russia is going through now. And in every country in such a condition we may expect to find the government falling into the hands of ruthless men, who have not by nature any love for freedom, and who will see little importance in hastening the transition from dictatorship to freedom. It is far more likely that such men will be tempted to embark upon new enterprises, requiring further concentration of forces, and postponing indefinitely the liberation of the populations which they use as their material.

For these reasons, equalization of wealth without equalization of power seems to me a rather small and unstable achievement. But equalization of power is not a thing that can be achieved in a day. It requires a considerable level of moral, intellectual, and technical education. It requires a long period without extreme crises, in order that habits of tolerance and good nature may become common. It requires vigour on the part of those who are acquiring power, without a too desperate resistance on the part of those whose share is diminishing. This is only possible if those who are acquiring power are not very fierce, and do not terrify their opponents by threats of ruin and death. It cannot be done quickly, because quick methods require that very mechanism and subordination of the individual which we should struggle to prevent.

But even equalization of power is not the whole of what is needed politically. The right grouping of men for different purposes is also essential. Self-government in industry, for example, is an indispensable condition of a good society. Those acts of an individual or a group which have no very great importance for outsiders ought to be freely decided by that individual or group. This is recognized as regards religion, but ought to be recognized over a much wider field.

Bolshevik theory seems to me to err by concentrating its attention upon one evil, namely inequality of wealth, which it believes to be at the bottom of all others. I do not believe any one evil can be thus isolated, but if I had to select one as the greatest of political evils, I should select inequality of power. And I should deny that this is likely to be cured by the class-war and the dictatorship of the Communist party. Only peace and a long period of gradual improvement can bring it about.

Good relations between individuals, freedom from hatred and violence and oppression, general diffusion of education, leisure rationally employed, the progress of art and science—these seem to me among the most important ends that a political theory ought to have in view. I do not believe that they can be furthered, except very rarely, by revolution and war; and I am convinced that at the present moment they can only be promoted by a diminution in the spirit of ruthlessness generated by the war. For these reasons, while admitting the necessity and even utility of Bolshevism in Russia, I do not wish to see it spread, or to encourage the adoption of its philosophy by advanced parties in the Western nations.

VI

WHY RUSSIAN COMMUNISM HAS FAILED

The civilized world seems almost certain, sooner or later, to follow the example of Russia in attempting a Communist organization of society. I believe that the attempt is essential to the progress and happiness of mankind during the next few centuries, but I believe also that the transition has appalling dangers. I believe that, if the Bolshevik theory as to the method of transition is adopted by Communists in Western nations, the result will be a prolonged chaos, leading neither to Communism nor to any other civilized system, but to a relapse into the barbarism of the Dark Ages. In the interests of Communism, no less than in the interests of civilization, I think it imperative that the Russian failure should be admitted and analysed. For this reason, if for no other, I cannot enter into the conspiracy of concealment which many Western Socialists who have visited Russia consider necessary.

I shall try first to recapitulate the facts which make me regard the Russian experiment as a failure, and then to seek out the causes of failure.

The most elementary failure in Russia is in regard to food. In a country which formerly produced a vast exportable surplus of cereals and other agricultural produce, and in which the non-agricultural population is only 15 per cent. of the total, it ought to be possible, without great difficulty, to provide enough food for the towns. Yet the Government has failed badly in this respect. The rations are inadequate and irregular, so that it is impossible to preserve health and vigour without the help of food purchased illicitly in the markets at speculative prices. I have given reasons for thinking that the breakdown of transport, though a contributory cause, is not the main reason for the shortage. The main reason is the hostility of the peasants, which, in turn, is due to the collapse of industry and to the policy of forced requisitions. In regard to corn and flour, the Government requisitions all that the peasant produces above a certain minimum required for himself and his family. If, instead, it exacted a fixed amount as rent, it would not destroy his incentive to production, and would not provide nearly such a strong motive for concealment. But this plan would have enabled the peasants to grow rich,and would have involved a confessed abandonment of Communist theory. It has therefore been thought better to employ forcible methods, which led to disaster, as they were bound to do.

The collapse of industry was the chief cause of the food difficulties, and has in turn been aggravated by them. Owing to the fact that there is abundant food in the country, industrial and urban workers are perpetually attempting to abandon their employment for agriculture. This is illegal, and is severely punished, by imprisonment or convict labour. Nevertheless, it continues, and in so vast a country as Russia it is not possible to prevent it. Thus, the ranks of industry become still further depleted.

Except as regards munitions of war, the collapse of industry in Russia is extraordinarily complete. The resolutions passed by the Ninth Congress of the Communist Party (April, 1920) speak of "the incredible catastrophes of public economy." This language is not too strong, though the recovery of the Baku oil has done something to produce a revival along the Volga basin.

The failure of the whole industrial side of the national economy, including transport, is at the bottom of the other failures of the Soviet Government. It is, to begin with, the main cause of the unpopularity of the Communists both in town and country: in town, because the people are hungry; in the country, because food is taken with no return except paper. If industry had been prosperous, the peasants could have had clothes and agricultural machinery, for which they would have willingly parted with enough food for the needs of the towns. The town population could then have subsisted in tolerable comfort; disease could have been coped with, and the general lowering of vitality averted. It would not have been necessary, as it has been in many cases, for men of scientific or artistic capacity to abandon the pursuits in which they were skilled for unskilled manual labour. The Communist Republic might have been agreeable to live in—at least for those who had been very poor before.

The unpopularity of the Bolsheviks, which is primarily due to the collapse of industry, has in turn been accentuated by the measures which it has driven the Government to adopt. In view of the fact that it was impossible to give adequate food to the ordinary population of Petrograd and Moscow, the Government decided that at any rate the men employed on important public work should be sufficiently nourished to preserve their efficiency. It is a gross libel to say that the Communists, or even the leading People's Commissaries, live luxurious lives according to our standards; but it is a fact that they are not exposed, like their subjects, to acute hunger and the weakening of energy that accompanies it. No tone can blame them for this, since the work of government must be carried on; but it is one of the ways in which class distinctions have reappeared where it was intended that they should be banished. I talked to an obviously hungry working man in Moscow, who pointed to the Kremlin and remarked: "In there they have enough to eat." He was expressing a widespread feeling which is fatal to the idealistic appeal that Communism attempts to make.

Owing to unpopularity, the Bolsheviks have had to rely upon the army and the Extraordinary Commission, and have been compelled to reduce the Soviet system to an empty form. More and more the pretence of representing the proletariat has grown threadbare. Amid official demonstrations and processions and meetings the genuine proletarian looks on, apathetic and disillusioned, unless he is possessed of unusual energy and fire, in which case he looks to the ideas of syndicalism or the I.W.W. to liberate him from a slavery far more complete than that of capitalism. A sweated wage, long hours, industrial conscription, prohibition of strikes, prison for slackers, diminution of the already insufficient rations in factories where the production falls below what the authorities expect, an army of spies ready to report any tendency to political disaffection and to procure imprisonment for its promoters—this is the reality of a system which still professes to govern in the name of the proletariat.

At the same time the internal and external peril has necessitated the creation of a vast army recruited by conscription, except as regards a Communist nucleus, from among a population utterly weary of war, who put the Bolsheviks in power because they alone promised peace. Militarism has produced its inevitable result in the way of a harsh and dictatorial spirit: the men in power go through their day's work with the consciousness that they command three million armed men, and that civilian opposition to their will can be easily crushed.

Out of all this has grown a system painfully like the old government of the Tsar—a system which is Asiatic in its centralized bureaucracy, its secret service, its atmosphere of governmental mystery and submissive terror. In many ways it resembles our Government of India. Like that Government, it stands for civilization, for education, sanitation, and Western ideas of progress; it is composed in the main of honest and hard-working men, who despise those whom they govern, but believe themselves possessed of something valuable which they must communicate to the population, however little it may be desired. Like our Government in India, they live in terror of popular risings, and are compelled to resort to cruel repressions in order to preserve their power. Like it, they represent an alien philosophy of life, which cannot be forced upon the people without a change of instinct, habit, and tradition so profound as to dry up the vital springs of action, producing listlessness and despair among the ignorant victims of militant enlightenment. It may be that Russia needs sternness and discipline more than anything else; it may be that a revival of Peter the Great's methods is essential to progress. From this point of view, much of what it is natural to criticize in the Bolsheviks becomes defensible; but this point of view has little affinity to Communism. Bolshevism may be defended, possibly, as a dire discipline through which a backward nation is to be rapidly industrialized; but as an experiment in Communism it has failed.

There are two things that a defender of the Bolsheviks may say against the argument that they have failed because the present state of Russia is bad. It may be said that it is too soon to judge, and it may be urged that whatever failure there has been is attributable to the hostility of the outside world.

As to the contention that it is too soon to judge,that is of course undeniable in a sense. But in a sense it is always too soon to judge of any historical movement, because its effects and developments go on for ever. Bolshevism has, no doubt, great changes ahead of it. But the last three years have afforded material for some judgments, though more definitive judgments will be possible later. And, for reasons which I have given in earlier chapters, I find it impossible to believe that later developments will realize more fully the Communist ideal. If trade is opened with the outer world, there will be an almost irresistible tendency to resumption of private enterprise. If trade is not re-opened, the plans of Asiatic conquest will mature, leading to a revival of Yenghis Khan and Timur. In neither case is the purity of the Communist faith likely to survive.

As for the hostility of the Entente, it is of course true that Bolshevism might have developed very differently if it had been treated in a friendly spirit. But in view of its desire to promote world-revolution, no one could expect—and the Bolsheviks certainly did not expect—that capitalist Governments would be friendly. If Germany had won the war, Germany would have shown a hostility more effective than that of the Entente. However we may blame Western Governments for their policy, we must realize that, according to the deterministic economic theory of the Bolsheviks, no other policy was to be expected from them. Other men might have been excused for not foreseeing the attitude of Churchill, Clemenceau and Millerand; but Marxians could not be excused, since this attitude was in exact accord with their own formula.

We have seen the symptoms of Bolshevik failure; I come now to the question of its profounder causes.

Everything that is worst in Russia we found traceable to the collapse of industry. Why has industry collapsed so utterly? And would it collapse equally if a Communist revolution were to occur in a Western country?

Russian industry was never highly developed, and depended always upon outside aid for much of its plant. The hostility of the world, as embodied in the blockade, left Russia powerless to replace the machinery and locomotives worn out during the war. The need of self-defence compelled the Bolsheviks to send their best workmen to the front, because they were the most reliable Communists, and the loss of them rendered their factories even more inefficient than they were under Kerensky. In this respect, and in the laziness and incapacity of the Russian workman, the Bolsheviks have had to face special difficulties which would be less in other countries. On the other hand, they have had special advantages in the fact that Russia is self-supporting in the matter of food; no other country could have endured the collapse of industry so long, and no other Great Power except the United States could have survived years of blockade.

The hostility of the world was in no way a surprise to those who made the October revolution; it was in accordance with their general theory, and its consequences should have been taken into account in making the revolution.

Other hostilities besides those of the outside world have been incurred by the Bolsheviks with open eyes, notably the hostility of the peasants and that of a great part of the industrial population. They have attempted, in accordance with their usual contempt for conciliatory methods, to substitute terror for reward as the incentive to work. Some amiable Socialists have imagined that, when the private capitalist had been eliminated, men would work from a sense of obligation to the community. The Bolsheviks will have none of such sentimentalism. In one of the resolutions of the ninth Communist Congress they say:

Every social system, whether based on slavery, feudalism, or capitalism, had its ways and means of labour compulsion and labour education in the interests of the exploiters.

The Soviet system is faced with the task of developing its own methods of labour compulsion to attain an increase of the intensity and wholesomeness of labour; this method is to be based on the socialization of public economy in the interests of the whole nation.

In addition to the propaganda by which the people are to be influenced and the repressions which are to be applied to all idlers, parasites and disorganizers who strive to undermine public zeal—the principal method for the increase of production will become the introduction of the system of compulsory labour.

In capitalist society rivalry assumed the character of competition and led to the exploitation of man by man. In a society where the means of production are nationalized, labour rivalry is to increase the products of labour without infringing its solidarity.

Rivalry between factories, regions, guilds, workshops, and individual workers should become the subject of careful organization and of close study on the side of the Trade Unions and the economic organs.

The system of premiums which is to be introduced should become one of the most powerful means of exciting rivalry. The system of rationing of food supply is to get into line with it; so long as Soviet Russia suffers from insufficiency of provisions, it is only just that the industrious and conscientious worker receives more than the careless worker.

It must be remembered that even the "industrious and conscientious worker" receives less food than is required to maintain efficiency.

Over the whole development of Russia and of Bolshevism since the October revolution there broods a tragic fatality. In spite of outward success the inner failure has proceeded by inevitable stages—stages which could, by sufficient acumen, have been foreseen from the first. By provoking the hostility of the outside world the Bolsheviks were forced to provoke the hostility of the peasants, and finally the hostility or utter apathy of the urban and industrial population. These various hostilities brought material disaster, and material disaster brought spiritual collapse. The ultimate source of the whole train of evils lies in the Bolshevik outlook on life: in its dogmatism of hatred and its belief that human nature can be completely transformed by force. To injure capitalists is not the ultimate goal of Communism, though among men dominated by hatred it is the part that gives zest to their activities. To face the hostility of the world may show heroism, but it is a heroism for which the country, not its rulers, has to pay the price. In the principles of Bolshevism there is more desire to destroy ancient evils than to build up new goods; it is for this reason that success in destruction has been so much greater than in construction. The desire to destroy is inspired by hatred, which is not a constructive principle. From this essential characteristic of Bolshevik mentality has sprung the willingness to subject Russia to its present martyrdom. It is only out of a quite different mentality that a happier world can be created.

And from this follows a further conclusion. The Bolshevik outlook is the outcome of the cruelty of the Tsarist régime and the ferocity of the years of the Great War, operating upon a ruined and starving nation maddened into universal hatred. If a different mentality is needed for the establishment of a successful Communism, then a quite different conjuncture must see its inauguration; men must be persuaded to the attempt by hope, not driven to it by despair. To bring this about should be the aim of every Communist who desires the happiness of mankind more than the punishment of capitalists and their governmental satellites.

VII

CONDITIONS FOR THE SUCCESS OF COMMUNISM

The fundamental ideas of Communism are by no means impracticable, and would, if realized, add immeasurably to the well-being of mankind. The difficulties which have to be faced are not in regard to the fundamental ideas, but in regard to the transition from capitalism. It must be assumed that those who profit by the existing system will fight to preserve it, and their fight may be sufficiently severe to destroy all that is best in Communism during the struggle, as well as everything else that has value in modern civilization. The seriousness of this problem of transition is illustrated by Russia, and cannot be met by the methods of the Third International. The Soviet Government, at the present moment, is anxious to obtain manufactured goods from capitalist countries, but the Third International is meanwhile endeavouring to promote revolutions which, if they occurred, would paralyse the industries of the countries concerned, and leave them incapable of supplying Russian needs.

The supreme condition of success in a Communist revolution is that it should not paralyse industry. If industry is paralysed, the evils which exist in modern Russia, or others just as great, seem practically unavoidable. There will be the problem of town and country, there will be hunger, there will be fierceness and revolts and military tyranny. All these things follow in a fatal sequence; and the end of them is almost certain to be something quite different from what genuine Communists desire.

If industry is to survive throughout a Communist revolution, a number of conditions must be fulfilled which are not, at present, fulfilled anywhere. Consider, for the sake of definiteness, what would happen if a Communist revolution were to occur in England to-morrow. Immediately America would place an embargo on all trade with us. The cotton industry would collapse, leaving about five million of the most productive portions of the population idle. The food supply would become inadequate, and would fail disastrously if, as is to be expected, the Navy were hostile or disorganized by the sabotage of the officers. The result would be that, unless there were a counter-revolution, about half the population would die within the first twelve months. On such a basis it would evidently be impossible to erect a successful Communist State.

What applies to England applies, in one form or another, to the remaining countries of Europe. Italian and German Socialists are, many of them, in a revolutionary frame of mind and could, if they chose, raise formidable revolts. They are urged by Moscow to do so, but they realize that, if they did, England

and America would starve them. France, for many reasons, dare not offend England and America beyond a point. Thus, in every country except America, a successful Communist revolution is impossible for economico-political reasons. America, being self-contained and strong, would be capable, so far as material conditions go, of achieving a successful revolution; but in America the psychological conditions are as yet adverse. There is no other civilized country where capitalism is so strong and revolutionary Socialism so weak as in America. At the present moment, therefore, though it is by no means impossible that Communist revolutions may occur all over the Continent, it is nearly certain that they cannot be successful in any real sense. They will have to begin by a war against America, and possibly England, by a paralysis of industry, by starvation, militarism and the whole attendant train of evils with which Russia has made us familiar.

That Communism, whenever and wherever it is adopted, will have to begin by fighting the bourgeoisie, is highly probable. The important question is not whether there is to be fighting, but how long and severe it is to be. A short war, in which Communism won a rapid and easy victory, would do little harm. It is long, bitter and doubtful wars that must be avoided if anything of what makes Communism desirable is to survive.

Two practical consequences flow from this conclusion: first, that nothing can succeed until America is either converted to Communism, or at any rate willing to remain neutral; secondly, that it is a mistake to attempt to inaugurate Communism in a country where the majority are hostile, or rather, where the active opponents are as strong as the active supporters, because in such a state of opinion a very severe civil war is likely to result. It is necessary to have a great body of opinion favourable to Communism, and a rather weak opposition, before a really successful Communist state can be introduced either by revolution or by more or less constitutional methods.

It may be assumed that when Communism is first introduced, the higher technical and business staff will side with the capitalists and attempt sabotage unless they have no hopes of a counter-revolution. For this reason, it is very necessary that among wage-earners there should be as wide a diffusion as possible of technical and business education, so that they may be able immediately to take control of big complex industries. In this respect Russia was very badly off, whereas England and America would be much more fortunate.

Self-government in industry is, I believe, the road by which England can best approach Communism. I do not doubt that the railways and the mines, after a little practice, could be run more efficiently by the workers, from the point of view of production, than they are at present by the capitalists. The Bolsheviks oppose self-government in industry everywhere, because it has failed in Russia, and their national self-esteem prevents them from admitting that this is due to the backwardness of Russia. This is one of the respects in which they are misled by the assumption that Russia must be in all ways a model to the rest of the world. I would go so far as to say that the winning of self-government in such industries as railways and mining is an essential preliminary to complete Communism. In England, especially, this is the case. The Unions can command whatever technical skill they may require; they are politically powerful; the demand for self-government is one for which there is widespread sympathy,and could be much more with adequate propaganda; moreover (what is important with the British temperament) self-government can be brought about gradually, by stages in each trade, and by extension from one trade to another. Capitalists value two things, their power and their money; many individuals among them value only the money. It is wiser to concentrate first on the power, as is done by seeking self-government in industry without confiscation of capitalist incomes. By this means the capitalists are gradually turned into obvious drones, their active functions in industry become nil, and they can be ultimately dispossessed without dislocation and without the possibility of any successful struggle on their parts.

Another advantage of proceeding by way of self-government is that it tends to prevent the Communist régime, when it comes, from having that truly terrible degree of centralization which now exists in Russia. The Russians have been forced to centralize, partly by the problems of the war, but more by the shortage of all kinds of skill. This has compelled the few competent men to attempt each to do the work of ten men, which has not proved satisfactory in spite of heroic efforts. The idea of democracy has become discredited as the result first of syndicalism, and then of Bolshevism. But there are two different things that may be meant by democracy: we may mean the system of Parliamentary government, or we may mean the participation of the people in affairs. The discredit of the former is largely deserved, and I have no desire to uphold Parliament as an ideal institution. But it is a great misfortune if, from a confusion of ideas, men come to think that, because Parliaments are imperfect, there is no reason why there should be self-government. The grounds for advocating self-government are very familiar: first, that no benevolent despot can be trusted to know or pursue the interests of his subjects; second, that the practice of self-government is the only effective method of political education; third, that it tends to place the preponderance of force on the side of the constitution, and thus to promote order and stable government. Other reasons could be found, but I think these are the chief. In Russia self-government has disappeared, except within the Communist Party. If it is not to disappear elsewhere during a Communist revolution, it is very desirable that there should exist already important industries competently administered by the workers themselves.

The Bolshevik philosophy is promoted very largely by despair of more gradual methods. But this despair is a mark of impatience, and is not really warranted by the facts. It is by no means impossible,in the near future, to secure self-government in British railways and mines by constitutional means. This is not the sort of measure which would bring into operation an American blockade or a civil war or any of the other catastrophic dangers that are to be feared from a full-fledged Communist revolution in the present international situation. Self-government in industry is feasible, and would be a great step towards Communism. It would both afford many of the advantages of Communism and also make the transition far easier without a technical break-down of production.

There is another defect in the methods advocated by the Third International. The sort of revolution which is recommended is never practically feasible except in a time of national misfortune; in fact, defeat in war seems to be an indispensable condition. Consequently, by this method, Communism will only be inaugurated where the conditions of life are difficult, where demoralization and disorganization make success almost impossible, and where men are in a mood of fierce despair very inimical to industrial construction. If Communism is to have a fair chance, it must be inaugurated in a prosperous country. But a prosperous country will not be readily moved by the arguments of hatred and universal upheaval which are employed by the Third International. It is necessary, in appealing to a prosperous country, to lay stress on hope rather than despair, and to show how the transition can be effected without a calamitous loss of prosperity. All this requires less violence and subversiveness, more patience and constructive propaganda, less appeal to the armed might of a determined minority.

The attitude of uncompromising heroism is attractive, and appeals especially to the dramatic instinct. But the purpose of the serious revolutionary is not personal heroism, nor martyrdom, but the creation of a happier world. Those who have the happiness of the world at heart will shrink from attitudes and the facile hysteria of "no parley with the enemy." They will not embark upon enterprises, however arduous and austere, which are likely to involve the martyrdom of their country and the discrediting of their ideals. It is by slower and less showy methods that the new world must be built: by industrial efforts after self-government, by proletarian training in technique and business administration, by careful study of the international situation, by a prolonged and devoted propaganda of ideas rather than tactics, especially among the wage-earners of the United States. It is not true that no gradual approaches to Communism are possible: self-government in industry is an important instance to the contrary. It is not true that any isolated European country, or even the whole of the Continent in unison, can, after the exhaustion produced by the war, introduce a successful form of Communism at the present moment, owing to the hostility and economic supremacy of America. To find fault with those who urge these considerations, or to accuse them of faint-heartedness, is mere sentimental self-indulgence, sacrificing the good we can do to the satisfaction of our own emotions.

Even under present conditions in Russia, it is possible still to feel the inspiration of the essential spirit of Communism, the spirit of creative hope, seeking to sweep away the incumbrances of injustice and tyranny and rapacity which obstruct the growth of the human spirit, to replace individual competition by collective action, the relation of master and slave by free co-operation. This hope has helped the best of the Communists to bear the harsh years through which Russia has been passing, and has become an inspiration to the world. The hope is not chimerical, but it can only be realized through a more patient labour, a more objective study of facts, and above all a longer propaganda, to make the necessity of the transition obvious to the great majority of wage-earners. Russian Communism may fail and go under, but Communism itself will not die. And if hope rather than hatred inspires its advocates, it can be brought about without the universal cataclysm preached by Moscow. The war and its sequel have proved the destructiveness of capitalism; let us see to it that the next epoch does not prove the still greater destructiveness of Communism, but rather its power to heal the wounds which the old evil system has inflicted upon the human spirit.

THE ANALYSIS OF MIND

PREFACE

This book has grown out of an attempt to harmonize two different tendencies, one in psychology, the other in physics, with both of which I find myself in sympathy, although at first sight they might seem inconsistent. On the one hand, many psychologists, especially those of the behaviourist school, tend to adopt what is essentially a materialistic position, as a matter of method if not of metaphysics. They make psychology increasingly dependent on physiology and external observation, and tend to think of matter as something much more solid and indubitable than mind. Meanwhile the physicists, especially Einstein and other exponents of the theory of relativity, have been making "matter" less and less material. Their world consists of "events," from which "matter" is derived by a logical construction. Whoever reads, for example, Professor Eddington's "Space, Time and Gravitation" (Cambridge University Press, 1920), will see that an old-fashioned materialism can receive no support from modern physics. I think that what has permanent value in the outlook of the behaviourists is the feeling that physics is the most fundamental science at present in existence. But this position cannot be called materialistic, if, as seems to be the case, physics does not assume the existence of matter.

The view that seems to me to reconcile the materialistic tendency of psychology with the anti-materialistic tendency of physics is the view of William James and the American new realists, according to which the "stuff" of the world is neither mental nor material, but a "neutral stuff," out of which both are constructed. I have endeavoured in this work to develop this view in some detail as regards the phenomena with which psychology is concerned.

My thanks are due to Professor John B. Watson and to Dr. T. P. Nunn for reading my MSS. at an early stage and helping me with many valuable suggestions; also to Mr. A. Wohlgemuth for much very useful information as regards important literature. I have also to acknowledge the help of the editor of this Library of Philosophy, Professor Muirhead, for several suggestions by which I have profited.

The work has been given in the form of lectures both in London and Peking, and one lecture, that on Desire, has been published in the Athenaeum.

There are a few allusions to China in this book, all of which were written before I had been in China, and are not intended to be taken by the reader as geographically accurate. I have used "China" merely as a synonym for "a distant country," when I wanted illustrations of unfamiliar things.

Peking, January 1921.

LECTURE I.

RECENT CRITICISMS OF "CONSCIOUSNESS"

There are certain occurrences which we are in the habit of calling "mental." Among these we may take as typical BELIEVING and DESIRING. The exact definition of the word "mental" will, I hope, emerge as the lectures proceed; for the present, I shall mean by it whatever occurrences would commonly be called mental.

I wish in these lectures to analyse as fully as I can what it is that really takes place when we, e.g. believe or desire. In this first lecture I shall be concerned to refute a theory which is widely held, and which I formerly held myself: the theory that the essence of everything mental is a certain quite peculiar something called "consciousness," conceived either as a relation to objects, or as a pervading quality of psychical phenomena.

The reasons which I shall give against this theory will be mainly derived from previous authors. There are two sorts of reasons, which will divide my lecture into two parts:

(1) Direct reasons, derived from analysis and its difficulties;

(2) Indirect reasons, derived from observation of animals (comparative psychology) and of the insane and hysterical (psycho-analysis).

Few things are more firmly established in popular philosophy than the distinction between mind and matter. Those who are not professional metaphysicians are willing to confess that they do not know what mind actually is, or how matter is constituted; but they remain convinced that there is an impassable gulf between the two, and that both belong to what actually exists in the world. Philosophers, on the other hand, have maintained often that matter is a mere fiction imagined by mind, and sometimes that mind is a mere property of a certain kind of matter. Those who maintain that mind is the reality and matter an evil dream are called "idealists"—a word which has a different meaning in philosophy from that which it bears in ordinary life. Those who argue that matter is the reality and mind a mere property of protoplasm are called "materialists." They have been rare among philosophers, but common, at certain periods, among men of science. Idealists, materialists, and ordinary mortals have been in agreement on one point: that they knew sufficiently what they meant by the words "mind" and "matter" to be able to conduct their debate intelligently. Yet it was just in this point, as to which they were at one, that they seem to me to have been all alike in error.

The stuff of which the world of our experience is composed is, in my belief, neither mind nor matter, but something more primitive than either. Both mind and matter seem to be composite, and the stuff of which they are compounded lies in a sense between the two, in a sense above them both, like a common ancestor. As regards matter, I have set forth my reasons for this view on former occasions, and I shall not now repeat them. But the question of mind is more difficult, and it is this question that I propose to discuss in these lectures. A great deal of what I shall have to say is not original; indeed, much recent work, in various fields, has tended to show the necessity of such theories as those which I shall be advocating. Accordingly in this first lecture I shall try to give a brief description of the systems of ideas within which our investigation is to be carried on.

If there is one thing that may be said, in the popular estimation, to characterize mind, that one thing is "consciousness." We say that we are "conscious" of what we see and hear, of what we remember, and of our own thoughts and feelings. Most of us believe that tables and chairs are not "conscious." We think that when we sit in a chair, we are aware of sitting in it, but it is not aware of being sat in. It cannot for a moment be doubted that we are right in believing that there is SOME difference between us and the chair in this respect: so much may be taken as fact, and as a datum for our inquiry. But as soon as we try to say what exactly the difference is, we become involved in perplexities. Is "consciousness" ultimate and simple, something to be merely accepted and contemplated? Or is it something complex, perhaps consisting in our way of behaving in the presence of objects, or, alternatively, in the existence in us of things called "ideas," having a certain relation to objects, though different from them, and only symbolically representative of them? Such questions are not easy to answer; but until they are answered we cannot profess to know what we mean by saying that we are possessed of "consciousness."

Before considering modern theories, let us look first at consciousness from the standpoint of conventional psychology, since this embodies views which naturally occur when we begin to reflect upon the subject. For this purpose, let us as a preliminary consider different ways of being conscious.

First, there is the way of PERCEPTION. We "perceive" tables and chairs, horses and dogs, our friends, traffic passing in the street—in short, anything which we recognize through the senses. I leave on one side for the present the question whether pure sensation is to be regarded as a form of consciousness: what I am speaking of now is perception, where, according to conventional psychology, we go beyond the sensation to the "thing" which it represents. When you hear a donkey bray, you not only hear a noise, but realize that it comes from a donkey. When you see a table, you not only see a coloured surface, but realize that it is hard. The addition of these elements that go beyond crude sensation is said to constitute perception. We shall have more to say about this at a later stage. For the moment, I am merely concerned to note that perception of objects is one of the most obvious examples of what is called "consciousness." We are "conscious" of anything that we perceive.

We may take next the way of MEMORY. If I set to work to recall what I did this morning, that is a form of consciousness different from perception, since it is concerned with the past. There are various problems as to how we can be conscious now of what no longer exists. These will be dealt with incidentally when we come to the analysis of memory.

From memory it is an easy step to what are called "ideas"—not in the Platonic sense, but in that of Locke, Berkeley and Hume, in which they are opposed to "impressions." You may be conscious of a friend either by seeing him or by "thinking" of him; and by "thought" you can be conscious of objects which cannot be seen, such as the human race, or physiology. "Thought" in the narrower sense is that form of consciousness which consists in "ideas" as opposed to impressions or mere memories.

We may end our preliminary catalogue with BELIEF, by which I mean that way of being conscious which may be either true or false. We say that a man is "conscious of looking a fool," by which we mean that he believes he looks a fool, and is not mistaken in this belief. This is a different form of consciousness from any of the earlier ones. It is the form which gives "knowledge" in the strict sense, and also error. It is, at least apparently, more complex than our previous forms of consciousness; though we shall find that they are not so separable from it as they might appear to be.

Besides ways of being conscious there are other things that would ordinarily be called "mental," such as desire and pleasure and pain. These raise problems of their own, which we shall reach in Lecture III. But the hardest problems are those that arise concerning ways of being "conscious." These ways, taken together, are called the "cognitive" elements in mind, and it is these that will occupy us most during the following lectures.

There is one element which SEEMS obviously in common among the different ways of being conscious, and that is, that they are all directed to OBJECTS. We are conscious "of" something. The consciousness, it seems, is one thing, and that of which we are conscious is another thing. Unless we are to acquiesce in the view that we can never be conscious of anything outside our own minds, we must say that the object of consciousness need not be mental, though the consciousness must be. (I am speaking within the circle of conventional doctrines, not expressing my own beliefs.) This direction towards an object is commonly regarded as typical of every form of cognition, and sometimes of mental life altogether. We may distinguish two different tendencies in traditional psychology. There are those who take mental phenomena naively, just as they would physical phenomena. This school of psychologists tends not to emphasize the object. On the other hand, there are those whose primary interest is in the apparent fact that we have KNOWLEDGE, that there is a world surrounding us of which we are aware. These men are interested in the mind because of its relation to the world, because knowledge, if it is a fact, is a very mysterious one. Their interest in psychology is naturally centred in the relation of consciousness to its object, a problem which, properly, belongs rather to theory of knowledge. We may take as one of the best and most typical representatives of this school the Austrian psychologist Brentano, whose "Psychology from the Empirical Standpoint," though published in 1874, is still influential and was the starting-point of a great deal of interesting work. He says :

"Every psychical phenomenon is characterized by what the scholastics of the Middle Ages called the intentional (also the mental) inexistence of an object, and what we, although with not quite unambiguous expressions, would call relation to a content, direction towards an object (which is not here to be understood as a reality), or immanent objectivity. Each contains something in itself as an object, though not each in the same way. In presentation something is presented, in judgment something is acknowledged or rejected, in love something is loved, in hatred hated, in desire desired, and so on.

"This intentional inexistence is exclusively peculiar to psychical phenomena. No physical phenomenon shows anything similar. And so we can define psychical phenomena by saying that they are phenomena which intentionally contain an object in themselves."

The view here expressed, that relation to an object is an ultimate irreducible characteristic of mental phenomena, is one which I shall be concerned to combat. Like Brentano, I am interested in psychology, not so much for its own sake, as for the light that it may throw on the problem of knowledge. Until very lately I believed, as he did, that mental phenomena have essential reference to objects, except possibly in the case of pleasure and pain. Now I no longer believe this, even in the case of knowledge. I shall try to make my reasons for this rejection clear as we proceed. It must be evident at first glance that the analysis of knowledge is rendered more difficult by the rejection; but the apparent simplicity of Brentano's view of knowledge will be found, if I am not mistaken, incapable of maintaining itself either against an analytic scrutiny or against a host of facts in psycho-analysis and animal psychology. I do not wish to minimize the problems. I will merely observe, in mitigation of our prospective labours, that thinking, however it is to be analysed, is in itself a delightful occupation, and that there is no enemy to thinking so deadly as a false simplicity. Travelling, whether in the mental or the physical world, is a joy, and it is good to know that, in the mental world at least, there are vast countries still very imperfectly explored.

The view expressed by Brentano has been held very generally, and developed by many writers. Among these we may take as an example his Austrian successor Meinong. According to him there are three elements involved in the thought of an object. These three he calls the act, the content and the object. The act is the same in any two cases of the same kind of consciousness; for instance, if I think of Smith or think of Brown, the act of thinking, in itself, is exactly similar on both occasions. But the content of my thought, the particular event that is happening in my mind, is different when I think of Smith and when I think of Brown. The content, Meinong argues, must not be confounded with the object, since the content must exist in my mind at the moment when I have the thought, whereas the object need not do so. The object may be something past or future; it may be physical, not mental; it may be something abstract, like equality for example; it may be something imaginary, like a golden mountain; or it may even be something self-contradictory, like a round square. But in all these cases, so he contends, the content exists when the thought exists, and is what distinguishes it, as an occurrence, from other thoughts.

To make this theory concrete, let us suppose that you are thinking of St. Paul's. Then, according to Meinong, we have to distinguish three elements which are necessarily combined in constituting the one thought. First, there is the act of thinking, which would be just the same whatever you were thinking about. Then there is what makes the character of the thought as contrasted with other thoughts; this is the content. And finally there is St. Paul's, which is the object of your thought. There must be a difference between the content of a thought and what it is about, since the thought is here and now, whereas what it is about may not be; hence it is clear that the thought is not identical with St. Paul's. This seems to show that we must distinguish between content and object. But if Meinong is right, there can be no thought without an object: the connection of the two is essential. The object might exist without the thought, but not the thought without the object: the three elements of act, content and object are all required to constitute the one single occurrence called "thinking of St. Paul's."

The above analysis of a thought, though I believe it to be mistaken, is very useful as affording a schema in terms of which other theories can be stated. In the remainder of the present lecture I shall state in outline the view which I advocate, and show how various other views out of which mine has grown result from modifications of the threefold analysis into act, content and object.

The first criticism I have to make is that the ACT seems unnecessary and fictitious. The occurrence of the content of a thought constitutes the occurrence of the thought. Empirically, I cannot discover anything corresponding to the supposed act; and theoretically I cannot see that it is indispensable. We say: "*I* think so-and-so," and this word "I" suggests that thinking is the act of a person. Meinong's "act" is the ghost of the subject, or what once was the full-blooded soul. It is supposed that thoughts cannot just come and go, but need a person to think them. Now, of course it is true that thoughts can be collected into bundles, so that one bundle is my thoughts, another is your thoughts, and a third is the thoughts of Mr. Jones. But I think the person is not an ingredient in the single thought: he is rather constituted by relations of the thoughts to each other and to the body. This is a large question, which need not, in its entirety, concern us at present. All that I am concerned with for the moment is that the grammatical forms "I think," "you think," and "Mr. Jones thinks," are misleading if regarded as indicating an analysis of a single thought. It would be better to say "it thinks in me," like "it rains here"; or better still, "there is a thought in me." This is simply on the ground that what Meinong calls the act in thinking is not empirically discoverable, or logically deducible from what we can observe.

The next point of criticism concerns the relation of content and object. The reference of thoughts to objects is not, I believe, the simple direct essential thing that Brentano and Meinong represent it as being. It seems to me to be derivative, and to consist largely in BELIEFS: beliefs that what constitutes the thought is connected with various other elements which together make up the object. You have, say, an image of St. Paul's, or merely the word "St. Paul's" in your head. You believe, however vaguely and dimly, that this is connected with what you would see if you went to St. Paul's, or what you would feel if you touched its walls; it is further connected with what other people see and feel, with services and the Dean and Chapter and Sir Christopher Wren. These things are not mere thoughts of yours, but your thought stands in a relation to them of which you are more or less aware. The awareness of this relation is a further thought, and constitutes your feeling that the original thought had an "object." But in pure imagination you can get very similar thoughts without these accompanying beliefs; and in this case your thoughts do not have objects or seem to have them. Thus in such instances you have content without object. On the other hand, in seeing or hearing it would be less misleading to say that you have object without content, since what you see or hear is actually part of the physical world, though not matter in the sense of physics. Thus the whole question of the relation of mental occurrences to objects grows very complicated, and cannot be settled by regarding reference to objects as of the essence of thoughts. All the above remarks are merely preliminary, and will be expanded later.

Speaking in popular and unphilosophical terms, we may say that the content of a thought is supposed to be something in your head when you think the thought, while the object is usually something in the outer world. It is held that knowledge of the outer world is constituted by the relation to the object, while the fact that knowledge is different from what it knows is due to the fact that knowledge comes by way of contents. We can begin to state the difference between realism and idealism in terms of this opposition of contents and objects. Speaking quite roughly and approximately, we may say that idealism tends to suppress the object, while realism tends to suppress the content. Idealism, accordingly, says that nothing can be known except thoughts, and all the reality that we know is mental; while realism maintains that we know objects directly, in sensation certainly, and perhaps also in memory and thought. Idealism does not say that nothing can be known beyond the present thought, but it maintains that the context of vague belief, which we spoke of in connection with the thought of St. Paul's, only takes you to other thoughts, never to anything radically different from thoughts. The difficulty of this view is in regard to sensation, where it seems as if we came into direct contact with the outer world. But the Berkeleian way of meeting this difficulty is so familiar that I need not enlarge upon it now. I shall return to it in a later lecture, and will only observe, for the present, that there seem to me no valid grounds for regarding what we see and hear as not part of the physical world.

Realists, on the other hand, as a rule, suppress the content, and maintain that a thought consists either of act and object alone, or of object alone. I have been in the past a realist, and I remain a realist as regards sensation, but not as regards memory or thought. I will try to explain what seem to me to be the reasons for and against various kinds of realism.

Modern idealism professes to be by no means confined to the present thought or the present thinker in regard to its knowledge; indeed, it contends that the world is so organic, so dove-tailed, that from any one portion the whole can be inferred, as the complete skeleton of an extinct animal can be inferred from one bone. But the logic by which this supposed organic nature of the world is nominally demonstrated appears to realists, as it does to me, to be faulty. They argue that, if we cannot know the physical world directly, we cannot really know any thing outside our own minds: the rest of the world may be merely our dream. This is a dreary view, and they there fore seek ways of escaping from it. Accordingly they maintain that in knowledge we are in direct contact with objects, which may be, and usually are, outside our own minds. No doubt they are prompted to this view, in the first place, by bias, namely, by the desire to think that they can know of the existence of a world outside themselves. But we have to consider, not what led them to desire the view, but whether their arguments for it are valid.

There are two different kinds of realism, according as we make a thought consist of act and object, or of object alone. Their difficulties are different, but neither seems tenable all through. Take, for the sake of definiteness, the remembering of a past event. The remembering occurs now, and is therefore necessarily not identical with the past event. So long as we retain the act, this need cause no difficulty. The act of remembering occurs now, and has on this view a certain essential relation to the past event which it remembers. There is no LOGICAL objection to this theory, but there is the objection, which we spoke of earlier, that the act seems mythical, and is not to be found by observation. If, on the other hand, we try to constitute memory without the act, we are driven to a content, since we must have something that happens NOW, as opposed to the event which happened in the past. Thus, when we reject the act, which I think we must, we are driven to a theory of memory which is more akin to idealism. These arguments, however, do not apply to sensation. It is especially sensation, I think, which is considered by those realists who retain only the object. Their views, which are chiefly held in America, are in large measure derived from William James, and before going further it will be well to consider the revolutionary doctrine which he advocated. I believe this doctrine contains important new truth, and what I shall have to say will be in a considerable measure inspired by it.

William James's view was first set forth in an essay called "Does 'consciousness' exist?" In this essay he explains how what used to be the soul has gradually been refined down to the "transcendental ego," which, he says, "attenuates itself to a thoroughly ghostly condition, being only a name for the fact that the 'content' of experience IS KNOWN. It loses personal form and activity—these passing over to the content—and becomes a bare Bewusstheit or Bewusstsein uberhaupt, of which in its own right absolutely nothing can be said. I believe (he continues) that 'consciousness,' when once it has evaporated to this estate of pure diaphaneity, is on the point of disappearing altogether. It is the name of a nonentity, and has no right to a place among first principles. Those who still cling to it are clinging to a mere echo, the faint rumour left behind by the disappearing 'soul' upon the air of philosophy".

He explains that this is no sudden change in his opinions. "For twenty years past," he says, "I have mistrusted 'consciousness' as an entity; for seven or eight years past I have suggested its non-existence to my students, and tried to give them its pragmatic equivalent in realities of experience. It seems to me that the hour is ripe for it to be openly and universally discarded"

His next concern is to explain away the air of paradox, for James was never wilfully paradoxical. "Undeniably," he says, "'thoughts' do exist." "I mean only to deny that the word stands for an entity, but to insist most emphatically that it does stand for a function. There is, I mean, no aboriginal stuff or quality of being, contrasted with that of which material objects are made, out of which our thoughts of them are made; but there is a function in experience which thoughts perform, and for the performance of which this quality of being is invoked. That function is KNOWING"

James's view is that the raw material out of which the world is built up is not of two sorts, one matter and the other mind, but that it is arranged in different patterns by its inter-relations, and that some arrangements may be called mental, while others may be called physical.

"My thesis is," he says, "that if we start with the supposition that there is only one primal stuff or material in the world, a stuff of which everything is composed, and if we call that stuff 'pure experience,' then knowing can easily be explained as a particular sort of relation towards one another into which portions of pure experience may enter. The relation itself is a part of pure experience; one of its 'terms' becomes the subject or bearer of the knowledge, the knower, the other becomes the object known".

After mentioning the duality of subject and object, which is supposed to constitute consciousness, he proceeds in italics: "EXPERIENCE, I BELIEVE, HAS NO SUCH INNER DUPLICITY; AND THE SEPARATION OF IT INTO CONSCIOUSNESS AND CONTENT COMES, NOT BY WAY OF SUBTRACTION, BUT BY WAY OF ADDITION".

He illustrates his meaning by the analogy of paint as it appears in a paint-shop and as it appears in a picture: in the one case it is just "saleable matter," while in the other it "performs a spiritual function. Just so, I maintain (he continues), does a given undivided portion of experience, taken in one context of associates, play the part of a knower, of a state of mind, of 'consciousness'; while in a different context the same undivided bit of experience plays the part of a thing known, of an objective 'content.' In a word, in one group it figures as a thought, in another group as a thing"

He does not believe in the supposed immediate certainty of thought. "Let the case be what it may in others," he says, "I am as confident as I am of anything that, in myself, the stream of thinking (which I recognize emphatically as a phenomenon) is only a careless name for what, when scrutinized, reveals itself to consist chiefly of the stream of my breathing. The 'I think' which Kant said must be able to accompany all my objects, is the 'I breathe' which actually does accompany them"

The same view of "consciousness" is set forth in the succeeding essay, "A World of Pure Experience". The use of the phrase "pure experience" in both essays points to a lingering influence of idealism. "Experience," like "consciousness," must be a product, not part of the primary stuff of the world. It must be possible, if James is right in his main contentions, that roughly the same stuff, differently arranged, would not give rise to anything that could be called "experience." This word has been dropped by the American realists, among whom we may mention specially Professor R. B. Perry of Harvard and Mr. Edwin B. Holt. The interests of this school are in general philosophy and the philosophy of the sciences, rather than in psychology; they have derived a strong impulsion from James, but have more interest than he had in logic and mathematics and the abstract part of philosophy. They speak of "neutral" entities as the stuff out of which both mind and matter are constructed. Thus Holt says: "If the terms and propositions of logic must be substantialized, they are all strictly of one substance, for which perhaps the least dangerous name is neutral-stuff. The relation of neutral-stuff to matter and mind we shall have presently to consider at considerable length."

My own belief—for which the reasons will appear in subsequent lectures—is that James is right in rejecting consciousness as an entity, and that the American realists are partly right, though not wholly, in considering that both mind and matter are composed of a neutral-stuff which, in isolation, is neither mental nor material. I should admit this view as regards sensations: what is heard or seen belongs equally to psychology and to physics. But I should say that images belong only to the mental world, while those occurrences (if any) which do not form part of any "experience" belong only to the physical world. There are, it seems to me, prima facie different kinds of causal laws, one belonging to physics and the other to psychology. The law of gravitation, for example, is a physical law, while the law of association is a psychological law. Sensations are subject to both kinds of laws, and are therefore truly "neutral" in Holt's sense. But entities subject only to physical laws, or only to psychological laws, are not neutral, and may be called respectively purely material and purely mental. Even those, however, which are purely mental will not have that intrinsic reference to objects which Brentano assigns to them and which constitutes the essence of "consciousness" as ordinarily understood. But it is now time to pass on to other modern tendencies, also hostile to "consciousness."

There is a psychological school called "Behaviourists," of whom the protagonist is Professor John B. Watson, formerly of the Johns Hopkins University. To them also, on the whole, belongs Professor John Dewey, who, with James and Dr. Schiller, was one of the three founders of pragmatism. The view of the "behaviourists" is that nothing can be known except by external observation. They deny altogether that there is a separate source of knowledge called "introspection," by which we can know things about ourselves which we could never observe in others. They do not by any means deny that all sorts of things MAY go on in our minds: they only say that such things, if they occur, are not susceptible of scientific observation, and do not therefore concern psychology as a science. Psychology as a science, they say, is only concerned with BEHAVIOUR, i.e. with what we DO; this alone, they contend, can be accurately observed. Whether we think meanwhile, they tell us, cannot be known; in their observation of the behaviour of human beings, they have not so far found any evidence of thought. True, we talk a great deal, and imagine that in so doing we are showing that we can think; but behaviourists say that the talk they have to listen to can be explained without supposing that people think. Where you might expect a chapter on "thought processes" you come instead upon a chapter on "The Language Habit." It is humiliating to find how terribly adequate this hypothesis turns out to be.

Behaviourism has not, however, sprung from observing the folly of men. It is the wisdom of animals that has suggested the view. It has always been a common topic of popular discussion whether animals "think." On this topic people are prepared to take sides without having the vaguest idea what they mean by "thinking." Those who desired to investigate such questions were led to observe the behaviour of animals, in the hope that their behaviour would throw some light on their mental faculties. At first sight, it might seem that this is so. People say that a dog "knows" its name because it comes when it is called, and that it "remembers" its master, because it looks sad in his absence, but wags its tail and barks when he returns. That the dog behaves in this way is matter of observation, but that it "knows" or "remembers" anything is an inference, and in fact a very doubtful one. The more such inferences are examined, the more precarious they are seen to be. Hence the study of animal behaviour has been gradually led to abandon all attempt at mental interpretation. And it can hardly be doubted that, in many cases of complicated behaviour very well adapted to its ends, there can be no prevision of those ends. The first time a bird builds a nest, we can hardly suppose it knows that there will be eggs to be laid in it, or that it will sit on the eggs, or that they will hatch into young birds. It does what it does at each stage because instinct gives it an impulse to do just that, not because it foresees and desires the result of its actions.

Careful observers of animals, being anxious to avoid precarious inferences, have gradually discovered more and more how to give an account of the actions of animals without assuming what we call "consciousness." It has seemed to the behaviourists that similar methods can be applied to human behaviour, without assuming anything not open to external observation. Let us give a crude illustration, too crude for the authors in question, but capable of affording a rough insight into their meaning. Suppose two children in a school, both of whom are asked "What is six times nine?" One says fifty-four, the other says fifty-six. The one, we say, "knows" what six times nine is, the other does not. But all that we can observe is a certain language-habit. The one child has acquired the habit of saying "six times nine is fifty-four"; the other has not. There is no more need of "thought" in this than there is when a horse turns into his accustomed stable; there are merely more numerous and complicated habits. There is obviously an observable fact called "knowing" such-and-such a thing; examinations are experiments for discovering such facts. But all that is observed or discovered is a certain set of habits in the use of words. The thoughts (if any) in the mind of the examinee are of no interest to the examiner; nor has the examiner any reason to suppose even the most successful examinee capable of even the smallest amount of thought.

Thus what is called "knowing," in the sense in which we can ascertain what other people "know," is a phenomenon exemplified in their physical behaviour, including spoken and written words. There is no reason—so Watson argues—to suppose that their knowledge IS anything beyond the habits shown in this behaviour: the inference that other people have something nonphysical called "mind" or "thought" is therefore unwarranted.

So far, there is nothing particularly repugnant to our prejudices in the conclusions of the behaviourists. We are all willing to admit that other people are thoughtless. But when it comes to ourselves, we feel convinced that we can actually perceive our own thinking. "Cogito, ergo sum" would be regarded by most people as having a true premiss. This, however, the behaviourist denies. He maintains that our knowledge of ourselves is no different in kind from our knowledge of other people. We may see MORE, because our own body is easier to observe than that of other people; but we do not see anything radically unlike what we see of others. Introspection, as a separate source of knowledge, is entirely denied by psychologists of this school. I shall discuss this question at length in a later lecture; for the present I will only observe that it is by no means simple, and that, though I believe the behaviourists somewhat overstate their case, yet there is an important element of truth in their contention, since the things which we can discover by introspection do not seem to differ in any very fundamental way from the things which we discover by external observation.

So far, we have been principally concerned with knowing. But it might well be maintained that desiring is what is really most characteristic of mind. Human beings are constantly engaged in achieving some end they feel pleasure in success and pain in failure. In a purely material world, it may be said, there would be no opposition of pleasant and unpleasant, good and bad, what is desired and what is feared. A man's acts are governed by purposes. He decides, let us suppose, to go to a certain place, whereupon he proceeds to the station, takes his ticket and enters the train. If the usual route is blocked by an accident, he goes by some other route. All that he does is determined—or so it seems—by the end he has in view, by what lies in front of him, rather than by what lies behind. With dead matter, this is not the case. A stone at the top of a hill may start rolling, but it shows no pertinacity in trying to get to the bottom. Any ledge or obstacle will stop it, and it will exhibit no signs of discontent if this happens. It is not attracted by the pleasantness of the valley, as a sheep or cow might be, but propelled by the steepness of the hill at the place where it is. In all this we have characteristic differences between the behaviour of animals and the behaviour of matter as studied by physics.

Desire, like knowledge, is, of course, in one sense an observable phenomenon. An elephant will eat a bun, but not a mutton chop; a duck will go into the water, but a hen will not. But when we think of our own desires, most people believe that we can know them by an immediate self-knowledge which does not depend upon observation of our actions. Yet if this were the case, it would be odd that people are so often mistaken as to what they desire. It is matter of common observation that "so-and-so does not know his own motives," or that "A is envious of B and malicious about him, but quite unconscious of being so." Such people are called self-deceivers, and are supposed to have had to go through some more or less elaborate process of concealing from themselves what would otherwise have been obvious. I believe that this is an entire mistake. I believe that the discovery of our own motives can only be made by the same process by which we discover other people's, namely, the process of observing our actions and inferring the desire which could prompt them. A desire is "conscious" when we have told ourselves that we have it. A hungry man may say to himself: "Oh, I do want my lunch." Then his desire is "conscious." But it only differs from an "unconscious" desire by the presence of appropriate words, which is by no means a fundamental difference.

The belief that a motive is normally conscious makes it easier to be mistaken as to our own motives than as to other people's. When some desire that we should be ashamed of is attributed to us, we notice that we have never had it consciously, in the sense of saying to ourselves, "I wish that would happen." We therefore look for some other interpretation of our actions, and regard our friends as very unjust when they refuse to be convinced by our repudiation of what we hold to be a calumny. Moral considerations greatly increase the difficulty of clear thinking in this matter. It is commonly argued that people are not to blame for unconscious motives, but only for conscious ones. In order, therefore, to be wholly virtuous it is only necessary to repeat virtuous formulas. We say: "I desire to be kind to my friends, honourable in business, philanthropic towards the poor, public-spirited in politics." So long as we refuse to allow ourselves, even in the watches of the night, to avow any contrary desires, we may be bullies at home, shady in the City, skinflints in paying wages and profiteers in dealing with the public; yet, if only conscious motives are to count in moral valuation, we shall remain model characters. This is an agreeable doctrine, and it is not surprising that men are un willing to abandon it. But moral considerations are the worst enemies of the scientific spirit and we must dismiss them from our minds if we wish to arrive at truth.

I believe—as I shall try to prove in a later lecture—that desire, like force in mechanics, is of the nature of a convenient fiction for describing shortly certain laws of behaviour. A hungry animal is restless until it finds food; then it becomes quiescent. The thing which will bring a restless condition to an end is said to be what is desired. But only experience can show what will have this sedative effect, and it is easy to make mistakes. We feel dissatisfaction, and think that such and-such a thing would remove it; but in thinking this, we are theorizing, not observing a patent fact. Our theorizing is often mistaken, and when it is mistaken there is a difference between what we think we desire and what in fact will bring satisfaction. This is such a common phenomenon that any theory of desire which fails to account for it must be wrong.

What have been called "unconscious" desires have been brought very much to the fore in recent years by psycho-analysis. Psycho-analysis, as every one knows, is primarily a method of understanding hysteria and certain forms of insanity; but it has been found that there is much in the lives of ordinary men and women which bears a humiliating resemblance to the delusions of the insane. The connection of dreams, irrational beliefs and foolish actions with unconscious wishes has been brought to light, though with some exaggeration, by Freud and Jung and their followers. As regards the nature of these unconscious wishes, it seems to me—though as a layman I speak with diffidence—that many psycho-analysts are unduly narrow; no doubt the wishes they emphasize exist, but others, e.g. for honour and power, are equally operative and equally liable to concealment. This, however, does not affect the value of their general theories from the point of view of theoretic psychology, and it is from this point of view that their results are important for the analysis of mind.

What, I think, is clearly established, is that a man's actions and beliefs may be wholly dominated by a desire of which he is quite unconscious, and which he indignantly repudiates when it is suggested to him. Such a desire is generally, in morbid cases, of a sort which the patient would consider wicked; if he had to admit that he had the desire, he would loathe himself. Yet it is so strong that it must force an outlet for itself; hence it becomes necessary to entertain whole systems of false beliefs in order to hide the nature of what is desired. The resulting delusions in very many cases disappear if the hysteric or lunatic can be made to face the facts about himself. The consequence of this is that the treatment of many forms of insanity has grown more psychological and less physiological than it used to be. Instead of looking for a physical defect in the brain, those who treat delusions look for the repressed desire which has found this contorted mode of expression. For those who do not wish to plunge into the somewhat repulsive and often rather wild theories of psychoanalytic pioneers, it will be worth while to read a little book by Dr. Bernard Hart on "The Psychology of Insanity." On this question of the mental as opposed to the physiological study of the causes of insanity, Dr. Hart says:

"The psychological conception [of insanity] is based on the view that mental processes can be directly studied without any reference to the accompanying changes which are presumed to take place in the brain, and that insanity may therefore be properly attacked from the standpoint of psychology".

This illustrates a point which I am anxious to make clear from the outset. Any attempt to classify modern views, such as I propose to advocate, from the old standpoint of materialism and idealism, is only misleading. In certain respects, the views which I shall be setting forth approximate to materialism; in certain others, they approximate to its opposite. On this question of the study of delusions, the practical effect of the modern theories, as Dr. Hart points out, is emancipation from the materialist method. On the other hand, as he also points out, imbecility and dementia still have to be considered physiologically, as caused by defects in the brain. There is no inconsistency in this If, as we maintain, mind and matter are neither of them the actual stuff of reality, but different convenient groupings of an underlying material, then, clearly, the question whether, in regard to a given phenomenon, we are to seek a physical or a mental cause, is merely one to be decided by trial. Metaphysicians have argued endlessly as to the interaction of mind and matter. The followers of Descartes held that mind and matter are so different as to make any action of the one on the other impossible. When I will to move my arm, they said, it is not my will that operates on my arm, but God, who, by His omnipotence, moves my arm whenever I want it moved. The modern doctrine of psychophysical parallelism is not appreciably different from this theory of the Cartesian school. Psycho-physical parallelism is the theory that mental and physical events each have causes in their own sphere, but run on side by side owing to the fact that every state of the brain coexists with a definite state of the mind, and vice versa. This view of the reciprocal causal independence of mind and matter has no basis except in metaphysical theory.For us, there is no necessity to make any such assumption, which is very difficult to harmonize with obvious facts. I receive a letter inviting me to dinner: the letter is a physical fact, but my apprehension of its meaning is mental. Here we have an effect of matter on mind. In consequence of my apprehension of the meaning of the letter, I go to the right place at the right time; here we have an effect of mind on matter. I shall try to persuade you, in the course of these lectures, that matter is not so material and mind not so mental as is generally supposed. When we are speaking of matter, it will seem as if we were inclining to idealism; when we are speaking of mind, it will seem as if we were inclining to materialism. Neither is the truth. Our world is to be constructed out of what the American realists call "neutral" entities, which have neither the hardness and indestructibility of matter, nor the reference to objects which is supposed to characterize mind.

There is, it is true, one objection which might be felt, not indeed to the action of matter on mind, but to the action of mind on matter. The laws of physics, it may be urged, are apparently adequate to explain everything that happens to matter, even when it is matter in a man's brain. This, however, is only a hypothesis, not an established theory. There is no cogent empirical reason for supposing that the laws determining the motions of living bodies are exactly the same as those that apply to dead matter. Sometimes, of course, they are clearly the same. When a man falls from a precipice or slips on a piece of orange peel, his body behaves as if it were devoid of life. These are the occasions that make Bergson laugh. But when a man's bodily movements are what we call "voluntary," they are, at any rate prima facie, very different in their laws from the movements of what is devoid of life. I do not wish to say dogmatically that the difference is irreducible; I think it highly probable that it is not. I say only that the study of the behaviour of living bodies, in the present state of our knowledge, is distinct from physics. The study of gases was originally quite distinct from that of rigid bodies, and would never have advanced to its present state if it had not been independently pursued. Nowadays both the gas and the rigid body are manufactured out of a more primitive and universal kind of matter. In like manner, as a question of methodology, the laws of living bodies are to be studied, in the first place, without any undue haste to subordinate them to the laws of physics. Boyle's law and the rest had to be discovered before the kinetic theory of gases became possible. But in psychology we are hardly yet at the stage of Boyle's law. Meanwhile we need not be held up by the bogey of the universal rigid exactness of physics. This is, as yet, a mere hypothesis, to be tested empirically without any preconceptions. It may be true, or it may not. So far, that is all we can say.

Returning from this digression to our main topic, namely, the criticism of "consciousness," we observe that Freud and his followers, though they have demonstrated beyond dispute the immense importance of "unconscious" desires in determining our actions and beliefs, have not attempted the task of telling us what an "unconscious" desire actually is, and have thus invested their doctrine with an air of mystery and mythology which forms a large part of its popular attractiveness. They speak always as though it were more normal for a desire to be conscious, and as though a positive cause had to be assigned for its being unconscious. Thus "the unconscious" becomes a sort of underground prisoner, living in a dungeon, breaking in at long intervals upon our daylight respectability with dark groans and maledictions and strange atavistic lusts. The ordinary reader, almost inevitably, thinks of this underground person as another consciousness, prevented by what Freud calls the "censor" from making his voice heard in company, except on rare and dreadful occasions when he shouts so loud that every one hears him and there is a scandal. Most of us like the idea that we could be desperately wicked if only we let ourselves go. For this reason, the Freudian "unconscious" has been a consolation to many quiet and well-behaved persons.

I do not think the truth is quite so picturesque as this. I believe an "unconscious" desire is merely a causal law of our behaviour, namely, that we remain restlessly active until a certain state of affairs is realized, when we achieve temporary equilibrium If we know beforehand what this state of affairs is, our desire is conscious; if not, unconscious. The unconscious desire is not something actually existing, but merely a tendency to a certain behaviour; it has exactly the same status as a force in dynamics. The unconscious desire is in no way mysterious; it is the natural primitive form of desire, from which the other has developed through our habit of observing and theorizing (often wrongly). It is not necessary to suppose, as Freud seems to do, that every unconscious wish was once conscious, and was then, in his terminology, "repressed" because we disapproved of it. On the contrary, we shall suppose that, although Freudian "repression" undoubtedly occurs and is important, it is not the usual reason for unconsciousness of our wishes. The usual reason is merely that wishes are all, to begin with, unconscious, and only become known when they are actively noticed. Usually, from laziness, people do not notice, but accept the theory of human nature which they find current, and attribute to themselves whatever wishes this theory would lead them to expect. We used to be full of virtuous wishes, but since Freud our wishes have become, in the words of the Prophet Jeremiah, "deceitful above all things and desperately wicked." Both these views, in most of those who have held them, are the product of theory rather than observation, for observation requires effort, whereas repeating phrases does not.

The interpretation of unconscious wishes which I have been advocating has been set forth briefly by Professor John B. Watson in an article called "The Psychology of Wish Fulfilment," which appeared in "The Scientific Monthly" in November, 1916. Two quotations will serve to show his point of view:

"The Freudians (he says) have made more or less of a 'metaphysical entity' out of the censor. They suppose that when wishes are repressed they are repressed into the 'unconscious,' and that this mysterious censor stands at the trapdoor lying between the conscious and the unconscious. Many of us do not believe in a world of the unconscious (a few of us even have grave doubts about the usefulness of the term consciousness), hence we try to explain censorship along ordinary biological lines. We believe that one group of habits can 'down' another group of habits—or instincts. In this case our ordinary system of habits—those which we call expressive of our 'real selves'—inhibit or quench (keep inactive or partially inactive) those habits and instinctive tendencies which belong largely in the past".

Again, after speaking of the frustration of some impulses which is involved in acquiring the habits of a civilized adult, he continues:

"It is among these frustrated impulses that I would find the biological basis of the unfulfilled wish. Such 'wishes' need never have been 'conscious,' and NEED NEVER HAVE BEEN SUPPRESSED INTO FREUD'S REALM OF THE UNCONSCIOUS. It may be inferred from this that there is no particular reason for applying the term 'wish' to such tendencies".

One of the merits of the general analysis of mind which we shall be concerned with in the following lectures is that it removes the atmosphere of mystery from the phenomena brought to light by the psycho-analysts. Mystery is delightful, but unscientific, since it depends upon ignorance. Man has developed out of the animals, and there is no serious gap between him and the amoeba. Something closely analogous to knowledge and desire, as regards its effects on behaviour, exists among animals, even where what we call "consciousness" is hard to believe in; something equally analogous exists in ourselves in cases where no trace of "consciousness" can be found. It is therefore natural to suppose that, what ever may be the correct definition of "consciousness," "consciousness" is not the essence of life or mind. In the following lectures, accordingly, this term will disappear until we have dealt with words, when it will re-emerge as mainly a trivial and unimportant outcome of linguistic habits.

LECTURE II.

INSTINCT AND HABIT

In attempting to understand the elements out of which mental phenomena are compounded, it is of the greatest importance to remember that from the protozoa to man there is nowhere a very wide gap either in structure or in behaviour. From this fact it is a highly probable inference that there is also nowhere a very wide mental gap. It is, of course, POSSIBLE that there may be, at certain stages in evolution, elements which are entirely new from the standpoint of analysis, though in their nascent form they have little influence on behaviour and no very marked correlatives in structure. But the hypothesis of continuity in mental development is clearly preferable if no psychological facts make it impossible. We shall find, if I am not mistaken, that there are no facts which refute the hypothesis of mental continuity, and that, on the other hand, this hypothesis affords a useful test of suggested theories as to the nature of mind.

The hypothesis of mental continuity throughout organic evolution may be used in two different ways. On the one hand, it may be held that we have more knowledge of our own minds than those of animals, and that we should use this knowledge to infer the existence of something similar to our own mental processes in animals and even in plants. On the other hand, it may be held that animals and plants present simpler phenomena, more easily analysed than those of human minds; on this ground it may be urged that explanations which are adequate in the case of animals ought not to be lightly rejected in the case of man. The practical effects of these two views are diametrically opposite: the first leads us to level up animal intelligence with what we believe ourselves to know about our own intelligence, while the second leads us to attempt a levelling down of our own intelligence to something not too remote from what we can observe in animals. It is therefore important to consider the relative justification of the two ways of applying the principle of continuity.

It is clear that the question turns upon another, namely, which can we know best, the psychology of animals or that of human beings? If we can know most about animals, we shall use this knowledge as a basis for inference about human beings; if we can know most about human beings, we shall adopt the opposite procedure. And the question whether we can know most about the psychology of human beings or about that of animals turns upon yet another, namely: Is introspection or external observation the surer method in psychology? This is a question which I propose to discuss at length in Lecture VI; I shall therefore content myself now with a statement of the conclusions to be arrived at.

We know a great many things concerning ourselves which we cannot know nearly so directly concerning animals or even other people. We know when we have a toothache, what we are thinking of, what dreams we have when we are asleep, and a host of other occurrences which we only know about others when they tell us of them, or otherwise make them inferable by their behaviour. Thus, so far as knowledge of detached facts is concerned, the advantage is on the side of self-knowledge as against external observation.

But when we come to the analysis and scientific understanding of the facts, the advantages on the side of self-knowledge become far less clear. We know, for example, that we have desires and beliefs, but we do not know what constitutes a desire or a belief. The phenomena are so familiar that it is difficult to realize how little we really know about them. We see in animals, and to a lesser extent in plants, behaviour more or less similar to that which, in us, is prompted by desires and beliefs, and we find that, as we descend in the scale of evolution, behaviour becomes simpler, more easily reducible to rule, more scientifically analysable and predictable. And just because we are not misled by familiarity we find it easier to be cautious in interpreting behaviour when we are dealing with phenomena remote from those of our own minds: Moreover, introspection, as psychoanalysis has demonstrated, is extraordinarily fallible even in cases where we feel a high degree of certainty. The net result seems to be that, though self-knowledge has a definite and important contribution to make to psychology, it is exceedingly misleading unless it is constantly checked and controlled by the test of external observation, and by the theories which such observation suggests when applied to animal behaviour. On the whole, therefore, there is probably more to be learnt about human psychology from animals than about animal psychology from human beings; but this conclusion is one of degree, and must not be pressed beyond a point.

It is only bodily phenomena that can be directly observed in animals, or even, strictly speaking, in other human beings. We can observe such things as their movements, their physiological processes, and the sounds they emit. Such things as desires and beliefs, which seem obvious to introspection, are not visible directly to external observation. Accordingly, if we begin our study of psychology by external observation, we must not begin by assuming such things as desires and beliefs, but only such things as external observation can reveal, which will be characteristics of the movements and physiological processes of animals. Some animals, for example, always run away from light and hide themselves in dark places. If you pick up a mossy stone which is lightly embedded in the earth, you will see a number of small animals scuttling away from the unwonted daylight and seeking again the darkness of which you have deprived them. Such animals are sensitive to light, in the sense that their movements are affected by it; but it would be rash to infer that they have sensations in any way analogous to our sensations of sight. Such inferences, which go beyond the observable facts, are to be avoided with the utmost care.

It is customary to divide human movements into three classes, voluntary, reflex and mechanical. We may illustrate the distinction by a quotation from William James :

"If I hear the conductor calling 'all aboard' as I enter the depot, my heart first stops, then palpitates, and my legs respond to the air-waves falling on my tympanum by quickening their movements. If I stumble as I run, the sensation of falling provokes a movement of the hands towards the direction of the fall, the effect of which is to shield the body from too sudden a shock. If a cinder enter my eye, its lids close forcibly and a copious flow of tears tends to wash it out.

"These three responses to a sensational stimulus differ, however, in many respects. The closure of the eye and the lachrymation are quite involuntary, and so is the disturbance of the heart. Such involuntary responses we know as 'reflex' acts. The motion of the arms to break the shock of falling may also be called reflex, since it occurs too quickly to be deliberately intended. Whether it be instinctive or whether it result from the pedestrian education of childhood may be doubtful; it is, at any rate, less automatic than the previous acts, for a man might by conscious effort learn to perform it more skilfully, or even to suppress it altogether. Actions of this kind, with which instinct and volition enter upon equal terms, have been called 'semi-reflex.' The act of running towards the train, on the other hand, has no instinctive element about it. It is purely the result of education, and is preceded by a consciousness of the purpose to be attained and a distinct mandate of the will. It is a 'voluntary act.' Thus the animal's reflex and voluntary performances shade into each other gradually, being connected by acts which may often occur automatically, but may also be modified by conscious intelligence.

"An outside observer, unable to perceive the accompanying consciousness, might be wholly at a loss to discriminate between the automatic acts and those which volition escorted. But if the criterion of mind's existence be the choice of the proper means for the attainment of a supposed end, all the acts alike seem to be inspired by intelligence, for APPROPRIATENESS characterizes them all alike."

There is one movement, among those that James mentions at first, which is not subsequently classified, namely, the stumbling. This is the kind of movement which may be called "mechanical"; it is evidently of a different kind from either reflex or voluntary movements, and more akin to the movements of dead matter. We may define a movement of an animal's body as "mechanical" when it proceeds as if only dead matter were involved. For example, if you fall over a cliff, you move under the influence of gravitation, and your centre of gravity describes just as correct a parabola as if you were already dead. Mechanical movements have not the characteristic of appropriateness, unless by accident, as when a drunken man falls into a waterbutt and is sobered. But reflex and voluntary movements are not ALWAYS appropriate, unless in some very recondite sense. A moth flying into a lamp is not acting sensibly; no more is a man who is in such a hurry to get his ticket that he cannot remember the name of his destination. Appropriateness is a complicated and merely approximate idea, and for the present we shall do well to dismiss it from our thoughts.

As James states, there is no difference, from the point of view of the outside observer, between voluntary and reflex movements. The physiologist can discover that both depend upon the nervous system, and he may find that the movements which we call voluntary depend upon higher centres in the brain than those that are reflex. But he cannot discover anything as to the presence or absence of "will" or "consciousness," for these things can only be seen from within, if at all. For the present, we wish to place ourselves resolutely in the position of outside observers; we will therefore ignore the distinction between voluntary and reflex movements. We will call the two together "vital" movements. We may then distinguish "vital" from mechanical movements by the fact that vital movements depend for their causation upon the special properties of the nervous system, while mechanical movements depend only upon the properties which animal bodies share with matter in general.

There is need for some care if the distinction between mechanical and vital movements is to be made precise. It is quite likely that, if we knew more about animal bodies, we could deduce all their movements from the laws of chemistry and physics. It is already fairly easy to see how chemistry reduces to physics, i.e. how the differences between different chemical elements can be accounted for by differences of physical structure, the constituents of the structure being electrons which are exactly alike in all kinds of matter. We only know in part how to reduce physiology to chemistry, but we know enough to make it likely that the reduction is possible. If we suppose it effected, what would become of the difference between vital and mechanical movements?

Some analogies will make the difference clear. A shock to a mass of dynamite produces quite different effects from an equal shock to a mass of steel: in the one case there is a vast explosion, while in the other case there is hardly any noticeable disturbance. Similarly, you may sometimes find on a mountain-side a large rock poised so delicately that a touch will set it crashing down into the valley, while the rocks all round are so firm that only a considerable force can dislodge them What is analogous in these two cases is the existence of a great store of energy in unstable equilibrium ready to burst into violent motion by the addition of a very slight disturbance. Similarly, it requires only a very slight expenditure of energy to send a post-card with the words "All is discovered; fly!" but the effect in generating kinetic energy is said to be amazing. A human body, like a mass of dynamite, contains a store of energy in unstable equilibrium, ready to be directed in this direction or that by a disturbance which is physically very small, such as a spoken word. In all such cases the reduction of behaviour to physical laws can only be effected by entering into great minuteness; so long as we confine ourselves to the observation of comparatively large masses, the way in which the equilibrium will be upset cannot be determined. Physicists distinguish between macroscopic and microscopic equations: the former determine the visible movements of bodies of ordinary size, the latter the minute occurrences in the smallest parts. It is only the microscopic equations that are supposed to be the same for all sorts of matter. The macroscopic equations result from a process of averaging out, and may be different in different cases. So, in our instance, the laws of macroscopic phenomena are different for mechanical and vital movements, though the laws of microscopic phenomena may be the same.

We may say, speaking somewhat roughly, that a stimulus applied to the nervous system, like a spark to dynamite, is able to take advantage of the stored energy in unstable equilibrium, and thus to produce movements out of proportion to the proximate cause. Movements produced in this way are vital movements, while mechanical movements are those in which the stored energy of a living body is not involved. Similarly dynamite may be exploded, thereby displaying its characteristic properties, or may (with due precautions) be carted about like any other mineral. The explosion is analogous to vital movements, the carting about to mechanical movements.

Mechanical movements are of no interest to the psychologist, and it has only been necessary to define them in order to be able to exclude them. When a psychologist studies behaviour, it is only vital movements that concern him. We shall, therefore, proceed to ignore mechanical movements, and study only the properties of the remainder.

The next point is to distinguish between movements that are instinctive and movements that are acquired by experience. This distinction also is to some extent one of degree. Professor Lloyd Morgan gives the following definition of "instinctive behaviour":

"That which is, on its first occurrence, independent of prior experience; which tends to the well-being of the individual and the preservation of the race; which is similarly performed by all members of the same more or less restricted group of animals; and which may be subject to subsequent modification under the guidance of experience."

This definition is framed for the purposes of biology, and is in some respects unsuited to the needs of psychology. Though perhaps unavoidable, allusion to "the same more or less restricted group of animals" makes it impossible to judge what is instinctive in the behaviour of an isolated individual. Moreover, "the well-being of the individual and the preservation of the race" is only a usual characteristic, not a universal one, of the sort of movements that, from our point of view, are to be called instinctive; instances of harmful instincts will be given shortly. The essential point of the definition, from our point of view, is that an instinctive movement is in dependent of prior experience.

We may say that an "instinctive" movement is a vital movement performed by an animal the first time that it finds itself in a novel situation; or, more correctly, one which it would perform if the situation were novel. The instincts of an animal are different at different periods of its growth, and this fact may cause changes of behaviour which are not due to learning. The maturing and seasonal fluctuation of the sex-instinct affords a good illustration. When the sex-instinct first matures, the behaviour of an animal in the presence of a mate is different from its previous behaviour in similar circumstances, but is not learnt, since it is just the same if the animal has never previously been in the presence of a mate.

On the other hand, a movement is "learnt," or embodies a "habit," if it is due to previous experience of similar situations, and is not what it would be if the animal had had no such experience.

There are various complications which blur the sharpness of this distinction in practice. To begin with, many instincts mature gradually, and while they are immature an animal may act in a fumbling manner which is very difficult to distinguish from learning. James ("Psychology," ii, 407) maintains that children walk by instinct, and that the awkwardness of their first attempts is only due to the fact that the instinct has not yet ripened. He hopes that "some scientific widower, left alone with his offspring at the critical moment, may ere long test this suggestion on the living subject." However this may be, he quotes evidence to show that "birds do not LEARN to fly," but fly by instinct when they reach the appropriate age . In the second place, instinct often gives only a rough outline of the sort of thing to do, in which case learning is necessary in order to acquire certainty and precision in action. In the third place, even in the clearest cases of acquired habit, such as speaking, some instinct is required to set in motion the process of learning. In the case of speaking, the chief instinct involved is commonly supposed to be that of imitation, but this may be questioned.

In spite of these qualifications, the broad distinction between instinct and habit is undeniable. To take extreme cases, every animal at birth can take food by instinct, before it has had opportunity to learn; on the other hand, no one can ride a bicycle by instinct, though, after learning, the necessary movements become just as automatic as if they were instinctive.

The process of learning, which consists in the acquisition of habits, has been much studied in various animals. For example: you put a hungry animal, say a cat, in a cage which has a door that can be opened by lifting a latch; outside the cage you put food. The cat at first dashes all round the cage, making frantic efforts to force a way out. At last, by accident, the latch is lifted and the cat pounces on the food. Next day you repeat the experiment, and you find that the cat gets out much more quickly than the first time, although it still makes some random movements. The third day it gets out still more quickly, and before long it goes straight to the latch and lifts it at once. Or you make a model of the Hampton Court maze, and put a rat in the middle, assaulted by the smell of food on the outside. The rat starts running down the passages, and is constantly stopped by blind alleys, but at last, by persistent attempts, it gets out. You repeat this experiment day after day; you measure the time taken by the rat in reaching the food; you find that the time rapidly diminishes, and that after a while the rat ceases to make any wrong turnings. It is by essentially similar processes that we learn speaking, writing, mathematics, or the government of an empire.

Professor Watson has an ingenious theory as to the way in which habit arises out of random movements. I think there is a reason why his theory cannot be regarded as alone sufficient, but it seems not unlikely that it is partly correct. Suppose, for the sake of simplicity, that there are just ten random movements which may be made by the animal—say, ten paths down which it may go—and that only one of these leads to food, or whatever else represents success in the case in question. Then the successful movement always occurs during the animal's attempts, whereas each of the others, on the average, occurs in only half the attempts. Thus the tendency to repeat a previous performance (which is easily explicable without the intervention of "consciousness") leads to a greater emphasis on the successful movement than on any other, and in time causes it alone to be performed. The objection to this view, if taken as the sole explanation, is that on improvement ought to set in till after the SECOND trial, whereas experiment shows that already at the second attempt the animal does better than the first time. Something further is, therefore, required to account for the genesis of habit from random movements; but I see no reason to suppose that what is further required involves "consciousness."

Mr. Thorndike formulates two "provisional laws of acquired behaviour or learning," as follows:

"The Law of Effect is that: Of several responses made to the same situation, those which are accompanied or closely followed by satisfaction to the animal will, other things being equal, be more firmly connected with the situation, so that, when it recurs, they will be more likely to recur; those which are accompanied or closely followed by discomfort to the animal will, other things being equal, have their connections with that situation weakened, so that, when it recurs, they will be less likely to occur. The greater the satisfaction or discomfort, the greater the strengthening or weakening of the bond.

"The Law of Exercise is that: Any response to a situation will, other things being equal, be more strongly connected with the situation in proportion to the number of times it has been connected with that situation and to the average vigour and duration of the connections."

With the explanation to be presently given of the meaning of "satisfaction" and "discomfort," there seems every reason to accept these two laws.

What is true of animals, as regards instinct and habit, is equally true of men. But the higher we rise in the evolutionary scale, broadly speaking, the greater becomes the power of learning, and the fewer are the occasions when pure instinct is exhibited unmodified in adult life. This applies with great force to man, so much so that some have thought instinct less important in the life of man than in that of animals. This, however, would be a mistake. Learning is only possible when instinct supplies the driving-force. The animals in cages, which gradually learn to get out, perform random movements at first, which are purely instinctive. But for these random movements, they would never acquire the experience which afterwards enables them to produce the right movement. (This is partly questioned by Hobhouse—wrongly, I think.) Similarly, children learning to talk make all sorts of sounds, until one day the right sound comes by accident. It is clear that the original making of random sounds, without which speech would never be learnt, is instinctive. I think we may say the same of all the habits and aptitudes that we acquire in all of them there has been present throughout some instinctive activity, prompting at first rather inefficient movements, but supplying the driving force while more and more effective methods are being acquired. A cat which is hungry smells fish, and goes to the larder. This is a thoroughly efficient method when there is fish in the larder, and it is often successfully practised by children. But in later life it is found that merely going to the larder does not cause fish to be there; after a series of random movements it is found that this result is to be caused by going to the City in the morning and coming back in the evening. No one would have guessed a priori that this movement of a middle-aged man's body would cause fish to come out of the sea into his larder, but experience shows that it does, and the middle-aged man therefore continues to go to the City, just as the cat in the cage continues to lift the latch when it has once found it. Of course, in actual fact, human learning is rendered easier, though psychologically more complex, through language; but at bottom language does not alter the essential character of learning, or of the part played by instinct in promoting learning. Language, however, is a subject upon which I do not wish to speak until a later lecture.

The popular conception of instinct errs by imagining it to be infallible and preternaturally wise, as well as incapable of modification. This is a complete delusion. Instinct, as a rule, is very rough and ready, able to achieve its result under ordinary circumstances, but easily misled by anything unusual. Chicks follow their mother by instinct, but when they are quite young they will follow with equal readiness any moving object remotely resembling their mother, or even a human being ergson, quoting Fabre, has made play with the supposed extraordinary accuracy of the solitary wasp Ammophila, which lays its eggs in a caterpillar. On this subject I will quote from Drever's "Instinct in Man,":

"According to Fabre's observations, which Bergson accepts, the Ammophila stings its prey EXACTLY and UNERRINGLY in EACH of the nervous centres. The result is that the caterpillar is paralyzed, but not immediately killed, the advantage of this being that the larva cannot be injured by any movement of the caterpillar, upon which the egg is deposited, and is provided with fresh meat when the time comes.

"Now Dr. and Mrs. Peckham have shown that the sting of the wasp is NOT UNERRING, as Fabre alleges, that the number of stings is NOT CONSTANT, that sometimes the caterpillar is NOT PARALYZED, and sometimes it is KILLED OUTRIGHT, and that THE DIFFERENT CIRCUMSTANCES DO NOT APPARENTLY MAKE ANY DIFFERENCE TO THE LARVA, which is not injured by slight movements of the caterpillar, nor by consuming food decomposed rather than fresh caterpillar."

This illustrates how love of the marvellous may mislead even so careful an observer as Fabre and so eminent a philosopher as Bergson.

In the same chapter of Dr. Drever's book there are some interesting examples of the mistakes made by instinct. I will quote one as a sample:

"The larva of the Lomechusa beetle eats the young of the ants, in whose nest it is reared. Nevertheless, the ants tend the Lomechusa larvae with the same care they bestow on their own young. Not only so, but they apparently discover that the methods of feeding, which suit their own larvae, would prove fatal to the guests, and accordingly they change their whole system of nursing".

Semon gives a good illustration of an instinct growing wiser through experience. He relates how hunters attract stags by imitating the sounds of other members of their species, male or female, but find that the older a stag becomes the more difficult it is to deceive him, and the more accurate the imitation has to be. The literature of instinct is vast, and illustrations might be multiplied indefinitely. The main points as regards instinct, which need to be emphasized as against the popular conceptions of it, are:

(1) That instinct requires no prevision of the biological end which it serves;

(2) That instinct is only adapted to achieve this end in the usual circumstances of the animal in question, and has no more precision than is necessary for success AS A RULE;

(3) That processes initiated by instinct often come to be performed better after experience;

(4) That instinct supplies the impulses to experimental movements which are required for the process of learning;

(5) That instincts in their nascent stages are easily modifiable, and capable of being attached to various sorts of objects.

All the above characteristics of instinct can be established by purely external observation, except the fact that instinct does not require prevision. This, though not strictly capable of being PROVED by observation, is irresistibly suggested by the most obvious phenomena. Who can believe, for example, that a new-born baby is aware of the necessity of food for preserving life? Or that insects, in laying eggs, are concerned for the preservation of their species? The essence of instinct, one might say, is that it provides a mechanism for acting without foresight in a manner which is usually advantageous biologically. It is partly for this reason that it is so important to understand the fundamental position of instinct in prompting both animal and human behaviour.

LECTURE III.

DESIRE AND FEELING

Desire is a subject upon which, if I am not mistaken, true views can only be arrived at by an almost complete reversal of the ordinary unreflecting opinion. It is natural to regard desire as in its essence an attitude towards something which is imagined, not actual; this something is called the END or OBJECT of the desire, and is said to be the PURPOSE of any action resulting from the desire. We think of the content of the desire as being just like the content of a belief, while the attitude taken up towards the content is different. According to this theory, when we say: "I hope it will rain," or "I expect it will rain," we express, in the first case, a desire, and in the second, a belief, with an identical content, namely, the image of rain. It would be easy to say that, just as belief is one kind of feeling in relation to this content, so desire is another kind. According to this view, what comes first in desire is something imagined, with a specific feeling related to it, namely, that specific feeling which we call "desiring" it. The discomfort associated with unsatisfied desire, and the actions which aim at satisfying desire, are, in this view, both of them effects of the desire. I think it is fair to say that this is a view against which common sense would not rebel; nevertheless, I believe it to be radically mistaken. It cannot be refuted logically, but various facts can be adduced which make it gradually less simple and plausible, until at last it turns out to be easier to abandon it wholly and look at the matter in a totally different way.

The first set of facts to be adduced against the common sense view of desire are those studied by psycho-analysis. In all human beings, but most markedly in those suffering from hysteria and certain forms of insanity, we find what are called "unconscious" desires, which are commonly regarded as showing self-deception. Most psycho-analysts pay little attention to the analysis of desire, being interested in discovering by observation what it is that people desire, rather than in discovering what actually constitutes desire. I think the strangeness of what they report would be greatly diminished if it were expressed in the language of a behaviourist theory of desire, rather than in the language of every-day beliefs. The general description of the sort of phenomena that bear on our present question is as follows: A person states that his desires are so-and-so, and that it is these desires that inspire his actions; but the outside observer perceives that his actions are such as to realize quite different ends from those which he avows, and that these different ends are such as he might be expected to desire. Generally they are less virtuous than his professed desires, and are therefore less agreeable to profess than these are. It is accordingly supposed that they really exist as desires for ends, but in a subconscious part of the mind, which the patient refuses to admit into consciousness for fear of having to think ill of himself. There are no doubt many cases to which such a supposition is applicable without obvious artificiality. But the deeper the Freudians delve into the underground regions of instinct, the further they travel from anything resembling conscious desire, and the less possible it becomes to believe that only positive self-deception conceals from us that we really wish for things which are abhorrent to our explicit life.

In the cases in question we have a conflict between the outside observer and the patient's consciousness. The whole tendency of psycho-analysis is to trust the outside observer rather than the testimony of introspection. I believe this tendency to be entirely right, but to demand a re-statement of what constitutes desire, exhibiting it as a causal law of our actions, not as something actually existing in our minds.

But let us first get a clearer statement of the essential characteristic of the phenomena.

A person, we find, states that he desires a certain end A, and that he is acting with a view to achieving it. We observe, however, that his actions are such as are likely to achieve a quite different end B, and that B is the sort of end that often seems to be aimed at by animals and savages, though civilized people are supposed to have discarded it. We sometimes find also a whole set of false beliefs, of such a kind as to persuade the patient that his actions are really a means to A, when in fact they are a means to B. For example, we have an impulse to inflict pain upon those whom we hate; we therefore believe that they are wicked, and that punishment will reform them. This belief enables us to act upon the impulse to inflict pain, while believing that we are acting upon the desire to lead sinners to repentance. It is for this reason that the criminal law has been in all ages more severe than it would have been if the impulse to ameliorate the criminal had been what really inspired it. It seems simple to explain such a state of affairs as due to "self-deception," but this explanation is often mythical. Most people, in thinking about punishment, have had no more need to hide their vindictive impulses from themselves than they have had to hide the exponential theorem. Our impulses are not patent to a casual observation, but are only to be discovered by a scientific study of our actions, in the course of which we must regard ourselves as objectively as we should the motions of the planets or the chemical reactions of a new element.

The study of animals reinforces this conclusion, and is in many ways the best preparation for the analysis of desire. In animals we are not troubled by the disturbing influence of ethical considerations. In dealing with human beings, we are perpetually distracted by being told that such-and-such a view is gloomy or cynical or pessimistic: ages of human conceit have built up such a vast myth as to our wisdom and virtue that any intrusion of the mere scientific desire to know the facts is instantly resented by those who cling to comfortable illusions. But no one cares whether animals are virtuous or not, and no one is under the delusion that they are rational. Moreover, we do not expect them to be so "conscious," and are prepared to admit that their instincts prompt useful actions without any prevision of the ends which they achieve. For all these reasons, there is much in the analysis of mind which is more easily discovered by the study of animals than by the observation of human beings.

We all think that, by watching the behaviour of animals, we can discover more or less what they desire. If this is the case—and I fully agree that it is—desire must be capable of being exhibited in actions, for it is only the actions of animals that we can observe. They MAY have minds in which all sorts of things take place, but we can know nothing about their minds except by means of inferences from their actions; and the more such inferences are examined, the more dubious they appear. It would seem, therefore, that actions alone must be the test of the desires of animals. From this it is an easy step to the conclusion that an animal's desire is nothing but a characteristic of a certain series of actions, namely, those which would be commonly regarded as inspired by the desire in question. And when it has been shown that this view affords a satisfactory account of animal desires, it is not difficult to see that the same explanation is applicable to the desires of human beings.

We judge easily from the behaviour of an animal of a familiar kind whether it is hungry or thirsty, or pleased or displeased, or inquisitive or terrified. The verification of our judgment, so far as verification is possible, must be derived from the immediately succeeding actions of the animal. Most people would say that they infer first something about the animal's state of mind—whether it is hungry or thirsty and so on—and thence derive their expectations as to its subsequent conduct. But this detour through the animal's supposed mind is wholly unnecessary. We can say simply: The animal's behaviour during the last minute has had those characteristics which distinguish what is called "hunger," and it is likely that its actions during the next minute will be similar in this respect, unless it finds food, or is interrupted by a stronger impulse, such as fear. An animal which is hungry is restless, it goes to the places where food is often to be found, it sniffs with its nose or peers with its eyes or otherwise increases the sensitiveness of its sense-organs; as soon as it is near enough to food for its sense-organs to be affected, it goes to it with all speed and proceeds to eat; after which, if the quantity of food has been sufficient, its whole demeanour changes it may very likely lie down and go to sleep. These things and others like them are observable phenomena distinguishing a hungry animal from one which is not hungry. The characteristic mark by which we recognize a series of actions which display hunger is not the animal's mental state, which we cannot observe, but something in its bodily behaviour; it is this observable trait in the bodily behaviour that I am proposing to call "hunger," not some possibly mythical and certainly unknowable ingredient of the animal's mind.

Generalizing what occurs in the case of hunger, we may say that what we call a desire in an animal is always displayed in a cycle of actions having certain fairly well marked characteristics. There is first a state of activity, consisting, with qualifications to be mentioned presently, of movements likely to have a certain result; these movements, unless interrupted, continue until the result is achieved, after which there is usually a period of comparative quiescence. A cycle of actions of this sort has marks by which it is broadly distinguished from the motions of dead matter. The most notable of these marks are—(1) the appropriateness of the actions for the realization of a certain result; (2) the continuance of action until that result has been achieved. Neither of these can be pressed beyond a point. Either may be (a) to some extent present in dead matter, and (b) to a considerable extent absent in animals, while vegetable are intermediate, and display only a much fainter form of the behaviour which leads us to attribute desire to animals. (a) One might say rivers "desire" the sea water, roughly speaking, remains in restless motion until it reaches either the sea or a place from which it cannot issue without going uphill, and therefore we might say that this is what it wishes while it is flowing. We do not say so, because we can account for the behaviour of water by the laws of physics; and if we knew more about animals, we might equally cease to attribute desires to them, since we might find physical and chemical reactions sufficient to account for their behaviour. (b) Many of the movements of animals do not exhibit the characteristics of the cycles which seem to embody desire. There are first of all the movements which are "mechanical," such as slipping and falling, where ordinary physical forces operate upon the animal's body almost as if it were dead matter. An animal which falls over a cliff may make a number of desperate struggles while it is in the air, but its centre of gravity will move exactly as it would if the animal were dead. In this case, if the animal is killed at the end of the fall, we have, at first sight, just the characteristics of a cycle of actions embodying desire, namely, restless movement until the ground is reached, and then quiescence. Nevertheless, we feel no temptation to say that the animal desired what occurred, partly because of the obviously mechanical nature of the whole occurrence, partly because, when an animal survives a fall, it tends not to repeat the experience.

There may be other reasons also, but of them I do not wish to speak yet. Besides mechanical movements, there are interrupted movements, as when a bird, on its way to eat your best peas, is frightened away by the boy whom you are employing for that purpose. If interruptions are frequent and completion of cycles rare, the characteristics by which cycles are observed may become so blurred as to be almost unrecognizable. The result of these various considerations is that the differences between animals and dead matter, when we confine ourselves to external unscientific observation of integral behaviour, are a matter of degree and not very precise. It is for this reason that it has always been possible for fanciful people to maintain that even stocks and stones have some vague kind of soul. The evidence that animals have souls is so very shaky that, if it is assumed to be conclusive, one might just as well go a step further and extend the argument by analogy to all matter. Nevertheless, in spite of vagueness and doubtful cases, the existence of cycles in the behaviour of animals is a broad characteristic by which they are prima facie distinguished from ordinary matter; and I think it is this characteristic which leads us to attribute desires to animals, since it makes their behaviour resemble what we do when (as we say) we are acting from desire.

I shall adopt the following definitions for describing the behaviour of animals:

A "behaviour-cycle" is a series of voluntary or reflex movements of an animal, tending to cause a certain result, and continuing until that result is caused, unless they are interrupted by death, accident, or some new behaviour-cycle. (Here "accident" may be defined as the intervention of purely physical laws causing mechanical movements.)

The "purpose" of a behaviour-cycle is the result which brings it to an end, normally by a condition of temporary quiescence-provided there is no interruption.

An animal is said to "desire" the purpose of a behaviour cycle while the behaviour-cycle is in progress.

I believe these definitions to be adequate also to human purposes and desires, but for the present I am only occupied with animals and with what can be learnt by external observation. I am very anxious that no ideas should be attached to the words "purpose" and "desire" beyond those involved in the above definitions.

We have not so far considered what is the nature of the initial stimulus to a behaviour-cycle. Yet it is here that the usual view of desire seems on the strongest ground. The hungry animal goes on making movements until it gets food; it seems natural, therefore, to suppose that the idea of food is present throughout the process, and that the thought of the end to be achieved sets the whole process in motion. Such a view, however, is obviously untenable in many cases, especially where instinct is concerned. Take, for example, reproduction and the rearing of the young. Birds mate, build a nest, lay eggs in it, sit on the eggs, feed the young birds, and care for them until they are fully grown. It is totally impossible to suppose that this series of actions, which constitutes one behaviour-cycle, is inspired by any prevision of the end, at any rate the first time it is performed. We must suppose that the stimulus to the performance of each act is an impulsion from behind, not an attraction from the future. The bird does what it does, at each stage, because it has an impulse to that particular action, not because it perceives that the whole cycle of actions will contribute to the preservation of the species. The same considerations apply to other instincts. A hungry animal feels restless, and is led by instinctive impulses to perform the movements which give it nourishment; but the act of seeking food is not sufficient evidence from which to conclude that the animal has the thought of food in its "mind."

Coming now to human beings, and to what we know about our own actions, it seems clear that what, with us, sets a behaviour-cycle in motion is some sensation of the sort which we call disagreeable. Take the case of hunger: we have first an uncomfortable feeling inside, producing a disinclination to sit still, a sensitiveness to savoury smells, and an attraction towards any food that there may be in our neighbourhood. At any moment during this process we may become aware that we are hungry, in the sense of saying to ourselves, "I am hungry"; but we may have been acting with reference to food for some time before this moment. While we are talking or reading, we may eat in complete unconsciousness; but we perform the actions of eating just as we should if we were conscious, and they cease when our hunger is appeased. What we call "consciousness" seems to be a mere spectator of the process; even when it issues orders, they are usually, like those of a wise parent, just such as would have been obeyed even if they had not been given. This view may seem at first exaggerated, but the more our so-called volitions and their causes are examined, the more it is forced upon us. The part played by words in all this is complicated, and a potent source of confusions; I shall return to it later. For the present, I am still concerned with primitive desire, as it exists in man, but in the form in which man shows his affinity to his animal ancestors.

Conscious desire is made up partly of what is essential to desire, partly of beliefs as to what we want. It is important to be clear as to the part which does not consist of beliefs.

The primitive non-cognitive element in desire seems to be a push, not a pull, an impulsion away from the actual, rather than an attraction towards the ideal. Certain sensations and other mental occurrences have a property which we call discomfort; these cause such bodily movements as are likely to lead to their cessation. When the discomfort ceases, or even when it appreciably diminishes, we have sensations possessing a property which we call PLEASURE. Pleasurable sensations either stimulate no action at all, or at most stimulate such action as is likely to prolong them. I shall return shortly to the consideration of what discomfort and pleasure are in themselves; for the present, it is their connection with action and desire that concerns us. Abandoning momentarily the standpoint of behaviourism, we may presume that hungry animals experience sensations involving discomfort, and stimulating such movements as seem likely to bring them to the food which is outside the cages. When they have reached the food and eaten it, their discomfort ceases and their sensations become pleasurable. It SEEMS, mistakenly, as if the animals had had this situation in mind throughout, when in fact they have been continually pushed by discomfort. And when an animal is reflective, like some men, it comes to think that it had the final situation in mind throughout; sometimes it comes to know what situation will bring satisfaction, so that in fact the discomfort does bring the thought of what will allay it. Nevertheless the sensation involving discomfort remains the prime mover.

This brings us to the question of the nature of discomfort and pleasure. Since Kant it has been customary to recognize three great divisions of mental phenomena, which are typified by knowledge, desire and feeling, where "feeling" is used to mean pleasure and discomfort. Of course, "knowledge" is too definite a word: the states of mind concerned are grouped together as "cognitive," and are to embrace not only beliefs, but perceptions, doubts, and the understanding of concepts. "Desire," also, is narrower than what is intended: for example, WILL is to be included in this category, and in fact every thing that involves any kind of striving, or "conation" as it is technically called. I do not myself believe that there is any value in this threefold division of the contents of mind. I believe that sensations (including images) supply all the "stuff" of the mind, and that everything else can be analysed into groups of sensations related in various ways, or characteristics of sensations or of groups of sensations. As regards belief, I shall give grounds for this view in later lectures. As regards desires, I have given some grounds in this lecture. For the present, it is pleasure and discomfort that concern us. There are broadly three theories that might be held in regard to them. We may regard them as separate existing items in those who experience them, or we may regard them as intrinsic qualities of sensations and other mental occurrences, or we may regard them as mere names for the causal characteristics of the occurrences which are uncomfortable or pleasant. The first of these theories, namely, that which regards discomfort and pleasure as actual contents in those who experience them, has, I think, nothing conclusive to be said in its favour. It is suggested chiefly by an ambiguity in the word "pain," which has misled many people, including Berkeley, whom it supplied with one of his arguments for subjective idealism. We may use "pain" as the opposite of "pleasure," and "painful" as the opposite of "pleasant," or we may use "pain" to mean a certain sort of sensation, on a level with the sensations of heat and cold and touch. The latter use of the word has prevailed in psychological literature, and it is now no longer used as the opposite of "pleasure." Dr. H. Head, in a recent publication, has stated this distinction as follows:

"It is necessary at the outset to distinguish clearly between 'discomfort' and 'pain.' Pain is a distinct sensory quality equivalent to heat and cold, and its intensity can be roughly graded according to the force expended in stimulation. Discomfort, on the other hand, is that feeling-tone which is directly opposed to pleasure. It may accompany sensations not in themselves essentially painful; as for instance that produced by tickling the sole of the foot. The reaction produced by repeated pricking contains both these elements; for it evokes that sensory quality known as pain, accompanied by a disagreeable feeling-tone, which we have called discomfort. On the other hand, excessive pressure, except when applied directly over some nerve-trunk, tends to excite more discomfort than pain."

The confusion between discomfort and pain has made people regard discomfort as a more substantial thing than it is, and this in turn has reacted upon the view taken of pleasure, since discomfort and pleasure are evidently on a level in this respect. As soon as discomfort is clearly distinguished from the sensation of pain, it becomes more natural to regard discomfort and pleasure as properties of mental occurrences than to regard them as separate mental occurrences on their own account. I shall therefore dismiss the view that they are separate mental occurrences, and regard them as properties of such experiences as would be called respectively uncomfortable and pleasant.

It remains to be examined whether they are actual qualities of such occurrences, or are merely differences as to causal properties. I do not myself see any way of deciding this question; either view seems equally capable of accounting for the facts. If this is true, it is safer to avoid the assumption that there are such intrinsic qualities of mental occurrences as are in question, and to assume only the causal differences which are undeniable. Without condemning the intrinsic theory, we can define discomfort and pleasure as consisting in causal properties, and say only what will hold on either of the two theories. Following this course, we shall say:

"Discomfort" is a property of a sensation or other mental occurrence, consisting in the fact that the occurrence in question stimulates voluntary or reflex movements tending to produce some more or less definite change involving the cessation of the occurrence.

"Pleasure" is a property of a sensation or other mental occurrence, consisting in the fact that the occurrence in question either does not stimulate any voluntary or reflex movement, or, if it does, stimulates only such as tend to prolong the occurrence in question.

Conscious" desire, which we have now to consider, consists of desire in the sense hitherto discussed, together with a true belief as to its "purpose," i.e. as to the state of affairs that will bring quiescence with cessation of the discomfort. If our theory of desire is correct, a belief as to its purpose may very well be erroneous, since only experience can show what causes a discomfort to cease. When the experience needed is common and simple, as in the case of hunger, a mistake is not very probable. But in other cases—e.g. erotic desire in those who have had little or no experience of its satisfaction—mistakes are to be expected, and do in fact very often occur. The practice of inhibiting impulses, which is to a great extent necessary to civilized life, makes mistakes easier, by preventing experience of the actions to which a desire would otherwise lead, and by often causing the inhibited impulses themselves to be unnoticed or quickly forgotten. The perfectly natural mistakes which thus arise constitute a large proportion of what is, mistakenly in part, called self-deception, and attributed by Freud to the "censor."

But there is a further point which needs emphasizing, namely, that a belief that something is desired has often a tendency to cause the very desire that is believed in. It is this fact that makes the effect of "consciousness" on desire so complicated.

When we believe that we desire a certain state of affairs, that often tends to cause a real desire for it. This is due partly to the influence of words upon our emotions, in rhetoric for example, and partly to the general fact that discomfort normally belongs to the belief that we desire such-and-such a thing that we do not possess. Thus what was originally a false opinion as to the object of a desire acquires a certain truth: the false opinion generates a secondary subsidiary desire, which nevertheless becomes real. Let us take an illustration. Suppose you have been jilted in a way which wounds your vanity. Your natural impulsive desire will be of the sort expressed in Donne's poem:

When by thy scorn, O Murderess, I am dead,

in which he explains how he will haunt the poor lady as a ghost, and prevent her from enjoying a moment's peace. But two things stand in the way of your expressing yourself so naturally: on the one hand, your vanity, which will not acknowledge how hard you are hit; on the other hand, your conviction that you are a civilized and humane person, who could not possibly indulge so crude a desire as revenge. You will therefore experience a restlessness which will at first seem quite aimless, but will finally resolve itself in a conscious desire to change your profession, or go round the world, or conceal your identity and live in Putney, like Arnold Bennett's hero. Although the prime cause of this desire is a false judgment as to your previous unconscious desire, yet the new conscious desire has its own derivative genuineness, and may influence your actions to the extent of sending you round the world. The initial mistake, however, will have effects of two kinds. First, in uncontrolled moments, under the influence of sleepiness or drink or delirium, you will say things calculated to injure the faithless deceiver. Secondly, you will find travel disappointing, and the East less fascinating than you had hoped—unless, some day, you hear that the wicked one has in turn been jilted. If this happens, you will believe that you feel sincere sympathy, but you will suddenly be much more delighted than before with the beauties of tropical islands or the wonders of Chinese art. A secondary desire, derived from a false judgment as to a primary desire, has its own power of influencing action, and is therefore a real desire according to our definition. But it has not the same power as a primary desire of bringing thorough satisfaction when it is realized; so long as the primary desire remains unsatisfied, restlessness continues in spite of the secondary desire's success. Hence arises a belief in the vanity of human wishes: the vain wishes are those that are secondary, but mistaken beliefs prevent us from realizing that they are secondary.

What may, with some propriety, be called self-deception arises through the operation of desires for beliefs. We desire many things which it is not in our power to achieve: that we should be universally popular and admired, that our work should be the wonder of the age, and that the universe should be so ordered as to bring ultimate happiness to all, though not to our enemies until they have repented and been purified by suffering. Such desires are too large to be achieved through our own efforts. But it is found that a considerable portion of the satisfaction which these things would bring us if they were realized is to be achieved by the much easier operation of believing that they are or will be realized. This desire for beliefs, as opposed to desire for the actual facts, is a particular case of secondary desire, and, like all secondary desire its satisfaction does not lead to a complete cessation of the initial discomfort. Nevertheless, desire for beliefs, as opposed to desire for facts, is exceedingly potent both individually and socially. According to the form of belief desired, it is called vanity, optimism, or religion. Those who have sufficient power usually imprison or put to death any one who tries to shake their faith in their own excellence or in that of the universe; it is for this reason that seditious libel and blasphemy have always been, and still are, criminal offences.

It is very largely through desires for beliefs that the primitive nature of desire has become so hidden, and that the part played by consciousness has been so confusing and so exaggerated.

We may now summarize our analysis of desire and feeling.

A mental occurrence of any kind—sensation, image, belief, or emotion—may be a cause of a series of actions, continuing, unless interrupted, until some more or less definite state of affairs is realized. Such a series of actions we call a "behaviour-cycle." The degree of definiteness may vary greatly: hunger requires only food in general, whereas the sight of a particular piece of food raises a desire which requires the eating of that piece of food. The property of causing such a cycle of occurrences is called "discomfort"; the property of the mental occurrences in which the cycle ends is called "pleasure." The actions constituting the cycle must not be purely mechanical, i.e. they must be bodily movements in whose causation the special properties of nervous tissue are involved. The cycle ends in a condition of quiescence, or of such action as tends only to preserve the status quo. The state of affairs in which this condition of quiescence is achieved is called the "purpose" of the cycle, and the initial mental occurrence involving discomfort is called a "desire" for the state of affairs that brings quiescence. A desire is called "conscious" when it is accompanied by a true belief as to the state of affairs that will bring quiescence; otherwise it is called "unconscious." All primitive desire is unconscious, and in human beings beliefs as to the purposes of desires are often mistaken. These mistaken beliefs generate secondary desires, which cause various interesting complications in the psychology of human desire, without fundamentally altering the character which it shares with animal desire.

LECTURE IV.

INFLUENCE OF PAST HISTORY ON PRESENT OCCURRENCES IN LIVING

ORGANISMS

In this lecture we shall be concerned with a very general characteristic which broadly, though not absolutely, distinguishes the behaviour of living organisms from that of dead matter. The characteristic in question is this:

The response of an organism to a given stimulus is very often dependent upon the past history of the organism, and not merely upon the stimulus and the HITHERTO DISCOVERABLE present state of the organism.

This characteristic is embodied in the saying "a burnt child fears the fire." The burn may have left no visible traces, yet it modifies the reaction of the child in the presence of fire. It is customary to assume that, in such cases, the past operates by modifying the structure of the brain, not directly. I have no wish to suggest that this hypothesis is false; I wish only to point out that it is a hypothesis. At the end of the present lecture I shall examine the grounds in its favour. If we confine ourselves to facts which have been actually observed, we must say that past occurrences, in addition to the present stimulus and the present ascertainable condition of the organism, enter into the causation of the response.

The characteristic is not wholly confined to living organisms. For example, magnetized steel looks just like steel which has not been magnetized, but its behaviour is in some ways different. In the case of dead matter, however, such phenomena are less frequent and important than in the case of living organisms, and it is far less difficult to invent satisfactory hypotheses as to the microscopic changes of structure which mediate between the past occurrence and the present changed response. In the case of living organisms, practically everything that is distinctive both of their physical and of their mental behaviour is bound up with this persistent influence of the past. Further, speaking broadly, the change in response is usually of a kind that is biologically advantageous to the organism.

Following a suggestion derived from Semon ("Die Mneme," Leipzig, 1904; 2nd edition, 1908, English translation, Allen & Unwin, 1921; "Die mnemischen Empfindungen," Leipzig, 1909), we will give the name of "mnemic phenomena" to those responses of an organism which, so far as hitherto observed facts are concerned, can only be brought under causal laws by including past occurrences in the history of the organism as part of the causes of the present response. I do not mean merely—what would always be the case—that past occurrences are part of a CHAIN of causes leading to the present event. I mean that, in attempting to state the PROXIMATE cause of the present event, some past event or events must be included, unless we take refuge in hypothetical modifications of brain structure. For example: you smell peat-smoke, and you recall some occasion when you smelt it before. The cause of your recollection, so far as hitherto observable phenomena are concerned, consists both of the peat smoke (present stimulus) and of the former occasion (past experience). The same stimulus will not produce the same recollection in another man who did not share your former experience, although the former experience left no OBSERVABLE traces in the structure of the brain. According to the maxim "same cause, same effect," we cannot therefore regard the peat-smoke alone as the cause of your recollection, since it does not have the same effect in other cases. The cause of your recollection must be both the peat-smoke and the past occurrence. Accordingly your recollection is an instance of what we are calling "mnemic phenomena."

Before going further, it will be well to give illustrations of different classes of mnemic phenomena.

(a) ACQUIRED HABITS.—In Lecture II we saw how animals can learn by experience how to get out of cages or mazes, or perform other actions which are useful to them but not provided for by their instincts alone. A cat which is put into a cage of which it has had experience behaves differently from the way in which it behaved at first. We can easily invent hypotheses, which are quite likely to be true, as to connections in the brain caused by past experience, and themselves causing the different response. But the observable fact is that the stimulus of being in the cage produces differing results with repetition, and that the ascertainable cause of the cat's behaviour is not merely the cage and its own ascertainable organization, but also its past history in regard to the cage. From our present point of view, the matter is independent of the question whether the cat's behaviour is due to some mental fact called "knowledge," or displays a merely bodily habit. Our habitual knowledge is not always in our minds, but is called up by the appropriate stimuli. If we are asked "What is the capital of France?" we answer "Paris," because of past experience; the past experience is as essential as the present question in the causation of our response. Thus all our habitual knowledge consists of acquired habits, and comes under the head of mnemic phenomena.

(b) IMAGES.—I shall have much to say about images in a later lecture; for the present I am merely concerned with them in so far as they are "copies" of past sensations. When you hear New York spoken of, some image probably comes into your mind, either of the place itself (if you have been there), or of some picture of it (if you have not). The image is due to your past experience, as well as to the present stimulus of the words "New York." Similarly, the images you have in dreams are all dependent upon your past experience, as well as upon the present stimulus to dreaming. It is generally believed that all images, in their simpler parts, are copies of sensations; if so, their mnemic character is evident. This is important, not only on its own account, but also because, as we shall see later, images play an essential part in what is called "thinking."

(c) ASSOCIATION.—The broad fact of association, on the mental side, is that when we experience something which we have experienced before, it tends to call up the context of the former experience. The smell of peat-smoke recalling a former scene is an instance which we discussed a moment ago. This is obviously a mnemic phenomenon. There is also a more purely physical association, which is indistinguishable from physical habit. This is the kind studied by Mr. Thorndike in animals, where a certain stimulus is associated with a certain act. This is the sort which is taught to soldiers in drilling, for example. In such a case there need not be anything mental, but merely a habit of the body. There is no essential distinction between association and habit, and the observations which we made concerning habit as a mnemic phenomenon are equally applicable to association.

(d) NON-SENSATIONAL ELEMENTS IN PERCEPTION.—When we perceive any object of a familiar kind, much of what appears subjectively to be immediately given is really derived from past experience. When we see an object, say a penny, we seem to be aware of its "real" shape we have the impression of something circular, not of something elliptical. In learning to draw, it is necessary to acquire the art of representing things according to the sensation, not according to the perception. And the visual appearance is filled out with feeling of what the object would be like to touch, and so on. This filling out and supplying of the "real" shape and so on consists of the most usual correlates of the sensational core in our perception. It may happen that, in the particular case, the real correlates are unusual; for example, if what we are seeing is a carpet made to look like tiles. If so, the non-sensational part of our perception will be illusory, i.e. it will supply qualities which the object in question does not in fact have. But as a rule objects do have the qualities added by perception, which is to be expected, since experience of what is usual is the cause of the addition. If our experience had been different, we should not fill out sensation in the same way, except in so far as the filling out is instinctive, not acquired. It would seem that, in man, all that makes up space perception, including the correlation of sight and touch and so on, is almost entirely acquired. In that case there is a large mnemic element in all the common perceptions by means of which we handle common objects. And, to take another kind of instance, imagine what our astonishment would be if we were to hear a cat bark or a dog mew. This emotion would be dependent upon past experience, and would therefore be a mnemic phenomenon according to the definition.

(e) MEMORY AS KNOWLEDGE.—The kind of memory of which I am now speaking is definite knowledge of some past event in one's own experience. From time to time we remember things that have happened to us, because something in the present reminds us of them. Exactly the same present fact would not call up the same memory if our past experience had been different. Thus our remembering is caused by—

(1) The present stimulus,

(2) The past occurrence.

It is therefore a mnemic phenomenon according to our definition. A definition of "mnemic phenomena" which did not include memory would, of course, be a bad one. The point of the definition is not that it includes memory, but that it includes it as one of a class of phenomena which embrace all that is characteristic in the subject matter of psychology.

(f) EXPERIENCE.—The word "experience" is often used very vaguely. James, as we saw, uses it to cover the whole primal stuff of the world, but this usage seems objection able, since, in a purely physical world, things would happen without there being any experience. It is only mnemic phenomena that embody experience. We may say that an animal "experiences" an occurrence when this occurrence modifies the animal's subsequent behaviour, i.e. when it is the mnemic portion of the cause of future occurrences in the animal's life. The burnt child that fears the fire has "experienced" the fire, whereas a stick that has been thrown on and taken off again has not "experienced" anything, since it offers no more resistance than before to being thrown on. The essence of "experience" is the modification of behaviour produced by what is experienced. We might, in fact, define one chain of experience, or one biography, as a series of occurrences linked by mnemic causation. I think it is this characteristic, more than any other, that distinguishes sciences dealing with living organisms from physics.

The best writer on mnemic phenomena known to me is Richard Semon, the fundamental part of whose theory I shall endeavour to summarize before going further:

When an organism, either animal or plant, is subjected to a stimulus, producing in it some state of excitement, the removal of the stimulus allows it to return to a condition of equilibrium. But the new state of equilibrium is different from the old, as may be seen by the changed capacity for reaction. The state of equilibrium before the stimulus may be called the "primary indifference-state"; that after the cessation of the stimulus, the "secondary indifference-state." We define the "engraphic effect" of a stimulus as the effect in making a difference between the primary and secondary indifference-states, and this difference itself we define as the "engram" due to the stimulus. "Mnemic phenomena" are defined as those due to engrams; in animals, they are specially associated with the nervous system, but not exclusively, even in man.

When two stimuli occur together, one of them, occurring afterwards, may call out the reaction for the other also. We call this an "ekphoric influence," and stimuli having this character are called "ekphoric stimuli." In such a case we call the engrams of the two stimuli "associated." All simultaneously generated engrams are associated; there is also association of successively aroused engrams, though this is reducible to simultaneous association. In fact, it is not an isolated stimulus that leaves an engram, but the totality of the stimuli at any moment; consequently any portion of this totality tends, if it recurs, to arouse the whole reaction which was aroused before. Semon holds that engrams can be inherited, and that an animal's innate habits may be due to the experience of its ancestors; on this subject he refers to Samuel Butler.

Semon formulates two "mnemic principles." The first, or "Law of Engraphy," is as follows: "All simultaneous excitements in an organism form a connected simultaneous excitement-complex, which as such works engraphically, i.e. leaves behind a connected engram-complex, which in so far forms a whole" ("Die mnemischen Empfindungen,"). The second mnemic principle, or "Law of Ekphory," is as follows: "The partial return of the energetic situation which formerly worked engraphically operates ekphorically on a simultaneous engram-complex". These two laws together represent in part a hypothesis (the engram), and in part an observable fact. The observable fact is that, when a certain complex of stimuli has originally caused a certain complex of reactions, the recurrence of part of the stimuli tends to cause the recurrence of the whole of the reactions.

Semon's applications of his fundamental ideas in various directions are interesting and ingenious. Some of them will concern us later, but for the present it is the fundamental character of mnemic phenomena that is in question.

Concerning the nature of an engram, Semon confesses that at present it is impossible to say more than that it must consist in some material alteration in the body of the organism It is, in fact, hypothetical, invoked for theoretical uses, and not an outcome of direct observation. No doubt physiology, especially the disturbances of memory through lesions in the brain, affords grounds for this hypothesis; nevertheless it does remain a hypothesis, the validity of which will be discussed at the end of this lecture.

I am inclined to think that, in the present state of physiology, the introduction of the engram does not serve to simplify the account of mnemic phenomena. We can, I think, formulate the known laws of such phenomena in terms, wholly, of observable facts, by recognizing provisionally what we may call "mnemic causation." By this I mean that kind of causation of which I spoke at the beginning of this lecture, that kind, namely, in which the proximate cause consists not merely of a present event, but of this together with a past event. I do not wish to urge that this form of causation is ultimate, but that, in the present state of our knowledge, it affords a simplification, and enables us to state laws of behaviour in less hypothetical terms than we should otherwise have to employ.

The clearest instance of what I mean is recollection of a past event. What we observe is that certain present stimuli lead us to recollect certain occurrences, but that at times when we are not recollecting them, there is nothing discoverable in our minds that could be called memory of them. Memories, as mental facts, arise from time to time, but do not, so far as we can see, exist in any shape while they are "latent." In fact, when we say that they are "latent," we mean merely that they will exist under certain circumstances. If, then, there is to be some standing difference between the person who can remember a certain fact and the person who cannot, that standing difference must be, not in anything mental, but in the brain. It is quite probable that there is such a difference in the brain, but its nature is unknown and it remains hypothetical. Everything that has, so far, been made matter of observation as regards this question can be put together in the statement: When a certain complex of sensations has occurred to a man, the recurrence of part of the complex tends to arouse the recollection of the whole. In like manner, we can collect all mnemic phenomena in living organisms under a single law, which contains what is hitherto verifiable in Semon's two laws. This single law is:

IF A COMPLEX STIMULUS A HAS CAUSED A COMPLEX REACTION B IN AN ORGANISM, THE OCCURRENCE OF A PART OF A ON A FUTURE OCCASION TENDS TO CAUSE THE WHOLE REACTION B.

This law would need to be supplemented by some account of the influence of frequency, and so on; but it seems to contain the essential characteristic of mnemic phenomena, without admixture of anything hypothetical.

Whenever the effect resulting from a stimulus to an organism differs according to the past history of the organism, without our being able actually to detect any relevant difference in its present structure, we will speak of "mnemic causation," provided we can discover laws embodying the influence of the past. In ordinary physical causation, as it appears to common sense, we have approximate uniformities of sequence, such as "lightning is followed by thunder," "drunkenness is followed by headache," and so on. None of these sequences are theoretically invariable, since something may intervene to disturb them. In order to obtain invariable physical laws, we have to proceed to differential equations, showing the direction of change at each moment, not the integral change after a finite interval, however short. But for the purposes of daily life many sequences are to all in tents and purposes invariable. With the behaviour of human beings, however, this is by no means the case. If you say to an Englishman, "You have a smut on your nose," he will proceed to remove it, but there will be no such effect if you say the same thing to a Frenchman who knows no English. The effect of words upon the hearer is a mnemic phenomena, since it depends upon the past experience which gave him understanding of the words. If there are to be purely psychological causal laws, taking no account of the brain and the rest of the body, they will have to be of the form, not "X now causes Y now," but—

"A, B, C,... in the past, together with X now, cause Y now." For it cannot be successfully maintained that our understanding of a word, for example, is an actual existent content of the mind at times when we are not thinking of the word. It is merely what may be called a "disposition," i.e. it is capable of being aroused whenever we hear the word or happen to think of it. A "disposition" is not something actual, but merely the mnemic portion of a mnemic causal law.

In such a law as "A, B, C,... in the past, together with X now, cause Y now," we will call A, B, C,... the mnemic cause, X the occasion or stimulus, and Y the reaction. All cases in which experience influences behaviour are instances of mnemic causation.

Believers in psycho-physical parallelism hold that psychology can theoretically be freed entirely from all dependence on physiology or physics. That is to say, they believe that every psychical event has a psychical cause and a physical concomitant. If there is to be parallelism, it is easy to prove by mathematical logic that the causation in physical and psychical matters must be of the same sort, and it is impossible that mnemic causation should exist in psychology but not in physics. But if psychology is to be independent of physiology, and if physiology can be reduced to physics, it would seem that mnemic causation is essential in psychology. Otherwise we shall be compelled to believe that all our knowledge, all our store of images and memories, all our mental habits, are at all times existing in some latent mental form, and are not merely aroused by the stimuli which lead to their display. This is a very difficult hypothesis. It seems to me that if, as a matter of method rather than metaphysics, we desire to obtain as much independence for psychology as is practically feasible, we shall do better to accept mnemic causation in psychology protem, and therefore reject parallelism, since there is no good ground for admitting mnemic causation in physics.

It is perhaps worth while to observe that mnemic causation is what led Bergson to deny that there is causation at all in the psychical sphere. He points out, very truly, that the same stimulus, repeated, does not have the same consequences, and he argues that this is contrary to the maxim, "same cause, same effect." It is only necessary, however, to take account of past occurrences and include them with the cause, in order to re-establish the maxim, and the possibility of psychological causal laws. The metaphysical conception of a cause lingers in our manner of viewing causal laws: we want to be able to FEEL a connection between cause and effect, and to be able to imagine the cause as "operating." This makes us unwilling to regard causal laws as MERELY observed uniformities of sequence; yet that is all that science has to offer. To ask why such-and-such a kind of sequence occurs is either to ask a meaningless question, or to demand some more general kind of sequence which includes the one in question. The widest empirical laws of sequence known at any time can only be "explained" in the sense of being subsumed by later discoveries under wider laws; but these wider laws, until they in turn are subsumed, will remain brute facts, resting solely upon observation, not upon some supposed inherent rationality.

There is therefore no a priori objection to a causal law in which part of the cause has ceased to exist. To argue against such a law on the ground that what is past cannot operate now, is to introduce the old metaphysical notion of cause, for which science can find no place. The only reason that could be validly alleged against mnemic causation would be that, in fact, all the phenomena can be explained without it. They are explained without it by Semon's "engram," or by any theory which regards the results of experience as embodied in modifications of the brain and nerves. But they are not explained, unless with extreme artificiality, by any theory which regards the latent effects of experience as psychical rather than physical. Those who desire to make psychology as far as possible independent of physiology would do well, it seems to me, if they adopted mnemic causation. For my part, however, I have no such desire, and I shall therefore endeavour to state the grounds which occur to me in favour of some such view as that of the "engram."

One of the first points to be urged is that mnemic phenomena are just as much to be found in physiology as in psychology. They are even to be found in plants, as Sir Francis Darwin pointed out. Habit is a characteristic of the body at least as much as of the mind. We should, therefore, be compelled to allow the intrusion of mnemic causation, if admitted at all, into non-psychological regions, which ought, one feels, to be subject only to causation of the ordinary physical sort. The fact is that a great deal of what, at first sight, distinguishes psychology from physics is found, on examination, to be common to psychology and physiology; this whole question of the influence of experience is a case in point. Now it is possible, of course, to take the view advocated by Professor J. S. Haldane, who contends that physiology is not theoretically reducible to physics and chemistry. But the weight of opinion among physiologists appears to be against him on this point; and we ought certainly to require very strong evidence before admitting any such breach of continuity as between living and dead matter. The argument from the existence of mnemic phenomena in physiology must therefore be allowed a certain weight against the hypothesis that mnemic causation is ultimate.

The argument from the connection of brain-lesions with loss of memory is not so strong as it looks, though it has also, some weight. What we know is that memory, and mnemic phenomena generally, can be disturbed or destroyed by changes in the brain. This certainly proves that the brain plays an essential part in the causation of memory, but does not prove that a certain state of the brain is, by itself, a sufficient condition for the existence of memory. Yet it is this last that has to be proved. The theory of the engram, or any similar theory, has to maintain that, given a body and brain in a suitable state, a man will have a certain memory, without the need of any further conditions. What is known, however, is only that he will not have memories if his body and brain are not in a suitable state. That is to say, the appropriate state of body and brain is proved to be necessary for memory, but not to be sufficient. So far, therefore, as our definite knowledge goes, memory may require for its causation a past occurrence as well as a certain present state of the brain.

In order to prove conclusively that mnemic phenomena arise whenever certain physiological conditions are fulfilled, we ought to be able actually to see differences between the brain of a man who speaks English and that of a man who speaks French, between the brain of a man who has seen New York and can recall it, and that of a man who has never seen that city. It may be that the time will come when this will be possible, but at present we are very far removed from it. At present, there is, so far as I am aware, no good evidence that every difference between the knowledge possessed by A and that possessed by B is paralleled by some difference in their brains. We may believe that this is the case, but if we do, our belief is based upon analogies and general scientific maxims, not upon any foundation of detailed observation. I am myself inclined, as a working hypothesis, to adopt the belief in question, and to hold that past experience only affects present behaviour through modifications of physiological structure. But the evidence seems not quite conclusive, so that I do not think we ought to forget the other hypothesis, or to reject entirely the possibility that mnemic causation may be the ultimate explanation of mnemic phenomena. I say this, not because I think it LIKELY that mnemic causation is ultimate, but merely because I think it POSSIBLE, and because it often turns out important to the progress of science to remember hypotheses which have previously seemed improbable.

LECTURE V.

PSYCHOLOGICAL AND PHYSICAL CAUSAL LAWS

The traditional conception of cause and effect is one which modern science shows to be fundamentally erroneous, and requiring to be replaced by a quite different notion, that of LAWS OF CHANGE. In the traditional conception, a particular event A caused a particular event B, and by this it was implied that, given any event B, some earlier event A could be discovered which had a relation to it, such that—

(1) Whenever A occurred, it was followed by B;

(2) In this sequence, there was something "necessary," not a mere de facto occurrence of A first and then B.

The second point is illustrated by the old discussion as to whether it can be said that day causes night, on the ground that day is always followed by night. The orthodox answer was that day could not be called the cause of night, because it would not be followed by night if the earth's rotation were to cease, or rather to grow so slow that one complete rotation would take a year. A cause, it was held, must be such that under no conceivable circumstances could it fail to be followed by its effect.

As a matter of fact, such sequences as were sought by believers in the traditional form of causation have not so far been found in nature. Everything in nature is apparently in a state of continuous change, so that what we call one "event" turns out to be really a process. If this event is to cause another event, the two will have to be contiguous in time; for if there is any interval between them, something may happen during that interval to prevent the expected effect. Cause and effect, therefore, will have to be temporally contiguous processes. It is difficult to believe, at any rate where physical laws are concerned, that the earlier part of the process which is the cause can make any difference to the effect, so long as the later part of the process which is the cause remains unchanged. Suppose, for example, that a man dies of arsenic poisoning, we say that his taking arsenic was the cause of death. But clearly the process by which he acquired the arsenic is irrelevant: everything that happened before he swallowed it may be ignored, since it cannot alter the effect except in so far as it alters his condition at the moment of taking the dose. But we may go further: swallowing arsenic is not really the proximate cause of death, since a man might be shot through the head immediately after taking the dose, and then it would not be of arsenic that he would die. The arsenic produces certain physiological changes, which take a finite time before they end in death. The earlier parts of these changes can be ruled out in the same way as we can rule out the process by which the arsenic was acquired. Proceeding in this way, we can shorten the process which we are calling the cause more and more. Similarly we shall have to shorten the effect. It may happen that immediately after the man's death his body is blown to pieces by a bomb. We cannot say what will happen after the man's death, through merely knowing that he has died as the result of arsenic poisoning. Thus, if we are to take the cause as one event and the effect as another, both must be shortened indefinitely. The result is that we merely have, as the embodiment of our causal law, a certain direction of change at each moment. Hence we are brought to differential equations as embodying causal laws. A physical law does not say "A will be followed by B," but tells us what acceleration a particle will have under given circumstances, i.e. it tells us how the particle's motion is changing at each moment, not where the particle will be at some future moment.

Laws embodied in differential equations may possibly be exact, but cannot be known to be so. All that we can know empirically is approximate and liable to exceptions; the exact laws that are assumed in physics are known to be somewhere near the truth, but are not known to be true just as they stand. The laws that we actually know empirically have the form of the traditional causal laws, except that they are not to be regarded as universal or necessary. "Taking arsenic is followed by death" is a good empirical generalization; it may have exceptions, but they will be rare. As against the professedly exact laws of physics, such empirical generalizations have the advantage that they deal with observable phenomena. We cannot observe infinitesimals, whether in time or space; we do not even know whether time and space are infinitely divisible. Therefore rough empirical generalizations have a definite place in science, in spite of not being exact of universal. They are the data for more exact laws, and the grounds for believing that they are USUALLY true are stronger than the grounds for believing that the more exact laws are ALWAYS true.

Science starts, therefore, from generalizations of the form, "A is usually followed by B." This is the nearest approach that can be made to a causal law of the traditional sort. It may happen in any particular instance that A is ALWAYS followed by B, but we cannot know this, since we cannot foresee all the perfectly possible circumstances that might make the sequence fail, or know that none of them will actually occur. If, however, we know of a very large number of cases in which A is followed by B, and few or none in which the sequence fails, we shall in PRACTICE be justified in saying "A causes B," provided we do not attach to the notion of cause any of the metaphysical superstitions that have gathered about the word.

There is another point, besides lack of universality and necessity, which it is important to realize as regards causes in the above sense, and that is the lack of uniqueness. It is generally assumed that, given any event, there is some one phenomenon which is THE cause of the event in question. This seems to be a mere mistake. Cause, in the only sense in which it can be practically applied, means "nearly invariable antecedent." We cannot in practice obtain an antecedent which is QUITE invariable, for this would require us to take account of the whole universe, since something not taken account of may prevent the expected effect. We cannot distinguish, among nearly invariable antecedents, one as THE cause, and the others as merely its concomitants: the attempt to do this depends upon a notion of cause which is derived from will, and will (as we shall see later) is not at all the sort of thing that it is generally supposed to be, nor is there any reason to think that in the physical world there is anything even remotely analogous to what will is supposed to be. If we could find one antecedent, and only one, that was QUITE invariable, we could call that one THE cause without introducing any notion derived from mistaken ideas about will. But in fact we cannot find any antecedent that we know to be quite invariable, and we can find many that are nearly so. For example, men leave a factory for dinner when the hooter sounds at twelve o'clock. You may say the hooter is THE cause of their leaving. But innumerable other hooters in other factories, which also always sound at twelve o'clock, have just as good a right to be called the cause. Thus every event has many nearly invariable antecedents, and therefore many antecedents which may be called its cause.

The laws of traditional physics, in the form in which they deal with movements of matter or electricity, have an apparent simplicity which somewhat conceals the empirical character of what they assert. A piece of matter, as it is known empirically, is not a single existing thing, but a system of existing things. When several people simultaneously see the same table, they all see something different; therefore "the" table, which they are supposed all to see, must be either a hypothesis or a construction. "The" table is to be neutral as between different observers: it does not favour the aspect seen by one man at the expense of that seen by another. It was natural, though to my mind mistaken, to regard the "real" table as the common cause of all the appearances which the table presents (as we say) to different observers. But why should we suppose that there is some one common cause of all these appearances? As we have just seen, the notion of "cause" is not so reliable as to allow us to infer the existence of something that, by its very nature, can never be observed.

Instead of looking for an impartial source, we can secure neutrality by the equal representation of all parties. Instead of supposing that there is some unknown cause, the "real" table, behind the different sensations of those who are said to be looking at the table, we may take the whole set of these sensations (together possibly with certain other particulars) as actually BEING the table. That is to say, the table which is neutral as between different observers (actual and possible) is the set of all those particulars which would naturally be called "aspects" of the table from different points of view. (This is a first approximation, modified later.)

It may be said: If there is no single existent which is the source of all these "aspects," how are they collected together? The answer is simple: Just as they would be if there were such a single existent. The supposed "real" table underlying its appearances is, in any case, not itself perceived, but inferred, and the question whether such-and-such a particular is an "aspect" of this table is only to be settled by the connection of the particular in question with the one or more particulars by which the table is defined. That is to say, even if we assume a "real" table, the particulars which are its aspects have to be collected together by their relations to each other, not to it, since it is merely inferred from them. We have only, therefore, to notice how they are collected together, and we can then keep the collection without assuming any "real" table as distinct from the collection. When different people see what they call the same table, they see things which are not exactly the same, owing to difference of point of view, but which are sufficiently alike to be described in the same words, so long as no great accuracy or minuteness is sought. These closely similar particulars are collected together by their similarity primarily and, more correctly, by the fact that they are related to each other approximately according to the laws of perspective and of reflection and diffraction of light. I suggest, as a first approximation, that these particulars, together with such correlated others as are unperceived, jointly ARE the table; and that a similar definition applies to all physical objects.

In order to eliminate the reference to our perceptions, which introduces an irrelevant psychological suggestion, I will take a different illustration, namely, stellar photography. A photographic plate exposed on a clear night reproduces the appearance of the portion of the sky concerned, with more or fewer stars according to the power of the telescope that is being used. Each separate star which is photographed produces its separate effect on the plate, just as it would upon ourselves if we were looking at the sky. If we assume, as science normally does, the continuity of physical processes, we are forced to conclude that, at the place where the plate is, and at all places between it and a star which it photographs, SOMETHING is happening which is specially connected with that star. In the days when the aether was less in doubt, we should have said that what was happening was a certain kind of transverse vibration in the aether. But it is not necessary or desirable to be so explicit: all that we need say is that SOMETHING happens which is specially connected with the star in question. It must be something specially connected with that star, since that star produces its own special effect upon the plate. Whatever it is must be the end of a process which starts from the star and radiates outwards, partly on general grounds of continuity, partly to account for the fact that light is transmitted with a certain definite velocity. We thus arrive at the conclusion that, if a certain star is visible at a certain place, or could be photographed by a sufficiently sensitive plate at that place, something is happening there which is specially connected with that star. Therefore in every place at all times a vast multitude of things must be happening, namely, at least one for every physical object which can be seen or photographed from that place. We can classify such happenings on either of two principles:

(1) We can collect together all the happenings in one place, as is done by photography so far as light is concerned;

(2) We can collect together all the happenings, in different places, which are connected in the way that common sense regards as being due to their emanating from one object.

Thus, to return to the stars, we can collect together either—

(1) All the appearances of different stars in a given place, or,

(2) All the appearances of a given star in different places.

But when I speak of "appearances," I do so only for brevity: I do not mean anything that must "appear" to somebody, but only that happening, whatever it may be, which is connected, at the place in question, with a given physical object—according to the old orthodox theory, it would be a transverse vibration in the aether. Like the different appearances of the table to a number of simultaneous observers, the different particulars that belong to one physical object are to be collected together by continuity and inherent laws of correlation, not by their supposed causal connection with an unknown assumed existent called a piece of matter, which would be a mere unnecessary metaphysical thing in itself. A piece of matter, according to the definition that I propose, is, as a first approximation,the collection of all those correlated particulars which would normally be regarded as its appearances or effects in different places. Some further elaborations are desirable, but we can ignore them for the present. I shall return to them at the end of this lecture.

According to the view that I am suggesting, a physical object or piece of matter is the collection of all those correlated particulars which would be regarded by common sense as its effects or appearances in different places. On the other hand, all the happenings in a given place represent what common sense would regard as the appearances of a number of different objects as viewed from that place. All the happenings in one place may be regarded as the view of the world from that place. I shall call the view of the world from a given place a "perspective." A photograph represents a perspective. On the other hand, if photographs of the stars were taken in all points throughout space, and in all such photographs a certain star, say Sirius, were picked out whenever it appeared, all the different appearances of Sirius, taken together, would represent Sirius. For the understanding of the difference between psychology and physics it is vital to understand these two ways of classifying particulars, namely:

(1) According to the place where they occur;

(2) According to the system of correlated particulars in different places to which they belong, such system being defined as a physical object.

Given a system of particulars which is a physical object, I shall define that one of the system which is in a given place (if any) as the "appearance of that object in that place."

When the appearance of an object in a given place changes, it is found that one or other of two things occurs. The two possibilities may be illustrated by an example. You are in a room with a man, whom you see: you may cease to see him either by shutting your eyes or by his going out of the room. In the first case, his appearance to other people remains unchanged; in the second, his appearance changes from all places. In the first case, you say that it is not he who has changed, but your eyes; in the second, you say that he has changed. Generalizing, we distinguish—

(1) Cases in which only certain appearances of the object change, while others, and especially appearances from places very near to the object, do not change;

(2) Cases where all, or almost all, the appearances of the object undergo a connected change.

In the first case, the change is attributed to the medium between the object and the place; in the second, it is attributed to the object itself.

It is the frequency of the latter kind of change, and the comparatively simple nature of the laws governing the simultaneous alterations of appearances in such cases, that have made it possible to treat a physical object as one thing, and to overlook the fact that it is a system of particulars. When a number of people at a theatre watch an actor, the changes in their several perspectives are so similar and so closely correlated that all are popularly regarded as identical with each other and with the changes of the actor himself. So long as all the changes in the appearances of a body are thus correlated there is no pressing prima facie need to break up the system of appearances, or to realize that the body in question is not really one thing but a set of correlated particulars. It is especially and primarily such changes that physics deals with, i.e. it deals primarily with processes in which the unity of a physical object need not be broken up because all its appearances change simultaneously according to the same law—or, if not all, at any rate all from places sufficiently near to the object, with increasing accuracy as we approach the object.

The changes in appearances of an object which are due to changes in the intervening medium will not affect, or will affect only very slightly, the appearances from places close to the object. If the appearances from sufficiently neighbouring places are either wholly unchanged, or changed to a diminishing extent which has zero for its limit, it is usually found that the changes can be accounted for by changes in objects which are between the object in question and the places from which its appearance has changed appreciably. Thus physics is able to reduce the laws of most changes with which it deals to changes in physical objects, and to state most of its fundamental laws in terms of matter. It is only in those cases in which the unity of the system of appearances constituting a piece of matter has to be broken up, that the statement of what is happening cannot be made exclusively in terms of matter. The whole of psychology, we shall find, is included among such cases; hence their importance for our purposes.

We can now begin to understand one of the fundamental differences between physics and psychology. Physics treats as a unit the whole system of appearances of a piece of matter, whereas psychology is interested in certain of these appearances themselves. Confining ourselves for the moment to the psychology of perceptions, we observe that perceptions are certain of the appearances of physical objects. From the point of view that we have been hitherto adopting, we might define them as the appearances of objects at places from which sense-organs and the suitable parts of the nervous system form part of the intervening medium. Just as a photographic plate receives a different impression of a cluster of stars when a telescope is part of the intervening medium, so a brain receives a different impression when an eye and an optic nerve are part of the intervening medium. An impression due to this sort of intervening medium is called a perception, and is interesting to psychology on its own account, not merely as one of the set of correlated particulars which is the physical object of which (as we say) we are having a perception.

We spoke earlier of two ways of classifying particulars. One way collects together the appearances commonly regarded as a given object from different places; this is, broadly speaking, the way of physics, leading to the construction of physical objects as sets of such appearances. The other way collects together the appearances of different objects from a given place, the result being what we call a perspective. In the particular case where the place concerned is a human brain, the perspective belonging to the place consists of all the perceptions of a certain man at a given time. Thus classification by perspectives is relevant to psychology, and is essential in defining what we mean by one mind.

I do not wish to suggest that the way in which I have been defining perceptions is the only possible way, or even the best way. It is the way that arose naturally out of our present topic. But when we approach psychology from a more introspective standpoint, we have to distinguish sensations and perceptions, if possible, from other mental occurrences, if any. We have also to consider the psychological effects of sensations, as opposed to their physical causes and correlates. These problems are quite distinct from those with which we have been concerned in the present lecture, and I shall not deal with them until a later stage.

It is clear that psychology is concerned essentially with actual particulars, not merely with systems of particulars. In this it differs from physics, which, broadly speaking, is concerned with the cases in which all the particulars which make up one physical object can be treated as a single causal unit, or rather the particulars which are sufficiently near to the object of which they are appearances can be so treated. The laws which physics seeks can, broadly speaking, be stated by treating such systems of particulars as causal units. The laws which psychology seeks cannot be so stated, since the particulars themselves are what interests the psychologist. This is one of the fundamental differences between physics and psychology; and to make it clear has been the main purpose of this lecture.

I will conclude with an attempt to give a more precise definition of a piece of matter. The appearances of a piece of matter from different places change partly according to intrinsic laws (the laws of perspective, in the case of visual shape), partly according to the nature of the intervening medium—fog, blue spectacles, telescopes, microscopes, sense-organs, etc. As we approach nearer to the object, the effect of the intervening medium grows less. In a generalized sense, all the intrinsic laws of change of appearance may be called "laws of perspective." Given any appearance of an object, we can construct hypothetically a certain system of appearances to which the appearance in question would belong if the laws of perspective alone were concerned. If we construct this hypothetical system for each appearance of the object in turn, the system corresponding to a given appearance x will be independent of any distortion due to the medium beyond x, and will only embody such distortion as is due to the medium between x and the object. Thus, as the appearance by which our hypothetical system is defined is moved nearer and nearer to the object, the hypothetical system of appearances defined by its means embodies less and less of the effect of the medium. The different sets of appearances resulting from moving x nearer and nearer to the object will approach to a limiting set, and this limiting set will be that system of appearances which the object would present if the laws of perspective alone were operative and the medium exercised no distorting effect. This limiting set of appearances may be defined, for purposes of physics, as the piece of matter concerned.

LECTURE VI.

INTROSPECTION

One of the main purposes of these lectures is to give grounds for the belief that the distinction between mind and matter is not so fundamental as is commonly supposed. In the preceding lecture I dealt in outline with the physical side of this problem. I attempted to show that what we call a material object is not itself a substance, but is a system of particulars analogous in their nature to sensations, and in fact often including actual sensations among their number. In this way the stuff of which physical objects are composed is brought into relation with the stuff of which part, at least, of our mental life is composed.

There is, however, a converse task which is equally necessary for our thesis, and that is, to show that the stuff of our mental life is devoid of many qualities which it is commonly supposed to have, and is not possessed of any attributes which make it incapable of forming part of the world of matter. In the present lecture I shall begin the arguments for this view.

Corresponding to the supposed duality of matter and mind, there are, in orthodox psychology, two ways of knowing what exists. One of these, the way of sensation and external perception, is supposed to furnish data for our knowledge of matter, the other, called "introspection," is supposed to furnish data for knowledge of our mental processes. To common sense, this distinction seems clear and easy. When you see a friend coming along the street, you acquire knowledge of an external, physical fact; when you realize that you are glad to meet him, you acquire knowledge of a mental fact. Your dreams and memories and thoughts, of which you are often conscious, are mental facts, and the process by which you become aware of them SEEMS to be different from sensation. Kant calls it the "inner sense"; sometimes it is spoken of as "consciousness of self"; but its commonest name in modern English psychology is "introspection." It is this supposed method of acquiring knowledge of our mental processes that I wish to analyse and examine in this lecture.

I will state at the outset the view which I shall aim at establishing. I believe that the stuff of our mental life, as opposed to its relations and structure, consists wholly of sensations and images. Sensations are connected with matter in the way that I tried to explain in Lecture V, i.e. each is a member of a system which is a certain physical object. Images, though they USUALLY have certain characteristics, especially lack of vividness, that distinguish them from sensations, are not INVARIABLY so distinguished, and cannot therefore be defined by these characteristics. Images, as opposed to sensations, can only be defined by their different causation: they are caused by association with a sensation, not by a stimulus external to the nervous system—or perhaps one should say external to the brain, where the higher animals are concerned. The occurrence of a sensation or image does not in itself constitute knowledge but any sensation or image may come to be known if the conditions are suitable. When a sensation—like the hearing of a clap of thunder—is normally correlated with closely similar sensations in our neighbours, we regard it as giving knowledge of the external world, since we regard the whole set of similar sensations as due to a common external cause. But images and bodily sensations are not so correlated. Bodily sensations can be brought into a correlation by physiology, and thus take their place ultimately among sources of knowledge of the physical world. But images cannot be made to fit in with the simultaneous sensations and images of others. Apart from their hypothetical causes in the brain, they have a causal connection with physical objects, through the fact that they are copies of past sensations; but the physical objects with which they are thus connected are in the past, not in the present. These images remain private in a sense in which sensations are not. A sensation SEEMS to give us knowledge of a present physical object, while an image does not, except when it amounts to a hallucination, and in this case the seeming is deceptive. Thus the whole context of the two occurrences is different. But in themselves they do not differ profoundly, and there is no reason to invoke two different ways of knowing for the one and for the other. Consequently introspection as a separate kind of knowledge disappears.

The criticism of introspection has been in the main the work of American psychologists. I will begin by summarizing an article which seems to me to afford a good specimen of their arguments, namely, "The Case against Introspection," by Knight Dunlap After a few historical quotations, he comes to two modern defenders of introspection, Stout and James. He quotes from Stout such statements as the following: "Psychical states as such become objects only when we attend to them in an introspective way. Otherwise they are not themselves objects, but only constituents of the process by which objects are recognized" ("Manual," 2nd edition. The word "recognized" in Dunlap's quotation should be "cognized.") "The object itself can never be identified with the present modification of the individual's consciousness by which it is cognized" his is to be true even when we are thinking about modifications of our own consciousness; such modifications are to be always at least partially distinct from the conscious experience in which we think of them.

At this point I wish to interrupt the account of Knight Dunlap's article in order to make some observations on my own account with reference to the above quotations from Stout. In the first place, the conception of "psychical states" seems to me one which demands analysis of a somewhat destructive character. This analysis I shall give in later lectures as regards cognition; I have already given it as regards desire. In the second place, the conception of "objects" depends upon a certain view as to cognition which I believe to be wholly mistaken, namely, the view which I discussed in my first lecture in connection with Brentano. In this view a single cognitive occurrence contains both content and object, the content being essentially mental, while the object is physical except in introspection and abstract thought. I have already criticized this view, and will not dwell upon it now, beyond saying that "the process by which objects are cognized" appears to be a very slippery phrase. When we "see a table," as common sense would say, the table as a physical object is not the "object" (in the psychological sense) of our perception. Our perception is made up of sensations, images and beliefs, but the supposed "object" is something inferential, externally related, not logically bound up with what is occurring in us. This question of the nature of the object also affects the view we take of self-consciousness. Obviously, a "conscious experience" is different from a physical object; therefore it is natural to assume that a thought or perception whose object is a conscious experience must be different from a thought or perception whose object is a physical object. But if the relation to the object is inferential and external, as I maintain, the difference between two thoughts may bear very little relation to the difference between their objects. And to speak of "the present modification of the individual's consciousness by which an object is cognized" is to suggest that the cognition of objects is a far more direct process, far more intimately bound up with the objects, than I believe it to be. All these points will be amplified when we come to the analysis of knowledge, but it is necessary briefly to state them now in order to suggest the atmosphere in which our analysis of "introspection" is to be carried on.

Another point in which Stout's remarks seem to me to suggest what I regard as mistakes is his use of "consciousness." There is a view which is prevalent among psychologists, to the effect that one can speak of "a conscious experience" in a curious dual sense, meaning, on the one hand, an experience which is conscious of something, and, on the other hand, an experience which has some intrinsic nature characteristic of what is called "consciousness." That is to say, a "conscious experience" is characterized on the one hand by relation to its object and on the other hand by being composed of a certain peculiar stuff, the stuff of "consciousness." And in many authors there is yet a third confusion: a "conscious experience," in this third sense, is an experience of which we are conscious. All these, it seems to me, need to be clearly separated. To say that one occurrence is "conscious" of another is, to my mind, to assert an external and rather remote relation between them. I might illustrate it by the relation of uncle and nephew a man becomes an uncle through no effort of his own, merely through an occurrence elsewhere. Similarly, when you are said to be "conscious" of a table, the question whether this is really the case cannot be decided by examining only your state of mind: it is necessary also to ascertain whether your sensation is having those correlates which past experience causes you to assume, or whether the table happens, in this case, to be a mirage. And, as I explained in my first lecture, I do not believe that there is any "stuff" of consciousness, so that there is no intrinsic character by which a "conscious" experience could be distinguished from any other.

After these preliminaries, we can return to Knight Dunlap's article. His criticism of Stout turns on the difficulty of giving any empirical meaning to such notions as the "mind" or the "subject"; he quotes from Stout the sentence: "The most important drawback is that the mind, in watching its own workings, must necessarily have its attention divided between two objects," and he concludes: "Without question, Stout is bringing in here illicitly the concept of a single observer, and his introspection does not provide for the observation of this observer; for the process observed and the observer are distinct" The objections to any theory which brings in the single observer were considered in Lecture I, and were acknowledged to be cogent. In so far, therefore, as Stout's theory of introspection rests upon this assumption, we are compelled to reject it. But it is perfectly possible to believe in introspection without supposing that there is a single observer.

William James's theory of introspection, which Dunlap next examines, does not assume a single observer. It changed after the publication of his "Psychology," in consequence of his abandoning the dualism of thought and things. Dunlap summarizes his theory as follows:

"The essential points in James's scheme of consciousness are SUBJECT, OBJECT, and a KNOWING of the object by the subject. The difference between James's scheme and other schemes involving the same terms is that James considers subject and object to be the same thing, but at different times In order to satisfy this requirement James supposes a realm of existence which he at first called 'states of consciousness' or 'thoughts,' and later, 'pure experience,' the latter term including both the 'thoughts' and the 'knowing.' This scheme, with all its magnificent artificiality, James held on to until the end, simply dropping the term consciousness and the dualism between the thought and an external reality".

He adds: "All that James's system really amounts to is the acknowledgment that a succession of things are known, and that they are known by something. This is all any one can claim, except for the fact that the things are known together, and that the knower for the different items is one and the same" .

In this statement, to my mind, Dunlap concedes far more than James did in his later theory. I see no reason to suppose that "the knower for different items is one and the same," and I am convinced that this proposition could not possibly be ascertained except by introspection of the sort that Dunlap rejects. The first of these points must wait until we come to the analysis of belief: the second must be considered now. Dunlap's view is that there is a dualism of subject and object, but that the subject can never become object, and therefore there is no awareness of an awareness. He says in discussing the view that introspection reveals the occurrence of knowledge: "There can be no denial of the existence of the thing (knowing) which is alleged to be known or observed in this sort of 'introspection.' The allegation that the knowing is observed is that which may be denied. Knowing there certainly is; known, the knowing certainly is not". And again: "I am never aware of an awareness" (ib.). And on the next page: "It may sound paradoxical to say that one cannot observe the process (or relation) of observation, and yet may be certain that there is such a process: but there is really no inconsistency in the saying. How do I know that there is awareness? By being aware of something. There is no meaning in the term 'awareness' which is not expressed in the statement 'I am aware of a colour (or what-not).'"

But the paradox cannot be so lightly disposed of. The statement "I am aware of a colour" is assumed by Knight Dunlap to be known to be true, but he does not explain how it comes to be known. The argument against him is not conclusive, since he may be able to show some valid way of inferring our awareness. But he does not suggest any such way. There is nothing odd in the hypothesis of beings which are aware of objects, but not of their own awareness; it is, indeed, highly probable that young children and the higher animals are such beings. But such beings cannot make the statement "I am aware of a colour," which WE can make. We have, therefore, some knowledge which they lack. It is necessary to Knight Dunlap's position to maintain that this additional knowledge is purely inferential, but he makes no attempt to show how the inference is possible. It may, of course, be possible, but I cannot see how. To my mind the fact (which he admits) that we know there is awareness, is ALL BUT decisive against his theory, and in favour of the view that we can be aware of an awareness.

Dunlap asserts (to return to James) that the real ground for James's original belief in introspection was his belief in two sorts of objects, namely, thoughts and things. He suggests that it was a mere inconsistency on James's part to adhere to introspection after abandoning the dualism of thoughts and things. I do not wholly agree with this view, but it is difficult to disentangle the difference as to introspection from the difference as to the nature of knowing. Dunlap suggests that what is called introspection really consists of awareness of "images," visceral sensations, and so on. This view, in essence, seems to me sound. But then I hold that knowing itself consists of such constituents suitably related, and that in being aware of them we are sometimes being aware of instances of knowing. For this reason, much as I agree with his view as to what are the objects of which there is awareness, I cannot wholly agree with his conclusion as to the impossibility of introspection.

The behaviourists have challenged introspection even more vigorously than Knight Dunlap, and have gone so far as to deny the existence of images. But I think that they have confused various things which are very commonly confused, and that it is necessary to make several distinctions before we can arrive at what is true and what false in the criticism of introspection.

I wish to distinguish three distinct questions, any one of which may be meant when we ask whether introspection is a source of knowledge. The three questions are as follows:

(1) Can we observe anything about ourselves which we cannot observe about other people, or is everything we can observe PUBLIC, in the sense that another could also observe it if suitably placed?

(2) Does everything that we can observe obey the laws of physics and form part of the physical world, or can we observe certain things that lie outside physics?

(3) Can we observe anything which differs in its intrinsic nature from the constituents of the physical world, or is everything that we can observe composed of elements intrinsically similar to the constituents of what is called matter?

Any one of these three questions may be used to define introspection. I should favour introspection in the sense of the first question, i.e. I think that some of the things we observe cannot, even theoretically, be observed by any one else. The second question, tentatively and for the present, I should answer in favour of introspection; I think that images, in the actual condition of science, cannot be brought under the causal laws of physics, though perhaps ultimately they may be. The third question I should answer adversely to introspection I think that observation shows us nothing that is not composed of sensations and images, and that images differ from sensations in their causal laws, not intrinsically. I shall deal with the three questions successively.

(1) PUBLICITY OR PRIVACY OF WHAT IS OBSERVED. Confining ourselves, for the moment, to sensations, we find that there are different degrees of publicity attaching to different sorts of sensations. If you feel a toothache when the other people in the room do not, you are in no way surprised; but if you hear a clap of thunder when they do not, you begin to be alarmed as to your mental condition. Sight and hearing are the most public of the senses; smell only a trifle less so; touch, again, a trifle less, since two people can only touch the same spot successively, not simultaneously. Taste has a sort of semi-publicity, since people seem to experience similar taste-sensations when they eat similar foods; but the publicity is incomplete, since two people cannot eat actually the same piece of food.

But when we pass on to bodily sensations—headache, toothache, hunger, thirst, the feeling of fatigue, and so on—we get quite away from publicity, into a region where other people can tell us what they feel, but we cannot directly observe their feeling. As a natural result of this state of affairs, it has come to be thought that the public senses give us knowledge of the outer world, while the private senses only give us knowledge as to our own bodies. As regards privacy, all images, of whatever sort, belong with the sensations which only give knowledge of our own bodies, i.e. each is only observable by one observer. This is the reason why images of sight and hearing are more obviously different from sensations of sight and hearing than images of bodily sensations are from bodily sensations; and that is why the argument in favour of images is more conclusive in such cases as sight and hearing than in such cases as inner speech.

The whole distinction of privacy and publicity, however, so long as we confine ourselves to sensations, is one of degree, not of kind. No two people, there is good empirical reason to think, ever have exactly similar sensations related to the same physical object at the same moment; on the other hand, even the most private sensation has correlations which would theoretically enable another observer to infer it.

That no sensation is ever completely public, results from differences of point of view. Two people looking at the same table do not get the same sensation, because of perspective and the way the light falls. They get only correlated sensations. Two people listening to the same sound do not hear exactly the same thing, because one is nearer to the source of the sound than the other, one has better hearing than the other, and so on. Thus publicity in sensations consists, not in having PRECISELY similar sensations, but in having more or less similar sensations correlated according to ascertainable laws. The sensations which strike us as public are those where the correlated sensations are very similar and the correlations are very easy to discover. But even the most private sensations have correlations with things that others can observe. The dentist does not observe your ache, but he can see the cavity which causes it, and could guess that you are suffering even if you did not tell him. This fact, however, cannot be used, as Watson would apparently wish, to extrude from science observations which are private to one observer, since it is by means of many such observations that correlations are established, e.g. between toothaches and cavities. Privacy, therefore does not by itself make a datum unamenable to scientific treatment. On this point, the argument against introspection must be rejected.

(2) DOES EVERYTHING OBSERVABLE OBEY THE LAWS OF PHYSICS? We come now to the second ground of objection to introspection, namely, that its data do not obey the laws of physics. This, though less emphasized, is, I think, an objection which is really more strongly felt than the objection of privacy. And we obtain a definition of introspection more in harmony with usage if we define it as observation of data not subject to physical laws than if we define it by means of privacy. No one would regard a man as introspective because he was conscious

of having a stomach ache. Opponents of introspection do not mean to deny the obvious fact that we can observe bodily sensations which others cannot observe. For example, Knight Dunlap contends that images are really muscular contractions,and evidently regards our awareness of muscular contractions as not coming under the head of introspection. I think it will be found that the essential characteristic of introspective data, in the sense which now concerns us, has to do with LOCALIZATION: either they are not localized at all, or they are localized, like visual images, in a place already physically occupied by something which would be inconsistent with them if they were regarded as part of the physical world. If you have a visual image of your friend sitting in a chair which in fact is empty, you cannot locate the image in your body, because it is visual, nor (as a physical phenomenon) in the chair, because the chair, as a physical object, is empty. Thus it seems to follow that the physical world does not include all that we are aware of, and that images, which are introspective data, have to be regarded, for the present, as not obeying the laws of physics; this is, I think, one of the chief reasons why an attempt is made to reject them. I shall try to show in Lecture VIII that the purely empirical reasons for accepting images are overwhelming. But we cannot be nearly so certain that they will not ultimately be brought under the laws of physics. Even if this should happen, however, they would still be distinguishable from sensations by their proximate causal laws, as gases remain distinguishable from solids.

(3) CAN WE OBSERVE ANYTHING INTRINSICALLY DIFFERENT FROM SENSATIONS? We come now to our third question concerning introspection. It is commonly thought that by looking within we can observe all sorts of things that are radically different from the constituents of the physical world, e.g. thoughts, beliefs, desires, pleasures, pains and emotions. The difference between mind and matter is increased partly by emphasizing these supposed introspective data, partly by the supposition that matter is composed of atoms or electrons or whatever units physics may at the moment prefer. As against this latter supposition, I contend that the ultimate constituents of matter are not atoms or electrons, but sensations, and other things similar to sensations as regards extent and duration. As against the view that introspection reveals a mental world radically different from sensations, I propose to argue that thoughts, beliefs, desires, pleasures, pains and emotions are all built up out of sensations and images alone, and that there is reason to think that images do not differ from sensations in their intrinsic character. We thus effect a mutual rapprochement of mind and matter, and reduce the ultimate data of introspection (in our second sense) to images alone. On this third view of the meaning of introspection, therefore, our decision is wholly against it.

There remain two points to be considered concerning introspection. The first is as to how far it is trustworthy; the second is as to whether, even granting that it reveals no radically different STUFF from that revealed by what might be called external perception, it may not reveal different RELATIONS, and thus acquire almost as much importance as is traditionally assigned to it.

To begin with the trustworthiness of introspection. It is common among certain schools to regard the knowledge of our own mental processes as incomparably more certain than our knowledge of the "external" world; this view is to be found in the British philosophy which descends from Hume, and is present, somewhat veiled, in Kant and his followers. There seems no reason whatever to accept this view. Our spontaneous, unsophisticated beliefs, whether as to ourselves or as to the outer world, are always extremely rash and very liable to error. The acquisition of caution is equally necessary and equally difficult in both directions. Not only are we often un aware of entertaining a belief or desire which exists in us; we are often actually mistaken. The fallibility of introspection as regards what we desire is made evident by psycho-analysis; its fallibility as to what we know is easily demonstrated. An autobiography, when confronted by a careful editor with documentary evidence, is usually found to be full of obviously inadvertent errors. Any of us confronted by a forgotten letter written some years ago will be astonished to find how much more foolish our opinions were than we had remembered them as being. And as to the analysis of our mental operations—believing, desiring, willing, or what not—introspection unaided gives very little help: it is necessary to construct hypotheses and test them by their consequences, just as we do in physical science. Introspection, therefore, though it is one among our sources of knowledge, is not, in isolation, in any degree more trustworthy than "external" perception.

I come now to our second question: Does introspection give us materials for the knowledge of relations other than those arrived at by reflecting upon external perception? It might be contended that the essence of what is "mental" consists of relations, such as knowing for example, and that our knowledge concerning these essentially mental relations is entirely derived from introspection. If "knowing" were an unanalysable relation, this view would be incontrovertible, since clearly no such relation forms part of the subject matter of physics. But it would seem that "knowing" is really various relations, all of them complex. Therefore, until they have been analysed, our present question must remain unanswered I shall return to it at the end of the present course of lectures.

LECTURE VII.

THE DEFINITION OF PERCEPTION

In Lecture V we found reason to think that the ultimate constituents of the world do not have the characteristics of either mind or matter as ordinarily understood: they are not solid persistent objects moving through space, nor are they fragments of "consciousness." But we found two ways of grouping particulars, one into "things" or "pieces of matter," the other into series of "perspectives," each series being what may be called a "biography." Before we can define either sensations or images, it is necessary to consider this twofold classification in somewhat greater detail, and to derive from it a definition of perception. It should be said that, in so far as the classification assumes the whole world of physics (including its unperceived portions), it contains hypothetical elements. But we will not linger on the grounds for admitting these, which belong to the philosophy of physics rather than of psychology.

The physical classification of particulars collects together all those that are aspects of one "thing." Given any one particular, it is found often (we do not say always) that there are a number of other particulars differing from this one in gradually increasing degrees. Those (or some of those) that differ from it only very slightly will be found to differ approximately according to certain laws which may be called, in a generalized sense, the laws of "perspective"; they include the ordinary laws of perspective as a special case. This approximation grows more and more nearly exact as the difference grows less; in technical language, the laws of perspective account for the differences to the first order of small quantities, and other laws are only required to account for second-order differences. That is to say, as the difference diminishes, the part of the difference which is not according to the laws of perspective diminishes much more rapidly, and bears to the total difference a ratio which tends towards zero as both are made smaller and smaller. By this means we can theoretically collect together a number of particulars which may be defined as the "aspects" or "appearances" of one thing at one time. If the laws of perspective were sufficiently known, the connection between different aspects would be expressed in differential equations.

This gives us, so far, only those particulars which constitute one thing at one time. This set of particulars may be called a "momentary thing." To define that series of "momentary things" that constitute the successive states of one thing is a problem involving the laws of dynamics. These give the laws governing the changes of aspects from one time to a slightly later time, with the same sort of differential approximation to exactness as we obtained for spatially neighbouring aspects through the laws of perspective. Thus a momentary thing is a set of particulars, while a thing (which may be identified with the whole history of the thing) is a series of such sets of particulars. The particulars in one set are collected together by the laws of perspective; the successive sets are collected together by the laws of dynamics. This is the view of the world which is appropriate to traditional physics.

The definition of a "momentary thing" involves problems concerning time, since the particulars constituting a momentary thing will not be all simultaneous, but will travel outward from the thing with the velocity of light (in case the thing is in vacuo). There are complications connected with relativity, but for our present purpose they are not vital, and I shall ignore them.

Instead of first collecting together all the particulars constituting a momentary thing, and then forming the series of successive sets, we might have first collected together a series of successive aspects related by the laws of dynamics, and then have formed the set of such series related by the laws of perspective. To illustrate by the case of an actor on the stage: our first plan was to collect together all the aspects which he presents to different spectators at one time, and then to form the series of such sets. Our second plan is first to collect together all the aspects which he presents successively to a given spectator, and then to do the same thing for the other spectators, thus forming a set of series instead of a series of sets. The first plan tells us what he does; the second the impressions he produces. This second way of classifying particulars is one which obviously has more relevance to psychology than the other. It is partly by this second method of classification that we obtain definitions of one "experience" or "biography" or "person." This method of classification is also essential to the definition of sensations and images, as I shall endeavour to prove later on. But we must first amplify the definition of perspectives and biographies.

In our illustration of the actor, we spoke, for the moment, as though each spectator's mind were wholly occupied by the one actor. If this were the case, it might be possible to define the biography of one spectator as a series of successive aspects of the actor related according to the laws of dynamics. But in fact this is not the case. We are at all times during our waking life receiving a variety of impressions, which are aspects of a variety of things. We have to consider what binds together two simultaneous sensations in one person, or, more generally, any two occurrences which forte part of one experience. We might say, adhering to the standpoint of physics, that two aspects of different things belong to the same perspective when they are in the same place. But this would not really help us, since a "place" has not yet been defined. Can we define what is meant by saying that two aspects are "in the same place," without introducing anything beyond the laws of perspective and dynamics?

I do not feel sure whether it is possible to frame such a definition or not; accordingly I shall not assume that it is possible, but shall seek other characteristics by which a perspective or biography may be defined.

When (for example) we see one man and hear another speaking at the same time, what we see and what we hear have a relation which we can perceive, which makes the two together form, in some sense, one experience. It is when this relation exists that two occurrences become associated. Semon's "engram" is formed by all that we experience at one time. He speaks of two parts of this total as having the relation of "Nebeneinander" (M. 118; M.E. 33 ff.), which is reminiscent of Herbart's "Zusammen." I think the relation may be called simply "simultaneity." It might be said that at any moment all sorts of things that are not part of my experience are happening in the world, and that therefore the relation we are seeking to define cannot be merely simultaneity. This, however, would be an error—the sort of error that the theory of relativity avoids. There is not one universal time, except by an elaborate construction; there are only local times, each of which may be taken to be the time within one biography. Accordingly, if I am (say) hearing a sound, the only occurrences that are, in any simple sense, simultaneous with my sensation are events in my private world, i.e. in my biography. We may therefore define the "perspective" to which the sensation in question belongs as the set of particulars that are simultaneous with this sensation. And similarly we may define the "biography" to which the sensation belongs as the set of particulars that are earlier or later than, or simultaneous with, the given sensation. Moreover, the very same definitions can be applied to particulars which are not sensations. They are actually required for the theory of relativity, if we are to give a philosophical explanation of what is meant by "local time" in that theory The relations of simultaneity and succession are known to us in our own experience; they may be analysable, but that does not affect their suitability for defining perspectives and biographies. Such time-relations as can be constructed between events in different biographies are of a different kind: they are not experienced, and are merely logical, being designed to afford convenient ways of stating the correlations between different biographies.

It is not only by time-relations that the parts of one biography are collected together in the case of living beings. In this case there are the mnemic phenomena which constitute the unity of one "experience," and transform mere occurrences into "experiences." I have already dwelt upon the importance of mnemic phenomena for psychology, and shall not enlarge upon them now, beyond observing that they are what transforms a biography (in our technical sense) into a life. It is they that give the continuity of a "person" or a "mind." But there is no reason to suppose that mnemic phenomena are associated with biographies except in the case of animals and plants.

Our two-fold classification of particulars gives rise to the dualism of body and biography in regard to everything in the universe, and not only in regard to living things. This arises as follows. Every particular of the sort considered by physics is a member of two groups (1) The group of particulars constituting the other aspects of the same physical object; (2) The group of particulars that have direct time-relations to the given particular.

Each of these is associated with a place. When I look at a star, my sensation is (1) A member of the group of particulars which is the star, and which is associated with the place where the star is; (2) A member of the group of particulars which is my biography, and which is associated with the place where I am.

The result is that every particular of the kind relevant to physics is associated with TWO places; e.g. my sensation of the star is associated with the place where I am and with the place where the star is. This dualism has nothing to do with any "mind" that I may be supposed to possess; it exists in exactly the same sense if I am replaced by a photographic plate. We may call the two places the active and passive places respectively. Thus in the case of a perception or photograph of a star, the active place is the place where the star is, while the passive place is the place where the percipient or photographic plate is.

We can thus, without departing from physics, collect together all the particulars actively at a given place, or all the particulars passively at a given place. In our own case, the one group is our body (or our brain), while the other is our mind, in so far as it consists of perceptions. In the case of the photographic plate, the first group is the plate as dealt with by physics, the second the aspect of the heavens which it photographs. (For the sake of schematic simplicity, I am ignoring various complications connected with time, which require some tedious but perfectly feasible elaborations.) Thus what may be called subjectivity in the point of view is not a distinctive peculiarity of mind: it is present just as much in the photographic plate. And the photographic plate has its biography as well as its "matter." But this biography is an affair of physics, and has none of the peculiar characteristics by which "mental" phenomena are distinguished, with the sole exception of subjectivity.

Adhering, for the moment, to the standpoint of physics, we may define a "perception" of an object as the appearance of the object from a place where there is a brain (or, in lower animals, some suitable nervous structure), with sense-organs and nerves forming part of the intervening medium. Such appearances of objects are distinguished from appearances in other places by certain peculiarities, namely:

(1) They give rise to mnemic phenomena;

(2) They are themselves affected by mnemic phenomena.

That is to say, they may be remembered and associated or influence our habits, or give rise to images, etc., and they are themselves different from what they would have been if our past experience had been different—for example, the effect of a spoken sentence upon the hearer depends upon whether the hearer knows the language or not, which is a question of past experience. It is these two characteristics, both connected with mnemic phenomena, that distinguish perceptions from the appearances of objects in places where there is no living being.

Theoretically, though often not practically, we can, in our perception of an object, separate the part which is due to past experience from the part which proceeds without mnemic influences out of the character of the object. We may define as "sensation" that part which proceeds in this way, while the remainder, which is a mnemic phenomenon, will have to be added to the sensation to make up what is called the "perception." According to this definition, the sensation is a theoretical core in the actual experience; the actual experience is the perception. It is obvious that there are grave difficulties in carrying out these definitions, but we will not linger over them. We have to pass, as soon as we can, from the physical standpoint, which we have been hitherto adopting, to the standpoint of psychology, in which we make more use of introspection in the first of the three senses discussed in the preceding lecture.

But before making the transition, there are two points which must be made clear. First: Everything outside my own personal biography is outside my experience; therefore if anything can be known by me outside my biography, it can only be known in one of two ways:

(1) By inference from things within my biography, or

(2) By some a priori principle independent of experience.

I do not myself believe that anything approaching certainty is to be attained by either of these methods, and therefore whatever lies outside my personal biography must be regarded, theoretically, as hypothesis. The theoretical argument for adopting the hypothesis is that it simplifies the statement of the laws according to which events happen in our experience. But there is no very good ground for supposing that a simple law is more likely to be true than a complicated law, though there is good ground for assuming a simple law in scientific practice, as a working hypothesis, if it explains the facts as well as another which is less simple. Belief in the existence of things outside my own biography exists antecedently to evidence, and can only be destroyed, if at all, by a long course of philosophic doubt. For purposes of science, it is justified practically by the simplification which it introduces into the laws of physics. But from the standpoint of theoretical logic it must be regarded as a prejudice, not as a well-grounded theory. With this proviso, I propose to continue yielding to the prejudice.

The second point concerns the relating of our point of view to that which regards sensations as caused by stimuli external to the nervous system (or at least to the brain), and distinguishes images as "centrally excited," i.e. due to causes in the brain which cannot be traced back to anything affecting the sense-organs. It is clear that, if our analysis of physical objects has been valid, this way of defining sensations needs reinterpretation. It is also clear that we must be able to find such a new interpretation if our theory is to be admissible.

To make the matter clear, we will take the simplest possible illustration. Consider a certain star, and suppose for the moment that its size is negligible. That is to say, we will regard it as, for practical purposes, a luminous point. Let us further suppose that it exists only for a very brief time, say a second. Then, according to physics, what happens is that a spherical wave of light travels outward from the star through space, just as, when you drop a stone into a stagnant pond, ripples travel outward from the place where the stone hit the water. The wave of light travels with a certain very nearly constant velocity, roughly 300,000 kilometres per second. This velocity may be ascertained by sending a flash of light to a mirror, and observing how long it takes before the reflected flash reaches you, just as the velocity of sound may be ascertained by means of an echo.

What it is that happens when a wave of light reaches a given place we cannot tell, except in the sole case when the place in question is a brain connected with an eye which is turned in the right direction. In this one very special case we know what happens: we have the sensation called "seeing the star." In all other cases, though we know (more or less hypothetically) some of the correlations and abstract properties of the appearance of the star, we do not know the appearance itself. Now you may, for the sake of illustration, compare the different appearances of the star to the conjugation of a Greek verb, except that the number of its parts is really infinite, and not only apparently so to the despairing schoolboy. In vacuo, the parts are regular, and can be derived from the (imaginary) root according to the laws of grammar, i.e. of perspective. The star being situated in empty space, it may be defined, for purposes of physics, as consisting of all those appearances which it presents in vacuo, together with those which, according to the laws of perspective, it would present elsewhere if its appearances elsewhere were regular. This is merely the adaptation of the definition of matter which I gave in an earlier lecture. The appearance of a star at a certain place, if it is regular, does not require any cause or explanation beyond the existence of the star. Every regular appearance is an actual member of the system which is the star, and its causation is entirely internal to that system. We may express this by saying that a regular appearance is due to the star alone, and is actually part of the star, in the sense in which a man is part of the human race.

But presently the light of the star reaches our atmosphere. It begins to be refracted, and dimmed by mist, and its velocity is slightly diminished. At last it reaches a human eye, where a complicated process takes place, ending in a sensation which gives us our grounds for believing in all that has gone before. Now, the irregular appearances of the star are not, strictly speaking, members of the system which is the star, according to our definition of matter. The irregular appearances, however, are not merely irregular: they proceed according to laws which can be stated in terms of the matter through which the light has passed on its way. The sources of an irregular appearance are therefore twofold:

(1) The object which is appearing irregularly;

2) The intervening medium.

It should be observed that, while the conception of a regular appearance is perfectly precise, the conception of an irregular appearance is one capable of any degree of vagueness. When the distorting influence of the medium is sufficiently great, the resulting particular can no longer be regarded as an appearance of an object, but must be treated on its own account. This happens especially when the particular in question cannot be traced back to one object, but is a blend of two or more. This case is normal in perception: we see as one what the microscope or telescope reveals to be many different objects. The notion of perception is therefore not a precise one: we perceive things more or less, but always with a very considerable amount of vagueness and confusion.

In considering irregular appearances, there are certain very natural mistakes which must be avoided. In order that a particular may count as an irregular appearance of a certain object, it is not necessary that it should bear any resemblance to the regular appearances as regard its intrinsic qualities. All that is necessary is that it should be derivable from the regular appearances by the laws which express the distorting influence of the medium. When it is so derivable, the particular in question may be regarded as caused by the regular appearances, and therefore by the object itself, together with the modifications resulting from the medium. In other cases, the particular in question may, in the same sense, be regarded as caused by several objects together with the medium; in this case, it may be called a confused appearance of several objects. If it happens to be in a brain, it may be called a confused perception of these objects. All actual perception is confused to a greater or less extent.

We can now interpret in terms of our theory the distinction between those mental occurrences which are said to have an external stimulus, and those which are said to be "centrally excited," i.e. to have no stimulus external to the brain. When a mental occurrence can be regarded as an appearance of an object external to the brain, however irregular, or even as a confused appearance of several such objects, then we may regard it as having for its stimulus the object or objects in question, or their appearances at the sense-organ concerned. When, on the other hand, a mental occurrence has not sufficient connection with objects external to the brain to be regarded as an appearance of such objects, then its physical causation (if any) will have to be sought in the brain. In the former case it can be called a perception; in the latter it cannot be so called. But the distinction is one of degree, not of kind. Until this is realized, no satisfactory theory of perception, sensation, or imagination is possible.

LECTURE VIII.

SENSATIONS AND IMAGES

The dualism of mind and matter, if we have been right so far, cannot be allowed as metaphysically valid. Nevertheless, we seem to find a certain dualism, perhaps not ultimate, within the world as we observe it. The dualism is not primarily as to the stuff of the world, but as to causal laws. On this subject we may again quote William James. He points out that when, as we say, we merely "imagine" things, there are no such effects as would ensue if the things were what we call "real." He takes the case of imagining a fire.

"I make for myself an experience of blazing fire; I place it near my body; but it does not warm me in the least. I lay a stick upon it and the stick either burns or remains green, as I please. I call up water, and pour it on the fire, and absolutely no difference ensues. I account for all such facts by calling this whole train of experiences unreal, a mental train. Mental fire is what won't burn real sticks; mental water is what won't necessarily (though of course it may) put out even a mental fire.... With 'real' objects, on the contrary, consequences always accrue; and thus the real experiences get sifted from the mental ones, the things from our thoughts of them, fanciful or true, and precipitated together as the stable part of the whole experience—chaos, under the name of the physical world."

In this passage James speaks, by mere inadvertence, as though the phenomena which he is describing as "mental" had NO effects. This is, of course, not the case: they have their effects, just as much as physical phenomena do, but their effects follow different laws. For example, dreams, as Freud has shown, are just as much subject to laws as are the motions of the planets. But the laws are different: in a dream you may be transported from one place to another in a moment, or one person may turn into another under your eyes. Such differences compel you to distinguish the world of dreams from the physical world.

If the two sorts of causal laws could be sharply distinguished, we could call an occurrence "physical" when it obeys causal laws appropriate to the physical world, and "mental" when it obeys causal laws appropriate to the mental world. Since the mental world and the physical world interact, there would be a boundary between the two: there would be events which would have physical causes and mental effects, while there would be others which would have mental causes and physical effects. Those that have physical causes and mental effects we should define as "sensations." Those that have mental causes and physical effects might perhaps be identified with what we call voluntary movements; but they do not concern us at present.

These definitions would have all the precision that could be desired if the distinction between physical and psychological causation were clear and sharp. As a matter of fact, however, this distinction is, as yet, by no means sharp. It is possible that, with fuller knowledge, it will be found to be no more ultimate than the distinction between the laws of gases and the laws of rigid bodies. It also suffers from the fact that an event may be an effect of several causes according to several causal laws we cannot, in general, point to anything unique as THE cause of such-and-such an event. And finally it is by no means certain that the peculiar causal laws which govern mental events are not really physiological. The law of habit, which is one of the most distinctive, may be fully explicable in terms of the peculiarities of nervous tissue, and these peculiarities, in turn, may be explicable by the laws of physics. It seems, therefore, that we are driven to a different kind of definition. It is for this reason that it was necessary to develop the definition of perception. With this definition, we can define a sensation as the non-mnemic elements in a perception.

When, following our definition, we try to decide what elements in our experience are of the nature of sensations, we find more difficulty than might have been expected. Prima facie, everything is sensation that comes to us through the senses: the sights we see, the sounds we hear, the smells we smell, and so on; also such things as headache or the feeling of muscular strain. But in actual fact so much interpretation, so much of habitual correlation, is mixed with all such experiences, that the core of pure sensation is only to be extracted by careful investigation. To take a simple illustration: if you go to the theatre in your own country, you seem to hear equally well in the stalls or the dress circle; in either case you think you miss nothing. But if you go in a foreign country where you have a fair knowledge of the language, you will seem to have grown partially deaf, and you will find it necessary to be much nearer the stage than you would need to be in your own country. The reason is that, in hearing our own language spoken, we quickly and unconsciously fill out what we really hear with inferences to what the man must be saying, and we never realize that we have not heard the words we have merely inferred. In a foreign language, these inferences are more difficult, and we are more dependent upon actual sensation. If we found ourselves in a foreign world, where tables looked like cushions and cushions like tables, we should similarly discover how much of what we think we see is really inference. Every fairly familiar sensation is to us a sign of the things that usually go with it, and many of these things will seem to form part of the sensation. I remember in the early days of motor-cars being with a friend when a tyre burst with a loud report. He thought it was a pistol, and supported his opinion by maintaining that he had seen the flash. But of course there had been no flash. Nowadays no one sees a flash when a tyre bursts.

In order, therefore, to arrive at what really is sensation in an occurrence which, at first sight, seems to contain nothing else, we have to pare away all that is due to habit or expectation or interpretation. This is a matter for the psychologist, and by no means an easy matter. For our purposes, it is not important to determine what exactly is the sensational core in any case; it is only important to notice that there certainly is a sensational core, since habit, expectation and interpretation are diversely aroused on diverse occasions, and the diversity is clearly due to differences in what is presented to the senses. When you open your newspaper in the morning, the actual sensations of seeing the print form a very minute part of what goes on in you, but they are the starting-point of all the rest, and it is through them that the newspaper is a means of information or mis-information. Thus, although it may be difficult to determine what exactly is sensation in any given experience, it is clear that there is sensation, unless, like Leibniz, we deny all action of the outer world upon us.

Sensations are obviously the source of our knowledge of the world, including our own body. It might seem natural to regard a sensation as itself a cognition, and until lately I did so regard it. When, say, I see a person I know coming towards me in the street, it SEEMS as though the mere seeing were knowledge. It is of course undeniable that knowledge comes THROUGH the seeing, but I think it is a mistake to regard the mere seeing itself as knowledge. If we are so to regard it, we must distinguish the seeing from what is seen: we must say that, when we see a patch of colour of a certain shape, the patch of colour is one thing and our seeing of it is another. This view, however, demands the admission of the subject, or act, in the sense discussed in our first lecture. If there is a subject, it can have a relation to the patch of colour, namely, the sort of relation which we might call awareness. In that case the sensation, as a mental event, will consist of awareness of the colour, while the colour itself will remain wholly physical, and may be called the sense-datum, to distinguish it from the sensation. The subject, however, appears to be a logical fiction, like mathematical points and instants. It is introduced, not because observation reveals it, but because it is linguistically convenient and apparently demanded by grammar. Nominal entities of this sort may or may not exist, but there is no good ground for assuming that they do. The functions that they appear to perform can always be performed by classes or series or other logical constructions, consisting of less dubious entities. If we are to avoid a perfectly gratuitous assumption, we must dispense with the subject as one of the actual ingredients of the world. But when we do this, the possibility of distinguishing the sensation from the sense-datum vanishes; at least I see no way of preserving the distinction. Accordingly the sensation that we have when we see a patch of colour simply is that patch of colour, an actual constituent of the physical world, and part of what physics is concerned with. A patch of colour is certainly not knowledge, and therefore we cannot say that pure sensation is cognitive. Through its psychological effects, it is the cause of cognitions, partly by being itself a sign of things that are correlated with it, as e.g. sensations of sight and touch are correlated, and partly by giving rise to images and memories after the sensation is faded. But in itself the pure sensation is not cognitive.

In the first lecture we considered the view of Brentano, that "we may define psychical phenomena by saying that they are phenomena which intentionally contain an object." We saw reasons to reject this view in general; we are now concerned to show that it must be rejected in the particular case of sensations. The kind of argument which formerly made me accept Brentano's view in this case was exceedingly simple. When I see a patch of colour, it seemed to me that the colour is not psychical, but physical, while my seeing is not physical, but psychical. Hence I concluded that the colour is something other than my seeing of the colour. This argument, to me historically, was directed against idealism: the emphatic part of it was the assertion that the colour is physical, not psychical. I shall not trouble you now with the grounds for holding as against Berkeley that the patch of colour is physical; I have set them forth before, and I see no reason to modify them. But it does not follow that the patch of colour is not also psychical, unless we assume that the physical and the psychical cannot overlap, which I no longer consider a valid assumption. If we admit—as I think we should—that the patch of colour may be both physical and psychical, the reason for distinguishing the sense-datum from the sensation disappears, and we may say that the patch of colour and our sensation in seeing it are identical.

This is the view of William James, Professor Dewey, and the American realists. Perceptions, says Professor Dewey, are not per se cases of knowledge, but simply natural events with no more knowledge status than (say) a shower. "Let them [the realists] try the experiment of conceiving perceptions as pure natural events, not cases of awareness or apprehension, and they will be surprised to see how little they miss." I think he is right in this, except in supposing that the realists will be surprised. Many of them already hold the view he is advocating, and others are very sympathetic to it. At any rate, it is the view which I shall adopt in these lectures.

The stuff of the world, so far as we have experience of it, consists, on the view that I am advocating, of innumerable transient particulars such as occur in seeing, hearing, etc., together with images more or less resembling these, of which I shall speak shortly. If physics is true, there are, besides the particulars that we experience, others, probably equally (or almost equally) transient, which make up that part of the material world that does not come into the sort of contact with a living body that is required to turn it into a sensation. But this topic belongs to the philosophy of physics, and need not concern us in our present inquiry.

Sensations are what is common to the mental and physical worlds; they may be defined as the intersection of mind and matter. This is by no means a new view;

it is advocated, not only by the American authors I have mentioned, but by Mach in his Analysis of Sensations, which was published in 1886. The essence of sensation, according to the view I am advocating, is its independence of past experience. It is a core in our actual experiences, never existing in isolation except possibly in very young infants. It is not itself knowledge, but it supplies the data for our knowledge of the physical world, including our own bodies.

There are some who believe that our mental life is built up out of sensations alone. This may be true; but in any case I think the only ingredients required in addition to sensations are images. What images are, and how they are to be defined, we have now to inquire.

The distinction between images and sensations might seem at first sight by no means difficult. When we shut our eyes and call up pictures of familiar scenes, we usually have no difficulty, so long as we remain awake, in discriminating between what we are imagining and what is really seen. If we imagine some piece of music that we know, we can go through it in our mind from beginning to end without any discoverable tendency to suppose that we are really hearing it. But although such cases are so clear that no confusion seems possible, there are many others that are far more difficult, and the definition of images is by no means an easy problem.

To begin with: we do not always know whether what we are experiencing is a sensation or an image. The things we see in dreams when our eyes are shut must count as images, yet while we are dreaming they seem like sensations. Hallucinations often begin as persistent images, and only gradually acquire that influence over belief that makes the patient regard them as sensations. When we are listening for a faint sound—the striking of a distant clock, or a horse's hoofs on the road—we think we hear it many times before we really do, because expectation brings us the image, and we mistake it for sensation. The distinction between images and sensations is, therefore, by no means always obvious to inspection.

We may consider three different ways in which it has been sought to distinguish images from sensations, namely:

(1) By the less degree of vividness in images;

(2) By our absence of belief in their "physical reality";

(3) By the fact that their causes and effects are different from those of sensations.

I believe the third of these to be the only universally applicable criterion. The other two are applicable in very many cases, but cannot be used for purposes of definition because they are liable to exceptions. Nevertheless, they both deserve to be carefully considered.

(1) Hume, who gives the names "impressions" and "ideas" to what may, for present purposes, be identified with our "sensations" and "images," speaks of impressions as "those perceptions which enter with most force and violence" while he defines ideas as "the faint images of these (i.e. of impressions) in thinking and reasoning." His immediately following observations, however, show the inadequacy of his criteria of "force" and "faintness." He says:

"I believe it will not be very necessary to employ many words in explaining this distinction. Every one of himself will readily perceive the difference betwixt feeling and thinking. The common degrees of these are easily distinguished, though it is not impossible but in particular instances they may very nearly approach to each other. Thus in sleep, in a fever, in madness, or in any very violent emotions of soul, our ideas may approach to our impressions; as, on the other hand, it sometimes happens, that our impressions are so faint and low that we cannot distinguish them from our ideas. But notwithstanding this near resemblance in a few instances, they are in general so very different, that no one can make a scruple to rank them under distinct heads, and assign to each a peculiar name to mark the difference" ("Treatise of Human Nature," Part I, Section I).

I think Hume is right in holding that they should be ranked under distinct heads, with a peculiar name for each. But by his own confession in the above passage, his criterion for distinguishing them is not always adequate. A definition is not sound if it only applies in cases where the difference is glaring: the essential purpose of a definition is to provide a mark which is applicable even in marginal cases—except, of course, when we are dealing with a conception, like, e.g. baldness, which is one of degree and has no sharp boundaries. But so far we have seen no reason to think that the difference between sensations and images is only one of degree.

Professor Stout, in his "Manual of Psychology," after discussing various ways of distinguishing sensations and images, arrives at a view which is a modification of Hume's. He says (I quote from the second edition):

"Our conclusion is that at bottom the distinction between image and percept, as respectively faint and vivid states, is based on a difference of quality. The percept has an aggressiveness which does not belong to the image. It strikes the mind with varying degrees of force or liveliness according to the varying intensity of the stimulus. This degree of force or liveliness is part of what we ordinarily mean by the intensity of a sensation. But this constituent of the intensity of sensations is absent in mental imagery".

This view allows for the fact that sensations may reach any degree of faintness—e.g. in the case of a just visible star or a just audible sound—without becoming images, and that therefore mere faintness cannot be the characteristic mark of images. After explaining the sudden shock of a flash of lightning or a steam-whistle, Stout says that "no mere image ever does strike the mind in this manner". But I believe that this criterion fails in very much the same instances as those in which Hume's criterion fails in its original form. Macbeth speaks of—

that suggestion

Whose horrid image doth unfix my hair

And make my seated heart knock at my ribs

Against the use of nature.

The whistle of a steam-engine could hardly have a stronger effect than this. A very intense emotion will often bring with it—especially where some future action or some undecided issue is involved—powerful compelling images which may determine the whole course of life, sweeping aside all contrary solicitations to the will by their capacity for exclusively possessing the mind. And in all cases where images, originally recognized as such, gradually pass into hallucinations, there must be just that "force or liveliness" which is supposed to be always absent from images. The cases of dreams and fever-delirium are as hard to adjust to Professor Stout's modified criterion as to Hume's. I conclude therefore that the test of liveliness, however applicable in ordinary instances, cannot be used to define the differences between sensations and images.

(2) We might attempt to distinguish images from sensations by our absence of belief in the "physical reality" of images. When we are aware that what we are experiencing is an image, we do not give it the kind of belief that we should give to a sensation: we do not think that it has the same power of producing knowledge of the "external world." Images are "imaginary"; in SOME sense they are "unreal." But this difference is hard to analyse or state correctly. What we call the "unreality" of images requires interpretation it cannot mean what would be expressed by saying "there's no such thing." Images are just as truly part of the actual world as sensations are. All that we really mean by calling an image "unreal" is that it does not have the concomitants which it would have if it were a sensation. When we call up a visual image of a chair, we do not attempt to sit in it, because we know that, like Macbeth's dagger, it is not "sensible to feeling as to sight"—i.e. it does not have the correlations with tactile sensations which it would have if it were a visual sensation and not merely a visual image. But this means that the so-called "unreality" of images consists merely in their not obeying the laws of physics, and thus brings us back to the causal distinction between images and sensations.

This view is confirmed by the fact that we only feel images to be "unreal" when we already know them to be images. Images cannot be defined by the FEELING of unreality, because when we falsely believe an image to be a sensation, as in the case of dreams, it FEELS just as real as if it were a sensation. Our feeling of unreality results from our having already realized that we are dealing with an image, and cannot therefore be the definition of what we mean by an image. As soon as an image begins to deceive us as to its status, it also deceives us as to its correlations, which are what we mean by its "reality."

(3) This brings us to the third mode of distinguishing images from sensations, namely, by their causes and effects. I believe this to be the only valid ground of distinction. James, in the passage about the mental fire which won't burn real sticks, distinguishes images by their effects, but I think the more reliable distinction is by their causes. Professor Stout says: "One characteristic mark of what we agree in calling sensation is its mode of production. It is caused by what we call a STIMULUS. A stimulus is always some condition external to the nervous system itself and operating upon it." I think that this is the correct view, and that the distinction between images and sensations can only be made by taking account of their causation. Sensations come through sense-organs, while images do not. We cannot have visual sensations in the dark, or with our eyes shut, but we can very well have visual images under these circumstances. Accordingly images have been defined as "centrally excited sensations," i.e. sensations which have their physiological cause in the brain only, not also in the sense-organs and the nerves that run from the sense-organs to the brain. I think the phrase "centrally excited sensations" assumes more than is necessary, since it takes it for granted that an image must have a proximate physiological cause. This is probably true, but it is an hypothesis, and for our purposes an unnecessary one. It would seem to fit better with what we can immediately observe if we were to say that an image is occasioned, through association, by a sensation or another image, in other words that it has a mnemic cause—which does not prevent it from also having a physical cause. And I think it will be found that the causation of an image always proceeds according to mnemic laws, i.e. that it is governed by habit and past experience. If you listen to a man playing the pianola without looking at him, you will have images of his hands on the keys as if he were playing the piano; if you suddenly look at him while you are absorbed in the music, you will experience a shock of surprise when you notice that his hands are not touching the notes. Your image of his hands is due to the many times that you have heard similar sounds and at the same time seen the player's hands on the piano. When habit and past experience play this part, we are in the region of mnemic as opposed to ordinary physical causation. And I think that, if we could regard as ultimately valid the difference between physical and mnemic causation, we could distinguish images from sensations as having mnemic causes, though they may also have physical causes. Sensations, on the other hand, will only have physical causes.

However this may be, the practically effective distinction between sensations and images is that in the causation of sensations, but not of images, the stimulation of nerves carrying an effect into the brain, usually from the surface of the body, plays an essential part. And this accounts for the fact that images and sensations cannot always be distinguished by their intrinsic nature.

Images also differ from sensations as regards their effects. Sensations, as a rule, have both physical and mental effects. As you watch the train you meant to catch leaving the station, there are both the successive positions of the train (physical effects) and the successive waves of fury and disappointment (mental effects). Images, on the contrary, though they MAY produce bodily movements, do so according to mnemic laws, not according to the laws of physics. All their effects, of whatever nature, follow mnemic laws. But this difference is less suitable for definition than the difference as to causes.

Professor Watson, as a logical carrying-out of his behaviourist theory, denies altogether that there are any observable phenomena such as images are supposed to be. He replaces them all by faint sensations, and especially by pronunciation of words sotto voce. When we "think" of a table (say), as opposed to seeing it, what happens, according to him, is usually that we are making small movements of the throat and tongue such as would lead to our uttering the word "table" if they were more pronounced. I shall consider his view again in connection with words; for the present I am only concerned to combat his denial of images. This denial is set forth both in his book on "Behavior" and in an article called "Image and Affection in Behavior" in the "Journal of Philosophy, Psychology and Scientific Methods," vol. x (July, 1913). It seems to me that in this matter he has been betrayed into denying plain facts in the interests of a theory, namely, the supposed impossibility of introspection. I dealt with the theory in Lecture VI; for the present I wish to reinforce the view that the facts are undeniable.

Images are of various sorts, according to the nature of the sensations which they copy. Images of bodily movements, such as we have when we imagine moving an arm or, on a smaller scale, pronouncing a word, might possibly be explained away on Professor Watson's lines, as really consisting in small incipient movements such as, if magnified and prolonged, would be the movements we are said to be imagining. Whether this is the case or not might even be decided experimentally. If there were a delicate instrument for recording small movements in the mouth and throat, we might place such an instrument in a person's mouth and then tell him to recite a poem to himself, as far as possible only in imagination. I should not be at all surprised if it were found that actual small movements take place while he is "mentally" saying over the verses. The point is important, because what is called "thought" consists mainly (though I think not wholly) of inner speech. If Professor Watson is right as regards inner speech, this whole region is transferred from imagination to sensation. But since the question is capable of experimental decision, it would be gratuitous rashness to offer an opinion while that decision is lacking.

But visual and auditory images are much more difficult to deal with in this way, because they lack the connection with physical events in the outer world which belongs to visual and auditory sensations. Suppose, for example, that I am sitting in my room, in which there is an empty arm-chair. I shut my eyes, and call up a visual image of a friend sitting in the arm-chair. If I thrust my image into the world of physics, it contradicts all the usual physical laws. My friend reached the chair without coming in at the door in the usual way; subsequent inquiry will show that he was somewhere else at the moment. If regarded as a sensation, my image has all the marks of the supernatural. My image, therefore, is regarded as an event in me, not as having that position in the orderly happenings of the public world that belongs to sensations. By saying that it is an event in me, we leave it possible that it may be PHYSIOLOGICALLY caused: its privacy may be only due to its connection with my body. But in any case it is not a public event, like an actual person walking in at the door and sitting down in my chair. And it cannot, like inner speech, be regarded as a SMALL sensation, since it occupies just as large an area in my visual field as the actual sensation would do.

Professor Watson says: "I should throw out imagery altogether and attempt to show that all natural thought goes on in terms of sensori-motor processes in the larynx." This view seems to me flatly to contradict experience. If you try to persuade any uneducated person that she cannot call up a visual picture of a friend sitting in a chair, but can only use words describing what such an occurrence would be like, she will conclude that you are mad. (This statement is based upon experiment.) Galton, as every one knows, investigated visual imagery, and found that education tends to kill it: the Fellows of the Royal Society turned out to have much less of it than their wives. I see no reason to doubt his conclusion that the habit of abstract pursuits makes learned men much inferior to the average in power of visualizing, and much more exclusively occupied with words in their "thinking." And Professor Watson is a very learned man.

I shall henceforth assume that the existence of images is admitted, and that they are to be distinguished from sensations by their causes, as well as, in a lesser degree, by their effects. In their intrinsic nature, though they often differ from sensations by being more dim or vague or faint, yet they do not always or universally differ from sensations in any way that can be used for defining them. Their privacy need form no bar to the scientific study of them, any more than the privacy of bodily sensations does. Bodily sensations are admitted by even the most severe critics of introspection, although, like images, they can only be observed by one observer. It must be admitted, however, that the laws of the appearance and disappearance of images are little known and difficult to discover, because we are not assisted, as in the case of sensations, by our knowledge of the physical world.

There remains one very important point concerning images, which will occupy us much hereafter, and that is, their resemblance to previous sensations. They are said to be "copies" of sensations, always as regards the simple qualities that enter into them, though not always as regards the manner in which these are put together. It is generally believed that we cannot imagine a shade of colour that we have never seen, or a sound that we have never heard. On this subject Hume is the classic. He says, in the definitions already quoted:

"Those perceptions, which enter with most force and violence, we may name IMPRESSIONS; and under this name I comprehend all our sensations, passions and emotions, as they make their first appearance in the soul. By IDEAS I mean the faint images of these in thinking and reasoning."

He next explains the difference between simple and complex ideas, and explains that a complex idea may occur without any similar complex impression. But as regards simple ideas, he states that "every simple idea has a simple impression, which resembles it, and every simple impression a correspondent idea." He goes on to enunciate the general principle "that all our simple ideas in their first appearance are derived from simple impressions, which are correspondent to them, and which they exactly represent" ("Treatise of Human Nature," Part I, Section I).

It is this fact, that images resemble antecedent sensations, which enables us to call them images "of" this or that. For the understanding of memory, and of knowledge generally, the recognizable resemblance of images and sensations is of fundamental importance.

There are difficulties in establishing Hume's principles, and doubts as to whether it is exactly true. Indeed, he himself signalized an exception immediately after stating his maxim. Nevertheless, it is impossible to doubt that in the main simple images are copies of similar simple sensations which have occurred earlier, and that the same is true of complex images in all cases of memory as opposed to mere imagination. Our power of acting with reference to what is sensibly absent is largely due to this characteristic of images, although, as education advances, images tend to be more and more replaced by words. We shall have much to say in the next two lectures on the subject of images as copies of sensations. What has been said now is merely by way of reminder that this is their most notable characteristic.

I am by no means confident that the distinction between images and sensations is ultimately valid, and I should be glad to be convinced that images can be reduced to sensations of a peculiar kind. I think it is clear, however, that, at any rate in the case of auditory and visual images, they do differ from ordinary auditory and visual sensations, and therefore form a recognizable class of occurrences, even if it should prove that they can be regarded as a sub-class of sensations. This is all that is necessary to validate the use of images to be made in the sequel.

LECTURE IX.

MEMORY

Memory, which we are to consider to-day, introduces us to knowledge in one of its forms. The analysis of knowledge will occupy us until the end of the thirteenth lecture, and is the most difficult part of our whole enterprise.

I do not myself believe that the analysis of knowledge can be effected entirely by means of purely external observation, such as behaviourists employ. I shall discuss this question in later lectures. In the present lecture I shall attempt the analysis of memory-knowledge, both as an introduction to the problem of knowledge in general, and because memory, in some form, is presupposed in almost all other knowledge. Sensation, we decided, is not a form of knowledge. It might, however, have been expected that we should begin our discussion of knowledge with PERCEPTION, i.e. with that integral experience of things in the environment, out of which sensation is extracted by psychological analysis. What is called perception differs from sensation by the fact that the sensational ingredients bring up habitual associates—images and expectations of their usual correlates—all of which are subjectively indistinguishable from the sensation. The FACT of past experience is essential in producing this filling-out of sensation, but not the RECOLLECTION of past experience. The non-sensational elements in perception can be wholly explained as the result of habit, produced by frequent correlations. Perception, according to our definition in Lecture VII, is no more a form of knowledge than sensation is, except in so far as it involves expectations. The purely psychological problems which it raises are not very difficult, though they have sometimes been rendered artificially obscure by unwillingness to admit the fallibility of the non-sensational elements of perception. On the other hand, memory raises many difficult and very important problems, which it is necessary to consider at the first possible moment.

One reason for treating memory at this early stage is that it seems to be involved in the fact that images are recognized as "copies" of past sensible experience. In the preceding lecture I alluded to Hume's principle "that all our simple ideas in their first appearance are derived from simple impressions, which are correspondent to them, and which they exactly represent." Whether or not this principle is liable to exceptions, everyone would agree that is has a broad measure of truth, though the word "exactly" might seem an overstatement, and it might seem more correct to say that ideas APPROXIMATELY represent impressions. Such modifications of Hume's principle, however, do not affect the problem which I wish to present for your consideration, namely: Why do we believe that images are, sometimes or always, approximately or exactly, copies of sensations? What sort of evidence is there? And what sort of evidence is logically possible? The difficulty of this question arises through the fact that the sensation which an image is supposed to copy is in the past when the image exists, and can therefore only be known by memory, while, on the other hand, memory of past sensations seems only possible by means of present images. How, then, are we to find any way of comparing the present image and the past sensation? The problem is just as acute if we say that images differ from their prototypes as if we say that they resemble them; it is the very possibility of comparison that is hard to understand. We think we can know that they are alike or different, but we cannot bring them together in one experience and compare them. To deal with this problem, we must have a theory of memory. In this way the whole status of images as "copies" is bound up with the analysis of memory.

In investigating memory-beliefs, there are certain points which must be borne in mind. In the first place, everything constituting a memory-belief is happening now, not in that past time to which the belief is said to refer. It is not logically necessary to the existence of a memory-belief that the event remembered should have occurred, or even that the past should have existed at all. There is no logical impossibility in the hypothesis that the world sprang into being five minutes ago, exactly as it then was, with a population that "remembered" a wholly unreal past. There is no logically necessary connection between events at different times; therefore nothing that is happening now or will happen in the future can disprove the hypothesis that the world began five minutes ago. Hence the occurrences which are CALLED knowledge of the past are logically independent of the past; they are wholly analysable into present contents, which might, theoretically, be just what they are even if no past had existed.

I am not suggesting that the non-existence of the past should be entertained as a serious hypothesis. Like all sceptical hypotheses, it is logically tenable, but uninteresting. All that I am doing is to use its logical tenability as a help in the analysis of what occurs when we remember.

In the second place, images without beliefs are insufficient to constitute memory; and habits are still more insufficient. The behaviourist, who attempts to make psychology a record of behaviour, has to trust his memory in making the record. "Habit" is a concept involving the occurrence of similar events at different times; if the behaviourist feels confident that there is such a phenomenon as habit, that can only be because he trusts his memory, when it assures him that there have been other times. And the same applies to images. If we are to know as it is supposed we do—that images are "copies," accurate or inaccurate, of past events, something more than the mere occurrence of images must go to constitute this knowledge. For their mere occurrence, by itself, would not suggest any connection with anything that had happened before.

Can we constitute memory out of images together with suitable beliefs? We may take it that memory-images, when they occur in true memory, are (a) known to be copies, (b) sometimes known to be imperfect copies (cf. footnote on previous page). How is it possible to know that a memory-image is an imperfect copy, without having a more accurate copy by which to replace it? This would SEEM to suggest that we have a way of knowing the past which is independent of images, by means of which we can criticize image-memories. But I do not think such an inference is warranted.

What results, formally, from our knowledge of the past through images of which we recognize the inaccuracy, is that such images must have two characteristics by which we can arrange them in two series, of which one corresponds to the more or less remote period in the past to which they refer, and the other to our greater or less confidence in their accuracy. We will take the second of these points first.

Our confidence or lack of confidence in the accuracy of a memory-image must, in fundamental cases, be based upon a characteristic of the image itself, since we cannot evoke the past bodily and compare it with the present image. It might be suggested that vagueness is the required characteristic, but I do not think this is the case. We sometimes have images that are by no means peculiarly vague, which yet we do not trust—for example, under the influence of fatigue we may see a friend's face vividly and clearly, but horribly distorted. In such a case we distrust our image in spite of its being unusually clear. I think the characteristic by which we distinguish the images we trust is the feeling of FAMILIARITY that accompanies them. Some images, like some sensations, feel very familiar, while others feel strange. Familiarity is a feeling capable of degrees. In an image of a well-known face, for example, some parts may feel more familiar than others; when this happens, we have more belief in the accuracy of the familiar parts than in that of the unfamiliar parts. I think it is by this means that we become critical of images, not by some imageless memory with which we compare them. I shall return to the consideration of familiarity shortly.

I come now to the other characteristic which memory-images must have in order to account for our knowledge of the past. They must have some characteristic which makes us regard them as referring to more or less remote portions of the past. That is to say if we suppose that A is the event remembered, B the remembering, and t the interval of time between A and B, there must be some characteristic of B which is capable of degrees, and which, in accurately dated memories, varies as t varies. It may increase as t increases, or diminish as t increases. The question which of these occurs is not of any importance for the theoretic serviceability of the characteristic in question.

In actual fact, there are doubtless various factors that concur in giving us the feeling of greater or less remoteness in some remembered event. There may be a specific feeling which could be called the feeling of "pastness," especially where immediate memory is concerned. But apart from this, there are other marks. One of these is context. A recent memory has, usually, more context than a more distant one. When a remembered event has a remembered context, this may occur in two ways, either (a) by successive images in the same order as their prototypes, or (b) by remembering a whole process simultaneously, in the same way in which a present process may be apprehended, through akoluthic sensations which, by fading, acquire the mark of just-pastness in an increasing degree as they fade, and are thus placed in a series while all sensibly present. It will be context in this second sense, more specially, that will give us a sense of the nearness or remoteness of a remembered event.

There is, of course, a difference between knowing the temporal relation of a remembered event to the present, and knowing the time-order of two remembered events. Very often our knowledge of the temporal relation of a remembered event to the present is inferred from its temporal relations to other remembered events. It would seem that only rather recent events can be placed at all accurately by means of feelings giving their temporal relation to the present, but it is clear that such feelings must play an essential part in the process of dating remembered events.

We may say, then, that images are regarded by us as more or less accurate copies of past occurrences because they come to us with two sorts of feelings: (1) Those that may be called feelings of familiarity; (2) those that may be collected together as feelings giving a sense of pastness. The first lead us to trust our memories, the second to assign places to them in the time-order.

We have now to analyse the memory-belief, as opposed to the characteristics of images which lead us to base memory-beliefs upon them.

If we had retained the "subject" or "act" in knowledge, the whole problem of memory would have been comparatively simple. We could then have said that remembering is a direct relation between the present act or subject and the past occurrence remembered: the act of remembering is present, though its object is past. But the rejection of the subject renders some more complicated theory necessary. Remembering has to be a present occurrence in some way resembling, or related to, what is remembered. And it is difficult to find any ground, except a pragmatic one, for supposing that memory is not sheer delusion, if, as seems to be the case, there is not, apart from memory, any way of ascertaining that there really was a past occurrence having the required relation to our present remembering. What, if we followed Meinong's terminology, we should call the "object" in memory, i.e. the past event which we are said to be remembering, is unpleasantly remote from the "content," i.e. the present mental occurrence in remembering. There is an awkward gulf between the two, which raises difficulties for the theory of knowledge. But we must not falsify observation to avoid theoretical difficulties. For the present, therefore, let us forget these problems, and try to discover what actually occurs in memory.

Some points may be taken as fixed, and such as any theory of memory must arrive at. In this case, as in most others, what may be taken as certain in advance is rather vague. The study of any topic is like the continued observation of an object which is approaching us along a road: what is certain to begin with is the quite vague knowledge that there is SOME object on the road. If you attempt to be less vague, and to assert that the object is an elephant, or a man, or a mad dog, you run a risk of error; but the purpose of continued observation is to enable you to arrive at such more precise knowledge. In like manner, in the study of memory, the certainties with which you begin are very vague, and the more precise propositions at which you try to arrive are less certain than the hazy data from which you set out. Nevertheless, in spite of the risk of error, precision is the goal at which we must aim.

The first of our vague but indubitable data is that there is knowledge of the past. We do not yet know with any precision what we mean by "knowledge," and we must admit that in any given instance our memory may be at fault. Nevertheless, whatever a sceptic might urge in theory, we cannot practically doubt that we got up this morning, that we did various things yesterday, that a great war has been taking place, and so on. How far our knowledge of the past is due to memory, and how far to other sources, is of course a matter to be investigated, but there can be no doubt that memory forms an indispensable part of our knowledge of the past.

The second datum is that we certainly have more capacity for knowing the past than for knowing the future. We know some things about the future, for example what eclipses there will be; but this knowledge is a matter of elaborate calculation and inference, whereas some of our knowledge of the past comes to us without effort, in the same sort of immediate way in which we acquire knowledge of occurrences in our present environment. We might provisionally, though perhaps not quite correctly, define "memory" as that way of knowing about the past which has no analogue in our knowledge of the future; such a definition would at least serve to mark the problem with which we are concerned, though some expectations may deserve to rank with memory as regards immediacy.

A third point, perhaps not quite so certain as our previous two, is that the truth of memory cannot be wholly practical, as pragmatists wish all truth to be. It seems clear that some of the things I remember are trivial and without any visible importance for the future, but that my memory is true (or false) in virtue of a past event, not in virtue of any future consequences of my belief. The definition of truth as the correspondence between beliefs and facts seems peculiarly evident in the case of memory, as against not only the pragmatist definition but also the idealist definition by means of coherence. These considerations, however, are taking us away from psychology, to which we must now return.

It is important not to confuse the two forms of memory which Bergson distinguishes in the second chapter of his "Matter and Memory," namely the sort that consists of habit, and the sort that consists of independent recollection. He gives the instance of learning a lesson by heart: when I know it by heart I am said to "remember" it, but this merely means that I have acquired certain habits; on the other hand, my recollection of (say) the second time I read the lesson while I was learning it is the recollection of a unique event, which occurred only once. The recollection of a unique event cannot, so Bergson contends, be wholly constituted by habit, and is in fact something radically different from the memory which is habit. The recollection alone is true memory. This distinction is vital to the understanding of memory. But it is not so easy to carry out in practice as it is to draw in theory. Habit is a very intrusive feature of our mental life, and is often present where at first sight it seems not to be. There is, for example, a habit of remembering a unique event. When we have once described the event, the words we have used easily become habitual. We may even have used words to describe it to ourselves while it was happening; in that case, the habit of these words may fulfil the function of Bergson's true memory, while in reality it is nothing but habit-memory. A gramophone, by the help of suitable records, might relate to us the incidents of its past; and people are not so different from gramophones as they like to believe.

In spite, however, of a difficulty in distinguishing the two forms of memory in practice, there can be no doubt that both forms exist. I can set to work now to remember things I never remembered before, such as what I had to eat for breakfast this morning, and it can hardly be wholly habit that enables me to do this. It is this sort of occurrence that constitutes the essence of memory Until we have analysed what happens in such a case as this, we have not succeeded in understanding memory.

The sort of memory with which we are here concerned is the sort which is a form of knowledge. Whether knowledge itself is reducible to habit is a question to which I shall return in a later lecture; for the present I am only anxious to point out that, whatever the true analysis of knowledge may be, knowledge of past occurrences is not proved by behaviour which is due to past experience. The fact that a man can recite a poem does not show that he remembers any previous occasion on which he has recited or read it. Similarly, the performances of animals in getting out of cages or mazes to which they are accustomed do not prove that they remember having been in the same situation before. Arguments in favour of (for example) memory in plants are only arguments in favour of habit-memory, not of knowledge-memory. Samuel Butler's arguments in favour of the view that an animal remembers something of the lives of its ancestors are, when examined, only arguments in favour of habit-memory. Semon's two books, mentioned in an earlier lecture, do not touch knowledge-memory at all closely. They give laws according to which images of past occurrences come into our minds, but do not discuss our belief that these images refer to past occurrences, which is what constitutes knowledge-memory. It is this that is of interest to theory of knowledge. I shall speak of it as "true" memory, to distinguish it from mere habit acquired through past experience. Before considering true memory, it will be well to consider two things which are on the way towards memory, namely the feeling of familiarity and recognition.

We often feel that something in our sensible environment is familiar, without having any definite recollection of previous occasions on which we have seen it. We have this feeling normally in places where we have often been before—at home, or in well-known streets. Most people and animals find it essential to their happiness to spend a good deal of their time in familiar surroundings, which are especially comforting when any danger threatens. The feeling of familiarity has all sorts of degrees, down to the stage where we dimly feel that we have seen a person before. It is by no means always reliable; almost everybody has at some time experienced the well-known illusion that all that is happening now happened before at some time. There are occasions when familiarity does not attach itself to any definite object, when there is merely a vague feeling that SOMETHING is familiar. This is illustrated by Turgenev's "Smoke," where the hero is long puzzled by a haunting sense that something in his present is recalling something in his past, and at last traces it to the smell of heliotrope. Whenever the sense of familiarity occurs without a definite object, it leads us to search the environment until we are satisfied that we have found the appropriate object, which leads us to the judgment: "THIS is familiar." I think we may regard familiarity as a definite feeling, capable of existing without an object, but normally standing in a specific relation to some feature of the environment, the relation being that which we express in words by saying that the feature in question is familiar. The judgment that what is familiar has been experienced before is a product of reflection, and is no part of the feeling of familiarity, such as a horse may be supposed to have when he returns to his stable. Thus no knowledge as to the past is to be derived from the feeling of familiarity alone.

A further stage is RECOGNITION. This may be taken in two senses, the first when a thing not merely feels familiar, but we know it is such-and-such. We recognize our friend Jones, we know cats and dogs when we see them, and so on. Here we have a definite influence of past experience, but not necessarily any actual knowledge of the past. When we see a cat, we know it is a cat because of previous cats we have seen, but we do not, as a rule, recollect at the moment any particular occasion when we have seen a cat. Recognition in this sense does not necessarily involve more than a habit of association: the kind of object we are seeing at the moment is associated with the word "cat," or with an auditory image of purring, or whatever other characteristic we may happen to recognize in the cat of the moment. We are, of course, in fact able to judge, when we recognize an object, that we have seen it before, but this judgment is something over and above recognition in this first sense, and may very probably be impossible to animals that nevertheless have the experience of recognition in this first sense of the word.

There is, however, another sense of the word, in which we mean by recognition, not knowing the name of a thing or some other property of it, but knowing that we have seen it before In this sense recognition does involve knowledge about the Fast. This knowledge is memory in one sense, though in another it is not. It does not involve a definite memory of a definite past event, but only the knowledge that something happening now is similar to something that happened before. It differs from the sense of familiarity by being cognitive; it is a belief or judgment, which the sense of familiarity is not. I do not wish to undertake the analysis of belief at present, since it will be the subject of the twelfth lecture; for the present I merely wish to emphasize the fact that recognition, in our second sense, consists in a belief, which we may express approximately in the words: "This has existed before."

There are, however, several points in which such an account of recognition is inadequate. To begin with, it might seem at first sight more correct to define recognition as "I have seen this before" than as "this has existed before." We recognize a thing (it may be urged) as having been in our experience before, whatever that may mean; we do not recognize it as merely having been in the world before. I am not sure that there is anything substantial in this point. The definition of "my experience" is difficult; broadly speaking, it is everything that is connected with what I am experiencing now by certain links, of which the various forms of memory are among the most important. Thus, if I recognize a thing, the occasion of its previous existence in virtue of which I recognize it forms part of "my experience" by DEFINITION: recognition will be one of the marks by which my experience is singled out from the rest of the world. Of course, the words "this has existed before" are a very inadequate translation of what actually happens when we form a judgment of recognition, but that is unavoidable: words are framed to express a level of thought which is by no means primitive, and are quite incapable of expressing such an elementary occurrence as recognition. I shall return to what is virtually the same question in connection with true memory, which raises exactly similar problems.

A second point is that, when we recognize something, it was not in fact the very same thing, but only something similar, that we experienced on a former occasion. Suppose the object in question is a friend's face. A person's face is always changing, and is not exactly the same on any two occasions. Common sense treats it as one face with varying expressions; but the varying expressions actually exist, each at its proper time, while the one face is merely a logical construction. We regard two objects as the same, for common-sense purposes, when the reaction they call for is practically the same. Two visual appearances, to both of which it is appropriate to say: "Hullo, Jones!" are treated as appearances of one identical object, namely Jones. The name "Jones" is applicable to both, and it is only reflection that shows us that many diverse particulars are collected together to form the meaning of the name "Jones." What we see on any one occasion is not the whole series of particulars that make up Jones, but only one of them (or a few in quick succession). On another occasion we see another member of the series, but it is sufficiently similar to count as the same from the standpoint of common sense. Accordingly, when we judge "I have seen THIS before," we judge falsely if "this" is taken as applying to the actual constituent of the world that we are seeing at the moment. The word "this" must be interpreted vaguely so as to include anything sufficiently like what we are seeing at the moment. Here, again, we shall find a similar point as regards true memory; and in connection with true memory we will consider the point again. It is sometimes suggested, by those who favour behaviourist views, that recognition consists in behaving in the same way when a stimulus is repeated as we behaved on the first occasion when it occurred. This seems to be the exact opposite of the truth. The essence of recognition is in the DIFFERENCE between a repeated stimulus and a new one. On the first occasion there is no recognition; on the second occasion there is. In fact, recognition is another instance of the peculiarity of causal laws in psychology, namely, that the causal unit is not a single event, but two or more events Habit is the great instance of this, but recognition is another. A stimulus occurring once has a certain effect; occurring twice, it has the further effect of recognition. Thus the phenomenon of recognition has as its cause the two occasions when the stimulus has occurred; either alone is insufficient. This complexity of causes in psychology might be connected with Bergson's arguments against repetition in the mental world. It does not prove that there are no causal laws in psychology, as Bergson suggests; but it does prove that the causal laws of psychology are Prima facie very different from those of physics. On the possibility of explaining away the difference as due to the peculiarities of nervous tissue I have spoken before, but this possibility must not be forgotten if we are tempted to draw unwarranted metaphysical deductions.

True memory, which we must now endeavour to understand, consists of knowledge of past events, but not of all such knowledge. Some knowledge of past events, for example what we learn through reading history, is on a par with the knowledge we can acquire concerning the future: it is obtained by inference, not (so to speak) spontaneously. There is a similar distinction in our knowledge of the present: some of it is obtained through the senses, some in more indirect ways. I know that there are at this moment a number of people in the streets of New York, but I do not know this in the immediate way in which I know of the people whom I see by looking out of my window. It is not easy to state precisely wherein the difference between these two sorts of knowledge consists, but it is easy to feel the difference. For the moment, I shall not stop to analyse it, but shall content myself with saying that, in this respect, memory resembles the knowledge derived from the senses. It is immediate, not inferred, not abstract; it differs from perception mainly by being referred to the past.

In regard to memory, as throughout the analysis of knowledge, there are two very distinct problems, namely (1) as to the nature of the present occurrence in knowing; (2) as to the relation of this occurrence to what is known. When we remember, the knowing is now, while what is known is in the past. Our two questions are, in the case of memory:

(1) What is the present occurrence when we remember?

(2) What is the relation of this present occurrence to the past event which is remembered?

Of these two questions, only the first concerns the psychologist; the second belongs to theory of knowledge. At the same time, if we accept the vague datum with which we began, to the effect that, in some sense, there is knowledge of the past, we shall have to find, if we can, such an account of the present occurrence in remembering as will make it not impossible for remembering to give us knowledge of the past. For the present, however, we shall do well to forget the problems concerning theory of knowledge, and concentrate upon the purely psychological problem of memory.

Between memory-image and sensation there is an intermediate experience concerning the immediate past. For example, a sound that we have just heard is present to us in a way which differs both from the sensation while we are hearing the sound and from the memory-image of something heard days or weeks ago. James states that it is this way of apprehending the immediate past that is "the ORIGINAL of our experience of pastness, from whence we get the meaning of the term"Everyone knows the experience of noticing (say) that the clock HAS BEEN striking, when we did not notice it while it was striking. And when we hear a remark spoken, we are conscious of the earlier words while the later ones are being uttered, and this retention feels different from recollection of something definitely past. A sensation fades gradually, passing by continuous gradations to the status of an image. This retention of the immediate past in a condition intermediate between sensation and image may be called "immediate memory." Everything belonging to it is included with sensation in what is called the "specious present." The specious present includes elements at all stages on the journey from sensation to image. It is this fact that enables us to apprehend such things as movements, or the order of the words in a spoken sentence. Succession can occur within the specious present, of which we can distinguish some parts as earlier and others as later. It is to be supposed that the earliest parts are those that have faded most from their original force, while the latest parts are those that retain their full sensational character. At the beginning of a stimulus we have a sensation; then a gradual transition; and at the end an image. Sensations while they are fading are called "akoluthic" sensations When the process of fading is completed (which happens very quickly), we arrive at the image, which is capable of being revived on subsequent occasions with very little change. True memory, as opposed to "immediate memory," applies only to events sufficiently distant to have come to an end of the period of fading. Such events, if they are represented by anything present, can only be represented by images, not by those intermediate stages, between sensations and images, which occur during the period of fading.

Immediate memory is important both because it provides experience of succession, and because it bridges the gulf between sensations and the images which are their copies. But it is now time to resume the consideration of true memory.

Suppose you ask me what I ate for breakfast this morning. Suppose, further, that I have not thought about my breakfast in the meantime, and that I did not, while I was eating it, put into words what it consisted of. In this case my recollection will be true memory, not habit-memory. The process of remembering will consist of calling up images of my breakfast, which will come to me with a feeling of belief such as distinguishes memory-images from mere imagination-images. Or sometimes words may come without the intermediary of images; but in this case equally the feeling of belief is essential.

Let us omit from our consideration, for the present, the memories in which words replace images. These are always, I think, really habit-memories, the memories that use images being the typical true memories.

Memory-images and imagination-images do not differ in their intrinsic qualities, so far as we can discover. They differ by the fact that the images that constitute memories, unlike those that constitute imagination, are accompanied by a feeling of belief which may be expressed in the words "this happened." The mere occurrence of images, without this feeling of belief, constitutes imagination; it is the element of belief that is the distinctive thing in memory.

There are, if I am not mistaken, at least three different kinds of belief-feeling, which we may call respectively memory, expectation and bare assent. In what I call bare assent, there is no time-element in the feeling of belief, though there may be in the content of what is believed. If I believe that Caesar landed in Britain in B.C. 55, the time-determination lies, not in the feeling of belief, but in what is believed. I do not remember the occurrence, but have the same feeling towards it as towards the announcement of an eclipse next year. But when I have seen a flash of lightning and am waiting for the thunder, I have a belief-feeling analogous to memory, except that it refers to the future: I have an image of thunder, combined with a feeling which may be expressed in the words: "this will happen." So, in memory, the pastness lies, not in the content of what is believed, but in the nature of the belief-feeling. I might have just the same images and expect their realization; I might entertain them without any belief, as in reading a novel; or I might entertain them together with a time-determination, and give bare assent, as in reading history. I shall return to this subject in a later lecture, when we come to the analysis of belief. For the present, I wish to make it clear that a certain special kind of belief is the distinctive characteristic of memory.

The problem as to whether memory can be explained as habit or association requires to be considered afresh in connection with the causes of our remembering something. Let us take again the case of my being asked what I had for breakfast this morning. In this case the question leads to my setting to work to recollect. It is a little strange that the question should instruct me as to what it is that I am to recall. This has to do with understanding words, which will be the topic of the next lecture; but something must be said about it now. Our understanding of the words "breakfast this morning" is a habit, in spite of the fact that on each fresh day they point to a different occasion. "This morning" does not, whenever it is used, mean the same thing, as "John" or "St. Paul's" does; it means a different period of time on each different day. It follows that the habit which constitutes our understanding of the words "this morning" is not the habit of associating the words with a fixed object, but the habit of associating them with something having a fixed time-relation to our present. This morning has, to-day, the same time-relation to my present that yesterday morning had yesterday. In order to understand the phrase "this morning" it is necessary that we should have a way of feeling time-intervals, and that this feeling should give what is constant in the meaning of the words "this morning." This appreciation of time-intervals is, however, obviously a product of memory, not a presupposition of it. It will be better, therefore, if we wish to analyse the causation of memory by something not presupposing memory, to take some other instance than that of a question about "this morning."

Let us take the case of coming into a familiar room where something has been changed—say a new picture hung on the wall. We may at first have only a sense that SOMETHING is unfamiliar, but presently we shall remember, and say "that picture was not on the wall before." In order to make the case definite, we will suppose that we were only in the room on one former occasion. In this case it seems fairly clear what happens. The other objects in the room are associated, through the former occasion, with a blank space of wall where now there is a picture. They call up an image of a blank wall, which clashes with perception of the picture. The image is associated with the belief-feeling which we found to be distinctive of memory, since it can neither be abolished nor harmonized with perception. If the room had remained unchanged, we might have had only the feeling of familiarity without the definite remembering; it is the change that drives us from the present to memory of the past.

We may generalize this instance so as to cover the causes of many memories. Some present feature of the environment is associated, through past experiences, with something now absent; this absent something comes before us as an image, and is contrasted with present sensation. In cases of this sort, habit (or association) explains why the present feature of the environment brings up the memory-image, but it does not explain the memory-belief. Perhaps a more complete analysis could explain the memory-belief also on lines of association and habit, but the causes of beliefs are obscure, and we cannot investigate them yet. For the present we must content ourselves with the fact that the memory-image can be explained by habit. As regards the memory-belief, we must, at least provisionally, accept Bergson's view that it cannot be brought under the head of habit, at any rate when it first occurs, i.e. when we remember something we never remembered before.

We must now consider somewhat more closely the content of a memory-belief. The memory-belief confers upon the memory-image something which we may call "meaning;" it makes us feel that the image points to an object which existed in the past. In order to deal with this topic we must consider the verbal expression of the memory-belief. We might be tempted to put the memory-belief into the words: "Something like this image occurred." But such words would be very far from an accurate translation of the simplest kind of memory-belief. "Something like this image" is a very complicated conception. In the simplest kind of memory we are not aware of the difference between an image and the sensation which it copies, which may be called its "prototype." When the image is before us, we judge rather "this occurred." The image is not distinguished from the object which existed in the past: the word "this" covers both, and enables us to have a memory-belief which does not introduce the complicated notion "something like this."

It might be objected that, if we judge "this occurred" when in fact "this" is a present image, we judge falsely, and the memory-belief, so interpreted, becomes deceptive. This, however, would be a mistake, produced by attempting to give to words a precision which they do not possess when used by unsophisticated people. It is true that the image is not absolutely identical with its prototype, and if the word "this" meant the image to the exclusion of everything else, the judgment "this occurred" would be false. But identity is a precise conception, and no word, in ordinary speech, stands for anything precise. Ordinary speech does not distinguish between identity and close similarity. A word always applies, not only to one particular, but to a group of associated particulars, which are not recognized as multiple in common thought or speech. Thus primitive memory, when it judges that "this occurred," is vague, but not false.

Vague identity, which is really close similarity, has been a source of many of the confusions by which philosophy has lived. Of a vague subject, such as a "this," which is both an image and its prototype, contradictory predicates are true simultaneously: this existed and does not exist, since it is a thing remembered, but also this exists and did not exist, since it is a present image. Hence Bergson's interpenetration of the present by the past, Hegelian continuity and identity-in-diversity, and a host of other notions which are thought to be profound because they are obscure and confused. The contradictions resulting from confounding image and prototype in memory force us to precision. But when we become precise, our remembering becomes different from that of ordinary life, and if we forget this we shall go wrong in the analysis of ordinary memory.

Vagueness and accuracy are important notions, which it is very necessary to understand. Both are a matter of degree. All thinking is vague to some extent, and complete accuracy is a theoretical ideal not practically attainable. To understand what is meant by accuracy, it will be well to consider first instruments of measurement, such as a balance or a thermometer. These are said to be accurate when they give different results for very slightly different stimuli A clinical thermometer is accurate when it enables us to detect very slight differences in the temperature of the blood. We may say generally that an instrument is accurate in proportion as it reacts differently to very slightly different stimuli. When a small difference of stimulus produces a great difference of reaction, the instrument is accurate; in the contrary case it is not.

Exactly the same thing applies in defining accuracy of thought or perception. A musician will respond differently to very minute differences in playing which would be quite imperceptible to the ordinary mortal. A negro can see the difference between one negro and another one is his friend, another his enemy. But to us such different responses are impossible: we can merely apply the word "negro" indiscriminately. Accuracy of response in regard to any particular kind of stimulus is improved by practice. Understanding a language is a case in point. Few Frenchmen can hear any difference between the sounds "hall" and "hole," which produce quite different impressions upon us. The two statements "the hall is full of water" and "the hole is full of water" call for different responses, and a hearing which cannot distinguish between them is inaccurate or vague in this respect.

Precision and vagueness in thought, as in perception, depend upon the degree of difference between responses to more or less similar stimuli. In the case of thought, the response does not follow immediately upon the sensational stimulus, but that makes no difference as regards our present question. Thus to revert to memory: A memory is "vague" when it is appropriate to many different occurrences: for instance, "I met a man" is vague, since any man would verify it. A memory is "precise" when the occurrences that would verify it are narrowly circumscribed: for instance, "I met Jones" is precise as compared to "I met a man." A memory is "accurate" when it is both precise and true, i.e. in the above instance, if it was Jones I met. It is precise even if it is false, provided some very definite occurrence would have been required to make it true.

It follows from what has been said that a vague thought has more likelihood of being true than a precise one. To try and hit an object with a vague thought is like trying to hit the bull's eye with a lump of putty: when the putty reaches the target, it flattens out all over it, and probably covers the bull's eye along with the rest. To try and hit an object with a precise thought is like trying to hit the bull's eye with a bullet. The advantage of the precise thought is that it distinguishes between the bull's eye and the rest of the target. For example, if the whole target is represented by the fungus family and the bull's eye by mushrooms, a vague thought which can only hit the target as a whole is not much use from a culinary point of view. And when I merely remember that I met a man, my memory may be very inadequate to my practical requirements, since it may make a great difference whether I met Brown or Jones. The memory "I met Jones" is relatively precise. It is accurate if I met Jones, inaccurate if I met Brown, but precise in either case as against the mere recollection that I met a man.

The distinction between accuracy and precision is however, not fundamental. We may omit precision from out thoughts and confine ourselves to the distinction between accuracy and vagueness. We may then set up the following definitions:

An instrument is "reliable" with respect to a given set of stimuli when to stimuli which are not relevantly different it gives always responses which are not relevantly different.

An instrument is a "measure" of a set of stimuli which are serially ordered when its responses, in all cases where they are relevantly different, are arranged in a series in the same order.

The "degree of accuracy" of an instrument which is a reliable measurer is the ratio of the difference of response to the difference of stimulus in cases where the difference of stimulus is small. That is to say, if a small difference of stimulus produces a great difference of response, the instrument is very accurate; in the contrary case, very inaccurate.

A mental response is called "vague" in proportion to its lack of accuracy, or rather precision.

These definitions will be found useful, not only in the case of memory, but in almost all questions concerned with knowledge.

It should be observed that vague beliefs, so far from being necessarily false, have a better chance of truth than precise ones, though their truth is less valuable than that of precise beliefs, since they do not distinguish between occurrences which may differ in important ways.

The whole of the above discussion of vagueness and accuracy was occasioned by the attempt to interpret the word "this" when we judge in verbal memory that "this occurred." The word "this," in such a judgment, is a vague word, equally applicable to the present memory-image and to the past occurrence which is its prototype. A vague word is not to be identified with a general word, though in practice the distinction may often be blurred. A word is general when it is understood to be applicable to a number of different objects in virtue of some common property. A word is vague when it is in fact applicable to a number of different objects because, in virtue of some common property, they have not appeared, to the person using the word, to be distinct. I emphatically do not mean that he has judged them to be identical, but merely that he has made the same response to them all and has not judged them to be different. We may compare a vague word to a jelly and a general word to a heap of shot. Vague words precede judgments of identity and difference; both general and particular words are subsequent to such judgments. The word "this" in the primitive memory-belief is a vague word, not a general word; it covers both the image and its prototype because the two are not distinguished.

But we have not yet finished our analysis of the memory-belief. The tense in the belief that "this occurred" is provided by the nature of the belief-feeling involved in memory; the word "this," as we have seen, has a vagueness which we have tried to describe. But we must still ask what we mean by "occurred." The image is, in one sense, occurring now; and therefore we must find some other sense in which the past event occurred but the image does not occur.

There are two distinct questions to be asked: (1) What causes us to say that a thing occurs? (2) What are we feeling when we say this? As to the first question, in the crude use of the word, which is what concerns us, memory-images would not be said to occur; they would not be noticed in themselves, but merely used as signs of the past event. Images are "merely imaginary"; they have not, in crude thought, the sort of reality that belongs to outside bodies. Roughly speaking, "real" things would be those that can cause sensations, those that have correlations of the sort that constitute physical objects. A thing is said to be "real" or to "occur" when it fits into a context of such correlations. The prototype of our memory-image did fit into a physical context, while our memory-image does not. This causes us to feel that the prototype was "real," while the image is "imaginary."

But the answer to our second question, namely as to what we are feeling when we say a thing "occurs" or is "real," must be somewhat different. We do not, unless we are unusually reflective, think about the presence or absence of correlations: we merely have different feelings which, intellectualized, may be represented as expectations of the presence or absence of correlations. A thing which "feels real" inspires us with hopes or fears, expectations or curiosities, which are wholly absent when a thing "feels imaginary." The feeling of reality is a feeling akin to respect: it belongs PRIMARILY to whatever can do things to us without our voluntary co-operation. This feeling of reality, related to the memory-image, and referred to the past by the specific kind of belief-feeling that is characteristic of memory, seems to be what constitutes the act of remembering in its pure form.

We may now summarize our analysis of pure memory.

Memory demands (a) an image, (b) a belief in past existence. The belief may be expressed in the words "this existed."

The belief, like every other, may be analysed into (1) the believing, (2) what is believed. The believing is a specific feeling or sensation or complex of sensations, different from expectation or bare assent in a way that makes the belief refer to the past; the reference to the past lies in the belief-feeling, not in the content believed. There is a relation between the belief-feeling and the content, making the belief-feeling refer to the content, and expressed by saying that the content is what is believed.

The content believed may or may not be expressed in words. Let us take first the case when it is not. In that case, if we are merely remembering that something of which we now have an image occurred, the content consists of (a) the image, (b) the feeling, analogous to respect, which we translate by saying that something is "real" as opposed to "imaginary," (c) a relation between the image and the feeling of reality, of the sort expressed when we say that the feeling refers to the image. This content does not contain in itself any time-determination.

The time-determination lies in the nature of the belief feeling, which is that called "remembering" or (better) "recollecting." It is only subsequent reflection upon this reference to the past that makes us realize the distinction between the image and the event recollected. When we have made this distinction, we can say that the image "means" the past event.

The content expressed in words is best represented by the words "the existence of this," since these words do not involve tense, which belongs to the belief-feeling, not to the content. Here "this" is a vague term, covering the memory-image and anything very like it, including its prototype. "Existence" expresses the feeling of a "reality" aroused primarily by whatever can have effects upon us without our voluntary co-operation. The word "of" in the phrase "the existence of this" represents the relation which subsists between the feeling of reality and the "this."

This analysis of memory is probably extremely faulty, but I do not know how to improve it.

NOTE.-When I speak of a FEELING of belief, I use the word "feeling" in a popular sense, to cover a sensation or an image or a complex of sensations or images or both; I use this word because I do not wish to commit myself to any special analysis of the belief-feeling.

LECTURE X.

WORDS AND MEANING

The problem with which we shall be concerned in this lecture is the problem of determining what is the relation called "meaning." The word "Napoleon," we say, "means" a certain person. In saying this, we are asserting a relation between the word "Napoleon" and the person so designated. It is this relation that we must now investigate.

Let us first consider what sort of object a word is when considered simply as a physical thing, apart from its meaning. To begin with, there are many instances of a word, namely all the different occasions when it is employed. Thus a word is not something unique and particular, but a set of occurrences. If we confine ourselves to spoken words, a word has two aspects, according as we regard it from the point of view of the speaker or from that of the hearer. From the point of view of the speaker, a single instance of the use of a word consists of a certain set of movements in the throat and mouth, combined with breath. From the point of view of the hearer, a single instance of the use of a word consists of a certain series of sounds, each being approximately represented by a single letter in writing, though in practice a letter may represent several sounds, or several letters may represent one sound. The connection between the spoken word and the word as it reaches the hearer is causal. Let us confine ourselves to the spoken word, which is the more important for the analysis of what is called "thought." Then we may say that a single instance of the spoken word consists of a series of movements, and the word consists of a whole set of such series, each member of the set being very similar to each other member. That is to say, any two instances of the word "Napoleon" are very similar, and each instance consists of a series of movements in the mouth.

A single word, accordingly, is by no means simple it is a class of similar series of movements (confining ourselves still to the spoken word). The degree of similarity required cannot be precisely defined: a man may pronounce the word "Napoleon" so badly that it can hardly be determined whether he has really pronounced it or not. The instances of a word shade off into other movements by imperceptible degrees. And exactly analogous observations apply to words heard or written or read. But in what has been said so far we have not even broached the question of the DEFINITION of a word, since "meaning" is clearly what distinguishes a word from other sets of similar movements, and "meaning" remains to be defined.

It is natural to think of the meaning of a word as something conventional. This, however, is only true with great limitations. A new word can be added to an existing language by a mere convention, as is done, for instance, with new scientific terms. But the basis of a language is not conventional, either from the point of view of the individual or from that of the community. A child learning to speak is learning habits and associations which are just as much determined by the environment as the habit of expecting dogs to bark and cocks to crow. The community that speaks a language has learnt it, and modified it by processes almost all of which are not deliberate, but the results of causes operating according to more or less ascertainable laws. If we trace any Indo-European language back far enough, we arrive hypothetically (at any rate according to some authorities) at the stage when language consisted only of the roots out of which subsequent words have grown. How these roots acquired their meanings is not known, but a conventional origin is clearly just as mythical as the social contract by which Hobbes and Rousseau supposed civil government to have been established. We can hardly suppose a parliament of hitherto speechless elders meeting together and agreeing to call a cow a cow and a wolf a wolf. The association of words with their meanings must have grown up by some natural process, though at present the nature of the process is unknown.

Spoken and written words are, of course, not the only way of conveying meaning. A large part of one of Wundt's two vast volumes on language in his "Volkerpsychologie" is concerned with gesture-language. Ants appear to be able to communicate a certain amount of information by means of their antennae. Probably writing itself, which we now regard as merely a way of representing speech, was originally an independent language, as it has remained to this day in China. Writing seems to have consisted originally of pictures, which gradually became conventionalized, coming in time to represent syllables, and finally letters on the telephone principle of "T for Tommy." But it would seem that writing nowhere began as an attempt to represent speech it began as a direct pictorial representation of what was to be expressed. The essence of language lies, not in the use of this or that special means of communication, but in the employment of fixed associations (however these may have originated) in order that something now sensible—a spoken word, a picture, a gesture, or what not—may call up the "idea" of something else. Whenever this is done, what is now sensible may be called a "sign" or "symbol," and that of which it is intended to call up the "idea" may be called its "meaning." This is a rough outline of what constitutes "meaning." But we must fill in the outline in various ways. And, since we are concerned with what is called "thought," we must pay more attention than we otherwise should do to the private as opposed to the social use of language. Language profoundly affects our thoughts, and it is this aspect of language that is of most importance to us in our present inquiry. We are almost more concerned with the internal speech that is never uttered than we are with the things said out loud to other people.

When we ask what constitutes meaning, we are not asking what is the meaning of this or that particular word. The word "Napoleon" means a certain individual; but we are asking, not who is the individual meant, but what is the relation of the word to the individual which makes the one mean the other. But just as it is useful to realize the nature of a word as part of the physical world, so it is useful to realize the sort of thing that a word may mean. When we are clear both as to what a word is in its physical aspect, and as to what sort of thing it can mean, we are in a better position to discover the relation of the two which is meaning.

The things that words mean differ more than words do. There are different sorts of words, distinguished by the grammarians; and there are logical distinctions, which are connected to some extent, though not so closely as was formerly supposed, with the grammatical distinctions of parts of speech. It is easy, however, to be misled by grammar, particularly if all the languages we know belong to one family. In some languages, according to some authorities, the distinction of parts of speech does not exist; in many languages it is widely different from that to which we are accustomed in the Indo-European languages. These facts have to be borne in mind if we are to avoid giving metaphysical importance to mere accidents of our own speech.

In considering what words mean, it is natural to start with proper names, and we will again take "Napoleon" as our instance. We commonly imagine, when we use a proper name, that we mean one definite entity, the particular individual who was called "Napoleon." But what we know as a person is not simple. There MAY be a single simple ego which was Napoleon, and remained strictly identical from his birth to his death. There is no way of proving that this cannot be the case, but there is also not the slightest reason to suppose that it is the case. Napoleon as he was empirically known consisted of a series of gradually changing appearances: first a squalling baby, then a boy, then a slim and beautiful youth, then a fat and slothful person very magnificently dressed This series of appearances, and various occurrences having certain kinds of causal connections with them, constitute Napoleon as empirically known, and therefore are Napoleon in so far as he forms part of the experienced world. Napoleon is a complicated series of occurrences, bound together by causal laws, not, like instances of a word, by similarities. For although a person changes gradually, and presents similar appearances on two nearly contemporaneous occasions, it is not these similarities that constitute the person, as appears from the "Comedy of Errors" for example.

Thus in the case of a proper name, while the word is a set of similar series of movements, what it means is a series of occurrences bound together by causal laws of that special kind that makes the occurrences taken together constitute what we call one person, or one animal or thing, in case the name applies to an animal or thing instead of to a person. Neither the word nor what it names is one of the ultimate indivisible constituents of the world. In language there is no direct way of designating one of the ultimate brief existents that go to make up the collections we call things or persons. If we want to speak of such existents—which hardly happens except in philosophy—we have to do it by means of some elaborate phrase, such as "the visual sensation which occupied the centre of my field of vision at noon on January 1, 1919." Such ultimate simples I call "particulars." Particulars MIGHT have proper names, and no doubt would have if language had been invented by scientifically trained observers for purposes of philosophy and logic. But as language was invented for practical ends, particulars have remained one and all without a name.

We are not, in practice, much concerned with the actual particulars that come into our experience in sensation; we are concerned rather with whole systems to which the particulars belong and of which they are signs. What we see makes us say "Hullo, there's Jones," and the fact that what we see is a sign of Jones (which is the case because it is one of the particulars that make up Jones) is more interesting to us than the actual particular itself. Hence we give the name "Jones" to the whole set of particulars, but do not trouble to give separate names to the separate particulars that make up the set.

Passing on from proper names, we come next to general names, such as "man," "cat," "triangle." A word such as "man" means a whole class of such collections of particulars as have proper names. The several members of the class are assembled together in virtue of some similarity or common property. All men resemble each other in certain important respects; hence we want a word which shall be equally applicable to all of them. We only give proper names to the individuals of a species when they differ inter se in practically important respects. In other cases we do not do this. A poker, for instance, is just a poker; we do not call one "John" and another "Peter."

There is a large class of words, such as "eating," "walking," "speaking," which mean a set of similar occurrences. Two instances of walking have the same name because they resemble each other, whereas two instances of Jones have the same name because they are causally connected. In practice, however, it is difficult to make any precise distinction between a word such as "walking" and a general name such as "man." One instance of walking cannot be concentrated into an instant: it is a process in time, in which there is a causal connection between the earlier and later parts, as between the earlier and later parts of Jones. Thus an instance of walking differs from an instance of man solely by the fact that it has a shorter life. There is a notion that an instance of walking, as compared with Jones, is unsubstantial, but this seems to be a mistake. We think that Jones walks, and that there could not be any walking unless there were somebody like Jones to perform the walking. But it is equally true that there could be no Jones unless there were something like walking for him to do. The notion that actions are performed by an agent is liable to the same kind of criticism as the notion that thinking needs a subject or ego, which we rejected in Lecture I. To say that it is Jones who is walking is merely to say that the walking in question is part of the whole series of occurrences which is Jones. There is no LOGICAL impossibility in walking occurring as an isolated phenomenon, not forming part of any such series as we call a "person."

We may therefore class with "eating," "walking," "speaking" words such as "rain," "sunrise," "lightning," which do not denote what would commonly be called actions. These words illustrate, incidentally, how little we can trust to the grammatical distinction of parts of speech, since the substantive "rain" and the verb "to rain" denote precisely the same class of meteorological occurrences. The distinction between the class of objects denoted by such a word and the class of objects denoted by a general name such as "man," "vegetable," or "planet," is that the sort of object which is an instance of (say) "lightning" is much simpler than (say) an individual man. (I am speaking of lightning as a sensible phenomenon, not as it is described in physics.) The distinction is one of degree, not of kind. But there is, from the point of view of ordinary thought, a great difference between a process which, like a flash of lightning, can be wholly comprised within one specious present and a process which, like the life of a man, has to be pieced together by observation and memory and the apprehension of causal connections. We may say broadly, therefore, that a word of the kind we have been discussing denotes a set of similar occurrences, each (as a rule) much more brief and less complex than a person or thing. Words themselves, as we have seen, are sets of similar occurrences of this kind. Thus there is more logical affinity between a word and what it means in the case of words of our present sort than in any other case.

There is no very great difference between such words as we have just been considering and words denoting qualities, such as "white" or "round." The chief difference is that words of this latter sort do not denote processes, however brief, but static features of the world. Snow falls, and is white; the falling is a process, the whiteness is not. Whether there is a universal, called "whiteness," or whether white things are to be defined as those having a certain kind of similarity to a standard thing, say freshly fallen snow, is a question which need not concern us, and which I believe to be strictly insoluble. For our purposes, we may take the word "white" as denoting a certain set of similar particulars or collections of particulars, the similarity being in respect of a static quality, not of a process.

From the logical point of view, a very important class of words are those that express relations, such as "in," "above," "before," "greater," and so on. The meaning of one of these words differs very fundamentally from the meaning of one of any of our previous classes, being more abstract and logically simpler than any of them. If our business were logic, we should have to spend much time on these words. But as it is psychology that concerns us, we will merely note their special character and pass on, since the logical classification of words is not our main business.

We will consider next the question what is implied by saying that a person "understands" a word, in the sense in which one understands a word in one's own language, but not in a language of which one is ignorant. We may say that a person understands a word when (a) suitable circumstances make him use it, (b) the hearing of it causes suitable behaviour in him. We may call these two active and passive understanding respectively. Dogs often have passive understanding of some words, but not active understanding, since they cannot use words.

It is not necessary, in order that a man should "understand" a word, that he should "know what it means," in the sense of being able to say "this word means so-and-so." Understanding words does not consist in knowing their dictionary definitions, or in being able to specify the objects to which they are appropriate. Such understanding as this may belong to lexicographers and students, but not to ordinary mortals in ordinary life. Understanding language is more like understanding cricket: it is a matter of habits, acquired in oneself and rightly presumed in others. To say that a word has a meaning is not to say that those who use the word correctly have ever thought out what the meaning is: the use of the word comes first, and the meaning is to be distilled out of it by observation and analysis. Moreover, the meaning of a word is not absolutely definite: there is always a greater or less degree of vagueness. The meaning is an area, like a target: it may have a bull's eye, but the outlying parts of the target are still more or less within the meaning, in a gradually diminishing degree as we travel further from the bull's eye. As language grows more precise, there is less and less of the target outside the bull's eye, and the bull's eye itself grows smaller and smaller; but the bull's eye never shrinks to a point, and there is always a doubtful region, however small, surrounding it.

A word is used "correctly" when the average hearer will be affected by it in the way intended. This is a psychological, not a literary, definition of "correctness." The literary definition would substitute, for the average hearer, a person of high education living a long time ago; the purpose of this definition is to make it difficult to speak or write correctly.

The relation of a word to its meaning is of the nature of a causal law governing our use of the word and our actions when we hear it used. There is no more reason why a person who uses a word correctly should be able to tell what it means than there is why a planet which is moving correctly should know Kepler's laws.

To illustrate what is meant by "understanding" words and sentences, let us take instances of various situations.

Suppose you are walking in London with an absent-minded friend, and while crossing a street you say, "Look out, there's a motor coming." He will glance round and jump aside without the need of any "mental" intermediary. There need be no "ideas," but only a stiffening of the muscles, followed quickly by action. He "understands" the words, because he does the right thing. Such "understanding" may be taken to belong to the nerves and brain, being habits which they have acquired while the language was being learnt. Thus understanding in this sense may be reduced to mere physiological causal laws.

If you say the same thing to a Frenchman with a slight knowledge of English he will go through some inner speech which may be represented by "Que dit-il? Ah, oui, une automobile!" After this, the rest follows as with the Englishman. Watson would contend that the inner speech must be incipiently pronounced; we should argue that it MIGHT be merely imaged. But this point is not important in the present connection.

If you say the same thing to a child who does not yet know the word "motor," but does know the other words you are using, you produce a feeling of anxiety and doubt you will have to point and say, "There, that's a motor." After that the child will roughly understand the word "motor," though he may include trains and steam-rollers If this is the first time the child has heard the word "motor," he may for a long time continue to recall this scene when he hears the word.

So far we have found four ways of understanding words:

(1) On suitable occasions you use the word properly.

(2) When you hear it you act appropriately.

(3) You associate the word with another word (say in a different language) which has the appropriate effect on behaviour.

(4) When the word is being first learnt, you may associate it with an object, which is what it "means," or a representative of various objects that it "means."

In the fourth case, the word acquires, through association, some of the same causal efficacy as the object. The word "motor" can make you leap aside, just as the motor can, but it cannot break your bones. The effects which a word can share with its object are those which proceed according to laws other than the general laws of physics, i.e. those which, according to our terminology, involve vital movements as opposed to merely mechanical movements. The effects of a word that we understand are always mnemic phenomena in the sense explained in Lecture IV, in so far as they are identical with, or similar to, the effects which the object itself might have.

So far, all the uses of words that we have considered can be accounted for on the lines of behaviourism.

But so far we have only considered what may be called the "demonstrative" use of language, to point out some feature in the present environment. This is only one of the ways in which language may be used. There are also its narrative and imaginative uses, as in history and novels. Let us take as an instance the telling of some remembered event.

We spoke a moment ago of a child who hears the word "motor" for the first time when crossing a street along which a motor-car is approaching. On a later occasion, we will suppose, the child remembers the incident and relates it to someone else. In this case, both the active and passive understanding of words is different from what it is when words are used demonstratively. The child is not seeing a motor, but only remembering one; the hearer does not look round in expectation of seeing a motor coming, but "understands" that a motor came at some earlier time. The whole of this occurrence is much more difficult to account for on behaviourist lines. It is clear that, in so far as the child is genuinely remembering, he has a picture of the past occurrence, and his words are chosen so as to describe the picture; and in so far as the hearer is genuinely apprehending what is said, the hearer is acquiring a picture more or less like that of the child. It is true that this process may be telescoped through the operation of the word-habit. The child may not genuinely remember the incident, but only have the habit of the appropriate words, as in the case of a poem which we know by heart, though we cannot remember learning it. And the hearer also may only pay attention to the words, and not call up any corresponding picture. But it is, nevertheless, the possibility of a memory-image in the child and an imagination-image in the hearer that makes the essence of the narrative "meaning" of the words. In so far as this is absent, the words are mere counters, capable of meaning, but not at the moment possessing it.

Yet this might perhaps be regarded as something of an overstatement. The words alone, without the use of images, may cause appropriate emotions and appropriate behaviour. The words have been used in an environment which produced certain emotions; by a telescoped process, the words alone are now capable of producing similar emotions. On these lines it might be sought to show that images are unnecessary. I do not believe, however, that we could account on these lines for the entirely different response produced by a narrative and by a description of present facts. Images, as contrasted with sensations, are the response expected during a narrative; it is understood that present action is not called for. Thus it seems that we must maintain our distinction words used demonstratively describe and are intended to lead to sensations, while the same words used in narrative describe and are only intended to lead to images.

We have thus, in addition to our four previous ways in which words can mean, two new ways, namely the way of memory and the way of imagination. That is to say:

(5) Words may be used to describe or recall a memory-image: to describe it when it already exists, or to recall it when the words exist as a habit and are known to be descriptive of some past experience.

(6) Words may be used to describe or create an imagination-image: to describe it, for example, in the case of a poet or novelist, or to create it in the ordinary case for giving information-though, in the latter case, it is intended that the imagination-image, when created, shall be accompanied by belief that something of the sort occurred.

These two ways of using words, including their occurrence in inner speech, may be spoken of together as the use of words in "thinking." If we are right, the use of words in thinking depends, at least in its origin, upon images, and cannot be fully dealt with on behaviourist lines. And this is really the most essential function of words, namely that, originally through their connection with images, they bring us into touch with what is remote in time or space. When they operate without the medium of images, this seems to be a telescoped process. Thus the problem of the meaning of words is brought into connection with the problem of the meaning of images.

To understand the function that words perform in what is called "thinking," we must understand both the causes and the effects of their occurrence. The causes of the occurrence of words require somewhat different treatment according as the object designated by the word is sensibly present or absent. When the object is present, it may itself be taken as the cause of the word, through association. But when it is absent there is more difficulty in obtaining a behaviourist theory of the occurrence of the word. The language-habit consists not merely in the use of words demonstratively, but also in their use to express narrative or desire. Professor Watson, in his account of the acquisition of the language-habit, pays very little attention to the use of words in narrative and desire. He says :

"The stimulus (object) to which the child often responds, a box, e.g. by movements such as opening and closing and putting objects into it, may serve to illustrate our argument. The nurse, observing that the child reacts with his hands, feet, etc., to the box, begins to say 'box' when the child is handed the box, 'open box' when the child opens it, 'close box' when he closes it, and 'put doll in box' when that act is executed. This is repeated over and over again. In the process of time it comes about that without any other stimulus than that of the box which originally called out the bodily habits, he begins to say 'box' when he sees it, 'open box' when he opens it, etc. The visible box now becomes a stimulus capable of releasing either the bodily habits or the word-habit, i.e. development has brought about two things: (1) a series of functional connections among arcs which run from visual receptor to muscles of throat, and (2) a series of already earlier connected arcs which run from the same receptor to the bodily muscles....

The object meets the child's vision. He runs to it and tries to reach it and says 'box.'... Finally the word is uttered without the movement of going towards the box being executed.... Habits are formed of going to the box when the arms are full of toys. The child has been taught to deposit them there. When his arms are laden with toys and no box is there, the word-habit arises and he calls 'box'; it is handed to him, and he opens it and deposits the toys therein. This roughly marks what we would call the genesis of a true language-habit."

We need not linger over what is said in the above passage as to the use of the word "box" in the presence of the box. But as to its use in the absence of the box, there is only one brief sentence, namely: "When his arms are laden with toys and no box is there, the word-habit arises and he calls 'box.'" This is inadequate as it stands, since the habit has been to use the word when the box is present, and we have to explain its extension to cases in which the box is absent.

Having admitted images, we may say that the word "box," in the absence of the box, is caused by an image of the box. This may or may not be true—in fact, it is true in some cases but not in others. Even, however, if it were true in all cases, it would only slightly shift our problem: we should now have to ask what causes an image of the box to arise. We might be inclined to say that desire for the box is the cause. But when this view is investigated, it is found that it compels us to suppose that the box can be desired without the child's having either an image of the box or the word "box." This will require a theory of desire which may be, and I think is, in the main true, but which removes desire from among things that actually occur, and makes it merely a convenient fiction, like force in mechanics. With such a view, desire is no longer a true cause, but merely a short way of describing certain processes.

In order to explain the occurrence of either the word or the image in the absence of the box, we have to assume that there is something, either in the environment or in our own sensations, which has frequently occurred at about the same time as the word "box." One of the laws which distinguish psychology (or nerve-physiology?) from physics is the law that, when two things have frequently existed in close temporal contiguity, either comes in time to cause the other This is the basis both of habit and of association. Thus, in our case, the arms full of toys have frequently been followed quickly by the box, and the box in turn by the word "box." The box itself is subject to physical laws, and does not tend to be caused by the arms full of toys, however often it may in the past have followed them—always provided that, in the case in question, its physical position is such that voluntary movements cannot lead to it. But the word "box" and the image of the box are subject to the law of habit; hence it is possible for either to be caused by the arms full of toys. And we may lay it down generally that, whenever we use a word, either aloud or in inner speech, there is some sensation or image (either of which may be itself a word) which has frequently occurred at about the same time as the word, and now, through habit, causes the word. It follows that the law of habit is adequate to account for the use of words in the absence of their objects; moreover, it would be adequate even without introducing images. Although, therefore, images seem undeniable, we cannot derive an additional argument in their favour from the use of words, which could, theoretically, be explained without introducing images.

When we understand a word, there is a reciprocal association between it and the images of what it "means." Images may cause us to use words which mean them, and these words, heard or read, may in turn cause the appropriate images. Thus speech is a means of producing in our hearers the images which are in us. Also, by a telescoped process, words come in time to produce directly the effects which would have been produced by the images with which they were associated. The general law of telescoped processes is that, if A causes B and B causes C, it will happen in time that A will cause C directly, without the intermediary of B. This is a characteristic of psychological and neural causation. In virtue of this law, the effects of images upon our actions come to be produced by words, even when the words do not call up appropriate images. The more familiar we are with words, the more our "thinking" goes on in words instead of images. We may, for example, be able to describe a person's appearance correctly without having at any time had any image of him, provided, when we saw him, we thought of words which fitted him; the words alone may remain with us as a habit, and enable us to speak as if we could recall a visual image of the man. In this and other ways the understanding of a word often comes to be quite free from imagery; but in first learning the use of language it would seem that imagery always plays a very important part.

Images as well as words may be said to have "meaning"; indeed, the meaning of images seems more primitive than the meaning of words. What we call (say) an image of St. Paul's may be said to "mean" St. Paul's. But it is not at all easy to say exactly what constitutes the meaning of an image. A memory-image of a particular occurrence, when accompanied by a memory-belief, may be said to mean the occurrence of which it is an image. But most actual images do not have this degree of definiteness. If we call up an image of a dog, we are very likely to have a vague image, which is not representative of some one special dog, but of dogs in general. When we call up an image of a friend's face, we are not likely to reproduce the expression he had on some one particular occasion, but rather a compromise expression derived from many occasions. And there is hardly any limit to the vagueness of which images are capable. In such cases, the meaning of the image, if defined by relation to the prototype, is vague: there is not one definite prototype, but a number, none of which is copied exactly.

There is, however, another way of approaching the meaning of images, namely through their causal efficacy. What is called an image "of" some definite object, say St. Paul's, has some of the effects which the object would have. This applies especially to the effects that depend upon association. The emotional effects, also, are often similar: images may stimulate desire almost as strongly as do the objects they represent. And conversely desire may cause images: a hungry man will have images of food, and so on. In all these ways the causal laws concerning images are connected with the causal laws concerning the objects which the images "mean." An image may thus come to fulfil the function of a general idea. The vague image of a dog, which we spoke of a moment ago, will have effects which are only connected with dogs in general, not the more special effects which would be produced by some dogs but not by others. Berkeley and Hume, in their attack on general ideas, do not allow for the vagueness of images: they assume that every image has the definiteness that a physical object would have This is not the case, and a vague image may well have a meaning which is general.

In order to define the "meaning" of an image, we have to take account both of its resemblance to one or more prototypes, and of its causal efficacy. If there were such a thing as a pure imagination-image, without any prototype whatever, it would be destitute of meaning. But according to Hume's principle, the simple elements in an image, at least, are derived from prototypes-except possibly in very rare exceptional cases. Often, in such instances as our image of a friend's face or of a nondescript dog, an image is not derived from one prototype, but from many; when this happens, the image is vague, and blurs the features in which the various prototypes differ. To arrive at the meaning of the image in such a case, we observe that there are certain respects, notably associations, in which the effects of images resemble those of their prototypes. If we find, in a given case, that our vague image, say, of a nondescript dog, has those associative effects which all dogs would have, but not those belonging to any special dog or kind of dog, we may say that our image means "dog" in general. If it has all the associations appropriate to spaniels but no others, we shall say it means "spaniel"; while if it has all the associations appropriate to one particular dog, it will mean that dog, however vague it may be as a picture. The meaning of an image, according to this analysis, is constituted by a combination of likeness and associations. It is not a sharp or definite conception, and in many cases it will be impossible to decide with any certainty what an image means. I think this lies in the nature of things, and not in defective analysis.

We may give somewhat more precision to the above account of the meaning of images, and extend it to meaning in general. We find sometimes that, IN MNEMIC CAUSATION, an image or word, as stimulus, has the same effect (or very nearly the same effect) as would belong to some object, say, a certain dog. In that case we say that the image or word means that object. In other cases the mnemic effects are not all those of one object, but only those shared by objects of a certain kind, e.g. by all dogs. In this case the meaning of the image or word is general: it means the whole kind. Generality and particularity are a matter of degree. If two particulars differ sufficiently little, their mnemic effects will be the same; therefore no image or word can mean the one as opposed to the other; this sets a bound to the particularity of meaning. On the other hand, the mnemic effects of a number of sufficiently dissimilar objects will have nothing discoverable in common; hence a word which aims at complete generality, such as "entity" for example, will have to be devoid of mnemic effects, and therefore of meaning. In practice, this is not the case: such words have VERBAL associations, the learning of which constitutes the study of metaphysics.

The meaning of a word, unlike that of an image, is wholly constituted by mnemic causal laws, and not in any degree by likeness (except in exceptional cases). The word "dog" bears no resemblance to a dog, but its effects, like those of an image of a dog, resemble the effects of an actual dog in certain respects. It is much easier to say definitely what a word means than what an image means, since words, however they originated, have been framed in later times for the purpose of having meaning, and men have been engaged for ages in giving increased precision to the meanings of words. But although it is easier to say what a word means than what an image means, the relation which constitutes meaning is much the same in both cases. A word, like an image, has the same associations as its meaning has. In addition to other associations, it is associated with images of its meaning, so that the word tends to call up the image and the image tends to call up the word., But this association is not essential to the intelligent use of words. If a word has the right associations with other objects, we shall be able to use it correctly, and understand its use by others, even if it evokes no image. The theoretical understanding of words involves only the power of associating them correctly with other words; the practical understanding involves associations with other bodily movements.

The use of words is, of course, primarily social, for the purpose of suggesting to others ideas which we entertain or at least wish them to entertain. But the aspect of words that specially concerns us is their power of promoting our own thought. Almost all higher intellectual activity is a matter of words, to the nearly total exclusion of everything else. The advantages of words for purposes of thought are so great that I should never end if I were to enumerate them. But a few of them deserve to be mentioned.

In the first place, there is no difficulty in producing a word, whereas an image cannot always be brought into existence at will, and when it comes it often contains much irrelevant detail. In the second place, much of our thinking is concerned with abstract matters which do not readily lend themselves to imagery, and are apt to be falsely conceived if we insist upon finding images that may be supposed to represent them. The word is always concrete and sensible, however abstract its meaning may be, and thus by the help of words we are able to dwell on abstractions in a way which would otherwise be impossible. In the third place, two instances of the same word are so similar that neither has associations not capable of being shared by the other. Two instances of the word "dog" are much more alike than (say) a pug and a great dane; hence the word "dog" makes it much easier to think about dogs in general. When a number of objects have a common property which is important but not obvious, the invention of a name for the common property helps us to remember it and to think of the whole set of objects that possess it. But it is unnecessary to prolong the catalogue of the uses of language in thought.

At the same time, it is possible to conduct rudimentary thought by means of images, and it is important, sometimes, to check purely verbal thought by reference to what it means. In philosophy especially the tyranny of traditional words is dangerous, and we have to be on our guard against assuming that grammar is the key to metaphysics, or that the structure of a sentence corresponds at all accurately with the structure of the fact that it asserts. Sayce maintained that all European philosophy since Aristotle has been dominated by the fact that the philosophers spoke Indo-European languages, and therefore supposed the world, like the sentences they were used to, necessarily divisible into subjects and predicates. When we come to the consideration of truth and falsehood, we shall see how necessary it is to avoid assuming too close a parallelism between facts and the sentences which assert them. Against such errors, the only safeguard is to be able, once in a way, to discard words for a moment and contemplate facts more directly through images. Most serious advances in philosophic thought result from some such comparatively direct contemplation of facts. But the outcome has to be expressed in words if it is to be communicable. Those who have a relatively direct vision of facts are often incapable of translating their vision into words, while those who possess the words have usually lost the vision. It is partly for this reason that the highest philosophical capacity is so rare: it requires a combination of vision with abstract words which is hard to achieve, and too quickly lost in the few who have for a moment achieved it.

LECTURE XI.

GENERAL IDEAS AND THOUGHT

It is said to be one of the merits of the human mind that it is capable of framing abstract ideas, and of conducting nonsensational thought. In this it is supposed to differ from the mind of animals. From Plato onward the "idea" has played a great part in the systems of idealizing philosophers. The "idea" has been, in their hands, always something noble and abstract, the apprehension and use of which by man confers upon him a quite special dignity.

The thing we have to consider to-day is this: seeing that there certainly are words of which the meaning is abstract, and seeing that we can use these words intelligently, what must be assumed or inferred, or what can be discovered by observation, in the way of mental content to account for the intelligent use of abstract words?

Taken as a problem in logic, the answer is, of course, that absolutely nothing in the way of abstract mental content is inferable from the mere fact that we can use intelligently words of which the meaning is abstract. It is clear that a sufficiently ingenious person could manufacture a machine moved by olfactory stimuli which, whenever a dog appeared in its neighbourhood, would say, "There is a dog," and when a cat appeared would throw stones at it. The act of saying "There is a dog," and the act of throwing stones, would in such a case be equally mechanical. Correct speech does not of itself afford any better evidence of mental content than the performance of any other set of biologically useful movements, such as those of flight or combat. All that is inferable from language is that two instances of a universal, even when they differ very greatly, may cause the utterance of two instances of the same word which only differ very slightly. As we saw in the preceding lecture, the word "dog" is useful, partly, because two instances of this word are much more similar than (say) a pug and a great dane. The use of words is thus a method of substituting for two particulars which differ widely, in spite of being instances of the same universal, two other particulars which differ very little, and which are also instances of a universal, namely the name of the previous universal. Thus, so far as logic is concerned, we are entirely free to adopt any theory as to general ideas which empirical observation may recommend.

Berkeley and Hume made a vigorous onslaught on "abstract ideas." They meant by an idea approximately what we should call an image. Locke having maintained that he could form an idea of triangle in general, without deciding what sort of triangle it was to be, Berkeley contended that this was impossible. He says:

"Whether others, have this wonderful faculty of abstracting their ideas, they best can tell: for myself, I dare be confident I have it not. I find, indeed, I have indeed a faculty of imagining, or representing to myself, the ideas of those particular things I have perceived, and of variously compounding and dividing them. I can imagine a man with two heads, or the upper parts of a man joined to the body of a horse. I can consider the hand, the eye, the nose, each by itself abstracted or separated from the rest of the body. But, then, whatever hand or eye I imagine, it must have some particular shape and colour. Likewise the idea of a man that I frame to myself must be either of a white, or a black, or a tawny, a straight, or a crooked, a tall, or a low, or a middle-sized man. I cannot by any effort of thought conceive the abstract idea above described. And it is equally impossible for me to form the abstract idea of motion distinct from the body moving, and which is neither swift nor slow, curvilinear nor rectilinear; and the like may be said of all other abstract general ideas whatsoever. To be plain, I own myself able to abstract in one sense, as when I consider some particular parts of qualities separated from others, with which, though they are united in some object, yet it is possible they may really exist without them. But I deny that I can abstract from one another, or conceive separately, those qualities which it is impossible should exist so separated; or that I can frame a general notion, by abstracting from particulars in the manner aforesaid—which last are the two proper acceptations of ABSTRACTION. And there is ground to think most men will acknowledge themselves to be in my case. The generality of men which are simple and illiterate never pretend to ABSTRACT NOTIONS. It is said they are difficult and not to be attained without pains and study; we may therefore reasonably conclude that, if such there be, they are confined only to the learned.

"I proceed to examine what can be alleged in defence of the doctrine of abstraction, and try if I can discover what it is that inclines the men of speculation to embrace an opinion so remote from common sense as that seems to be. There has been a late excellent and deservedly esteemed philosopher who, no doubt, has given it very much countenance, by seeming to think the having abstract general ideas is what puts the widest difference in point of understanding betwixt man and beast. 'The having of general ideas,' saith he, 'is that which puts a perfect distinction betwixt man and brutes, and is an excellency which the faculties of brutes do by no means attain unto. For, it is evident we observe no footsteps in them of making use of general signs for universal ideas; from which we have reason to imagine that they have not the faculty of abstracting, or making general ideas, since they have no use of words or any other general signs.' And a little after: 'Therefore, I think, we may suppose that it is in this that the species of brutes are discriminated from men, and it is that proper difference wherein they are wholly separated, and which at last widens to so wide a distance. For, if they have any ideas at all, and are not bare machines (as some would have them), we cannot deny them to have some reason. It seems as evident to me that they do, some of them, in certain instances reason as that they have sense; but it is only in particular ideas, just as they receive them from their senses. They are the best of them tied up within those narrow bounds, and have not (as I think) the faculty to enlarge them by any kind of abstraction. I readily agree with this learned author, that the faculties of brutes can by no means attain to abstraction. But, then, if this be made the distinguishing property of that sort of animals, I fear a great many of those that pass for men must be reckoned into their number. The reason that is here assigned why we have no grounds to think brutes have abstract general ideas is, that we observe in them no use of words or any other general signs; which is built on this supposition-that the making use of words implies the having general ideas. From which it follows that men who use language are able to abstract or generalize their ideas. That this is the sense and arguing of the author will further appear by his answering the question he in another place puts: 'Since all things that exist are only particulars, how come we by general terms?' His answer is: 'Words become general by being made the signs of general ideas.' ("Essay on Human Understanding," Bk. III, chap. III, paragraph 6.) But it seems that a word becomes general by being made the sign, not of an abstract general idea, but of several particular ideas, any one of which it indifferently suggests to the mind. For example, when it is said 'the change of motion is proportional to the impressed force,' or that 'whatever has extension is divisible,' these propositions are to be understood of motion and extension in general; and nevertheless it will not follow that they suggest to my thoughts an idea of motion without a body moved, or any determinate direction and velocity, or that I must conceive an abstract general idea of extension, which is neither line, surface, nor solid, neither great nor small, black, white, nor red, nor of any other determinate colour. It is only implied that whatever particular motion I consider, whether it be swift or slow, perpendicular, horizontal, or oblique, or in whatever object, the axiom concerning it holds equally true. As does the other of every particular extension, it matters not whether line, surface, or solid, whether of this or that magnitude or figure.

"By observing how ideas become general, we may the better judge how words are made so. And here it is to be noted that I do not deny absolutely there are general ideas, but only that there are any ABSTRACT general ideas; for, in the passages we have quoted wherein there is mention of general ideas, it is always supposed that they are formed by abstraction, after the manner set forth in sections 8 and 9. Now, if we will annex a meaning to our words, and speak only of what we can conceive, I believe we shall acknowledge that an idea which, considered in itself, is particular, becomes general by being made to represent or stand for all other particular ideas of the same sort. To make this plain by an example, suppose a geometrician is demonstrating the method of cutting a line in two equal parts. He draws, for instance, a black line of an inch in length: this, which in itself is a particular line, is nevertheless with regard to its signification general, since, as it is there used, it represents all particular lines whatsoever; so that what is demonstrated of it is demonstrated of all lines, or, in other words, of a line in general. And, as THAT PARTICULAR LINE becomes general by being made a sign, so the NAME 'line,' which taken absolutely is particular, by being a sign is made general. And as the former owes its generality not to its being the sign of an abstract or general line, but of all particular right lines that may possibly exist, so the latter must be thought to derive its generality from the same cause, namely, the various particular lines which it indifferently denotes."

Berkeley's view in the above passage, which is essentially the same as Hume's, does not wholly agree with modern psychology, although it comes nearer to agreement than does the view of those who believe that there are in the mind single contents which can be called abstract ideas. The way in which Berkeley's view is inadequate is chiefly in the fact that images are as a rule not of one definite prototype, but of a number of related similar prototypes. On this subject Semon has written well. In "Die Mneme," discussing the effect of repeated similar stimuli in producing and modifying our images, he says: "We choose a case of mnemic excitement whose existence we can perceive for ourselves by introspection, and seek to ekphore the bodily picture of our nearest relation in his absence, and have thus a pure mnemic excitement before us. At first it may seem to us that a determinate quite concrete picture becomes manifest in us, but just when we are concerned with a person with whom we are in constant contact, we shall find that the ekphored picture has something so to speak generalized. It is something like those American photographs which seek to display what is general about a type by combining a great number of photographs of different heads over each other on one plate. In our opinion, the generalizations happen by the homophonic working of different pictures of the same face which we have come across in the most different conditions and situations, once pale, once reddened, once cheerful, once earnest, once in this light, and once in that. As soon as we do not let the whole series of repetitions resound in us uniformly, but give our attention to one particular moment out of the many... this particular mnemic stimulus at once overbalances its simultaneously roused predecessors and successors, and we perceive the face in question with concrete definiteness in that particular situation." A little later he says: "The result is—at least in man, but probably also in the higher animals—the development of a sort of PHYSIOLOGICAL abstraction. Mnemic homophony gives us, without the addition of other processes of thought, a picture of our friend X which is in a certain sense abstract, not the concrete in any one situation, but X cut loose from any particular point of time. If the circle of ekphored engrams is drawn even more widely, abstract pictures of a higher order appear: for instance, a white man or a negro. In my opinion, the first form of abstract concepts in general is based upon such abstract pictures. The physiological abstraction which takes place in the above described manner is a predecessor of purely logical abstraction. It is by no means a monopoly of the human race, but shows itself in various ways also among the more highly organized animals." The same subject is treated in more detail in Chapter xvi of "Die mnemischen Empfindungen," but what is said there adds nothing vital to what is contained in the above quotations.

It is necessary, however, to distinguish between the vague and the general. So long as we are content with Semon's composite image, we MAY get no farther than the vague. The question whether this image takes us to the general or not depends, I think, upon the question whether, in addition to the generalized image, we have also particular images of some of the instances out of which it is compounded. Suppose, for example, that on a number of occasions you had seen one negro, and that you did not know whether this one was the same or different on the different occasions. Suppose that in the end you had an abstract memory-image of the different appearances presented by the negro on different occasions, but no memory-image of any one of the single appearances. In that case your image would be vague. If, on the other hand, you have, in addition to the generalized image, particular images of the several appearances, sufficiently clear to be recognized as different, and as instances of the generalized picture, you will then not feel the generalized picture to be adequate to any one particular appearance, and you will be able to make it function as a general idea rather than a vague idea. If this view is correct, no new general content needs to be added to the generalized image. What needs to be added is particular images compared and contrasted with the generalized image. So far as I can judge by introspection, this does occur in practice. Take for example Semon's instance of a friend's face. Unless we make some special effort of recollection, the face is likely to come before us with an average expression, very blurred and vague, but we can at will recall how our friend looked on some special occasion when he was pleased or angry or unhappy, and this enables us to realize the generalized character of the vague image.

There is, however, another way of distinguishing between the vague, the particular and the general, and this is not by their content, but by the reaction which they produce. A word, for example, may be said to be vague when it is applicable to a number of different individuals, but to each as individuals; the name Smith, for example, is vague: it is always meant to apply to one man, but there are many men to each of whom it applies The word "man," on the other hand, is general. We say, "This is Smith," but we do not say "This is man," but "This is a man." Thus we may say that a word embodies a vague idea when its effects are appropriate to an individual, but are the same for various similar individuals, while a word embodies a general idea when its effects are different from those appropriate to individuals. In what this difference consists it is, however, not easy to say. I am inclined to think that it consists merely in the knowledge that no one individual is represented, so that what distinguishes a general idea from a vague idea is merely the presence of a certain accompanying belief. If this view is correct, a general idea differs from a vague one in a way analogous to that in which a memory-image differs from an imagination-image. There also we found that the difference consists merely of the fact that a memory-image is accompanied by a belief, in this case as to the past.

It should also be said that our images even of quite particular occurrences have always a greater or a less degree of vagueness. That is to say, the occurrence might have varied within certain limits without causing our image to vary recognizably. To arrive at the general it is necessary that we should be able to contrast it with a number of relatively precise images or words for particular occurrences; so long as all our images and words are vague, we cannot arrive at the contrast by which the general is defined. This is the justification for the view which I quoted on from Ribot viz. that intelligence progresses from the indefinite to the definite, and that the vague appears earlier than either the particular or the general.

I think the view which I have been advocating, to the effect that a general idea is distinguished from a vague one by the presence of a judgment, is also that intended by Ribot when he says "The generic image is never, the concept is always, a judgment. We know that for logicians (formerly at any rate) the concept is the simple and primitive element; next comes the judgment, uniting two or several concepts; then ratiocination, combining two or several judgments. For the psychologists, on the contrary, affirmation is the fundamental act; the concept is the result of judgment (explicit or implicit), of similarities with exclusion of differences."

A great deal of work professing to be experimental has been done in recent years on the psychology of thought. A good summary of such work up to the year agog is contained in Titchener's "Lectures on the Experimental Psychology of the Thought Processes" (1909). Three articles in the "Archiv fur die gesammte Psychologie" by Watt, Messer and Buhler contain a great deal of the material amassed by the methods which Titchener calls experimental.

For my part I am unable to attach as much importance to this work as many psychologists do. The method employed appears to me hardly to fulfil the conditions of scientific experiment. Broadly speaking, what is done is, that a set of questions are asked of various people, their answers are recorded, and likewise their own accounts, based upon introspection, of the processes of thought which led them to give those answers. Much too much reliance seems to me to be placed upon the correctness of their introspection. On introspection as a method I have spoken earlier (Lecture VI). I am not prepared, like Professor Watson, to reject it wholly, but I do consider that it is exceedingly fallible and quite peculiarly liable to falsification in accordance with preconceived theory. It is like depending upon the report of a shortsighted person as to whom he sees coming along the road at a moment when he is firmly convinced that Jones is sure to come. If everybody were shortsighted and obsessed with beliefs as to what was going to be visible, we might have to make the best of such testimony, but we should need to correct its errors by taking care to collect the simultaneous evidence of people with the most divergent expectations. There is no evidence that this was done in the experiments in question, nor indeed that the influence of theory in falsifying the introspection was at all adequately recognized. I feel convinced that if Professor Watson had been one of the subjects of the questionnaires, he would have given answers totally different from those recorded in the articles in question. Titchener quotes an opinion of Wundt on these investigations, which appears to me thoroughly justified. "These experiments," he says, "are not experiments at all in the sense of a scientific methodology; they are counterfeit experiments, that seem methodical simply because they are ordinarily performed in a psychological laboratory, and involve the co-operation of two persons, who purport to be experimenter and observer. In reality, they are as unmethodical as possible; they possess none of the special features by which we distinguish the introspections of experimental psychology from the casual introspections of everyday life." Titchener, of course, dissents from this opinion, but I cannot see that his reasons for dissent are adequate. My doubts are only increased by the fact that Buhler at any rate used trained psychologists as his subjects. A trained psychologist is, of course, supposed to have acquired the habit of observation, but he is at least equally likely to have acquired a habit of seeing what his theories require. We may take Buhler's "Uber Gedanken" to illustrate the kind of results arrived at by such methods. Buhler says : "We ask ourselves the general question: 'WHAT DO WE EXPERIENCE WHEN WE THINK?' Then we do not at all attempt a preliminary determination of the concept 'thought,' but choose for analysis only such processes as everyone would describe as processes of thought." The most important thing in thinking, he says, is "awareness that..." (Bewusstheit dass), which he calls a thought. It is, he says, thoughts in this sense that are essential to thinking. Thinking, he maintains, does not need language or sensuous presentations. "I assert rather that in principle every object can be thought (meant) distinctly, without any help from sensuous presentation (Anschauungshilfen). Every individual shade of blue colour on the picture that hangs in my room I can think with complete distinctness unsensuously (unanschaulich), provided it is possible that the object should be given to me in another manner than by the help of sensations. How that is possible we shall see later." What he calls a thought (Gedanke) cannot be reduced, according to him, to other psychic occurrences. He maintains that thoughts consist for the most part of known rules It is clearly essential to the interest of this theory that the thought or rule alluded to by Buhler should not need to be expressed in words, for if it is expressed in words it is immediately capable of being dealt with on the lines with which the behaviourists have familiarized us. It is clear also that the supposed absence of words rests solely upon the introspective testimony of the persons experimented upon. I cannot think that there is sufficient certainty of their reliability in this negative observation to make us accept a difficult and revolutionary view of thought, merely because they have failed to observe the presence of words or their equivalent in their thinking. I think it far more likely, especially in view of the fact that the persons concerned were highly educated, that we are concerned with telescoped processes, in which habit has caused a great many intermediate terms to be elided or to be passed over so quickly as to escape observation.

I am inclined to think that similar remarks apply to the general idea of "imageless thinking," concerning which there has been much controversy. The advocates of imageless thinking are not contending merely that there can be thinking which is purely verbal; they are contending that there can be thinking which proceeds neither in words nor in images. My own feeling is that they have rashly assumed the presence of thinking in cases where habit has rendered thinking unnecessary. When Thorndike experimented with animals in cages, he found that the associations established were between a sensory stimulus and a bodily movement (not the idea of it), without the need of supposing any non-physiological intermediary The same thing, it seems to me, applies to ourselves. A certain sensory situation produces in us a certain bodily movement. Sometimes this movement consists in uttering words. Prejudice leads us to suppose that between the sensory stimulus and the utterance of the words a process of thought must have intervened, but there seems no good reason for such a supposition. Any habitual action, such as eating or dressing, may be performed on the appropriate occasion, without any need of thought, and the same seems to be true of a painfully large proportion of our talk. What applies to uttered speech applies of course equally to the internal speech which is not uttered. I remain, therefore, entirely unconvinced that there is any such phenomenon as thinking which consists neither of images nor of words, or that "ideas" have to be added to sensations and images as part of the material out of which mental phenomena are built.

The question of the nature of our consciousness of the universal is much affected by our view as to the general nature of the relation of consciousness to its object. If we adopt the view of Brentano, according to which all mental content has essential reference to an object, it is then natural to suppose that there is some peculiar kind of mental content of which the object is a universal, as oppose to a particular. According to this view, a particular cat can be PERceived or imagined, while the universal "cat" is CONceived. But this whole manner of viewing our dealings with universals has to be abandoned when the relation of a mental occurrence to its "object" is regarded as merely indirect and causal, which is the view that we have adopted. The mental content is, of course, always particular, and the question as to what it "means" (in case it means anything) is one which cannot be settled by merely examining the intrinsic character of the mental content, but only by knowing its causal connections in the case of the person concerned. To say that a certain thought "means" a universal as opposed to either a vague or a particular, is to say something exceedingly complex. A horse will behave in a certain manner whenever he smells a bear, even if the smell is derived from a bearskin. That is to say, any environment containing an instance of the universal "smell of a bear" produces closely similar behaviour in the horse, but we do not say that the horse is conscious of this universal. There is equally little reason to regard a man as conscious of the same universal, because under the same circumstances he can react by saying, "I smell a bear." This reaction, like that of the horse, is merely closely similar on different occasions where the environment affords instances of the same universal. Words of which the logical meaning is universal can therefore be employed correctly, without anything that could be called consciousness of universals. Such consciousness in the only sense in which it can be said to exist is a matter of reflective judgment consisting in the observation of similarities and differences. A universal never appears before the mind as a single object in the sort of way in which something perceived appears. I THINK a logical argument could be produced to show that universals are part of the structure of the world, but they are an inferred part, not a part of our data. What exists in us consists of various factors, some open to external observation, others only visible to introspection. The factors open to external observation are primarily habits, having the peculiarity that very similar reactions are produced by stimuli which are in many respects very different from each other. Of this the reaction of the horse to the smell of the bear is an instance, and so is the reaction of the man who says "bear" under the same circumstances. The verbal reaction is, of course, the most important from the point of view of what may be called knowledge of universals. A man who can always use the word "dog" when he sees a dog may be said, in a certain sense, to know the meaning of the word "dog," and IN THAT SENSE to have knowledge of the universal "dog." But there is, of course, a further stage reached by the logician in which he not merely reacts with the word "dog," but sets to work to discover what it is in the environment that causes in him this almost identical reaction on different occasions. This further stage consists in knowledge of similarities and differences: similarities which are necessary to the applicability of the word "dog," and differences which are compatible with it. Our knowledge of these similarities and differences is never exhaustive, and therefore our knowledge of the meaning of a universal is never complete.

In addition to external observable habits (including the habit of words), there is also the generic image produced by the superposition, or, in Semon's phrase, homophony, of a number of similar perceptions. This image is vague so long as the multiplicity of its prototypes is not recognized, but becomes universal when it exists alongside of the more specific images of its instances, and is knowingly contrasted with them. In this case we find again, as we found when we were discussing words in general in the preceding lecture, that images are not logically necessary in order to account for observable behaviour, i.e. in this case intelligent speech. Intelligent speech could exist as a motor habit, without any accompaniment of images, and this conclusion applies to words of which the meaning is universal, just as much as to words of which the meaning is relatively particular. If this conclusion is valid, it follows that behaviourist psychology, which eschews introspective data, is capable of being an independent science, and of accounting for all that part of the behaviour of other people which is commonly regarded as evidence that they think. It must be admitted that this conclusion considerably weakens the reliance which can be placed upon introspective data. They must be accepted simply on account of the fact that we seem to perceive them, not on account of their supposed necessity for explaining the data of external observation.

This, at any rate, is the conclusion to which we are forced, so long as, with the behaviourists, we accept common-sense views of the physical world. But if, as I have urged, the physical world itself, as known, is infected through and through with subjectivity, if, as the theory of relativity suggests, the physical universe contains the diversity of points of view which we have been accustomed to regard as distinctively psychological, then we are brought back by this different road to the necessity for trusting observations which are in an important sense private. And it is the privacy of introspective data which causes much of the behaviourists' objection to them.

This is an example of the difficulty of constructing an adequate philosophy of any one science without taking account of other sciences. The behaviourist philosophy of psychology, though in many respects admirable from the point of view of method, appears to me to fail in the last analysis because it is based upon an inadequate philosophy of physics. In spite, therefore, of the fact that the evidence for images, whether generic or particular, is merely introspective, I cannot admit that images should be rejected, or that we should minimize their function in our knowledge of what is remote in time or space.

LECTURE XII.

BELIEF

Belief, which is our subject to-day, is the central problem in the analysis of mind. Believing seems the most "mental" thing we do, the thing most remote from what is done by mere matter. The whole intellectual life consists of beliefs, and of the passage from one belief to another by what is called "reasoning." Beliefs give knowledge and error; they are the vehicles of truth and falsehood. Psychology, theory of knowledge and metaphysics revolve about belief, and on the view we take of belief our philosophical outlook largely depends.

Before embarking upon the detailed analysis of belief, we shall do well to note certain requisites which any theory must fulfil.

(1) Just as words are characterized by meaning, so beliefs are characterized by truth or falsehood. And just as meaning consists in relation to the object meant, so truth and falsehood consist in relation to something that lies outside the belief. You may believe that such-and-such a horse will win the Derby. The time comes, and your horse wins or does not win; according to the outcome, your belief was true or false. You may believe that six times nine is fifty-six; in this case also there is a fact which makes your belief false. You may believe that America was discovered in 1492, or that it was discovered in 1066. In the one case your belief is true, in the other false; in either case its truth or falsehood depends upon the actions of Columbus, not upon anything present or under your control. What makes a belief true or false I call a "fact." The particular fact that makes a given belief true or false I call its "objective," and the relation of the belief to its objective I call the "reference" or the "objective reference" of the belief. Thus, if I believe that Columbus crossed the Atlantic in 1492, the "objective" of my belief is Columbus's actual voyage, and the "reference" of my belief is the relation between my belief and the voyage—that relation, namely, in virtue of which the voyage makes my belief true (or, in another case, false). "Reference" of beliefs differs from "meaning" of words in various ways, but especially in the fact that it is of two kinds, "true" reference and "false" reference. The truth or falsehood of a belief does not depend upon anything intrinsic to the belief, but upon the nature of its relation to its objective. The intrinsic nature of belief can be treated without reference to what makes it true or false. In the remainder of the present lecture I shall ignore truth and falsehood, which will be the subject of Lecture XIII. It is the intrinsic nature of belief that will concern us to-day.

(2) We must distinguish between believing and what is believed. I may believe that Columbus crossed the Atlantic, that all Cretans are liars, that two and two are four, or that nine times six is fifty-six; in all these cases the believing is just

the same, and only the contents believed are different. I may remember my breakfast this morning, my lecture last week, or my first sight of New York. In all these cases the feeling of memory-belief is just the same, and only what is remembered differs. Exactly similar remarks apply to expectations. Bare assent, memory and expectation are forms of belief; all three are different from what is believed, and each has a constant character which is independent of what is believed.

In Lecture I we criticized the analysis of a presentation into act, content and object. But our analysis of belief contains three very similar elements, namely the believing, what is believed and the objective. The objections to the act (in the case of presentations) are not valid against the believing in the case of beliefs, because the believing is an actual experienced feeling, not something postulated, like the act. But it is necessary first to complete our preliminary requisites, and then to examine the content of a belief. After that, we shall be in a position to return to the question as to what constitutes believing.

(3) What is believed, and the believing, must both consist of present occurrences in the believer, no matter what may be the objective of the belief. Suppose I believe, for example, "that Caesar crossed the Rubicon." The objective of my belief is an event which happened long ago, which I never saw and do not remember. This event itself is not in my mind when I believe that it happened. It is not correct to say that I am believing the actual event; what I am believing is something now in my mind, something related to the event (in a way which we shall investigate in Lecture XIII), but obviously not to be confounded with the event, since the event is not occurring now but the believing is. What a man is believing at a given moment is wholly determinate if we know the contents of his mind at that moment; but Caesar's crossing of the Rubicon was an historical physical event, which is distinct from the present contents of every present mind. What is believed, however true it may be, is not the actual fact that makes the belief true, but a present event related to the fact. This present event, which is what is believed, I shall call the "content" of the belief. We have already had occasion to notice the distinction between content and objective in the case of memory-beliefs, where the content is "this occurred" and the objective is the past event.

(4) Between content and objective there is sometimes a very wide gulf, for example in the case of "Caesar crossed the Rubicon." This gulf may, when it is first perceived, give us a feeling that we cannot really "know" anything about the outer world. All we can "know," it may be said, is what is now in our thoughts. If Caesar and the Rubicon cannot be bodily in our thoughts, it might seem as though we must remain cut off from knowledge of them. I shall not now deal at length with this feeling, since it is necessary first to define "knowing," which cannot be done yet. But I will say, as a preliminary answer, that the feeling

assumes an ideal of knowing which I believe to be quite mistaken. It assumes, if it is thought out, something like the mystic unity of knower and known. These two are often said to be combined into a unity by the fact of cognition; hence when this unity is plainly absent, it may seem as if there were no genuine cognition. For my part, I think such theories and feelings wholly mistaken: I believe knowing to be a very external and complicated relation, incapable of exact definition, dependent upon causal laws, and involving no more unity than there is between a signpost and the town to which it points. I shall return to this question on a later occasion; for the moment these provisional remarks must suffice.

(5) The objective reference of a belief is connected with the fact that all or some of the constituents of its content have meaning. If I say "Caesar conquered Gaul," a person who knows the meaning of the three words composing my statement knows as much as can be known about the nature of the objective which would make my statement true. It is clear that the objective reference of a belief is, in general, in some way derivative from the meanings of the words or images that occur in its content. There are, however, certain complications which must be borne in mind. In the first place, it might be contended that a memory-image acquires meaning only through the memory-belief, which would seem, at least in the case of memory, to make belief more primitive than the meaning of images. In the second place, it is a very singular thing that meaning, which is single, should generate objective reference, which is dual, namely true and false. This is one of the facts which any theory of belief must explain if it is to be satisfactory.

It is now time to leave these preliminary requisites, and attempt the analysis of the contents of beliefs.

The first thing to notice about what is believed, i.e. about the content of a belief, is that it is always complex: We believe that a certain thing has a certain property, or a certain relation to something else, or that it occurred or will occur (in the sense discussed at the end of Lecture IX); or we may believe that all the members of a certain class have a certain property, or that a certain property sometimes occurs among the members of a class; or we may believe that if one thing happens, another will happen (for example, "if it rains I shall bring my umbrella"), or we may believe that something does not happen, or did not or will not happen (for example, "it won't rain"); or that one of two things must happen (for example, "either you withdraw your accusation, or I shall bring a libel action"). The catalogue of the sorts of things we may believe is infinite, but all of them are complex.

Language sometimes conceals the complexity of a belief. We say that a person believes in God, and it might seem as if God formed the whole content of the belief. But what is really believed is that God exists, which is very far from being simple. Similarly, when a person has a memory-image with a memory-belief, the belief is "this occurred," in the sense explained in Lecture IX; and "this occurred" is not simple. In like manner all cases where the content of a belief seems simple at first sight will be found, on examination, to confirm the view that the content is always complex.

The content of a belief involves not merely a plurality of constituents, but definite relations between them; it is not determinate when its constituents alone are given. For example, "Plato preceded Aristotle" and "Aristotle preceded Plato" are both contents which may be believed, but, although they consist of exactly the same constituents, they are different, and even incompatible.

The content of a belief may consist of words only, or of images only, or of a mixture of the two, or of either or both together with one or more sensations. It must contain at least one constituent which is a word or an image, and it may or may not contain one or more sensations as constituents. Some examples will make these various possibilities clear.

We may take first recognition, in either of the forms "this is of such-and-such a kind" or "this has occurred before." In either case, present sensation is a constituent. For example, you hear a noise, and you say to yourself "tram." Here the noise and the word "tram" are both constituents of your belief; there is also a relation between them, expressed by "is" in the proposition "that is a tram." As soon as your act of recognition is completed by the occurrence of the word "tram," your actions are affected: you hurry if you want the tram, or cease to hurry if you want a bus. In this case the content of your belief is a sensation (the noise) and a word ("tram") related in a way which may be called predication.

The same noise may bring into your mind the visual image of a tram, instead of the word "tram." In this case your belief consists of a sensation and an image suitable related. Beliefs of this class are what are called "judgments of perception." As we saw in Lecture VIII, the images associated with a sensation often come with such spontaneity and force that the unsophisticated do not distinguish them from the sensation; it is only the psychologist or the skilled observer who is aware of the large mnemic element that is added to sensation to make perception. It may be objected that what is added consists merely of images without belief. This is no doubt sometimes the case, but is certainly sometimes not the case. That belief always occurs in perception as opposed to sensation it is not necessary for us to maintain; it is enough for our purposes to note that it sometimes occurs, and that when it does, the content of our belief consists of a sensation and an image suitably related.

In a PURE memory-belief only images occur. But a mixture of words and images is very common in memory. You have an image of the past occurrence, and you say to yourself: "Yes, that's how it was." Here the image and the words together make up the content of the belief. And when the remembering of an incident has become a habit, it may be purely verbal, and the memory-belief may consist of words alone.

The more complicated forms of belief tend to consist only of words. Often images of various kinds accompany them, but they are apt to be irrelevant, and to form no part of what is actually believed. For example, in thinking of the Solar System, you are likely to have vague images of pictures you have seen of the earth surrounded by clouds, Saturn and his rings, the sun during an eclipse, and so on; but none of these form part of your belief that the planets revolve round the sun in elliptical orbits. The only images that form an actual part of such beliefs are, as a rule, images of words. And images of words, for the reasons considered in Lecture VIII, cannot be distinguished with any certainty from sensations, when, as is often, if not usually, the case, they are kinaesthetic images of pronouncing the words.

It is impossible for a belief to consist of sensations alone, except when, as in the case of words, the sensations have associations which make them signs possessed of meaning. The reason is that objective reference is of the essence of belief, and objective reference is derived from meaning. When I speak of a belief consisting partly of sensations and partly of words, I do not mean to deny that the words, when they are not mere images, are sensational, but that they occur as signs, not (so to speak) in their own right. To revert to the noise of the tram, when you hear it and say "tram," the noise and the word are both sensations (if you actually pronounce the word), but the noise is part of the fact which makes your belief true, whereas the word is not part of this fact. It is the MEANING of the word "tram," not the actual word, that forms part of the fact which is the objective of your belief. Thus the word occurs in the belief as a symbol, in virtue of its meaning, whereas the noise enters into both the belief and its objective. It is this that distinguishes the occurrence of words as symbols from the occurrence of sensations in their own right: the objective contains the sensations that occur in their own right, but contains only the meanings of the words that occur as symbols.

For the sake of simplicity, we may ignore the cases in which sensations in their own right form part of the content of a belief, and confine ourselves to images and words. We may also omit the cases in which both images and words occur in the content of a belief. Thus we become confined to two cases: (a) when the content consists wholly of images, (b) when it consists wholly of words. The case of mixed images and words has no special importance, and its omission will do no harm.

Let us take in illustration a case of memory. Suppose you are thinking of some familiar room. You may call up an image of it, and in your image the window may be to the left of the door. Without any intrusion of words, you may believe in the correctness of your image. You then have a belief, consisting wholly of images, which becomes, when put into words, "the window is to the left of the door." You may yourself use these words and proceed to believe them. You thus pass from an image-content to the corresponding word-content. The content is different in the two cases, but its objective reference is the same. This shows the relation of image-beliefs to word-beliefs in a very simple case. In more elaborate cases the relation becomes much less simple.

It may be said that even in this very simple case the objective reference of the word-content is not quite the same as that of the image-content, that images have a wealth of concrete features which are lost when words are substituted, that the window in the image is not a mere window in the abstract, but a window of a certain shape and size, not merely to the left of the door, but a certain distance to the left, and so on. In reply, it may be admitted at once that there is, as a rule, a certain amount of truth in the objection. But two points may be urged to minimize its force. First, images do not, as a rule, have that wealth of concrete detail that would make it IMPOSSIBLE to express them fully in words. They are vague and fragmentary: a finite number of words, though perhaps a large number, would exhaust at least their SIGNIFICANT features. For—and this is our second point—images enter into the content of a belief through the fact that they are capable of meaning, and their meaning does not, as a rule, have as much complexity as they have: some of their characteristics are usually devoid of meaning. Thus it may well be possible to extract in words all that has meaning in an image-content; in that case the word-content and the image-content will have exactly the same objective reference.

The content of a belief, when expressed in words, is the same thing (or very nearly the same thing) as what in logic is called a "proposition." A proposition is a series of words (or sometimes a single word) expressing the kind of thing that can be asserted or denied. "That all men are mortal," "that Columbus discovered America," "that Charles I died in his bed," "that all philosophers are wise," are propositions. Not any series of words is a proposition, but only such series of words as have "meaning," or, in our phraseology, "objective reference." Given the meanings of separate words, and the rules of syntax, the meaning of a proposition is determinate. This is the reason why we can understand a sentence we never heard before. You probably never heard before the proposition "that the inhabitants of the Andaman Islands habitually eat stewed hippopotamus for dinner," but there is no difficulty in understanding the proposition. The question of the relation between the meaning of a sentence and the meanings of the separate words is difficult, and I shall not pursue it now; I brought it up solely as being illustrative of the nature of propositions.

We may extend the term "proposition" so as to cover the image-contents of beliefs consisting of images. Thus, in the case of remembering a room in which the window is to the left of the door, when we believe the image-content the proposition will consist of the image of the window on the left together with the image of the door on the right. We will distinguish propositions of this kind as "image-propositions" and propositions in words as "word-propositions." We may identify propositions in general with the contents of actual and possible beliefs, and we may say that it is propositions that are true or false. In logic we are concerned with propositions rather than beliefs, since logic is not interested in what people do in fact believe, but only in the conditions which determine the truth or falsehood of possible beliefs. Whenever possible, except when actual beliefs are in question, it is generally a simplification to deal with propositions.

It would seem that image-propositions are more primitive than word-propositions, and may well ante-date language. There is no reason why memory-images, accompanied by that very simple belief-feeling which we decided to be the essence of memory, should not have occurred before language arose; indeed, it would be rash to assert positively that memory of this sort does not occur among the higher animals. Our more elementary beliefs, notably those that are added to sensation to make perception, often remain at the level of images. For example, most of the visual objects in our neighbourhood rouse tactile images: we have a different feeling in looking at a sofa from what we have in looking at a block of marble, and the difference consists chiefly in different stimulation of our tactile imagination. It may be said that the tactile images are merely present, without any accompanying belief; but I think this view, though sometimes correct, derives its plausibility as a general proposition from our thinking of explicit conscious belief only. Most of our beliefs, like most of our wishes, are "unconscious," in the sense that we have never told ourselves that we have them. Such beliefs display themselves when the expectations that they arouse fail in any way. For example, if someone puts tea (without milk) into a glass, and you drink it under the impression that it is going to be beer; or if you walk on what appears to be a tiled floor, and it turns out to be a soft carpet made to look like tiles. The shock of surprise on an occasion of this kind makes us aware of the expectations that habitually enter into our perceptions; and such expectations must be classed as beliefs, in spite of the fact that we do not normally take note of them or put them into words. I remember once watching a cock pigeon running over and over again to the edge of a looking-glass to try to wreak vengeance on the particularly obnoxious bird whom he expected to find there, judging by what he saw in the glass. He must have experienced each time the sort of surprise on finding nothing, which is calculated to lead in time to the adoption of Berkeley's theory that objects of sense are only in the mind. His expectation, though not expressed in words, deserved, I think, to be called a belief.

I come now to the question what constitutes believing, as opposed to the content believed.

To begin with, there are various different attitudes that may be taken towards the same content. Let us suppose, for the sake of argument, that you have a visual image of your breakfast-table. You may expect it while you are dressing in the morning; remember it as you go to your work; feel doubt as to its correctness when questioned as to your powers of visualizing; merely entertain the image, without connecting it with anything external, when you are going to sleep; desire it if you are hungry, or feel aversion for it if you are ill. Suppose, for the sake of definiteness, that the content is "an egg for breakfast." Then you have the following attitudes "I expect there will be an egg for breakfast"; "I remember there was an egg for breakfast"; "Was there an egg for breakfast?" "An egg for breakfast: well, what of it?" "I hope there will be an egg for breakfast"; "I am afraid there will be an egg for breakfast and it is sure to be bad." I do not suggest that this is a list of all possible attitudes on the subject; I say only that they are different attitudes, all concerned with the one content "an egg for breakfast."

These attitudes are not all equally ultimate. Those that involve desire and aversion have occupied us in Lecture III. For the present, we are only concerned with such as are cognitive. In speaking of memory, we distinguished three kinds of belief directed towards the same content, namely memory, expectation and bare assent without any time-determination in the belief-feeling. But before developing this view, we must examine two other theories which might be held concerning belief, and which, in some ways, would be more in harmony with a behaviourist outlook than the theory I wish to advocate.

(1) The first theory to be examined is the view that the differentia of belief consists in its causal efficacy I do not wish to make any author responsible for this theory: I wish merely to develop it hypothetically so that we may judge of its tenability.

We defined the meaning of an image or word by causal efficacy, namely by associations: an image or word acquires meaning, we said, through having the same associations as what it means.

We propose hypothetically to define "belief" by a different kind of causal efficacy, namely efficacy in causing voluntary movements. (Voluntary movements are defined as those vital movements which are distinguished from reflex movements as involving the higher nervous centres. I do not like to distinguish them by means of such notions as "consciousness" or "will," because I do not think these notions, in any definable sense, are always applicable. Moreover, the purpose of the theory we are examining is to be, as far as possible, physiological and behaviourist, and this purpose is not achieved if we introduce such a conception as "consciousness" or "will." Nevertheless, it is necessary for our purpose to find some way of distinguishing between voluntary and reflex movements, since the results would be too paradoxical, if we were to say that reflex movements also involve beliefs.) According to this definition, a content is said to be "believed" when it causes us to move. The images aroused are the same if you say to me, "Suppose there were an escaped tiger coming along the street," and if you say to me, "There is an escaped tiger coming along the street." But my actions will be very different in the two cases: in the first, I shall remain calm; in the second, it is possible that I may not. It is suggested, by the theory we are considering, that this difference of effects constitutes what is meant by saying that in the second case I believe the proposition suggested, while in the first case I do not. According to this view, images or words are "believed" when they cause bodily movements.

I do not think this theory is adequate, but I think it is suggestive of truth, and not so easily refutable as it might appear to be at first sight.

It might be objected to the theory that many things which we certainly believe do not call for any bodily movements. I believe that Great Britain is an island, that whales are mammals, that Charles I was executed, and so on; and at first sight it seems obvious that such beliefs, as a rule, do not call for any action on my part. But when we investigate the matter more closely, it becomes more doubtful. To begin with, we must distinguish belief as a mere DISPOSITION from actual active belief. We speak as if we always believed that Charles I was executed, but that only means that we are always ready to believe it when the subject comes up. The phenomenon we are concerned to analyse is the active belief, not the permanent disposition. Now, what are the occasions when, we actively believe that Charles I was executed? Primarily: examinations, when we perform the bodily movement of writing it down; conversation, when we assert it to display our historical erudition; and political discourses, when we are engaged in showing what Soviet government leads to. In all these cases bodily movements (writing or speaking) result from our belief.

But there remains the belief which merely occurs in "thinking." One may set to work to recall some piece of history one has been reading, and what one recalls is believed, although it probably does not cause any bodily movement whatever. It is true that what we believe always MAY influence action. Suppose I am invited to become King of Georgia: I find the prospect attractive, and go to Cook's to buy a third-class ticket to my new realm. At the last moment I remember Charles I and all the other monarchs who have come to a bad end; I change my mind, and walk out without completing the transaction. But such incidents are rare, and cannot constitute the whole of my belief that Charles I was executed. The conclusion seems to be that, although a belief always MAY influence action if it becomes relevant to a practical issue, it often exists actively (not as a mere disposition) without producing any voluntary movement whatever. If this is true, we cannot define belief by the effect on voluntary movements.

There is another, more theoretical, ground for rejecting the view we are examining. It is clear that a proposition can be either believed or merely considered, and that the content is the same in both cases. We can expect an egg for breakfast, or merely entertain the supposition that there may be an egg for breakfast. A moment ago I considered the possibility of being invited to become King of Georgia, but I do not believe that this will happen. Now, it seems clear that, since believing and considering have different effects if one produces bodily movements while the other does not, there must be some intrinsic difference between believing and considering; for if they were precisely similar, their effects also would be precisely similar. We have seen that the difference between believing a given proposition and merely considering it does not lie in the content; therefore there must be, in one case or in both, something additional to the content which distinguishes the occurrence of a belief from the occurrence of a mere consideration of the same content. So far as the theoretical argument goes, this additional element may exist only in belief, or only in consideration, or there may be one sort of additional element in the case of belief, and another in the case of consideration. This brings us to the second view which we have to examine.

(1) The theory which we have now to consider regards belief as belonging to every idea which is entertained, except in so far as some positive counteracting force interferes. In this view belief is not a positive phenomenon, though doubt and disbelief are so. What we call belief, according to this hypothesis, involves only the appropriate content, which will have the effects characteristic of belief unless something else operating simultaneously inhibits them. James quotes with approval, though inaccurately, a passage from Spinoza embodying this view:

"Let us conceive a boy imagining to himself a horse, and taking note of nothing else. As this imagination involves the existence of the horse, AND THE BOY HAS NO PERCEPTION WHICH ANNULS ITS EXISTENCE [James's italics], he will necessarily contemplate the horse as present, nor will he be able to doubt of its existence, however little certain of it he may be. I deny that a man in so far as he imagines [percipit] affirms nothing. For what is it to imagine a winged horse but to affirm that the horse [that horse, namely] has wings? For if the mind had nothing before it but the winged horse, it would contemplate the same as present, would have no cause to doubt of its existence, nor any power of dissenting from its existence, unless the imagination of the winged horse were joined to an idea which contradicted [tollit] its existence"

To this doctrine James entirely assents, adding in italics:

"ANY OBJECT WHICH REMAINS UNCONTRADICTED IS IPSO FACTO BELIEVED AND POSITED AS ABSOLUTE REALITY."

If this view is correct, it follows (though James does not draw the inference) that there is no need of any specific feeling called "belief," and that the mere existence of images yields all that is required. The state of mind in which we merely consider a proposition, without believing or disbelieving it, will then appear as a sophisticated product, the result of some rival force adding to the image-proposition a positive feeling which may be called suspense or non-belief—a feeling which may be compared to that of a man about to run a race waiting for the signal. Such a man, though not moving, is in a very different condition from that of a man quietly at rest And so the man who is considering a proposition without believing it will be in a state of tension, restraining the natural tendency to act upon the proposition which he would display if nothing interfered. In this view belief primarily consists merely in the existence of the appropriate images without any counteracting forces.

There is a great deal to be said in favour of this view, and I have some hesitation in regarding it as inadequate. It fits admirably with the phenomena of dreams and hallucinatory images, and it is recommended by the way in which it accords with mental development. Doubt, suspense of judgment and disbelief all seem later and more complex than a wholly unreflecting assent. Belief as a positive phenomenon, if it exists, may be regarded, in this view, as a product of doubt, a decision after debate, an acceptance, not merely of THIS, but of THIS-RATHER-THAN-THAT. It is not difficult to suppose that a dog has images (possible olfactory) of his absent master, or of the rabbit that he dreams of hunting. But it is very difficult to suppose that he can entertain mere imagination-images to which no assent is given.

I think it must be conceded that a mere image, without the addition of any positive feeling that could be called "belief," is apt to have a certain dynamic power, and in this sense an uncombated image has the force of a belief. But although this may be true, it accounts only for some of the simplest phenomena in the region of belief. It will not, for example, explain memory. Nor can it explain beliefs which do not issue in any proximate action, such as those of mathematics. I conclude, therefore, that there must be belief-feelings of the same order as those of doubt or disbelief, although phenomena closely analogous to those of belief can be produced by mere uncontradicted images.

(3) I come now to the view of belief which I wish to advocate. It seems to me that there are at least three kinds of belief, namely memory, expectation and bare assent. Each of these I regard as constituted by a certain feeling or complex of sensations, attached to the content believed. We may illustrate by an example. Suppose I am believing, by means of images, not words, that it will rain. We have

here two interrelated elements, namely the content and the expectation. The content consists of images of (say) the visual appearance of rain, the feeling of wetness, the patter of drops, interrelated, roughly, as the sensations would be if it were raining. Thus the content is a complex fact composed of images. Exactly the same content may enter into the memory "it was raining" or the assent "rain occurs." The difference of these cases from each other and from expectation does not lie in the content. The difference lies in the nature of the belief-feeling. I, personally, do not profess to be able to analyse the sensations constituting respectively memory, expectation and assent; but I am not prepared to say that they cannot be analysed. There may be other belief-feelings, for example in disjunction and implication; also a disbelief-feeling.

It is not enough that the content and the belief-feeling should coexist: it is necessary that there should be a specific relation between them, of the sort expressed by saying that the content is what is believed. If this were not obvious, it could be made plain by an argument. If the mere co-existence of the content and the belief-feeling sufficed, whenever we were having (say) a memory-feeling we should be remembering any proposition which came into our minds at the same time. But this is not the case, since we may simultaneously remember one proposition and merely consider another.

We may sum up our analysis, in the case of bare assent to a proposition not expressed in words, as follows: (a) We have a proposition, consisting of interrelated images, and possibly partly of sensations; (b) we have the feeling of assent, which is presumably a complex sensation demanding analysis; (c) we have a relation, actually subsisting, between the assent and the proposition, such as is expressed by saying that the proposition in question is what is assented to. For other forms of belief-feeling or of content, we have only to make the necessary substitutions in this analysis.

If we are right in our analysis of belief, the use of words in expressing beliefs is apt to be misleading. There is no way of distinguishing, in words, between a memory and an assent to a proposition about the past: "I ate my breakfast" and "Caesar conquered Gaul" have the same verbal form, though (assuming that I remember my breakfast) they express occurrences which are psychologically very different. In the one case, what happens is that I remember the content "eating my breakfast"; in the other case, I assent to the content "Caesar's conquest of Gaul occurred." In the latter case, but not in the former, the pastness is part of the content believed. Exactly similar remarks apply to the difference between expectation, such as we have when waiting for the thunder after a flash of lightning, and assent to a proposition about the future, such as we have in all the usual cases of inferential knowledge as to what will occur. I think this difficulty in the verbal expression of the temporal aspects of beliefs is one among the causes which have hampered philosophy in the consideration of time.

The view of belief which I have been advocating contains little that is novel except the distinction of kinds of belief-feeling—such as memory and expectation. Thus James says: "Everyone knows the difference between imagining a thing and believing in its existence, between supposing a proposition and acquiescing in its truth...IN ITS INNER NATURE, BELIEF, OR THE SENSE OF REALITY, IS A SORT OF FEELING MORE ALLIED TO THE EMOTIONS THAN TO ANYTHING ELSE" He proceeds to point out that drunkenness, and, still more, nitrous-oxide intoxication, will heighten the sense of belief: in the latter case, he says, a man's very soul may sweat with conviction, and he be all the time utterly unable to say what he is convinced of. It would seem that, in such cases, the feeling of belief exists unattached, without its usual relation to a content believed, just as the feeling of familiarity may sometimes occur without being related to any definite familiar object. The feeling of belief, when it occurs in this separated heightened form, generally leads us to look for a content to which to attach it. Much of what passes for revelation or mystic insight probably comes in this way: the belief-feeling, in abnormal strength, attaches itself, more or less accidentally, to some content which we happen to think of at the appropriate moment. But this is only a speculation, upon which I do not wish to lay too much stress.

LECTURE XIII.

TRUTH AND FALSEHOOD

The definition of truth and falsehood, which is our topic to-day, lies strictly outside our general subject, namely the analysis of mind. From the psychological standpoint, there may be different kinds of belief, and different degrees of certainty, but there cannot be any purely psychological means of distinguishing between true and false beliefs. A belief is rendered true or false by relation to a fact, which may lie outside the experience of the person entertaining the belief. Truth and falsehood, except in the case of beliefs about our own minds, depend upon the relations of mental occurrences to outside things, and thus take us beyond the analysis of mental occurrences as they are in themselves. Nevertheless, we can hardly avoid the consideration of truth and falsehood. We wish to believe that our beliefs, sometimes at least, yield KNOWLEDGE, and a belief does not yield knowledge unless it is true. The question whether our minds are instruments of knowledge, and, if so, in what sense, is so vital that any suggested analysis of mind must be examined in relation to this question. To ignore this question would be like describing a chronometer without regard to its accuracy as a time-keeper, or a thermometer without mentioning the fact that it measures temperature.

Many difficult questions arise in connection with knowledge. It is difficult to define knowledge, difficult to decide whether we have any knowledge, and difficult, even if it is conceded that we sometimes have knowledge to discover whether we can ever know that we have knowledge in this or that particular case. I shall divide the discussion into four parts:

I. We may regard knowledge, from a behaviourist standpoint, as exhibited in a certain kind of response to the environment. This response must have some characteristics which it shares with those of scientific instruments, but must also have others that are peculiar to knowledge. We shall find that this point of view is important, but not exhaustive of the nature of knowledge.

II. We may hold that the beliefs that constitute knowledge are distinguished from such as are erroneous or uncertain by properties which are intrinsic either to single beliefs or to systems of beliefs, being in either case discoverable without reference to outside fact. Views of this kind have been widely held among philosophers, but we shall find no reason to accept them.

III. We believe that some beliefs are true, and some false. This raises the problem of VERIFIABILITY: are there any circumstances which can justifiably give us an unusual degree of certainty that such and such a belief is true? It is obvious that there are circumstances which in fact cause a certainty of this sort, and we wish to learn what we can from examining these circumstances.

IV. Finally, there is the formal problem of defining truth and falsehood, and deriving the objective reference of a proposition from the meanings of its component words.

We will consider these four problems in succession.

I. We may regard a human being as an instrument, which makes various responses to various stimuli. If we observe these responses from outside, we shall regard them as showing knowledge when they display two characteristics, ACCURACY and APPROPRIATENESS. These two are quite distinct, and even sometimes incompatible. If I am being pursued by a tiger, accuracy is furthered by turning round to look at him, but appropriateness by running away without making any search for further knowledge of the beast. I shall return to the question of appropriateness later; for the present it is accuracy that I wish to consider.

When we are viewing a man from the outside, it is not his beliefs, but his bodily movements, that we can observe. His knowledge must be inferred from his bodily movements, and especially from what he says and writes. For the present we may ignore beliefs, and regard a man's knowledge as actually consisting in what he says and does. That is to say, we will construct, as far as possible, a purely behaviouristic account of truth and falsehood.

If you ask a boy "What is twice two?" and the boy says "four," you take that as prima facie evidence that the boy knows what twice two is. But if you go on to ask what is twice three, twice four, twice five, and so on, and the boy always answers "four," you come to the conclusion that he knows nothing about it. Exactly similar remarks apply to scientific instruments. I know a certain weather-cock which has the pessimistic habit of always pointing to the north-east. If you were to see it first on a cold March day, you would think it an excellent weather-cock; but with the first warm day of spring your confidence would be shaken. The boy and the weather-cock have the same defect: they do not vary their response when the stimulus is varied. A good instrument, or a person with much knowledge, will give different responses to stimuli which differ in relevant ways. This is the first point in defining accuracy of response.

We will now assume another boy, who also, when you first question him, asserts that twice two is four. But with this boy, instead of asking him different questions, you make a practice of asking him the same question every day at breakfast. You find that he says five, or six, or seven, or any other number at random, and you conclude that he also does not know what twice two is, though by good luck he answered right the first time. This boy is like a weather-cock which, instead of being stuck fast, is always going round and round, changing without any change of wind. This boy and weather-cock have the opposite defect to that of the previous pair: they give different responses to stimuli which do not differ in any relevant way.

In connection with vagueness in memory, we already had occasion to consider the definition of accuracy. Omitting some of the niceties of our previous discussion, we may say that an instrument is ACCURATE when it avoids the defects of the two boys and weather-cocks, that is to say, when—

(a) It gives different responses to stimuli which differ in relevant ways;

(b) It gives the same response to stimuli which do not differ in relevant ways.

What are relevant ways depends upon the nature and purpose of the instrument. In the case of a weather-cock, the direction of the wind is relevant, but not its strength; in the case of the boy, the meaning of the words of your question is relevant, but not the loudness of your voice, or whether you are his father or his schoolmaster If, however, you were a boy of his own age, that would be relevant, and the appropriate response would be different.

It is clear that knowledge is displayed by accuracy of response to certain kinds of stimuli, e.g. examinations. Can we say, conversely, that it consists wholly of such accuracy of response? I do not think we can; but we can go a certain distance in this direction. For this purpose we must define more carefully the kind of accuracy and the kind of response that may be expected where there is knowledge.

From our present point of view, it is difficult to exclude perception from knowledge; at any rate, knowledge is displayed by actions based upon perception. A bird flying among trees avoids bumping into their branches; its avoidance is a response to visual sensations. This response has the characteristic of accuracy, in the main, and leads us to say that the bird "knows," by sight, what objects are in its neighbourhood. For a behaviourist, this must certainly count as knowledge, however it may be viewed by analytic psychology. In this case, what is known, roughly, is the stimulus; but in more advanced knowledge the stimulus and what is known become different. For example, you look in your calendar and find that Easter will be early next year. Here the stimulus is the calendar, whereas the response concerns the future. Even this can be paralleled among instruments: the behaviour of the barometer has a present stimulus but foretells the future, so that the barometer might be said, in a sense, to know the future. However that may be, the point I am emphasizing as regards knowledge is that what is known may be quite different from the stimulus, and no part of the cause of the knowledge-response. It is only in sense-knowledge that the stimulus and what is known are, with qualifications, identifiable. In knowledge of the future, it is obvious that they are totally distinct, since otherwise the response would precede the stimulus. In abstract knowledge also they are distinct, since abstract facts have no date. In knowledge of the past there are complications, which we must briefly examine.

Every form of memory will be, from our present point of view, in one sense a delayed response. But this phrase does not quite clearly express what is meant. If you light a fuse and connect it with a heap of dynamite, the explosion of the dynamite may be spoken of, in a sense, as a delayed response to your lighting of the fuse. But that only means that it is a somewhat late portion of a continuous process of which the earlier parts have less emotional interest. This is not the case with habit. A display of habit has two sorts of causes: (a) the past occurrences which generated the habit, (b) the present occurrence which brings it into play. When you drop a weight on your toe, and say what you do say, the habit has been caused by imitation of your undesirable associates, whereas it is brought into play by the dropping of the weight. The great bulk of our knowledge is a habit in this sense: whenever I am asked when I was born, I reply correctly by mere habit. It would hardly be correct to say that getting born was the stimulus, and that my reply is a delayed response But in cases of memory this way of speaking would have an element of truth. In an habitual memory, the event remembered was clearly an essential part of the stimulus to the formation of the habit. The present stimulus which brings the habit into play produces a different response from that which it would produce if the habit did not exist. Therefore the habit enters into the causation of the response, and so do, at one remove, the causes of the habit. It follows that an event remembered is an essential part of the causes of our remembering.

In spite, however, of the fact that what is known is SOMETIMES an indispensable part of the cause of the knowledge, this circumstance is, I think, irrelevant to the general question with which we are concerned, namely What sort of response to what sort of stimulus can be regarded as displaying knowledge? There is one characteristic which the response must have, namely, it must consist of voluntary movements. The need of this characteristic is connected with the characteristic of APPROPRIATENESS, which I do not wish to consider as yet. For the present I wish only to obtain a clearer idea of the sort of ACCURACY that a knowledge-response must have. It is clear from many instances that accuracy, in other cases, may be purely mechanical. The most complete form of accuracy consists in giving correct answers to questions, an achievement in which calculating machines far surpass human beings. In asking a question of a calculating machine, you must use its language: you must not address it in English, any more than you would address an Englishman in Chinese. But if you address it in the language it understands, it will tell you what is 34521 times 19987, without a moment's hesitation or a hint of inaccuracy. We do not say the machine KNOWS the answer, because it has no purpose of its own in giving the answer: it does not wish to impress you with its cleverness, or feel proud of being such a good machine. But as far as mere accuracy goes, the machine leaves nothing to be desired.

Accuracy of response is a perfectly clear notion in the case of answers to questions, but in other cases it is much more obscure. We may say generally that an object whether animate or inanimate, is "sensitive" to a certain feature of the environment if it behaves differently according to the presence or absence of that feature. Thus iron is sensitive to anything magnetic. But sensitiveness does not constitute knowledge, and knowledge of a fact which is not sensible is not sensitiveness to that fact, as we have seen in distinguishing the fact known from the stimulus. As soon as we pass beyond the simple case of question and answer, the definition of knowledge by means of behaviour demands the consideration of purpose. A carrier pigeon flies home, and so we say it "knows" the way. But if it merely flew to some place at random, we should not say that it "knew" the way to that place, any more than a stone rolling down hill knows the way to the valley.

On the features which distinguish knowledge from accuracy of response in general, not much can be said from a behaviourist point of view without referring to purpose. But the necessity of SOMETHING besides accuracy of response may be brought out by the following consideration: Suppose two persons, of whom one believed whatever the other disbelieved, and disbelieved whatever the other believed. So far as accuracy and sensitiveness of response alone are concerned, there would be nothing to choose between these two persons. A thermometer which went down for warm weather and up for cold might be just as accurate as the usual kind; and a person who always believes falsely is just as sensitive an instrument as a person who always believes truly. The observable and practical difference between them would be that the one who always believed falsely would quickly come to a bad end. This illustrates once more that accuracy of response to stimulus does not alone show knowledge, but must be reinforced by appropriateness, i.e. suitability for realizing one's purpose. This applies even in the apparently simple case of answering questions: if the purpose of the answers is to deceive, their falsehood, not their truth, will be evidence of knowledge. The proportion of the combination of appropriateness with accuracy in the definition of knowledge is difficult; it seems that both enter in, but that appropriateness is only required as regards the general type of response, not as regards each individual instance.

II. I have so far assumed as unquestionable the view that the truth or falsehood of a belief consists in a relation to a certain fact, namely the objective of the belief. This view has, however, been often questioned. Philosophers have sought some intrinsic criterion by which true and false beliefs could be distinguished. I am afraid their chief reason for this search has been the wish to feel more certainty than seems otherwise possible as to what is true and what is false. If we could discover the truth of a belief by examining its intrinsic characteristics, or those of some collection of beliefs of which it forms part, the pursuit of truth, it is thought, would be a less arduous business than it otherwise appears to be. But the attempts which have been made in this direction are not encouraging. I will take two criteria which have been suggested, namely, (1) self-evidence, (2) mutual coherence. If we can show that these are inadequate, we may feel fairly certain that no intrinsic criterion hitherto suggested will suffice to distinguish true from false beliefs.

1) Self-evidence.—Some of our beliefs seem to be peculiarly indubitable. One might instance the belief that two and two are four, that two things cannot be in the same place at the same time, nor one thing in two places, or that a particular buttercup that we are seeing is yellow. The suggestion we are to examine is that such: beliefs have some recognizable quality which secures their truth, and the truth of whatever is deduced from them according to self-evident principles of inference. This theory is set forth, for example, by Meinong in his book, "Ueber die Erfahrungsgrundlagen unseres Wissens."

If this theory is to be logically tenable, self-evidence must not consist merely in the fact that we believe a proposition. We believe that our beliefs are sometimes erroneous, and we wish to be able to select a certain class of beliefs which are never erroneous. If we are to do this, it must be by some mark which belongs only to certain beliefs, not to all; and among those to which it belongs there must be none that are mutually inconsistent. If, for example, two propositions p and q were self-evident, and it were also self-evident that p and q could not both be true, that would condemn self-evidence as a guarantee of truth. Again, self-evidence must not be the same thing as the absence of doubt or the presence of complete certainty. If we are completely certain of a proposition, we do not seek a ground to support our belief. If self-evidence is alleged as a ground of belief, that implies that doubt has crept in, and that our self-evident proposition has not wholly resisted the assaults of scepticism. To say that any given person believes some things so firmly that he cannot be made to doubt them is no doubt true. Such beliefs he will be willing to use as premisses in reasoning, and to him personally they will seem to have as much evidence as any belief can need. But among the propositions which one man finds indubitable there will be some that another man finds it quite possible to doubt. It used to seem self-evident that there could not be men at the Antipodes, because they would fall off, or at best grow giddy from standing on their heads. But New Zealanders find the falsehood of this proposition self-evident. Therefore, if self-evidence is a guarantee of truth, our ancestors must have been mistaken in thinking their beliefs about the Antipodes self-evident. Meinong meets this difficulty by saying that some beliefs are falsely thought to be self-evident, but in the case of others it is self-evident that they are self-evident, and these are wholly reliable. Even this, however, does not remove the practical risk of error, since we may mistakenly believe it self-evident that a certain belief is self-evident. To remove all risk of error, we shall need an endless series of more and more complicated self-evident beliefs, which cannot possibly be realized in practice. It would seem, therefore, that self-evidence is useless as a practical criterion for insuring truth.

The same result follows from examining instances. If we take the four instances mentioned at the beginning of this discussion, we shall find that three of them are logical, while the fourth is a judgment of perception. The proposition that two and two are four follows by purely logical deduction from definitions: that means that its truth results, not from the properties of objects, but from the meanings of symbols. Now symbols, in mathematics, mean what we choose; thus the feeling of self-evidence, in this case, seems explicable by the fact that the whole matter is within our control. I do not wish to assert that this is the whole truth about mathematical propositions, for the question is complicated, and I do not know what the whole truth is. But I do wish to suggest that the feeling of self-evidence in mathematical propositions has to do with the fact that they are concerned with the meanings of symbols, not with properties of the world such as external observation might reveal.

Similar considerations apply to the impossibility of a thing being in two places at once, or of two things being in one place at the same time. These impossibilities result logically, if I am not mistaken, from the definitions of one thing and one place. That is to say, they are not laws of physics, but only part of the intellectual apparatus which we have manufactured for manipulating physics. Their self-evidence, if this is so, lies merely in the fact that they represent our decision as to the use of words, not a property of physical objects.

Judgments of perception, such as "this buttercup is yellow," are in a quite different position from judgments of logic, and their self-evidence must have a different explanation. In order to arrive at the nucleus of such a judgment, we will eliminate, as far as possible, the use of words which take us beyond the present fact, such as "buttercup" and "yellow." The simplest kind of judgment underlying the perception that a buttercup is yellow would seem to be the perception of similarity in two colours seen simultaneously. Suppose we are seeing two buttercups, and we perceive that their colours are similar. This similarity is a physical fact, not a matter of symbols or words; and it certainly seems to be indubitable in a way that many judgments are not.

The first thing to observe, in regard to such judgments, is that as they stand they are vague. The word "similar" is a vague word, since there are degrees of similarity, and no one can say where similarity ends and dissimilarity begins. It is unlikely that our two buttercups have EXACTLY the same colour, and if we judged that they had we should have passed altogether outside the region of self-evidence. To make our proposition more precise, let us suppose that we are also seeing a red rose at the same time. Then we may judge that the colours of the buttercups are more similar to each other than to the colour of the rose. This judgment seems more complicated, but has certainly gained in precision. Even now, however, it falls short of complete precision, since similarity is not prima facie measurable, and it would require much discussion to decide what we mean by greater or less similarity. To this process of the pursuit of precision there is strictly no limit.

The next thing to observe (although I do not personally doubt that most of our judgments of perception are true) is that it is very difficult to define any class of such judgments which can be known, by its intrinsic quality, to be always exempt from error. Most of our judgments of perception involve correlations, as when we judge that a certain noise is that of a passing cart. Such judgments are all obviously liable to error, since there is no correlation of which we have a right to be certain that it is invariable. Other judgments of perception are derived from recognition, as when we say "this is a buttercup," or even merely "this is yellow." All such judgments entail some risk of error, though sometimes perhaps a very small one; some flowers that look like buttercups are marigolds, and colours that some would call yellow others might call orange. Our subjective certainty is usually a result of habit, and may lead us astray in circumstances which are unusual in ways of which we are unaware.

For such reasons, no form of self-evidence seems to afford an absolute criterion of truth. Nevertheless, it is perhaps true that judgments having a high degree of subjective certainty are more apt to be true than other judgments. But if this be the case, it is a result to be demonstrated, not a premiss from which to start in defining truth and falsehood. As an initial guarantee, therefore, neither self-evidence nor subjective certainty can be accepted as adequate.

(2) Coherence.—Coherence as the definition of truth is advocated by idealists, particularly by those who in the main follow Hegel. It is set forth ably in Mr. Joachim's book, "The Nature of Truth" (Oxford, 1906). According to this view, any set of propositions other than the whole of truth can be condemned on purely logical grounds, as internally inconsistent; a single proposition, if it is what we should ordinarily call false, contradicts itself irremediably, while if it is what we should ordinarily call true, it has implications which compel us to admit other propositions, which in turn lead to others, and so on, until we find ourselves committed to the whole of truth. One might illustrate by a very simple example: if I say "so-and-so is a married man," that is not a self-subsistent proposition. We cannot logically conceive of a universe in which this proposition constituted the whole of truth. There must be also someone who is a married woman, and who is married to the particular man in question. The view we are considering regards everything that can be said about any one object as relative in the same sort of way as "so-and-so is a married man." But everything, according to this view, is

relative, not to one or two other things, but to all other things, so that from one bit of truth the whole can be inferred.

The fundamental objection to this view is logical, and consists in a criticism of its doctrine as to relations. I shall omit this line of argument, which I have developed elsewhere. For the moment I will content myself with saying that the powers of logic seem to me very much less than this theory supposes. If it were taken seriously, its advocates ought to profess that any one truth is logically inferable from any other, and that, for example, the fact that Caesar conquered Gaul, if adequately considered, would enable us to discover what the weather will be to-morrow. No such claim is put forward in practice, and the necessity of empirical observation is not denied; but according to the theory it ought to be.

Another objection is that no endeavour is made to show that we cannot form a consistent whole composed partly or wholly of false propositions, as in a novel. Leibniz's conception of many possible worlds seems to accord much better with modern logic and with the practical empiricism which is now universal. The attempt to deduce the world by pure thought is attractive, and in former times was largely supposed capable of success. But nowadays most men admit that beliefs must be tested by observation, and not merely by the fact that they harmonize with other beliefs. A consistent fairy-tale is a different thing from truth, however elaborate it may be. But to pursue this topic would lead us into difficult technicalities; I shall therefore assume, without further argument, that coherence is not sufficient as a definition of truth.

III. Many difficult problems arise as regards the verifiability of beliefs. We believe various things, and while we believe them we think we know them. But it sometimes turns out that we were mistaken, or at any rate we come to think we were. We must be mistaken either in our previous opinion or in our subsequent recantation; therefore our beliefs are not all correct, and there are cases of belief which are not cases of knowledge. The question of verifiability is in essence this: can we discover any set of beliefs which are never mistaken or any test which, when applicable, will always enable us to discriminate between true and false beliefs? Put thus broadly and abstractly, the answer must be negative. There is no way hitherto discovered of wholly eliminating the risk of error, and no infallible criterion. If we believe we have found a criterion, this belief itself may be mistaken; we should be begging the question if we tried to test the criterion by applying the criterion to itself.

But although the notion of an absolute criterion is chimerical, there may be relative criteria, which increase the probability of truth. Common sense and science hold that there are. Let us see what they have to say.

One of the plainest cases of verification, perhaps ultimately the only case, consists in the happening of something expected. You go to the station believing that there will be a train at a certain time; you find the train, you get into it, and it starts at the expected time This constitutes verification, and is a perfectly definite experience. It is, in a sense, the converse of memory instead of having first sensations and then images accompanied by belief, we have first images accompanied by belief and then sensations. Apart from differences as to the time-order and the accompanying feelings, the relation between image and sensation is closely similar in the two cases of memory and expectation; it is a relation of similarity, with difference as to causal efficacy—broadly, the image has the psychological but not the physical effects that the sensation would have. When an image accompanied by an expectation-belief is thus succeeded by a sensation which is the "meaning" of the image, we say that the expectation-belief has been verified. The experience of verification in this sense is exceedingly familiar; it happens every time that accustomed activities have results that are not surprising, in eating and walking and talking and all our daily pursuits.

But although the experience in question is common, it is not wholly easy to give a theoretical account of it. How do we know that the sensation resembles the previous image? Does the image persist in presence of the sensation, so that we can compare the two? And even if SOME image does persist, how do we know that it is the previous image unchanged? It does not seem as if this line of inquiry offered much hope of a successful issue. It is better, I think, to take a more external and causal view of the relation of expectation to expected occurrence. If the occurrence, when it comes, gives us the feeling of expectedness, and if the expectation, beforehand, enabled us to act in a way which proves appropriate to the occurrence, that must be held to constitute the maximum of verification. We have first an expectation, then a sensation with the feeling of expectedness related to memory of the expectation. This whole experience, when it occurs, may be defined as verification, and as constituting the truth of the expectation. Appropriate action, during the period of expectation, may be regarded as additional verification, but is not essential. The whole process may be illustrated by looking up a familiar quotation, finding it in the expected words, and in the expected part of the book. In this case we can strengthen the verification by writing down beforehand the words which we expect to find.

I think all verification is ultimately of the above sort. We verify a scientific hypothesis indirectly, by deducing consequences as to the future, which subsequent experience confirms. If somebody were to doubt whether Caesar had crossed the Rubicon, verification could only be obtained from the future. We could proceed to display manuscripts to our historical sceptic, in which it was said that Caesar had behaved in this way. We could advance arguments, verifiable by future experience, to prove the antiquity of the manuscript from its texture, colour, etc. We could find inscriptions agreeing with the historian on other points, and tending to show his general accuracy. The causal laws which our arguments would assume could be verified by the future occurrence of events inferred by means of them. The existence and persistence of causal laws, it is true, must be regarded as a fortunate accident, and how long it will continue we cannot tell. Meanwhile verification remains often practically possible. And since it is sometimes possible, we can gradually discover what kinds of beliefs tend to be verified by experience, and what kinds tend to be falsified; to the former kinds we give an increased degree of assent, to the latter kinds a diminished degree. The process is not absolute or infallible, but it has been found capable of sifting beliefs and building up science. It affords no theoretical refutation of the sceptic, whose position must remain logically unassailable; but if complete scepticism is rejected, it gives the practical method by which the system of our beliefs grows gradually towards the unattainable ideal of impeccable knowledge.

IV. I come now to the purely formal definition of the truth or falsehood of a belief. For this definition it is necessary first of all to consider the derivation of the objective reference of a proposition from the meanings of its component words or images.

Just as a word has meaning, so a proposition has an objective reference. The objective reference of a proposition is a function (in the mathematical sense) of the meanings of its component words. But the objective reference differs from the meaning of a word through the duality of truth and falsehood. You may believe the proposition "to-day is Tuesday" both when, in fact, to-day is Tuesday, and when to-day is not Tuesday. If to-day is not Tuesday, this fact is the objective of your belief that to-day is Tuesday. But obviously the relation of your belief to the fact is different in this case from what it is in the case when to-day is Tuesday. We may say, metaphorically, that when to-day is Tuesday, your belief that it is Tuesday points TOWARDS the fact, whereas when to-day is not Tuesday your belief points AWAY FROM the fact. Thus the objective reference of a belief is not determined by the fact alone, but by the direction of the belief towards or away from the fact. If, on a Tuesday, one man believes that it is Tuesday while another believes that it is not Tuesday, their beliefs have the same objective, namely the fact that it is Tuesday but the true belief points towards the fact while the false one points away from it. Thus, in order to define the reference of a proposition we have to take account not only of the objective, but also of the direction of pointing, towards the objective in the case of a true proposition and away from it in the case of a false one.

This mode of stating the nature of the objective reference of a proposition is necessitated by the circumstance that there are true and false propositions, but not true and false facts. If to-day is Tuesday, there is not a false objective "to-day is not Tuesday," which could be the objective of the false belief "to-day is not Tuesday." This is the reason why two beliefs which are each other's contradictories have the same objective. There is, however, a practical inconvenience, namely that we cannot determine the objective reference of a proposition, according to this definition, unless we know whether the proposition is true or false. To avoid this inconvenience, it is better to adopt a slightly different phraseology, and say: The "meaning" of the proposition "to-day is Tuesday" consists in pointing to the fact "to-day is Tuesday" if that is a fact, or away from the fact "to-day is not Tuesday" if that is a fact. The "meaning" of the proposition "to-day is not Tuesday" will be exactly the opposite. By this hypothetical form we are able to speak of the meaning of a proposition without knowing whether it is true or false. According to this definition, we know the meaning of a proposition when we know what would make it true and what would make it false, even if we do not know whether it is in fact true or false.

The meaning of a proposition is derivative from the meanings of its constituent words. Propositions occur in pairs, distinguished (in simple cases) by the absence or presence of the word "not." Two such propositions have the same objective, but opposite meanings: when one is true, the other is false, and when one is false, the other is true.

The purely formal definition of truth and falsehood offers little difficulty. What is required is a formal expression of the fact that a proposition is true when it points towards its objective, and false when it points away from it, In very simple cases we can give a very simple account of this: we can say that true propositions actually resemble their objectives in a way in which false propositions do not. But for this purpose it is necessary to revert to image-propositions instead of word-propositions. Let us take again the illustration of a memory-image of a familiar room, and let us suppose that in the image the window is to the left of the door. If in fact the window is to the left of the door, there is a correspondence between the image and the objective; there is the same relation between the window and the door as between the images of them. The image-memory consists of the image of the window to the left of the image of the door. When this is true, the very same relation relates the terms of the objective (namely the window and the door) as relates the images which mean them. In this case the correspondence which constitutes truth is very simple.

In the case we have just been considering the objective consists of two parts with a certain relation (that of left-to-right), and the proposition consists of images of these parts with the very same relation. The same proposition, if it were false, would have a less simple formal relation to its objective. If the image-proposition consists of an image of the window to the left of an image of the door, while in fact the window is not to the left of the door, the proposition does not result from the objective by the mere substitution of images for their prototypes. Thus in this unusually simple case we can say that a true proposition "corresponds" to its objective in a formal sense in which a false proposition does not. Perhaps it may be possible to modify this notion of formal correspondence in such a way as to be more widely applicable, but if so, the modifications required will be by no means slight. The reasons for this must now be considered.

To begin with, the simple type of correspondence we have been exhibiting can hardly occur when words are substituted for images, because, in word-propositions, relations are usually expressed by words, which are not themselves relations. Take such a proposition as "Socrates precedes Plato." Here the word "precedes" is just as solid as the words "Socrates" and "Plato"; it MEANS a relation, but is not a relation. Thus the objective which makes our proposition true consists of TWO terms with a relation between them, whereas our proposition consists of THREE terms with a relation of order between them. Of course, it would be perfectly possible, theoretically, to indicate a few chosen relations, not by words, but by relations between the other words. "Socrates-Plato" might be used to mean "Socrates precedes Plato"; "Plato-Socrates" might be used to mean "Plato was born before Socrates and died after him"; and so on. But the possibilities of such a method would be very limited. For aught I know, there may be languages that use it, but they are not among the languages with which I am acquainted. And in any case, in view of the multiplicity of relations that we wish to express, no language could advance far without words for relations. But as soon as we have words for relations, word-propositions have necessarily more terms than the facts to which they refer, and cannot therefore correspond so simply with their objectives as some image-propositions can.

The consideration of negative propositions and negative facts introduces further complications. An image-proposition is necessarily positive: we can image the window to the left of the door, or to the right of the door, but we can form no image of the bare negative "the window not to the left of the door." We can DISBELIEVE the image-proposition expressed by "the window to the left of the door," and our disbelief will be true if the window is not to the left of the door. But we can form no image of the fact that the window is not to the left of the door. Attempts have often been made to deny such negative facts, but, for reasons which I have given elsewhere, I believe these attempts to be mistaken, and I shall assume that there are negative facts.

Word-propositions, like image-propositions, are always positive facts. The fact that Socrates precedes Plato is symbolized in English by the fact that the word "precedes" occurs between the words "Socrates" and "Plato." But we cannot symbolize the fact that Plato does not precede Socrates by not putting the word "precedes" between "Plato" and "Socrates." A negative fact is not sensible, and language, being intended for communication, has to be sensible. Therefore we symbolize the fact that Plato does not precede Socrates by putting the words "does not precede" between "Plato" and "Socrates." We thus obtain a series of words which is just as positive a fact as the series "Socrates precedes Plato." The propositions asserting negative facts are themselves positive facts; they are merely different positive facts from those asserting positive facts.

We have thus, as regards the opposition of positive and negative, three different sorts of duality, according as we are dealing with facts, image-propositions, or word-propositions. We have, namely:

(1) Positive and negative facts;

(2) Image-propositions, which may be believed or disbelieved, but do not allow any duality of content corresponding to positive and negative facts;

(3) Word-propositions, which are always positive facts, but are of two kinds: one verified by a positive objective, the other by a negative objective.

Owing to these complications, the simplest type of correspondence is impossible when either negative facts or negative propositions are involved.

Even when we confine ourselves to relations between two terms which are both imaged, it may be impossible to form an image-proposition in which the relation of the terms is represented by the same relation of the images. Suppose we say "Caesar was 2,000 years before Foch," we express a certain temporal relation between Caesar and Foch; but we cannot allow 2,000 years to elapse between our image of Caesar and our image of Foch. This is perhaps not a fair example, since "2,000 years before" is not a direct relation. But take a case where the relation is direct, say, "the sun is brighter than the moon." We can form visual images of sunshine and moonshine, and it may happen that our image of the sunshine is the brighter of the two, but this is by no means either necessary or sufficient. The act of comparison, implied in our judgment, is something more than the mere coexistence of two images, one of which is in fact brighter than the other. It would take us too far from our main topic if we were to go into the question what actually occurs when we make this judgment. Enough has been said to show that the correspondence between the belief and its objective is more complicated in this case than in that of the window to the left of the door, and this was all that had to be proved.

In spite of these complications, the general nature of the formal correspondence which makes truth is clear from our instances. In the case of the simpler kind of propositions, namely those that I call "atomic" propositions, where there is only one word expressing a relation, the objective which would verify our proposition, assuming that the word "not" is absent, is obtained by replacing each word by what it means, the word meaning a relation being replaced by this relation among the meanings of the other words. For example, if the proposition is "Socrates precedes Plato," the objective which verifies it results from replacing the word "Socrates" by Socrates, the word "Plato" by Plato, and the word "precedes" by the relation of preceding between Socrates and Plato. If the result of this process is a fact, the proposition is true; if not, it is false. When our proposition is "Socrates does not precede Plato," the conditions of truth and falsehood are exactly reversed. More complicated propositions can be dealt with on the same lines. In fact, the purely formal question, which has occupied us in this last section, offers no very formidable difficulties.

I do not believe that the above formal theory is untrue, but I do believe that it is inadequate. It does not, for example, throw any light upon our preference for true beliefs rather than false ones. This preference is only explicable by taking account of the causal efficacy of beliefs, and of the greater appropriateness of the responses resulting from true beliefs. But appropriateness depends upon purpose, and purpose thus becomes a vital part of theory of knowledge.

LECTURE XIV.

EMOTIONS AND WILL

On the two subjects of the present lecture, I have nothing original to say, and I am treating them only in order to complete the discussion of my main thesis, namely that all psychic phenomena are built up out of sensations and images alone.

Emotions are traditionally regarded by psychologists as a separate class of mental occurrences: I am, of course, not concerned to deny the obvious fact that they have characteristics which make a special investigation of them necessary. What I am concerned with is the analysis of emotions. It is clear that an emotion is essentially complex, and we have to inquire whether it ever contains any non-physiological material not reducible to sensations and images and their relations.

Although what specially concerns us is the analysis of emotions, we shall find that the more important topic is the physiological causation of emotions. This is a subject upon which much valuable and exceedingly interesting work has been done, whereas the bare analysis of emotions has proved somewhat barren. In view of the fact that we have defined perceptions, sensations, and images by their physiological causation, it is evident that our problem of the analysis of the emotions is bound up with the problem of their physiological causation.

Modern views on the causation of emotions begin with what is called the James-Lange theory. James states this view in the following terms :

"Our natural way of thinking about these coarser emotions, grief, fear, rage, love, is that the mental perception of some fact excites the mental affection called the emotion, and that this latter state of mind gives rise to the bodily expression. My theory, on the contrary, is that THE BODILY CHANGES FOLLOW DIRECTLY THE PERCEPTION OF THE EXCITING FACT, AND THAT OUR FEELING OF THE SAME CHANGES AS THEY OCCUR *IS* THE EMOTION (James's italics). Common sense says: we lose our fortune, are sorry and weep; we meet a bear, are frightened and run; we are insulted by a rival, are angry and strike. The hypothesis here to be defended says that this order of sequence is incorrect, that the one mental state is not immediately induced by the other, that the bodily manifestations must first be interposed between, and that the more rational statement is that we feel sorry because we cry, angry because we strike, afraid because we tremble, and not that we cry, strike, or tremble, because we are sorry, angry, or fearful, as the case may be. Without the bodily states following on the perception, the latter would be purely cognitive in form, pale, colourless, destitute of emotional warmth."

Round this hypothesis a very voluminous literature has grown up. The history of its victory over earlier criticism, and its difficulties with the modern experimental work of Sherrington and Cannon, is well told by James R. Angell in an article called "A Reconsideration of James's Theory of Emotion in the Light of Recent Criticisms." In this article Angell defends James's theory and to me—though I speak with diffidence on a question as to which I have little competence—it appears that his defence is on the whole successful.

Sherrington, by experiments on dogs, showed that many of the usual marks of emotion were present in their behaviour even when, by severing the spinal cord in the lower cervical region, the viscera were cut off from all communication with the brain, except that existing through certain cranial nerves. He mentions the various signs which "contributed to indicate the existence of an emotion as lively as the animal had ever shown us before the spinal operation had been made." He infers that the physiological condition of the viscera cannot be the cause of the emotion displayed under such circumstances, and concludes: "We are forced back toward the likelihood that the visceral expression of emotion is SECONDARY to the cerebral action occurring with the psychical state.... We may with James accept visceral and organic sensations and the memories and associations of them as contributory to primitive emotion, but we must regard them as re-enforcing rather than as initiating the psychosis."

Angell suggests that the display of emotion in such cases may be due to past experience, generating habits which would require only the stimulation of cerebral reflex arcs. Rage and some forms of fear, however, may, he thinks, gain expression without the brain. Rage and fear have been especially studied by Cannon, whose work is of the greatest importance. His results are given in his book, "Bodily Changes in Pain, Hunger, Fear and Rage" (D. Appleton and Co., 1916).

The most interesting part of Cannon's book consists in the investigation of the effects produced by secretion of adrenin. Adrenin is a substance secreted into the blood by the adrenal glands. These are among the ductless glands, the functions of which, both in physiology and in connection with the emotions, have only come to be known during recent years. Cannon found that pain, fear and rage occurred in circumstances which affected the supply of adrenin, and that an artificial injection of adrenin could, for example, produce all the symptoms of fear. He studied the effects of adrenin on various parts of the body; he found that it causes the pupils to dilate, hairs to stand erect, blood vessels to be constricted, and so on. These effects were still produced if the parts in question were removed from the body and kept alive artificially.

Cannon's chief argument against James is, if I understand him rightly, that similar affections of the viscera may accompany dissimilar emotions, especially fear and rage. Various different emotions make us cry, and therefore it cannot be true to say, as James does, that we "feel sorry because we cry," since sometimes we cry when we feel glad. This argument, however, is by no means conclusive against James, because it cannot be shown that there are no visceral differences for different emotions, and indeed it is unlikely that this is the case.

As Angell says (loc. cit.): "Fear and joy may both cause cardiac palpitation, but in one case we find high tonus of the skeletal muscles, in the other case relaxation and the general sense of weakness."

Angell's conclusion, after discussing the experiments of Sherrington and Cannon, is: "I would therefore submit that, so far as concerns the critical suggestions by these two psychologists, James's essential contentions are not materially affected." If it were necessary for me to take sides on this question, I should agree with this conclusion; but I think my thesis as to the analysis of emotion can be maintained without coming to a probably premature conclusion upon the doubtful parts of the physiological problem.

According to our definitions, if James is right, an emotion may be regarded as involving a confused perception of the viscera concerned in its causation, while if Cannon and Sherrington are right, an emotion involves a confused perception of its external stimulus. This follows from what was said in Lecture VII. We there defined a perception as an appearance, however irregular, of one or more objects external to the brain. And in order to be an appearance of one or more objects, it is only necessary that the occurrence in question should be connected with them by a continuous chain, and should vary when they are varied sufficiently. Thus the question whether a mental occurrence can be called a perception turns upon the question whether anything can be inferred from it as to its causes outside the brain: if such inference is possible, the occurrence in question will come within our definition of a perception. And in that case, according to the definition in Lecture VIII, its non-mnemic elements will be sensations. Accordingly, whether emotions are caused by changes in the viscera or by sensible objects, they contain elements which are sensations according to our definition.

An emotion in its entirety is, of course, something much more complex than a perception. An emotion is essentially a process, and it will be only what one may call a cross-section of the emotion that will be a perception, of a bodily condition according to James, or (in certain cases) of an external object according to his opponents. An emotion in its entirety contains dynamic elements, such as motor impulses, desires, pleasures and pains. Desires and pleasures and pains, according to the theory adopted in Lecture III, are characteristics of processes, not separate ingredients. An emotion—rage, for example—will be a certain kind of process, consisting of perceptions and (in general) bodily movements. The desires and pleasures and pains involved are properties of this process, not separate items in the stuff of which the emotion is composed. The dynamic elements in an emotion, if we are right in our analysis, contain, from our point of view, no ingredients beyond those contained in the processes considered in Lecture III. The ingredients of an emotion are only sensations and images and bodily movements succeeding each other according to a certain pattern. With this conclusion we may leave the emotions and pass to the consideration of the will.

The first thing to be defined when we are dealing with Will is a VOLUNTARY MOVEMENT. We have already defined vital movements, and we have maintained that, from a behaviourist standpoint, it is impossible to distinguish which among such movements are reflex and which voluntary. Nevertheless, there certainly is a distinction. When we decide in the morning that it is time to get up, our consequent movement is voluntary. The beating of the heart, on the other hand, is involuntary: we can neither cause it nor prevent it by any decision of our own, except indirectly, as e.g. by drugs. Breathing is intermediate between the two: we normally breathe without the help of the will, but we can alter or stop our breathing if we choose.

James maintains that the only distinctive characteristic of a voluntary act is that it involves an idea of the movement to be performed, made up of memory-images of the kinaesthetic sensations which we had when the same movement occurred on some former occasion. He points out that, on this view, no movement can be made voluntarily unless it has previously occurred involuntarily.

I see no reason to doubt the correctness of this view. We shall say, then, that movements which are accompanied by kinaesthetic sensations tend to be caused by the images of those sensations, and when so caused are called VOLUNTARY.

Volition, in the emphatic sense, involves something more than voluntary movement. The sort of case I am thinking of is decision after deliberation. Voluntary movements are a part of this, but not the whole. There is, in addition to them, a judgment: "This is what I shall do"; there is also a sensation of tension during doubt, followed by a different sensation at the moment of deciding. I see no reason whatever to suppose that there is any specifically new ingredient; sensations and images, with their relations and causal laws, yield all that seems to be wanted for the analysis of the will, together with the fact that kinaesthetic images tend to cause the movements with which they are connected. Conflict of desires is of course essential in the causation of the emphatic kind of will: there will be for a time kinaesthetic images of incompatible movements, followed by the exclusive image of the movement which is said to be willed. Thus will seems to add no new irreducible ingredient to the analysis of the mind.

LECTURE XV.

CHARACTERISTICS OF MENTAL PHENOMENA

At the end of our journey, it is time to return to the question from which we set out, namely: What is it that characterizes mind as opposed to matter? Or, to state the same question in other terms: How is psychology to be distinguished from physics? The answer provisionally suggested at the outset of our inquiry was that psychology and physics are distinguished by the nature of their causal laws, not by their subject matter. At the same time we held that there is a certain subject matter, namely images, to which only psychological causal laws are applicable; this subject matter, therefore, we assigned exclusively to psychology. But we found no way of defining images except through their causation; in their intrinsic character they appeared to have no universal mark by which they could be distinguished from sensations.

In this last lecture I propose to pass in review various suggested methods of distinguishing mind from matter. I shall then briefly sketch the nature of that fundamental science which I believe to be the true metaphysic, in which mind and matter alike are seen to be constructed out of a neutral stuff, whose causal laws have no such duality as that of psychology, but form the basis upon which both physics and psychology are built.

In search for the definition of "mental phenomena," let us begin with "consciousness," which is often thought to be the essence of mind. In the first lecture I gave various arguments against the view that consciousness is fundamental, but I did not attempt to say what consciousness is. We must find a definition of it, if we are to feel secure in deciding that it is not fundamental. It is for the sake of the proof that it is not fundamental that we must now endeavour to decide what it is.

"Consciousness," by those who regard it as fundamental, is taken to be a character diffused throughout our mental life, distinct from sensations and images, memories, beliefs and desires, but present in all of them. Dr. Henry Head, in an article which I quoted in Lecture III, distinguishing sensations from purely physiological occurrences, says: "Sensation, in the strict sense of the term, demands the existence of consciousness." This statement, at first sight, is one to which we feel inclined to assent, but I believe we are mistaken if we do so. Sensation is the sort of thing of which we MAY be conscious, but not a thing of which we MUST be conscious. We have been led, in the course of our inquiry, to admit unconscious beliefs and unconscious desires. There is, so far as I can see, no class of mental or other occurrences of which we are always conscious whenever they happen.

The first thing to notice is that consciousness must be of something. In view of this, I should define "consciousness" in terms of that relation of an image of a word to an object which we defined, in Lecture XI, as "meaning." When a sensation is followed by an image which is a "copy" of it, I think it may be said that the existence of the image constitutes consciousness of the sensation, provided it is accompanied by that sort of belief which, when we reflect upon it, makes us feel that the image is a "sign" of something other than itself. This is the sort of belief which, in the case of memory, we expressed in the words "this occurred"; or which, in the case of a judgment of perception, makes us believe in qualities correlated with present sensations, as e.g., tactile and visual qualities are correlated. The addition of some element of belief seems required, since mere imagination does not involve consciousness of anything, and there can be no consciousness which is not of something. If images alone constituted consciousness of their prototypes, such imagination-images as in fact have prototypes would involve consciousness of them; since this is not the case, an element of belief must be added to the images in defining consciousness. The belief must be of that sort that constitutes objective reference, past or present. An image, together with a belief of this sort concerning it, constitutes, according to our definition, consciousness of the prototype of the image.

But when we pass from consciousness of sensations to consciousness of objects of perception, certain further points arise which demand an addition to our definition. A judgment of perception, we may say, consists of a core of sensation, together with associated images, with belief in the present existence of an object to which sensation and images are referred in a way which is difficult to analyse. Perhaps we might say that the belief is not fundamentally in any PRESENT existence, but is of the nature of an expectation: for example, when we see an object, we expect certain sensations to result if we proceed to touch it. Perception, then, will consist of a present sensation together with expectations of future sensations. (This, of course, is a reflective analysis, not an account of the way perception appears to unchecked introspection.) But all such expectations are liable to be erroneous, since they are based upon correlations which are usual but not invariable. Any such correlation may mislead us in a particular case, for example, if we try to touch a reflection in a looking-glass under the impression that it is "real." Since memory is fallible, a similar difficulty arises as regards consciousness of past objects. It would seem odd to say that we can be "conscious" of a thing which does not or did not exist. The only way to avoid this awkwardness is to add to our definition the proviso that the beliefs involved in consciousness must be TRUE.

In the second place, the question arises as to whether we can be conscious of images. If we apply our definition to this case, it seems to demand images of images. In order, for example, to be conscious of an image of a cat, we shall require, according to the letter of the definition, an image which is a copy of our image of the cat, and has this image for its prototype. Now, it hardly seems probable, as a matter of observation, that there are images of images, as opposed to images of sensations. We may meet this difficulty in two ways, either by boldly denying consciousness of images, or by finding a sense in which, by means of a different accompanying belief, an image, instead of meaning its prototype, can mean another image of the same prototype.

The first alternative, which denies consciousness of images, has already been discussed when we were dealing with Introspection in Lecture VI. We then decided that there must be, in some sense, consciousness of images. We are therefore left with the second suggested way of dealing with knowledge of images. According to this second hypothesis, there may be two images of the same prototype, such that one of them means the other, instead of meaning the prototype. It will be remembered that we defined meaning by association a word or image means an object, we said, when it has the same associations as the object. But this definition must not be interpreted too absolutely: a word or image will not have ALL the same associations as the object which it means. The word "cat" may be associated with the word "mat," but it would not happen except by accident that a cat would be associated with a mat. And in like manner an image may have certain associations which its prototype will not have, e.g. an association with the word "image." When these associations are active, an image means an image, instead of meaning its prototype. If I have had images of a given prototype many times, I can mean one of these, as opposed to the rest, by recollecting the time and place or any other distinctive association of that one occasion. This happens, for example, when a place recalls to us some thought we previously had in that place, so that we remember a thought as opposed to the occurrence to which it referred. Thus we may say that we think of an image A when we have a similar image B associated with recollections of circumstances connected with A, but not with its prototype or with other images of the same prototype. In this way we become aware of images without the need of any new store of mental contents, merely by the help of new associations. This theory, so far as I can see, solves the problems of introspective knowledge, without requiring heroic measures such as those proposed by Knight Dunlap, whose views we discussed in Lecture VI.

According to what we have been saying, sensation itself is not an instance of consciousness, though the immediate memory by which it is apt to be succeeded is so. A sensation which is remembered becomes an object of consciousness as soon as it begins to be remembered, which will normally be almost immediately after its occurrence (if at all); but while it exists it is not an object of consciousness. If, however, it is part of a perception, say of some familiar person, we may say that the person perceived is an object of consciousness. For in this case the sensation is a SIGN of the perceived object in much the same way in which a memory-image is a sign of a remembered object. The essential practical function of "consciousness" and "thought" is that they enable us to act with reference to what is distant in time or space, even though it is not at present stimulating our senses. This reference to absent objects is possible through association and habit. Actual sensations, in themselves, are not cases of consciousness, because they do not bring in this reference to what is absent. But their connection with consciousness is very close, both through immediate memory, and through the correlations which turn sensations into perceptions.

Enough has, I hope, been said to show that consciousness is far too complex and accidental to be taken as the fundamental characteristic of mind. We have seen that belief and images both enter into it. Belief itself, as we saw in an earlier lecture, is complex. Therefore, if any definition of mind is suggested by our analysis of consciousness, images are what would naturally suggest themselves. But since we found that images can only be defined causally, we cannot deal with this suggestion, except in connection with the difference between physical and psychological causal laws.

I come next to those characteristics of mental phenomena which arise out of mnemic causation. The possibility of action with reference to what is not sensibly present is one of the things that might be held to characterize mind. Let us take first a very elementary example. Suppose you are in a familiar room at night, and suddenly the light goes out. You will be able to find your way to the door without much difficulty by means of the picture of the room which you have in your mind. In this case visual images serve, somewhat imperfectly it is true, the purpose which visual sensations would otherwise serve. The stimulus to the production of visual images is the desire to get out of the room, which, according to what we found in Lecture III, consists essentially of present sensations and motor impulses caused by them. Again, words heard or read enable you to act with reference to the matters about which they give information; here, again, a present sensible stimulus, in virtue of habits formed in the past, enables you to act in a manner appropriate to an object which is not sensibly present. The whole essence of the practical efficiency of "thought" consists in sensitiveness to signs: the sensible presence of A, which is a sign of the present or future existence of B, enables us to act in a manner appropriate to B. Of this, words are the supreme example, since their effects as signs are prodigious, while their intrinsic interest as sensible occurrences on their own account is usually very slight. The operation of signs may or may not be accompanied by consciousness. If a sensible stimulus A calls up an image of B, and we then act with reference to B, we have what may be called consciousness of B. But habit may enable us to act in a manner appropriate to B as soon as A appears, without ever having an image of B. In that case, although A operates as a sign, it operates without the help of consciousness. Broadly speaking, a very familiar sign tends to operate directly in this manner, and the intervention of consciousness marks an imperfectly established habit.

The power of acquiring experience, which characterizes men and animals, is an example of the general law that, in mnemic causation, the causal unit is not one event at one time, but two or more events at two or more times.& A burnt child fears the fire, that is to say, the neighbourhood of fire has a different effect upon a child which has had the sensations of burning than upon one which has not. More correctly, the observed effect, when a child which has been burnt is put near a fire, has for its cause, not merely the neighbourhood of the fire, but this together with the previous burning. The general formula, when an animal has acquired experience through some event A, is that, when B occurs at some future time, the animal to which A has happened acts differently from an animal which A has not happened. Thus A and B together, not either separately, must be regarded as the cause of the animal's behaviour, unless we take account of the effect which A has had in altering the animal's nervous tissue, which is a matter not patent to external observation except under very special circumstances. With this possibility, we are brought back to causal laws, and to the suggestion that many things which seem essentially mental are really neural. Perhaps it is the nerves that acquire experience rather than the mind. If so, the possibility of acquiring experience cannot be used to define mind.

Very similar considerations apply to memory, if taken as the essence of mind. A recollection is aroused by something which is happening now, but is different from the effect which the present occurrence would have produced if the recollected event had not occurred. This may be accounted for by the physical effect of the past event on the brain, making it a different instrument from that which would have resulted from a different experience. The causal peculiarities of memory may, therefore, have a physiological explanation. With every special class of mental phenomena this possibility meets us afresh. If psychology is to be a separate science at all, we must seek a wider ground for its separateness than any that we have been considering hitherto.

We have found that "consciousness" is too narrow to characterize mental phenomena, and that mnemic causation is too wide. I come now to a characteristic which, though difficult to define, comes much nearer to what we require, namely subjectivity.

Subjectivity, as a characteristic of mental phenomena, was considered in Lecture VII, in connection with the definition of perception. We there decided that those particulars which constitute the physical world can be collected into sets in two ways, one of which makes a bundle of all those particulars that are appearances of a given thing from different places, while the other makes a bundle of all those particulars which are appearances of different things from a given place. A bundle of this latter sort, at a given time, is called a "perspective"; taken throughout a period of time, it is called a "biography." Subjectivity is the characteristic of perspectives and biographies, the characteristic of giving the view of the world from a certain place. We saw in Lecture VII that this characteristic involves none of the other characteristics that are commonly associated with mental phenomena, such as consciousness, experience and memory. We found in fact that it is exhibited by a photographic plate, and, strictly speaking, by any particular taken in conjunction with those which have the same "passive" place in the sense defined in Lecture VII. The particulars forming one perspective are connected together primarily by simultaneity; those forming one biography, primarily by the existence of direct time-relations between them. To these are to be added relations derivable from the laws of perspective. In all this we are clearly not in the region of psychology, as commonly understood; yet we are also hardly in the region of physics. And the definition of perspectives and biographies, though it does not yet yield anything that would be commonly called "mental," is presupposed in mental phenomena, for example in mnemic causation: the causal unit in mnemic causation, which gives rise to Semon's engram, is the whole of one perspective—not of any perspective, but of a perspective in a place where there is nervous tissue, or at any rate living tissue of some sort. Perception also, as we saw, can only be defined in terms of perspectives. Thus the conception of subjectivity, i.e. of the "passive" place of a particular, though not alone sufficient to define mind, is clearly an essential element in the definition.

I have maintained throughout these lectures that the data of psychology do not differ in, their intrinsic character from the data of physics. I have maintained that sensations are data for psychology and physics equally, while images, which may be in some sense exclusively psychological data, can only be distinguished from sensations by their correlations, not by what they are in themselves. It is now necessary, however, to examine the notion of a "datum," and to obtain, if possible, a definition of this notion.

The notion of "data" is familiar throughout science, and is usually treated by men of science as though it were perfectly clear. Psychologists, on the other hand, find great difficulty in the conception. "Data" are naturally defined in terms of theory of knowledge: they are those propositions of which the truth is known without demonstration, so that they may be used as premisses in proving other propositions. Further, when a proposition which is a datum asserts the existence of something, we say that the something is a datum, as well as the proposition asserting its existence. Thus those objects of whose existence we become certain through perception are said to be data.

There is some difficulty in connecting this epistemological definition of "data" with our psychological analysis of knowledge; but until such a connection has been effected, we have no right to use the conception "data."

It is clear, in the first place, that there can be no datum apart from a belief. A sensation which merely comes and goes is not a datum; it only becomes a datum when it is remembered. Similarly, in perception, we do not have a datum unless we have a JUDGMENT of perception. In the sense in which objects (as opposed to propositions) are data, it would seem natural to say that those objects of which we are conscious are data. But consciousness, as we have seen, is a complex notion, involving beliefs, as well as mnemic phenomena such as are required for perception and memory. It follows that no datum is theoretically indubitable, since no belief is infallible; it follows also that every datum has a greater or less degree of vagueness, since there is always some vagueness in memory and the meaning of images.

Data are not those things of which our consciousness is earliest in time. At every period of life, after we have become capable of thought, some of our beliefs are obtained by inference, while others are not. A belief may pass from either of these classes into the other, and may therefore become, or cease to be, a belief giving a datum. When, in what follows, I speak of data, I do not mean the things of which we feel sure before scientific study begins, but the things which, when a science is well advanced, appear as affording grounds for other parts of the science, without themselves being believed on any ground except observation. I assume, that is to say, a trained observer, with an analytic attention, knowing the sort of thing to look for, and the sort of thing that will be important. What he observes is, at the stage of science which he has reached, a datum for his science. It is just as sophisticated and elaborate as the theories which he bases upon it, since only trained habits and much practice enable a man to make the kind of observation that will be scientifically illuminating. Nevertheless, when once it has been observed, belief in it is not based on inference and reasoning, but merely upon its having been seen. In this way its logical status differs from that of the theories which are proved by its means.

In any science other than psychology the datum is primarily a perception, in which only the sensational core is ultimately and theoretically a datum, though some such accretions as turn the sensation into a perception are practically unavoidable. But if we postulate an ideal observer, he will be able to isolate the sensation, and treat this alone as datum. There is, therefore, an important sense in which we may say that, if we analyse as much as we ought, our data, outside psychology, consist of sensations, which include within themselves certain spatial and temporal relations.

Applying this remark to physiology, we see that the nerves and brain as physical objects are not truly data; they are to be replaced, in the ideal structure of science, by the sensations through which the physiologist is said to perceive them. The passage from these sensations to nerves and brain as physical objects belongs really to the initial stage in the theory of physics, and ought to be placed in the reasoned part, not in the part supposed to be observed. To say we see the nerves is like saying we hear the nightingale; both are convenient but inaccurate expressions. We hear a sound which we believe to be causally connected with the nightingale, and we see a sight which we believe to be causally connected with a nerve. But in each case it is only the sensation that ought, in strictness, to be called a datum. Now, sensations are certainly among the data of psychology. Therefore, all the data of the physical sciences are also psychological data. It remains to inquire whether all the data of psychology are also data of physical science, and especially of physiology.

If we have been right in our analysis of mind, the ultimate data of psychology are only sensations and images and their relations. Beliefs, desires, volitions, and so on, appeared to us to be complex phenomena consisting of sensations and images variously interrelated. Thus (apart from certain relations) the occurrences which seem most distinctively mental, and furthest removed from physics, are, like physical objects, constructed or inferred, not part of the original stock of data in the perfected science. From both ends, therefore, the difference between physical and psychological data is diminished. Is there ultimately no difference, or do images remain as irreducibly and exclusively psychological? In view of the causal definition of the difference between images and sensations, this brings us to a new question, namely: Are the causal laws of psychology different from those of any other science, or are they really physiological?

Certain ambiguities must be removed before this question can be adequately discussed.

First, there is the distinction between rough approximate laws and such as appear to be precise and general. I shall return to the former presently; it is the latter that I wish to discuss now.

Matter, as defined at the end of Lecture V, is a logical fiction, invented because it gives a convenient way of stating causal laws. Except in cases of perfect regularity in appearances (of which we can have no experience), the actual appearances of a piece of matter are not members of that ideal system of regular appearances which is defined as being the matter in question. But the matter is, after all, inferred from its appearances, which are used to VERIFY physical laws. Thus, in so far as physics is an empirical and verifiable science, it must assume or prove that the inference from appearances to matter is, in general, legitimate, and it must be able to tell us, more or less, what appearances to expect. It is through this question of verifiability and empirical applicability to experience that we are led to a theory of matter such as I advocate. From the consideration of this question, it results that physics, in so far as it is an empirical science, not a logical phantasy, is concerned with particulars of just the same sort as those which psychology considers under the name of sensations. The causal laws of physics, so interpreted, differ from those of psychology only by the fact that they connect a particular with other appearances in the same piece of matter, rather than with other appearances in the same perspective. That is to say, they group together particulars having the same "active" place, while psychology groups together those having the same "passive" place. Some particulars, such as images, have no "active" place, and therefore belong exclusively to psychology.

We can now understand the distinction between physics and psychology. The nerves and brain are matter: our visual sensations when we look at them may be, and I think are, members of the system constituting irregular appearances of this matter, but are not the whole of the system. Psychology is concerned, inter alia, with our sensations when we see a piece of matter, as opposed to the matter which we see. Assuming, as we must, that our sensations have physical causes, their causal laws are nevertheless radically different from the laws of physics, since the consideration of a single sensation requires the breaking up of the group of which it is a member. When a sensation is used to verify physics, it is used merely as a sign of a certain material phenomenon, i.e. of a group of particulars of which it is a member. But when it is studied by psychology, it is taken away from that group and put into quite a different context, where it causes images or voluntary movements. It is primarily this different grouping that is characteristic of psychology as opposed to all the physical sciences, including physiology; a secondary difference is that images, which belong to psychology, are not easily to be included among the aspects which constitute a physical thing or piece of matter.

There remains, however, an important question, namely: Are mental events causally dependent upon physical events in a sense in which the converse dependence does not hold? Before we can discuss the answer to this question, we must first be clear as to what our question means.

When, given A, it is possible to infer B, but given B, it is not possible to infer A, we say that B is dependent upon A in a sense in which A is not dependent upon B. Stated in logical terms, this amounts to saying that, when we know a many-one relation of A to B, B is dependent upon A in respect of this relation. If the relation is a causal law, we say that B is causally dependent upon A. The illustration that chiefly concerns us is the system of appearances of a physical object. We can, broadly speaking, infer distant appearances from near ones, but not vice versa. All men look alike when they are a mile away, hence when we see a man a mile off we cannot tell what he will look like when he is only a yard away. But when we see him a yard away, we can tell what he will look like a mile away. Thus the nearer view gives us more valuable information, and the distant view is causally dependent upon it in a sense in which it is not causally dependent upon the distant view.

It is this greater causal potency of the near appearance that leads physics to state its causal laws in terms of that system of regular appearances to which the nearest appearances increasingly approximate, and that makes it value information derived from the microscope or telescope. It is clear that our sensations, considered as irregular appearances of physical objects, share the causal dependence belonging to comparatively distant appearances; therefore in our sensational life we are in causal dependence upon physical laws.

This, however, is not the most important or interesting part of our question. It is the causation of images that is the vital problem. We have seen that they are subject to mnenic causation, and that mnenic causation may be reducible to ordinary physical causation in nervous tissue. This is the question upon which our attitude must turn towards what may be called materialism. One sense of materialism is the view that all mental phenomena are causally dependent upon physical phenomena in the above-defined sense of causal dependence. Whether this is the case or not, I do not profess to know. The question seems to me the same as the question whether mnemic causation is ultimate, which we considered without deciding in Lecture IV. But I think the bulk of the evidence points to the materialistic answer as the more probable.

In considering the causal laws of psychology, the distinction between rough generalizations and exact laws is important. There are many rough generalizations in psychology, not only of the sort by which we govern our ordinary behaviour to each other, but also of a more nearly scientific kind. Habit and association belong among such laws. I will give an illustration of the kind of law that can be obtained. Suppose a person has frequently experienced A and B in close temporal contiguity, an association will be established, so that A, or an image of A, tends to cause an image of B. The question arises: will the association work in either direction, or only from the one which has occurred earlier to the one which has occurred later? In an article by Mr. Wohlgemuth, called "The Direction of Associations" ("British Journal of Psychology," vol. v, part iv, March, 1913), it is claimed to be proved by experiment that, in so far as motor memory (i.e. memory of movements) is concerned, association works only from earlier to later, while in visual and auditory memory this is not the case, but the later of two neighbouring experiences may recall the earlier as well as the earlier the later. It is suggested that motor memory is physiological, while visual and auditory memory are more truly psychological. But that is not the point which concerns us in the illustration. The point which concerns us is that a law of association, established by purely psychological observation, is a purely psychological law, and may serve as a sample of what is possible in the way of discovering such laws. It is, however, still no more than a rough generalization, a statistical average. It cannot tell us what will result from a given cause on a given occasion. It is a law of tendency, not a precise and invariable law such as those of physics aim at being.

If we wish to pass from the law of habit, stated as a tendency or average, to something more precise and invariable, we seem driven to the nervous system. We can more or less guess how an occurrence produces a change in the brain, and how its repetition gradually produces something analogous to the channel of a river, along which currents flow more easily than in neighbouring paths. We can perceive that in this way, if we had more knowledge, the tendency to habit through repetition might be replaced by a precise account of the effect of each occurrence in bringing about a modification of the sort from which habit would ultimately result. It is such considerations that make students of psychophysiology materialistic in their methods, whatever they may be in their metaphysics. There are, of course, exceptions, such as Professor J. S. Haldane, who maintains that it is theoretically impossible to obtain physiological explanations of psychical phenomena, or physical explanations of physiological phenomena. But I think the bulk of expert opinion, in practice, is on the other side.

The question whether it is possible to obtain precise causal laws in which the causes are psychological, not material, is one of detailed investigation. I have done what I could to make clear the nature of the question, but I do not believe that it is possible as yet to answer it with any confidence. It seems to be by no means an insoluble question, and we may hope that science will be able to produce sufficient grounds for regarding one answer as much more probable than the other. But for the moment I do not see how we can come to a decision.

I think, however, on grounds of the theory of matter explained in Lectures V and VII, that an ultimate scientific account of what goes on in the world, if it were ascertainable, would resemble psychology rather than physics in what we found to be the decisive difference between them. I think, that is to say, that such an account would not be content to speak, even formally, as though matter, which is a logical fiction, were the ultimate reality. I think that, if our scientific knowledge were adequate to the task, which it neither is nor is likely to become, it would exhibit the laws of correlation of the particulars constituting a momentary condition of a material unit, and would state the causal laws of the world in terms of these particulars, not in terms of matter. Causal laws so stated would, I believe, be applicable to psychology and physics equally; the science in which they were stated would succeed in achieving what metaphysics has vainly attempted, namely a unified account of what really happens, wholly true even if not the whole of truth, and free from all convenient fictions or unwarrantable assumptions of metaphysical entities. A causal law applicable to particulars would count as a law of physics if it could be stated in terms of those fictitious systems of regular appearances which are matter; if this were not the case, it would count as a law of psychology if one of the particulars were a sensation or an image, i.e. were subject to mnemic causation. I believe that the realization of the complexity of a material unit, and its analysis into constituents analogous to sensations, is of the utmost importance to philosophy, and vital for any understanding of the relations between mind and matter, between our perceptions and the world which they perceive. It is in this direction, I am convinced, that we must look for the solution of many ancient perplexities.

It is probable that the whole science of mental occurrences, especially where its initial definitions are concerned, could be simplified by the development of the fundamental unifying science in which the causal laws of particulars are sought, rather than the causal laws of those systems of particulars that constitute the material units of physics. This fundamental science would cause physics to become derivative, in the sort of way in which theories of the constitution of the atom make chemistry derivative from physics; it would also cause psychology to appear less singular and isolated among sciences. If we are right in this, it is a wrong philosophy of matter which has caused many of the difficulties in the philosophy of mind—difficulties which a right philosophy of matter would cause to disappear.

The conclusions at which we have arrived may be summed up as follows:

I. Physics and psychology are not distinguished by their material. Mind and matter alike are logical constructions; the particulars out of which they are constructed, or from which they are inferred, have various relations, some of which are studied by physics, others by psychology. Broadly speaking, physics group particulars by their active places, psychology by their passive places.

II. The two most essential characteristics of the causal laws which would naturally be called psychological are SUBJECTIVITY and MNEMIC CAUSATION; these are not unconnected, since the causal unit in mnemic causation is the group of particulars having a given passive place at a given time, and it is by this manner of grouping that subjectivity is defined.

III. Habit, memory and thought are all developments of mnemic causation. It is probable, though not certain, that mnemic causation is derivative from ordinary physical causation in nervous (and other) tissue.

IV. Consciousness is a complex and far from universal characteristic of mental phenomena.

V. Mind is a matter of degree, chiefly exemplified in number and complexity of habits.

VI. All our data, both in physics and psychology, are subject to psychological causal laws; but physical causal laws, at least in traditional physics, can only be stated in terms of matter, which is both inferred and constructed, never a datum. In this respect psychology is nearer to what actually exists.

Printed at Repro India Ltd.